Noncommutative Geometry

Noncommutative Geometry

Alain Connes
Collège de France
Institut des Hautes Etudes Scientifiques
Paris, France

ACADEMIC PRESS
An Imprint of Elsevier
San Diego New York Boston
London Sydney Tokyo Toronto

This book is printed on acid-free paper. ♾

An English translation of the original French edition of the
author's GÉOMÉTRIE NON COMMUTATIVE (© 1990, InterEditions, Paris)
was provided by Sterling K. Berberian, Professor (Emeritus) of
Mathematics at the University of Texas at Austin.

Scientific editor for the translations: Marc A. Rieffel, Professor
of Mathematics, University of California, Berkeley.

Although this book is based on Alain Connes, GÉOMÉTRIE NON
COMMUTATIVE, InterEditions, 1990, and its English translation by
Sterling Berberian, more than two thirds of the material is new
with this work.

Front cover image by A. Connes. For back cover image identification
and credits see pages 4, 23, 34, and 54.

Academic Press
An Imprint of Elsevier
525 B Street, Suite 1900, San Diego, California 92101-4495

United Kingdom Edition published by
Academic Press Limited
24-28 Oval Road, London, NW1 7DX

Library of Congress Cataloging-in-Publication Data

Connes, Alain.
[Geometrie non commutative. English]
Noncommutative geometry / by Alain Connes.
p. cm.
Includes index.
ISBN-13: 978-0-12-185860-5 ISBN-10: 0-12-185860-X
1. Geometry, Algebraic. 2. Noncommutative rings I. Title.
QA564.C6713 1994 94-26550
516.3'5—dc20 CIP

ISBN-13: 978-0-12-185860-5
ISBN-10: 0-12-185860-X

Printed and bound in Great Britain by
CPI Antony Rowe, Chippenham and Eastbourne

Transferred to Digital Printing, 2010

TABLE OF CONTENTS

PREFACE

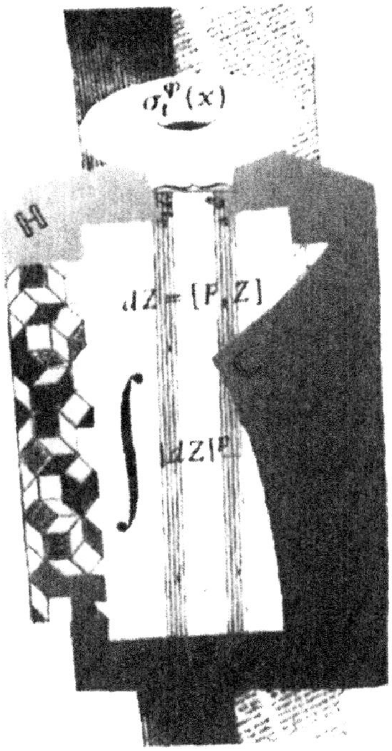

This book is the English version of the French "Géométrie non commutative" published by InterEditions Paris (1990). After the excellent initial translation by S.K. Berberian, a considerable amount of rewriting was done and many additions made, multiplying by 3.8 the size of the original manuscript. In particular the present text contains several unpublished results.

My thanks go first of all to Cécile whose patience and care for the manuscript have been essential to its completion. This second version of the book greatly benefited from the modifications suggested by many people: foremost was Marc Rieffel, but important contributions were made by D. Sullivan, J.-L. Loday, J. Lott, J. Bellissard, P. B. Cohen, R. Coquereaux, J. Dixmier, M. Karoubi, P. Krée, H. Bacry, P. de la Harpe, A. Hof, G. Kasparov, J. Cuntz, D. Testard, D. Kastler, T. Loring, J. Pradines, V. Nistor, R. Plymen, R. Brown, C. Kassel, and M. Gerstenhaber, with several of whom I have shared the pleasure of collaboration.

Patrick Ion and Arthur Greenspoon played a decisive rôle in the finalisation of the book, clearing up many mathematical imprecisions and considerably smoothing the initial manuscript. I wish to express my deep gratitude for their generous help and their insight.

Finally, my thanks go to Marie Claude for her help in creating the picture on the cover of the book, to Gilles who took the photograph, and to Bonnie Ion and Françoise for their help with the bibliography. Many thanks go also to Peter Renz who orchestrated the whole thing.

Alain Connes
30 June 1994
Paris

INTRODUCTION

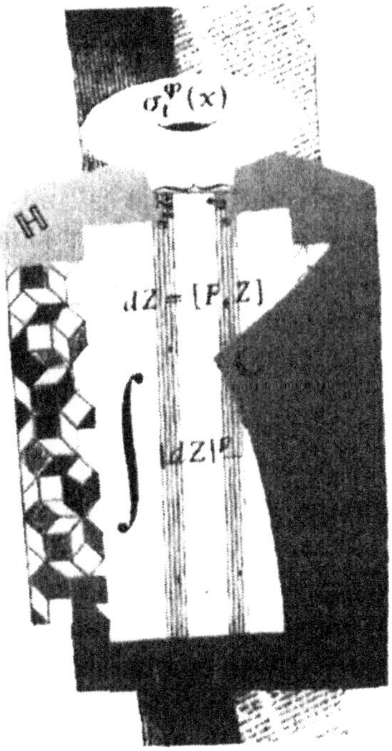

The correspondence between geometric spaces and commutative algebras is a familiar and basic idea of algebraic geometry. The purpose of this book is to extend this correspondence to the noncommutative case in the framework of real analysis. The theory, called noncommutative geometry, rests on two essential points:

1. The existence of many natural spaces for which the classical set-theoretic tools of analysis, such as measure theory, topology, calculus, and metric ideas lose their pertinence, but which correspond very naturally to a noncommutative algebra. Such spaces arise both in mathematics and in quantum physics and we shall discuss them in more detail below; examples include:

a) The space of Penrose tilings

b) The space of leaves of a foliation

c) The space of irreducible unitary representations of a discrete group

d) The phase space in quantum mechanics

e) The Brillouin zone in the quantum Hall effect

f) Space-time.

Moreover, even for classical spaces, which correspond to commutative algebras, the new point of view will give new tools and results, for instance for the Julia sets of iteration theory.

2. The extension of the classical tools, such as measure theory, topology, differential calculus and Riemannian geometry, to the noncommutative situation. This extension involves, of course, an algebraic reformulation of the above tools, but passing from the commutative to the noncommutative case is never straightforward. On the one hand, completely new phenomena arise in the noncommutative case, such as the existence of a canonical time evolution for a

noncommutative measure space. On the other hand, the constraint of developing the theory in the noncommutative framework leads to a new point of view and new tools even in the commutative case, such as cyclic cohomology and the quantized differential calculus which, unlike the theory of distributions, is perfectly adapted to products and gives meaning and uses expressions like $\int f(Z)|dZ|^p$ where Z is not differentiable (and p not necessarily an integer).

Let us now discuss in more detail the extension of the classical tools of analysis to the noncommutative case.

A. Measure theory (Chapters I and V)

It has long been known to operator algebraists that the theory of von Neumann algebras and weights constitutes a far reaching generalization of classical measure theory. Given a countably generated measure space X, the linear space of square-integrable (classes of) measurable functions on X forms a Hilbert space. It is one of the great virtues of the Lebesgue theory that every element of the latter Hilbert space is represented by a measurable function, a fact which easily implies the Radon-Nikodým theorem, for instance. There is, up to isomorphism, only one Hilbert space with a countable basis, and in the above construction the original measure space is encoded by the representation (by multiplication operators) of its algebra of bounded measurable functions. This algebra turns out to be the prototype of a *commutative* von Neumann algebra, which is dual to an (essentially unique) measure space X.

In general a construction of a Hilbert space with a countable basis provides one with specific automorphisms (unitary operators) of that space. The algebra of operators in the Hilbert space which commute with these particular automorphisms is a *von Neumann algebra*, and all von Neumann algebras are obtained in that manner. The theory of not necessarily commutative von Neumann algebras was initiated by Murray and von Neumann and is considerably more difficult than the commutative case.

The center of a von Neumann algebra is a commutative von Neumann algebra, and, as such, dual to an essentially unique measure space. The general case thus decomposes over the center as a direct integral of so-called *factors*, i.e. von Neumann algebras with trivial center.

In increasing degree of complexity the factors were initially classified by Murray and von Neumann into three types, I, II, and III.

The type I factors and more generally the type I von Neumann algebras, (i.e. direct integrals of type I factors) are isomorphic to commutants of *commutative* von Neumann algebras. Thus, up to the notion of multiplicity they correspond to classical measure theory.

The type II factors exhibit a completely new phenomenon, that of *continuous dimension*. Thus, whereas a type I factor corresponds to the geometry of lines, planes, ..., k-dimensional complex subspaces of a given Hilbert space, the subspaces that belong to a type II factor are no longer classified by a dimension which is an integer but by a dimension which is a positive *real number* and will span a continuum of values (an interval). Moreover, crucial properties such as the equality

$$\dim(E \wedge F) + \dim(E \vee F) = \dim(E) + \dim(F)$$

remain true in this continuous geometry ($E \wedge F$ is the intersection of the subspaces and $E \vee F$ the closure of the linear span of E and F).

The type III factors are those which remain after the type I and type II cases have been considered. They appear at first sight to be singular and intractable. Relying on Tomita's theory of modular Hilbert algebras and on the earlier work of Powers, Araki, Woods and Krieger, I showed in my thesis that type III is subdivided into types III_λ, $\lambda \in [0, 1]$ and that a factor of type III_λ, $\lambda \neq 1$, can be reconstructed uniquely as a crossed product of a type II von Neumann algebra by an automorphism contracting the trace. This result was then extended by M. Takesaki to cover the III_1 case as well, using a one-parameter group of automorphisms instead of a single automorphism.

These results thus reduce the understanding of type III factors to that of type II factors and their automorphisms, a task which was completed in the hyperfinite case and culminates in the complete classification of hyperfinite von Neumann algebras presented briefly in Chapter I Section 3 and in great detail in Chapter V.

The reduction from type III to type II has some resemblance to the reduction of arbitrary locally compact groups to unimodular ones by a semidirect product construction. There is one essential difference, however, which is that the range of the module, which is a closed subgroup of $\mathbb{R}_+^*$ in the locally compact group case, has to be replaced for type III_0 factors by an ergodic action of $\mathbb{R}_+^*$: the flow of weights of the type III factor. This flow is an invariant of the factor and can, by Krieger's theorem (Chapter V) be any ergodic flow, thus exhibiting an intrinsic relation between type III factors and ergodic theory and lending support to the ideas of G. Mackey on virtual subgroups. Indeed, in Mackey's terminology, a virtual subgroup of $\mathbb{R}_+^*$ corresponds exactly to an ergodic action of $\mathbb{R}_+^*$.

Since general von Neumann algebras have such an unexpected and powerful structure theory it is natural to look for them in more common parts of mathematics and to start using them as tools. After some earlier work by Singer, Coburn, Douglas, and Schaeffer, and by Shubin (whose work is the first application of type II techniques to the spectral theory of operators), a decisive step in this direction was taken up by M.F. Atiyah and I. M. Singer. They showed that the type II von Neumann algebra generated by the regular representation of a discrete group (already considered by Murray and von Neumann) provides, thanks to the *continuous dimension*, the necessary tool to measure the multiplicity of the kernel of an invariant elliptic differential operator on a Galois covering space. Moreover, they showed that the type II index on the covering equals the ordinary (type I) index on the quotient manifold. Atiyah then went on, with W. Schmid, to show the power of this result in the geometric construction of discrete series representations of semisimple Lie groups.

Motivated by this result and by the second construction of factors by Murray and von Neumann, namely the group-measure-space construction, I then showed that a foliated manifold gives rise in a canonical manner to a von Neumann algebra (Chapter I Section 4). A general element of this algebra is just a random operator in the L^2 space of the generic leaf of the foliation and can thus be seen as an operator-valued function on the badly behaved leaf space X of the foliation. As in the case of covering spaces the generic leaf is in general not compact even if the ambient manifold is compact. A notable first difference from the case of discrete groups and covering spaces is that in general the von Neumann algebra of a foliation is not of type II. In fact every type can occur and, for instance, very standard foliations such as the Anosov foliation of the unit sphere bundle of a compact Riemann surface of genus > 1 give a factor of type III_1. This allows one to illustrate by concrete geometric examples the meaning of type I, type II, type III_λ ... and we shall do that as early as Section 4 of Chapter I. Geometrically the reduction from type III to type II amounts to the replacement of the initial noncommutative space by the total space of an $\mathbb{R}_+^*$ principal bundle over it.

We shall see much later (Chapter III Section 6) the deep relation between the flow of weights and the Godbillon-Vey class for codimension-one foliations.

The second notable difference is in the formulation of the index theorem, which, as in Atiyah's case, uses the type II continuous dimensions as the key tool. For foliations one needs first to realize that the type II Radon measures, i.e. the traces on the C^*-algebra of the foliation (cf. below) correspond exactly to the holonomy invariant measures. Such measures are characterized (cf. Chapter I Section 5) by a de Rham current, the Ruelle-Sullivan current, and the index formula for the type II index of a longitudinal elliptic differential operator now involves the homology class of the Ruelle -Sullivan current. In contrast to the case of covering spaces this homology class is in general not even rational; the continuous dimensions involved can now assume arbitrary real values, and the index is not related to an integer-valued index.

In the case of measured foliations the continuous dimensions acquire a very clear geometric meaning. First, a general projection belonging to the von Neumann algebra of the foliation yields a random Hilbert space, i.e. a measurable bundle of Hilbert spaces over the badly behaved space X of leaves of the foliation. Next, any such random Hilbert space is isomorphic to one associated to a transversal as follows: the transversal intersected with a generic leaf yields a countable set; the fiber over the leaf is then the Hilbert space with basis this countable set. Finally, in the above isomorphism, the transverse measure of the transversal is independent of any choices and gives the *continuous dimension* of the original projection. One can then formulate the index theorem independently of von Neumann algebras, which we do in Section 5 of Chapter I. In simple cases where the ergodic theorem applies, one recovers the transverse measure of a transversal as the density of the corresponding discrete subset of the generic leaf, i.e. as the limit of the number of points of this subset over increasingly large volumes. Thus the Murray and von Neumann continuous dimensions bear the same relation to ordinary dimensions as (continuous) densities bear to the counting of finite sets.

In order to get some intuitive idea of what the generic leaf of a foliation can be like, as well as its space X of leaves (at this level of measure theory) one can consider the space of Penrose tilings of the plane (or Penrose universes). After the discovery of aperiodic tilings of the plane, the number of required tiles was gradually reduced to two.

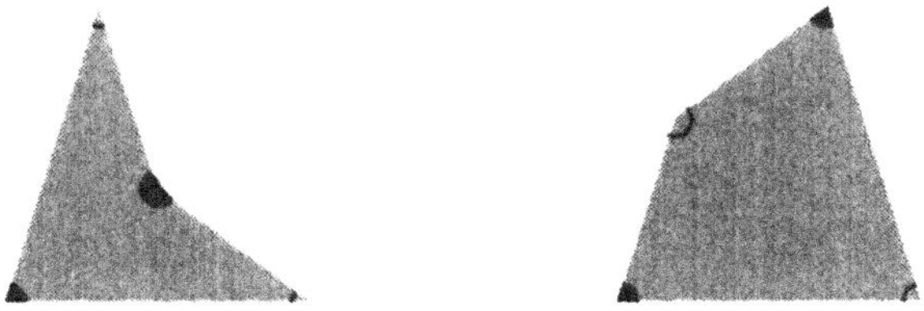

Figure 0.1. A dart and a kite

Given the two basic tiles: the Penrose kites and darts of Figure 1, one can tile the plane with these two tiles (with a matching condition on the colors of the vertices) but no such tiling is periodic. Two tilings are called identical if they are carried into each other by an isometry of the plane. Examples of nonidentical tilings are given by the star tiling of Figure 2 and the cartwheel tiling of Figure 3. The set X of all nonidentical tilings of the plane by the above two tiles is a very strange set because of the following:

> "Every finite patch of tiles in a tiling by kites and darts does occur, and infinitely many times, in any other tiling by the same tiles".

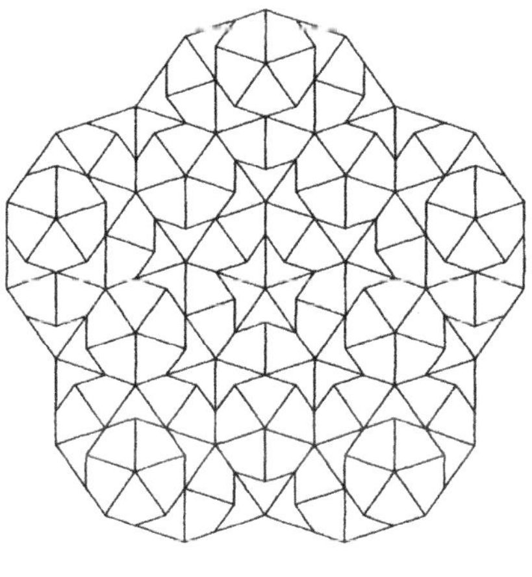

Figure 0.2. A star patch in a tiling ([Gru-S])

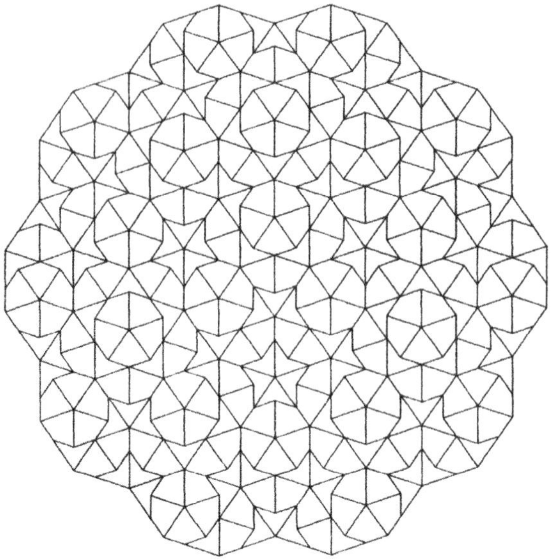

Figure 0.3. A cartwheel patch in a tiling ([Gru-S])

This means that it is impossible to decide locally with which tiling one is dealing. Any pair of tilings, such as those of Figures 2 and 3, can be matched on arbitrarily large patches. When analyzed with classical tools the space X appears pathological, and the usual tools of measure theory or topology give poor results: the only natural topology on the set X is the coarse topology with $\emptyset$ and X as the only open sets, and a similar result holds in measure theory. In fact one can go even further and show that the *effective* cardinality of the set X is strictly larger than the continuum (cf. Appendix C of Chapter I). The natural first reaction to such a space X is to dismiss it as pathological. Our thesis in this book is that X only looks pathological because one tries to understand it with classical tools whereas, even as a *set*, X is inherently of quantum mechanical nature.

What we mean by this, and we shall fully justify it in Chapters I and II, is that all the pathological properties of X disappear if we analyse it using, instead of the usual complex-valued functions, q-number or operator-valued functions. Equivalently one can write (in many ways) the space X as a quotient of an ordinary well behaved space by an equivalence relation. For instance the space Y of pairs (tiling, tile belonging to the tiling), is well behaved and possesses an obvious equivalence relation with quotient X. One could also use in a similar manner the group of isometries of the plane. All these constructions yield *equivalence relations*, or better, *groupoids* or pre-equivalence relations in the sense of Grothendieck. The noncommutative algebra is then the convolution algebra of the groupoid.

It is fashionable among mathematicians to despise groupoids and to consider that only groups have an authentic mathematical status, probably because of the pejorative suffix oid. To remove this prejudice we start Chapter I with Heisenberg's discovery of quantum mechanics. The reader will, we hope, realise there how the experimental results of spectroscopy forced Heisenberg to replace the classical frequency group of the system by the groupoid of quantum transitions. Imitating for this groupoid the construction of a group convolution algebra, Heisenberg rediscovered matrix multiplication and invented quantum mechanics.

In the case of the space X of Penrose tilings the noncommutative C^*-algebra which replaces the commutative C^*-algebra of continuous functions on X has trivial center and a unique trace. The corresponding factor of type II which describes X from the measure theoretic point of view has a natural subfactor of Jones index $(2\cos\pi/5)^2$ (cf. Chapter V Section 10).

Before we pass to the natural extension of topology to noncommutative spaces we need to describe its role in noncommutative measure theory. One crucial use of the (local) compactness of an ordinary space X is the Riesz representation theorem. It extends the construction of the Lebesgue integral, starting from an arbitrary *positive* linear form on the algebra of continuous functions over the space X. This theorem extends as follows to the noncommutative case. First, the involutive algebras (over $\mathbb{C}$) of complex-valued functions over compact spaces are, by Gel'fand's fundamental result, exactly the commutative C^*-algebras with unit. Moreover, this establishes a perfect duality between the category of compact (resp. locally compact) spaces and continuous (resp. proper and continuous) maps and the category of unital C^*-algebras and $*$-homomorphisms (resp. not necessarily unital).

The algebraic definition of a C^*-algebra turns out to be remarkably simple and to make no use of commutativity. One is used to introducing C^*-algebras as involutive Banach algebras for which the following equality holds:

$$\|x^*x\| = \|x\|^2$$

for any element x. But this hides an absolutely crucial feature by letting one believe that, as in a Banach algebra, there is freedom in the choice of the norm. In fact if an involutive algebra is a C^*-algebra it is so for a *unique* norm, given for any x by

$$\|x\| = (\text{Spectral radius of } x^*x)^{1/2}.$$

The general tools of functional analysis show that C^*-algebras constitute the natural framework for noncommutative Radon measure theory. Thus, for instance, the elements of a C^*-algebra which are of the form x^*x constitute a closed convex cone, the cone of *positive* elements of the C^*-algebra. Any element of the dual cone, i.e. any *positive linear form*, yields by the Gel'fand-Naimark-Segal construction a Hilbert space representation of the C^*-algebra. This bridges the gap with noncommutative measure theory, i.e. von Neumann algebras. Indeed, the positive linear form extends by continuity to the von Neumann algebra generated in the above Hilbert space representation. Moreover, the remarkable up-down theorem of G. Pedersen asserts that any selfadjoint element of the von Neumann algebra is obtained by monotone limits (iterated twice) of "continuous functions" i.e. of elements of the C^*-algebra.

This construction of measures from positive linear forms on C^*-algebras plays an important role in the case of foliated manifolds where, as we already mentioned, the traces on the C^*-algebra of the foliation correspond exactly to the holonomy invariant transverse measures. It also plays a crucial role in quantum statistical mechanics in formulating what a state of the system is in the thermodynamic limit: exactly a positive normalized linear form on the C^*-algebra of observables. The equilibrium or Gibbs states are then characterized by the Kubo-Martin-Schwinger condition as revealed by Haag, Hugenholtz, and Winnink (cf. Chapter I Section 2). The relation between this "KMS" condition and the modular operator of Tomita's Hilbert algebras, discovered by M. Takesaki and M. Winnink, remains one of the deepest points of contact between physics and pure mathematics.

B. Topology and K-theory (Chapter II)

In the above use of C^*-algebras as a tool to construct measures (in the commutative or the noncommutative case) the fine topological features of the spaces under consideration are not relevant and do not show up. But, by Gel'fand's theorem any homeomorphism invariant of a compact space X is an algebraic invariant of the C^*-algebra $C(X)$ of continuous functions on X so that one should be able to recover it purely algebraically.

The first invariant for which this was done is the Atiyah-Hirzebruch topological K-theory. Indeed, the abelian group $K(X)$ generated by stable isomorphism classes of complex vector bundles over the compact topological space X has a very simple and natural description in terms of the C^*-algebra $C(X)$. By a result of Serre and Swan, it is the abelian group generated by stable isomorphism classes of finite projective modules over $C(X)$, a purely algebraic notion, which moreover makes no use of the commutativity of $C(X)$. The key result of topological K-theory is the periodicity theorem of R. Bott. The original proof of Bott relied on Morse theory applied to loop spaces of Lie groups. Thanks to the work of Atiyah and Bott, the result, once formulated in the algebraic context, has a very simple proof and holds for any (not necessarily commutative) Banach algebra and in particular for C^*-algebras.

The second invariant which was naturally extended to the noncommutative and algebraic framework is K-homology, which appeared as a result of the influence of the work of Atiyah and Singer on the index theorem for elliptic operators on a compact manifold and was developed by Atiyah, Kasparov, Brown, Douglas, and Fillmore. The pseudodifferential calculus on a compact manifold, as used in the proof of the index theorem, gives rise to a short exact sequence of C^*-algebras which encodes in an algebraic way the information given by the index map. The last term of the exact sequence is the commutative algebra of continuous functions on the unit sphere of the cotangent space of the manifold, and its K-theory group is the natural receptacle for the symbol of the elliptic operator. The first term of the exact sequence is a noncommutative C^*-algebra, independent of the context: the algebra of compact operators in a Hilbert space with a countable basis. This unique C^*-algebra is called the *elementary* C^*-algebra and is the only separable infinite-dimensional C^*-algebra which admits (up to unitary equivalence) only one irreducible unitary representation. Its K-theory group is equal to the additive group of relative integers and is the natural receptacle for indices of Fredholm operators. The connecting map of the exact sequence of pseudodifferential calculus is the index map of Atiyah and Singer.

In their work on extension theory, L. Brown, R. Douglas, and P. Fillmore showed how to associate to any compact space X an abelian group classifying short exact sequences of the above type, called extensions of $C(X)$ by the elementary C^*-algebra. They proved that the invariant obtained is K-homology, i.e. the homology theory associated by duality to K-theory in the odd case. The even case of that theory was treated by M.F. Atiyah and G. G. Kasparov.

The resulting theory was then extended to the noncommutative case, i.e. with $C(X)$ replaced by a noncommutative C^*-algebra, thanks to the remarkable generalisation by D. Voiculescu of the Weyl-von Neumann perturbation theorem. This allowed the proof that classes of extensions (by the elementary C^*-algebra) form a group, provided the original C^*-algebra is *nuclear*. The class of nuclear C^*-algebras was introduced by M. Takesaki in his work on tensor products of C^*-algebras. A C^*-algebra is nuclear if and only if its associated von Neumann algebra is hyperfinite, in any unitary representation. This class of C^*-algebras covers many, though not all, interesting examples.

In the meantime considerable progress had been made by topologists concerning the use, in surgery theory of non-simply-connected manifolds M, of algebraic invariants of the group ring of the fundamental group. In 1965, S. Novikov conjectured the homotopy invariance of numbers of the form $\langle Lx, [M]\rangle$ where L is the characteristic class of the Hirzebruch signature theorem, and x a product of one-dimensional cohomology classes. His conjecture was proved independently by Farrell-Hsiang and Kasparov using geometric methods.

The higher signatures of manifolds M with fundamental group Γ are the numbers of the form $\langle L\ \psi^*(y), [M]\rangle$, where y is a cohomology class on $B\Gamma$ and $\psi : M \to B\Gamma$ the classifying map. The original conjecture of Novikov is the special case Γ abelian. In 1970 Mishchenko constructed the equivariant signature of non-simply-connected manifolds as an element of the Wall group of the group ring. He proved that this signature is a homotopy invariant of the manifold. Lusztig gave a new proof of the result of Farrell-Hsiang and Kasparov based on families of elliptic operators and extended it to the symplectic groups. Following this line, and by a crucial use of C^*-algebras, Mishchenko was able to prove the homotopy invariance of higher signatures under the assumption that the classifying space $B\Gamma$ of the fundamental group Γ is a compact manifold with a Riemannian metric of non-positive curvature.

The reason why C^*-algebras play a key role at this point is the following: The Wall group of an involutive algebra (such as a group ring) classifies Hermitian quadratic forms over that algebra, and is in general far more difficult to compute than the K-group which classifies finite projective modules. However, as shown by Gel'fand and Mishchenko, the two groups coincide when the involutive algebra is a C^*-algebra. This equality *does not hold* for general Banach involutive algebras. As the group ring of a discrete group can be completed canonically to a C^*-algebra one thus obtains the equivariant signature of a non-simply-connected manifold as an element of the K-group of this C^*-algebra.

Mishchenko went on and used the dual theory, namely K-homology, in the guise of Fredholm representations of the fundamental group, to obtain numerical invariants by pairing with K-theory.

Thus the K-theory of the highly noncommutative C^*-algebra of the fundamental group played a key role in the solution of a classical problem in the theory of non-simply-connected manifolds.

The K-theory of C^*-algebras, the extension theory of Brown, Douglas, and Fillmore, the L-theory of Atiyah, and the Fredholm representations of Mishchenko are all special cases of the bivariant KK-theory of Kasparov. With his theory, whose main tool is the intersection product, Kasparov proved the homotopy invariance of the extension theory and solved the Novikov conjecture for discrete subgroups of arbitrary Lie groups. Together with the breakthrough of Pimsner and Voiculescu in the computation of K-groups of crossed products the Kasparov theory played a decisive role in the understanding of K-groups of noncommutative C^*-algebras.

In Chapter II we shall use a variant of Kasparov's theory, the deformation theory (Chapter II Appendix B) due to N. Higson and myself, which succeeds in defining the intersection product in full generality for extensions. It also has the advantage of having an easily defined product (cf. Appendix B) and of being exact in full generality. Nevertheless we shall need KK-theory later and have gathered its main results in Appendix A of Chapter IV.

The importance of the K-group as an invariant of noncommutative C^*-algebras was already clear from the work of Bratteli and Elliott on the classification of C^*-algebras which are inductive limits of finite-dimensional algebras. They showed that the pointed K-group, with its natural order, is a complete invariant in the above class of algebras, and Effros, Chen, and Handelman completely characterized the pointed ordered groups thus obtained. As a simple example where this applies consider (as above) the space X of Penrose tilings and the noncommutative C^*-algebra which replaces, in this singular example, the algebra $C(X)$. This C^*-algebra turns out to be an inductive limit of finite-dimensional algebras and the computation of its K-group (Chapter II Section 3) gives the abelian group $\mathbb{Z}^2$ ordered by the half-plane determined by the golden ratio. The space X is thus well understood as a "0-dimensional" noncommutative space.

In order to be able to apply the above tools of noncommutative topology to spaces such as the space of leaves of a foliation we need to describe more carefully how the topology of such spaces give rise to a noncommutative C^*-algebra.

This is done in great detail in Chapter II with a lot of examples. The general principle is, instead of taking the quotient of a space by an equivalence relation, to retain this equivalence relation as the basic information. An important intermediate notion which emerges is that of smooth groupoid. It plays the same role as the pre-equivalence relations of Grothendieck. The same noncommutative space can be presented by several equivalent smooth groupoids. The corresponding C^*-algebras are then strongly Morita equivalent in the sense of M. Rieffel and have consequently the same K-theory invariants (cf. Chapter II Appendix A).

Even in the context of ordinary manifolds smooth groupoids are quite pertinent. To illustrate this point we give in Chapter II Section 5 a proof of the Atiyah-Singer index theorem. It follows directly from the construction of the tangent groupoid of an arbitrary manifold, a geometric construction which encodes the naive interpretation of a tangent vector as a pair of points whose distance is comparable to a given infinitesimal number ε.

Smooth groupoids contain manifolds, Lie groups, and discrete groups as special cases. The smooth groupoid which corresponds to the space of leaves of a foliation is the holonomy groupoid or graph of the foliation, first considered by R. Thom. The smooth groupoid which corresponds to the quotient of a manifold by a group of diffeomorphisms is the semidirect product of the manifold by the action of the group.

Smooth groupoids are special cases of locally compact groupoids and J. Renault has shown how to associate a C^*-algebra to the latter. The advantage of smoothness, however, is that, as in the original case of foliations, the use of $\frac{1}{2}$-densities removes any artificial choices in the construction.

We define the K-theory of spaces, such as the space of leaves of a foliation, as the K-theory of the associated C^*-algebra, i.e. here the convolution algebra of the holonomy groupoid. In the special case of fibrations the leaf space is a manifold and the above definition of its K-theory coincides with the usual one. The first role of the K-group is as an invariant of the leaf space, unaffected by modifications of the foliation such as leafwise surgery, which do not modify the space of leaves. Thus, for instance, for the Kronecker foliations $dy = \theta dx$ of the two-torus one gets back the class of θ modulo $PSL(2,\mathbb{Z})$ from the K-theory of the leaf space. But the main role of the K-group is as a receptacle for the index of families.

Given a space X such as the leaf space of a foliation, an example of a family $(D_\ell)_{\ell \in X}$ of elliptic operators parametrized by X is given by a longitudinal elliptic differential operator D on the ambient manifold: it restricts to each leaf ℓ as an elliptic operator D_ℓ. In terms of the holonomy groupoid of the foliation this provides us with a functor to the category of manifolds and elliptic operators. The general notion of family is defined in these terms. It is a general principle of great relevance that the analytical index of such families $(D_x)_{x \in X}$ makes sense as an element of the K-group of the parameter space (defined as above through the associated C^*-algebra). Thus the index of a longitudinal elliptic operator now makes sense, irrespective of the existence of a transverse measure, as an element of the K-group of the leaf space. Similarly, the index of an invariant elliptic operator on a Galois covering of a manifold makes sense as an element of the K-group of the C^*-algebra of the covering group. Moreover, the invariance properties, such as the homotopy invariance of the equivariant signature (due to Mishchenko and Kasparov) or the vanishing of the Dirac index in the presence of positive scalar curvature (due to Gromov, Lawson, and Rosenberg) do hold at the level of the K-group.

The obvious problem then is to compute these K-groups in geometric terms for the above spaces. Motivated by the case of foliations where closed transversals give idempotents of the C^*-algebra, I was led with P. Baum and G. Skandalis to the construction, in the above generality of smooth groupoids G, of a geometrically defined group $K^*_{\text{top}}(G)$ and of an additive map μ from this group to the K-group of the C^*-algebra. So far each new computation of K-groups confirms the validity of the general conjecture according to which the map μ is an isomorphism.

Roughly speaking, the injectivity of μ is a generalized form of the Novikov higher signature conjecture, while the surjectivity of μ is a general form of the Selberg principle on the vanishing of orbital integrals of non-elliptic elements in the theory of semisimple Lie groups. It also has deep connections with the zero divisor conjecture of discrete group theory. Chapter II contains a detailed account of the construction of the geometric group, of the map μ and their properties. Besides the cases of discrete groups, quotients of manifolds by group actions, and foliations, we also treat carefully the case of Lie groups where the conjecture, intimately related to the work of Atiyah and Schmid, has been proved, in the semisimple case, by A. Wassermann.

The general problem of injectivity of the analytic assembly map μ is an important reason for developing the analogue of de Rham homology for the above spaces, which is done in Chapter III.

C. Cyclic cohomology (Chapter III)

In 1981 I discovered cyclic cohomology and the spectral sequence relating it to Hochschild cohomology ([Co$_{15}$]). My original motivation came from the trace formulas of Helton-Howe and Carey-Pincus for operators with trace-class commutators. These formulae together with the operator theoretic definition of K-homology discussed above, lead naturally to cyclic cohomology as the natural receptacle for the Chern character (not the usual one but the Chern character in

K-homology). Motivated by my work, J.-L. Loday, C. Kassel and D. Quillen developed the dual theory, cyclic homology, and related it to the homology of the Lie algebra of matrices. This result was obtained independently by B.Tsygan who, not having access to my work, discovered cyclic homology from a completely different motivation, that of additive K-theory.

We shall come back in Chapter IV to cyclic cohomology as the natural receptacle for the Chern character of K-homology classes and to the quantized differential calculus. In Chapter III we develop both the algebraic and analytic properties of cyclic cohomology with as motivation the construction of K-theory invariants, generalizing to the noncommutative case the Chern-Weil theory.

It is worthwhile to consider the most elementary example of that theory, namely the Gauss-Bonnet theorem. Thus, let $\Sigma \subset \mathbb{R}^3$ be a smooth closed surface embedded in three-space. Let us recall the notion of curvature.

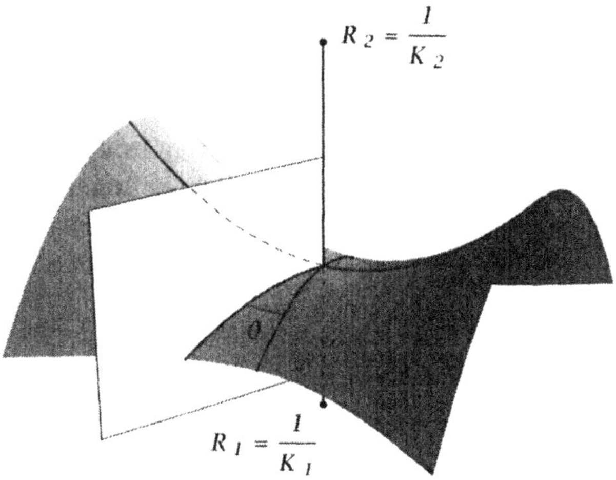

Figure 0.4. Curvature

Through a point P of the surface, one can draw the normal to the surface and, to reduce dimension by 1, cut the surface by a plane containing the normal. One obtains in this way a curve which, at the point P, has a *curvature:* the inverse of the radius of the circle, with center on the normal, which best fits the curve at the point P. Clearly the curvature of the above curve depends upon the plane containing the normal, and an old result of Euler asserts that there are two extreme values K_1 and K_2 for this curvature, attained for two perpendicular planes. Moreover, given any plane containing the normal, and making an angle θ with the plane corresponding to K_1, the curvature of its intersection with the surface is

$$K_\theta = K_1 \cos^2 \theta + K_2 \sin^2 \theta.$$

The Gauss-Bonnet theorem then asserts that, provided Σ is oriented, the integral over Σ of the product $K_1 K_2$ of the principal curvatures is equal to $2\pi(2-2g)$, where g is an integer depending only upon the topology of Σ, and called the genus.

One remarkable feature of this result is that the above integral has an extraordinary property of stability. One has indeed at one's disposal an infinite number of parameters to modify the embedding of Σ in $\mathbb{R}^3$, say by making an arbitrary small bump on the surface. The theorem nevertheless asserts that, in doing so, one will introduce equal amounts of positive and negative curvature (here on the top of the bump and the sides of it, respectively).

One can of course give a rather straightforward proof of the above theorem by invoking the concept of degree of a map in differential topology, but we claim that, once considered from the algebraic point of view, the above idea of stability has considerable potential, as we shall now see. By the algebraic point of view I mean that we let $\mathcal{A}$ be the algebra of all smooth functions on Σ, i.e. $\mathcal{A} = C^\infty(\Sigma)$, and we consider on $\mathcal{A}$ the following trilinear form: $\tau(f^0, f^1, f^2) = \int_\Sigma f^0 df^1 \wedge df^2$. In more geometric terms this trilinear form associates to every map $f : \Sigma \to \mathbb{R}^3$ given by the three functions f^0, f^1, f^2 on Σ, the oriented volume bounded by the image. This trilinear form possesses the following compatibility, reminiscent of the properties of a trace, with the algebra structure of $\mathcal{A}$:

1) $\tau(f^1, f^2, f^0) = \tau(f^0, f^1, f^2) \quad \forall f^j \in \mathcal{A}$

2) $\tau(f^0 f^1, f^2, f^3) - \tau(f^0, f^1 f^2, f^3) + \tau(f^0, f^1, f^2 f^3) - \tau(f^3 f^0, f^1, f^2) = 0 \quad \forall f^j \in \mathcal{A}.$

We can now see the remarkable stability of the integral formula of the Gauss-Bonnet theorem as a special case of a general algebraic lemma, valid for *noncommutative* algebras. It simply asserts that if $\mathcal{A}$ is an algebra (over $\mathbb{R}$ or $\mathbb{C}$) and τ a trilinear form on $\mathcal{A}$ with properties 1) and 2), then for each idempotent $E = E^2 \in \mathcal{A}$ the scalar $\tau(E, E, E)$ remains constant when E is deformed among the idempotents of $\mathcal{A}$.

The proof of this statement is simple; the idea is that since the deformation of E is isospectral (all idempotents have spectrum $\subset \{0, 1\}$) one can find $X \in \mathcal{A}$ with $\dot{E} = [X, E]$ (where $\dot{E}$ is the time derivative of E) and algebraic manipulations using 1) and 2) then show that the time derivative of $\tau(E, E, E)$ is zero.

It is quite important that this lemma does not make use of *commutativity*, even when we apply it in the above case. Indeed, if we apply it to $\mathcal{A} = C^\infty(\Sigma)$ we get nothing of interest since as Σ is connected there are no nontrivial (i.e. different from 0 and 1) idempotents in $\mathcal{A}$. We may however, use $M_2(\mathcal{A}) = \mathcal{A} \otimes M_2(\mathbb{C})$, the algebra of 2×2 matrices with entries in $\mathcal{A}$. We need first to extend the functional τ to $M_2(\mathcal{A})$ and this is done canonically as follows:

$$\tilde{\tau}(f^0 \otimes \mu^0, f^1 \otimes \mu^1, f^2 \otimes \mu^2) = \tau(f^0, f^1, f^2) \text{ Trace } (\mu^0 \mu^1 \mu^2)$$

where $f^j \in \mathcal{A}$, $\mu^j \in M_2(\mathbb{C})$ and Trace is the ordinary trace on $M_2(\mathbb{C})$. One then checks easily that properties 1) and 2) still hold for the extended $\tilde{\tau}$. (Note however, at this point, that the original τ did fulfill a stronger property, namely $\tau(f^{\sigma(0)}, f^{\sigma(1)}, f^{\sigma(2)}) = \varepsilon(\sigma)\, \tau(f^0, f^1, f^2)$ for any permutation σ, but that due to the cyclicity of the trace *only* the property 1) survives after passing to $\tilde{\tau}$.) The next step is to find an interesting idempotent $E \in M_2(\mathcal{A})$. For this one uses the fundamental but straightforward fact: An idempotent of $M_2(C^\infty(\Sigma))$ is exactly a smooth map from Σ to the Grassmannian of idempotents of $M_2(\mathbb{C})$. Since the latter Grassmannian is, using selfadjoint idempotents to simplify, exactly the two sphere S^2, we see that the normal map of the embedding of Σ in $\mathbb{R}^3$ provides us with a specific idempotent $E \in M_2(\mathcal{A})$ to which the above stability lemma applies.

In fact the above lemma shows that, given a trilinear form τ with properties 1) and 2) on an algebra $\mathcal{A}$, the formula

$$E \to \tau(E, E, E)$$

defines an invariant of K-theory, i.e. an additive map of $K_0(\mathcal{A})$ to the complex numbers. This follows because any finite projective module over an algebra $\mathcal{A}$ is the image of an idempotent E belonging to a matrix algebra over $\mathcal{A}$. The homotopy invariance provided by the lemma allows one to eliminate the ambiguity of the choice of E in the stable isomorphism class of finite projective modules $x \in K_0(\mathcal{A})$.

The above simple algebraic manipulations extend easily to higher dimensions, where conditions 1) and 2) have obvious analogues. The analogue of 1) in dimension n is

$$\tau(f^1, f^2, \ldots, f^n, f^0) = (-1)^n\, \tau(f^0, \ldots, f^n) \quad \forall f^j \in \mathcal{A}$$

where the $-$ sign for n odd accounts for the oddness of the cyclic permutation in that case.

The analogue of 2) is just the vanishing of $b\tau$, where

$$(b\tau)(f^0, \ldots, f^{n+1}) = \sum_{j=0}^{n} (-1)^j\, \tau(f^0, \ldots, f^j f^{j+1}, \ldots, f^{n+1}) + (-1)^{n+1}\, \tau(f^{n+1} f^0, f^1, \ldots, f^n) \quad \forall f^j \in \mathcal{A}.$$

One checks that for $n = 0$ this vanishing characterizes the *traces* on $\mathcal{A}$. An $(n+1)$-linear form on an algebra $\mathcal{A}$ satisfying 1) and 2) is called a *cyclic cocycle* of dimension n. The above lemma and its proof easily extend to the general case: an *even-dimensional* cyclic cocycle τ gives an invariant of K-theory by the formula

$$E \to \tau(E, E, \ldots, E).$$

This immediately implies the construction of the usual Chern character of vector bundles for manifolds. Indeed, any homology class is represented by a closed de Rham current, i.e. by a continuous linear form C on differential forms, which vanishes on coboundaries. One checks as above that the following formula then defines a cyclic cocycle (of the same dimension as the current) on the algebra $\mathcal{A}$ of smooth functions on the manifold:

$$\tau(f^0, \ldots, f^n) = \langle C, f^0\, df^1 \wedge \cdots \wedge df^n \rangle \quad \forall f^j \in \mathcal{A}.$$

Up to a normalization factor the above pairing of τ with the K-group of $\mathcal{A}$ gives the usual Chern character for a vector bundle E, the scalar $\langle \mathrm{ch}(E), C \rangle$.

So far we have just reformulated algebraically the Chern-Weil construction, dispensing completely with any commutativity assumption. The remarkable fact is that the highly noncommutative group rings, i.e. the convolution algebras $\mathbb{C}\Gamma$ of discrete groups Γ, possess very nontrivial cyclic cocycles associated to the ordinary group cocycles on Γ. The group cohomology $H^*(\Gamma, \mathbb{C})$ is the cohomology of the classifying space $B\Gamma$ of Γ, the (unique up to homotopy) quotient of a contractible topological space on which Γ acts freely and properly. It may be described purely algebraically in terms of group cocycles on Γ: Given an integer n, an n-group cocycle on Γ (with complex coefficients) is a function of n variables $g_i \in \Gamma$ which fulfills the condition

$$\sum_{j=0}^{n+1} (-1)^j\, c(g_1, \ldots, g_j\, g_{j+1}, \ldots, g_{n+1}) = 0$$

where for $j = 0$ one takes $c(g_2, g_3, \ldots, g_{n+1})$ and for $j = n+1$ one takes $(-1)^{n+1}\, c(g_1, \ldots, g_n)$ as the corresponding terms in the sum. Moreover, such cocycles can be normalized so as to vanish if any of the g_i's or their product $g_1 \cdots g_n$ is the unit element of Γ.

Now the key observation is that such a normalized group cocycle uniquely defines a corresponding *cyclic cocycle* of the same dimension n, on the group ring $\mathbb{C}\Gamma$. Since the latter is obtained by linearizing Γ it is enough to give the value of the cyclic cocycle on $(n+1)$-tuples $(g_0, g_1, \ldots, g_n) \in \Gamma^{n+1}$. The rule is the following:

$$\tau(g_0, g_1, \ldots, g_n) = 0 \ \text{ if } \ g_0 g_1 \cdots g_n \neq 1$$
$$\tau(g_0, g_1, \ldots, g_n) = c(g_1, \ldots, g_n) \ \text{ if } \ g_0 g_1 \cdots g_n = 1.$$

Observe that the conditions on the right are invariant under cyclic permutations, since these do not alter the conjugacy class of the product $g_0 \cdots g_n$.

This formula was initially found in [Co_{17}] and later extended by M. Karoubi and D. Burghelea, who computed the cyclic cohomology of group rings (cf. Chapter III Section 2).

We have used in a crucial manner the above cyclic cocycles on group rings, in collaboration with H. Moscovici, to prove the Novikov conjecture for the class of all hyperbolic groups in the sense of Gromov (cf. Chapter III Section 5). They play, in the noncommutative case, the role that de Rham currents play in the commutative case on the Pontryagin dual of the discrete group Γ.

More precisely, when Γ is abelian, Lusztig, in his proof of the Novikov conjecture, has shown how to obtain it from the Atiyah-Singer index theorem for *families* of elliptic operators, with parameter space X the Pontryagin dual of Γ. The idea is that each element $\chi \in X$, being a *character* of the group Γ, determines a flat line bundle on any manifold M with fundamental group Γ. Twisting the signature operator of the compact oriented manifold M by such flat bundles, one obtains a family $(D_\chi)_{\chi \in X}$ of elliptic operators, whose Atiyah-Singer index (an element of $K(X)$) is a homotopy invariant of M.

Next, the Atiyah-Singer index theorem for families, as formulated in terms of the *cohomology* of X rather than K-theory, gives the homotopy invariance of Novikov's higher signatures. Note that it is crucial here to use the Chern character, $\mathrm{ch} : K(X) \to H^*(X)$, to express the index formula in cohomology. In fact, even when X is a single point, the K-theory formulation of the Atiyah-Singer index theorem only becomes easy to use after translation, using the Chern character, into cohomological terms.

The role of cyclic cohomology in the context of the Novikov conjecture is exactly the same. The space X, the Pontryagin dual of Γ, no longer exists when Γ is noncommutative, but the commutative C^*-algebra $C(X)$ is replaced by the group C^*-algebra $C^*(\Gamma)$ which makes perfectly good sense. This C^*-algebra contains, as a dense subalgebra of fairly smooth elements, the group ring $\mathbb{C}\Gamma$. The index of the above family of elliptic operators still makes perfectly good sense and is the element of the K-group of $C^*(\Gamma)$ given by the Mishchenko-Kasparov signature of the covering space of M. In fact it can even be obtained as an element of the K-group of the group ring $\mathcal{R}\Gamma$ of Γ over the ground ring $\mathcal{R}$ of infinite matrices of rapid decay.

Moreover, when paired with the above cyclic cocycle on $\mathbb{C}\Gamma$ (extended to $\mathcal{R}\Gamma$) associated to a given group cocycle, the index of the signature family gives exactly the Novikov higher signature associated to the group cohomology class of the cocycle. This follows from a general index theorem due to Moscovici and myself (Section 4 of Chapter III) for elliptic operators on covering spaces. With such a result it would appear at first sight that one has proved the Novikov conjecture in full generality. The difficulty is that the homotopy invariance of the signature index is only known at the level of the K-group of the C^*-algebra $C^*(\Gamma)$ but not for the K-group of the ring $\mathcal{R}\Gamma$.

One way to overcome this is to construct an intermediate algebra playing the role of the algebra of smooth functions on X, which on the one hand is large enough to have the same K-theory as $C^*(\Gamma)$ and on the other small enough so that the cyclic cocycle still makes sense. This is achieved, by the analogue of the Harish-Chandra-Schwartz space, for hyperbolic groups (Section 5 of Chapter III).

We shall give in Section 6 of Chapter III another striking application of cyclic cohomology and K-theory invariants to the Godbillon-Vey class, along the lines of earlier work of S. Hurder. Using cyclic cohomology as a key tool, we shall show that the Godbillon-Vey class (the simplest instance of Gel'fand-Fuchs cohomology class) of a codimension-one foliation yields a pairing with complex values between the K-group of the leaf space (defined as above using the C^*-algebra of the foliation) and the flow of weights $\text{mod}(M)$ of the von Neumann algebra of the foliation.

Together with the construction of the analytic assembly map of Chapter II this result immediately implies that if the Godbillon-Vey class of the foliation does not vanish then the flow of weights $\text{mod}(M)$ admits a finite invariant probability measure. In other words the (virtual) modular spectrum is a (virtual) subgroup of finite covolume in $\mathbb{R}_+^*$ and one is in particular in the type III situation, the previous result of Hurder.

This shows that in the noncommutative framework there is a very intricate relation between differential geometry and measure theory, where the nonvanishing of differential geometric quantities implies the type III behaviour at the measure theoretic level. In fact, as we shall see, one can, thanks to cyclic cohomology, give the following formula for the Godbillon-Vey class as a 2-dimensional closed current GV on the leaf space (i.e. in our algebraic terms as a 2-dimensional cyclic cocycle on the noncommutative algebra of the foliation)

$$GV = i_H\,[\dot{X}].$$

It comes from the interplay between the transverse fundamental class $[X]$ of the leaf space X and the canonical time evolution σ_t of the von Neumann algebra of the foliation. The transverse fundamental class $[X]$ of the leaf space X is given by a one-dimensional cyclic cocycle which corresponds to integration of transverse 1-forms. The time evolution σ_t of the von Neumann algebra of the foliation is the algebraic counterpart of its noncommutativity and is one of the great surprises of noncommutative measure theory (cf. Chapter I).

Now these two pieces of data, $[X]$ and σ_t, are in general slightly incompatible in that $[X]$ is not in general static, i.e. invariant under σ_t. This is not too bad, however, since one has

$$\frac{d^2}{(dt)^2}\,\sigma_t[X] = 0.$$

It follows then, as in ordinary differential geometry, that if one contracts the closed current

$$[\dot{X}] = \frac{d}{dt}\,\sigma_t[X]$$

with the derivation H generating the one-parameter group σ_t, then one obtains a 2-*dimensional closed current.* The fundamental equation is then

$$GV = i_H\,[\dot{X}].$$

It gives a simple example of a general fact: that the leaf space of a foliation can have higher cohomological dimension than the naive value, the codimension of the foliation.

Using a lot more analysis all this is extended to higher Gel'fand-Fuchs cohomology, and many geometric applications are given. They all show that provided one uses the tools of noncommutative geometry one can indeed think of a foliated manifold as a bundle of leaves over the noncommutative leaf space. For instance, we show that if the $\hat{A}$-genus of a manifold V does not vanish, then V does not admit a foliation with leaves of strictly positive scalar curvature, a result which is easy for fibrations by a Fubini-type argument and a well known result of A. Lichnerowicz.

D. The quantized calculus (Chapter IV)

The basic new idea of noncommutative differential geometry is a new calculus which replaces the usual differential and integral calculus.

This new calculus can be succinctly described by the following dictionary. We fix a pair $(\mathcal{H}, F)$, where $\mathcal{H}$ is an infinite-dimensional separable Hilbert space and F is a selfadjoint operator of square 1 in $\mathcal{H}$. Giving F is the same as giving the decomposition of $\mathcal{H}$ as the direct sum of the two orthogonal closed subspaces

$$\{\xi \in \mathcal{H}\,;\; F\xi = \pm\xi\}.$$

Assuming, as we shall, that both subspaces are infinite-dimensional, we see that all such pairs $(\mathcal{H}, F)$ are unitarily equivalent. The dictionary is then the following:

CLASSICAL	QUANTUM
Complex variable	Operator in $\mathcal{H}$
Real variable	Selfadjoint operator in $\mathcal{H}$
Infinitesimal	Compact operator in $\mathcal{H}$
Infinitesimal of order α	Compact operator in $\mathcal{H}$ whose characteristic values μ_n satisfy $\mu_n = O(n^{-\alpha})$, $n \to \infty$
Differential of real or complex variable	$df = [F, f] = Ff - fF$
Integral of infinitesimal of order 1	Dixmier trace $\mathrm{Tr}_\omega(T)$

Let us comment in some detail on each entry of the dictionary.

The range of a complex variable corresponds to the *spectrum* $\mathrm{Sp}(T)$ of an operator. The holomorphic functional calculus for operators in Hilbert space gives meaning to $f(T)$ for any holomorphic function f defined on $\mathrm{Sp}(T)$ and only holomorphic functions act in that generality. This reflects the need for holomorphy in the theory of complex variables. For real variables the situation is quite different. Indeed, when the operator T is selfadjoint, $f(T)$ makes sense for any Borel function f on the line.

The role of infinitesimal variables is played by the compact operators T in $\mathcal{H}$. First $\mathcal{K} = \{T \in \mathcal{L}(\mathcal{H})\ ;\ T \text{ compact}\}$ is a two-sided ideal in the algebra $\mathcal{L}(\mathcal{H})$ of bounded operators in $\mathcal{H}$, and it is the largest nontrivial ideal. An operator T in $\mathcal{H}$ is compact iff for any $\varepsilon > 0$ the size of T is smaller than ε except for a finite-dimensional subspace of $\mathcal{H}$. More precisely, one lets, for $n \in \mathbb{N}$,

$$\mu_n(T) = \mathrm{Inf}\{\|T - R\|\ ;\ R \text{ operator of rank } \le n\}$$

where the rank of an operator is the dimension of its range. Then T is compact iff $\mu_n(T) \to 0$ when $n \to \infty$. Moreover, the $\mu_n(T)$ are the eigenvalues, ordered by decreasing size, of the absolute value $|T| = (T^*T)^{1/2}$ of T. The rate of decay of the $\mu_n(T)$ as $n \to \infty$ is a precise measure of the *size* of the infinitesimal T.

In particular, for each positive real α the condition

$$\mu_n(T) = O(n^{-\alpha}) \quad n \to \infty$$

(i.e. there exists $C < \infty$ such that $\mu_n(T) \le Cn^{-\alpha} \quad \forall n \ge 1$) defines the infinitesimals of order α. They form a two-sided ideal, as is easily checked using the above formula for $\mu_n(T)$. Moreover, if T_1 is of order α_1 and T_2 of order α_2, then T_1T_2 is of order $\alpha_1 + \alpha_2$.

The differential df of a real or complex variable, usually given by the differential geometric expression

$$df = \sum \frac{\partial f}{\partial x^i}\, dx^i$$

is replaced in the new calculus by the commutator

$$df = [F, f].$$

The passage from the classical formula to the above operator-theoretic one is analogous to the quantization of the Poisson brackets $\{f, g\}$ of classical mechanics as commutators $[f, g]$. This is at the origin of the name "quantized calculus". The Leibniz rule $d(fg) = (df)g + f\, dg$ still holds. The equality $F^2 = 1$ is used to show that the differential df has vanishing anticommutator with F.

The next key ingredient of our calculus is the analogue of integration; it is given by the Dixmier trace. The Dixmier trace is a general tool designed to treat in a classical manner data of quantum mechanical nature. It is given as a positive linear form Tr_ω on the ideal of infinitesimals of order 1, and is a *trace:*

$$\mathrm{Tr}_\omega(ST) = \mathrm{Tr}_\omega(TS) \quad \forall T \text{ of order } 1,\ S \text{ bounded.}$$

In the classical differential calculus it is an important fact that one can neglect all infinitesimals of order > 1. Similarly, the Dixmier trace does neglect (i.e. vanishes on) any infinitesimal of order > 1, i.e.

$$\mathrm{Tr}_\omega(T) = 0 \quad \text{if} \quad \mu_n(T) = o(n^{-1})$$

where the little o means, as usual, that $n\mu_n \to 0$ as $n \to \infty$. This vanishing allows considerable simplification to occur, as in the symbolic calculus, for expressions to which the Dixmier trace is applied.

The value of $\mathrm{Tr}_\omega(T)$ is given for $T \geq 0$ by a suitable limit of the bounded sequence

$$\frac{1}{\log N} \sum_{n=0}^{N} \mu_n(T).$$

It is then extended by linearity to all compact operators of order 1.

In general the above sequence does not converge, so that Tr_ω a priori depends on a limiting procedure ω. As we shall see, however, in all the applications one can prove the independence of $\mathrm{Tr}_\omega(T)$ on ω. Such operators T will be called measurable. For instance, when T is a pseudodifferential operator on a manifold it is measurable and its Dixmier trace coincides with the Manin-Wodzicki-Guillemin residue computed by a local formula. In general the term residue(T) for the common value of $\mathrm{Tr}_\omega(T)$, T measurable, would be appropriate since for $T > 0$ it coincides with the residue at $s = 1$ of the Dirichlet series $\zeta(s) = \mathrm{Trace}\,(T^s)$, $s \in \mathbb{C}$, $\mathrm{Re}(s) > 1$.

We have now completed our description of the framework of the quantized calculus. To use it for a given noncommutative space X we need a representation of the algebra $\mathcal{A}$ of functions on X in the Hilbert space $\mathcal{H}$. The compatibility of this representation with the operator F is simply that all operators f in $\mathcal{H}$ coming from $\mathcal{A}$ have infinitesimal differential

$$[F, f] \in \mathcal{K} \quad \forall f \in \mathcal{A}.$$

Such a representation is called a Fredholm module, and these are the basic cycles for the K-homology of A when A is a C^*-algebra.

To see how the new calculus works and allows operations not possible in distribution theory we shall start with a simple example. There is a unique way to quantize, in the above sense, the calculus of functions of one real variable, i.e. for $X = \mathbb{R}$, in a translation and scale invariant manner. It is given by the representation of functions as multiplication operators in $L^2(\mathbb{R})$, while F is the Hilbert transform. Similarly, for $X = \mathbb{S}^1$ one lets $L^\infty(\mathbb{S}^1)$ act on $L^2(\mathbb{S}^1)$ by multiplication, while F is again the Hilbert transform, given by the multiplication by the sign of n in the Fourier basis $(e_n)_{n\in\mathbb{Z}}$ of $L^2(\mathbb{S}^1)$, with $e_n(\theta) = \exp(in\theta) \quad \forall \theta \in \mathbb{S}^1$.

The first virtue of the new calculus is that df continues to make sense, as an operator in $L^2(\mathbb{S}^1)$, for an arbitrary measurable $f \in L^\infty(\mathbb{S}^1)$. This of course would also hold if we were to define df using distribution theory, but the essential difference is the following. A distribution is defined as an element of the topological dual of the locally convex vector space of smooth functions, here $C^\infty(\mathbb{S}^1)$. Thus only the linear structure on $C^\infty(\mathbb{S}^1)$ is used, not the *algebra* structure of $C^\infty(\mathbb{S}^1)$. It is consequently not surprising that distributions are incompatible with pointwise product or absolute value. Thus while, with f nondifferentiable, df makes sense as a distribution, we cannot make any sense of $|df|$ or powers $|df|^p$ as distributions on $\mathbb{S}^1$.

Let us give a concrete example where one would like to use such an expression for nondifferentiable f. Let c be a complex number and let J be the Julia set given by the complex dynamical system $z \to z^2 + c = \varphi(z)$. More specifically J is here the boundary of the set $B = \{z \in \mathbb{C}; \sup_{n\in\mathbb{N}} |\varphi^n(z)| < \infty\}$. For small values of c like the one chosen in Figure 5, the Julia set J is a Jordan curve and B is the bounded component of its complement. Now the Riemann mapping theorem provides us with a conformal equivalence Z of the unit disk, $D = \{z \in \mathbb{C}; |z| < 1\}$ with the inside of B, and by a result of Carathéodory, the conformal mapping Z extends continuously to the boundary $\mathbb{S}^1$ of D as a homeomorphism, which we still denote by Z, from $\mathbb{S}^1$ to J. By a known result of D. Sullivan, the Hausdorff dimension p of the Julia set is strictly bigger than 1, $1 < p < 2$, and is close to 2, for instance, in the example of Figure 5. This shows that the function Z is nowhere of bounded variation on $\mathbb{S}^1$ and forbids a distribution interpretation of the naive expression

$$\int f(Z)|dZ|^p \quad \forall f \in C(J)$$

that would be the natural candidate for the Hausdorff measure on J.

Figure 0.5. A Julia set

We shall show that the above expression makes sense in the quantized calculus and that it does give the Hausdorff measure on the Julia set J. The first essential fact is that as $dZ = [F, Z]$ is now an operator in Hilbert space one can, irrespective of the regularity of Z, talk about $|dZ|$: it is the absolute value $|T| = (T^*T)^{1/2}$ of the operator $T = [F, Z]$. This gives meaning to any function $h(|dZ|)$ where h is a bounded measurable function on $\mathbb{R}_+$ and in particular to $|dZ|^p$. The next essential step is to give meaning to the integral of $f(Z)|dZ|^p$. The latter expression is an operator in $L^2(\mathbb{S}^1)$, and we shall use a result of hard analysis due to V.V. Peller, together with the homogeneity properties of the Julia set, to show that the operator $f(Z)|dZ|^p$ belongs to the domain of definition of the Dixmier trace Tr_ω, i.e. is an infinitesimal of order 1. Moreover, if one works modulo infinitesimals of order > 1 the rules of the usual differential calculus such as

$$|d\varphi(Z)|^p = |\varphi'(Z)|^p \; |dZ|^p$$

are valid and show that the measure

$$f \to \mathrm{Tr}_\omega(f(Z)|dZ|^p) \quad \forall f \in C(J)$$

has the right conformal weight and is a non-zero multiple of the Hausdorff measure. The corresponding constant governs the asymptotic expansion for the distance, in the sup norm on $\mathbb{S}^1$, between the function Z and restrictions to $\mathbb{S}^1$ of rational functions with at most n poles outside the unit disk.

This example of quantized calculus will be explained in Section 3 of Chapter IV. The first two sections of this chapter deal with the properties of the usual trace (Section 1) and the Dixmier trace Tr_ω (Section 2).

In Section 1 we shall use the ordinary trace to evaluate differential expressions such as $f^0\, df^1 \dots df^n$, where $f^j \in \mathcal{A}$ and $(\mathcal{H}, F)$ is a Fredholm module over $\mathcal{A}$.

The *dimension* of such modules is measured by the size of the differentials $df = [F, f]$ for $f \in \mathcal{A}$, i.e. by the growth of the characteristic values of these operators. More specifically, a Fredholm module $(\mathcal{H}, F)$ over an algebra $\mathcal{A}$ (commutative or not) is called p-summable iff the operators df, $f \in \mathcal{A}$, all belong to the Schatten ideal $\mathcal{L}^p$ of operators: $T \in \mathcal{L}^p$ iff $\sum \mu_n(T)^p < \infty$. We shall show in Section 1 that the Chern character of a p-summable Fredholm module over an algebra $\mathcal{A}$ can be defined by a trace formula (in line with earlier works of Helton-Howe and Carey-Pincus) and gives a cyclic cocycle of dimension n for any given integer n larger than $p - 1$ and of the same parity as the Fredholm module. (For a module to be even one requires that the Hilbert space be $\mathbb{Z}/2$-graded, with F anticommuting with the grading, as required by general K-homology definitions.) Moreover, the various cyclic cocycles thus obtained are all images of a single one by the periodicity operator S of cyclic cohomology. The cyclic cohomology class $\mathrm{ch}_*(\mathcal{H}, F)$ thus obtained has a remarkable property of integrality, displaying the *quantum* nature of our calculus. Thus, when pairing this class $\mathrm{ch}_*(\mathcal{H}, F)$, the Chern character of a Fredholm module over an algebra $\mathcal{A}$, with the K-group of $\mathcal{A}$ one only gets integers:

$$\langle \mathrm{ch}_*(\mathcal{H}, F), K(\mathcal{A})\rangle \subset \mathbb{Z}.$$

This follows from a simple formula computing the value of the pairing as the index of a Fredholm operator (Section 1).

In Section 2 we introduce the Dixmier trace and give a general formula, in terms of the Dixmier trace, for the Hochschild class of the character $\mathrm{ch}_*(\mathcal{H}, F)$ defined above.

The relevance of this class is that, although it is not the whole cohomological information contained in the character, it is exactly the obstruction to writing the latter as the image, by the periodicity operator S of cyclic cohomology, of a lower-dimensional cyclic cocycle (as follows from general results of Chapter III Section 1.)

The formula obtained for the Hochschild cocycle uses an unbounded selfadjoint operator D whose sign is equal to F and which controls uniformly the size of the $da = [F, a]$, $a \in \mathcal{A}$ by the conditions

$$\alpha)\ [D, a] \text{ bounded } \quad \forall a \in \mathcal{A} \ , \quad \beta)\ |D|^{-n} \text{ of order } 1.$$

The formula obtained gives a Hochschild n-cocycle as the Dixmier trace applied to the product $a^0[D, a^1] \dots [D, a^n]\ |D|^{-n}$, where $[D, a^j]$ is the commutator of D with the element a^j of $\mathcal{A}$. The proof that this Hochschild cocycle is cohomologous, in Hochschild cohomology, to the character of $(\mathcal{H}, D)$ is very involved because it relates the original cyclic cocycle, computed in terms of the ordinary trace, to an expression involving the logarithmic or Dixmier trace. It implies in particular that if the obstruction to lowering the dimension does not vanish then the Dixmier trace of D^{-n} cannot vanish and the eigenvalues of D^{-1} cannot be $o(k^{-1/n})$.

At the end of Section 2 we show, using results of D. Voiculescu, that the existence of such a D controls the size of abelian subalgebras of $\mathcal{A}$, i.e. that the number of independent commuting quantities in the algebra $\mathcal{A}$ is bounded above by n. This is another justification of the depth of the relation between the degree of summability and dimension.

While Section 3 is devoted to the quantized calculus in one real variable we show in Section 4 that on a conformal manifold the calculus can be canonically quantized. This is related to the work of Donaldson and Sullivan on quasiconformal 4-manifolds. We show that the Polyakov action of two-dimensional conformal field theory is given, in terms of the components X^μ of a map $X : \Sigma \to \mathbb{R}^d$ of a Riemann surface to d-dimensional space, by the formula

$$I(X) = \mathrm{Tr}_\omega(\eta_{\mu\nu}\, \mathrm{d}X^\mu\, \mathrm{d}X^\nu) \;\;,\;\; dX^\mu = [F, X^\mu].$$

The remarkable fact is that the right-hand side continues to make sense when Σ is a 4-dimensional conformal manifold. This is due to the discovery by M. Wodzicki of the residue of pseudodifferential operators, a unique tracial extension of the Dixmier trace. We compute this new 4-dimensional action at the end of Section 4 and relate it to the Paneitz Laplacian. In Section 5 we move to highly noncommutative examples coming from group rings and return to the origin of the quantized calculus as a substitute of differential topology in the noncommutative context. We exploit the *integrality* of the Chern character $\mathrm{ch}_*(\mathcal{H}, F)$.

We show the power of this method by giving an elegant proof, in the spirit of differential topology, of the Pimsner-Voiculescu theorem solving the Kadison conjecture for the C^*-algebras of free groups. As another remarkable application of this integrality result we describe in detail in Section 6 of Chapter IV the work of J. Bellissard on the quantum Hall effect. Experimental results, of von Klitzing, Pepper, and Dorda, in 1980, showed the existence of plateaux for the transverse conductivity as a function of the natural parameters of the experiment. This unexpected finding gave a precise meaning to the numerical value of the conductivity on these plateaux and yielded precision measurements of the fine structure constant which were independent of quantum field theory. The numbers obtained have the same unexpected *stability* property as the integral occurring above in the Gauss-Bonnet theorem.

Bellissard constructed a natural cyclic 2-cocycle on the *noncommutative* algebra of "functions on the Brillouin zone" and expressed the Hall conductivity as the pairing between this cyclic 2-cocycle and an idempotent of the algebra, the spectral projection of the Hamiltonian. This accounts for the stability part. He then went on and showed that his 2-cocycle is in fact the Chern character of a Fredholm module, from which the remarkable *integrality* of the conductivity in units $e^2/\hbar$ does follow. In doing so he defined the analogue of the Brillouin zone as a noncommutative space and the framework to apply the above quantized calculus to this example. All this is explained in great detail in Section 6. We also show there how the ordinary pseudodifferential calculus adapts to the noncommutative torus, the simplest and probably the first example of a smooth noncommutative space. We then apply it to prove an index theorem for finite difference-differential equations on the real line.

The rest of this Chapter IV is devoted to the infinite-dimensional case, i.e. to the construction of the Chern character for Fredholm modules which are not finitely summable. This is motivated by the following two classes of infinite-dimensional noncommutative spaces: the dual spaces of non-amenable discrete groups and the phase space in quantum field theory (cf. Section 9 of Chapter IV). The first step in order to adapt the above tools to the infinite-dimensional case is to develop entire cyclic cohomology, which has the same relation to ordinary cyclic cohomology as entire functions have to polynomials.

From the very beginning of cyclic cohomology, the possibility of defining it by means of cocycles with finite support in the (b, B) fundamental bicomplex did suggest the existence of infinite-dimensional cocycles, not reducible to finite-dimensional ones, by considering cocycles with arbitrary support in the (b, B) bicomplex. However, a key algebraic result, the vanishing of the E_1 term of the first spectral sequence of the bicomplex (cf. Chapter III Section 1) shows that the cohomology of cochains with arbitrary supports is always trivial. It is thus necessary to impose a nontrivial limitation on the growth of the components (φ_n) of a cochain with infinite support, in order to obtain a nontrivial cohomology. This growth condition eluded me for a long time but turned out to be dictated by the pairing with K-theory.

In Section 7 of Chapter IV we shall adapt it to arbitrary locally convex algebras, which shows in particular that entire cyclic cohomology applies to any algebra over $\mathbb{C}$ (endowing it with the fine locally convex topology). We give in that section the equivalence between three points of view on entire cyclic cocycles, which can be viewed as (normalized) cocycles in the (b, B) bicomplex, characters of infinite-dimensional cycles, or traces on the Cuntz algebra of $\mathcal{A}$, i.e. the free product of $\mathcal{A}$ with the group with 2 elements. The growth condition for entire cocycles then means that they extend to a suitable completion of the universal differential algebra $\Omega\mathcal{A}$ (resp. of the Cuntz algebra), whose quasi-nilpotency is analyzed. We end this section with the computation of the entire cyclic cohomology of the algebra of Laurent polynomials and its relation with a remarkable deformation of the Cuntz algebra in that case (also noticed independently by Cuntz and Quillen).

In Section 8 of Chapter IV we extend the previous finite-dimensional construction of the Chern character in K-homology to the infinite-dimensional case. For a Fredholm module $(\mathcal{H}, F)$ over an algebra $\mathcal{A}$ the infinite-dimensional summability condition is now that the commutators $da = [F, a]$ for $a \in \mathcal{A}$ all belong to the two-sided ideal of compact operators T such that $\mu_n(T)$, the n-th characteristic value of T, is of the order of $(\log n)^{-1/2}$ when n goes to ∞.

We first show that given such a Fredholm module over $\mathcal{A}$ one can find a selfadjoint unbounded operator D whose sign is equal to F (i.e. $D = F|D|$ is the polar decomposition of D) and which satisfies the following two conditions:

1) The commutator $[D, a]$ is *bounded* for any $a \in \mathcal{A}$

2) Trace $(e^{-D^2}) < \infty$.

Such θ-summable modules $(\mathcal{H}, D)$ over an algebra will turn out, as we shall see in Chapter VI, to be an excellent point of departure for the metric aspect in noncommutative geometry, i.e. the analogue of Riemannian geometry in our framework. This fact is already visible in the proof of the existence of the operator D, given F. The heuristic formula for D^{-2} is indeed

$$D^{-2} = \sum (dx^\mu)^* \, g_{\mu\nu} \, (dx^\nu)$$

where the x^μ are generators of the algebra $\mathcal{A}$, the $g_{\mu\nu}$ are the entries of a positive matrix, and the operator dx for $x \in \mathcal{A}$ is the quantum differential $[F, x]$.

Such modules $(\mathcal{H}, D)$ will be called (θ-summable) K-cycles for short. We showed how to construct the Chern character of such a K-homology class as an entire cyclic cocycle. Our formula for this character was quite complicated and Jaffe, Lesniewski, and Osterwalder, motivated by supersymmetric quantum field theory, then obtained a much simpler formula for the same cohomology class.

All this is explained in detail in Section 8, where the role of the quasinilpotent algebra of convolution of distributions with support in $\mathbb{R}_+$ is put into evidence.

In Section 9 we describe the two basic examples of θ-summable K-cycles not reducible to finitely summable ones. The first example combines the construction of Mishchenko and Kasparov of Fredholm representations of discrete subgroups of semisimple Lie groups with the twisted de Rham operator of Witten in Morse theory. We leave open the problem of completing the computation of the character of this K-cycle for discrete subgroups of higher rank Lie groups. The second example is the supersymmetric Wess-Zumino model of quantum field theory. We prove there the new unexpected result that, even for the free theory, the K-theory of the algebra $\mathcal{A}(\mathcal{O})$ of observables localized in an open subset $\mathcal{O}$ of space-time is very nontrivial, of infinite rank as an abelian group, and pairs nontrivially with the K-homology class given by the supercharge operator Q of the theory.

We have already explained that Chapter V is a detailed account of the theory of von Neumann algebras. In its last section we explain our joint work with J.-B. Bost, motivated by earlier work of B. Julia, on a natural system of quantum statistical mechanics related to the arithmetic of $\mathbb{Z}$ and the Riemann zeta function. We show that our system, formulated as a C^*-dynamical system, undergoes a phase transition with spontaneous symmetry breaking with symmetry group the Galois group of the field of roots of unity.

E. The metric aspect of noncommutative geometry

In the last chapter of this book we shall develop operator theoretic ideas about the metric aspect of geometry and then apply these ideas to a fundamental example: space-time.

Given a (not necessarily commutative) $*$-algebra $\mathcal{A}$ corresponding to the functions on a "space X", our basic data in order to develop geometry on X is simply a K-cycle $(\mathcal{H}, D)$ over $\mathcal{A}$ in the above sense (cf. D). A first example is provided by ordinary Riemannian geometry. Given a Riemannian manifold (K-oriented for convenience) we let its algebra $\mathcal{A}$ of functions act by multiplication in the Hilbert space $\mathcal{H}$ of L^2 spinors on X, and we let D be the Dirac operator in $\mathcal{H}$. One thus obtains a K-cycle over $\mathcal{A}$ which represents the fundamental class of X in K-homology. We shall give (Section 1 of Chapter VI) four simple formulas showing how to recover, from the K-cycle $(\mathcal{H}, D)$ the following classical geometric notions on X:

1) The geodesic distance d on X
2) The integration against the volume form $\det(g_{\mu\nu})^{1/2}\, dx^1 \wedge \ldots \wedge dx^n$
3) The affine space of gauge potentials
4) The Yang-Mills action functional.

The essential fact is that all these notions will continue to make sense in our general framework, and in particular will apply nontrivially even to the case of finite spaces X.

The operator theoretic formula 1) for the geodesic distance is in essence the dual of the usual formula in terms of the length of paths γ in X. Recall that given two points p and q in a Riemannian manifold X, their geodesic distance $d(p,q)$ is given as the infimum of the lengths of paths γ from p to q. The latter length is computed by means of an integral along γ of the square root of $g_{\mu\nu}\, dx^\mu\, dx^\nu$. Our formula for the same quantity $d(p,q)$ is different in nature in that it is a supremum (instead of an infimum) and it invokes as a variable a *function* f on X instead of a path γ in X. It is the following:

$$d(p,q) = \operatorname{Sup}\{|f(p) - f(q)|\,;\ f \in \mathcal{A}\,,\ \|[D,f]\| \le 1\}\,.$$

The norm involved in the right-hand side is the norm of operators in Hilbert space. Note also that the points p and q of X are used to convert the element f of $\mathcal{A}$ into a scalar, namely they are used, in conformity with Gel'fand's theorem, as characters of the algebra $\mathcal{A}$.

There are two important features of this formula. The first is that since it does not invoke paths but functions it continues to be meaningful and gives a finite nonzero result when the space X is no longer arcwise connected. Indeed, one can take a space X with two points $a \neq b$: the algebra $\mathcal{A}$ of functions on X is then simply $\mathcal{A} = \mathbb{C} \oplus \mathbb{C}$, the direct sum of two copies of $\mathbb{C}$, and a natural K-cycle is given by the diagonal action of $\mathcal{A}$ in $\ell^2(X)$ with operator $D = \begin{bmatrix} 0 & \mu \\ \mu & 0 \end{bmatrix}$. One then checks that with our formula the distance between a and b is given by $\ell = \mu^{-1}$.

An equally important feature of our formula is that, since it uses functions instead of paths, it is directly compatible with the formalism of quantum mechanics in which particles are described by wave functions rather than by any classical path.

The next step is to recover the tools of Riemannian geometry which go beyond the mere "metric space" attributes of such spaces such as 3) and 4). First, the volume form and the corresponding integral are recovered by the Dixmier trace, already mentioned in the discussion of Chapter IV. In fact the case already covered in Chapter IV of the Hausdorff measure on Julia sets in terms of the Dixmier trace was far more difficult. In Section 1 we shall develop the formalism of gauge theory (commutative or not, of course) and the Yang-Mills action functional from our operator theoretic data. The main mathematical result of Section 2 is the analogue of the basic inequality between the second Chern number of a Hermitian vector bundle and the minimum of the Yang-Mills functional on compatible connections on this bundle. This is proved in Section 2 using the main theorem of Chapter IV Section 2. In Section 3 we determine the Yang-Mills connections on the noncommutative torus, a joint result of M. Rieffel and myself, after defining the metric structure of that simple noncommutative space in great detail. As it turns out, even though the space we start with is non commutative, the moduli space of Yang-Mills connections is nicely behaved and finite-dimensional.

The end of the chapter (and the book) is occupied by the analysis, with the above new geometric tool, of the structure of space-time. The idea can be simply stated as follows: It was originally through the Maxwell equations that, by the well-known steps due to Lorentz, Michelson, Morley, Poincaré and Einstein, the model of Minkowski's space-time was elaborated. In more modern terms the Maxwell-Dirac Lagrangian accounts perfectly for all phenomena involving only the electromagnetic interaction. However, a century has passed since then and essential discoveries, both experimental and theoretical, have shown that in order to account for weak and strong forces a number of modifications in the above Lagrangian were necessary. (See the small chronological table below for the early development of weak forces.)

Chronological table

1896	H. Becquerel discovers radioactivity
1898	M. Curie shows that radioactivity is an intrinsically atomic property
1901	E. Rutherford shows the inhomogeneity of uranic rays and isolates the β rays (i.e. the weak interaction part)
1902	W. Kaufman shows that the β rays are formed of electrons
1907-14	J. Chadwick shows that the spectrum of β rays is continuous
1930	W. Pauli introduces the neutrino to account for the above continuous spectrum
1932	W. Heisenberg introduces the concept of isospin (in the other context of strong forces)
1934	E. Fermi gives a phenomenological theory of weak interactions
1935	H. Yukawa introduces the idea of heavy bosons mediating short range forces.

Among the next steps which led finally to the Glashow-Weinberg-Salam Lagrangian of the standard model let us mention the discovery of the vector-axial form of the weak currents, and the intermediate boson hypothesis which replaces the 4-fermion interaction of the Fermi theory by a renormalizable interaction. This requires a considerable mass for the intermediate-boson in order to get a possible unification of the weak and electromagnetic forces using the ideas of nonabelian gauge theory of Yang and Mills. There was a major difficulty at this point since a nonabelian pure Yang-Mills field is necessarily massless. This difficulty was resolved by the theoretical discovery of the Higgs mechanism, allowing for the generation, by spontaneous symmetry breaking, of nontrivial masses for fermions and intermediate bosons. The Higgs field thus introduced appears in three of the five natural terms of the Lagrangian and completely spoils their geometric significance.

Now our idea is quite simple. We have at our disposal a more flexible notion of geometric space in which the continuum (i.e. manifolds say) and the discrete (cf. the two-point space example above) are treated on the same footing, and in which the Maxwell-Dirac Lagrangian does make sense: its bosonic part is the already treated Yang-Mills action and the fermionic part is easy to get since we are given the Dirac operator to start with. Thus we can keep track of the above modifications of the Lagrangian as modifications of the geometry of space-time. In other words, we look for a geometry in our sense, i.e. a triple $(\mathcal{A}, \mathcal{H}, D)$ as above, such that the associated Maxwell-Dirac action functional produces the standard model of electroweak and strong interactions with all its refinements dictated by experimental results.

The result that we obtain is canonically derived from the standard model considered as a *phenomenological* model. What we find is a geometric space which is neither a continuum nor a discrete space but a mixture of both. This space is the product of the ordinary continuum by a discrete space with only two points, which, for reasons which will become clear later, we shall call L and R. The geometry, in our sense, of the finite space $\{L, R\}$ is given by a finite-dimensional Hilbert space representation $\mathcal{H}$ and a selfadjoint operator D in $\mathcal{H}$. These have direct physical meaning since the operator D is given by the Yukawa coupling matrix which encodes the masses of the elementary fermions as well as their mixing, i.e. the Kobayashi-Maskawa mixing parameters. Computing the distance between the two points L and R thus gives a number of the order of 10^{-16}cm, i.e. the inverse mass of the heaviest fermion.

The naive picture that emerges is that of a double space-time, i.e. the product of ordinary space-time by a very tiny discrete two-point space. By construction, purely left-handed particles such as neutrinos live on the left-handed copy X_L while electrons involve both X_L and X_R in $X = X_L \cup X_R$. Note that each point p of X_L is extremely close to a corresponding point p' of X_R, and when computing the differential of a function f on $X = X_L \cup X_R$ we shall find that it invokes not only the ordinary differential of f on X_L or on X_R but also the finite difference $f(p) - f(p')$ of the values of f on corresponding points of X_L and X_R. It is this finite difference, occurring as a new "transverse" component of the gauge potential, which yields the famous Higgs fields which appear in the three non-geometric terms of the standard model Lagrangian (cf. Chapter VI) $\mathcal{L} = \mathcal{L}_G + \mathcal{L}_f + \mathcal{L}_\varphi + \mathcal{L}_Y + \mathcal{L}_V$.

Note that there is no symmetry whatsoever between the points L and R, and in fact the natural vector bundle used over that space has fiber $\mathbb{C}^2$ over L and $\mathbb{C}$ over R. All this works very well for the electroweak sector of the standard model, but in order to account for the (already essentially geometric) strong structure much more work was necessary. It has been done in collaboration with J. Lott and is described in the last two Sections 4 and 5. It hinges on the fundamental problem of defining correctly what is a "manifold" in the setup of noncommutative geometry. Our proposed answer, which is essentially dictated by the $SU(2) \times SU(3)$ structure of up-down quark iso-doublets, is given by Poincaré duality discussed in Section 4. It fits remarkably well with the work of Sullivan on the fundamental role of Poincaré duality in K-homology for ordinary manifolds (cf. Section 4).

In Section 5 we discuss the full standard model, and also account for the hypercharge assignment of particles from a general unimodularity condition. One should understand this work as pure phenomenology, at the classical level, interpreting the detailed structure of the best phenomenological model, as the *fine structure* of space-time rather than as a long list of new particles. Applications of this point of view in quantum field theory, say as a model building device, are still ahead of us.

I

NONCOMMUTATIVE SPACES AND MEASURE THEORY

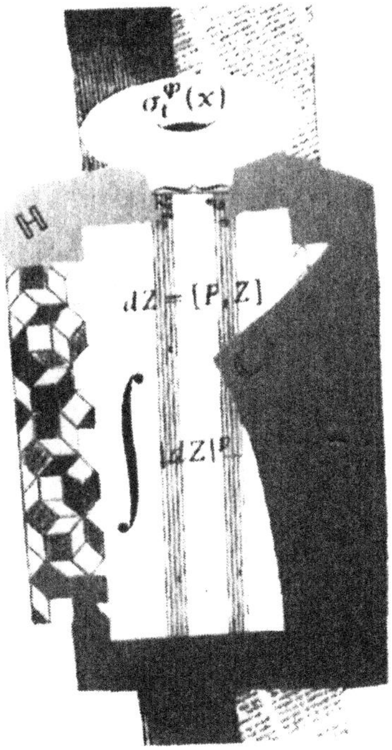

My first goal in this chapter is to show the extent to which Heisenberg's discovery of matrix mechanics, or quantum mechanics, was guided by the experimental results of spectroscopy. The resulting replacement of the phase space by a "noncommutative space", more precisely the replacement of the algebra of functions on the phase space by the algebra of matrices, is one of the most important conceptual steps, on which it is important to dwell. I will then attempt to place into evidence, thanks to quantum statistical mechanics, the interaction between theoretical physics and pure mathematics in the realm of operator algebras, and to show how this interaction was at the origin of the classification of factors summarized below. Next I will show how the theory of operator algebras replaces ordinary measure theory for the "noncommutative spaces" that one encounters in mathematics, such as the space of leaves of a foliation. This will first of all give a detailed illustration of the classification of factors by geometric examples. It will also allow us to extend the Atiyah-Singer index theorem to the non-compact leaves of a measured foliation, using the full force of the Murray and von Neumann theory of continuous dimensions.

I.1. Heisenberg and the Noncommutative Algebra of Physical Quantities Associated to a Microscopic System

The classification of the simple chemical elements into Mendeleev's periodic table is without doubt the most striking result in chemistry in the 19th century.

The theoretical explanation of this classification, by Schrödinger's equation and Pauli's exclusion principle, is an equivalent success of physics in the 20th century, and, more precisely, of quantum mechanics. One can look at this theory from very diverse points of view.

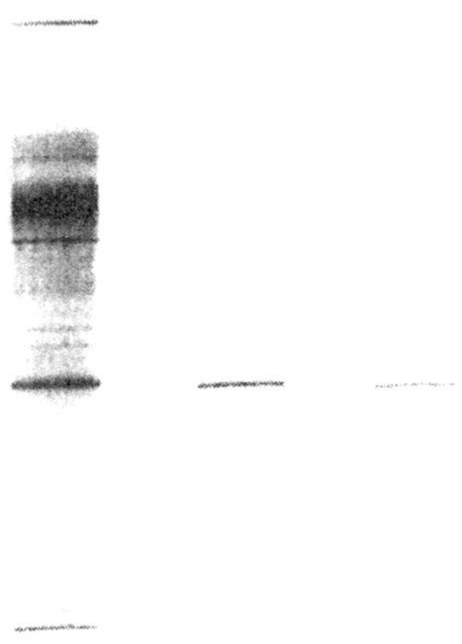

Figure I.1. The visible part of the spectrum of the hydrogen atom.©CNRS. Paris Observatory photo. L. Pamia.

With Planck, it has its origins in thermodynamics and manifests itself in the discretization of the energy levels of oscillators. With Bohr, it is the discretization of angular momentum. For de Broglie and Schrödinger, it is the wave nature of matter. These diverse points of view are all corollaries of that of Heisenberg: *physical quantities are governed by noncommutative algebra.* My first goal will be to show how close the latter point of view is to experimental reality. Towards the end of the 19th century, numerous experiments permitted determining with precision the lines of the emission spectra of the atoms that make up the elements. One considers a Geissler tube filled with a gas such as hydrogen. The light emitted by the tube is analyzed with a spectrometer, the simplest being a prism, and one obtains a certain number of lines, indexed by their wavelengths. The configuration thus obtained is the most direct source of information on the atomic structure and constitutes, as it were, the signature of the element under consideration. It depends only on the element considered, and characterizes it. It is thus essential to find the regularities that appear in these configurations or *atomic spectra.* It is hydrogen that, in conformity with Mendeleev's table, has the simplest spectrum (Figure 1).

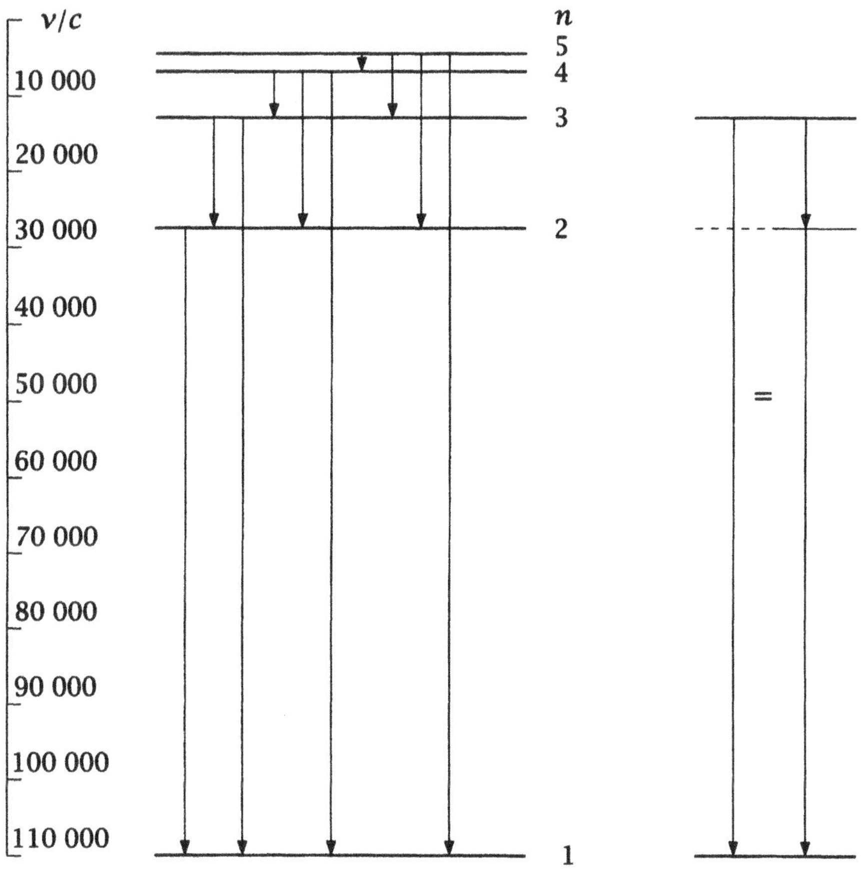

Figure I.2. The vertical arrows represent the transitions. They are indexed by a pair of indices (i, j).

The numerical expression of the regularity of the lines $H_\alpha, H_\beta, H_\gamma, \ldots$ was obtained by Balmer in 1885 in the form

$$H_\alpha = \frac{9}{5}L, \; H_\beta = \frac{16}{12}L, \; H_\gamma = \frac{25}{21}L, \; H_\delta = \frac{36}{32}L,$$

where the value of the length L is approximately 3645.6×10^{-8} cm. In other words, the wavelengths of the lines in Figure 1 are of the form

$$\lambda = \frac{n^2}{n^2 - 4} L, \text{ where } n \text{ is an integer equal to 3, 4, 5 or 6.}$$

Around 1890, Rydberg showed that for a complex atom the lines of the spectrum can be classified into series, each of them being of the form

$$\frac{1}{\lambda} = \frac{R}{m^2} - \frac{R}{n^2} \text{ with } n \text{ and } m \text{ integers, } m \text{ fixed.}$$

Here $R = 4/L$ is Rydberg's constant. From this experimental discovery one deduces that, on the one hand, the frequency $\nu = c/\lambda$ is a more natural parameter than the wavelength λ for indexing the lines of the spectrum and, on the other hand, the spectrum is a set of differences of frequencies, that is, there exists a set I of frequencies such that the spectrum is the set of differences $\nu_{ij} = \nu_i - \nu_j$ of arbitrary pairs of elements of I. This property shows that one can combine two frequencies ν_{ij} and ν_{jk} to obtain a third, $\nu_{ik} = \nu_{ij} + \nu_{jk}$. This important corollary is the Ritz-Rydberg combination principle: the spectrum is naturally endowed with a partially defined law of composition; the sum of certain frequencies of the spectrum is again a frequency of the spectrum (Figure 2).

Now, these experimental results could not be explained in the framework of the theoretical physics of the 19th century, based as it was on Newton's mechanics and Maxwell's electromagnetism. If one applies the classical conception of mechanics to the microscopic level, then an atom is described mathematically by the phase space and the Hamiltonian, and its interaction with radiation is described by Maxwell's theory. As we shall see, this theory predicts that the set of frequencies forms a subgroup Γ of $\mathbb{R}$.

Let us begin with the phase space. In the model of classical mechanics, to determine the later trajectory of a particle it is necessary to know both its initial position and its velocity. The initial data thus forms a set with six parameters, which are the three coordinates of position and the three coordinates of velocity, or rather of the momentum $p = m\dot{q}$. If one is interested in a number n of particles, it is necessary to know the position q and momentum of each of them. One is thus dealing with a set with $6n$ parameters, called the phase space of the mechanical system under consideration.

Starting with a function on this space which is called the Hamiltonian and which measures energy, classical mechanics prescribes the differential equations that determine the trajectory from the initial data. The natural structure of the phase space is that of a symplectic manifold X whose points are the "states" of the system. The Hamiltonian H is a function on X that intervenes to specify the evolution of every observable physical quantity, that is, of every function f on X, by the equation

$$\dot{f} = \{H, f\},$$

where $\{\ ,\ \}$ denotes the Poisson bracket and $\dot{f} = \frac{d}{dt} f$.

In the good cases such as, for example, the planetary model of the hydrogen atom, the dynamical system obtained is totally integrable. This means that there are sufficiently many "constants of motion" so that, on specifying them, the system is reduced to an almost periodic motion. The description of such a system is very simple. For, on the one hand, the algebra of observable quantities is the commutative algebra of almost periodic series

$$q(t) = \sum q_{n_1 \ldots n_k} \exp(2\pi i \langle n, \nu \rangle t),$$

where the n_i are integers, the ν_i are positive real numbers called the fundamental frequencies, and $\langle n, \nu \rangle = \sum n_i \nu_i$. On the other hand, the evolution in time is given by translation of the variable t.

The interaction between a classical atom and the electromagnetic field is described by Maxwell's theory. Such an atom emits an electromagnetic wave whose radiative part is calculated by superposing the plane waves W_n, $n = (n_1, \ldots, n_k)$, with frequencies $\langle n, \nu \rangle = \sum n_i \nu_i$, and whose amplitude and polarization are calculated simply from the fundamental observable that is the dipole moment. The dipole moment Q has three components Q_x, Q_y and Q_z, each of which is an observable quantity,

$$Q_x(t) = \sum q_{x,n} \exp(2\pi i \langle n, \nu \rangle t),$$

and which give the intensity of the emitted radiation of frequency $\langle n, \nu \rangle$ by the equation

$$I = \frac{dE}{dt} = \frac{2}{3c^3}(2\pi \langle n, \nu \rangle)^4 (|q_{x,n}|^2 + |q_{y,n}|^2 + |q_{z,n}|^2),$$

where c denotes the speed of light. It follows, in particular, that the set of frequencies of the emitted radiations is an additive subgroup, $\Gamma \subset \mathbb{R}$, of the real numbers. Thus, to each frequency emitted there are associated all of its integral multiples or harmonics.

In fact, spectroscopy and its numerous experimental results show that this last theoretical result is contradicted by experiment. The set of frequencies emitted by an atom does not form a group, and it is false that the sum of two frequencies of the spectrum is again one. What experiment dictates is the Ritz-Rydberg combination principle, which permits indexing the spectral lines by the set Δ of all pairs (i, j) of elements of a set I of indices. The frequencies ν_{ij} and $\nu_{k\ell}$ only combine when $j = k$ to yield $\nu_{i\ell} = \nu_{ij} + \nu_{j\ell}$. The theory of Bohr, by artificially discretizing the angular momentum of the electron, succeeds in predicting the frequencies of the radiations emitted by the hydrogen atom, but it is unable to predict their intensities and polarizations.

It is by a fundamental calling into question of classical mechanics that Heisenberg arrived at this goal and went well beyond his predecessors. This questioning of classical mechanics runs approximately as follows: in the classical model, the algebra of observable physical quantities can be directly read from the *group* Γ of emitted frequencies; it is the convolution algebra of this group of frequencies. Since Γ is a commutative group, the convolution algebra is commutative. Now, in reality one is not dealing with a group of frequencies but rather, due to the Ritz-Rydberg combination principle, with a groupoid $\Delta = \{(i,j);\ i,j \in I\}$ having the composition rule $(i,j)\cdot(j,k) = (i,k)$. The convolution algebra still has meaning when one passes from a group to a groupoid, and the convolution algebra of the groupoid Δ is none other than *the algebra of matrices* since the convolution product may be written

$$(ab)_{(i,k)} = \sum_j a_{(i,j)} b_{(j,k)},$$

which is identical with the product rule for matrices.

On replacing the commutative convolution algebra of the group Γ by the noncommutative convolution algebra of the groupoid Δ dictated by experimental results, Heisenberg replaced classical mechanics, in which the observable quantities commute pairwise, by *matrix mechanics*, in which observable quantities as important as position and momentum no longer commute. Heisenberg's rules of algebraic calculation were imposed on him by the experimental results of spectroscopy. However, Heisenberg did not understand right away that the algebra he was working with was already known to mathematicians and was called the algebra of matrices. It was Jordan and Born who noticed this. In fact, Jordan had remarked that the conditions which, in Heisenberg's formalism, correspond to the Bohr-Sommerfeld quantization rules, signified that the diagonal elements of the matrix $[p,q]$ were equal to $-i\hbar$. In Heisenberg's matrix mechanics, an observable physical quantity is given by its coefficients $q_{(i,j)}$, indexed by elements $(i,j) \in \Delta$ of the groupoid Δ. The time evolution of an observable is given by the homomorphism $(i,j) \in \Delta \mapsto \nu_{ij} \in \mathbb{R}$ of Δ into $\mathbb{R}$ which associates with each spectral line its frequency. One has

$$q_{(i,j)}(t) = q_{(i,j)} \exp(2\pi i \nu_{ij} t). \tag{1}$$

This formula is the analogue of the classical formula

$$q_{n_1\dots n_k}(t) = q_{n_1\dots n_k} \exp(2\pi i \langle n, \nu\rangle t).$$

To obtain the analogue of Hamilton's law of evolution

$$\frac{d}{dt} q = \{H, q\},$$

one defines a particular physical quantity H that plays the role of the classical energy and is given by its coefficients $H_{(i,j)}$, with

$$H_{(i,j)} = 0 \text{ if } i \neq j,\ H_{(i,i)} = h\nu_i, \text{ where } \nu_i - \nu_j = \nu_{ij}\ \forall\ i,j \in I,$$

where h is Planck's constant, a factor that converts frequencies into energies. One sees that H is defined uniquely, up to the addition of a multiple of the identity matrix. Moreover, the above formula (1) is equivalent to

$$\frac{d}{dt} q = \frac{2\pi i}{h}(Hq - qH). \tag{2}$$

This equation is similar to the one of Hamilton that uses the Poisson brackets. It is in fact simpler, since it only uses the product of the observables, and more precisely the commutator $[A, B] = AB - BA$, which plays the role that the Poisson bracket plays in Hamiltonian mechanics. By analogy with classical mechanics, one requires the observables q of position and p of momentum to satisfy $[p, q] = -i\hbar$, where $\hbar = h/2\pi$. The simple algebraic form of the classical energy as a function of p and q then gives the equation of Schrödinger for determining the set $\{v_i;\ i \in I\}$, or the spectrum of H.

The quantum system thus described is much simpler and more rigid than its classical analogue. One thus obtains a nonnegligible payoff for abandoning the commutativity of classical mechanics. Though less intuitive, quantum mechanics is more directly accessible by virtue of its simplicity and its contact with spectroscopy.

In fact, the results of the theory of $*$-products show that a *symplectic* structure on a manifold such as the phase space is none other than the indication of the existence of a deformation with one parameter (h here) of the algebra of functions into a noncommutative algebra. I refer the reader to the literature ([Arn-C-M-P], [B-F-F-L-S_1], [B-F-F-L-S_2], [DeW-L], [Dr], [Ger-S], [Li], [Ri_3], [Vey]) for a description of the results of this theory.

I.2. Statistical State of a Macroscopic System and Quantum Statistical Mechanics

A cubic centimeter of water contains on the order of $N = 3 \times 10^{22}$ molecules of water agitated by an incessant movement. The detailed description of the motion of each molecule is not necessary, any more than is the precise knowledge of the microscopic state of the system, to determine the results of macroscopic observations. In classical statistical mechanics, a microscopic state of the system is represented by a point of the phase space, which is of dimension $6N$ for N point molecules. A statistical state is described not by a point of the phase space but by a measure μ on that space, a measure that associates with each observable f its mean value

$$\int f \, d\mu .$$

For a system that is maintained at fixed temperature by means of a thermostat, the measure μ is called the Gibbs canonical ensemble; it is given by a formula that involves the Hamiltonian H of the system and the Liouville measure that arises from the symplectic structure of the phase space. One sets

$$d\mu = \frac{1}{Z} e^{-\beta H} \cdot \text{Liouville measure}, \tag{3}$$

where $\beta = 1/kT$, T being the absolute temperature and k the Boltzmann constant, whose value is approximately 1.38×10^{-23} joules per degree Kelvin, and where Z is a normalization factor.

The thermodynamic quantities, such as the entropy or the free energy, are calculated as functions of β and a small number of macroscopic parameters introduced in the formula that gives the Hamiltonian H. For a finite system, the free energy is an analytic function of these parameters. For an infinite system, some discontinuities appear that correspond to the phenomenon of phase transition. The rigorous proof, starting with the mathematical formula that specifies H, of the absence or existence of these discontinuities is a difficult branch of mathematical analysis ([Rue$_2$]).

However, as we have seen, the microscopic description of matter cannot be carried out without quantum mechanics. Let us consider, to fix our ideas, a solid having an atom at each vertex of a crystal lattice $\mathbb{Z}^3$. The algebra of observable physical quantities associated with each atom $x = (x_1, x_2, x_3)$ is a matrix algebra Q_x, and if we assume for simplicity that these atoms are of the same nature and can only occupy a finite number n of quantum states, then $Q_x = M_n(\mathbb{C})$ for every x. Now let Λ be a finite subset of the lattice. The algebra Q_Λ of observable physical quantities for the system formed by the atoms contained in Λ is given by the tensor product $Q_\Lambda = \bigotimes_{x\in\Lambda} Q_x$.

The Hamiltonian H_Λ of this finite system is a self-adjoint matrix that is typically of the form

$$H_\Lambda = \sum_{x\in\Lambda} H_x + \lambda H_{\text{int}},$$

where the first term corresponds to the absence of interactions between distinct atoms, and where λ is a coupling constant that governs the intensity of the interaction. A statistical state of the finite system Λ is given by a linear form φ that associates with each observable $A \in Q_\Lambda$ its mean value $\varphi(A)$ and which has the same positivity and normalization properties as a probability measure μ, namely,

a) Positivity: $\varphi(A^*A) \geq 0 \quad \forall\, A \in Q_\Lambda$;

b) Normalization: $\varphi(1) = 1$.

If the system is maintained at fixed temperature T, the equilibrium state is given by the quantum analogue of the above formula (3)

$$\varphi(A) = \frac{1}{Z} \text{trace}(e^{-\beta H_\Lambda} A) \;\; \forall\, A \in Q_\Lambda, \tag{4}$$

where the unique trace on the algebra Q_Λ replaces the Liouville measure.

As in classical statistical mechanics, the interesting phenomena appear when one passes to the thermodynamic limit, that is, when $\Lambda \to \mathbb{Z}^3$. A state of the infinite system being given by the family (φ_Λ) of its restrictions to the finite systems indexed by Λ, one obtains in this way all the families such that

a) for every Λ, φ_Λ is a state on Q_Λ;

b) if $\Lambda_1 \subset \Lambda_2$ then the restriction of φ_{Λ_2} to Q_{Λ_1} is equal to φ_{Λ_1}.

In general, the family φ_Λ defined above by means of $\exp(-\beta H_\Lambda)$ does not satisfy the condition b) and it is necessary to understand better the concept of state of an infinite system. This is where C^*-algebras make their appearance. In fact, if one takes the inductive limit Q of the finite-dimensional C^*-algebras Q_Λ, one obtains a C^*-algebra that has the following property:

An arbitrary state φ on Q is given by a family (φ_Λ) satisfying the conditions a) *and* b).

Thus, the families (φ_Λ) satisfying a) and b), that is the states of the infinite system, are in natural bijective correspondence with the states of the C^*-algebra Q. Moreover, the family (H_Λ) uniquely determines a one-parameter group (α_t) of automorphisms of the C^*-algebra Q by the equation

$$\frac{d}{dt}\alpha_t(A) = \lim_{\Lambda \to \mathbb{Z}^3} \frac{2\pi i}{h}[H_\Lambda, A].$$

This one-parameter group gives the time evolution of the observables of the infinite system that are given by the elements A of Q, and is calculated by passing to the limit, starting from Heisenberg's formula. For a finite system, maintained at temperature T, the formula (4) gives the equilibrium state in a unique manner as a function of H_Λ, but in the thermodynamic limit one cannot have a simple correspondence between the Hamiltonian of the system, or, if one prefers, the group of time evolution, and the equilibrium state of the system. Indeed, during phase transitions, distinct states can coexist, which precludes uniqueness of the equilibrium state as a function of the group (α_t). It is impossible to give a simple formula that would define in a unique manner the equilibrium state as a function of the one-parameter group (α_t). In compensation, there does exist a relation between a state φ on Q and the one-parameter group (α_t), that does not always uniquely specify φ from the knowledge of α_t, but which is the analogue of the formula (4). This relation is the *Kubo-Martin-Schwinger condition* ([Ku], [Mart-S]) as formulated by R. Haag, N. Hugenholtz and M. Winnink [H-H-W]:

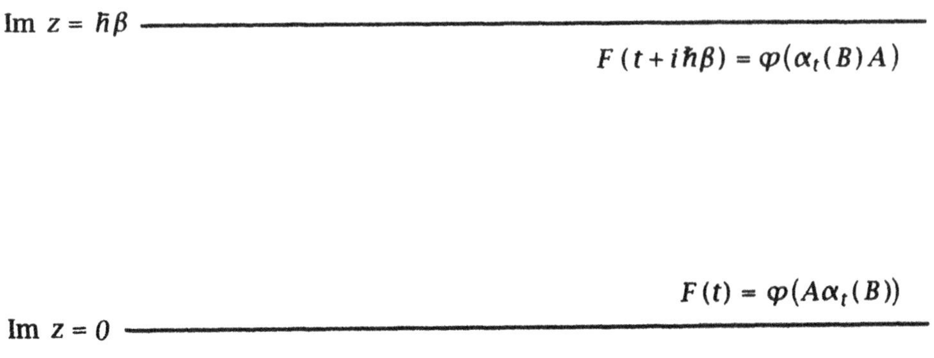

Figure I.3. The function F is holomorphic in the unshaded strip and connects $\varphi(A\alpha_t(B))$ and $\varphi(\alpha_t(B)A)$.

Given T, a state φ on Q, and the one-parameter group (α_t) of automorphisms of Q satisfy the KMS-condition if and only if for every pair A, B of elements of Q there exists a function $F(z)$ holomorphic in the strip $\{z \in \mathbb{C};\ \text{Im } z \in [0, \hbar\beta]\}$ such that (Figure 3)

$$F(t) = \varphi(A\alpha_t(B)), \quad F(t + i\hbar\beta) = \varphi(\alpha_t(B)A) \quad (\forall\, t \in \mathbb{R}).$$

Here t is a time parameter, as is $\hbar\beta = \hbar/kT$ which, for $T = 1°$ K, has value approximately 10^{-11}s.

This condition allows us to formulate mathematically, in quantum statistical mechanics, the problem of the coexistence of distinct phases at given temperature T, that is, the problem of the uniqueness of φ, given (α_t) and β. We shall give, in Chapter V Section 11, an explicit example of a phase transition with spontaneous symmetry breaking coming from the statistical theory of prime numbers.

This same condition has played an essential role in the modular theory of operator algebras. It has thus become an indisputable point of interaction between theoretical physics and pure mathematics.

I.3. Modular Theory and the Classification of Factors

Between 1957 and 1967, a Japanese mathematician, Minoru Tomita, who was motivated in particular by the harmonic analysis of nonunimodular locally compact groups, proved a theorem of considerable importance for the theory of von Neumann algebras. His original manuscript was very hard to decipher, and his results would have remained unknown without the lecture notes of M. Takesaki [T2], who also contributed greatly to the theory.

Before giving the technical definition of a von Neumann algebra, it must be explained that the theory of commutative von Neumann algebras is equivalent to Lebesgue's measure theory and to the spectral theorem for self-adjoint operators. The noncommutative theory was elaborated at the outset by Murray and von Neumann, quantum mechanics being one of their motivations. The theory of noncommutative von Neumann algebras only achieved its maturity with the modular theory; it now constitutes an indispensable tool in the analysis of noncommutative spaces.

A von Neumann algebra is an involutive subalgebra M of the algebra of operators on a Hilbert space $\mathcal{H}$ that has the property of being the commutant of its commutant: $(M')' = M$.

This property is equivalent to saying that M is an involutive algebra of operators that is closed under weak limits. To see intuitively what the equality $(M')' = M$ means, it suffices to say that it characterizes the algebras of operators on Hilbert space that are invariant under a group of unitary operators: The commutant of any subgroup of the unitary group of the Hilbert space is a von Neumann algebra, and they are all of that form (given M take as a subgroup, the unitary group of M'). In the general noncommutative case, the classical notion of probability measure is replaced by the notion of state. A typical state on the algebra M is given by a linear form $\varphi(A) = \langle A\xi, \xi\rangle$, where ξ is a vector of length 1 in the Hilbert space. Tomita's theory, which has as an ancestor the notion of quasi-Hilbert algebra ([Di2]), consists in analyzing, given a von Neumann algebra M on the Hilbert space $\mathcal{H}$ and a vector $\xi \subset \mathcal{H}$ such that $M\xi$ and $M'\xi$ are dense in $\mathcal{H}$, the following unbounded operator S:

$$Sx\xi = x^*\xi \quad \forall x \in M.$$

This is an operator with dense domain in $\mathcal{H}$ that is conjugate-linear; the results of the theory are as follows:

1) S is closable and equal to its inverse.
2) The phase $J = S|S|^{-1}$ of S satisfies $JMJ = M'$.
3) The modulus squared $\Delta = |S|^2 = S^*S$ of S satisfies $\Delta^{it}M\Delta^{-it} = M$ for every $t \in \mathbb{R}$.

Thus, to every state φ on M one associates a one-parameter group (σ_t^{φ}) of automorphisms of M, given by $\sigma_t^{\varphi}(x) = \Delta^{it}x\Delta^{-it}$ $(\forall x \in M)$ $(\forall t \in \mathbb{R})$, the group of modular automorphisms of φ. It is precisely at this point that the interaction between theoretical physics and pure mathematics takes place. Indeed, Takesaki and Winnink showed simultaneously [T$_2$][Winn] that the connection between the state φ and the one-parameter group (σ_{-t}^{φ}) of Tomita's theorem is exactly the KMS condition for $\hbar\beta = 1$.

These results, as well as the work of R. Powers [Pow$_1$], and of H. Araki and E. J. Woods [Ar-W] on factors that are infinite tensor products, proved to be of considerable importance in setting in motion the classification of factors.

The point of departure of my work on the classification of factors was the discovery of the relation between the Araki-Woods invariants and Tomita's theory. For this it had to be shown that the evolution group associated with a state by that theory harbored properties of the algebra M independent of the particular choice of the state φ.

A von Neumann algebra is far from having just one state φ, which has as consequence that only the properties of σ_t^{φ} that do not depend on the choice of φ have real significance for M. The crucial result that allowed me to get the classification of factors going is the following analogue of the Radon-Nikodým theorem [Co$_4$]:

For every pair φ, ψ of states on M, there exists a canonical 1-cocycle

$$t \to u_t, \qquad u_{t_1+t_2} = u_{t_1}\sigma_{t_1}^{\varphi}(u_{t_2}) \qquad \forall t_1, t_2 \in \mathbb{R}$$

with values in the unitary group of M, such that

$$\sigma_t^{\psi}(x) = u_t \sigma_t^{\varphi}(x) u_t^* \qquad \forall x \in M, \ \forall t \in \mathbb{R}.$$

Moreover, $\sqrt{-1}\left(\frac{d}{dt}u_t\right)_{t=0}$ coincides

1) in the commutative case, with the logarithm of the Radon-Nikodým derivative $(d\psi/d\varphi)$;
2) in the case of statistical mechanics, with the difference of the Hamiltonians corresponding to two equilibrium states, or the relative Hamiltonian of Araki [Ar$_2$].

It follows that, given a von Neumann algebra M, there exists a canonical homomorphism δ of $\mathbb{R}$ into the group $\operatorname{Out} M = \operatorname{Aut} M / \operatorname{Inn} M$ (the quotient of the automorphism group by the normal subgroup of inner automorphisms), given by the class of σ_t^{φ}, independently of the choice of φ. Thus, $\ker \delta = T(M)$ is an invariant of M, as is $\operatorname{Sp} \delta = S(M) = \bigcap_{\varphi} \operatorname{Sp} \Delta_{\varphi}$.

Thus von Neumann algebras are *dynamical* objects. Such an algebra possesses a group of automorphism classes parametrized by $\mathbb{R}$. This group, which is completely canonical, is a manifestation of the noncommutativity of the algebra M. It has no counterpart in the commutative case and attests to the originality of noncommutative measure theory with respect to the usual theory.

Twenty years after Tomita's theorem, and after considerable work (for more details see Chapter V), we now have at our disposal a complete classification of all the hyperfinite von Neumann algebras. Rather than give a definition of this class, let us simply note that:

1) If G is a connected Lie group and $\pi \in \operatorname{Rep} G$ is a unitary representation of G, then its commutant $\pi(G)'$ is hyperfinite.

2) If Γ is an amenable discrete group and $\pi \in \operatorname{Rep}\Gamma$, then $\pi(\Gamma)'$ is hyperfinite.

3) If a C^*-algebra A is an inductive limit of finite-dimensional algebras and if $\pi \in \operatorname{Rep} A$, then $\pi(A)''$ is hyperfinite.

Moreover, the classification of hyperfinite von Neumann algebras reduces to that of the hyperfinite factors on writing $M = \int M_t d\mu(t)$, where each M_t is a factor, that is, has center equal to $\mathbb{C}$. Finally, the list of hyperfinite factors is as follows:

I_n $M = M_n(\mathbb{C})$.

I_∞ $M = \mathcal{L}(\mathcal{H})$, the algebra of all operators on an infinite-dimensional Hilbert space.

II_1 $R = \operatorname{Cliff}_{\mathbb{C}}(E)$, the Clifford algebra of an infinite-dimensional Euclidean space E.

II_∞ $R_{0,1} = R \otimes \mathrm{I}_\infty$.

III_λ $R_\lambda =$ the Powers factors ($\lambda \in]0,1[$).

III_1 $R_\infty = R_{\lambda_1} \otimes R_{\lambda_2}$ ($\forall \lambda_1, \lambda_2$, $\lambda_1/\lambda_2 \notin \mathbb{Q}$), the Araki–Woods factor.

III_0 R_W, the Krieger factor associated with an ergodic flow W.

After my own work, case III_1 was the only one that remained to be elucidated. U. Haagerup has since shown that all the hyperfinite factors of type III_1 are isomorphic. All these results are explained in great detail in Chapter V.

I.4. Geometric Examples of von Neumann Algebras: Measure Theory of Noncommutative Spaces

My aim in this section is to show by means of examples that the theory of von Neumann algebras replaces ordinary measure theory when one has to deal with noncommutative spaces. It will allow us to analyse such spaces even though they appear singular when considered from the classical point of view, i.e. when investigated using measurable real-valued functions.

We shall first briefly review the classical Lebesgue measure theory and explain in particular its intimate relation with *commutative* von Neumann algebras. We then give the general construction of the von Neumann algebras of the noncommutative spaces X which arise naturally in differential geometry, namely the spaces of leaves of foliations. Our first use of this construction will be to illustrate the classification expounded above by numerous geometric examples.

Figure I.4A. The Lady in Blue. Gainsborough's portrait of the Duchess of Beaufort. Photograph by Giraudon of the original at the Hermitage, St. Petersburg, used by their permission.

4.α Classical Lebesgue measure theory

H. Lebesgue was the first to succeed in defining the integral $\int_a^b f(x)dx$ of a bounded function of a real variable x without imposing any serious restrictions on f. At a technical level, for the definition to make sense it is necessary to require that the function f be measurable. However, this measurability condition is so little restrictive that one has to use the uncountable axiom of choice to prove the existence of nonmeasurable functions. In fact, a very instructive debate took place in 1905 between Borel, Baire, and Lebesgue on the one hand, and Hadamard (and Zermelo) on the other, as to the "existence" of a well ordering on the real line (see Lebesgue's letter in Appendix C). A result of the logician Solovay shows that (modulo the existence of strongly inaccessible cardinals) a nonmeasurable function cannot be constructed using only the axiom of conditional choice.

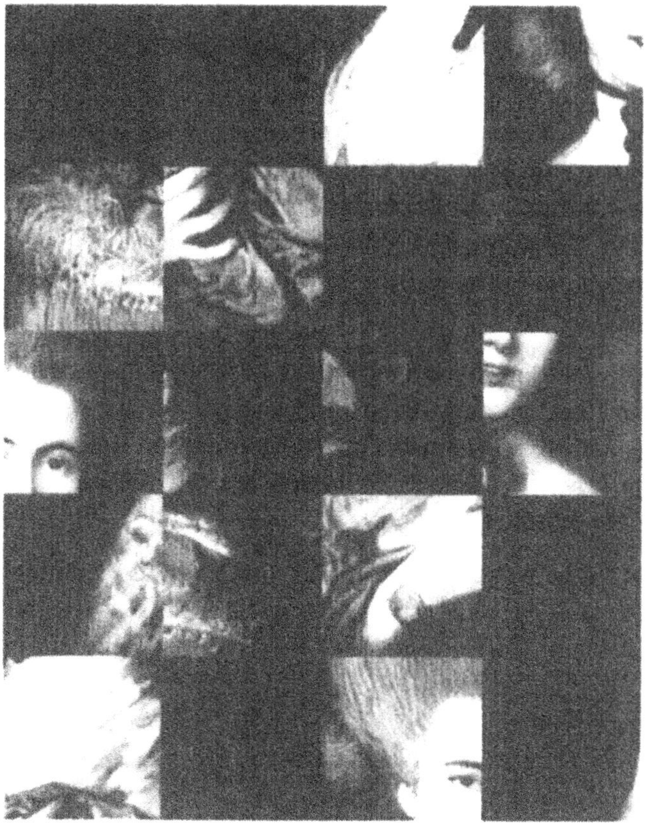

Figure I.4B. The Permuted Lady in Blue. See Figure I.4A.

This result on measurability still holds if the interval $[a, b]$ is replaced by a standard Borel space X. It shows, in particular, that among the classical structures on a set X obtained by specifying a class of functions, $f : X \to \mathbb{R}$, such as continuous or smooth functions, the *measure space* structure, obtained by specifying the measurable functions, is the coarsest possible.

In particular, a measure space X is not modified by a transformation T such as the one indicated in Figure 4, which consists in making the picture explode like a jigsaw puzzle whose pieces are scattered. This transformation has the effect of destroying all shape; nevertheless, a particular subset A of X does not change in measure even though it may be scattered into several pieces.

This wealth of possible transformations of X is bound up with the existence, up to isomorphism, of only one interesting measure space: an interval equipped with Lebesgue measure. Even spaces as complex in appearance as a functional space of distributions equipped with a functional measure are in fact isomorphic to it as measure spaces. In particular, the notion of dimension has no intrinsic meaning in measure theory. The key tools of this theory are positivity and the completeness of the Hilbert space $L^2(X, \mu)$ of square-integrable functions f, with $\int |f|^2 d\mu < \infty$ on X. A crucial result is the Radon-Nikodým theorem, by which the derivative $d\mu/d\nu$ of one measure with respect to another may be defined as a function on X.

The topology of *compact* metrizable spaces X has a remarkable compatibility with measure theory. In fact, the finite Borel measures on X correspond exactly to the continuous linear forms on the Banach space $C(X)$ of continuous functions on X, equipped with the norm $\|f\| = \sup_{x\in X} |f(x)|$. The positive measures correspond to the positive linear forms, i.e. the linear forms φ such that $\varphi\left(\bar{f}f\right) \geq 0$ for all $f \in C(X)$.

To go into matters more deeply, it is necessary to understand how this theory arises naturally from the spectral analysis of selfadjoint operators in Hilbert space, and thus becomes a special commutative case of the theory of von Neumann algebras (Chapter 5), which is itself the natural extension of *linear algebra* to infinite dimensions.

Thus, let $\mathcal{H}$ be a (separable) Hilbert space and T a bounded selfadjoint operator on $\mathcal{H}$, i.e. $\langle T\xi, \eta\rangle = \langle \xi, T\eta\rangle \ \forall \xi, \eta \in \mathcal{H}$. Then if $p(u)$ is a polynomial of the real variable u, $p(u) = a_0 u^n + \cdots + a_n$, one can define the operator $p(T)$ as

$$p(T) = \sum_{k=0}^{n} a_k \, T^{n-k}.$$

It is rather amazing that this definition of $p(T)$ extends by continuity to all bounded Borel functions f of a real variable u. This extension is uniquely given by the condition

$$\langle f_n(T)\xi, \eta\rangle \to \langle f(T)\xi, \eta\rangle \qquad \forall \xi, \eta \in \mathcal{H},$$

if the sequence f_n converges *simply* to f, i.e. $f_n(u) \to f(u)$ for all $u \in \mathbb{R}$.

There exists then a unique *measure class* μ on $\mathbb{R}$, carried by the compact interval $I = [-\|T\|, \|T\|]$, such that

$$f(T) = 0 \Longleftrightarrow \int |f| d\mu = 0.$$

Moreover, the algebra M of operators on $\mathcal{H}$ of the form $f(T)$ for some bounded Borel function f is a *von Neumann algebra*, and is the von Neumann algebra generated by T. In other words if S is a bounded operator on $\mathcal{H}$ which has the same symmetries as T (i.e. which commutes with all unitary operators U, $U^*U = UU^* = 1$ on $\mathcal{H}$ which fix T, so $UTU^* = T$), then there exists a bounded Borel function f with $S = f(T)$. This commutative von Neumann algebra is naturally isomorphic to $L^\infty(I, \mu)$, the algebra of classes modulo equality almost everywhere, of bounded measurable functions on the interval I.

4.β Foliations

We shall now describe a large class of noncommutative spaces arising from differential geometry, see why the Lebesgue theory is not able to analyse such spaces, and replace it by the theory of noncommutative von Neumann algebras.

The spaces X considered are spaces of manifolds which are solutions of a differential equation, i.e. spaces of leaves of a foliation.

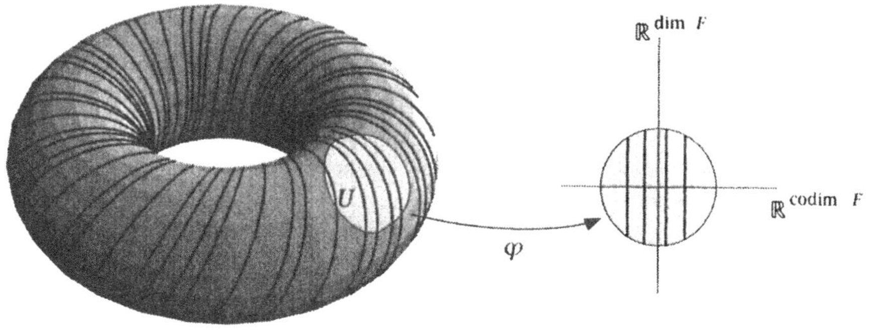

Figure I.5. Foliation

Let V be a smooth manifold and TV its tangent bundle, so that for each $x \in V$, T_xV is the tangent space of V at x. A smooth subbundle F of TV is called *integrable* iff one of the following equivalent conditions is satisfied:

a) Every $x \in V$ is contained in a submanifold W of V such that

$$T_y(W) = F_y \qquad \forall y \in W.$$

b) Every $x \in V$ is in the domain $U \subset V$ of a submersion $p : U \to \mathbb{R}^q$ ($q = \operatorname{codim} F$) with $F_y = \ker(p_*)_y$ $\forall y \in U$.

c) $C^\infty(V,F) = \{X \in C^\infty(V,TV); X_x \in F_x \ \forall x \in V\}$ is a Lie subalgebra of the Lie algebra of vector fields on V.

d) The ideal $J(F)$ of smooth differential forms which vanish on F is stable under differentiation: $dJ \subset J$.

Any 1-dimensional subbundle F of TV is integrable, but for $\dim F \geq 2$ the condition is nontrivial; for instance, if $P \xrightarrow{p} B$ is a principal H-bundle, with compact structure group H, the bundle of horizontal vectors for a given connection is integrable iff this connection is flat.

A foliation of V is given by an integrable subbundle F of TV. The leaves of the foliation (V,F) are the maximal connected submanifolds L of V with $T_x(L) = F_x$ $\forall x \in L$, and the partition of V into leaves $V = \bigcup L_\alpha, \alpha \in A$, is characterized geometrically by its "local triviality": every point $x \in V$ has a neighborhood U and a system of local coordinates $(x^j)_{j=1,\ldots,\dim V}$, which is called a *foliation chart*, so that the partition of U into connected components of leaves, called *plaques* (they are the leaves of the restriction of the foliation to the open set U), corresponds to the partition of $\mathbb{R}^{\dim V} = \mathbb{R}^{\dim F} \times \mathbb{R}^{\operatorname{codim} F}$ into the parallel affine subspaces $\mathbb{R}^{\dim F} \times \mathrm{pt}$ (cf. Figure 5).

The manifolds L which are the leaves of the foliation are defined in a fairly implicit manner, and the simplest examples show that:

1) Even though the ambient manifold V is compact, the leaves L can fail to be compact.

2) The space X of leaves can fail to be Hausdorff for the quotient topology.

For instance, let V be the two-dimensional torus $V = \mathbb{R}^2/\mathbb{Z}^2$, with the Kronecker foliation associated to a real number θ, i.e. given by the differential equation

$$dy = \theta dx.$$

Then if θ is an irrational number, the leaves L are diffeomorphic to $\mathbb{R}$, while the quotient topology on the space X of leaves is the same as the quotient topology of $S^1 = \mathbb{R}/\mathbb{Z}$ divided by the partition into orbits of the rotation given by $\alpha \in S^1 \overset{R_\theta}{\mapsto} \alpha + \theta$, and is thus the coarse topology. Thus there are no open sets in X except $\emptyset$ and X. Similarly, the ergodicity of the rotation R_θ on S^1 shows that if we endow X with the quotient measure class of the Lebesgue measure class on V we get the following "pathological" behaviour: any measurable function $f : X \to \mathbb{R}$ is almost everywhere equal to a constant. This implies that classical measure theory does not distinguish between X and a one-point space, and, in particular, the L^p-spaces of analysis, $L^p(X)$, are one-dimensional, $L^p(X) = \mathbb{C}$, and essentially useless.

4.γ The von Neumann algebra of a foliation

Let (V, F) be a foliated manifold. We shall now construct a von Neumann algebra $W(V, F)$ canonically associated to (V, F) and depending only on the Lebesgue measure class on the space X of leaves of the foliation. The classical point of view, $L^\infty(X)$, will only give the *center* $Z(W)$ of W.

The basic idea of the construction is that while in general we cannot find on X interesting scalar-valued functions, there are always, as we shall see, plenty of operator-valued functions on X. In other words we just replace c-numbers by q-numbers. To be more precise, let us denote, for each leaf ℓ of (V, F), by $L^2(\ell)$ the canonical Hilbert space of square-integrable half-densities ([Gu-S]) on ℓ.

We assume here to simplify the discussion that the set of leaves with nontrivial holonomy is Lebesgue negligible. In general (cf. Chapter II) one just replaces ℓ by $\tilde{\ell}$, its holonomy covering.

Figure I.6. Kronecker foliation: $dy = \theta dx$

Definition 1. *A random operator* $q = (q_\ell)_{\ell \in X}$ *is a measurable map* $\ell \to q_\ell$ *which associates to each leaf* $\ell \in X$ *a bounded operator* q_ℓ *in the Hilbert space* $L^2(\ell)$.

To define the *measurability* of random operators we note the following simple facts:

a) Let $\lambda : V \to X$ be the canonical projection which to each $x \in V$ assigns the unique leaf $\ell = \lambda(x)$ passing through x. Then the bundle $(L^2(\lambda(x)))_{x \in V}$ of Hilbert spaces over V is *measurable.* More specifically, consider the Borel subset of $V \times V$, $G = \{(x, y) \in V \times V \; ; \; y \in \lambda(x)\}$. Then, modulo the irrelevant choice of a $\frac{1}{2}$-density $|dy|^{1/2}$ along the leaves, the sections $(\xi_x)_{x \in V}$ of the bundle $(L^2(\lambda(x)))_{x \in V}$ are just scalar functions on G and measurability has its ordinary meaning.

b) A map $\ell \to q_\ell$ as in Definition 1 defines by composition with λ an endomorphism $(q_{\lambda(x)})_{x \in V}$ of the bundle $(L^2(\lambda(x)))_{x \in V}$. We shall then say that $q = (q_\ell)_{\ell \in X}$ is *measurable* iff the corresponding endomorphism $(q_{\lambda(x)})_{x \in V}$ is measurable in the usual sense, i.e. iff for any pair of measurable sections $(\xi_x)_{x \in V}$, $(\eta_x)_{x \in V}$ of $(L^2(\lambda(x)))_{x \in V}$ the following function on V is Lebesgue measurable:

$$x \in V \to \langle q_{\lambda(x)}\, \xi_x\, ,\, \eta_x \rangle \in \mathbb{C}.$$

There are many equivalent ways of defining measurability of random operators. As we already noted in Section α), any natural construction of families $(q_\ell)_{\ell \in X}$ will automatically satisfy the above measurability condition.

Let us give some examples of random operators.

1) Let f be a bounded Borel function on V. Then for each leaf $\ell \in X$ let $\tilde{f}_\ell$ be the multiplication operator

$$(\tilde{f}_\ell\, \xi)(x) = f(x)\, \xi(x) \qquad \forall x \in \ell\, ,\ \forall \xi \in L^2(\ell).$$

This defines a random operator $\tilde{f} = (\tilde{f}_\ell)_{\ell \in X}$.

2) Let $X \in C^\infty(V,F)$ be a real vector field on V tangent to the leaves of the foliation. Then let $\psi_t = \exp(tX)$ be the associated group of diffeomorphisms of V. (Assume for instance that V is compact to ensure the existence of ψ_t.) By construction, $\psi_t(x) \in \lambda(x)$ for any $x \in V$ so that it defines for each ℓ a diffeomorphism of ℓ. Let $(U_\ell)_{\ell \in X}$ be the corresponding family of unitaries (the map $\ell \to L^2(\ell)$ is functorial). It is a random operator $U = (U_\ell)_{\ell \in X}$.

Let $q = (q_\ell)_{\ell \in X}$ be a random operator. The operator norm $\|q_\ell\|$ of q_ℓ in $L^2(\ell)$ defines a measurable function on V by composition with $\lambda : V \to X$ which gives meaning to the norm

$$\|q\| = \text{Essential Supremum of } \|q_\ell\|\, ,\ \ell \in X. \qquad (*)$$

We say that q is zero almost everywhere iff $q \circ \lambda$ is so.

The von Neumann algebra $W(V,F)$ of a foliation is obtained as follows:

Proposition 2. *The classes of bounded random operators $(q_\ell)_{\ell \in X}$ modulo equality almost everywhere, endowed with the following algebraic rules, form a von Neumann algebra $W(V,F)$:*

$$\begin{aligned} (p+q)_\ell &= p_\ell + q_\ell & \forall \ell \in X \\ (pq)_\ell &= p_\ell q_\ell & \forall \ell \in X \\ (p^*)_\ell &= (p_\ell)^* & \forall \ell \in X. \end{aligned}$$

The norm, uniquely defined by the involutive algebra structure, is given by $(*)$. We have used here the possibility of defining a von Neumann algebra without a specific representation in Hilbert space, cf. Chapter V. If one wishes to realise $W(V,F)$ concretely as operators on a Hilbert space, one can, for instance, let random operators act by

$$(q\xi)_s = q_{\lambda(s)} \xi_s \quad \forall s \in V$$

in the Hilbert space $\mathcal{H}$ of square-integrable sections $(\xi_s)_{s \in V}$ of the bundle $\lambda^* L^2$ of Hilbert spaces on V. One can also use the restriction of $\lambda^* L^2$ to a sufficiently large transverse submanifold T of V. The invariants of von Neumann algebras are independent of the choice of a specific representation in Hilbert space and depend only upon the algebraic structure.

The von Neumann algebra $W(V,F)$ only depends upon the space X of leaves with its Lebesgue measure class; one has

Proposition 3. *Two foliations (V_i, F_i) with the same leaf space have isomorphic von Neumann algebras: $W(V_1,F_1) \simeq W(V_2,F_2)$.*

To be more precise one has to require in Proposition 3 that $\dim F_i > 0$ and that the assumed isomorphism $V_1/F_1 \simeq V_2/F_2$ be given by a bijection $\psi : V_1/F_1 \to V_2/F_2$ which preserves the Lebesgue measure class and is Borel in the sense that $\{(x,y) \in V_1 \times V_2\ ;\ \psi(\lambda_1(x)) = \lambda_2(y)\}$ is a Borel subset of $V_1 \times V_2$, with $\lambda_i : V_i \to V_i/F_i$ the quotient map.

For any von Neumann algebra M, its center

$$Z(M) = \{x \in M\ ;\ xy = yx \quad \forall y \in M\}$$

is a commutative von Neumann subalgebra of M.

In the above context one has the easily proved

Proposition 4. *Let (V,F) be a foliated manifold; then the construction 1) of random operators yields an isomorphism*

$$L^\infty(X) \simeq Z(W(V,F))$$

of the algebra of bounded measurable (classes of) functions on $X = V/F$ with the center of $W(V,F)$.

In particular $W(V,F)$ is a factor iff the foliation is ergodic.

We shall now illustrate the type classification of von Neumann algebras by geometric examples given by foliations. This will allow a non-algebraically minded reader to form a mental picture of these notions.

Type I von Neumann algebras

By definition a von Neumann algebra M is of type I iff it is algebraically isomorphic to the commutant of a *commutative* von Neumann algebra. Thus type I is the stable form of the hypothesis of commutativity. We refer to Chapter V for the easy classification of type I von Neumann algebras (with separable preduals) as direct integrals of matrix algebras $M_n(\mathbb{C})$, $n \in \{1,2,\ldots,\infty\}$.

A simple criterion ensuring a von Neumann algebra M be of type I is that it contains an abelian projection e with central support equal to 1. A projection $e = e^* = e^2 \in M$ is called *abelian* iff the reduced von Neumann algebra

$$M_e = \{x \in M\ ;\ xe = ex = x\}$$

is commutative. The central support of e is equal to 1 iff one has $ex \neq 0$ for any $x \in Z(M)$, $x \neq 0$.

Using this criterion one easily gets that the von Neumann algebra of a foliation (V, F) is of type I iff the leaf space X is an ordinary measure space. More precisely:

Proposition 5. *Let (V, F) be a foliated manifold. Then the associated von Neumann algebra is of type I iff one of the following equivalent conditions is satisfied.*

α) The quotient map $\lambda : V \to X$ admits a Lebesgue measurable section.

β) The space of leaves X, with its quotient structure, is, up to a null set, a standard Borel space.

A measurable section α of λ is a measurable map from V to V constant along the leaves and such that

$$\alpha(x) \in \lambda(x) \qquad \forall x \in V.$$

In general, an abelian projection $e \in W(V, F)$ is given by a measurable section $\ell \to \xi_\ell \in L^2(\ell)$ of the bundle $(L^2(\ell))_{\ell \in X}$ such that $\|\xi_\ell\| \in \{0, 1\} \quad \forall \ell \in X$. The formula for the random operator $e = (e_\ell)_{\ell \in X}$ is then

$$e_\ell \, \eta = \langle \eta, \xi_\ell \rangle \xi_\ell \qquad \forall \eta \in L^2(\ell) \, , \; \forall \ell \in X.$$

The central support of e is equal to 1 iff $\|\xi_\ell\| = 1$ a.e. The above type I property rarely holds for a foliation. It does, of course, for fibrations or if all leaves are compact; but it also holds for some foliations whose leaves are not compact, such as the Reeb foliations (Figure 7). In this last example the map α, which assigns to each leaf ℓ the point $\alpha(\ell) \in \ell$ where the extrinsic curvature is maximal, yields a measurable section of $\lambda : V \to X$.

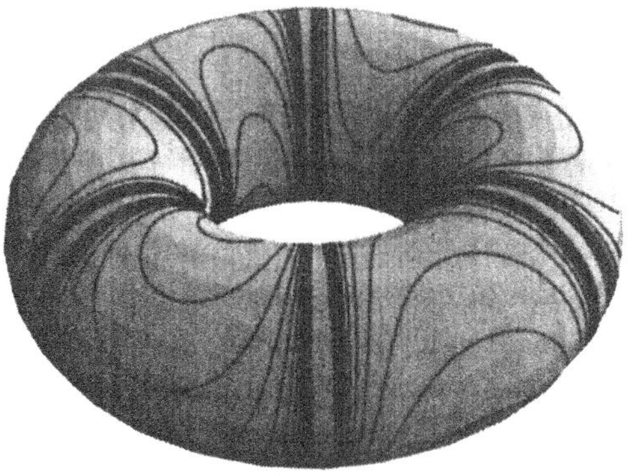

Figure I.7. Example of type I foliation

Type II von Neumann algebras

A finite trace τ on an algebra is a linear form τ such that $\tau(xy) = \tau(yx)$ for any pair x, y of elements of that algebra. Let $\mathcal{H}$ be an infinite-dimensional Hilbert space and $M_\infty(\mathbb{C}) = \mathcal{L}(\mathcal{H})$ be the von Neumann algebra of all bounded operators on $\mathcal{H}$. It is the only factor of type I_∞. This factor does not possess any finite trace, but the usual trace of operators, extended with the value $+\infty$ to positive operators which are not in $\mathcal{L}^1(\mathcal{H})$, is a positive semi-finite faithful normal trace on $M_\infty(\mathbb{C})$. We refer to Chapter V Section 3 for the technical definitions.

By definition, a von Neumann algebra M is semi-finite iff it has a positive semi-finite faithful normal trace. Any type I von Neumann algebra is semi-finite. Any semi-finite von Neumann algebra M is uniquely decomposable as the direct sum $M = M_{\mathrm{I}} \oplus M_{\mathrm{II}}$ of a type I von Neumann algebra M_{I} and a semi-finite von Neumann algebra M_{II} with no nonzero abelian projection. One says that M_{II} is of type II.

Let (V, F) be a foliated manifold and $M = W(V, F)$ the associated von Neumann algebra. We shall see now that the positive faithful semi-finite normal traces τ on M correspond exactly to the geometric notion of *positive invariant transverse densities* on V, which we first describe.

Given a smooth manifold Y, a *positive density* ρ on Y is given by a section of the following canonical principal bundle P on Y. At each point $y \in Y$ the fiber P_y is the space of nonzero homogeneous maps $\rho : \wedge^d T_y \to \mathbb{R}_+^*$, $\rho(\lambda v) = |\lambda|\, \rho(v) \quad \forall \lambda \in \mathbb{R}, \forall v \in \wedge^d T_y$, where $T_y = T_y(Y)$ is the tangent space of Y at $y \in Y$, and d is the dimension of Y.

By construction P is a principal $\mathbb{R}_+^*$-bundle over Y. The choice of a measurable density determines a measure in the Lebesgue class, and all measures in that class are obtained in this manner.

Let us now pass to the case of foliations (V, F). We want to define the notion of a density on the leaf space X. Given a leaf $\ell \in X$, the tangent space $T_\ell(X)$ can be naively thought of as the q-dimensional real vector space, where $q = \dim V - \dim F$, obtained as follows. For each $x \in V$ the quotient $N_x = T_x(V)/F_x$ is a good candidate for $T_\ell(X)$; it remains to identify canonically the vector spaces N_x and N_y for $x, y \in \ell$.

Let $U \subset V$ be the domain of a foliation chart so that x and y belong to the same plaque of U, i.e. to the same leaf p of the restriction of F to U (Figure 8). The leaves of (U, F) are the fibres of a submersion $\pi : U \to \mathbb{R}^q$ whose tangent map at x, y gives the desired identification: $N_x \simeq N_y$. This identification is independent of any choices when the leaf ℓ has no holonomy and is transitive. Since we assume that the set of leaves with nontrivial holonomy is Lebesgue negligible we thus get a well defined real measurable bundle $T(X)$ over X, with fiber $T_\ell(X) \simeq N_x$ for any $x \in \ell$. As above in the case of manifolds we let P be the associated principal $\mathbb{R}_+^*$-bundle of *positive densities*. The fiber P_ℓ of P at $\ell \in X$ is the space of nonzero homogeneous maps $\rho : \wedge^q T_\ell \to \mathbb{R}_+^*$, $\rho(\lambda v) = |\lambda|\, \rho(v) \quad \forall \lambda \in \mathbb{R}, \forall v \in \wedge^q T_\ell$. Here $q = \dim T_\ell$ is the codimension of F.

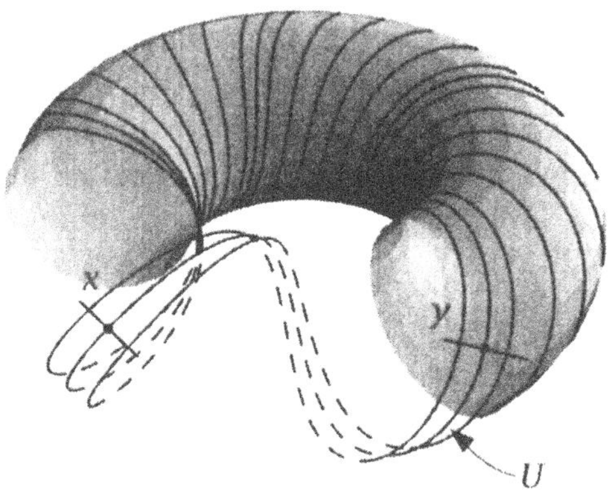

Figure I.8. Holonomy: x and y belong to the same leaf, U is a domain of foliation chart containing x, y and x, y belong to the same plaque.

We shall now see that the positive densities on X, i.e. the measurable sections ρ of P on X, correspond exactly to the positive semi-finite faithful normal traces on the von Neumann algebra $W(V, F)$:

Proposition 6. *Let (V, F) be a foliated manifold, X its leaf space.*

1) *Let ρ be a measurable section of P on X, and $T = (T_\ell)_{\ell \in X}$ be a positive random operator. Let $\mathcal{U} = (U_\alpha)_{\alpha \in I}$ be a locally finite open covering of V by domains of foliation charts, and $(\xi_\alpha)_{\alpha \in I}$ a smooth partition of unity associated to this covering. For each plaque p of U_α, let $\mathrm{Trace}\,((\xi_\alpha T) | p)$ be the trace of the operator $\xi_\alpha^{1/2}\, T_\ell\, \xi_\alpha^{1/2}$ in $L^2(p) \subset L^2(\ell)$, where ℓ is the leaf of p. Then the following number $\tau_\rho(T) \in [0, +\infty]$ is independent of the choices of U_α, ξ_α:*

$$\tau_\rho(T) = \sum_{\alpha \in I} \int \mathrm{Trace}\,((\xi_\alpha T) | p)\; \rho(p).$$

2) *The functional $\tau_\rho : M^+ \to [0, +\infty]$ thus obtained is a positive faithful semi-finite normal trace on M, and all such traces on M are obtained in this way.*

In particular we get a simple criterion for $W(V, F)$ to be semi-finite:

Corollary 7. *$W(V, F)$ is semi-finite iff the principal $\mathbb{R}_+^*$-bundle P over X admits some measurable section.*

For $\theta \notin \mathbb{Q}$ the Kronecker foliation $dy = \theta dx$ of the 2-torus fulfills this condition; its von Neumann algebra $W(\mathbb{T}^2, F_\theta)$ is the unique hyperfinite factor of type II_∞, namely $R_{0,1}$ (cf. Section 3). Similarly, any flow which is ergodic for an invariant measure in the Lebesgue measure class gives the same hyperfinite factor of type II_∞. Any product $(V_1 \times V_2, F_1 \times F_2)$ of the above examples will also yield this factor $R_{0,1}$.

Let us now describe a type II foliation giving rise to a non-hyperfinite factor. Let $\Gamma = \pi_1(S)$ be the fundamental group of a compact Riemann surface S of genus > 1 and $\alpha : \Gamma \to SO(N)$ be an orthogonal faithful representation of Γ. Let V be the total space of the flat principal $SO(N)$-bundle over S associated to α. Then the horizontal foliation of V is a two-dimensional foliation F whose associated von Neumann algebra is a type II_∞ non-hyperfinite factor.

Type III von Neumann algebras

Any von Neumann algebra M is canonically the direct sum $M = M_{\mathrm{I}} \oplus M_{\mathrm{II}} \oplus M_{\mathrm{III}}$ of von Neumann algebras of the respective types. A type III von Neumann algebra M is such that the reduced algebra M_e, for any $e = e^2 = e^*$, $e \neq 0$ is never semi-finite. Of course this is a negative statement, but the modular theory (cf. Section 3) yields a fundamental invariant, the flow of weights $\mathrm{mod}(M)$, which is an action of the group $\mathbb{R}_+^*$ by automorphisms of a Lebesgue measure space, as well as the canonical continuous decomposition of M as the crossed product of a type II von Neumann algebra by an action of $\mathbb{R}_+^*$. We refer to Chapter V Section 8 for the technical definitions, but we shall now spell out for the case of the von Neumann algebra M of a foliation (V,F) what these invariants are. Let (V,F) be a foliated manifold. We first need to come back to the construction of the principal $\mathbb{R}_+^*$-bundle P over the leaf space X and interpret the total space of P as the leaf space of a new foliation (V',F'). Let N, $N_x = T_x(V)/F_x$, be the transverse bundle of the foliation (V,F). It is a real vector bundle, of dimension $q = \operatorname{codim} F$, over the manifold V. We let Q be the associated principal $\mathbb{R}_+^*$-bundle of positive densities. Thus the fiber of Q at $x \in V$ is the $\mathbb{R}_+^*$ homogeneous space of nonzero maps:

$$\delta : \wedge^q N_x \to \mathbb{R}_+ \,, \; \delta(\lambda v) = |\lambda| \, \delta(v) \qquad \forall \lambda \in \mathbb{R} \,, \; \forall v \in \wedge^q N_x.$$

By construction, Q is a smooth principal bundle, and we denote by V' its total space. The restriction of Q to any leaf ℓ of the foliation has a *canonical flat connection* ∇. This follows from the above canonical identification of the transverse spaces N_x, $x \in \ell$. Since it is a local statement it does not use any holonomy hypothesis. Using the flat connection ∇ we define a foliation F' of V', $\dim F' = \dim F$, as the horizontal lifts of F. Thus for $y \in V' = Q$ sitting over $x \in V$, the space $F'_y \subset T_y(V')$ is the horizontal lift of F_x. The flatness of ∇ ensures the integrability of F'. One then checks that:

Proposition 8. 1) *(V',F') is a foliated manifold canonically associated to (V,F).*

2) *The group $\mathbb{R}_+^*$ acts by automorphisms of the foliation (V',F').*

3) *With the above holonomy hypothesis the leaf space of (V',F') is the total space of the principal $\mathbb{R}_+^*$-bundle P over the leaf space X of (V,F).*

In 2) the action is of course that of the $\mathbb{R}_+^*$-bundle $Q = V'$, we shall denote it by $(\theta_\lambda)_{\lambda \in \mathbb{R}_+^*}$, $\theta_\lambda \in \mathrm{Aut}(V', F')$. With the above notation we now have

Proposition 9. a) *The von Neumann algebra $W(V', F')$ is always semi-finite.*

b) *$W(V, F)$ is of type III iff $W(V', F')$ is of type II.*

c) *The flow of weights of $W(V, F)$ is given by the action θ of $\mathbb{R}_+^*$ on the commutative von Neumann algebra*

$$L^\infty(P) = Z(W(V', F')).$$

d) *The continuous decomposition of $W(V, F)$ is given by the crossed product of $W(V', F')$ by the action θ of $\mathbb{R}_+^*$.*

With this result we can now illustrate the classification of Section 3 by geometric examples.

All invariants discussed in Section 3, $S(M)$, $T(M)$... are computed in terms of the flow of weights $\mathrm{mod}(M)$, thus, for instance, (for factors)

$S(M) = \{\lambda \in \mathbb{R}_+^* \,;\, \theta_\lambda = \mathrm{id}\}$

$T(M) =$ Point spectrum of $\theta = \{t \in \mathbb{R};\ \exists u \neq 0,\ \theta_\lambda(u) = \lambda^{it} u \ \ \forall \lambda \in \mathbb{R}_+^*\}$.

In particular M is of type III_1 iff $\mathrm{mod}(M)$ is the constant flow. Thus an ergodic foliation $W(V, F)$ is of type III_1 iff (V', F') is ergodic.

A simple example of this situation is given by the Anosov foliation F of the unit sphere bundle V of a compact Riemann surface S of genus $g > 1$ endowed with its Riemannian metric of constant curvature -1. Thus (cf. for instance [Mi$_4$]) the manifold V is the quotient $V = G/\Gamma$ of the semisimple Lie group $G = PSL(2, \mathbb{R})$ by the discrete cocompact subgroup $\Gamma = \pi_1(S)$, and the foliation F of V is given by the orbits of the action by left multiplication on $V = G/\Gamma$ of the subgroup $B \subset G$ of upper triangular matrices.

The von Neumann algebra $M = W(V, F)$ of this foliation is the (unique) hyperfinite factor of type $\mathrm{III}_1 : R_\infty$. One can indeed check that the associated type II foliation (V', F') is ergodic. It has the same space of leaves as the horocycle flow given by the action on $V = G/\Gamma$ of the group of upper triangular matrices of the form $\begin{bmatrix} 1 & t \\ 0 & 1 \end{bmatrix}$, $t \in \mathbb{R}$.

Let us give an example of a foliation whose associated von Neumann algebra is the hyperfinite factor R_λ of type III_λ, $\lambda \in \,]0,1[$. Let $G = PSL(2, \mathbb{R})$, $\Gamma \subset G$ as above, and B be the subgroup of G of upper triangular matrices $\begin{bmatrix} a & t \\ 0 & a^{-1} \end{bmatrix}$, $t \in \mathbb{R}$, $a \in \mathbb{R}_+^*$. Let V be the manifold $V = G/\Gamma \times \mathbb{T}$ where $\mathbb{T}$ is the one-dimensional torus, $\mathbb{T} = \mathbb{R}_+^*/\lambda^{\mathbb{Z}}$. The group B acts on G/Γ by left translations and on $\mathbb{T}$ by multiplication by a. The product action of B on V gives a two-dimensional foliation (V, F) and

$$W(V, F) \simeq R_\lambda.$$

The same construction with $\mathbb{T}$ and the action of $\mathbb{R}_+^*$ on $\mathbb{T}$ replaced by an arbitrary ergodic smooth flow Y yields a foliation (V, F) whose associated von Neumann algebra is the unique hyperfinite factor algebra of type III_0 with Y as flow of weights, namely the Krieger factor R_Y (cf. Chapter V). In fact the hyperfinite factors of type III_λ, $\lambda \in]0,1]$ and a large class of hyperfinite factors of type III_0 already arise from smooth foliations of the 2-dimensional torus ([Katz]).

We cannot end this section without mentioning the deep relation between the Godbillon-Vey class and the flow of weights which will be discussed in Chapter III Section 6.

I.5. The Index Theorem for Measured Foliations

In the previous section we used the geometric examples of noncommutative spaces coming from the spaces of leaves of foliations to illustrate the classification of factors. In this section we shall show how the theory of type II von Neumann algebras is used as an essential tool in measuring the continuous dimension of the random Hilbert spaces of L^2-solutions of a leafwise elliptic differential equation. An extraordinary property of factors of type II_1, such as the hyperfinite factor R, is the complete classification of the equivalence classes of projections $e \in R$ by a real number $\dim(e) \in [0,1]$ that can take on *any value* between 0 and 1. The Grassmannian of the projections $e \in R$ thus no longer describes the lines, planes, etc. of ordinary geometry but instead "spaces of dimension $\alpha \in [0,1]$", in other words a continuous geometry. The force of this discovery comes across very clearly as one reads the original texts of Murray and von Neumann ([Mur-N$_1$]). As in the usual case, one can speak of the intersection of "subspaces"; the corresponding projection $e \wedge f$ is the largest projection majorized by e and f. One can likewise speak of the subspace generated by e and f; the corresponding projection is denoted $e \vee f$. The fundamental equation is then

$$\dim(e \wedge f) + \dim(e \vee f) = \dim(e) + \dim(f) \quad \forall e, f.$$

These properties make it possible to extend the notion of continuous dimension to the representations of a type II_1 factor N in Hilbert space, so to speak to N-modules, so that these modules are classified exactly by their dimension $\dim_N(\mathcal{H})$, which can be any real number in $[0, +\infty]$ (cf. Chapter V, Section 10). We first explain in detail the notion of transverse measure for foliations and the Ruelle-Sullivan current associated to such a measure. We insist on this notion because it is the geometric counterpart of the notion of a trace on a C^*-algebra, and is at the heart of noncommutative measure theory. We then discuss the special case of the leafwise de Rham complex and show how the continuous dimensions of Murray and von Neumann allow one to define the real-valued Betti numbers. We finally describe the index theorem which replaces, for the noncompact leaves L of a measured foliation (V, F), the Atiyah-Singer index theorem for compact manifolds, and is directly along the lines of the index theorem for covering spaces due to Atiyah [At$_5$] and Singer [Sin$_2$].

5.α Transverse measures for foliations

To get acquainted with the notion of transverse measure for a foliation, we shall first describe it in the simplest case: $\dim F = 1$, i.e. when the leaves are one-dimensional. The foliation is then given by an arbitrary smooth one-dimensional subbundle F of TV, the integrability condition being automatically satisfied. To simplify even further, we assume that F is oriented, so that the complement of the zero section has two components F^+ and $F^- = -F^+$. Then using partitions of unity, one gets the existence of smooth sections of F^+, and any two such vector fields X and $X' \in C^\infty(V, F^+)$ are related by $X' = \phi X$ where $\phi \in C^\infty(V, \mathbb{R}_+^*)$. The leaves are the orbits of the flow $\exp tX$. The flows $H_t = \exp tX$ and $H'_t = \exp tX'$ have the same orbits and differ by a time change:

$$H'_t(p) = H_{T(t,p)}(p) \quad \forall t \in \mathbb{R}\,,\ p \in V.$$

The dependence in p of this time change $T(t, p)$ makes it clear that a measure μ on V which is invariant under the flow H (i.e. $H_t\mu = \mu$, $\forall t \in \mathbb{R}$) is not in general invariant under H' (take the simplest case $X = \frac{\partial}{\partial\theta}$ on S^1). To be more precise let us first translate the invariance $H_t\mu = \mu$ by a condition involving the vector field X rather than the flow $H_t = \exp tX$. Recall that a de Rham current C of dimension q on V is a continuous linear form on the complex topological vector space $C^\infty(V, \wedge^q T_{\mathbb{C}}^*)$ of smooth complex-valued differential forms on V of degree q. In particular a measure μ on V defines a 0-dimensional current by the equality $\langle \mu, \omega\rangle = \int \omega\, d\mu$, $\forall \omega \in C^\infty(V)$. All the usual operations, the Lie derivative ∂_X with respect to a vector field, the boundary d, and the contraction i_X with a vector field, are extended to currents by duality, and the equality $\partial_X = di_X + i_X d$ remains, of course, true. Now the condition $H_t\mu = \mu$, $\forall t \in \mathbb{R}$ is equivalent to $\partial_X \mu = 0$, and since, as μ is 0-dimensional its boundary $d\mu$ is 0, it is equivalent to $d(i_X\mu) = 0$. This condition is obviously not invariant if one changes X into $X' = \phi X$, $\phi \in C^\infty(V, \mathbb{R}_+^*)$. However, if we replace X by $X' = \phi X$ and μ by $\mu' = \phi^{-1}\mu$, the current $i_{X'}\mu'$ is equal to $i_X\mu$ and hence is closed, so that μ' is now invariant under $H'_t = \exp(tX')$.

So, while we do not have a single measure μ on V invariant under all possible flows defining the foliation, we can keep track of the invariant measures for each of these flows using the 1-dimensional current $i_X\mu = C$. To reconstruct μ from the current C and the vector field X, define

$$\langle \mu, f\rangle = \langle C, \omega\rangle\,,\ \forall \omega \in C^\infty(V, \wedge^1 T_{\mathbb{C}}^*)\,,\ \omega(X) = f.$$

Given a 1-dimensional current C on V and a vector field $X \in C^\infty(V, F^+)$, the above formula will define a positive invariant measure for $H_t = \exp tX$ iff C satisfies the following conditions:

1) C is closed, i.e. $dC = 0$

2) C is positive in the leaf direction, i.e. if ω is a smooth 1-form whose restriction to leaves is positive then $\langle C, \omega \rangle \geq 0$.

We could also replace condition 1) by any of the following:

1′) There exists a vector field $X \in C^\infty(V, F^+)$ such that $\exp tX$ leaves C invariant.

1″) Same as 1′), but for all $X \in C^\infty(V, F^+)$.

In fact if C satisfies 2) then $\langle C, \omega \rangle = 0$ for any ω whose restriction to F is 0, thus $i_X C = 0$, $\forall X \in C^\infty(V, F^+)$, and $\partial_X C = d i_X C + i_X dC = i_X dC$ is zero iff $dC = 0$.

From the above discussion we get two equivalent points of view on what the notion of an invariant measure should be for the one-dimensional foliation F:

a) An equivalence class of pairs (X, μ), where $X \in C^\infty(V, F^+)$, μ is an $\exp tX$ invariant measure on V, and $(X, \mu) \sim (X', \mu')$ when $X' = \phi X, \mu = \phi \mu'$ for $\phi \in C^\infty(V, \mathbb{R}_+^*)$.

b) A one-dimensional current C, positive in the leaf direction (cf. 2) and invariant under all, or equivalently some, flows $\exp tX$, $X \in C^\infty(V, F^+)$.

Before we proceed to describe a) and b) for arbitrary foliations we relate them to a third point of view c), that of holonomy invariant transverse measures. A submanifold N of V is called a transversal if, at each $p \in N$, $T_p(V)$ splits as the direct sum of the subspaces $T_p(N)$ and F_p. Thus the dimension of N is equal to the codimension of F. Let $p \in N$, and let U, $p \in U$, be the domain of a foliation chart. One can choose U small enough so that the plaques of U correspond bijectively to points of $N \cap U$, each plaque of U meeting N in one and only one point.

Starting from a pair (X, μ) as in a), one defines on each transversal N a positive measure as follows: the conditional measures of μ (restricted to U) on the plaques, are, since μ is invariant by $H_t = \exp tX$, proportional to the obvious Lebesgue measures determined by X, so the formula $\Lambda(B) = \lim_{\varepsilon \to 0} \frac{1}{\varepsilon} \mu(B_\varepsilon)$, $B_\varepsilon = \bigcup_{t \in [0,\varepsilon]} H_t(B)$ makes sense for any Borel subset B of N.

If one replaces (X, μ) by an equivalent pair (X', μ') it is obvious that Λ does not change, since $\mu' = \phi^{-1}\mu$ while $X' = \phi X$. By construction, Λ is invariant under any of the flows $H_t = \exp tX$, i.e. $\Lambda(H_t B) = \Lambda(B)$, $\forall t \in \mathbb{R}$, and any Borel subset B of a transversal. In fact much more is true:

Lemma 1. *Let N_1 and N_2 be two transversals, B_i a Borel subset of N_i, $i = 1,2$, and $\psi : B_1 \to B_2$, a Borel bijection such that, for each $x \in B_1$, $\psi(x)$ is on the leaf of x; then we have $\Lambda(B_1) = \Lambda(B_2)$.*

To prove this, note that if $p_1 \in N_1$ and $H_t(p_1) = p_2 \in N_2$ for some $t \in \mathbb{R}$, then there exists a smooth function ϕ defined in a neighborhood of p_1 and such that $\phi(p_1) = t$, $H_{\phi(p)}(p) \in N_2$. Thus there exists a sequence ϕ_n of smooth functions defined on open sets of N_1 and such that

$$\{(p,t) \in N_1 \times \mathbb{R} \; ; \; H_t(p) \in N_2\} = \bigcup \text{Graph } \phi_n.$$

Let then (P_n) be a Borel partition of B_1 such that for $p \in P_n$ one has $\psi(p) = H_{\phi_n(p)}(p)$. It is enough to show that $\Lambda(\psi(P_n)) = \Lambda(P_n)$ for all n. But since on P_n, ψ coincides with $p \mapsto H_{\phi_n(p)}(p)$ the result follows from the invariance of Λ under all flows $\exp tY$, $Y \in C^\infty(V, F^+)$.

Thus the transverse measure $\Lambda(B)$ depends in a certain sense only on the intersection number of the leaves L of the foliation, with the Borel set B. For instance, if the current C is carried by a single closed leaf L the transverse measure $\Lambda(B)$ only depends on the number of points of intersection of L and B and hence is proportional to $B \mapsto (B \cap L)^\#$.

By a Borel transversal B to (V,F) we mean a Borel subset B of V such that $B \cap L$ *is countable for any leaf* L. If there exists a Borel injection ψ of B into a transversal N with $\psi(x) \in \text{Leaf }(x)$ $\forall x \in B$, define $\Lambda(B)$ as $\Lambda(\psi(B))$ (which by Lemma 1, is independent of the choices of N and ψ). Then extend Λ to arbitrary Borel transversals by σ-additivity, after remarking that any Borel transversal is a countable union of the previous ones.

We thus obtain a transverse measure Λ for (V,F) in the following sense:

Definition 2. *A transverse measure Λ for the foliation (V,F) is a σ-additive map $B \to \Lambda(B)$ from Borel transversals (i.e. Borel sets in V with $V \cap L$ countable for any leaf L) to $[0, +\infty]$ such that*

1) *If $\psi : B_1 \to B_2$ is a Borel bijection and $\psi(x)$ is on the leaf of x for any $x \in B_1$, then $\Lambda(B_1) = \Lambda(B_2)$.*

2) *$\Lambda(K) < \infty$ if K is a compact subset of a smooth transversal.*

We have seen that the points of view a) and b) are equivalent and how to pass from a) to c). Given a transverse measure Λ as in Definition 2, we get for any distinguished open set U (where U is the domain of a foliation chart), a measure μ_U on the set of plaques π of U, such that for any transversal $B \subset U$ one has

$$\Lambda(B) = \int \text{Card } (B \cap \pi) \, d\mu_U(\pi).$$

Put $\langle C_U, \omega \rangle = \int (\int_\pi \omega) \, d\mu_U(\pi)$, where ω is a differential form on U and $\int_\pi \omega$ is its integral over the plaque π of U. Then on $U \cap U'$ the currents C_U and $C_{U'}$ agree so that one gets a current C on V which obviously satisfies the conditions b), 1), and 2). One thus gets the equivalence between the three points of view a), b), and c).

Let us now pass to the general notion of transverse measure for foliations. We first state how to modify a) and b) for arbitrary foliations ($\dim F \neq 1$). To simplify we assume that the bundle F is oriented. For a) we considered, in the case $\dim F = 1$, pairs (X, μ) up to the very simple equivalence relation saying that only $X \otimes_{C^\infty(V)} \mu$ matters. In the general case, since F is oriented we can talk of the positive part $(\wedge^{\dim F} F)^+$ of $\wedge^{\dim F} F$, and, using partitions of unity, construct sections ν of this bundle. These will play the role of the vector field X. Given a smooth section $\nu \in C^\infty(V, \wedge^{\dim F} F)^+$, we have on each leaf L of the foliation a corresponding volume element. For $k = \dim F$, it corresponds to the unique k-form ω on L such that $\omega(\nu) = 1$. In the case $\dim F = 1$, the measure μ had to be invariant under the flow $H_t = \exp(tX)$. This occurs iff *in each domain of a foliation chart U, the conditional measures of μ on the plaques of U are proportional to the measures determined by the volume element ν*. Thus we shall define the invariance of the pair (ν, μ) in general by this condition. So a) becomes classes of invariant pairs (ν, μ) where $\nu \in C^\infty(V, \wedge^{\dim F} F)^+$, while $(\nu, \mu) \sim (\nu', \mu')$ iff $\nu' = \phi\nu$ and $\mu = \phi\mu'$ for some $\phi \in C^\infty(V, \mathbb{R}_+^*)$. The condition of invariance of μ is of course local. If it is satisfied and $Y \in C^\infty(V, F)$ is a vector field leaving the volume element invariant, i.e. $\partial_Y(\nu) = 0$, then Y also leaves μ invariant. Moreover, since Lebesgue measure in $\mathbb{R}^k$ is characterized by its invariance under translation, one checks that if $\partial_Y \mu = 0\ \forall Y \in C^\infty(V, F)$ with $\partial_Y(\nu) = 0$, then (ν, μ) is invariant. To see this one can choose local coordinates in U transforming ν into the k-vector field $\nu = \frac{\partial}{\partial x^1} \wedge \cdots \wedge \frac{\partial}{\partial x^k} \in C^\infty(V, \wedge F)$. With the above trivialization of ν it is clear that the current $i_\nu \mu$ one gets by contracting ν with the 0-dimensional current μ is a closed current, which is locally of the form

$$\langle C, \omega \rangle = \int \left(\int_\pi \omega \right) d\mu_U(\pi)$$

where ω is a k-form with support in U, $\int_\pi \omega$ is its integral over a plaque π of U, and μ_U is the measure on the set of plaques coming from the disintegration of μ restricted to U with respect to the conditional measures associated to ν.

Clearly C satisfies conditions 1) and 2) of b) and we may also check that $\partial_Y C = 0$, $\forall Y \in C^\infty(V, F)$. Thus for b) in general we take the same object as in the case $k = 1$, namely a *closed current positive in the leaf direction*, the condition "closed" being equivalent to

$$\partial_Y C = 0 \quad \forall Y \in C^\infty(V, F).$$

To recover μ, given C and ν, one considers an arbitrary k-form ω on V such that $\omega(\nu) = 1$ and puts $\langle f, \mu \rangle = \langle C, f\omega \rangle\ \forall f \in C^\infty(V)$. One checks in this way that a) and b) are equivalent points of view. For c) one takes exactly the same definition as for $k = 1$. Given a transverse measure Λ as in Definition 2 one constructs a current exactly as for $k = 1$. Conversely, given a current C

satisfying b), one gets for each domain U of a foliation chart a measure μ_U on the set of plaques of U, such that the restriction of C to U is given by

$$\langle C, \omega \rangle = \int \left(\int_\pi \omega \right) d\mu_U(\pi).$$

Thus one can define $\Lambda(B)$ for each Borel transversal B: if B is contained in U then

$$\Lambda(B) = \int \text{Card } (B \cap \pi) d\mu_U(\pi).$$

One easily checks, as in the case of flows, that Λ satisfies Definition 2.

5.β The Ruelle-Sullivan cycle and the Euler number of a measured foliation

By a measured foliation we mean a foliation (V, F) equipped with a transverse measure Λ. We assume that F is oriented and we let C be the current defining Λ in the point of view b). As C is closed, $dC = 0$, it defines a cycle $[C] \in H_k(V, \mathbb{R})$, by looking at its de Rham homology class. The distinction here between cycles and cocycles is only a question of orientability. If one assumes that F is transversally oriented then the current becomes even and it defines a cohomology class (cf. [Rue-S]).

Now let $e(F) \in H^k(V, \mathbb{R})$ be the Euler class of the oriented bundle F on V (cf. [Mi$_4$]). Using the pairing which makes $H^k(V, \mathbb{R})$ the dual of the finite-dimensional vector space $H_k(V, \mathbb{R})$, we get a scalar $\chi(F, \Lambda) = \langle e(F), [C] \rangle \in \mathbb{R}$ which we shall first interpret in two ways as the average Euler characteristic of the leaves of the measured foliation. First recall that for an oriented compact manifold M the Euler characteristic $\chi(M)$ is given by the well known theorem of H. Poincaré and H. Hopf:

$$\chi(M) = \sum_{p \in \text{ Zero } X} \omega(X, p)$$

where X is a smooth vector field on M with only finitely many zeros, while $\omega(X, p)$ is the local degree of X around p. Given generically, p is a nondegenerate zero, i.e. in local coordinates, $X = \sum a^i \frac{\partial}{\partial x^i}$, the matrix $\frac{\partial a^i}{\partial x^j}(p)$ is nondegenerate and the local degree is the sign of its determinant.

Also, choosing arbitrarily on M a Riemannian metric, one has the generalized Gauss-Bonnet theorem which expresses the Euler characteristic as the integral over M of a form Ω on M, equal to the Pfaffian of $(2\pi)^{-1}K$ where K is the curvature form. These two interpretations of the Euler characteristic extend immediately to the case of measured foliations (V, F, Λ).

Figure I.9. Vector field on the generic leaf of a foliation

First, if $X \in C^\infty(V,F)$ is a generic vector field tangent to the foliation, its set of zeros, $T = \{p \in V; X(p) = 0\}$, defines a submanifold of V which is not in general everywhere transverse to the foliation F but is so Λ-almost everywhere. Thus Λ-almost everywhere the local degree $\omega(X,p) = \pm 1$ of the restriction of X to the leaf of p is well defined, and using the transverse measure we can thus form $\int_{p\in \text{Zero } X} \omega(X,p)\, d\Lambda(p)$. This scalar is again independent of the choice of X and equals $\chi(F,\Lambda) = \langle e(F),[C]\rangle$, as is easily seen from the geometric definition of the Euler class by taking the odd cycle associated to the zeros of a generic section.

Second, let $\| \; \|$ be a Euclidean structure on the bundle F. Then each leaf L is equipped with a Euclidean structure on its tangent bundle, i.e. with a Riemannian metric. So the curvature form of each leaf L allows us to define a form Ω_L, of maximal degree, on L, by taking as above the Pfaffian of $(2\pi)^{-1}$ times the curvature form. Now Ω_L is only defined on L, but one can easily define its integral, using the transverse measure Λ; it is formally

$$\int \left(\int_L \Omega_L \right) d\Lambda(L).$$

(In point of view a), this integral is $\mu(\phi)$, where $\phi(p) = \Omega_p(\nu_p)$; in b) it is $\langle \Omega', C\rangle$ where Ω' is any k-form on V whose restriction to each leaf L is Ω_L; and in c) it is the above integral computed using a covering by domains U_i of foliation charts and a partition of unity ϕ_i, the value of $\int (\int_L \phi_i \Omega)\, d\Lambda(L)$ being given by $\int (\int_\pi \phi_i \Omega)\, d\Lambda(\pi)$ where π varies over the set of plaques of U_i).

To show that the above integral is equal to $\chi(F,\Lambda)$, choose a connection ∇ for the bundle F on V compatible with the metric, and using its curvature form K, take $\Omega' = \mathrm{Pf}(K/2\pi)$. This gives a closed k-form on V, and by [Mi$_4$] p.311, the Euler class of F is represented by Ω' in $H^k(V,\mathbb{R})$. Now the restriction of

∇ to leaves is not necessarily equal to the Riemannian connection. However, both are compatible with the metric. It follows then that on each leaf L there is a canonical $(k-1)$-form ω_L with $\Omega_L - \Omega'_L = d\omega_L$, where Ω'_L is the restriction of Ω' to L. Since ω_L is canonical, it is the restriction to L of a $(k-1)$-form ω on V and hence

$$\int \left(\int_L \Omega_L \right) d\Lambda(L) = \langle \Omega' + d\omega, C \rangle = \langle e(F), [C] \rangle .$$

So in fact both interpretations of $\chi(F, \Lambda)$ follow from the general theory of characteristic classes. One is, however, missing the third interpretation of the Euler number $\chi(M)$ of a compact manifold

$$\chi(M) = \sum (-1)^i \beta_i ,$$

where the β_i are the Betti numbers

$$\beta_i = \dim \left(H^i(M, \mathbb{R}) \right) .$$

The first approach to defining the β_i in the foliation case is to consider the transverse measure Λ as a way of defining the density of discrete subsets of the generic leaf, and then to take β_i as the density of holes of dimension i. However the simplest examples show that one may very well have a foliation with all leaves diffeomorphic to $\mathbb{R}^2$ while $\chi(F, \Lambda) < 0$, so that $\beta_1 > 0$ cannot be defined in the above naive sense. Specifically, let ρ be a faithful orthogonal representation of the fundamental group Γ of a Riemann surface S of genus 2. Then the corresponding principal bundle V on S has a natural foliation: V is the quotient of $\tilde{S} \times SO(n)$ by the action of Γ, and the foliation with leaves $\tilde{S} \times$ points is globally invariant under Γ and hence drops down as a foliation on V. $\tilde{S}$ is the universal covering of S. The bundle F is the bundle of horizontal vectors for a flat connection on V, and each fiber $p^{-1}\{x\}$ is a closed transversal which intersects each leaf in exactly one orbit of Γ. So the Haar measure of $SO(n)$ defines a transverse measure Λ. Since the transverse measure of each fiber is one it is clear that $\chi(F, \Lambda) = -2$. Moreover, since the representation ρ of Γ in $SO(n)$ is faithful, each leaf L of the foliation is equal to the covering space $\tilde{S}$, i.e. is conformal to the unit disk in $\mathbb{C}$. So each leaf is simply-connected while "$\beta_0 - \beta_1 + \beta_2$"$< 0$. This clearly shows that one cannot obtain β_1 by counting in a naive way the handles of this surface. However, though the Poincaré disk (i.e. the unit disk of $\mathbb{C}$ considered as a complex curve) is simply connected it has plenty of nonzero harmonic 1-forms. For instance if f is a bounded holomorphic function in the disk D, then the form $\omega = f(z)dz$ is harmonic, and its L^2 norm $\int \omega \wedge *\omega$ is finite. Thus the space $H^1(D, \mathbb{C})$ of square-integrable harmonic 1-forms is infinite-dimensional and β_1 will be obtained by evaluating its "density of dimension", following the original idea of Atiyah [At_5], for covering spaces.

Given a compact foliated manifold we can, in many ways, choose a Euclidean metric $\| \ \|$ on F. However, since V is compact, two such metrics $\| \ \|$, $\| \ \|'$ always satisfy an inequality of the form $C^{-1}\| \ \|' \leq \| \ \| \leq C\| \ \|'$. So for each leaf L the two Riemannian structures defined by $\| \ \|$ and $\| \ \|'$ will be well related, the identity map from (L) to $(L)'$ being a quasi-isometry. Letting then $\mathcal{H}$ be the Hilbert space of square-integrable 1-forms on L with respect to $\| \ \|$ (resp. $\mathcal{H}'$ with respect to $\| \ \|'$) the above quasi-isometry determines a bounded invertible operator T from $\mathcal{H}$ to $\mathcal{H}'$. We let P (resp. P') be the orthogonal projection of $\mathcal{H}$ (resp. $\mathcal{H}'$) on the subspace of harmonic 1-forms. Then $P'T$ (resp. PT^{-1}) is a bounded operator from $H^1(L,\mathbb{C})$ to $H^1(L',\mathbb{C})$ (resp. from $H^1(L',\mathbb{C})$ to $H^1(L,\mathbb{C})$). These operators are inverses of each other, since, for instance, the form $PT^{-1}P'T\omega - \omega$ is harmonic on L, is in the closure of the range of the boundary operator, and hence is 0. (At this point, of course, one has to know precisely the domains of the unbounded operators used; the compactness of the ambient manifold V shows that each leaf with its quasi-isometric structure is complete; in particular, there is no boundary condition needed to define the Laplacian, since its minimal and maximal domains coincide as in [At$_5$]).

From the above discussion we get that the Hilbert space $H^j(L,\mathbb{C})$ of square integrable harmonic j-forms on the leaf L is well defined up to a quasi-isometry. Of course this fixes only its dimension, $\dim H^j(L,\mathbb{C}) \in \{0,1,\ldots,+\infty\}$. In the above example all leaves were equal to the Poincaré disk D so that for each L: $\dim H^1(L,\mathbb{C}) = +\infty$. However, in this example $\chi(F,\Lambda) = -2$ was finite, which is not compatible with a definition of β_1 as the constant value $+\infty$ of $\dim(H^1(L,\mathbb{C}))$. One can easily see that $H^0(L,\mathbb{C}) = 0$ and $H^2(L,\mathbb{C}) = 0$, thus we should have $\beta_1 = 2$.

Now note the quasi-isometry defined above from $H^j(L,\mathbb{C})$ to $H^j(L',\mathbb{C})$ $(j = 1)$ is canonical. This means that, on the space of leaves V/F of (V,F), the two "bundles" of Hilbert spaces are isomorphic as "bundles" and not only fiberwise. The point that there is more information in the "bundle" than in the individual fibers is well-known. However, it is also well-known that in classical measure theory all bundles are trivial. If (X,μ) is a Lebesgue measure space and $(H_x)_{x\in X}$, $(H'_x)_{x\in X}$ are two measurable fields of Hilbert spaces (cf. [Di$_2$]) with isomorphic fibers (i.e. $\dim H_x = \dim H'_x$ a.e.) then they are isomorphic as bundles.

Since in our example $\dim H^1(L,\mathbb{C}) = +\infty \ \forall L \in V/F$, one could think that this bundle of Hilbert spaces is measurably trivial. In fact it admits no measurable cross-section of norm one. This follows as in the examples of Section 4 Type I, using the ergodicity of the transverse measure Λ.

Thus we see that the measurable bundle $H^1(L,\mathbb{C})$ is not trivial. It is however, isomorphic to a much simpler measurable bundle, which we now describe. Let B be a Borel transversal; then to each leaf L of the foliation we associate the Hilbert space $H_L = \ell^2(L\cap B)$ with orthonormal basis (e_y) canonically parametrized by the discrete countable subset $B\cap L$ of the leaf L. To define the measurable structure of this bundle, note that its pull-back to V assigns to each $x\in V$ the space $\ell^2(L_x\cap B)$ where L_x is the leaf through x, so given a section $(s(x))_{x\in V}$ of this pull-back, we shall say that it is measurable iff the function $(x,y)\in V\times B \mapsto \langle s(x), e_y\rangle$ is measurable.

Lemma 3. *Let B and B' be Borel transversals. Then if the two bundles of Hilbert spaces $(H_L)_{L\in V/F}$ with $H_L = \ell^2(L\cap B)$, and $(H'_L)_{L\in V/F}$ with $H'_L = \ell^2(L\cap B')$, are measurably isomorphic, one has $\Lambda(B) = \Lambda(B')$.*

Proof. Let $(U_L)_{L\in V/F}$ be a measurable family of unitaries from H_L to H'_L. For each $x \in B$ let λ_x be the probability measure on B' given by $\lambda_x(\{y\}) = |\langle U_{\ell(x)}e_x, e_y\rangle|^2$. By construction, λ_x is carried by the intersection of the leaf of x with B'. From the existence of this map $x \mapsto \lambda_x$, which is obviously measurable, one concludes that $\Lambda(B) \leq \Lambda(B')$. Q.E.D.

From this lemma we get an unambiguous definition of the dimension for measurable bundles of Hilbert spaces of the form $H_L = \ell^2(L\cap B)$ by taking

$$\dim_\Lambda(H) = \Lambda(B).$$

We can now state the main result of this section. We shall assume that the set of leaves of (V,F) with nontrivial holonomy is Λ-negligible. This is not always true and we shall explain in Remark 1 how the statement has to be modified for the general case.

Theorem 4. a) *For each $j = 0,1,2,\ldots,\dim F$, there exists a Borel transversal B_j such that the bundle $(H^j(L,\mathbb{C}))_{L\in V/F}$ of the square-integrable harmonic forms on L is measurably isomorphic to $(\ell^2(L\cap B_j))_{L\in V/F}$.*

b) *The scalar $\beta_j = \Lambda(B_j)$ is finite, independent of the choice of B_j and of the choice of the Euclidean structure on F.*

c) *One has $\sum(-1)^j\beta_j = \chi(F,\Lambda)$.*

Of course if $F = TV$ so that there is only one leaf, a Borel transversal B is a finite subset of V and $\Lambda(B)$ is its cardinality, so one recovers the usual relationship between Euler characteristic and Betti numbers. Let us specialise now to two-dimensional leaves, i.e. $\dim F = 2$. Then we get $\beta_0 - \beta_1 + \beta_2 = \frac{1}{2\pi}\int K d\Lambda$ where K is the intrinsic Gaussian curvature of the leaves. Now β_0 is the dimension of the measurable bundle $H_L = \{$square-integrable harmonic 0-forms on $L\}$. Thus, as harmonic 0-forms are constant, there are two cases:

If L is not compact, one has $H_L = \{0\}$.

If L is compact, one has $H_L = \mathbb{C}$.

Using the $*$-operation as an isomorphism of $H^0(L,\mathbb{C})$ with $H^2(L,\mathbb{C})$ one gets the same result for $H^2(L,\mathbb{C})$, and hence

Corollary 5. *If the set of compact leaves of (V,F) is Λ-negligible, then the integral $\int K d\Lambda$ of the intrinsic Gaussian curvature of the leaves is ≤ 0.*

Proof.

$$\frac{1}{2\pi}\int K d\Lambda = \beta_0 - \beta_1 + \beta_2 = -\beta_1 \leq 0. \qquad \text{Q.E.D.}$$

Remark 6. 1) The above theorem was proven under the hypothesis: "The set of leaves with non-trivial holonomy is negligible". To state it in general, one has to replace the generic leaf L, wherever it appears, by its holonomy covering $\tilde{L}$. Thus for instance the measurable bundle $H^j(L,\mathbb{C})$ is replaced by $H^j(\tilde{L},\mathbb{C})$, and $\ell^2(L \cap B)$ is replaced by $\ell^2(L\tilde{\cap}B)$, where $L\tilde{\cap}B$ is a shorthand notation for the inverse image of $L \cap B$ in the covering space $\tilde{L}$. One has to be careful at one point, since the holonomy group of L acts naturally on both $H^j(\tilde{L},\mathbb{C})$ and $\ell^2(L\tilde{\cap}B)$, and the unitary equivalence $U_L : H^j(\tilde{L},\mathbb{C}) \mapsto \ell^2(B\tilde{\cap}L)$ *is supposed to commute with the action of the holonomy group.* Then with these precautions the above theorem holds in full generality. Now unless $\tilde{L}$ is compact one has $H^0(\tilde{L},\mathbb{C}) = 0$. Thus, unless L is compact with finite holonomy (which by the Reeb stability theorem implies that nearby leaves are also compact) one has $\beta_0 = \beta_2 = 0$. This of course strengthens the above corollary: to get $\int K d\Lambda \leq 0$ it is enough that the set of leaves isomorphic to S^2 be Λ-negligible. Of course, one may have $\int K d\Lambda > 0$, as occurs for foliations with S^2-leaves.

2) Using the analytical proof of the Morse inequalities for manifolds ([Wit]) one can prove their analogues for foliations ([Co-F]). The main point here is a result of Igusa [Ig] which shows that given a foliation (V,F), one can always find a smooth function $\phi \in C^\infty(V)$ such that the singularities associated to the critical points of the restriction of ϕ to the leaves are at most of degree 3. (As above, for vector fields $X(x) \in F_x \ \forall x \in V$, it is not possible in general to assume that the restriction of ϕ to the leaves will be a Morse function, since this would yield a closed transversal to the foliation.)

3) If $T \subset V$ is a closed transversal to a foliation F, then one can perform surgery along the leaves at the points of $T \cap L$. This will not affect the space of leaves or the transverse measure, but will, in general, modify the real Betti numbers β_i. By using this operation on products of Kronecker foliations $(\mathbb{T}^2, F_{\theta_j})$, one gets examples where the real Betti numbers β_i, $i = 1,\ldots,k/2$, $k = \dim F$ have given (irrational) preassigned values.

4) The homotopy invariance of the real Betti numbers β_i has been proved by J. Heitsch and C. Lazarov [Hei-La].

5.γ The index theorem for measured foliations

The above formula $\sum(-1)^j\beta_j = \chi(F,\Lambda)$, which relates the real Betti numbers of a measured foliation to the Euler characteristic of F evaluated on the Ruelle-Sullivan cycle, is a special case of a general theorem which extends to measured foliations the Atiyah-Singer index theorems for compact manifolds [At-Si$_1$] and for covering spaces [At$_5$] [Sin$_2$]. As we have already seen above, a typical feature of the leaves of foliations is that they fail to be compact even if the ambient manifold V is compact. However, the continuous dimensions due to Murray and von Neumann allow us to measure with a finite positive real number, $\dim_\Lambda(\ker D)$, the dimension of the random Hilbert space $(\ker D_L)_{L\in X}$ of L^2 solutions of a leafwise elliptic differential equation $D_L\xi = 0$. This continuous dimension vanishes iff the solution space $\ker D_L$ vanishes for Λ-almost all leaves.

More specifically, one starts with a pair of smooth vector bundles E_1, E_2 on V together with a differential operator D on V from sections of E_1 to sections of E_2 such that:

1) D restricts to leaves, i.e. $(D\xi)_x$ only depends upon the restriction of ξ to a neighborhood of x in the leaf of x (i.e. D only uses partial differentiation in the leaf direction).

2) D is elliptic when restricted to any leaf, i.e. the principal symbol given by $\sigma_D(x,\xi) \in \mathrm{Hom}(E_{1,x}, E_{2,x})$ is invertible for any $\xi \in F_x^*$, $\xi \neq 0$.

For each leaf L, let D_L be the restriction of D to L (replace L by the holonomy covering $\tilde{L}$ if L has holonomy). Then D_L is an ordinary elliptic operator on this manifold, and its L^2-kernel $\{\xi \in L^2(L, E_1); D_L(\xi) = 0\}$ is formed of smooth sections of E_1 on L. As in [At$_5$], one does not have problems of domains for the definition of D_L as an unbounded operator in L^2 since its minimal and maximal domains coincide. In this discussion we fix, once and for all, a 1-density α on the leaves. This choice determines $L^2(L, E_i)$, $i = 1, 2$, as well as the formal adjoint D_L^* of D_L, which coincides with its Hilbert space adjoint.

The principal symbol σ_D of D gives, as in [At-Si$_2$], an element $[\sigma_D]$ of the K-theory with compact supports, $K^*(F^*)$, of the total space F^* of the real vector bundle F^* over V. Using the Thom isomorphism in cohomology, as in [At-Si$_2$], yields the Chern character (for simplicity, F is assumed oriented) $\mathrm{ch}(\sigma_D) \in H^*(V, \mathbb{Q})$ as an element of the rational cohomology of the manifold V. We can now state the general result ([Co$_{10}$]).

Theorem 7. *Let (V, F) be a compact foliated manifold (1), D a longitudinal elliptic operator on V, and $X = V/F$ the space of leaves.*

a) *There exists a Borel transversal B to F such that the bundle $(\ker D_L)_{L \in X}$ is measurably isomorphic to $(\ell^2(L \cap B))_{L \in X}$, and the scalar $\Lambda(B)$ is finite and independent of the choice of B.*

b) $\dim_\Lambda(\ker(D)) - \dim_\Lambda(\ker(D^*)) = \varepsilon\langle \mathrm{ch}\,(\sigma_D)\ \mathrm{Td}(F_{\mathbb{C}}), [C]\rangle$,
$\varepsilon = (-1)^{\frac{k(k+1)}{2}}$, *and* $k = \dim F$.

In this formula $[C]$ is the Ruelle-Sullivan cycle, $\mathrm{Td}(F_{\mathbb{C}})$ is the Todd genus of the complexified bundle F. Using the flatness of F in the leaf direction together with the orthogonality of C to the ideal of forms vanishing on the leaves, one can replace $\mathrm{Td}(F_{\mathbb{C}})$ by the Todd genus of $T_{\mathbb{C}}V$.

Let us note that in b) the two sides of the formula are of very different natures. The left-hand side gives global information about the leaves by measuring the dimension of the space of global L^2 solutions. It obviously depends on the transverse measure Λ. The right-hand side only depends upon the homology class $[C] \in H^k(V, \mathbb{R})$ (a finite-dimensional vector space) of the Ruelle-Sullivan cycle, on the subbundle F of TV which defined the foliation, and on the symbol of D which is also a local datum. In particular, to compute the right-hand side it is not necessary to integrate the bundle F. This makes sense since the conditions on a current C to have it correspond to a transverse measure on (V,F) are meaningful without integrating the foliation (C should be closed and positive on k-forms ω whose restriction to F is positive).

For a more thorough discussion of this theorem and several applications we refer to the book [Mo-S] of C. Moore and C. Schochet. We shall just illustrate it by the following simple example, which shows how the usual Riemann-Roch theorem extends to the non-compact case.

Let (V,F) be a compact foliated manifold with two-dimensional leaves. Let us assume that F is oriented. Then every Euclidean structure on F determines canonically a complex structure on the leaves of (V,F). This is analogous to the canonical complex structure of an oriented Riemann surface. Next, let us assume that the foliation (V,F) has some leaf of subexponential growth (Appendix A) and let L_j be an associated averaging sequence, assumed to be regular ([Mo-S]), and Λ the associated transverse measure. Then the index theorem for measured foliations, applied to the longitudinal $\bar\partial$ operator with coefficients in a complex line bundle E, takes the following form ([Mo-S]):

$$\begin{aligned}\dim_\Lambda\left(\ker\bar\partial_E\right) - \dim_\Lambda\left(\ker(\bar\partial_E)^*\right) &= \lim_{j\to\infty}\frac{\text{Number of zeros of } E \text{ in } L_j}{\mathrm{Vol}(L_j)}\\ &\quad - \lim_{j\to\infty}\frac{\text{Number of poles of } E \text{ in } L_j}{\mathrm{Vol}(L_j)}\\ &\quad + \frac{1}{2}\,\text{Average Euler Characteristic}\end{aligned}$$

where the zeros and poles of the line bundle E are defined using a classifying map ([Mo-S]), but take on the usual concrete meaning when E is associated to a divisor. More explicitly, consider the case of a foliation of a 3-manifold V by surfaces. Let $\{\gamma_1,\ldots,\gamma_d\}$ be a collection of d embedded closed curves in V which are transverse to F, and $\{n_1,\ldots,n_d\}$ non-zero relative integers. These data define a complex line bundle E on V whose restriction to each leaf is the complex line bundle with divisor $\sum_{i=1}^{d} n_i(\gamma_i\cap L)$. Then the above formula becomes ([Mo-S])

$$\dim_\Lambda\left(\ker\bar\partial_E\right) - \dim_\Lambda\left(\ker(\bar\partial_E)^*\right) = \sum_{i=1}^{d} n_i\Lambda(\gamma_i) + \frac{1}{2}\langle C, e(F)\rangle$$

in perfect analogy with the usual Riemann-Roch theorem, with the counting of poles and zeros replaced by the counting of their densities.

Problem. Use the above "Riemann-Roch" theorem in conjunction with the method of construction of measures used in the proof of Szemerédi's theorem [Fu] to obtain results on noncompact manifolds of subexponential growth independent of any foliation with transverse measure. Important results in this direction have been obtained by J. Roe (cf. [Ro$_2$]).

I. Appendix A. Transverse Measures and Averaging Sequences

Let (V, F) be a compact foliated manifold. Let us fix a Euclidean structure $\| \ \|$ on the bundle F, tangent to the leaves, and endow the leaves with the corresponding Riemannian metric. Let L be a leaf and, for $x \in L$ and $r > 0$, let $B(x, r)$ be the ball of radius r and center x in the leaf L.

Definition 1. *A leaf L has non-exponential growth if*

$$\liminf_{r \to \infty} \frac{1}{r} \log(\mathrm{Vol}(B(x,r))) = 0.$$

This condition is independent of the choice of $x \in L$ and of the Euclidean structure.

For each leaf of non-exponential growth there is a sequence $r_j \to \infty$ for which the compact subsets $L_j = B(x, r_j)$ satisfy $\frac{\mathrm{Vol}(\partial L_j)}{\mathrm{Vol}(L_j)} \to 0$. Such a sequence of compact submanifolds with boundary is called an *averaging sequence* ([G-P]).

Proposition 2. [G-P] *Given an averaging sequence L_j the following equality defines a transverse measure C for (V, F)*

$$\langle C, \omega \rangle = \lim_{j \to \infty} \frac{1}{\mathrm{Vol}(L_j)} \int_{L_j} \omega$$

for any differential form ω on V of degree equal to the dimension of F.

This construction of transverse measures is rather general, but it is not true that all transverse measures can be obtained this way, as one can see using, for instance, the two-dimensional foliation discussed in Section 5.β). It shows, however, that any foliation of a compact manifold with some leaf of nonexponential growth does admit a nontrivial transverse measure.

I. Appendix B. Abstract Transverse Measure Theory

Let X be a standard Borel space. Then given a standard Borel space Y and a Borel map $Y \xrightarrow{p} X$ with countable fibers ($p^{-1}\{x\}$ countable for any $x \in X$), there exists a Borel bijection ψ of Y on the subgraph $\{(x, n); 0 \le n < F(x)\}$ of the integer-valued function F defined by $F(x) = \mathrm{card}(p^{-1}\{x\})$ on X. In particular if (Y_1, p_1) and (Y_2, p_2) are as above, then the two integer-valued functions $F_i(x) = \mathrm{Card}(p_i^{-1}\{x\})$ coincide iff there exists a Borel bijection $\psi : Y_1 \to Y_2$ with $p_2 \circ \psi = p_1$. Let X be a set, and Y_1 and Y_2 be standard Borel spaces with maps $p_i : Y_i \to X$ with countable fibers and such that $\{(y_i, y_j) \in Y_i \times Y_j; p_i(y_i) = p_j(y_j)\}$ is Borel in $Y_i \times Y_j$, $i, j = 1, 2$. Then we say that the "functions" from X to the integers defined by (Y_1, p_1) and (Y_2, p_2) are the same iff there exists a Borel bijection $\psi : Y_1 \to Y_2$ with $p_2 \circ \psi = p_1$. By the above, if $\forall x \in X$, $\operatorname{Card}\left(p_1^{-1}\{x\}\right) = \operatorname{Card}\left(p_2^{-1}\{x\}\right)$, and *if the quotient Borel structure on X is standard*, then the two "functions" are the same. However, in general we obtain a more refined notion of integer-valued function and of measure space. Let (V, F) be a foliation with transverse measure Λ. We shall now define, using Λ, such a generalized measure on the set X of leaves of (V, F). Each Borel transversal B is a standard Borel space endowed with a projection $p : x \in B \mapsto$ (leaf of x) $\in X$ with countable fiber. Clearly, for any two transversals we check the compatibility condition that in $B_1 \times B_2$, the set $\{(b_1, b_2); b_1$ on the leaf of $b_2\}$ is Borel.

Definition 1. *Let p be a map with countable fibers from a standard Borel space Y to the space X of leaves. We say that p is Borel iff* $\{(y,x); y \in Y\ ,\ x \in$ *leaf* $p(y)\}$ *is Borel in* $Y \times V$.

Equivalently, one could say that the pair (Y, p) is compatible with the pairs associated to Borel transversals. Given a Borel pair (Y, p) we shall define its transverse measure $\Lambda(Y, p)$ by observing that if (B, q) is a Borel transversal with $q(B) = X$, we can define on $B \cup Y$ the equivalence relation coming from the projection to X, and then find a Borel partition $Y = \bigcup_{n=1}^{\infty} Y_n$ and Borel maps ψ_n : $Y_n \to B$ with $q \circ \psi_n = p$, ψ_n injective. Thus $\sum_1^{\infty} \Lambda(\psi_n(Y_n))$ is an unambiguous definition of $\Lambda(Y, p)$.

We then obtain probably the most interesting example of the following abstract measure theory:

A measure space $(X, \mathcal{B})$ is a set X together with a collection $\mathcal{B}$ of pairs (Y, p), where Y is a standard Borel space and p a map with countable fibers from Y to X, with the only axiom:

A pair (Y, p) *belongs to* $\mathcal{B}$ *iff it is compatible with all other pairs of* $\mathcal{B}$ (i.e. iff for any (Z, q) in $\mathcal{B}$ one has $\{(y, z); p(y) = q(z)\}$ is a Borel subset in $Y \times Z$).

A measure Λ on $(X, \mathcal{B})$ is a map which assigns a real number, $\Lambda(Y, p) \in [0, +\infty]$, to any pair (Y, p) in $\mathcal{B}$ with the following axioms:

σ-additivity: $\Lambda\left(\sum(Y_n, p_n)\right) = \sum \Lambda(Y_n, p_n)$ where $\sum(Y_n, p_n)$ is the disjoint union Y of the Y_n with the obvious projection p.

Invariance. If $\psi : Y_1 \to Y_2$ is a Borel bijection with $p_2 \circ \psi = p_1$, then

$$\Lambda(Y_1, p_1) = \Lambda(Y_2, p_2).$$

Of course the measure theory obtained contains as a special case the usual measure theory on standard Borel spaces. It is however much more suitable for spaces like the space of leaves of a foliation, since giving a transverse measure for the foliation (V, F) is the same as giving a measure (in the above sense) on the space of leaves, which satisfies the following finiteness condition: $\Lambda(K, p) < \infty$ for any compact subset K of a smooth transversal.

The role of the abstract theory of transverse measures ([Co_{10}]) thus obtained is made clear by its *functorial property*: if h is a Borel map of the leaf space of (V_1, F_1) to the leaf space of (V_2, F_2) then $h_*(\Lambda)$ is a "measure" on V_2/F_2 for any "measure" Λ on V_1/F_1 (V/F is the space of leaves of (V, F)).

I. Appendix C. Noncommutative Spaces and Set Theory

We have seen in Appendix B that it is possible to formulate what is a transverse measure for a foliation using only the space X of its leaves, provided we take into account in a crucial manner the following principle: one only uses measurable maps between spaces. Thus, using this principle we saw that the notion of an integer-valued function on X automatically becomes more refined, and leads directly to transverse measures. In fact, the space X, viewed as a set, has, if one applies the above principle, an effective cardinality which is strictly larger than the cardinality of the continuum. Indeed, it is easy to construct a Borel injection of $[0, 1]$ into X, but in the ergodic case it is impossible to inject X into $[0, 1]$ by a measurable map since such maps are almost everywhere constant.

The impossibility of constructing effectively an injection of X into $[0, 1]$ is equivalent to the impossibility of distinguishing the elements of X from each other by means of a *denumerable* family of properties P_n each of which defines a measurable subset of X. The noncommutative sets are thus characterized by the effective indiscernability of their elements. In connection with the above "measurability" principle, let us reproduce the following letter of H. Lebesgue to E. Borel. (Taken from *Oeuvres de Jacques Hadamard*, © Publications CNRS, Paris, 1968. Authorized reproduction.)

"You ask me my opinion on the Note of Mr. Zermelo (*Math. Ann.*, v. 59), on the objections that you have made to him (*Math. Ann.*, v. 60) and on the letter of J. Hadamard that you communicated to me; here it is. Forgive me for being so lengthy, for I have tried to be clear.

First of all, I am in agreement with you in this: Mr. Zermelo has very ingeniously proved that one knows how to solve Problem A:

A. *Put a set* M *into well-ordered form,*

whenever one knows how to solve Problem B:

B. *Make correspond to each set* M′ *formed of elements of* M *a particular element* m' *of* M′.

Unfortunately, Problem B is not easy to solve, it seems, except for the sets one knows how to well-order; consequently, one does not have a general solution of Problem A.

I strongly doubt that a general solution of this problem can be given, at least if one accepts, along with Mr. Cantor, that to define a set M is to name a property P belonging to certain elements of a previously defined set N and characterizing, by definition, the elements of M. In effect, with this definition, nothing is known about the elements of M other than this: they possess all of the *unknown* properties of the elements of N and they are the only ones that have the *unknown* property P. Nothing in this permits distinguishing between two elements of M, still less classifying them as one would have to do in order to solve A.

This objection, made *a priori* to every attempt at solving A, obviously falls if one specializes N or P; the objection falls, for example, if N is the set of numbers. All that one can hope to do in general is to indicate problems, such as B, whose solution would imply that of A and which are possible in certain special but frequently encountered cases. Whence the interest, in my opinion, of Mr. Zermelo's reasoning.

I believe that Mr. Hadamard is more faithful than you to Mr. Zermelo's thinking, in interpreting that author's Note as an attempt, not at the effective solution of A, but at proving the existence of a solution. The question comes down to this, which is hardly new: *can one prove the existence of a mathematical entity without defining it*?

This is obviously a matter of convention; however, I believe that one can build solidly only *by accepting that one can prove the existence of an entity only by defining it*. From this point of view, close to that of Kronecker and of Mr. Drach, there is no distinction to be made between A and the problem C:

C. *Can every set be well-ordered*?

I would have nothing more to say if the convention I indicated were universally accepted; however, I must confess that one often uses, and that I myself have often used, the word *existence* in other senses. For example, when one interprets a well-known argument of Mr. Cantor by saying that *there exists a nondenumerable infinity of numbers*, one nevertheless does not give the means of naming such an infinity. One merely shows, as you have said before me, that whenever one has a denumerable infinity of numbers, one can define a number not belonging to this infinity. (The word *define* always has the meaning of *naming a characteristic property of what is defined*.) An existence of this nature may be used in an argument, and in the following way: a property is true if its denial leads to admitting that one can arrange all of the numbers into a denumerable sequence. I believe that it can only intervene in this way.

Mr. Zermelo makes use of the *existence* of a *correspondence* between the subsets of M and certain of their elements. You see, even if the existence of such correspondences were not in doubt, because of the manner in which this existence had been proved, it would not be evident that one had the right to use this existence in the way that Mr. Zermelo does.

I now come to the argument that you state as follows: "It is possible, in a particular set M′, to choose *ad libitum* the distinguished element m'; it being possible to make this choice for each of the sets M′, it is possible to make it for the set of these sets"; and from which the existence of the correspondences appears to result.

First of all, M′ being given, is it obvious that one can choose m'? This would be obvious if M′ existed in the nearly Kroneckerian sense I mentioned, since to say that M′ exists would then be to affirm that one knows how to name certain of its elements. But let us extend the meaning of the word *exists*. The set Γ of correspondences between the subsets M′ and the distinguished elements m' certainly *exists* for Messrs. Hadamard and Zermelo; the latter even represents the number of its elements by a transfinite product. Nevertheless, does one know how to choose an element of Γ? Obviously not, since this would yield a definite solution of B for M.

It is true that I use the word *choose* in the sense of *naming* and that it would perhaps suffice for Mr. Zermelo's argument that *to choose* mean *to think of*. But it would still be necessary to observe that it is not indicated which one is being thought of, and that it is nevertheless necessary to Mr. Zermelo's argument that one think of *a definite correspondence that is always the same.* Mr. Hadamard believes, it seems to me, that it is not necessary to prove that one can *determine* an element (and one only); this is, in my opinion, the source of the differences in appreciation.

To give you a better feeling for the difficulty that I see, I remind you that in my thesis I proved the existence (in the non-Kroneckerian sense and perhaps difficult to make precise) of measurable sets that are not measurable B, but it remains doubtful to me that anyone can ever name one. Under these circumstances, would I have had the right to base an argument on this hypothesis: *suppose chosen a measurable set that is not measurable* B, when I doubted that anyone could ever name one?

Thus, I already see a difficulty in this "in a definite M′, I can choose a definite m' ", since there exist sets (the set C for example, which one could regard as a set M′ coming from a more general set) in which it is perhaps impossible to choose an element. There is then the difficulty that you note relative to the infinity of choices, the result of which is that, if one wants to regard the argument of Mr. Zermelo as entirely general, it must be admitted that one speaks of an infinity of choices, perhaps an infinity of very large power; moreover, one gives neither the law of this infinity, nor the law of one of the choices; one does not know if it is possible to name a law defining a set of choices having the power of the set of the M′; one does not know whether it is possible, given an M′, to name an m'.

In summary, when I examine closely the argument of Mr. Zermelo, as indeed with several general arguments on sets, I find it insufficiently Kroneckerian to attribute a meaning to it (only as an existence theorem for C, of course).

You refer to the argument: "To well-order a set, it suffices to choose an element of it, then another, etc." It is certain that this argument presents enormous difficulties, even greater than, at least in appearance, that of Mr. Zermelo; and I am tempted to believe, along with Mr. Hadamard, that there is progress in having replaced an infinity of successive and mutually dependent choices by an infinity, unordered, of independent choices. Perhaps this is nothing more than an illusion and the apparent simplification comes only from the fact that one must replace an ordered infinity of choices by an unordered infinity, but of much larger power. So that the fact that one can reduce to a single difficulty, placed at the beginning of Mr. Zermelo's argument, all of the difficulties of the simplistic argument that you cite, perhaps proves simply that this single difficulty is very great. In any case, it does not seem to me to disappear because it involves an unordered set of independent choices. For example, if I believe in the existence of functions $y(x)$ such that, given any x, y is never bound to x by an algebraic equation with integer coefficients, it is because I believe, along with Mr. Hadamard, that it is possible to construct one; but, for me, it is not the immediate consequence of the existence, given any x, of numbers y that are not bound to x by any equation with integer coefficients.[1]

I am fully in agreement with Mr. Hadamard when he declares that the difficulty in speaking of an infinity of choices without giving the law for them is equally grave whether it is a matter of a denumerable infinity or not. When one says, as in the argument that you criticize, "it being possible to make this choice for each of the sets M′, it is possible to make it for the set of these sets", one hasn't said anything if one does not explain the terms employed. To make a choice could be the writing or the naming of the chosen element; to make an infinity of choices cannot be the writing or naming of the chosen elements one by one: life is too short. Thus one has to say what it means to do it. By this is meant, in general, giving the law that defines the chosen elements, but this law is for me, as for Mr. Hadamard, equally indispensable whether it is a matter of a denumerable infinity or not.

Perhaps, however, I am again in agreement with you on this point, because, if I do not establish theoretical distinctions between the two infinities, from a practical point of view I make a great distinction between them. When I hear someone speak of a law defining a transfinite infinity of choices, I am very suspicious, because I have never yet seen any such laws, whereas I do know of laws defining a denumerable infinity of choices. But perhaps this is only a matter of habit and, on reflection, I sometimes see difficulties equally grave, in my opinion, in arguments in which only a denumerable infinity of choices occur, as in arguments where there is a transfinity. For example, if I do not regard as established by the classical argument that every set of power greater than the denumerable contains a set whose power is that of the set of transfinite numbers of Mr. Cantor's class II, I attribute no greater value to the method by which one proves that a non-finite set contains a denumerable set. Although I strongly doubt that one ever names a set that is neither finite nor infinite, the impossibility of such a set does not seem to me to be proved. But I have already spoken to you of these questions."

---

[1] *In correcting proofs, I add that in fact the argument, whereby one usually legitimizes the statement A of Mr. Hadamard (p. 262), legitimizes at the same time the statement B. And, in my opinion, it is because it legitimizes B that it legitimizes A. {Translator's note: The footnote is Lebesgue's. Here A and B refer to the statements in Hadamard's letter to Borel, reproduced on pp. 335-337 of the previously cited Collected Works of Hadamard. Lebesgue's letter is the third of Five letters on the theory of sets (between Hadamard, Borel, Lebesgue and Baire), originally published together in the "Correspondence" section of the Bulletin of the French Mathematical Society [Bull. Soc. Math. France* 33 *(1905), 261-273].}*

II

TOPOLOGY AND K-THEORY

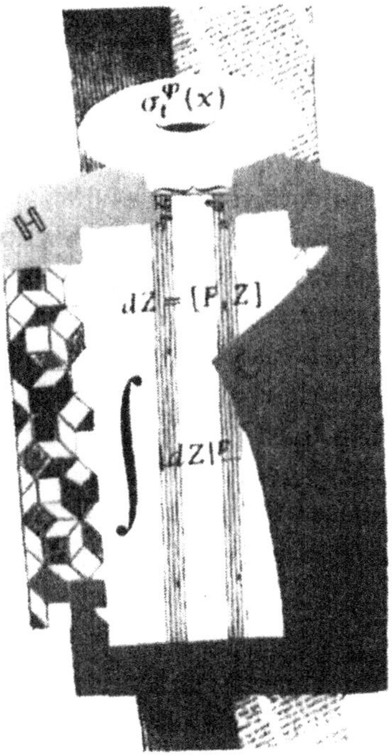

In this chapter we shall extend *topology* beyond its classical set theoretic framework in order to understand, from the topological point of view, the following spaces which are ill behaved as sets:

1) The space X of Penrose tilings of the plane.
2) The dual space $\hat{\Gamma}$ of a discrete group Γ.
3) The orbit space of a group action on a manifold.
4) The leaf space of a foliation.
5) The dual space $\hat{G}$ of a Lie group G.

One could base this extension of topology on the notion of topos due to Grothendieck. Our aim, however, is to establish contact with the powerful tools of functional analysis such as positivity and Hilbert space techniques, and with K-theory. The first general principle at work is that to any of the above spaces X there corresponds, in a natural manner, a *C^*-algebra* which plays the role of the involutive algebra $C(X)$ of continuous functions on X. To the non-classical nature of such spaces corresponds the *noncommutativity* of the associated C^*-algebra. When it happens that the space X is well behaved classically then its associated C^*-algebra is equivalent (in the sense of strong Morita equivalence, cf. Appendix A) to a commutative C^*-algebra $C(X)$.

The second general principle is that for the above spaces the K-theory of the associated C^*-algebra "$C(X)$" is the natural place for invariants of families of ordinary spaces $(Y_x)_{x\in X}$ parametrized by X. Thus, for instance, the signature $\mathrm{Sign}(L)$ of the generic leaf of a foliated compact manifold (V,F) is a specific element

$$\mathrm{Sign}(L) \in K(X)$$

of the K-theory group of the C^*-algebra of the foliation. Moreover, this element is an invariant of leafwise homotopy equivalence. This type of result extends the result of Mishchenko on the homotopy invariance of the Γ-equivariant signature of covering spaces ([Mis$_2$]) which corresponds to Example 2. There are several reasons, reviewed in Section 1, which make the K-theory of C^*-algebras both relevant and tractable. In fact, computing the K-groups of the C^*-algebra associated to one of the above spaces (of Examples 1 to 5) amounts to classifying the stable isomorphism classes of virtual "vector bundles" over such spaces and, as such, is a natural prerequisite for using them as geometric spaces.

The third principle at work in all the above examples is that one may construct a geometric group, denoted $K_*(BX)$, easily computable by standard techniques of algebraic topology, and a map μ, the analytic assembly map, from the geometric group to the K-group $K(X)$ of the C^*-algebra associated to X

$$\mu : K_*(BX) \to K(X).$$

The general conjecture ([Bau-C$_1$]) that this map μ is an isomorphism is a guiding principle of great relevance. The notations BX and $K_*(BX)$ will be explained later, but, essentially, BX stands for the *homotopy type* of an ordinary space which fibers over X with contractible fibers. Thus, for instance, if (V, F) is a compact foliated manifold with contractible leaves, then BX is homotopy equivalent to V, since V fibers over the leaf space X with contractible fibers, the leaves of the foliation. Also, in Example 2 if Γ is *torsion-free* then BX is the usual classifying space $B\Gamma$ of the discrete group Γ. In this last example K_* is K-homology and the map μ is the assembly map of Mishchenko and Kasparov.

The bivariant KK-theory of Kasparov [Kas$_1$] plays a crucial role in the construction of the map μ and in the computation of the K-theory of C^*-algebras. We shall use a variant of that theory, the E-theory or *deformation theory*, which is both easier to develop (Appendix B) and more appropriate for K-theory maps, since unlike KK it is half-exact in its two arguments. However we shall come back to KK later in Chapter IV and use explicitly the more precise geometric significance of the KK cycles. In this chapter we shall construct our K-theory maps from deformations of C^*-algebras (Section 6). In particular we shall get the Atiyah-Singer index theorem as a corollary of the Thom isomorphism theorem (Section 5).

II.1. C^*-algebras and their K-theory

Given an ordinary compact or locally compact space Y, Urysohn's lemma [Ru] shows that there exist sufficiently many continuous functions $f \in C(Y)$ on Y to determine the topology of Y uniquely. This is true for real-valued functions and *a fortiori* for complex-valued ones. Thus, the topology of Y determines, and is determined by, the algebra $A = C(Y)$ of continuous complex-valued functions on Y, equipped with the involution $f \to f^*$, where $f^*(y) = \overline{f(y)}$

($\forall y \in Y$). (The analogous statement would be false if $\mathbb{R}$ or $\mathbb{C}$ were replaced by a totally disconnected local field.)

Moreover, the class of involutive $\mathbb{C}$-algebras A obtained in this way can be characterized very simply: they are the *commutative* C^*-algebras with units. General C^*-algebras have a very simple axiomatic characterization that expresses exactly what we expect from functions of class C^0. The theory of C^*-algebras began in 1943 with a paper of Gel'fand and Naimark [G-N]. (See also M.H. Stone [Sto].)

Definition 1. *A C^*-algebra is a Banach algebra over $\mathbb{C}$ with a conjugate-linear involution $x \to x^*$ such that*

$$(xy)^* = y^*x^* \quad \textit{and} \quad \|x^*x\| = \|x\|^2 \quad \textit{for} \quad x, y \in A.$$

A C^*-algebra A is an involutive Banach algebra for the norm $x \to \|x\|$ uniquely determined from the involutive algebra structure by considering the spectral radius of x^*x:

$$\begin{aligned}\|x\|^2 &= \text{spectral radius } (x^*x)\\ &= \sup\{|\lambda|;\ x^*x - \lambda \text{ not invertible}\}.\end{aligned}$$

For an involutive algebra A to be a C^*-algebra, it is necessary and sufficient that it admit a $*$-representation π on a Hilbert space, such that

$$\begin{aligned}&1)\pi(x) = 0 \Longrightarrow x = 0;\\ &2)\pi(A) \text{ is norm closed.}\end{aligned} \tag{4}$$

A theorem of Gel'fand, which is based on complex analysis and the structure of the maximal ideals of A, shows that if A is a *commutative* C^*-algebra then there exists a compact topological space $Y = \mathrm{Sp}(A)$ such that $A = C(Y)$.

Let A be a *commutative* C^*-algebra, and suppose that A has a unit. The set $\mathrm{Sp}\,A$ of homomorphisms χ of A into $\mathbb{C}$ such that $\chi(1) = 1$, equipped with the topology of pointwise convergence on A, is *compact*; the compact space $\mathrm{Sp}\,A$ is called the *spectrum* of A.

Theorem 2. *Let A be a commutative C^*-algebra with unit and let $X = \mathrm{Sp}\,A$ be its spectrum. The Gel'fand transform*

$$x \in A \to \textit{the function } \hat{x}(\chi) = \chi(x) \quad (\chi \in \mathrm{Sp}\,A)$$

is an isomorphism of A onto the C^*-algebra $C(X)$ of continuous complex functions on X.

Thus, the contravariant functor C, that associates to every compact space X the C^*-algebra $C(X)$, effects an equivalence between the category of compact spaces with continuous mappings, and the opposite of the category of commutative unital C^*-algebras with unit-preserving homomorphisms. To a continuous mapping $f : X \to Y$ there corresponds the homomorphism $C(f)$:

$C(Y) \to C(X)$ that sends $h \in C(Y)$ to $h \circ f \in C(X)$. In particular, two commutative C^*-algebras are isomorphic if and only if their spectra are homeomorphic.

The basis for noncommutative geometry is the possibility of adapting most of the classical tools, such as Radon measures, K-theory, cohomology, etc., necessary for the study of a compact space Y, to the case where the C^*-algebra $A = C(Y)$ is replaced by a noncommutative C^*-algebra. In particular, the general theory is not limited to the spaces X given in Examples 1) to 5) above. It is, however, important to note that to a general C^*-algebra A there corresponds a space X, namely the space of equivalence classes of irreducible representations π of A on Hilbert space. One can still prove the existence of sufficiently many irreducible representations of A, and when A is *simple*, i.e. has no nontrivial two-sided ideals, the natural topology of the space X has the same triviality property as that of the set X of Penrose tilings encountered below (Section 3). Moreover, exactly as one can find any finite portion of a given tiling in any other tiling, two irreducible representations π_1 and π_2 of A can, by means of a unitary transformation, be made as similar to each other as one likes. This follows from [K] and [Vo_2].

The cohomology theory for compact spaces X which is the easiest to extend to the noncommutative case is the K-theory $K(X)$.

The group $K^0(X)$ associated to a compact space X has a very simple description in terms of the locally trivial, finite-dimensional complex vector bundles E over X. To such a bundle E there corresponds an element $[E]$ of $K^0(X)$ that depends only on the stable isomorphism class of E, where E_1 and E_2 are said to be stably isomorphic if there exists a vector bundle F such that the direct sum $E_1 \oplus F$ is isomorphic to $E_2 \oplus F$. Moreover, the classes $[E]$ generate the group $K^0(X)$ and the relations $[E_1 \oplus E_2] = [E_1] + [E_2]$ constitute a presentation of this group.

One obtains in this way a cohomology theory, topological K-theory, that is both simple to define and to calculate, thanks to the Bott periodicity theorem (cf. below).

More precisely, the functor K is a contravariant functor from the category of compact topological spaces and continuous mappings to the category of $\mathbb{Z}/2$-graded abelian groups: $K = K^0 \oplus K^1$. The group $K^1(X)$ is obtained as the group $\pi_0(GL_\infty)$ of connected components of the group $GL_\infty = \bigcup_{n \geq 1} GL_n(A)$, the increasing union of the groups $GL_n(A)$ of invertible elements of the algebras $M_n(A) = M_n(C(X))$ of $n \times n$ matrices with elements in the algebra $A = C(X)$. The inclusion mapping of GL_n into GL_{n+1} is given by $g \to \begin{bmatrix} g & 0 \\ 0 & 1 \end{bmatrix}$. It is because $GL_n(A)$ is an *open* set in $M_n(A)$ (a general property of Banach algebras A, commutative or not) that the group $K^1(X) = \pi_0(GL_\infty)$ is denumerable (if A is separable as a Banach space) and can be calculated by the methods of differential topology.

The fundamental tool of this theory is Bott's periodicity theorem, which remains valid for a not necessarily commutative Banach algebra A [At_3] [Woo] [Kar_2]. It can be stated as the periodicity of period two, $\pi_k(GL_\infty) \simeq \pi_{k+2}(GL_\infty)$, of the homotopy groups, but this periodicity has as its most important corollary the existence of a short exact sequence for the functor K, valid for every closed set $Y \subset X$:

$$\begin{array}{ccccc} K^0(X\backslash Y) & \longrightarrow & K^0(X) & \longrightarrow & K^0(Y) \\ \uparrow & & & & \downarrow \\ K^1(Y) & \longleftarrow & K^1(X) & \longleftarrow & K^1(X\backslash Y) \end{array} \qquad (*)$$

One defines the group K^i for a locally compact space such as $Z = X\backslash Y$ as the kernel of the restriction $K^i(\tilde{Z}) \to K^i(\{\infty\})$, where $\tilde{Z} = Z \cup \{\infty\}$ is the one-point compactification of Z.

It follows from the Bott periodicity theorem that the group $K^0(X)$ is the fundamental group $\pi_1(GL_\infty(A))$, where $A = C(X)$.

All of the features of the topology of compact spaces we have just described adapt remarkably well to the noncommutative case, where the algebra $C(X)$ of continuous complex functions on X is replaced by a not necessarily commutative C^*-algebra A. As for K-theory, the definitions of the groups K^i, in the form $K_i(A) = \pi_{2k+i+1}(GL_\infty(A))$, where $k \in \mathbb{N}$ is an arbitrary integer, and the Bott periodicity theorem are unchanged. The exact sequence $(*)$ remains valid for every closed two-sided ideal J of A in the form

$$\begin{array}{ccccc} K_0(J) & \longrightarrow & K_0(A) & \longrightarrow & K_0(B) \\ \uparrow & & & & \downarrow \\ K_1(B) & \longleftarrow & K_1(A) & \longleftarrow & K_1(J) \end{array} \qquad (*')$$

where $B = A/J$ is the C^*-algebra quotient of A by J. Also, as above, if A is not unital one defines $K_j(A)$ as the kernel of the augmentation map $K_j(\tilde{A}) \xrightarrow{\varepsilon} K_j(\mathbb{C})$, where $\tilde{A}$ is obtained from A by adjoining a unit, i.e.

$$\tilde{A} = \{a + \lambda 1 \,;\, a \in A\,,\ \lambda \in \mathbb{C}\}\,,\ \varepsilon(a + \lambda 1) = \lambda \in \mathbb{C}.$$

Moreover, the description of K^0 in terms of vector bundles is carried out in terms of *projective modules of finite type* over the algebra A. Indeed, by a theorem of Serre and Swan [Ser$_3$] [Sw] the locally trivial finite-dimensional complex vector bundles E over a compact space X correspond canonically to the finite projective modules over $A = C(X)$. To the vector bundle E one associates the $C(X)$-module $\mathcal{E} = C(X, E)$ of continuous sections of E. Conversely, if $\mathcal{E}$ is a finite projective module over $A = C(X)$, the fiber of the associated vector bundle E at a point $p \in X$ is the tensor product

$$E_p = \mathcal{E} \otimes_A \mathbb{C}$$

where A acts on $\mathbb{C}$ by the character χ, $\chi(f) = f(p)\ \forall f \in C(X)$.

The direct sum of vector bundles corresponds to the direct sum of the associated modules. Isomorphism and stable isomorphism thus have a meaning in general, and for a unital C^*-algebra A the group $K_0(A)$ is the group of stable isomorphism classes of finite projective modules over A (or, more pedantically, of formal differences of such classes). If $\mathcal{E}$ is a finite projective module over

A there exists a projection (or idempotent) $e = e^2$, $e \in M_n(A)$, and an isomorphism of $\mathcal{E}$ with the right A-module $eA^n = \{(\xi_i)_{i=1,\dots,n};\ \xi_i \in A,\ e\xi = \xi\}$. For non-unital C^*-algebras the group K_0 is the reduced group of the C^*-algebra $\tilde{A}$ obtained by adjoining a unit.

One important feature of the K-theory groups K_0, K_1 of a C^*-algebra is that if A is *separable* as a Banach space (which in the commutative case $A = C(X)$ is equivalent to X being metrisable), then these K-groups are *countable abelian groups*. This feature is shared by the K-theory of Banach algebras, but not by K-theory of general algebras. One can for instance show that the group K_0 of the convolution algebra $C_c^\infty(\mathbb{R})$ of smooth compactly supported functions on $\mathbb{R}$ is an abelian group with the cardinality of the continuum cf. Section 10.

There is one crucial feature of the K-theory of C^*-algebras which differs notably from the case of general Banach algebras. It was discovered in connection with the Novikov conjecture [Mis$_3$]. The point is that for C^*-algebras one has a canonical isomorphism

$$K_0(A) \simeq \mathrm{Witt}(A)$$

where the Witt group $\mathrm{Witt}(A)$ classifies the quadratic forms Q on finite projective modules over A. More precisely, let us first recall the definition of the Witt group of an involutive algebra A over $\mathbb{C}$. Given a finite projective (right) module $\mathcal{E}$ over A, a *Hermitian form* Q on $\mathcal{E}$ is a sesquilinear form $\mathcal{E} \times \mathcal{E} \to A$ such that

1) $Q(\xi a, \eta b) = a^* Q(\xi, \eta) b \quad \forall \xi, \eta \in \mathcal{E},\ \forall a, b \in A.$

2) $Q(\eta, \xi) = Q(\xi, \eta)^* \quad \forall \xi, \eta \in \mathcal{E}.$

To a Hermitian form Q on $\mathcal{E}$ corresponds a linear map $\tilde{Q}$ of $\mathcal{E}$ into $\mathcal{E}^* = \mathrm{Hom}_A(\mathcal{E}, A)$, given by

$$(\tilde{Q}\xi)(\eta) = Q(\xi, \eta) \quad \forall \xi, \eta \in \mathcal{E}.$$

The Hermitian form is called *invertible* when $\tilde{Q}$ is invertible. There are obvious notions of isomorphism and direct sum of Hermitian forms. The Witt group $\mathrm{Witt}(A)$ is the group generated by isomorphism classes $[Q]$ of invertible Hermitian forms Q on arbitrary finite projective modules over A with the relations

α) $[Q_1 \oplus Q_2] = [Q_1] + [Q_2]$

β) $[Q] + [-Q] = 0.$

When A is a C^*-algebra with unit, there exists on any given finite projective module $\mathcal{E}$ over A an invertible Hermitian form Q satisfying the following *positivity* condition:

3) $Q(\xi, \xi) \geq 0 \quad \forall \xi \in \mathcal{E}$

(cf. Chapter V for the notion of positive elements in a C^*-algebra). Moreover, given $\mathcal{E}$, all the positive invertible Hermitian forms on $\mathcal{E}$ are pairwise isomorphic ([Mis$_3$]) thus yielding a well defined map $\varphi : K_0(A) \to \mathrm{Witt}(A)$ such that $\varphi(\mathcal{E})$ is the class of $(\mathcal{E}, Q)$ with Q positive. This map is actually an isomorphism ([Mis$_3$]); this property uses in an essential manner the fact that the involutive Banach algebra A is a C^*-algebra, and in particular that

$$T = T^* \ , \ T \in A \Longrightarrow \text{Spectrum}(T) \subset \mathbb{R}.$$

(In other words, if T is a selfadjoint element of A and $\lambda \in \mathbb{C}$ is not real then $T-\lambda$ is invertible in A.) This property, as well as the isomorphism $K_0(A) \simeq \text{Witt}(A)$ fails for general involutive Banach algebras, but one should expect that for any involutive Banach algebra B one has

$$\text{Witt}(B) = \text{Witt}(C^*B)$$

where C^*B is the enveloping C^*-algebra of B, i.e. the completion of B for the seminorm $\|x\|_{C^*} = \sup\{\|\pi(x)\| \ ; \ \pi$ a unitary representation of B in a Hilbert space$\}$.

This equality was proved by J.-B. Bost for arbitrary commutative involutive Banach algebras [Bos$_2$].

II.2. Elementary Examples of Quotient Spaces

We shall formulate the algebraic counterpart of the geometric operation of forming the quotient of a space by an equivalence relation, and show how to handle non-Hausdorff quotients by using noncommutative C^*-algebras.

Let $Y = \{a, b\}$ be a set consisting in two elements a and b, so that the algebra $C(Y)$ of (complex-valued) functions on Y is the commutative algebra $\mathbb{C} \oplus \mathbb{C}$ of 2×2 diagonal matrices. There are two ways of declaring that the two points a and b of Y are identical, i.e. of quotienting Y by the equivalence relation $a \sim b$.

1) The first is to consider the *subalgebra* $A \subset C(Y)$ of functions f on Y which take the same values at a and b: $f(a) = f(b)$.

2) The second is to consider the *larger algebra* $B \supset C(Y)$ of all 2×2 matrices:

$$f = \begin{bmatrix} f_{aa} & f_{ab} \\ f_{ba} & f_{bb} \end{bmatrix}.$$

The relation between the two algebras is the notion of strong Morita equivalence of C^*-algebras, due to M. Rieffel [Ri$_2$] (cf. Appendix A). This relation preserves many invariants of C^*-algebras, such as K-theory and the topology of the space of irreducible representations. Thus, in our example the pure states ω_a and ω_b of B which come directly from the points a and b of Y by the formula

$$\omega_a(f) = f_{aa} \ , \ \omega_b(f) = f_{bb} \quad \forall f \in B$$

yield equivalent irreducible representations of $B = M_2(\mathbb{C})$, which corresponds to the identification $a \simeq b$ in the spectrum of B. We shall now work out many examples of the above algebraic operation of quotient, first (example α)) to give an alternate description of existing spaces, but mainly to give precise topological meaning to non-existent quotient spaces, for which the first method above yields a trivial result while the second yields a nontrivial noncommutative C^*-algebra.

2.α Open covers of manifolds.

Let X be a compact manifold obtained by pasting together some open pieces $U_i \subset X$ of Euclidean space along their intersections $U_i \cap U_j$. To obtain X from the Euclidean open set $V = \coprod U_j$, disjoint union of the U_j's, one must identify the pairs (z, z') of elements of V where the pasting takes place. To be more precise, there is an equivalence relation $\mathcal{R}$ on V generated by such pairs, and for $z, z' \in V$ one has $z \sim z'(\mathcal{R})$ iff $p(z) = p(z')$, where $p : V \to X$ is the obvious surjection. Since X is compact we shall assume that the above covering is finite.

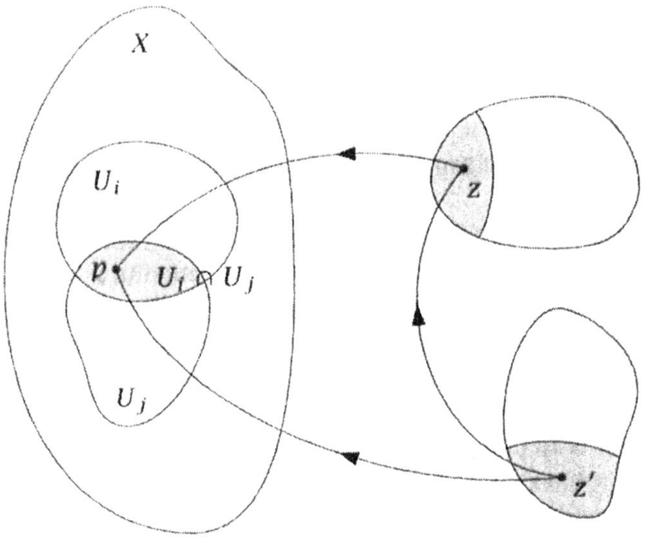

Figure II.1. One identifies z with z' using the algebra of matrices

To this description of X we associate the following C^*-algebra:

Proposition 1. *Let $\mathcal{R}$ be the graph, $\mathcal{R} = \{(z, z') \in V \times V; z \sim z'\}$, of the equivalence relation endowed with its locally compact topology. The following algebraic operations turn the vector space $C_0(\mathcal{R})$ of continuous functions vanishing at infinity on $\mathcal{R}$ into a C^*-algebra, $C^*(\mathcal{R})$, which is strongly Morita equivalent to $C(X)$:*

$$(f * g)(z, z'') = \sum_{z \sim z' \sim z''} f(z, z')g(z', z'')$$

$$(f^*)(z, z') = \overline{f(z', z)}.$$

Indeed, by construction, $C^*(\mathcal{R})$ is identical with the C^*-algebra of compact endomorphisms of the continuous field of Hilbert spaces $(\mathcal{H}_x)_{x\in X}$ over X with $\mathcal{H}_x = \ell^2(p^{-1}(x))$ $\forall x \in X$ (cf. Appendix A).

The locally compact space $\mathcal{R}$ is the disjoint union of the open sets $\mathcal{R}_{ij} = \{(z,z') \in U_i \times U_j;\, p(z) = p(z')\} \simeq U_i \cap U_j$. Thus a function $f \in C_0(\mathcal{R})$ can be viewed as a matrix (f_{ij}) of functions where each f_{ij} belongs to $C_0(U_i \cap U_j)$, while the algebraic rules are just the matrix rules. Since (U_i) is an open covering of X one can easily construct an idempotent $e \in C^*(\mathcal{R})$, $e = e^* = e^2$, $e = (e_{ij})$ which at each point $x \in X$ gives a matrix $(e_{ij}(x))$ of rank one, thus exhibiting the strong Morita equivalence

$$C^*(\mathcal{R}) \simeq C(X).$$

The C^*-algebra $C^*(\mathcal{R})$ is *not commutative* but is strongly Morita equivalent to a commutative C^*-algebra. The latter algebra is uniquely determined since strong Morita equivalence between two commutative C^*-algebras is the same as isomorphism.

In trading $C(X)$ for $C^*(\mathcal{R})$ we lose the commutativity, but we keep the same spectrum, i.e. the topological space of irreducible representations (canonically isomorphic to X), and the same topological invariants such as K-theory: $K(C^*(\mathcal{R})) = K(X)$. But the main gain is that we use the topology of the quotient space X nowhere in the construction of $C^*(\mathcal{R})$. The crucial ingredients were the topology of $\mathcal{R}$ and its composition law, $(z,z')\circ(z',z'') = (z,z'')$.

As we shall see later, these ingredients still exist in situations where the quotient topology of X is the coarse topology, and hence is useless.

2.β The dual of the infinite dihedral group $\Gamma = \mathbb{Z} \rtimes \mathbb{Z}/2$

Let us now describe another example in which the above two algebraic operations of quotient 1) and 2) yield obviously different (not strongly Morita equivalent) algebras. Thus, let $Y = I_1 \cup I_2$ be the disjoint union of two copies I_1 and I_2 of the interval $[0,1]$ and on Y let $\mathcal{R}$ be the equivalence relation which identifies (s,i) with (s,j) for any $i,j \in \{1,2\}$ provided $s \in \,]0,1[$. In other words the quotient space $X = Y/\mathcal{R}$ is obtained by gluing the two interiors of the intervals I_1 and I_2 but not the end points.

If we apply the operation 1) we get a subalgebra of $C(I_1 \cup I_2)$ and since two continuous functions $f \in C[0,1]$ which agree on a dense (open) set are equal, we see that this subalgebra is $C([0,1])$. In particular it is homotopic (Appendix A Definition 10) to $\mathbb{C}$ and its K-theory group is $K_0 = \mathbb{Z}$.

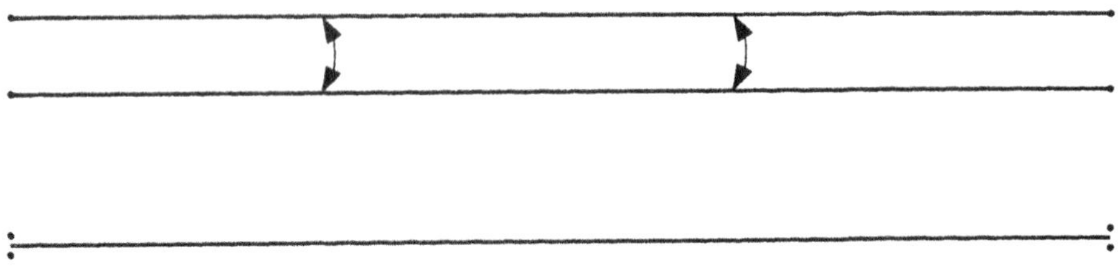

Figure II.2. The dual of the infinite dihedral group. The points inside the intervals are identified, but not the endpoints.

If we apply the operation 2) we get the C^*-subalgebra of the C^*-algebra $M_2(C([0,1])) = M_2(\mathbb{C}) \otimes C([0,1])$ given by

$$A = \left\{(x(t))_{t\in[0,1]} \in M_2(C([0,1]))\ ;\ x(0) \text{ and } x(1) \text{ are diagonal matrices}\right\}.$$

(We view a generic element $x \in M_2(C([0,1]))$ as a continuous map $t \mapsto x(t) \in M_2(\mathbb{C})$, $t \in [0,1]$.)

Obviously, the space of irreducible representations of A is the space $X = Y/\mathcal{R}$ with its non-Hausdorff topology. The K-theory of this C^*-algebra A is much less trivial than that of $C[0,1]$. Indeed, from the exact sequence of C^*-algebras given by evaluation at the end points

$$0 \to J \to A \to \mathbb{C}^4 \to 0$$

where $J = M_2(\mathbb{C}) \otimes C_0(]0,1[)$, one derives by using the six-term exact sequence of K-groups (Section 1) that $K_0(A)$ is isomorphic to $\mathbb{Z}^3$.

The above example shows clearly that the two operations 1) and 2) *do not* in general yield equivalent results, and 2) is plainly much finer.

II.3. The Space X of Penrose Tilings

I had the good fortune to attend a lecture by R. Penrose on quasiperiodic tilings of the plane (Figure 3). Penrose has constructed such tilings using two simple tiles A and B. The tiling T depends on successive choices made in the course of the construction. More precisely, to construct such a tiling one must choose a sequence $(z_n)_{n=0,1,2,\ldots}$ of 0's and 1's that satisfies the following coherence rule:

$$z_n = 1 \Longrightarrow z_{n+1} = 0. \tag{1}$$

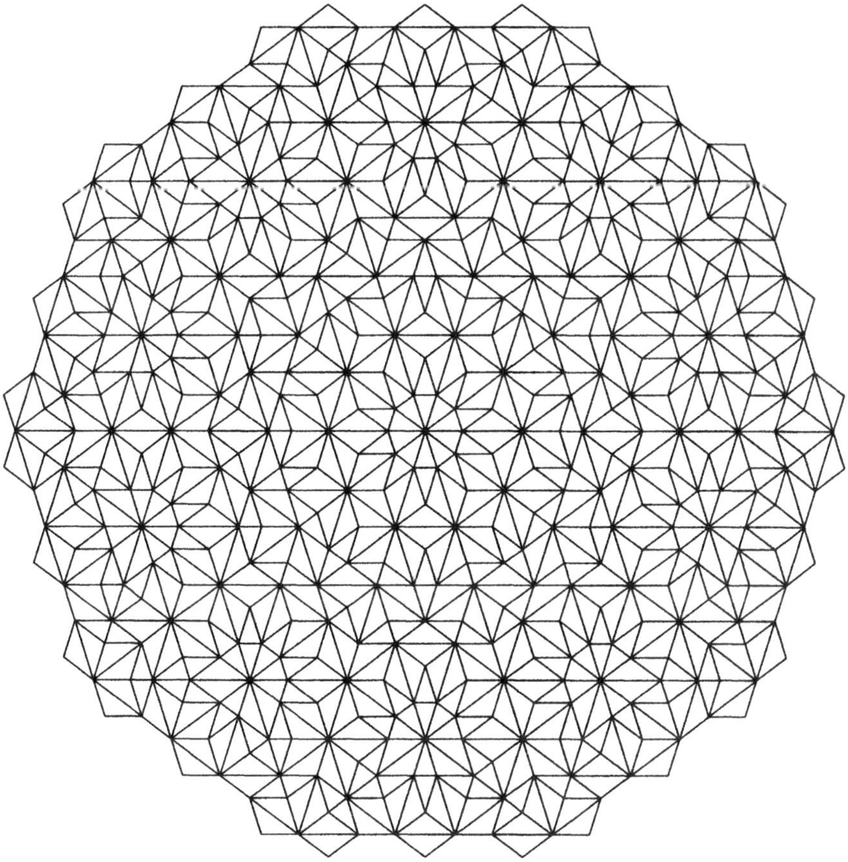

Figure II.3. Penrose tiling

We use the value of z_n as indicated in the construction of T in Appendix D. However, it can happen that two different sequences $z = (z_n)$ and $z' = (z'_n)$ lead to the same tiling; in fact,

$$T \text{ is identical to } T' \Longleftrightarrow \exists n \text{ such that } z_j = z'_j \;\; \forall j \geq n. \qquad (2)$$

Thus, the space X of tilings obtained is the quotient set $K/\mathcal{R}$ of a compact space K by an equivalence relation $\mathcal{R}$. The compact space K, homeomorphic to the Cantor set, is the set of sequences $z = (z_n)_{n \in \mathbb{N}}$ of 0's and 1's that satisfy the rule (1); it is by construction a closed subset of the product of an infinite number of two-element sets. The equivalence relation $\mathcal{R}$ is given by

$$z \sim z' \Longleftrightarrow \exists n \text{ such that } z_j = z'_j \;\; \forall j \geq n. \qquad (3)$$

Thus X is the quotient space $K/\mathcal{R}$. It is exactly in this form that Penrose presented his set of tilings in his exposition. If one tries to understand the space

X as an ordinary space, one sees very quickly that the classical tools do not work, and do not distinguish X from the space consisting of a single point. For example, given two tilings T_1 and T_2 and any finite portion P of T_1, one can find exactly the same configuration P occurring in T_2; thus no finite portion enables one to distinguish T_1 from T_2 ([Gru-S]). Thus any configuration which does occur in *some* tiling T, such as those of Figure 3, will occur (and infinitely many times) in any other tiling T'. This geometric property translates into the triviality of the topology of X. The natural topology on X has for closed sets $F \subset X$ the closed sets of K that are saturated for $\mathcal{R}$. But, every equivalence class for $\mathcal{R}$ is dense in K, and it follows that the only closed sets $F \subset X$ are $F = \emptyset$ and $F = X$. Thus the topology of X is trivial and does not distinguish X from a point; it is, of course, not Hausdorff and it contains no interesting information.

One possible attitude toward such an example would be to say that, up to fluctuations, there is *only one* tiling T of the plane and not to be disturbed by the distinction between X and a point. However, we shall see that X is a very interesting "noncommutative" space or "quantum" space, and that one of its topological invariants, the dimension group, is the subgroup of $\mathbb{R}$ generated by $\mathbb{Z}$ and the golden number $\frac{1+\sqrt{5}}{2}$. This will show, in particular, why the density, or frequency of appearance, of a motif in the tiling must, entirely on account of the topology of the set of all tilings, be an element of the group $\mathbb{Z} + \left(\frac{1+\sqrt{5}}{2}\right)\mathbb{Z}$.

Let us now give the construction of the C^*-algebra A of the space $X = K/\mathcal{R}$ of Penrose tilings (see also [Cu-K]). A general element $a \in A$ is given by a matrix $(a_{z,z'})$ of complex numbers, indexed by the pairs $(z, z') \in \mathcal{R}$. The product of two elements of A is given by the matrix product $(ab)_{z,z''} = \sum_{z'} a_{z,z'} b_{z',z''}$. To each element x of X there corresponds an equivalence class for $\mathcal{R}$, which is a denumerable subset of K. One can therefore associate to x the Hilbert space ℓ^2_x having this denumerable set for an orthonormal basis. Every element a of A defines an operator on ℓ^2_x by the equation

$$(a(x)\zeta)_z = \sum_{z'} a_{z,z'} \zeta_{z'} \quad \forall \zeta \in \ell^2_x.$$

For $a \in A$ the norm $\|a(x)\|$ of the operator $a(x)$ is finite and does not depend on $x \in X$; it is the C^*-algebra norm. Of course, to get more technical, one must give a precise definition of the class of matrices a that we are considering. Let us do it. The relation $\mathcal{R} \subset K \times K$ is the increasing union of the relations $\mathcal{R}_n = \{(z, z'); z_j = z'_j \ (\forall j \geq n)\}$, and as such $\mathcal{R}$ inherits a natural locally compact topology (not equal to its topology as a subset of $K \times K$). Then A is by definition the norm closure of $C_c(\mathcal{R})$ in the above norm. One can show that every element $a \in A$ of this norm closure comes from a matrix $(a_{z,z'})$, $(z, z') \in \mathcal{R}$.

We can summarize the above construction by saying that, while the space X cannot be described nontrivially by means of functions with values in $\mathbb{C}$, there exists a very rich class of operator-valued functions on X:

$$a(x) \in \mathcal{L}(\ell^2_x) \quad \forall x \in X.$$

The algebraic structure of A is dictated by this point of view, because one has

$$(\lambda a + \mu b)(x) = \lambda a(x) + \mu b(x)$$
$$(ab)(x) = a(x)b(x)$$

for all $a, b \in A$, $\lambda, \mu \in \mathbb{C}$ and $x \in X$.

To exhibit clearly the richness of the C^*-algebra A, I will now give an equivalent description of this C^*-algebra, as an inductive limit of finite-dimensional algebras. The Cantor set K is by construction the projective limit of the finite sets K_n, where K_n is the set of sequences of $n+1$ elements $(z_j)_{j=0,1,\dots,n}$ of 0's and 1's, satisfying the rule $z_j = 1 \Longrightarrow z_{j+1} = 0$. There is an obvious projection $K_{n+1} \to K_n$ that consists in 'forgetting' the final z_{n+1}. On the finite set K_n, consider the relation

$$\mathcal{R}^n = \{(z, z') \in K_n \times K_n \ ; \ z_n = z'_n\}.$$

Every function $a = a_{z,z'}$ on $\mathcal{R}^n$ defines an element $\tilde{a}$ of the C^*-algebra A by the equations

$$\begin{aligned} \tilde{a}_{z,z'} &= a_{(z_0,\dots,z_n),(z'_0,\dots,z'_n)} && \text{if } (z, z') \in \mathcal{R}^n \\ \tilde{a}_{z,z'} &= 0 && \text{if } (z, z') \notin \mathcal{R}^n. \end{aligned}$$

Moreover, the subalgebra of A obtained from the functions on $\mathcal{R}^n$ is easily calculated: it is the direct sum $A_n = M_{k_n}(\mathbb{C}) \oplus M_{k'_n}(\mathbb{C})$ of two matrix algebras, where k_n (resp. k'_n) is the number of elements of K_n that end with 0 (resp. with 1). Finally, the inclusion $A_n \to A_{n+1}$ is uniquely determined by the equalities $k_{n+1} = k_n + k'_n$ and $k'_{n+1} = k_n$, which permit embedding $M_{k_n} \oplus M_{k'_n}$ as block matrices into $M_{k_{n+1}}$ and by the homomorphism $(a, a') \to a$ into $M_{k'_{n+1}}$.

The C^*-algebra A is then the inductive limit of the finite-dimensional algebras A_n, and one can calculate its invariants for the classification of Bratteli, Elliott, Effros, Shen and Handelman ([Br], [El$_1$], [El$_2$], [E-H-S]) of these particular C^*-algebras. The invariant to be calculated, due to G. Elliott, is an ordered group, namely the group $K_0(A)$, described earlier, generated by the stable isomorphism classes of finite projective modules over A. In an equivalent way, it is generated by the equivalence classes of projections, as in the work of Murray and von Neumann on factors.

A projection $e \in M_k(A)$ is an element of $M_k(A)$ such that $e^2 = e$ and $e = e^*$. The projections e and f are equivalent if there exists $u \in M_k(A)$ such that $u^*u = e$ and $uu^* = f$. In order to be able to add equivalence classes of projections, one uses the matrices $M_n(A)$, with n arbitrary; this permits assigning a meaning to

$$e \oplus f = \begin{bmatrix} e & 0 \\ 0 & f \end{bmatrix}$$

for two projections $e, f \in M_k(A)$. The ordered group $(K_0(A), K_0(A)^+)$ is obtained canonically from the semigroup of equivalence classes of projections $e \in M_n(A)$ by the usual symmetrization operation by which one passes from the semigroup $\mathbb{N}$ of nonnegative integers to the ordered group $(\mathbb{Z}, \mathbb{Z}^+)$ of integers.

This ordered group is very easy to calculate for finite-dimensional algebras such as the algebras A_n, and for the algebra $A = \overline{\bigcup A_n}$ encountered earlier. Since A_n is the direct sum of two matrix algebras, we have

$$K_0(A_n) = \mathbb{Z}^2 \, , \; K_0(A_n)^+ = \mathbb{Z}^+ \oplus \mathbb{Z}^+ \subset \mathbb{Z} \oplus \mathbb{Z}.$$

The ordered group $(K_0(A), K_0(A)^+)$ is then the inductive limit of the ordered groups $(\mathbb{Z} \oplus \mathbb{Z} \, , \; \mathbb{Z}^+ \oplus \mathbb{Z}^+)$, the inclusion of the nth into the $(n+1)$st being given by the matrix

$$\begin{bmatrix} 1 & 1 \\ 1 & 0 \end{bmatrix}$$

which corresponds to the inclusion $A_n \subset A_{n+1}$ described earlier.

Since this matrix defines a bijection $(a, b) \to (a + b, a)$ of $\mathbb{Z}^2$ onto $\mathbb{Z}^2$, the desired inductive limit is the group $K_0(A) = \mathbb{Z}^2$. However, this bijection is not a bijection of $\mathbb{Z}^+ \oplus \mathbb{Z}^+$ onto $\mathbb{Z}^+ \oplus \mathbb{Z}^+$, and in the limit the semigroup $K_0(A)^+$ becomes $K_0(A)^+ = \{(a, b) \in \mathbb{Z}^2 ; \left(\frac{1+\sqrt{5}}{2}\right) a + b \geq 0\}$, as one sees on diagonalizing the matrix $\begin{bmatrix} 0 & 1 \\ 1 & 1 \end{bmatrix}$ (Figure 4).

It follows that, modulo the choice of basis of $\mathbb{Z}^2$, i.e. modulo $PSL(2, \mathbb{Z})$, the golden number appears as a topological invariant of the space X through the C^*-algebra A. In another formulation, one shows that this C^*-algebra has a unique trace τ. Thus there exists a unique linear form τ on A such that $\tau(xy) = \tau(yx)$ for all $x, y \in A$, and $\tau(1) = 1$. The values of τ on the projections form the intersection $\left(\mathbb{Z} + \frac{1+\sqrt{5}}{2} \mathbb{Z}\right) \cap \mathbb{R}_+$. This trace τ is *positive*, i.e., satisfies

$$\tau(a^* a) \geq 0 \quad (\forall a \in A)$$

and it may be calculated for $a = a_{(z,z')}$ directly as the integral of the diagonal entries of the matrix a:

$$\tau(a) = \int_K a_{(z,z)} \mathrm{d}\mu(z),$$

where the probability measure μ on K is uniquely determined by the condition $\tau(ab) = \tau(ba)$ $(\forall a, b \in A)$. In classical measure theory, given a Radon measure ρ on a compact space Y, the Hilbert space $L^2(Y, \rho)$ is obtained as the completion of $C(Y)$ for the scalar product $\langle f, g \rangle = \int f \bar{g} \, \mathrm{d}\rho$, and $L^\infty(Y, \rho)$ is the weak closure of the algebra $C(Y)$ acting by multiplication on $L^2(Y, \rho)$. In the case that we are interested in, the algebra A replaces $C(Y)$, the positive trace τ replaces the Radon measure ρ, and the weak closure of the action of A by left multiplication on $L^2(A, \tau)$ (the completion of A for the scalar product $\langle a, b \rangle = \tau(a^* b)$) is the hyperfinite factor of type II_1 of Murray and von Neumann [Mur-N], which we shall denote by R.

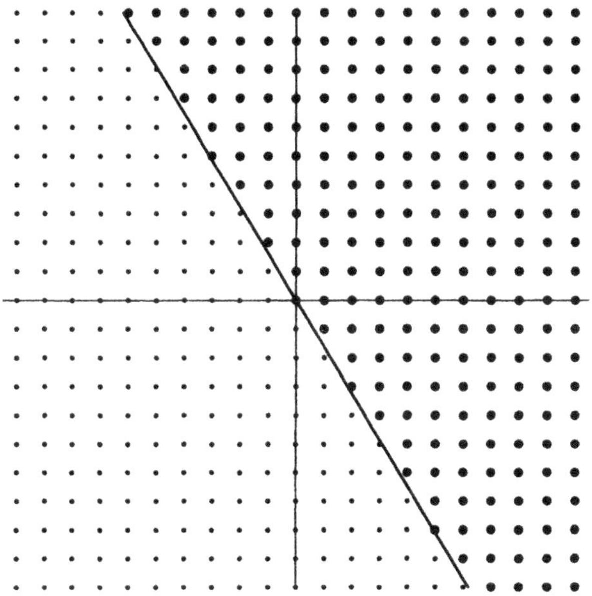

Figure II.4. $K_0^+(A) = \left\{(n,m) \in \mathbb{Z}^2 \; ; \; n\left(\frac{1+\sqrt{5}}{2}\right) + m \geq 0\right\}$

Whereas the continuous dimension of Murray and von Neumann can take on all the positive real values for the projections of R (or of the matrix rings $M_n(R)$) since $\dim(e) = \tau(e)$, for the projections that belong to A this dimension can only take values in the subgroup $\mathbb{Z} + \left(\frac{1+\sqrt{5}}{2}\right)\mathbb{Z}$, which accounts for the role of these numbers in measuring the densities of tiles, or of patterns of a given type in a generic Penrose tiling, in accordance with our interpretation in Chapter I of the continuous dimensions as densities. To summarize, we have shown in this example that the "topology" of X is far from being trivial, that it gives rise to the C^*-algebra A, which is a *simple* C^*-algebra, uniquely characterized as a C^*-algebra (up to Morita equivalence) by the properties

1) A is the inductive limit of finite-dimensional algebras; it is said to be approximately finite (or AF).

2) $(K_0(A), K_0(A)^+) = \left(\mathbb{Z}^2, \left\{(a,b); \left(\frac{1+\sqrt{5}}{2}\right)a + b \geq 0\right\}\right).$

Finally, we note that the C^*-algebra A is exactly the one that appears in the construction by Vaughan Jones [Jone$_2$] of subfactors of index less than 4, for index equal to the golden ratio (Chapter V, Section 10). This can be exploited to describe explicitly the factor in this geometric situation.

II.4. Duals of Discrete Groups and the Novikov Conjecture

We have seen in the above example of the space X of Penrose tilings how the pathologies that a non-Hausdorff space exhibits from the set theoretic point of view disappear when it is considered from the algebraic and noncommutative point of view. To a conservative mathematician this example might appear as rather special, and one could be tempted to stay away from such spaces by dealing exclusively with more central parts of mathematics. Since every finitely presented discrete group Γ appears as the fundamental group, $\Gamma = \pi_1(M)$, of a smooth compact 4-manifold, it would hardly be tenable to exclude discrete groups as well. As we shall see shortly, however, as soon as a discrete group Γ fails to be of type I (a finitely generated discrete group Γ is of type I only if it contains an abelian normal subgroup $\Gamma_0 \subset \Gamma$ of finite index, which is of course very rare), its dual space, i.e. the space $X = \hat{\Gamma}$ of irreducible representations of Γ, is of the same nature as the space of Penrose tilings. Fortunately, as in the example of Penrose tilings all the unpleasant properties of this space from the set theoretic point of view disappear when we treat it from the algebraic point of view.

When the group Γ is abelian, its dual space $\hat{\Gamma} = X$ is a compact space, the Pontryagin dual of Γ, whose topology is characterized by the commutative C^*-algebra $C(X)$ of continuous functions on X. This C^*-algebra has, thanks to the Fourier transform, an equivalent description as the norm closure $C^*(\Gamma)$ of the group ring $\mathbb{C}\Gamma$ of Γ in the regular representation of Γ in $\ell^2(\Gamma)$. More precisely, every element $a = (a_g)_{g\in\Gamma}$ of $\mathbb{C}\Gamma$ is a function with finite support on Γ, and it acts in the Hilbert space $\ell^2(\Gamma)$ as a convolution operator:

$$(a * \xi)_g = \sum_{g_1 g_2 = g} a_{g_1} \xi_{g_2} \quad \forall \xi \in \ell^2(\Gamma).$$

The C^*-algebra $C^*(\Gamma)$ is the norm closure of this algebra of operators. Let us now give a simple example of a non-type-I discrete group and investigate its dual from a set theoretic point of view. Let Γ be the semidirect product of the abelian group $\mathbb{Z}^2$ by the group $\mathbb{Z}$, acting on $\mathbb{Z}^2$ by the powers of the automorphism $\alpha \in \mathrm{Aut}(\mathbb{Z}^2)$, given by the two-by-two matrix $\alpha = \begin{bmatrix} 1 & 1 \\ 1 & 2 \end{bmatrix}$. By construction, Γ is a finitely generated solvable group. Let Y be the 2-torus which is the dual of the normal subgroup $\mathbb{Z}^2 \subset \Gamma$. The theory of induced representations [M₄] shows immediately that the dual space $\hat{\Gamma}$ of Γ contains the space $Y/\mathbb{Z}$ of orbits of the transformation $\hat{\alpha}$ of Y, since each such orbit defines an irreducible representation of Γ, and different orbits yield inequivalent representations. The quotient space $Y/\mathbb{Z}$ is of course not Hausdorff and is of the same nature as the space X of Penrose tilings, thus a fortiori the dual space $\hat{\Gamma}$ of Γ has, if one tries to understand it with the classical set theoretic tools, a pathological aspect. This aspect disappears if, as we did for the space of Penrose tilings, we analyse $\hat{\Gamma}$ by means of the associated noncommutative algebra. Thus for instance the *measure theory* of $\hat{\Gamma}$ for the Plancherel measure class is described by the von Neumann algebra $\lambda(\Gamma)''$, the weak closure of the left regular representation of Γ in $\ell^2(\Gamma)$ (cf. Chapter V). This remains true for arbitrary discrete groups. In our example $\lambda(\Gamma)''$ is the hyperfinite factor R of type II_1. For an arbitrary discrete group one always has a finite trace τ, the Plancherel measure, on the von Neumann algebra $\lambda(\Gamma)''$, and moreover the following equivalence holds:

$$\Gamma \text{ is amenable} \iff \lambda(\Gamma)'' \text{ is hyperfinite}$$

(cf. Chapter V).

The topology of $\hat{\Gamma}$ is of course described by the convolution C^*-algebra of Γ, but this requires some elaboration when the group Γ is not amenable. The point is that in this case there is a natural closed subset $\hat{\Gamma}_r$ of the dual $\hat{\Gamma}$, the support of the Plancherel measure, called the *reduced dual* of Γ, so that one has two natural convolution C^*-algebras of Γ, namely:

$$C^*(\Gamma) = \text{“}C(\hat{\Gamma})\text{”}\,,\ C_r^*(\Gamma) = \text{“}C(\hat{\Gamma}_r)\text{”}.$$

By definition $C^*(\Gamma)$ is the completion of the group ring $\mathbb{C}\Gamma$ for the norm

$$\|a\|_{\max} = \sup\{\|\pi(a)\|,\ \pi \text{ unitary representation of } \Gamma\}.$$

It is a C^*-algebra whose representations are exactly the unitary representations of Γ.

The reduced dual $\hat{\Gamma}_r$ of Γ is the space of irreducible representations of Γ which are weakly contained in the regular representation ([Di$_3$]). It is the space of irreducible representations of the C^*-algebra $C_r^*(\Gamma)$ which is the norm closure of $\mathbb{C}\Gamma$ in $\ell^2(\Gamma)$.

To the inclusion $\hat{\Gamma}_r \subset \hat{\Gamma}$ corresponds a surjection of C^*-algebras

$$C^*(\Gamma) \xrightarrow{r} C_r^*(\Gamma).$$

The K-theory of the C^*-algebra of a discrete group Γ plays a crucial role in the work of Mishchenko and Kasparov on the conjecture of Novikov on homotopy invariance of the higher signatures for non-simply-connected manifolds. Let us first recall the statement of this conjecture. Let Γ be a discrete group and $x \in H^*(B\Gamma, \mathbb{R}) = H^*(\Gamma, \mathbb{R})$ a group cocycle; then the pair (Γ, x) satisfies the Novikov conjecture iff the following number is a homotopy invariant for maps (M, Ψ) of compact oriented manifolds M to the classifying space $B\Gamma$:

$$\mathrm{Sign}_x(M, \Psi) = \langle L(M) \cup \Psi^*(x), [M] \rangle.$$

In other words what is asserted is that if (M', Ψ') is another pair consisting in a compact oriented manifold M' and a map $\Psi' : M' \to B\Gamma$, which is homotopic to the pair (M, Ψ), then

$$\mathrm{Sign}_x(M, \Psi) = \mathrm{Sign}_x(M', \Psi').$$

As a corollary it follows that one has a homotopy invariant for compact oriented manifolds M with fundamental group Γ; the map Ψ is then automatically given as the classifying map of the universal cover $\widetilde{M}$ of M.

The cohomology class $L(M)$ is the Hirzebruch L-genus which enters in the Hirzebruch signature theorem. The latter asserts that for a $4k$-manifold the signature of the intersection form on H^{2k} is given by a universal polynomial $L_k(p_1,\ldots,p_k)$ in the rational Pontryagin classes of the manifold:

$$\mathrm{Sign}(M) = \langle L_k(p_1,\ldots,p_k),[M]\rangle.$$

The Hirzebruch formula only involves the top-dimensional component of the L-genus but the above higher signature formula uses all its components (cf. [Mi$_4$]):

$$L(M) = \sum_k L_k(p_1,\ldots,p_k)$$

$$L_1(p_1) = \frac{1}{3}p_1\,,\ L_2(p_1,p_2) = \frac{1}{45}(7p_2 - p_1^2),\ldots.$$

One says that a discrete group Γ satisfies the Novikov conjecture if the above assertion holds for any group cocycle $x \in H^*(\Gamma,\mathbb{R})$. The natural generalization of the signature $\mathrm{Sign}(M)$ of an oriented compact $4k$-manifold to the non-simply-connected situation yields (cf. [Mis$_2$]) an element $\mathrm{Sign}_\Gamma(M)$, where $\Gamma = \pi_1(M)$, in the Wall group $L_{4k}(\mathbb{Q}[\Gamma])$. This follows from the theory of algebraic Poincaré complexes due to Mishchenko [Mis$_2$]. Here $\mathbb{Q}[\Gamma]$ is the group ring of Γ with rational coefficients, and the Wall group L_{4k} of a ring with involution is equal to the Witt group which classifies symmetric bilinear forms modulo the hyperbolic ones. The crucial reason for the role of the K-theory of the C^**-algebra* of the group Γ is the isomorphism ρ

$$\mathrm{Witt}(A) \simeq K_0(A)$$

valid for any C^*-algebra (but false for involutive Banach algebras in general), which allows one to extract from the homotopy invariant $\mathrm{Sign}_\Gamma(M) \in L_{4k}(\mathbb{Q}[\Gamma])$, a homotopy invariant belonging to a K-theory group, namely the image

$$\rho\,\mathrm{Sign}_\Gamma(M) \in K_0(C^*(\Gamma)).$$

Simple examples, such as the case of commutative Γ (cf.[Lus]) show that in such cases the relevant information is not lost in retaining only the K-theory signature, and that the latter, being in the topological K-theory of the Pontryagin dual $\hat{\Gamma}$ of Γ which is a torus, is much easier to use than the L-theory signature. This point is an essential motivation to study the K-theory of the C^*-algebras of discrete groups.

In L-theory there is a natural spectrum, the L-theory spectrum and a map, called the assembly map, $\alpha : h_*(\Gamma,L) \to L_*(\mathbb{Q}(\Gamma))$, whose range contains all the elements $\mathrm{Sign}_\Gamma(M)$ and whose rational injectivity implies the Novikov conjecture. Kasparov and Mishchenko have constructed an analytic assembly map from the K-homology $K_*(B\Gamma) = h_*(B\Gamma, \mathbf{BU})$ (where $\mathbf{BU}$ is the K-theory spectrum) of the classifying space $B\Gamma$ of Γ to the K-theory of the C^*-algebra of Γ

$$K_*(B\Gamma) \xrightarrow{\mu} K_*(C^*(\Gamma))$$

which makes the following diagram commute:

$$\begin{array}{ccc} h_*(B\Gamma, \mathbf{L}) & \xrightarrow{\alpha} & L_*(\mathbb{C}(\Gamma)) \\ \downarrow & & \downarrow \rho \\ K_*(B\Gamma) & \xrightarrow{\mu} & K_*(C^*(\Gamma)) \end{array} \qquad (*)$$

where the horizontal arrows are the assembly maps, the left vertical arrow comes from the natural map from the L-theory spectrum to the BU-spectrum, and the right vertical arrow comes from the equality for C^*-algebras of L-theory with K-theory.

The simplest description of the map $\mu : K_*(B\Gamma) \to K_*(C^*(\Gamma))$ is based on the existence for every compact subset K of $B\Gamma$ of a canonical "Mishchenko line bundle"

$$\ell_K \in K_0(C(K) \otimes C^*(\Gamma))$$

which is described as a finite projective C^*-module over $C(K) \otimes C^*(\Gamma)$ by the following proposition:

Proposition 1. *Let K be a compact space, $\varphi : K \to B\Gamma$ a continuous map and $\widetilde{K} \xrightarrow{p} K$ the principal Γ-bundle over K, the pull-back of $E\Gamma \to B\Gamma$. Let $\mathcal{E}$ be the completion of $C_c(\widetilde{K})$ for the norm $\|\xi\| = \|\langle\xi, \xi\rangle\|^{1/2}$, where for $\xi, \eta \in C_c(\widetilde{K})$ the element $\langle\xi, \eta\rangle$ of $C(K) \otimes C^*(\Gamma)$ is defined by*

$$\langle\xi, \eta\rangle = \sum_{g \in \Gamma} \langle\xi, \eta\rangle_g \otimes g$$

$$\langle\xi, \eta\rangle_g(x) = \sum_{p(\tilde{x})=x} \bar{\xi}(\tilde{x})\eta(g^{-1}\tilde{x}) \quad \forall x \in K.$$

Then $\mathcal{E}$ is a right C^*-module, finite and projective, over $C(K) \otimes C^*(\Gamma)$, for the right action uniquely defined by the equality

$$(\xi \cdot (f \otimes g))\,(\tilde{x}) = f(p(\tilde{x}))\xi(g\tilde{x})$$

$\forall \xi \in C_c(\widetilde{K}), f \in C(K), g \in \Gamma, \tilde{x} \in \widetilde{X}.$

The proof is straightforward. The local triviality of the principal Γ-bundle $\widetilde{X}$ over X yields in fact an explicit idempotent $e \in M_n(C(K) \otimes C^*(\Gamma))$ whose associated right module is $\mathcal{E}$.

The construction of ℓ_K is naturally compatible with the inductive system of compact subsets of $B\Gamma$. By definition the K-homology of $B\Gamma$ is the inductive limit

$$h_*(B\Gamma, \mathbf{BU}) = \pi_*(B\Gamma \wedge \mathbf{BU}) = \varinjlim \pi_*(K \wedge \mathbf{BU})$$

where K runs through finite subcomplexes of $B\Gamma$. To construct μ one uses the following formula which holds both in the Kasparov KK-theory and in E-theory:

$$\mu(z) = (z \otimes 1) \circ \ell_K \in E(\mathbb{C}, C^*(\Gamma)) = K(C^*(\Gamma)) \qquad (**)$$

$$\forall z \in h_*(K, \mathbf{BU}) = E(C(K), \mathbb{C}).$$

One has $z \otimes 1 \in E(C(K) \otimes C^*(\Gamma), C^*(\Gamma))$ and $\ell_K \in E(\mathbb{C}, C(K) \otimes C^*(\Gamma))$ so that $\mu(z) \in E(\mathbb{C}, C^*(\Gamma)) = K(C^*(\Gamma))$.

One then has the following index theorem of Kasparov and Mishchenko ([Kas$_7$] [Kas$_6$] [Mis$_3$] [Mis-S])

Theorem 2. *The analytic assembly map,* $\mu : K_*(B\Gamma) \to K_*(C^*(\Gamma))$ *defined by* $(**)$ *makes the diagram* $(*)$ *commute.*

In particular, let M be a compact oriented manifold of even dimension n and $\varphi : M \to B\Gamma$ a continuous map. Then there is a corresponding element $z \in K_*(B\Gamma)$ whose image $\mu(z) \in K_*(C^*(\Gamma))$ is the Γ-equivariant signature of M, i.e. the image $\rho(z')$ of the Mishchenko L-theory class $z' = \mathrm{Sign}_\Gamma(M)$. One has $z = \varphi_*(\sigma(M))$, where $\sigma(M) \in K_*(M)$ is the K-homology class of the signature operator on M. What matters here is that the Chern character of $\sigma(M)$, $\mathrm{ch}_*(\sigma(M)) \in H_*(M, \mathbb{Q})$, is given by

$$\mathrm{ch}_*(\sigma(M)) = 2^{n/2} \mathcal{L}(M) \cap [M],$$

where $\mathcal{L}_k(M) = 2^{-2k} L_k(M)$ for all k. Thus the Novikov conjecture for a discrete group Γ is equivalent to the homotopy invariance of the image $\varphi_*(\sigma(M))$ in the rational K-homology of $B\Gamma$. Since the Mishchenko L-theory class $z' \in L_n(\mathbb{C}\Gamma)$ is homotopy invariant, one gets as a corollary of Theorem 2,

Corollary 3. (loc. cit) *The Novikov conjecture for* Γ *is implied by the rational injectivity of the analytic assembly map*

$$\mu : K_*(B\Gamma) \to K_*(C^*(\Gamma)).$$

This so-called *strong Novikov conjecture* has been proved by Mishchenko for fundamental groups of negatively curved compact Riemannian manifolds ([Mis$_1$]) and by Kasparov for discrete subgroups of Lie groups ([Kas$_5$]).

Thanks to the work of Pimsner and Voiculescu [Pi-V$_3$], of Cuntz [Cu$_1$] for free groups, and of Kasparov [Kas$_4$], Fox and Haskell [F-H] and Kasparov and Julg [Ju-K] for discrete subgroups of $SO(n,1)$ and $SU(n,1)$, one knows for many torsion-free discrete groups Γ that the analytic assembly map μ is an *isomorphism*. However, there is no discrete group Γ, infinite and with Kazhdan's property T, for which $K(C_r^*(\Gamma))$ or $K(C^*(\Gamma))$ has actually been computed. One difficulty is that due to property T the natural homomorphism

$$C^*(\Gamma) \xrightarrow{r} C_r^*(\Gamma)$$

fails to be a K-theory isomorphism. Indeed, the idempotent corresponding by property T to the trivial representation of Γ defines a non-zero K-theory class $[e] \in K_0(C^*(\Gamma))$ whose image $r(e)$ is zero. Thus there are indeed two K-theories to compute. The map μ does not contain $[e]$ in its range and thus cannot yield an isomorphism with $K(C^*(\Gamma))$. The possibility that $r \circ \mu = \mu_r$ is an isomorphism is still open, but by a counterexample of G. Skandalis ([Sk$_4$]) it cannot be proved by a KK-theory or E-theory equivalence. When the group Γ has torsion the above map μ is too crude to be expected to yield the K-theory of $C^*(\Gamma)$. Indeed, even with Γ finite, say $\Gamma = \mathbb{Z}/2\mathbb{Z}$, one has $K_0(C^*(\Gamma)) = K_0(\mathbb{C} \oplus \mathbb{C}) = \mathbb{Z}^2$ while $K_0(B\Gamma)_{\mathbb{Q}} = \mathbb{Q}$. We shall see later how to modify $K_*(B\Gamma)$ and μ when Γ has torsion.

II.5. The Tangent Groupoid of a Manifold

Let M be a compact smooth manifold. The pseudodifferential calculus on M allows one to show that an elliptic pseudodifferential operator D on M is a Fredholm operator and hence has an index, $\mathrm{Ind}(D) \in \mathbb{Z}$. The Fredholm theory then shows that this index only depends upon the K-theory class of the symbol of D. One thus obtains ([At-Si$_1$]) an additive map, the analytic index map

$$\mathrm{Ind}_a : K_0(T^*M) \to \mathbb{Z}$$

from the K-theory of the locally compact space T^*M to $\mathbb{Z}$.

In this section we shall show how the same map Ind_a arises naturally from a geometric construction, that of the tangent groupoid of the manifold M. This groupoid encodes the deformation of T^*M to a single point, using the equivalence relation on $M \times [0,1]$ which identifies any pairs (x,ε) and (y,ε) provided $\varepsilon > 0$.

In the above section, as well as in Chapter I, we have met implicitly the notion of groupoid. All our algebra structures could be written in the following form:

$$(a * b)(\gamma) = \sum_{\gamma_1 \circ \gamma_2 = \gamma} a(\gamma_1) b(\gamma_2)$$

where the γ's vary in a *groupoid* G, i.e. in a small category with inverses, or more explicitly:

Definition 1. *A groupoid consists of a set G, a distinguished subset $G^{(0)} \subset G$, two maps $r, s : G \to G^{(0)}$ and a law of composition*

$$\circ : G^{(2)} = \{(\gamma_1, \gamma_2) \in G \times G \,;\, s(\gamma_1) = r(\gamma_2)\} \to G$$

such that

(1) $s(\gamma_1 \circ \gamma_2) = s(\gamma_2)$, $r(\gamma_1 \circ \gamma_2) = r(\gamma_1)$ $\forall (\gamma_1, \gamma_2) \in G^{(2)}$

(2) $s(x) = r(x) = x \quad \forall x \in G^{(0)}$

(3) $\gamma \circ s(\gamma) = \gamma$, $r(\gamma) \circ \gamma = \gamma \quad \forall \gamma \in G$

(4) $(\gamma_1 \circ \gamma_2) \circ \gamma_3 = \gamma_1 \circ (\gamma_2 \circ \gamma_3)$

(5) *Each* γ *has a two-sided inverse* γ^{-1}, *with* $\gamma\gamma^{-1} = r(\gamma)$, $\gamma^{-1}\gamma = s(\gamma)$.

The maps r, s are called the range and source maps.

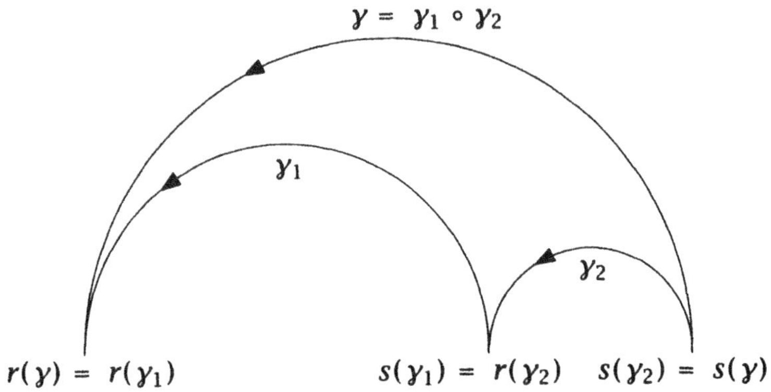

Figure II.5. Groupoid

Here are a few important examples of groupoids.

Equivalence relations.

Given an equivalence relation $\mathcal{R} \subset X \times X$ on a set X, one gets a groupoid in the following obvious way: $G = \mathcal{R}$, $G^{(0)}$ = diagonal of $X \times X \subset \mathcal{R}$, $r(x, y) = x$, $s(x, y) = y$ for any $\gamma = (x, y) \in \mathcal{R} \subset X \times X$ and

$$(x, y) \circ (y, z) = (x, z) \, , \; (x, y)^{-1} = (y, x).$$

Groups.

Given a group Γ one takes $G = \Gamma$, $G^{(0)} = \{e\}$, and the law of composition is the group law.

Group actions. (Section 7).

Given an action $X \times \Gamma \overset{\alpha}{\to} X$ of a group Γ on a set X, $\alpha(x, g) = xg$, so that $x(g_1 g_2) = (x g_1) g_2$ $\forall x \in X$, $g_i \in \Gamma$, one takes $G = X \times \Gamma$, $G^{(0)} = X \times \{e\}$, and

$$r(x, g) = x \, , \; s(x, g) = xg \quad \forall (x, g) \in X \times \Gamma$$

$$(x, g_1)(y, g_2) = (x, g_1 g_2) \text{ if } x g_1 = y$$

$$(x, g)^{-1} = (xg, g^{-1}) \quad \forall (x, g) \in X \times \Gamma.$$

This groupoid $G = X \rtimes \Gamma$ is called the semi-direct product of X by Γ.

In all the examples we have met so far, the groupoid G has a natural locally compact topology and the fibers $G^x = r^{-1}\{x\}$, $x \in G^{(0)}$, of the map r, are *discrete*. This is what allows us to define the convolution algebra very simply by

$$(a * b)(\gamma) = \sum_{\gamma_1 \circ \gamma_2 = \gamma} a(\gamma_1) b(\gamma_2).$$

We refer to [Ren$_1$] [Bro$_2$] for the general case of locally compact groupoids. Our next example of the tangent groupoid of a manifold will be easier to handle than the general case; though no longer discrete, it will be *smooth* in the following sense:

Definition 2. *A smooth groupoid G is a groupoid together with a differentiable structure on G and $G^{(0)}$ such that the maps r and s are submersions, and the object inclusion map $G^{(0)} \to G$ is smooth, as is the composition map $G^{(2)} \to G$.*

The general notion is due to Ehresmann [Ehr$_{1,2,3}$] and the specific definition here to Pradines [Pra] who proved that in a smooth groupoid G, all the maps $s : G^x \to G^{(0)}$ are subimmersions, where $G^x = \{\gamma \in G; r(\gamma) = x\}$.

The notion of a $\frac{1}{2}$-density on a smooth manifold allows one to define in a canonical manner the convolution algebra of a smooth groupoid G. More specifically, given G, we let $\Omega^{1/2}$ be the line bundle over G whose fiber $\Omega^{1/2}_\gamma$ at $\gamma \in G$, $r(\gamma) = x$, $s(\gamma) = y$, is the linear space of maps

$$\rho : \wedge^k T_\gamma(G^x) \otimes \wedge^k T_\gamma(G_y) \to \mathbb{C}$$

such that $\rho(\lambda v) = |\lambda|^{1/2} \rho(v)$ $\forall \lambda \in \mathbb{R}$. Here $G_y = \{\gamma \in G; s(\gamma) = y\}$ and $k = \dim T_\gamma(G^x) = \dim T_\gamma(G_y)$ is the dimension of the fibers of the submersions $r : G \to G^{(0)}$ and $s : G \to G^{(0)}$.

Then we endow the linear space $C_c^\infty(G, \Omega^{1/2})$ of smooth compactly supported sections of $\Omega^{1/2}$ with the convolution product

$$(a * b)(\gamma) = \int_{\gamma_1 \circ \gamma_2 = \gamma} a(\gamma_1) b(\gamma_2) \quad \forall a, b \in C_c^\infty(G, \Omega^{1/2})$$

where the integral on the right-hand side makes sense since it is the integral of a 1-density, namely $a(\gamma_1) b(\gamma_1^{-1}\gamma)$, on the manifold G^x, $x = r(\gamma)$.

As two easy examples of this construction one can take:

α) The groupoid $G = M \times M$ where M is a compact manifold, r and s are the two projections $G \to M = G^{(0)} = \{(x,x); x \in M\}$ and the composition is $(x,y) \circ (y,z) = (x,z)$ $\forall x,y,z \in M$.

The convolution algebra is then the algebra of smoothing kernels on the manifold M.

β) A Lie group G is, in a trivial way, a groupoid with $G^{(0)} = \{e\}$. One then gets the convolution algebra $C_c^\infty(G, \Omega^{1/2})$ of smooth 1-densities on G.

Coming back to the general case, one has:

Proposition 3. *Let G be a smooth groupoid, and let $C_c^\infty(G,\Omega^{1/2})$ be the convolution algebra of smooth compactly supported $\frac{1}{2}$-densities, with involution $*$, $f^*(\gamma) = \overline{f(\gamma^{-1})}$.*

Then for each $x \in G^{(0)}$ the following defines an involutive representation π_x of $C_c^\infty(G,\Omega^{1/2})$ in the Hilbert space $L^2(G_x)$:

$$(\pi_x(f)\xi)(\gamma) = \int f(\gamma_1)\xi(\gamma_1^{-1}\gamma) \qquad \forall \gamma \in G_x\,,\ \xi \in L^2(G_x).$$

The completion of $C_c^\infty(G,\Omega^{1/2})$ for the norm $\|f\| = \sup_{x\in G^{(0)}} \|\pi_x(f)\|$ is a C^*-algebra, denoted $C_r^*(G)$.

We refer to [Co$_{10}$][Ren$_1$] for the proof. As in the case of discrete groups (Section 4) one defines the C^*-algebra $C^*(G)$ as the completion of the involutive algebra $C_c^\infty(G,\Omega^{1/2})$ for the norm

$$\|f\|_{\max} = \sup\{\|\pi(f)\|;\pi \text{ involutive Hilbert space representation of } C_c^\infty(G,\Omega^{1/2})\}.$$

We let $r : C^*(G) \to C_r^*(G)$ be the canonical surjection. We refer to [Ren$_1$] for a discussion of the corresponding notion of amenability of G.

Let us now pass to an interesting example of smooth groupoid, namely we construct the *tangent groupoid* of a manifold M. Let us first describe G at the groupoid level; we shall then describe its smooth structure.

We let $G = (M \times M\times\,]0,1]) \cup (TM)$, where TM is the total space of the tangent bundle of M.

We let $G^{(0)} \subset G$ be $M \times [0,1]$ with inclusion given by

$$(x,\varepsilon) \to (x,x,\varepsilon) \in M \times M\times\,]0,1] \text{ for } x \in M, \varepsilon > 0.$$

$$(x,0) \to x \in M \subset TM \text{ as the 0-section, for } \varepsilon = 0.$$

The range and source maps are given respectively by

$$\begin{cases} r(x,y,\varepsilon) = (x,\varepsilon) & \text{for } \ x \in M, \varepsilon > 0 \\ r(x,X) = (x,0) & \text{for } \ x \in M, X \in T_x(M). \end{cases}$$

$$\begin{cases} s(x,y,\varepsilon) = (y,\varepsilon) & \text{for } \ y \in M, \varepsilon > 0 \\ s(x,X) = (x,0) & \text{for } \ x \in M, X \in T_x(M). \end{cases}$$

The composition is given by

$$\begin{array}{lll} (x,y,\varepsilon)\circ(y,z,\varepsilon) = (x,z,\varepsilon) & \text{for} & \varepsilon > 0 \text{ and } x,y,z \in M \\ (x,X)\circ(x,Y) = (x,X+Y) & \text{for} & x \in M \text{ and } X,Y \in T_x(M). \end{array}$$

Putting this in other words, the groupoid G is the union (a union of groupoids is again a groupoid) of the product G_1 of the groupoid $M \times M$ of example α) by $]0,1]$ (a set is a groupoid where all the elements belong to $G^{(0)}$) and of the groupoid $G_2 = TM$ which is a union of groups: the tangent spaces

$T_x(M)$. This decomposition $G = G_1 \cup G_2$ of G as a disjoint union is true set theoretically but not at the manifold level. Indeed, we shall now endow G with the manifold structure that it inherits from its identification with the space obtained by blowing up the diagonal $\Delta = M \subset M \times M$ in the cartesian square $M \times M$. More explicitly, the topology of G is such that G_1 is an open subset of G and a sequence $(x_n, y_n, \varepsilon_n)$ of elements of $G_1 = M \times M \times \,]0,1]$ with $\varepsilon_n \to 0$ converges to a tangent vector (x, X); $X \in T_x(M)$ iff the following holds:

$$x_n \to x \,,\; y_n \to x \,,\; \frac{x_n - y_n}{\varepsilon_n} \to X.$$

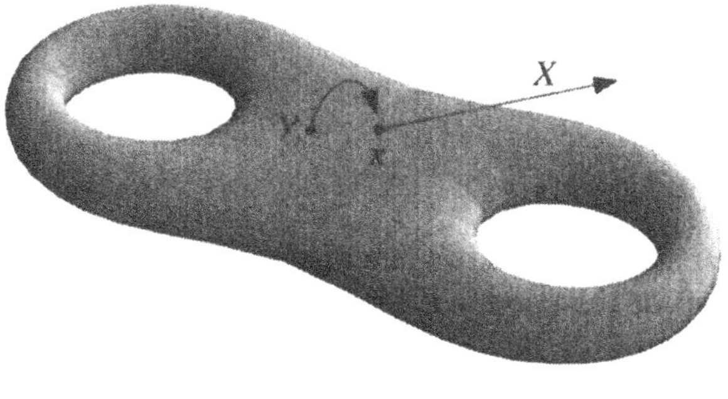

Figure II.6. The tangent groupoid of M

The last equality makes sense in any local chart around x independently of any choice. One obtains in this way a manifold with boundary, and a local chart around a boundary point $(x, X) \in TM$ is provided, for instance, by a choice of Riemannian metric on M and the following map of an open set of $TM \times [0,1]$ to G:

$$\begin{aligned} \psi(x, X, \varepsilon) &= (x, \exp_x(-\varepsilon X), \varepsilon) \in M \times M \times \,]0,1] \,, \text{ for } \varepsilon > 0 \\ \psi(x, X, 0) &= (x, X) \in TM. \end{aligned}$$

Proposition 4. *With the above structure G is a smooth groupoid.*

We shall call it the tangent groupoid of the manifold M and denote it by G_M. The structure of the C^*-algebra of this groupoid G_M is given by the following immediate translation of the inclusion of $G_2 = TM$ as a closed subgroupoid of G_M, with complement G_1.

Proposition 5. 1) *To the decomposition $G_M = G_1 \cup G_2$ of G_M as a union of an open and a closed subgroupoid corresponds the exact sequence of C^*-algebras*

$$0 \to C^*(G_1) \to C^*(G) \xrightarrow{\sigma} C^*(G_2) \to 0.$$

2) *The C^*-algebra $C^*(G_1)$ is isomorphic to $C_0(]0,1]) \otimes \mathcal{K}$, where $\mathcal{K}$ is the elementary C^*-algebra (all compact operators on Hilbert space).*

3) *The C^*-algebra $C^*(G_2)$ is isomorphic to $C_0(T^*M)$, the isomorphism being given by the Fourier transform: $C^*(T_xM) \simeq C_0(T_x^*M)$, for each $x \in M$.*

It follows from 2) that the C^*-algebra $C^*(G_1)$ is *contractible*: it admits a pointwise norm continuous family θ_λ of endomorphisms, $\lambda \in [0,1]$, such that $\theta_0 = \text{id}$ and $\theta_1 = 0$. (This is easy to check for $C_0(]0,1])$.) In particular, from the long exact sequence in K-theory we thus get isomorphisms

$$\sigma_* : K_i(C^*(G)) \simeq K_i(C^*(G_2)) = K^i(T^*M).$$

On the right-hand side $K^i(T^*M)$ is the K-theory with compact supports of the total space of the cotangent bundle. We now have the following geometric reformulation of the analytic index map Ind_a of Atiyah and Singer.

Lemma 6. *Let $\rho : C^*(G) \to \mathcal{K} = C^*(M \times M)$ be the transpose of the inclusion $M \times M \to G$: $(x,y) \to (x,y,1)$ $\forall x,y \in M$. Then the Atiyah-Singer analytic index is given by*

$$\text{Ind}_a = \rho_* \circ (\sigma_*)^{-1} : K^0(T^*M) \to \mathbb{Z} = K_0(\mathcal{K}).$$

The proof is straightforward. The map $\sigma : C^*(G) \to C^*(G_2) \simeq C_0(T^*M)$ is the symbol map of the pseudodifferential calculus for asymptotic pseudodifferential operators ([Gu-S]).

We shall end this Section by giving a proof of the index theorem, closely related to the proof of Atiyah and Singer ([At-Si$_1$]) but which can be adapted to many other situations. Lemma 6 above shows that the analytic index Ind_a has a simple interpretation in terms of the tangent groupoid $G_M = G$. If the smooth groupoids G, G_1, and G_2 involved in this interpretation were equivalent (in the sense of the equivalence of small categories) to ordinary spaces X_j (viewed as groupoids in a trivial way, i.e. $X_j = X_j^{(0)}$), then we would already have a geometric interpretation of Ind_a, i.e. an index formula. Now the groupoid $G_1 = M \times M \times]0,1]$ is equivalent to the space $]0,1]$ since $M \times M$ is equivalent to a single point. Thus the problem comes from G_2 which involves the *groups* T_xM and is not equivalent to a space. Given any smooth groupoid G and a (smooth) homomorphism h from G to the additive group $\mathbb{R}^N$ one can form the following smooth groupoid G_h:

$$G_h = G \times \mathbb{R}^N, \quad G_h^{(0)} = G^{(0)} \times \mathbb{R}^N$$

with $r(\gamma,X) = (r(\gamma),X)$, $s(\gamma,X) = (s(\gamma),X + h(\gamma))$ $\forall \gamma \in G$, $X \in \mathbb{R}^N$, and $(\gamma_1,X_1)\circ(\gamma_2,X_2) = (\gamma_1\circ\gamma_2,X_1)$ for any composable pair.

Heuristically, if G corresponds to a space X, then the homomorphism h fixes a principal $\mathbb{R}^N$-bundle over X and G_h corresponds to the total space of this principal bundle. At the level of the associated C^*-algebras one has the following:

Proposition 7. *Let G be a smooth groupoid, $h : G \to \mathbb{R}^N$ a homomorphism.*

1) *For each character $\chi \in \mathbb{R}_N$ of the group $\mathbb{R}^N$ the following formula defines an automorphism α_χ of $C^*(G)$:*

$$(\alpha_\chi(f))(\gamma) = \chi(h(\gamma))f(\gamma) \quad \forall f \in C_c^\infty(G,\Omega^{1/2}).$$

2) *The crossed product* $C^*(G) \rtimes_\alpha \hat{\mathbb{R}}_N$ *of* $C^*(G)$ *by the above action* α *of* $\hat{\mathbb{R}}_N = (\mathbb{R}^N)^\wedge$ *is the* C^**-algebra* $C^*(G_h)$.

Thus we see in particular that if N is even, the Thom isomorphism for C^*-algebras (Appendix C) gives us a natural isomorphism:

$$K_0(C^*(G)) \simeq K_0(C^*(G_h)).$$

In the case where G corresponds to a space X, the above isomorphism is of course the usual Bott periodicity isomorphism. We shall now see that for a suitable choice of homomorphism $G \xrightarrow{h} \mathbb{R}^N$, where $G = G_M$ is the tangent groupoid of M, the smooth groupoids G_h, $G_{1,h}$, and $G_{2,h}$ will be equivalent to spaces, thus yielding a geometric computation of Ind_a and the index theorem.

Let $M \xrightarrow{j} \mathbb{R}^N$ be an immersion of M in a Euclidean space $\mathbb{R}^N$. Then to j corresponds the following *homomorphism* h of the tangent groupoid G of M into the group $\mathbb{R}^N$:

$$h(x,y,\varepsilon) = \tfrac{j(x)-j(y)}{\varepsilon} \quad \varepsilon > 0$$

$$h(x,X) = j_*(X) \qquad \forall X \in T_x(M).$$

One checks immediately that $j(\gamma_1 \circ \gamma_2) = j(\gamma_1) + j(\gamma_2)$ whenever $(\gamma_1, \gamma_2) \in G^{(2)}$.

This homomorphism h defines a *free and proper* action of G, by translations, on the contractible space $\mathbb{R}^N$. This follows because j is an immersion, so that j_* is injective. The smooth groupoid G_h is thus equivalent to the following space BG, which is the quotient of $G^{(0)} \times \mathbb{R}^N$ by the equivalence relation

$$(x,X) \sim (y,Y) \text{ iff } \exists \gamma \in G \ \ r(\gamma) = x \ , \ s(\gamma) = y \ , \ X = Y + h(\gamma).$$

Since the action is free and proper the quotient makes good sense. Similar statements hold for G_1 and G_2. A straightforward computation yields

$$BG = \left(]0,1] \times \mathbb{R}^N\right) \cup \nu(M)$$

where $\nu(M)$ is the total space of the *normal bundle* of M in $\mathbb{R}^N$.

In this decomposition, $BG = BG_1 \cup BG_2$, one identifies BG_2, the quotient of $G_2^{(0)} \times \mathbb{R}^N = M \times \mathbb{R}^N$ by the action of $G_2 = TM$, with the total space of ν, $\nu_x = \mathbb{R}^N / T_x(M)$. The isomorphism α of $(]0,1] \times \mathbb{R}^N)$ with the quotient BG_1 of $G_1^{(0)} \times \mathbb{R}^N = (]0,1] \times M) \times \mathbb{R}^N$ by the action of G_1 depends upon the choice of a base point $x_0 \in M$, and to simplify the formulae we take $j(x_0) = 0 \in \mathbb{R}^N$. One then has

$$\alpha(\varepsilon, X) = ((x_0, \varepsilon), X) \ \ \forall \varepsilon > 0 \ , \ X \in \mathbb{R}^N.$$

With this notation the locally compact topology of BG is obtained by gluing $]0,1] \times \mathbb{R}^N$ to $\nu(M)$ by the following rule:

$$(\varepsilon_n, X_n) \to (x, Y) \ \text{ for } \ \varepsilon_n \to 0 \ , \ x \in M, Y \in \nu_x(M)$$

iff $X_n \to j(x) \in \mathbb{R}^N$ and $\frac{X_n - j(x)}{\varepsilon_n} \to Y$ in $\nu_x(M)$.

Using the Euclidean structure of $\mathbb{R}^N$ we can view $\nu_x(M)$ as the subspace orthogonal to $j_* T_x(M) \subset \mathbb{R}^N$ and use the following local chart around $(x, Y) \in \nu(M)$:

$$\varphi(x, Y, \varepsilon) = (\varepsilon, j(x) + \varepsilon Y) \in]0, 1] \times \mathbb{R}^N \quad \text{for } \varepsilon > 0.$$

To the decomposition of G_h as a union of the open groupoid $G_{1,h}$ and the closed groupoid $G_{2,h}$ corresponds the decomposition $BG = BG_1 \cup BG_2$. As in Proposition 5, BG_1 is properly contractible and thus we get a well-defined K-theory map

$$\psi : K^0(BG_2) \simeq K^0(BG) \to K^0(\mathbb{R}^N)$$

which corresponds to the analytic index $\mathrm{Ind}_a = \rho_* \circ (\sigma_*)^{-1}$ under the Thom isomorphisms $K_0(C^*(G_i)) \simeq K_0(C^*(G_{h,i})) = K^0(BG_i)$. Now, from the definition of the topology of BG it follows that ψ is the natural excision map

$$K^0(\nu(M)) \to K^0(\mathbb{R}^N)$$

of the normal bundle of M, viewed as an open set in $\mathbb{R}^N$. Moreover, the Thom isomorphism

$$K^0(\mathbb{R}^N) \overset{\beta}{\simeq} \mathbb{Z}$$

is the Bott periodicity, while the Thom isomorphism

$$K^0(T^*M) \simeq K_0(C^*(G_2)) \simeq K_0(C^*(G_{2,h})) \simeq K^0(BG_2)$$

is the usual Thom isomorphism $\tau : K^0(T^*M) \simeq K^0(\nu(M))$. Thus we have obtained the following formula:

$$\mathrm{Ind}_a = \beta \circ \psi \circ \tau$$

which is the Atiyah-Singer index theorem ([At-Si$_1$]), the right-hand side being the topological index Ind_t.

We used this proof to illustrate the general principle of first reformulating, as in Lemma 6, the analytical index problems in terms of smooth groupoids and their K-theory (through the associated C^*-algebras), and then of making use of free and proper actions of groupoids on contractible spaces to replace the groupoids involved by spaces, for which the computations become automatically geometric.

II.6. Wrong-way Functoriality in K-theory as a Deformation

Let X and Y be manifolds, and $f : X \to Y$ a K-oriented smooth map. Then the Gysin or wrong-way functoriality map

$$f! : K(X) \to K(Y)$$

can be described ([Co-S]) as an element of the Kasparov group $KK(X, Y)$. We shall give here the description of $f!$ as an element of $E(X, Y)$, i.e. as a deformation. The construction is a minor elaboration of the construction of the tangent groupoid of Section 5. The advantage of this construction is that, given a smooth map: $f : X \to Y$, it yields a *canonical* element of

$$E(T^*X \oplus f^*TY, Y).$$

Thus it adapts to the equivariant and to the noncommutative cases.

6.α The index groupoid of a linear map

Let E and F be finite-dimensional real vector spaces and $L : E \to F$ a linear map. Then the *group* E acts by translations on the *space* F, and we can thus form the groupoid $F \rtimes_L E = \mathrm{Ind}(L)$ which is the semidirect product of F by the action of E. More specifically, with $G = \mathrm{Ind}(L)$ we have: $G = F \times E$, $G^{(0)} = F \times \{0\}$, and for $(\eta, \xi) \in F \times E$, $r(\eta, \xi) = \eta$, $s(\eta, \xi) = \eta + L(\xi)$, while

$$(\eta, \xi) \circ (\eta', \xi') = (\eta, \xi + \xi') \ \text{ if } \ \eta + L(\xi) = \eta'.$$

By construction, $G = \mathrm{Ind}(L)$ is a smooth groupoid and is equivalent to the product of the *quotient space* $F/\mathrm{im}(L)$ by the *group* $\ker L$. Thus, using the Fourier transform, the C^*-algebra of $\mathrm{Ind}\, L$ is strongly Morita equivalent to the C^*-algebra $C_0((F/\mathrm{im}\, L) \times (\ker L)^*)$, which justifies the notation $\mathrm{Ind}\, L$.

Proposition 1. a) *Let* $L : E \to F$ *be a linear map. Then the family* $\mathrm{Ind}(\varepsilon L)$, $\varepsilon \in [0, 1]$, *gives a canonical deformation*

$$\delta_L \in E(C_0(F \times E^*), C^*(\mathrm{Ind}\, L)).$$

b) *Let* $L : E \to E'$ *and* $L' : E' \to E''$ *be linear maps. Then the family* $\begin{bmatrix} L & \varepsilon\, \mathrm{id}_{E'} \\ 0 & L' \end{bmatrix}$ *of linear maps from* $E \oplus E'$ *to* $E' \oplus E''$ *gives a canonical deformation of* $(\mathrm{Ind}\, L) \times (\mathrm{Ind}\, L')$ *to* $\mathrm{Ind}(L' \circ L)$.

Statement a) follows because $\mathrm{Ind}(0) = F \rtimes_0 E$ has $C_0(F \times E^*)$ as its associated C^*-algebra, while for $\varepsilon > 0$, $F \rtimes_{\varepsilon L} E$ is canonically isomorphic to $F \rtimes_L E = \mathrm{Ind}\, L$.

To get part b) one uses the equivalence of $\mathrm{Ind} \begin{bmatrix} L & \mathrm{id}_{E'} \\ 0 & L' \end{bmatrix}$ with the smooth groupoid $\mathrm{Ind}(L' \circ L)$.

All of the above discussion goes over for *families* of vector spaces and linear maps, i.e. for vector bundles E and F over a smooth base B and vector bundle maps $L : E \to F$. We shall still denote by $\mathrm{Ind}\, L = \bigcup_{x \in B} \mathrm{Ind}\, L_x$ the corresponding smooth groupoid. Proposition 1 then applies without any change, where in a) $F \times E^*$ denotes the *total space* of the vector bundle $F \times E^*$.

6.β Construction of $f! \in E(T^*M \oplus f^*TN, N)$

Let M and N be smooth manifolds and $f : M \to N$ a smooth map. The tangent map f_* of f is a vector bundle map

$$f_* : TM \to f^*(TN)$$

and thus by α) it yields a smooth groupoid,

$$\mathrm{Ind}(f_*) = \bigcup_{x \in M} \mathrm{Ind}(f_{*,x}).$$

As a manifold $\mathrm{Ind}(f_*)$ is the total space of the vector bundle $TM \oplus f^*(TN)$ over M, while its groupoid structure is as a union of the groupoids $\mathrm{Ind}(f_{*,x})$, with $f_{*,x} : T_xM \to T_{f(x)}N$. We shall construct a canonical deformation of the groupoid $\mathrm{Ind}(f_*)$ to the product of the *space* N (viewed in a trivial way as a groupoid, with $N = N^{(0)}$) by the trivial groupoid (equivalent to a point) $M \times M$. Thus, let $G_1 = \mathrm{Ind}\, f_*$, $G_2 = N \times (M \times M) \times\,]0,1]$ and let us endow the groupoid $G = G_1 \cup G_2$ with the relevant smooth structure. The topology of G is uniquely specified by the closedness of G_1 and the following convergence condition for sequences of elements of G_2: for $\varepsilon_n > 0$ with $\varepsilon_n \to 0$, a sequence $(t_n, (x_n, y_n), \varepsilon_n)$ of elements of $G_2 = N \times (M \times M) \times\,]0,1]$ converges to $(x, \eta, \xi) \in G_1$, with $x \in M$, $\eta \in T_{f(x)}(N)$ and $\xi \in T_x(M)$, iff one has

$$x_n \to x\,,\ y_n \to x\,,\ t_n \to f(x)$$

$$\frac{x_n - y_n}{\varepsilon_n} \to \xi\,,\ \frac{t_n - f(x_n)}{\varepsilon_n} \to \eta.$$

One checks that this topology is compatible with the groupoid structure of G, which becomes a smooth groupoid with the following local diffeomorphism $\varphi : G_1 \times [0,1] \to G_2$ around any point of $G_1 \times \{0\}$:

$$\varphi((x, \eta, \xi), \varepsilon) = (\exp_{f(x)}(\varepsilon\eta), (x, \exp_x(-\varepsilon\xi), \varepsilon)) \in G_2$$

where the exponential maps are relative to arbitrary Riemannian metrics on M and N.

Thus $C^*(G)$ gives a canonical deformation of $C^*(\mathrm{Ind}(f_*))$ to the C^*-algebra of $N \times (M \times M)$, and, combining it with the element δ of Proposition 1 a) and the strong Morita equivalence $C^*(N \times (M \times M)) = C(N) \otimes \mathcal{K} \simeq C(N)$, we get a canonical element

$$f! \in E(T^*M \oplus f^*(TN), N).$$

In order to convert $f!$ to an element of $E(M,N)$ we need a K-orientation of f as an element of

$$E(M, T^*M \oplus f^*(TN)).$$

6.y K-orientations of vector bundles and maps

Let M be a compact space and F a real finite-dimensional vector bundle over M, with $p : F \to M$ the corresponding projection of the total space of the bundle to the base M.

Let n be the fiber dimension of F and define $\mathrm{Spin}^c(n)$ as the Lie group $\mathrm{Spin}(n) \times_{\mathbb{Z}/2\mathbb{Z}} U(1)$, the quotient of the product group by the element $(-1, -1)$ (cf. [L-M] p.390). Then a Spin^c structure on the bundle F is a reduction of its structure group $GL_n(\mathbb{R})$ to $\mathrm{Spin}^c(n)$ (cf. loc. cit, p.391 for more detail).

Definition 2. *The vector bundle F is K-*oriented *if it is endowed with a* Spin^c *structure.*

In particular this implies a line bundle ℓ and a reduction of the structure group of F to $SO(n)$, i.e. an orientation and a Euclidean structure on F. The latter is in general *not invariant* in the Γ-equivariant case, in which case the Definition 2 has been adapted by P. Baum (cf. [Hi-S$_1$]), replacing the group $\mathrm{Spin}^c(n)$ by the group $ML_n^c = ML_n(\mathbb{R}) \times_{\mathbb{Z}/2} U(1)$, where $ML_n(\mathbb{R})$ is the nontrivial two-fold covering of $GL_n^+(\mathbb{R})$. The following proposition then still holds in the equivariant case ([Hi-S$_1$]).

Proposition 3. a) *A complex vector bundle has a canonical K-orientation given by the homomorphism $SU(n) \to \mathrm{Spin}^c(2n)$.*

b) *The dual F^* of a K-oriented bundle inherits a canonical K-orientation.*

c) *For any bundle F, $F \oplus F^*$ has a canonical K-orientation.*

d) *Let $0 \to F \to F' \to F'' \to 0$ be an exact sequence of vector bundles. If two of these bundles are K-oriented the third inherits a canonical K-orientation.*

e) *Let $M \xrightarrow{f} N$ be a continuous map, and F a K-oriented bundle over N. Then $f^*(F)$ is a K-oriented bundle over M.*

The group $H^2(M, \mathbb{Z})$ of complex line bundles on M acts transitively and freely on the set of K-orientations of F.

Definition 4. *Let M and N be smooth manifolds, $f : M \to N$ a smooth map. Then we define a K-orientation of f as a K-orientation of the real vector bundle $F = T^*M \oplus f^*(TN)$.*

As an immediate corollary of Proposition 3 one gets

Proposition 5. a) *If the tangent bundles of M and N are K-oriented then any $f : M \to N$ inherits a K-orientation.*

b) *The identity map $\mathrm{Id}_M : M \to M$ has a canonical K-orientation.*

c) *The composition of K-oriented maps is K-oriented.*

d) *A K-orientation of $f : M \to \mathrm{pt}$ is a* Spin^c *structure on M.*

We assume for simplicity that the fibers of F are even-dimensional, the odd case being treated similarly (with K^1 instead of K^0). There is a natural map $\sigma \to \tilde{\sigma}$ from $K^0(F)$ to $E(C(M), C_0(F))$, which we now describe. First let $\tilde{p}$: $F \to F \times M$ be the proper continuous map given by $\tilde{p}(\xi) = (\xi, p(\xi))$, and $\tilde{p}^* \in \mathrm{Hom}(C_0(F) \otimes C(M), C_0(F))$ be the corresponding morphism of C^*-algebras. Then for $\sigma \in K^0(F) = E(\mathbb{C}, C_0(F))$, let $\tilde{\sigma}$ be given by

$$\tilde{\sigma} = \tilde{p}^* \circ (\sigma \otimes \mathrm{id}_M) \in E(C(M), C_0(F))$$

(one has $\sigma \otimes \mathrm{id}_M \in E(C(M), C_0(F) \otimes C(M))$ and $\tilde{p}^* \in E(C_0(F) \otimes C(M), C_0(F))$ so that $\tilde{\sigma}$ is well defined).

Any given K-orientation on the bundle F yields canonically ([At-Si$_1$]) an element σ of $K^0(F)$ given by the spinor bundle $S^+ \oplus S^-$ with its natural $\mathbb{Z}/2$ grading, and the morphism of p^*S^+ to p^*S^- (over F) given by Clifford multiplication $\gamma(\xi)$, $\xi \in F$. The latter morphism is an isomorphism outside the 0-section $M \subset F$ and thus defines an element of $K^0(F)$ ([At-Si$_1$]). Moreover, the Thom isomorphism in K-theory ([At-Si$_1$]) then shows that the corresponding element $\tilde{\sigma} \in E(C(M), C_0(F))$ is invertible. In particular, σ is a generator of $K(F)$ as a $K(M)$-module. We shall denote by σ_f and $\tilde{\sigma}_f$ the corresponding elements of $K(F)$ and $E(C(M), C_0(F))$.

Remark 6. a) The above discussion of K-orientations of vector bundles and the construction of $\tilde{\sigma} \in E(C_0(M), C_0(F))$ extends (cf. [Co-S]) to the case of non-compact manifolds. The element σ, which we shall not use, is, however, no longer an element of $K(F)$.

b) The Todd genus $\mathrm{Td}(F)$ of a Spin^c vector bundle is defined (cf. [Bau-D] p.136 and [L-M]) by the formula $\mathrm{Td}(F) = e^{c/2}\, \hat{A}(F)$, where c is the first Chern class of the line bundle ℓ associated to the Spin^c structure.

6.δ Wrong-way functoriality for K-oriented maps

Let M and N be smooth manifolds and let $f : M \to N$ be a smooth map. We have constructed above (in β) a deformation $f! \in E(T^*M \oplus f^*(TN), N)$ canonically associated to f. Let us now assume that f is *K-oriented* and let $\sigma_f \in E(M, T^*M \oplus f^*(TN))$ be the corresponding invertible element. With a slight abuse of language we shall still denote by $f!$ the following element

$$f! = f! \circ \sigma_f \in E(M, N).$$

We can then formulate as follows the results of Section 2 of [Co-S]:

Theorem 7. *α) The element $f! \in E(M,N)$ only depends upon the K-oriented homotopy class of f.*

β) One has $(\mathrm{Id}_M)! = 1 \in E(M,M)$.

γ) For any composable smooth maps $f_1 : M_1 \to M_2$, and $f_2 : M_2 \to M_3$ one has, with the corresponding K-orientations,

$$(f_2 \circ f_1)! = f_2! \circ f_1! \in E(M_1, M_3).$$

This result can be deduced from [Co-S] and the natural functor from the KK-category to the E-category (Appendix B). However, the direct proof is easier due to the canonical construction of Section β) and we invite the reader to do it as an exercise.

As in [Co-S], one gets the following strengthening of Bott periodicity.

Corollary 8. *Let M, N be (not necessarily compact) smooth manifolds and $f : M \to N$ a (not necessarily proper) homotopy equivalence. Then $f!$ is invertible.*

In particular $f!$ yields natural isomorphisms, for arbitrary C^*-algebras A and B, between the groups

$$E(A, B \otimes C_0(N)) \simeq E(A, B \otimes C_0(M))$$

$$E(A \otimes C_0(M), B) \simeq E(A \otimes C_0(N), B).$$

The Bott periodicity is the special case with $M = \mathbb{R}^{2n}$ and N =pt; it also applies to any (non-compact) contractible manifold.

Remark 9. Let $f : M \to N$ be a K-oriented map and $f!$ the associated map from $K^0(M) = K_0(C_0(M))$ to $K^0(N) = K_0(C_0(N))$. The compatibility of $f!$ with the Chern character is the equality

$$\mathrm{ch}(f!(x)) = f!(\mathrm{ch}\, x\, \mathrm{Td}(f))$$

where $\mathrm{Td}(f)$ is the Todd genus, defined in Remark 6 b), of the *difference* bundle $TM \ominus f^*(TN)$, defined as the ratio

$$\mathrm{Td}(f) = \mathrm{Td}(T_{\mathbb{C}}M) / \mathrm{Td}(TM \oplus f^*(TN)).$$

II.7. The Orbit Space of a Group Action

In this section, as a preparation for the case of leaf spaces of foliations, we shall consider spaces of the same nature as the space of Penrose tilings, namely the spaces of orbits for an action of a discrete group Γ on a manifold V. We assume that Γ acts on V by diffeomorphisms and use the notation of a right action $(x, g) \in V \times \Gamma \to xg \in V$ with

$$x(g_1 g_2) = (x g_1) g_2 \quad \forall g_1, g_2 \in \Gamma\,,\ \forall x \in V.$$

Such an action is called *free* if each $g \in \Gamma$, $g \neq 1$, has no fixed point, and *proper* when the map $(x, g) \to (x, xg)$ is a proper map from $V \times \Gamma$ to $V \times V$.

When the action of Γ is free and proper the quotient space $X = V/\Gamma$ is Hausdorff and is a manifold of the same dimension as V. The next proposition shows that the space X is equivalently described, as a topological space, by the C^*-algebra crossed product $C_0(V) \rtimes \Gamma$ (Appendix C).

Proposition 1. [Ri_2] *Let Γ act freely and properly on V and let $X = V/\Gamma$. Then the C^*-algebra $C_0(X)$ is strongly Morita equivalent to the crossed product C^*-algebra $C_0(V) \rtimes \Gamma$.*

It is important to mention that in this case, and for any Γ amenable or not, there is no distinction between the reduced and unreduced crossed products. Also, the equivalence C^*-bimodule $\mathcal{E}$ is easy to describe; it is given by the bundle of Hilbert spaces $(\mathcal{H}_x)_{x \in X}$ over X, whose fiber at $x \in X = V/\Gamma$ is the ℓ^2-space of the orbit $x \in X$. This yields the required $(C_0(V) \rtimes \Gamma, C_0(X))$ C^*-bimodule.

When the action of Γ is free but not proper, for instance when V is compact and Γ infinite, the quotient space $X = V/\Gamma$ is no longer Hausdorff and its topology is of little use, so also is the algebra $C_0(X)$ of continuous functions on X. But then Proposition 1 no longer holds and the topology of the orbit space is much better encoded by the crossed product C^*-algebra $C_0(V) \rtimes \Gamma$. Our aim in this section is to construct K-theory classes in this C^*-algebra from geometric data. For that purpose we shall not need the hypothesis of the freeness of the action of Γ. In fact, the case of Γ acting on the space V reduced to a single point, already treated in Section 4, will also guide us. We shall assume that Γ is torsion-free and treat the general case later in Section 10.

The motivation for the construction of K-theory classes, due to P. Baum and myself ([Bau-C_1]) is the following. While there are in general very few continuous maps $f : X \to W$ from a noncommutative space such as V/Γ to an ordinary manifold, there are always plenty of smooth maps from ordinary manifolds W to such a space as $X = V/\Gamma$. Thus, since we have at our disposal, by Section 6, the wrong-way functoriality map $f!$ in K-theory, we should expect to construct elements of K-theory of the form

$$f!(z) \in K(C_0(V) \rtimes \Gamma) \quad \forall z \in K(C_0(W))$$

for any smooth K-oriented map $f : W \to V/\Gamma$ from an ordinary manifold to our space $X = V/\Gamma$.

Given a smooth map $\rho : W \to V$, one obtains by composition with $p : V \to V/\Gamma = X$ a "smooth" map to X. But, clearly, there are other natural "smooth" maps to X which do not factor through V. Indeed, if we are given an open cover $(W_i)_{i \in I}$ of W and smooth maps $\rho_i : W_i \to V$, such that on any non-empty intersection $W_i \cap W_j$ one has $p \circ \rho_i = p \circ \rho_j$, then $p \circ \rho : W \to V/\Gamma$ still makes sense. Since we do not want to assume that the action of Γ on V is free, we strengthen the equality $p \circ \rho_i = p \circ \rho_j$ on $W_i \cap W_j$ by specifying a Čech 1-cocycle

$$\rho_{ij} : W_i \cap W_j \to \Gamma$$

such that

$$\rho_j(x)\, \rho_{ij}(x) = \rho_i(x) \quad \forall x \in W_i \cap W_j.$$

To the Čech 1-cocycle (W_i, ρ_{ij}) with values in Γ there corresponds a principal Γ-bundle

$$\widetilde{W} \xrightarrow{q} W$$

and local sections $s_i : W_i \to \widetilde{W}$ such that

$$s_j(x)\,\rho_{ij}(x) = s_i(x) \quad \forall x \in W_i \cap W_j.$$

There exists then a unique Γ-equivariant smooth map from $\widetilde{W}$ to V such that

$$\rho(s_i(x)) = \rho_i(x) \quad \forall x \in W_i.$$

To summarize, we obtain the following notion:

Definition 2. *Let Γ be a discrete group acting by diffeomorphisms on a manifold V. Then a smooth map $W \to V/\Gamma$ from a manifold W to the orbit space, is given by a principal Γ-bundle $\widetilde{W} \xrightarrow{q} W$ and a Γ-equivariant smooth map $\rho : \widetilde{W} \to V$.*

The reason for the existence of sufficiently many such maps is that homotopy classes of smooth maps $W \mapsto V/\Gamma$ correspond exactly to homotopy classes of continuous maps

$$W \to V \times_\Gamma E\Gamma = V_\Gamma$$

where $E\Gamma \xrightarrow{\Gamma} B\Gamma$ is the universal principal Γ-bundle over the classifying space $B\Gamma$ of Γ.

First, given $\widetilde{W} \xrightarrow{q} W$ and $\rho : \widetilde{W} \to V$ as in Definition 2, we get a Γ-equivariant continuous map

$$\widetilde{W} \xrightarrow{(\rho,\varphi)} V \times E\Gamma$$

where φ is a classifying map $\widetilde{W} \to E\Gamma$ for the bundle $\widetilde{W}$. The map (ρ, φ) then yields a continuous map $\psi : W \to V \times_\Gamma E\Gamma$. Second, given a continuous map $\psi : W \to V \times_\Gamma E\Gamma$, one can pull back to W the principal Γ-bundle over $V \times_\Gamma E\Gamma$ given by

$$V \times E\Gamma \xrightarrow{p} V \times_\Gamma E\Gamma.$$

One then has a principal Γ-bundle $\widetilde{W} \xrightarrow{q} W$ over W and a continuous Γ-equivariant map $\tilde{\psi} : \widetilde{W} \to V \times E\Gamma$. The composition of $\tilde{\psi}$ with the projection $\mathrm{pr}_V : V \times E\Gamma \to V$ gives us a Γ-equivariant continuous map $\widetilde{W} \xrightarrow{\rho} V$. Since $V \times_\Gamma E\Gamma$ is a locally finite CW-complex, there is a natural notion of a smooth map $W \mapsto V \times_\Gamma E\Gamma$ and every continuous map from W to $V \times_\Gamma E\Gamma$ can be smoothed, thus yielding smooth maps $W \overset{\rho}{\mapsto} V/\Gamma$ in the sense of Definition 2. Thus:

Proposition 3. *Let Γ be a discrete group acting by diffeomorphisms on the manifold V, and let W be a manifold. Then homotopy classes of continuous maps $W \mapsto V_\Gamma = V \times_\Gamma E\Gamma$ correspond bijectively to homotopy classes of smooth maps $W \to V/\Gamma$ in the sense of Definition 2.*

Any Γ-equivariant bundle F on V is still Γ-equivariant on $V \times E\Gamma$, and hence drops down to a bundle on V_Γ. This applies in particular to the tangent bundle

TV of V, yielding a bundle τ on V_Γ. At a formal level the geometric group $K_{*,\tau}(V_\Gamma)$ is defined as follows:

Definition 4. *$K_{*,\tau}(V_\Gamma)$ is the K-homology of the pair $(B\tau, S\tau)$ consisting in the unit-ball and unit-sphere bundles of τ over V_Γ.*

Since $V_\Gamma = V \times_\Gamma E\Gamma$ is not in general a finite simplicial complex we have to be precise regarding the definition of K-homology for arbitrary simplicial complexes X. We take $K_*(X) = \varinjlim K_*(Y)$ where Y runs through compact subsets of X. In other words we choose K-homology *with compact supports* in the sense of [Sp] Axiom 11 p. 203.

Let $H_*^\tau(V_\Gamma, \mathbb{Q})$ be the ordinary singular homology of the pair $(B\tau, S\tau)$ over V_Γ, with coefficients in $\mathbb{Q}$. Since this is also a theory with compact supports, the Chern character

$$\mathrm{ch} : K_{*,\tau}(V_\Gamma) \to H_*^\tau(V_\Gamma, \mathbb{Q})$$

is a rational isomorphism.

Since we are interested mainly in the case of orientation preserving diffeomorphisms, let us assume that V is oriented and that Γ preserves this orientation. Then the bundle τ over $V_\Gamma = V \times_\Gamma E\Gamma$ is still oriented, and letting U be the orientation class of τ on V_Γ, we can use the Thom isomorphism ([Mi$_4$] Theorem 10, p. 259)

$$\phi : H_{q+n}^\tau(V_\Gamma, \mathbb{Q}) \to H_q(V_\Gamma, \mathbb{Q}) \quad (n = \dim \tau = \dim V)$$

where $\phi(z) = p_*(U \cap z)\ \forall z \in H_{q+n}((B\tau, S\tau, \mathbb{Q})$, and where p is the projection from $B\tau$ to the base V_Γ.

Thus $\phi \circ \mathrm{ch}$ is a rational isomorphism:

$$\phi \circ \mathrm{ch} : K_{*,\tau}(V_\Gamma) \to H_*(V_\Gamma, \mathbb{Q}).$$

To construct the analytic assembly map $\mu : K_{*,\tau}(V_\Gamma) \to K_*(C_0(V) \rtimes \Gamma)$ requires a better understanding of the K-homology of an arbitrary pair (here the pair is $(B\tau, S\tau)$ over V_Γ). This follows from:

Proposition 5. a) *Let M be a* Spin^c*-manifold with boundary, and assume that M is compact. Then one has a Poincaré duality isomorphism*

$$K^*(M) \cong K_*(M, \partial M).$$

b) *Let (X, A) be a topological pair of simplicial complexes and let $x \in K_*(X, A)$. Then there exists a compact* Spin^c*-manifold with boundary $(M, \partial M)$, a continuous map $f : (M, \partial M) \to (X, A)$ and an element y of $K_*(M, \partial M)$ with $f_*(y) = x$.*

The Chern character is then uniquely characterized by the properties:

1) $f_* \mathrm{ch}(y) = \mathrm{ch}(f_*(y))$.

2) If M is a compact Spin^c-manifold with boundary, and $z \in K_*(M, \partial M)$ is the image of $y \in K^*(M)$ under Poincaré duality, one has

$$\mathrm{ch}(z) = (\mathrm{ch}(y) \cdot \mathrm{Td}(M)) \cap [M, \partial M],$$

where $\mathrm{Td}(M)$ is the characteristic class associated to the Spin^c structure of M as in [Bau-D] p.136 and [L-M] p.399, $\mathrm{Td}(M) = e^{c/2}\,\hat{A}(M)$ where c is the first Chern class of the line bundle ℓ associated to the Spin^c structure.

Now let (N, F, g) be a triple where N is a compact manifold without boundary, $F \in K^*(N)$, and g is a continuous map from N to $V_\Gamma = V \times_\Gamma E\Gamma$, which is K-oriented, i.e. such that the bundle $TN \oplus g^*\tau$ is endowed with a Spin^c structure. To such a triple corresponds an element of $K_*(B\tau, S\tau)$ as follows. Let B and S be the unit-ball and unit-sphere bundles of $g^*\tau$ on N. Then B is a Spin^c-manifold with boundary so that the Poincaré duality isomorphism assigns a class $y \in K_*(B, S)$ to the pull-back of F to B. Then put $[(N, F, g)] = g_*(y) \in K_*(B\tau, S\tau)$. For convenience any triple (N, F, g) as above will be called a geometric cycle.

Proposition 6. a) *Any element of* $K_{*,\tau}(V_\Gamma) = K_*(B\tau, S\tau)$ *is of the form* $[(N, F, g)]$ *for some geometric cycle* (N, F, g).

b) *Let* (N, F, g) *be a geometric cycle*, N' *be a compact manifold and assume* $f : N' \to N$ *a continuous map which is K-oriented, i.e.* $TN' \oplus f^*TN$ *is endowed with a* Spin^c *structure. Then, for any* $F' \in K^*(N')$ *with* $f!(F') = F$ *one has*

$$[(N', F', g \circ f)] = [(N, F, g)] \text{ in } K_*(B\tau, S\tau).$$

Here $f! : K^*(N') \to K^*(N)$ is the push-forward map in K-theory (cf. [Bau-D]).

Proof. a) By Proposition 5 b) there exists a compact Spin^c-manifold with boundary $(M, \partial M)$, an element y of $K_*(M, \partial M)$, and a continuous map $f : (M, \partial M) \to (B\tau, S\tau)$ with $x = f_*(y)$. By transversality one may assume that the inverse image in M of the 0-section of τ is a submanifold N of M whose normal bundle ν is the restriction of $f^*(\tau)$ to N. Since the boundary of M maps to $S\tau$, the manifold N is closed without boundary. Let g be the restriction of f to N. Then g is a continuous map from N to V_Γ, and the bundle $TN \oplus g^*\tau = TN \oplus \nu$ has a Spin^c structure. Let B be the unit-ball bundle $B \xrightarrow{p} N$ of the bundle $\nu = g^*\tau$. Then, since $p^* : K^*(N) \to K^*(B)$ is an isomorphism, the assertion follows from Proposition 5 a).

b) Let (B, S) and (B', S') be the unit-ball and unit-sphere bundles of $g^*\tau$ and $f^*g^*\tau$ over N and N', and $\tilde{f} : (B', S') \to (B, S)$ be the natural extension of f. One has $\tilde{f}\,!(p'^*(F')) = p^*(F)$. Thus the conclusion follows since $\tilde{f}\,!$ is Poincaré dual to $\tilde{f}_* : K_*(B', S') \to K_*(B, S)$.

If we translate the Chern character $\phi \circ \mathrm{ch}$ in terms of geometric cycles we get:

Proposition 7. *Let (N, F, g) be a geometric cycle. Then*

$$\phi \circ \mathrm{ch}[(N, F, g)] = g_*(\mathrm{ch}(F) \cdot \mathrm{Td}(TN \oplus g^*\tau) \cap [N]) \in H_*(V_\Gamma, \mathbb{Q}).$$

Note that we assumed that τ was oriented, so that N is oriented since the bundle $TN \oplus g^*\tau$ is Spin^c, and hence oriented.

So far we have just described the general elements of the group $K_{*,\tau}(V_\Gamma)$ and computed their Chern characters. We shall now construct the analytic assembly map

$$\mu : K_{*,\tau}(V_\Gamma) \to K_*(C_0(V) \rtimes \Gamma).$$

Given a geometric cycle (N, F, g), by Proposition 3 we have $\tilde{N}$ the corresponding principal Γ-bundle over N and $\tilde{g} : \tilde{N} \to V$ the corresponding Γ-equivariant K-oriented map. By Proposition 3 we can assume that $\tilde{g}$ is smooth. Then by Section 6 we get a well-defined element $\tilde{g}! \in E(C_0(\tilde{N}), C_0(V))$, and, since the construction of $\tilde{g}!$ is natural, the corresponding deformation passes to the crossed products by Γ using Proposition 2 b) of Appendix C, hence yielding

$$\tilde{g}!_\Gamma \in E(C_0(\tilde{N}) \rtimes \Gamma \ , \ C_0(V) \rtimes \Gamma).$$

As the action of Γ on $\tilde{N}$ is free and proper, the crossed product $C_0(\tilde{N}) \rtimes \Gamma$ is strongly Morita equivalent to $C_0(\tilde{N}/\Gamma) = C_0(N)$ so that we can view $\tilde{g}!_\Gamma$ as an element of

$$E(C_0(N), C_0(V) \rtimes \Gamma).$$

Now *define* $\mu(N, F, g)$ as the image $\tilde{g}!_\Gamma(F)$ of the K-theory class $F \in K_*(C_0(N))$ under the element $\tilde{g}!_\Gamma$.

Theorem 8. *There exists an additive map μ of $K_{*,\tau}(V_\Gamma)$ to $K_*(C_0(V) \rtimes \Gamma)$ such that for any geometric cycle (N, F, g) as above $\mu(N, F, g) = \tilde{g}!_\Gamma(F)$, where $F \in K_*(C_0(N))$ is viewed as an element of $K(C_0(\tilde{N} \rtimes \Gamma))$ through the Morita equivalence, and $\tilde{g}!_\Gamma$ as an element of $E(C_0(\tilde{N}) \rtimes \Gamma, C_0(V) \rtimes \Gamma)$.*

We shall now apply this theorem to give examples of crossed products $A = C(V) \rtimes \Gamma$, with V a compact manifold, where the K-theory class of the unit 1 of A is trivial. We first need to find a geometric cycle (N, F, g) whose image $\mu(N, F, g)$ is the unit of A.

Lemma 9. *Let $F \in K_*(C_0(V))$, and consider the geometric cycle (V, F, p), where $p : V \to V/\Gamma$ is the quotient map with its tautological K-orientation. Then $\mu(V, F, p)$ is equal to $j_*(F)$, where $j : C_0(V) \to A = C_0(V) \rtimes \Gamma$ is the canonical homomorphism of $C_0(V)$ into the crossed product by Γ.*

The proof is straightforward ($[\mathrm{Co}_{14}]$).

With this we are ready to prove

Theorem 10. *Let V be a compact oriented manifold on which Γ acts by orientation preserving diffeomorphisms. Assume that in the induced fibration over $B\Gamma$ with fiber V, $V \times_\Gamma E\Gamma \to B\Gamma$, the fundamental class $[V]$ of the fiber becomes 0 in $H_n(V \times_\Gamma E\Gamma, \mathbb{Q})$. Then the unit of the C^*-algebra $A = C(V) \rtimes \Gamma$ is a torsion element of $K_0(A)$.*

Proof. By Theorem 8 we know that the map $\mu : K^*(V,\Gamma) \to K_*(C(V) \rtimes \Gamma)$ is well-defined. Thus it is enough to find $x \in K_\tau^*(V_\Gamma)$ with $\mu(x) = 1_A$ and $\mathrm{ch}(x) = 0$ (since the Chern character is a rational isomorphism (see above)). Now Lemma 9 shows that $\mu(x) = 1_A$, where x is the K-cycle $(V, 1_V, g)$ and 1_V stands for the trivial line bundle on V, g is the map from V to V_Γ given by $g(s) = (s \times t_0)/\Gamma$ for some $t_0 \in E\Gamma$, and where the K-orientation of the bundle $TV \oplus TV$ comes from its natural complex structure. By Proposition 7 the Chern character of this K-cycle is equal to $\mathrm{Td}(\tau_{\mathbb{C}})\; g_*([V]) = 0$. Thus x is a torsion element in $K_{*,\tau}(V_\Gamma)$ and also $\mu(x) = 1_A$ in $K_0(A)$.

As a nice example where this theorem applies let us mention:

Corollary 11. *Let $\Gamma \subset \mathrm{PSL}(2,\mathbb{R})$ be a torsion-free cocompact discrete subgroup. Let Γ act on $V = P_1(\mathbb{R})$ in the obvious way. Then in the C^*-algebra $A = C(V) \rtimes \Gamma$ the unit 1_A is a torsion element of $K_0(A)$.*

Proof. Let $H = \{z \in \mathbb{C}; \mathrm{im}\, z > 0\}$ be the Poincaré space, and $M = H/\Gamma$ be the quotient Riemann surface. Let us identify H with $E\Gamma$ since it is a contractible space on which Γ acts freely and properly. Then, as in [Mi$_4$] p.313, we can identify the induced bundle $V_\Gamma = P_1(\mathbb{R}) \times_\Gamma E\Gamma = P_1(\mathbb{R}) \times_\Gamma H$ over $B\Gamma = M$ with the unit-sphere bundle of M, denoted by η. It follows that the Euler class $e(\eta)$ of η is equal to $2 - 2g$ times the generator of $H^2(M,\mathbb{Z})$. Applying this and the Gysin sequence of the tangent vector bundle TM, (cf. [Mi$_4$] p.143),

$$H^0(M,\mathbb{Z}) \xrightarrow{\cup e} H^2(M,\mathbb{Z}) \xrightarrow{\pi^*} H^2(\eta,\mathbb{Z}) \to \cdots$$

shows that $\pi^*([M]^*)$ is a torsion element of $H^2(\eta,\mathbb{Z})$, where $\pi : \eta \to M$ is the projection and $[M]^*$ is the generator of $H^2(M,\mathbb{Z})$. But η is an oriented manifold, and the homology class of the fiber, $[P_1(\mathbb{R})]$, is Poincaré dual to $\pi^*([M])$; hence it is also a torsion element, so that $\mathrm{ch}(x) = 0$.

Remark 12. Let X be a simplicial complex and τ a real vector bundle over X. The Chern character $\mathrm{ch} : K_{*,\tau}(X) \to H_*^\tau(X,\mathbb{Q})$ is a rational isomorphism. It is convenient for computations to introduce the rational isomorphism ch_τ given by:

$$\mathrm{ch}_\tau(x) = \mathrm{Td}(\tau_{\mathbb{C}})^{-1}\,\mathrm{ch}(x) \in H_*^\tau(X,\mathbb{Q}) \qquad \forall x \in K_{*,\tau}(X).$$

II.8. The Leaf Space of a Foliation

8.α Construction of $C_r^*(V,F)$

Let (V,F) be a foliated manifold of codimension q. Given any $x \in V$ and a small enough open set $W \subset V$ containing x, the restriction of the foliation F to W has, as its leaf space, an open set of $\mathbb{R}^q$, which we shall call for short a transverse neighborhood of x. In other words this open set W/F is the set of *plaques* around x. Now, given a leaf L of (V,F) and two points $x, y \in L$ of this leaf, any simple path γ from x to y on the leaf L uniquely determines a germ $h(\gamma)$ of a diffeomorphism from a transverse neighborhood of x to a transverse neighborhood of y. One can obtain $h(\gamma)$, for instance, by restricting the foliation F to a neighborhood N of γ in V sufficiently small to be a transverse neighborhood of both x and y as well as of any $\gamma(t)$. The germ of diffeomorphism $h(\gamma)$ thus obtained only depends upon the homotopy class of γ in the fundamental groupoid of the leaf L, and is called the *holonomy* of the path γ. The *holonomy groupoid* of a leaf L is the quotient of its fundamental groupoid by the equivalence relation which identifies two paths γ and γ' from x to y (both in L) iff $h(\gamma) = h(\gamma')$. The holonomy covering $\tilde{L}$ of a leaf is the covering of L associated to the normal subgroup of its fundamental group $\pi_1(L)$ given by paths with trivial holonomy. The *holonomy groupoid* of the foliation is the union G of the holonomy groupoids of its leaves. Given an element γ of G, we denote by $x = s(\gamma)$ the origin of the path γ, by $y = r(\gamma)$ its end point, and r and s are called the range and source maps.

An element γ of G is thus given by two points $x = s(\gamma)$ and $y = r(\gamma)$ of V together with an equivalence class of smooth paths: the $\gamma(t)$, $t \in [0,1]$ with $\gamma(0) = x$ and $\gamma(1) = y$, tangent to the bundle F (i.e. with $\dot{\gamma}(t) \in F_{\gamma(t)}$, $\forall t \in \mathbb{R}$) identifying γ_1 and γ_2 as equivalent iff the *holonomy* of the path $\gamma_2 \cdot \gamma_1^{-1}$ at the point x is the *identity*. The graph G has an obvious composition law. For γ and $\gamma' \in G$, the composition $\gamma \circ \gamma'$ makes sense if $s(\gamma) = r(\gamma')$. The groupoid G is by construction a (not necessarily Hausdorff) manifold of dimension $\dim G = \dim V + \dim F$ (cf. [Ehr$_4$] for the original construction, and also [Win]).

If the leaf L which contains both x and y has no holonomy (this is generic in the topological sense of truth on a dense G_δ) then the class in G of the path $\gamma(t)$ depends only on the pair (y,x). In general, if one fixes $x = s(\gamma)$, the map from $G_x = \{\gamma; s(\gamma) = x\}$ to the leaf L through x, given by $\gamma \in G_x \mapsto y = r(\gamma)$, is the holonomy covering of L.

Both maps r and s from the manifold G to V are smooth submersions, and G is a smooth groupoid in the sense of Definition 5.1. The map (r,s) to $V \times V$ is an immersion whose image in $V \times V$ is the (often singular) subset $\{(y,x) \in V \times V;\ y$ and x are on the same leaf$\}$. The topology of G is *not* in general the same as its topology as a subset of $V \times V$, and the map $G \to V \times V$ is not a proper map. By construction, the C^*-algebra of the foliation is the C^*-algebra $C_r^*(G)$ of the smooth groupoid G constructed according to Proposition 5.3, but we shall describe it in detail. We shall assume for notational convenience that the manifold G is Hausdorff, but as this fails to be the case in very interesting examples we shall refer to [Co₁₁] for the removal of this hypothesis.

The basic elements of $C_r^*(V,F)$ are smooth half-densities with compact supports on G, $f \in C_c^\infty(G,\Omega^{1/2})$, where $\Omega_\gamma^{1/2}$ for $\gamma \in G$ is the one-dimensional complex vector space $\Omega_x^{1/2} \otimes \Omega_y^{1/2}$, where $s(\gamma) = x$, $r(\gamma) = y$, and $\Omega_x^{1/2}$ is the one-dimensional complex vector space of maps from the exterior power $\wedge^k F_x$, $k = \dim F$, to $\mathbb{C}$ such that

$$\rho(\lambda v) = |\lambda|^{1/2}\rho(v) \quad \forall v \in \wedge^k F_x\,,\ \forall \lambda \in \mathbb{R}.$$

Of course the bundle $(\Omega_x^{1/2})_{x\in V}$ is trivial, and we could choose once and for all a trivialization ν making elements of $C_c^\infty(G,\Omega^{1/2})$ into functions, but we want to have canonical algebraic operations.

For $f,g \in C_c^\infty(G,\Omega^{1/2})$, the convolution product $f * g$ is given by the equality

$$(f * g)(\gamma) = \int_{\gamma_1\circ\gamma_2=\gamma} f(\gamma_1)g(\gamma_2).$$

This makes sense because for fixed $\gamma : x \to y$, and with $v_x \in \wedge^k F_x$ and $v_y \in \wedge^k F_y$, the product $f(\gamma_1)g(\gamma_1^{-1}\gamma)$ defines a 1-density on $G^y = \{\gamma_1 \in G; r(\gamma_1) = y\}$, which is smooth with compact support (it vanishes if $\gamma_1 \notin$ support f), and hence can be integrated over G^y to give a scalar, $(f * g)(\gamma)$ evaluated on $v_x \otimes v_y$. One has to check that $f * g$ is still smooth with compact support, which is trivial here, and remains true in the non-Hausdorff case.

The $*$-operation is given by $f^*(\gamma) = \overline{f(\gamma^{-1})}$, i.e. if $\gamma : x \to y$ and $v_x \in \wedge^k F_x, v_y \in \wedge^k F_y$, then $f^*(\gamma)$ evaluated on $v_x \otimes v_y$ is equal to $\overline{f(\gamma^{-1})}$ evaluated on $v_y \otimes v_x$. We thus get a $*$-algebra $C_c^\infty(G,\Omega^{1/2})$. For each leaf L of (V,F) one has a natural representation π_L of this $*$-algebra on the L^2-space of the holonomy covering $\tilde{L}$ of L. Fixing a base point $x \in L$, one identifies $\tilde{L}$ with $G_x = \{\gamma; s(\gamma) = x\}$ and defines

$$(\pi_x(f)\xi)(\gamma) = \int_{\gamma_1\circ\gamma_2=\gamma} f(\gamma_1)\xi(\gamma_2) \quad \forall \xi \in L^2(G_x)$$

where ξ is a square-integrable half-density on G_x. Given $\gamma : x \to y$ one has a natural isometry of $L^2(G_x)$ onto $L^2(G_y)$ which transforms the representation π_x to π_y. Applying Proposition 5.3 we thus get

Definition 1. *$C_r^*(V,F)$ is the C^*-algebra completion of $C_c^\infty(G,\Omega^{1/2})$ with the norm* $\|f\| = \sup_{x\in V} \|\pi_x(f)\|$.

If the leaf L has trivial holonomy, then the representation π_x, $x \in L$, is irreducible. In general its commutant is generated by the action of the discrete holonomy group G_x^x in $L^2(G_x)$. If the foliation comes from a submersion $p : V \to B$, then its graph G is $\{(x,y) \in V \times V; p(x) = p(y)\}$, which is a submanifold of $V \times V$, and $C_r^*(V,F)$ is identical with the algebra of the continuous field of Hilbert spaces $L^2\left(p^{-1}\{x\}\right)_{x\in B}$. Thus, unless $\dim F = 0$, it is isomorphic to the tensor product of $C_0(B)$ with the elementary C^*-algebra of compact operators. It is always strongly Morita equivalent to $C_0(B)$. If the foliation comes from an action of a Lie group H in such a way that the graph is identical with $V \times H$, then $C_r^*(V,F)$ is identical with the reduced crossed product of $C_0(V)$ by H (cf. Appendix C). Moreover, the construction of $C_r^*(V,F)$ is local in the following sense:

Lemma 2. *If $V' \subset V$ is an open set and F' is the restriction of F to V', then the graph G' of (V',F') is an open set in the graph G of (V,F), and the inclusion $C_c^\infty(G',\Omega^{1/2}) \subset C_c^\infty(G,\Omega^{1/2})$ extends to an isometric $*$-homomorphism of $C_r^*(V',F')$ into $C_r^*(V,F)$.*

This lemma, which is still valid in the non-Hausdorff case ([Co$_{11}$]), allows one to reflect algebraically the local triviality of the foliation. Thus one can cover the manifold V by open sets W_i such that F restricted to W_i has a Hausdorff space of leaves, $B_i = W_i/F$, and hence such that the C^*-algebras $C_r^*(W_i,F)$ are strongly Morita equivalent to the commutative C^*-algebras $C_0(B_i)$. These subalgebras $C_r^*(W_i,F)$ generate $C_r^*(V,F)$, but of course they fit together in a very complicated way which is related to the global properties of the foliation.

By construction, $C_r^*(V,F) = C_r^*(G)$ where the smooth groupoid G is the graph of (V,F). Similarly, we let $C^*(V,F)$ be the maximal C^*-algebra $C^*(G)$, (Proposition 5.3) and r be the canonical surjection

$$r : C^*(V,F) \to C_r^*(V,F).$$

8.β Closed transversals and idempotents of $C_r^*(V,F)$

Let (V,F) be a foliated manifold, and $N \subset V$ a compact submanifold with $\dim N = \operatorname{codim} F$, everywhere transverse to the foliation. Then the restriction F' of F to a small enough tubular neighborhood V' of N defines on V' a fibration with compact base N. In other words, there exists a fibration $V' \xrightarrow{p} N$ with fibers $\mathbb{R}^k$, $k = \dim F$, such that for any $x \in V$ one has $F_x = \ker(p_*)_x$. The C^*-algebra $C_r^*(V',F')$ of the restriction of F to V' is thus strongly Morita equivalent to $C(N)$, and in fact isomorphic to $C(N) \otimes \mathcal{K}$ where $\mathcal{K}$ is the C^*-algebra of compact operators. In particular it contains an idempotent $e = e^2 = e^*$, $e = 1_N \otimes f \in C(N) \otimes \mathcal{K}$, where f is a minimal projection in $\mathcal{K}$. Using the inclusion $C_r^*(V',F') \subset C_r^*(V,F)$ given by the above Lemma 2, we thus get an idempotent of $C_r^*(V,F)$, which we shall now describe more concretely. The transversality of N to the foliation ensures the existence of a neighborhood $\mathcal{V}$ of $G^{(0)}$ in G such that

$$\gamma \in \mathcal{V} \circ \mathcal{V}^{-1}\ ,\ s(\gamma) \in N\ ,\ r(\gamma) \in N \Longrightarrow \gamma \in G^{(0)},$$

where $G^{(0)} = \{(x,x); x \in V\}$ is the set of units of G. Let then ξ be a smooth section on the submanifold $r^{-1}(N) \subset G$ of the bundle $s^*(\Omega^{1/2})$ of half-densities, $\xi \in C_c^\infty\left(r^{-1}(N), s^*(\Omega^{1/2})\right)$, such that

α) Support $\xi \subset \mathcal{V}$

β) $\int_{r(\gamma)=y} |\xi(\gamma)|^2 = 1 \quad \forall y \in N$.

The equality $e(\gamma) = \sum_{\substack{s(\gamma')=s(\gamma)\\ r(\gamma')\in N}} \bar{\xi}(\gamma'\gamma^{-1})\xi(\gamma')$ defines an idempotent $e \in C_c^\infty(G,\Omega^{1/2}) \subset C_r^*(V,F)$. Indeed, for each $x \in V$ the operator $\pi_x(e)$ in $L^2(G_x)$ is the orthogonal projection on the closed subspace of $L^2(G_x)$ spanned by a set of orthogonal vectors labelled by the countable set $I = r^{-1}(N) \cap G_x$, namely $(\eta_\gamma)_{\gamma\in I}$, where

$$\eta_\gamma(\gamma') = \xi\left(\gamma(\gamma')^{-1}\right) \quad \forall \gamma' \in G_x.$$

A more canonical way to define the (equivalence class of the) above idempotent $e \in C_r^*(V,F)$ is to show that the C^*-module eA, $A = C_r^*(V,F)$, is the same as the completion of $C_c^\infty\left(r^{-1}(N), s^*(\Omega^{1/2})\right)$ with the right action of $C_c^\infty(G,\Omega^{1/2})$ given by convolution, for $f \in C_c^\infty(G,\Omega^{1/2})$,

$$(\xi * f)(\gamma) = \int_{\gamma_1\circ\gamma_2=\gamma} \xi(\gamma_1)f(\gamma_2) \quad \forall \xi \in C_c^\infty\left(r^{-1}(N), s^*(\Omega^{1/2})\right)$$

and with the $C_c^\infty(G,\Omega^{1/2})$-valued inner product given by

$$\langle \xi,\eta\rangle(\gamma) = \sum_{\substack{s(\gamma')=s(\gamma)\\ r(\gamma')\in N}} \bar{\xi}(\gamma'\gamma^{-1})\eta(\gamma'), \quad \forall \xi,\eta \in C_c^\infty\left(r^{-1}(N), s^*(\Omega^{1/2})\right).$$

The construction of this C^*-module $\mathcal{E}_N$ over $C_r^*(V,F)$ makes sense whether N is compact or not, but in the former case the identity is a compact endomorphism of $\mathcal{E}_N$ and $\mathcal{E}_N = eC_r^*(V,F)$ for a suitable idempotent $e \in C_r^*(V,F)$ (the identity being of rank one).

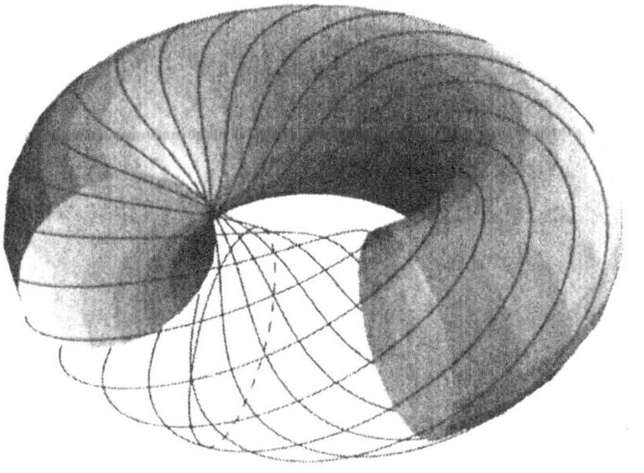

Figure II.7. Transversal

If N is compact and meets every leaf of (V,F), then $\mathcal{E}_N$ gives a strong Morita equivalence between $C_r^*(V,F)$ and the C^*-algebra of compact endomorphisms of $\mathcal{E}_N$ which, if we assume that N is compact, is *unital*. This C^*-algebra is the convolution C^*-algebra of the reduced groupoid G_N:

$$G_N = \{\gamma \in G; r(\gamma) \in N, s(\gamma) \in N\}.$$

The groupoid G_N is a manifold of dimension $q = \operatorname{codim} F$, and the convolution product in $C_c^\infty(G_N)$ is given by

$$(f * g)(\gamma) = \sum_{\gamma_1 \circ \gamma_2 = \gamma} f(\gamma_1) g(\gamma_2)$$
$$f^*(\gamma) = \overline{f(\gamma^{-1})}.$$

The C^*-algebra norm on $C_c^\infty(G_N)$ is given by the supremum of the norm $\|\pi_x(f)\|$ where for each $x \in N = G_N^{(0)}$, π_x is the representation of $C_c^\infty(G_N)$ in $\ell^2(G_{N,x})$ given by

$$(\pi_x(f)\xi)(\gamma) = \sum_{\gamma_1 \circ \gamma_2 = \gamma} f(\gamma_1)\xi(\gamma_2) \quad \forall \gamma \in G_N, \ s(\gamma) = x.$$

We shall denote this C^*-algebra by $C^*_{r,N}(V,F)$, and compute it for a simple example.

Thus, let (V,F) be the Kronecker foliation $dy = \theta dx$ of the 2-torus $V = \mathbb{T}^2 = \mathbb{R}^2/\mathbb{Z}^2$ with natural coordinates $(x,y) \in \mathbb{R}^2$. Here $\theta \in]0,1[$ is an irrational number.

The graph G of this foliation is the manifold $G = \mathbb{T}^2 \times \mathbb{R}$ with range and source maps $G \to \mathbb{T}^2$ given by

$$r((x,y),t) = (x+t, y+\theta t)$$
$$s((x,y),t) = (x,y)$$

and with composition given by $((x,y),t)((x',y'),t') = ((x',y'),t+t')$ for any pair of composable elements.

Every closed geodesic of the flat torus $\mathbb{T}^2$ yields a compact transversal. More precisely, for each pair (p,q) of relatively prime integers we let $N_{p,q}$ be the submanifold of $\mathbb{T}^2$ given by

$$N_{p,q} = \{(ps, qs)\ ;\ s \in \mathbb{R}/\mathbb{Z}\}.$$

The graph G reduced by $N = N_{p,q}$, i.e. $G_N = \{\gamma \in G; r(\gamma) \in N, s(\gamma) \in N\}$, is then the manifold $G_N = \mathbb{T} \times \mathbb{Z}$ with range and source maps given by:

$$r(u,n) = u + n\theta' \ , \ s(u,n) = u$$

where $\theta' \in \mathbb{R}/\mathbb{Z}$ is determined uniquely by any pair (p',q') of integers such that $pq' - p'q = 1$, $\theta' = \frac{p'\theta - q'}{p\theta - q}$.

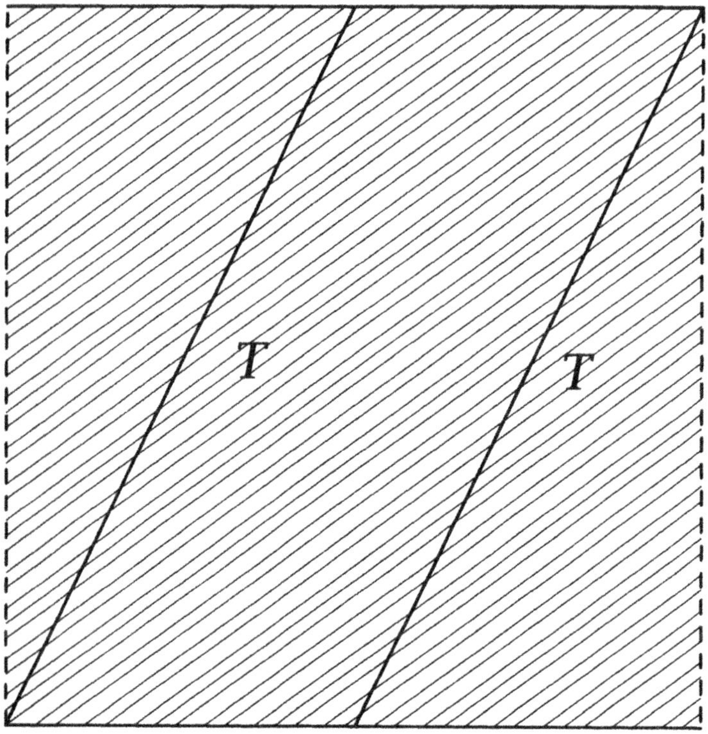

Figure II.8. The Kronecker foliation with a closed transversal T

The C^*-algebra $C^*_{r,N}(V,F)$ is the crossed product of $C(\mathbb{T})$ by the rotation of angle θ', i.e. it is the irrational rotation C^*-algebra ([Ri$_1$]) $A_{\theta'}$, generated by two unitaries U and V such that:

$$VU = \exp(2\pi i\theta')UV.$$

Since $N = N_{p,q}$ meets every leaf of the foliation, it follows that $C_r^*(V,F)$ is strongly Morita equivalent to $A_{\theta'}$ for any relatively prime pair (p,q). In particular for $p = 0$, $q = 1$ we get A_θ, and, by transitivity of strong Morita equivalence, we see that if θ and θ' are on the same orbit of the action of $PSL(2,\mathbb{Z})$ then A_θ is strongly Morita equivalent to $A_{\theta'}$, an important result due to M. Rieffel [Ri$_1$]. The finite projective C^*-module $\mathcal{E} = \mathcal{E}_{p,q}$ over A_θ which achieves the strong Morita equivalence with $A_{\theta'}$ (i.e. $A_{\theta'} \simeq \mathrm{End}_{A_\theta}(\mathcal{E})$) is obtained directly from the above geometric picture. One first determines the manifold

$$E_{p,q} = \left\{\gamma \in G \,;\, r(\gamma) \in N_{p,q} \,,\, s(\gamma) \in N_{0,1}\right\}.$$

Let us assume that $p > 0$ to avoid the trivial case $p = 0$. One finds then that $E_{p,q} = \{((0,y),t) \,;\, t \in \mathbb{R},\, py = t(q - p\theta) \text{ modulo } 1\}$. It is thus the disjoint union of p copies of the manifold $\mathbb{R}$, since the value of $y \in \mathbb{R}/\mathbb{Z}$ is not uniquely determined by the equality

$$py = t(q - p\theta) \text{ modulo } 1.$$

Working out the A_θ-valued inner product on $C_c^\infty(E_{p,q})$, one finds the following description of the C^*-module $\mathcal{E}_{p,q}$.

One lets $\varepsilon = \frac{q}{p} - \theta$, and lets W_1 and W_2 be unitary operators in $\mathbb{C}^p = K$ such that $W_1^p = W_2^p = 1$ and $W_1 W_2 = \exp\left(2\pi i \frac{q}{p}\right) W_2 W_1$. Let V_1 and V_2 be the operators on $C_c^\infty(\mathbb{R})$ given by

$$(V_1\xi)(s) = \xi(s - \varepsilon)\,,\ (V_2\xi)(s) = \exp(2\pi i s)\xi(s) \quad \forall s \in \mathbb{R}.$$

Then the action of A_θ on $\mathcal{E}_{p,q}$ is determined using the identification $C_c^\infty(E_{p,q}) = C_c^\infty(\mathbb{R}) \otimes K$ by the equalities

$$\xi U = (V_1 \otimes W_1)\xi\,,\ \xi V = (V_2 \otimes W_2)\xi \quad \forall \xi \in C_c^\infty(E_{p,q})$$

where U, V are the above generators of A_θ.

The A_θ-valued inner product which allows one to complete $C_c^\infty(E_{p,q})$ and get $\mathcal{E}_{p,q}$ is given by the value of the components of $\langle \xi, \eta\rangle = \sum_{\mathbb{Z}^2} \langle \xi, \eta\rangle_{m,n}\, U^m V^n$

$$\langle \xi, \eta\rangle_{m,n} = \int_{-\infty}^{\infty} \langle W_2^n W_1^m\, \xi(s - m\varepsilon), \eta(s)\rangle\ \exp(-2\pi i n s)ds.$$

We shall come back to these modules in later chapters.

Of course foliations can fail to have such a closed transversal N, and we shall show in an example that even $C_r^*(V,F)$ can fail to have any non-zero idempotent. We let Γ be a discrete cocompact subgroup of $SL(2,\mathbb{R})$. Then $V = SL(2,\mathbb{R})/\Gamma$ has a natural flow H_t, the horocycle flow, defined by the action by left translations of the subgroup

$$\left\{\begin{bmatrix} 1 & 0 \\ t & 1 \end{bmatrix} ;\ t \in \mathbb{R}\right\} \text{ of } SL(2,\mathbb{R}).$$

We let F be the foliation of V into orbits of the horocycle flow. First, the flow is minimal, so, $C_r^*(V,F)$ is a simple C^*-algebra ([F-S][Hi-S$_2$]). Then, letting μ be the measure on V associated to the Haar measure of $SL(2,\mathbb{R})$, we can associate to μ (which is H_t-invariant for all $t \in \mathbb{R}$) a transverse measure Λ for (V,F), and hence a trace τ on $C_r^*(V,F)$. By simplicity of $C_r^*(V,F)$ this trace τ is faithful. Thus for any idempotent $e \in C_r^*(V,F)$ one has

$$0 < \tau(e) < \infty \text{ if } e \neq 0.$$

Now let G_s, $s \in \mathbb{R}$, be the geodesic flow on V, defined by the action by left translation of the subgroup $\left\{\begin{bmatrix} e^s & 0 \\ 0 & e^{-s} \end{bmatrix} ;\ s \in \mathbb{R}\right\}$. For every s, G_s is an automorphism of (V,F), since the equality $G_s H_t G_s^{-1} = H_{e^{-2s}t}$ shows that for $y = G_s(x)$, $F_y = (G_s)_* F_x$. This follows from

$$\begin{bmatrix} 1 & 0 \\ te^{-2s} & 1 \end{bmatrix} = \begin{bmatrix} e^s & 0 \\ 0 & e^{-s} \end{bmatrix} \begin{bmatrix} 1 & 0 \\ t & 1 \end{bmatrix} \begin{bmatrix} e^{-s} & 0 \\ 0 & e^s \end{bmatrix}.$$

Let θ_s be the corresponding automorphism of $C_r^*(V,F)$. For every $x \in C_r^*(V,F)$ the map $s \mapsto \theta_s(x)$ from $\mathbb{R}$ to $C_r^*(V,F)$ is norm continuous, which shows that if e is a self-adjoint idempotent, then $\theta_s(e)$ is equivalent to e for all $s \in \mathbb{R}$, and hence $\tau(\theta_s(e)) = \tau(e)$, $\forall s \in \mathbb{R}$.

But though μ is obviously invariant under the geodesic flow, the transverse measure Λ is *not* invariant under G_s; indeed the equality $G_s H_t G_s^{-1} = H_{e^{-2s}t}$ shows that $G_s(\Lambda) = e^{-2s}\Lambda$ for all $s \in \mathbb{R}$. Thus $\tau \circ \theta_s = e^{-2s}\tau$, and so $\tau(e) = 0$ for any self-adjoint idempotent e. So $C_r^*(V,F)$ does not have any non-zero idempotent though it is simple (cf. [Bl$_2$] for the first example of such a C^*-algebra). We shall describe in Section 9 another example with a unital C^*-algebra.

8.γ The analytic assembly map $\mu : K_{*,\tau}(BG) \to K(C^*(V,F))$

In this section we shall show how the construction of K-theory classes for the orbit space of a group action, discussed in Section 7, adapts to the leaf spaces of foliations. We shall thus get a very general construction of elements of $K(C^*(V,F))$ from geometric cycles, i.e. K-oriented maps from compact manifolds to the leaf space. Our first task will thus be to define carefully what we mean by a smooth map

$$f : W \to V/F$$

where W is a manifold.

Any smooth map $W \xrightarrow{\rho} V$ gives, by composition with the canonical projection $p : V \to V/F$, a smooth map $f = p \circ \rho$ from W to V/F. But, as in the case of orbit spaces (Section 7), a general smooth map $f : W \to V/F$ does not factorize through V, and is given by a Čech cocycle (Ω_i, γ_{ij}) on W with values in the graph G of (V,F). More precisely, $(\Omega_i)_{i\in I}$ is an open cover of W and there is a collection of smooth maps $\gamma_{ij} : \Omega_i \cap \Omega_j \to G$ such that

$$\gamma_{ij}(x) \circ \gamma_{jk}(x) = \gamma_{ik}(x) \quad \forall x \in \Omega_i \cap \Omega_j \cap \Omega_k.$$

Thus the smooth maps $\gamma_{ii} : \Omega_i \to V$ patch together as maps $p \circ \gamma_{ii}$ from Ω_i to V/F.

Given such a Čech cocycle (Ω_i, γ_{ij}) one constructs as follows a principal right G-bundle over W which captures all the relevant information about the map f. Let G_f be the manifold obtained by gluing together the open sets

$$\tilde{\Omega}_i = \{(x,\gamma) \in \Omega_i \times G \; ; \; \gamma_{ii}(x) = r(\gamma)\}$$

with the maps $(x,\gamma) \to (x, \gamma_{ji}(x) \circ \gamma)$. Then let r_f and s_f be the smooth maps given by

$$r_f : G_f \to W \, , \; r_f(x,\gamma) = x \quad \forall (x,\gamma) \in \tilde{\Omega}_i$$

$$s_f : G_f \to V \, , \; s_f(x,\gamma) = s(\gamma) \quad \forall (x,\gamma) \in \tilde{\Omega}_i.$$

The groupoid G acts on the right on G_f; for $\alpha \in G_f$ and $\gamma' \in G$ such that $s_f(\alpha) = r(\gamma')$ the composition $\alpha \circ \gamma' \in G_f$ is given by

$$(x,\gamma) \circ \gamma' = (x, \gamma \circ \gamma') \quad \forall (x,\gamma) \in \tilde{\Omega}_i.$$

Given $\alpha_1, \alpha_2 \in G_f$ such that $r_f(\alpha_1) = r_f(\alpha_2)$ there exists a unique $\gamma' \in G$ such that $\alpha_2 = \alpha_1 \circ \gamma'$. Thus G_f is a principal right G-bundle over W in an obvious sense.

Definition 3. *α) A smooth map $f : W \to V/F$ is given by its graph, which is a principal right G-bundle G_f over W.*

β) The map f is called a submersion if the map s_f is a submersion: $G_f \to V$.

As in the case of orbit spaces, the reason for the existence of sufficiently many smooth maps $f : W \to V/F$ is that homotopy classes of such maps correspond exactly to homotopy classes of continuous maps:

$$W \to BG$$

where BG is the classifying space of the topological groupoid G. The space BG is only defined up to homotopy, as the quotient of a free and proper action of G on contractible spaces. More precisely, a right action of G on topological spaces is given by a topological space Y, a continuous map $s_Y : Y \to G^{(0)}$ and a continuous map $Y \times_s G \overset{\circ}{\to} Y$, where $Y \times_s G = \{(y, \gamma) \in Y \times G \, ; \, s_Y(y) = r(\gamma)\}$ such that $(y \circ \gamma_1) \circ \gamma_2 = y \circ (\gamma_1 \gamma_2)$ for any $y \in Y$, $\gamma_1 \in G$, $\gamma_2 \in G$ with $r(\gamma_1) = s_Y(y)$, $s(\gamma_1) = r(\gamma_2)$.

In other words the map which to each $t \in G^{(0)}$ assigns the topological space $Y_t = s_Y^{-1}\{t\}$ and to each $\gamma \in G$, $\gamma : t \to t'$, assigns the homeomorphism $y \to y \circ \gamma$ from $Y_{t'}$ to Y_t, is a contravariant functor from the small category G to the category of topological spaces. We shall say that such an action of G on Y is *free* iff for any $y \in Y$ the map $\gamma \in G^{s_Y(y)} \to y \circ \gamma \in Y$ is injective, and that it is *proper* iff the map $(y, \gamma) \to (y, y \circ \gamma)$ is proper. A free and proper action of G on Y is the same thing as a principal G-bundle on the quotient space $Z = Y/G$, quotient of Y by the equivalence relation

$$y_1 \sim y_2 \ \text{ iff } \ \exists \gamma \in G \, , \ y_1 \circ \gamma = y_2 .$$

We shall say that the action of G on Y has *contractible fibers* iff the fibers of $s_Y : Y \to G^{(0)}$ are contractible. Then exactly as for groups (and with the usual paracompactness conditions) the classifying space BG is the quotient EG/G of a principal G-bundle $Y = EG$ with contractible fibers. It is unique up to homotopy, and it classifies, when G is the graph of a foliation (V, F), the homotopy classes of smooth maps to V/F. In fact we have already seen in Sections 5 and 7 two other examples of classifying spaces for smooth groupoids. In Section 5 we computed BG, where G is the tangent groupoid of a manifold M, by using the action of G on $Y = G^{(0)} \times \mathbb{R}^N$ coming from an immersion $j : M \to \mathbb{R}^N$. In Section 7 the homotopy quotient $V \times_\Gamma E\Gamma$ is the classifying space of the semidirect product $G = V \rtimes \Gamma$, which is a smooth groupoid.

For the case of foliations one can, assuming that the holonomy groups G_x^x are torsion-free, construct ([Co$_{11}$]) EG as the space of measures with finite support on the leaves of (V, F). What matters is that BG is an *ordinary space* which maps to V/F with contractible fibers. More precisely any foliation (V',F') with the same leaf space as (V,F) determines a principal G-bundle, and if the holonomy coverings of the leaves of (V',F') are contractible then the total space V' is the classifying space BG.

We shall now, under the above assumption that G_x^x is torsion-free for all $x \in V$, construct the geometric group $K_{*,\tau}(BG)$ and the analytic assembly map ([Bau-C$_1$] [Co-S])

$$K_{*,\tau}(BG) \xrightarrow{\mu} K(C^*(V,F)).$$

First, the transverse bundle $\tau_x = T_x(V)/F_x$ of the foliation is a G-equivariant bundle, the action of G on τ being given by the differential of the holonomy, $h(y)_* : \tau_x \to \tau_y \ \forall y : x \to y$. To τ corresponds an induced real vector bundle, which we "abusively" still denote by τ, on $BG = EG/G$. Thus the group $K_{*,\tau}(BG)$ is well-defined, exactly as in Definition 7.4, as the twisted K-homology, with compact supports, of the space BG. (The twisting by τ is defined as in 7.4 from the pair $(B\tau, S\tau)$ of the unit-ball and unit-sphere bundles of τ.) We, moreover, have the exact analogue of Proposition 7.6: By a *geometric cycle* we mean a triple (W,y,g) where W is a compact manifold, $y \in K^*(W)$ a K-theory class, and g a smooth map $W \to V/F$ (Definition 3) which is K-oriented by a choice of a Spin^c structure on the real vector bundle $TW \oplus g^*\tau$.

Proposition 4. a) *Any element of $K_{*,\tau}(BG)$ is represented by a geometric cycle (W,y,g).*

b) *Let (W,y,g) be a geometric cycle and $f : W' \to W$ a K-oriented smooth map. Then for any $x \in K^*(W')$, the following geometric cycles represent the same element of $K_{*,\tau}(BG)$:*

$$(W',x,g\circ f) \sim (W,f!(x),g).$$

For any smooth K-oriented map $W \xrightarrow{h} V/F$ one can factorize h as $g \circ f$, where $f : W \to W_1$ is K-oriented and smooth, while $g : W_1 \to V/F$ is a submersion in the sense of Definition 3. β) ([Co$_{11}$]). Thus using Proposition 4.b we see that in order to define the analytic assembly map $\mu : K_{*,\tau}(BG) \to K(C^*(V,F))$ we just need to define $\mu(W,y,g) = g!(y)$ in the case where $g : W \to V/F$ is a *smooth K-oriented submersion.* Using F_W, the pull-back by g of the foliation F, which is a foliation of W because g is a submersion (cf. [Co-S]) one can then easily reduce the task to the construction of the wrong-way functoriality map

$$p!: \ K^*(F^*) \to K(C^*(V,F))$$

where F^* means the *total space* of the vector bundle F^* on V. (We refer to [Co-S] for the reduction to this case, which invokes the construction of a homomorphism of $C^*(W,F_W)$ to the C^*-algebra of compact endomorphisms of a C^*-module over $C^*(V,F)$, thus yielding an element $\varepsilon_g \in E(C^*(W,F_W),C^*(V,F))$.)

We shall construct $p! \in E(C_0(F^*), C^*(V,F))$ as a deformation of the convolution C^*-algebra of the vector bundle F to that of the smooth groupoid, G, of the foliation. In order to do so, we just need to deform the groupoid F (i.e. $F^{(0)} = V \subset F$ is the 0-section and F is the union of the *groups* F_x, $x \in V$) to the groupoid G. In fact we just need the relevant topology on the union

$$G' = F \cup (G \times \,]0,1])$$

where both terms are groupoids, $]0,1]$ being, as in Section 5, viewed as a space. Here F is a *closed* subset of G', and a sequence $(\gamma_n, \varepsilon_n)$, $\gamma_n \in G$, $\varepsilon_n > 0$, of elements of $G\times \,]0,1]$ converges to $(x,X) \in F_x \subset F$ iff $\varepsilon_n \to 0$, $x_n = s(\gamma_n) \to x$, $y_n = r(\gamma_n) \to x$, $\frac{y_n - x_n}{\varepsilon_n} \to X$ and $\text{length}(\gamma_n) \to 0$. The condition $\text{length}(\gamma) < \varepsilon$ defines a basis of neighborhoods of the diagonal $G^{(0)}$ in G. In fact, using a Euclidean structure on F, we can give the local diffeomorphism $F \times [0,1] \xrightarrow{\varphi} G'$ near $((x,X),0)$ by

$$\varphi((y,Y),\varepsilon) = ((y, \exp_y(-\varepsilon Y)), \varepsilon) \in G\times \,]0,1] \quad \text{for } \varepsilon > 0$$

where the latter pair of points y and $\exp_y(-\varepsilon Y) \in$ (Leaf of y) define the end points of the path $\gamma(t) = \exp_y(-\varepsilon t Y)$, $\gamma \in G$.

As in Section 5 one checks that with the above structure G' is a smooth groupoid, and this suffices, using the associated C^*-algebra, to get the required deformation

$$p! \in E(C_0(F^*), C^*(V,F))$$

where we identified $C^*(F)$ with $C_0(F^*)$ using the Fourier transform. Thus, given a smooth K-oriented submersion $g : W \to V/F$ we obtain an element $g! \in E(C(W), C^*(V,F))$ as the composition of $\varepsilon_g \circ (p_W)! \in E(C_0(F_W^*), C^*(V,F))$ with the Thom isomorphism $\beta \in E(C(W), C_0(F_W^*))$ given by the K-orientation of g. We can then state the main result of [Co-S] as

Theorem 5. a) *Let W be a compact manifold and let $g : W \to V/F$ be a smooth K-oriented map. Then the composition $f! \circ j! = g! \in E(C(W), C^*(V,F))$ is independent of the factorization of $g = f \circ j$ through a K-oriented* submersion $f : W' \to V/F$.

b) *The element $g!$ only depends upon the K-oriented homotopy class of g and one has $(g \circ h)! = g! \circ h!$ for any K-oriented smooth map $h : X \to W$.*

The construction of the analytic assembly map μ follows immediately from this theorem.

Corollary 6. *Let $x \in K_{*,\tau}(BG)$ and (W, y, g) be a geometric cycle representing x. The element $g!(y) \in K(C^*(V,F))$ only depends upon x, and μ is an additive map:*

$$\mu : K_{*,\tau}(BG) \to K(C^*(V,F)).$$

By composition with the canonical surjection $r : C^*(V,F) \to C_r^*(V,F)$ one obtains a corresponding map $\mu_r : K_{*,\tau}(BG) \to K(C_r^*(V,F))$.

Let us also note that this construction of K-theory classes of $C^*(V,F)$ contains, of course, the construction of the classes of idempotents $e_T \in C_r^*(V,F)$ associated to closed transversals T of (V,F). More precisely, the composition of the inclusion $T \subset V$ with the projection $p : V \to V/F$ yields a smooth map $f : T \to V/F$ which is by construction étale and hence K-oriented. One then has a geometric cycle $(T, 1_T, f)$, where 1_T is the class in $K_0(T)$ of the trivial one-dimensional vector bundle. One checks easily that $\mu_r(T, 1_T, f) = [e_T]$.

The construction of $g!$ for a smooth map $g : W \to V/F$ has been extended by M. Hilsum and G. Skandalis ([Hi-S$_1$]) to smooth maps of leaf spaces, after solving the very difficult case of the submersion $V/F \to \text{pt}$. We shall come back to this point in Chapter III.

Remarks 7. a) We have only defined the geometric group as $K_{*,\tau}(BG)$ when the holonomy groups G_x^x, $x \in V$, are torsion-free. As in the case of discrete groups, the general case requires more care and will be treated in Section 10 (cf. also [Bau-C-H]).

b) Exactly as in Section 7, the Chern character ch_* is a rational isomorphism of $K_{*,\tau}(BG)$ with $H_*(BG, \mathbb{Q})$, (if we assume to simplify that τ is oriented, i.e. that the foliation is transversally oriented). Thus, as in Section 7, any element $z \in K(C^*(V,F))$ of the form $\mu(x)$, where $x \in K_{*,\tau}(BG)$ and $\text{ch}_*(x) = 0$, is a torsion element.

II.9. The Longitudinal Index Theorem for Foliations

Let (V,F) be a compact foliated manifold. Let E_1 and E_2 be smooth complex vector bundles over V, and let $D : C^\infty(V, E_1) \to C^\infty(V, E_2)$ be a differential operator on V from sections of E_1 to sections of E_2. Let us make the following hypotheses:

1) *D restricts to leaves,* i.e. $(D\xi)_x$ depends only upon the restriction of ξ to a neighborhood of x in the leaf of x.

2) *D is elliptic when restricted to leaves,* so that, for any $\eta \in F_x^*$, $\eta \neq 0$, the principal symbol $\sigma_D(x, \eta) \in \text{Hom}(E_{1,x}, E_{2,x})$ is invertible.

In any domain of a foliation chart $U = T \times P$ the operator D appears as a family, indexed by $t \in T$, of elliptic operators on the plaques P_t. One can then use the local construction of a parametrix for families of elliptic operators and patch the resulting operators by using a partition of unity in V subordinate to a covering (U_j) by domains of foliation charts. What one obtains is an inverse Q for D modulo the algebra $C_c^\infty(G, \Omega^{1/2}) = J$ of the foliation (Section 8). To be more precise let us fix for convenience a smooth nonvanishing 1-density along the leaves and drop the Ω's from the notation. Then to D corresponds a distribution section $D \in C_c^{-\infty}(G, s^*(E_1^*) \otimes r^*(E_2))$ with support on $G^{(0)}$. To the quasi-inverse Q of D corresponds a section $Q \in C_c^{-\infty}(G, s^*(E_2^*) \otimes r^*(E_1))$. The quasi-inverse property is then the following

$$QD - 1_{E_1} \in C_c^\infty(G, s^*(E_1^*) \otimes r^*(E_1))$$

$$DQ - 1_{E_2} \in C_c^\infty(G, s^*(E_2^*) \otimes r^*(E_2))$$

where 1_{E_i} corresponds to the identity operator $C^\infty(V, E_i) \to C^\infty(V, E_i)$.

Now the algebra $J = C_c^\infty(G)$ is a two-sided ideal in the larger $\mathcal{A}$, the algebra under convolution of distributions $T \in C_c^{-\infty}(G)$ which are multipliers of $C_c^\infty(G)$, i.e. satisfy $T * f \in C_c^\infty(G)$ and $f * T \in C_c^\infty(G)$, for any $f \in C_c^\infty(G)$. Thus, neglecting the bundles E_i for a while, the existence of Q means that D yields an invertible element of $\mathcal{A}/J$. It is, however, not always possible to find a representative $D' \in \mathcal{A}$ of the same class, i.e. $D' \in D + J$, which is *invertible* in $\mathcal{A}$. The obstruction to doing so is an element of the K-theory group $K_0(J)$, as follows from elementary algebraic K-theory ([Mi$_1$]). We shall, however, recall in detail the construction of this obstruction $\mathrm{Ind}(D) \in K_0(J)$, since this allows us to take the bundles E_i into account and will also be useful in Chapter III.

9.α Construction of $\mathrm{Ind}(D) \in K_0(J)$

Let J be a non-unital algebra over $\mathbb{C}$, and define $K_0(J)$ as the kernel of the map

$$K_0(\tilde{J}) \overset{\varepsilon_*}{\to} K_0(\mathbb{C}) = \mathbb{Z}$$

where $\tilde{J}$ is the algebra obtained by adjoining a unit to J, that is,

$$\tilde{J} = \{(a, \lambda); a \in J, \lambda \in \mathbb{C}\}, \text{and } \varepsilon(a, \lambda) = \lambda \ \forall (a, \lambda) \in \tilde{J}.$$

(For a unital algebra K_0 is the group associated to stable isomorphism classes of finite projective modules viewed as a semigroup under direct sum.)

Let $\mathcal{A}$ be a unital algebra (over $\mathbb{C}$) containing J as a two-sided ideal, and let $j : \mathcal{A} \to \mathcal{A}/J = \Lambda$ be the quotient map. Recall that finite projective modules push forward under morphisms of algebras.

Definition 1. *Given $J \subset \mathcal{A}$ as above, a* quasi-isomorphism *is given by a triple $(\mathcal{E}_1, \mathcal{E}_2, h)$, where $\mathcal{E}_1$ and $\mathcal{E}_2$ are finite projective modules over $\mathcal{A}$ and h is an isomorphism*

$$h : j_* \mathcal{E}_1 \to j_* \mathcal{E}_2.$$

Any element D of $\mathcal{A}$ which is invertible modulo J determines the quasi-isomorphism $(\mathcal{A}, \mathcal{A}, j(D))$. A quasi-isomorphism is called *degenerate* when h comes from an isomorphism $T : \mathcal{E}_1 \to \mathcal{E}_2$. There is an obvious notion of direct sum of quasi-isomorphisms, and a simple but crucial lemma ([Mi$_1$]) shows that the direct sum $(\mathcal{E}_1, \mathcal{E}_2, h) \oplus (\mathcal{E}_2, \mathcal{E}_1, h^{-1})$ is always degenerate.

More explicitly, let $D \in \mathrm{Hom}_{\mathcal{A}}(\mathcal{E}_1, \mathcal{E}_2)$ and $Q \in \mathrm{Hom}_{\mathcal{A}}(\mathcal{E}_2, \mathcal{E}_1)$ be such that $j(D) = h$ and $j(Q) = h^{-1}$. Then the matrix

$$T = \begin{bmatrix} D + (1 - DQ)D & DQ - 1 \\ 1 - QD & Q \end{bmatrix}$$

defines an isomorphism of $\mathcal{E}_1 \oplus \mathcal{E}_2$ with $\mathcal{E}_2 \oplus \mathcal{E}_1$ such that $j(T) = \begin{bmatrix} h & 0 \\ 0 & h^{-1} \end{bmatrix}$.

It follows that quasi-isomorphisms modulo degenerate ones form a group which, as we shall see now, is canonically isomorphic to $K_0(J)$ independently of the choice of $\mathcal{A}$.

Let us first consider the special case $\mathcal{A} = \tilde{J}$. Then the exact sequence $0 \to J \to \tilde{J} \xrightarrow{\varepsilon} \mathbb{C} \to 0$ has a natural section $r : \mathbb{C} \to \tilde{J}$, $\varepsilon \circ r = \mathrm{id}$. Thus for any finite projective module $\mathcal{E}$ over $\tilde{J}$ the triple $(\mathcal{E}, (r \circ \varepsilon)_* \mathcal{E}, \mathrm{id})$ is a quasi-isomorphism, which we denote $\rho(\mathcal{E})$.

Now, let $\mathcal{A}$ be arbitrary, and let α be the homomorphism $\alpha : \tilde{J} \to \mathcal{A}$, $\alpha(a, \lambda) = a + \lambda 1$ $\forall a \in J, \lambda \in \mathbb{C}$.

Proposition 2. *Given $J \subset \mathcal{A}$ as above, the map $\alpha_* \circ \rho$ is an isomorphism from $K_0(J)$ to the group of classes of quasi-isomorphisms modulo degenerate ones.*

The proof follows from the computation ([Mi_1]) of the K-theory of the fibered product algebra

$$\{(a_1, a_2) \in \mathcal{A} \times \mathcal{A}; j(a_1) = j(a_2)\}.$$

Given a quasi-isomorphism $(\mathcal{E}_1, \mathcal{E}_2, h)$ we shall let $\mathrm{Ind}(h) \in K_0(J)$ be the associated element of $K_0(J)$ (Proposition 2). For instance, if D is an element of $\mathcal{A}$ which is invertible modulo J then $\mathrm{Ind}(D)$ is the element of $K_0(J)$ given by $[e] - [e_0]$, where the idempotents $e, e_0 \in M_2(\tilde{J})$ are $e_0 = \begin{bmatrix} 1 & 0 \\ 0 & 0 \end{bmatrix}$ and $e = T e_0 T^{-1}$ with, as above,

$$T = \begin{bmatrix} D + (1 - DQ)D & DQ - 1 \\ 1 - QD & Q \end{bmatrix}.$$

Thus, with $S_0 = 1 - QD$, $S_1 = 1 - DQ$ one gets

$$e = \begin{bmatrix} 1 - S_1^2 & (S_1 + S_1^2)D \\ S_0 Q & S_0^2 \end{bmatrix} \in M_2(\tilde{J}).$$

One has $\mathrm{Ind}(h_2 \circ h_1) = \mathrm{Ind}(h_1) + \mathrm{Ind}(h_2)$ for any pair $((\mathcal{E}_1, \mathcal{E}_2, h_1), (\mathcal{E}_2, \mathcal{E}_3, h_2))$ of quasi-isomorphisms.

Let (V, F) be a compact foliated manifold and let D be, as above, a longitudinal elliptic operator from the bundle E_1 to E_2. Then the inclusion $C^\infty(V) \subset \mathcal{A}$ of multiplication operators on V, as longitudinal differential operators of order 0 allows us to induce the vector bundles E_i to finite projective modules $\mathcal{E}_i$ over

$\mathcal{A}$. The above existence of an inverse for D modulo $C_c^\infty(G)$ is then precisely encoded in

Proposition 3. *The triple $(\mathcal{E}_1, \mathcal{E}_2, D)$ defines a quasi-isomorphism over the algebra $J = C_c^\infty(G) \subset \mathcal{A}$.*

We shall let $\mathrm{Ind}(D) \in K_0(C_c^\infty(G))$ be the index associated to $(\mathcal{E}_1, \mathcal{E}_2, D)$ by Proposition 2.

9.β Significance of the C^*-algebra index

By the construction of $C^*(V,F)$ as a completion of the algebra $C_c^\infty(G)$, one has a natural homomorphism

$$C_c^\infty(G) \xrightarrow{j} C^*(V,F)$$

and we shall denote by $\mathrm{Ind}_a(D)$ the image $j_* \mathrm{Ind}(D)$. In general we do not expect the map

$$j_* : K_0(C_c^\infty(G)) \to K_0(C^*(V,F))$$

to be an isomorphism, and thus we lose information in replacing $\mathrm{Ind}(D)$ by $\mathrm{Ind}_a(D) = j_*(\mathrm{Ind}(D))$. It is, however, only for the latter index that one has vanishing or homotopy invariance results of the following type:

Proposition 4. [Co$_{14}$] *Let (V,F) be a compact foliated manifold. Assume that the real vector bundle F is endowed with a Spin structure and a Euclidean structure whose leafwise scalar curvature is strictly positive. Let D be the leafwise Dirac operator. Then D is a longitudinal elliptic operator and* $\mathrm{Ind}_a(D) = 0$.

In other words, with F of even dimension and oriented by its Spin structure, one lets $S^\pm$ be the bundle of spinors and $D : C^\infty(V,S^+) \to C^\infty(V,S^-)$ be the partial differential operator on V which restricts to leaves as the leafwise Dirac operator.

The proof of the vanishing of $\mathrm{Ind}_a(D)$ is the same as the proof of J. Rosenberg for covering spaces (cf. [Gro-L]); using the Lichnerowicz formula for D^*D and DD^* one shows that these two operators are bounded from below by a strictly positive scalar. This shows the vanishing of $\mathrm{Ind}_a(D) \in K_0(C^*(V,F))$ because operator inequalities imply spectral properties in C^*-algebras. It is, however, not sufficient to prove the vanishing of $\mathrm{Ind}(D) \in K_0(C_c^\infty(G))$. As another example, let us consider the leafwise homotopy invariance of the *longitudinal signature*, i.e. of $\mathrm{Ind}_a(D)$, where D is the longitudinal signature operator. This question is the exact analogue of the question of the homotopy invariance of the Γ-invariant signature for covering spaces, which was proved by Mishchenko and Kasparov. One gets ([Bau-C$_2$] [Hi-S$_3$]):

Proposition 5. *Let (V,F) be a compact foliated manifold with F even-dimensional and oriented. Let D be the leafwise signature operator. Then its analytic index* $\mathrm{Ind}_a(D) \in K_0(C^*(V,F))$ *is preserved under leafwise oriented homotopy equivalences.*

9.γ The longitudinal index theorem

Let $V \overset{p}{\to} B$ be a *fibration*, where V and B are smooth compact manifolds, and let F be the vertical foliation, so that the leaves of (V, F) are the fibers of the fibration and the base B is the space V/F of leaves of the foliation. Then a longitudinal elliptic operator is the same thing as a family $(D_y)_{y\in B}$ of elliptic operators on the fibers in the sense of [At-Si$_2$]. Moreover, the C^*-algebra of the foliation (V, F) is strongly Morita equivalent to $C(B)$, and one has a canonical isomorphism

$$K(C^*(V,F)) \simeq K(C(B)) = K(B).$$

Under this isomorphism our analytic index, $\mathrm{Ind}_a(D) \in K(C^*(V,F))$ is the same as the Atiyah-Singer index for the family $D = (D_y)_{y\in B}$, $\mathrm{Ind}_a(D) \in K(B)$ (cf. [At-Si$_2$]). In this situation the Atiyah-Singer index theorem for families (loc. cit) gives a topological formula for $\mathrm{Ind}_a(D)$ as an equality

$$\mathrm{Ind}_a(D) = \mathrm{Ind}_t(D)$$

where the *topological* index $\mathrm{Ind}_t(D)$ only involves the K-theory class $\sigma_D \in K(F^*)$ of the principal symbol of D, and uses in its construction an auxiliary embedding of V in the Euclidean space $\mathbb{R}^N$. (cf. loc. cit).

We shall now explain the index theorem for foliations ([Co-S]) which extends the above result to the case of *arbitrary* foliations of compact manifolds and immediately implies the index theorem for *measured* foliations of Chapter I.

As in the Atiyah-Singer theorem we shall use an auxiliary embedding of V in $\mathbb{R}^n$ in order to define the topological index, $\mathrm{Ind}_t(D)$, and the theorem will be the equality $\mathrm{Ind}_a = \mathrm{Ind}_t$. This equality holds in $K(C^*(V,F))$ and thus we need an easy way to land in this group. Now, given a foliation (V', F') of a not necessarily compact manifold, and a not necessarily compact submanifold N of V' which is everywhere transverse to F', then Lemma 8.2 provides us with an easy map from $K(N) = K(C_0(N))$ to $K(C^*(V', F'))$. Indeed, for a suitable open neighborhood V'' of N in V' one has

$$C_0(N) \sim C^*(V'', F') \subset C^*(V', F')$$

(where the first equivalence is a strong Morita equivalence). Of course, the resulting map $K(N) \to K(C^*(V', F'))$ coincides with the map $e!$, where $e : N \to V'/F'$ is the obvious étale map, but we do not need this equality (except as notation) to define $e!$. The main point in the construction of the topological index Ind_t is that an embedding $i : V \to \mathbb{R}^n$ allows one to consider the *normal bundle* $\nu = i_*(F)^\perp$ of $i_*(F)$ in $\mathbb{R}^n$ as a manifold N, transversal to the foliation of $V' = V \times \mathbb{R}^n$ by the integrable bundle $F' = F \times \{0\} \subset TV'$. First note that the bundle ν over V has a total space of dimension $d = \dim V + n - \dim F$, which is the same as the codimension of F' in V'. Next consider the map φ from the total space ν to $V' = V \times \mathbb{R}^n$ given by

$$\varphi(x, \xi) = (x, i(x) + \xi) \quad \forall x \in V\,,\ \xi \in \nu_x = (i_*(F_x))^\perp,$$

and check that on a small enough neighborhood N of the 0-section in ν the map $\varphi : N \to V'$ is transverse to F'. It is enough to check this transversality on the 0-section $V \subset \nu$ where it is obvious, and only uses the injectivity of i_* on $F \subset TV$.

Theorem 6. [Co-S] *Let (V,F) be a compact foliated manifold, D a longitudinal elliptic differential operator. Let $i : V \to \mathbb{R}^n$ be an embedding, let $\nu = (i_*(F))^\perp$ be the normal bundle of i_*F in $\mathbb{R}^n$, and let $N \sim \nu$ be the corresponding transversal to the foliation of $V \times \mathbb{R}^n = V'$ by $F' = F \times \{0\}$. Then the analytic index,* $\mathrm{Ind}_a(D)$ *is equal to* $\mathrm{Ind}_t(\sigma(D))$, *where $\sigma(D) \in K(F^*)$ is the K-theory class of the principal symbol of D, while* $\mathrm{Ind}_t(\sigma(D))$ *is the image of $\sigma(D) \in K(F^*) \simeq K(N)$ (through the Thom isomorphism) by the map*

$$K(N) \xrightarrow{e!} K(C^*(V',F')) \simeq K(C^*(V,F))$$

(through the Bott periodicity isomorphism since $C^*(V',F') = S^n C^*(V,F) = C^*(V,F) \otimes C_0(\mathbb{R}^n)$.)

The proof easily follows from Theorem 8.3 and the following analogue of Lemma 5.6:

Lemma 7. *With the notation of Theorem 6, let $x = (F, \sigma(D), p \circ \pi)$ be the geometric cycle given by the total space of the bundle F, the K-theory class $\sigma(D) \in K^*(F)$ of the symbol of D, and the K-oriented map $F \to V \to V/F$. Then*

$$\mathrm{Ind}_a(D) = \mu(x) \in K(C^*(V,F)).$$

Note that a direct proof of Theorem 6 is possible using the Thom isomorphism as in Section 5.

Of course, the above theorem does not require the existence of a transverse measure for the foliation, and it implies the index theorem for measured foliations of Chapter I. A transverse measure Λ on (V,F) determines a positive semi-continuous semi-finite trace, trace_Λ, on $C^*(V,F)$ and hence an additive map

$$\mathrm{trace}_\Lambda : K(C^*(V,F)) \to \mathbb{R}.$$

The reason why $\mathrm{trace}_\Lambda \circ \mathrm{Ind}_t$ is much easier to compute than $\mathrm{trace}_\Lambda \circ \mathrm{Ind}_a$ is that the former computation localizes on the restriction of the foliation F' to a neighborhood V'' of N in V' on which the space of leaves V''/F' is an ordinary space N, so that all difficulties disappear.

Let us apply these results in the simplest example of a one-dimensional oriented foliation (with no stable compact leaves) and compute

$$\dim_\Lambda \; K(C^*(V,F)) \subset \mathbb{R}.$$

For such foliations the graph G is $V \times \mathbb{R}$ and the classifying space BG is thus homotopic to V. Thus, up to a shift of parity, the geometric group $K_{*,\tau}(BG)$ is the K-theory group $K^{*+1}(V)$. The projection

$$p : V \to V/F$$

is naturally K-oriented and the following maps from $K^*(V)$ to $K(C^*(V,F))$ coincide:

1) $p!$ (cf. Theorem 8.3) which, up to the Thom isomorphism, $K^*(F) \simeq K^{*+1}(V)$ is equal by Lemma 7 to $\mathrm{Ind}_a = \mathrm{Ind}_t$.

2) The Thom isomorphism (Appendix C)

$$\phi : K^*(V) = K(C(V)) \to K(C(V) \rtimes \mathbb{R}) = K(C^*(V,F))$$

where we used an arbitrary flow defining the foliation. In particular, by Appendix C, ϕ is an isomorphism and so are the map $p!$ and the index Ind_a : $K^*(F) \to K(C^*(V,F))$.

Taking for instance the horocycle foliation of $V = SL(2,\mathbb{R})/\Gamma$, where Γ is a discrete cocompact subgroup of $SL(2,\mathbb{R})$, by the left action of the subgroup of lower triangular matrices of the form $\begin{bmatrix} 1 & 0 \\ t & 1 \end{bmatrix}$, $t \in \mathbb{R}$, one sees that the analytic index gives a (degree-1) isomorphism of $K^*(V)$ onto $K(C^*(V,F))$, while for the only transverse measure Λ for (V,F) (the horocycle flow being strongly ergodic), the composition of the above analytic index with $\dim_\Lambda : K^*(C^*(V,F)) \mapsto \mathbb{R}$ is equal to 0, so in particular $\dim_\Lambda$ is identically 0. This example shows that even when the foliation (V,F) does have a non-trivial transverse measure, the K-theoretic formulation of the index theorem (Theorem 6) gives much more information than the index theorem for the measured foliation (F,Λ). As a corollary of the above we get:

Corollary 8. *Let (V,F) be one-dimensional as above, and let Λ be a transverse measure for (V,F). Then the image $\dim_\Lambda(K(C^*(V,F)))$ is equal to*

$$\{\langle \mathrm{ch}(E), [C]\rangle ; [E] \in K^*(V)\}.$$

Here ch is the usual Chern character, mapping $K^*(V)$ to $H^*(V,\mathbb{Q})$, and $[C]$ is the Ruelle-Sullivan cycle of Chapter I.

Corollary 9. *Let V be a compact smooth manifold, and let φ be a minimal diffeomorphism of V. Assume that the first cohomology group $H^1(V,\mathbb{Z})$ is equal to 0. Then the crossed product $A = C(V) \rtimes_\varphi \mathbb{Z}$ is a simple unital C^*-algebra without any non-trivial idempotents.*

As a very nice example where this corollary applies one can take the diffeomorphism φ given by left translation by $\begin{bmatrix} 1 & 0 \\ 1 & 1 \end{bmatrix} \in SL(2,\mathbb{R})$ of the manifold $V = SL(2,\mathbb{R})/\Gamma$, where the group Γ is chosen discrete and cocompact in such a way that V is a homology 3-sphere. For instance, one can take in the Poincaré disk a regular triangle T with its three angles equal to $\pi/4$ and as Γ the group formed by products of an even number of hyperbolic reflections along the sides of T.

In order to exploit the results on vanishing or homotopy invariance of the analytic index Ind_a we shall construct in Chapter III higher-dimensional generalizations of the above maps

$$\dim_\Lambda : K(C^*(V,F)) \to \mathbb{R}$$

associated to higher-dimensional "currents" on the space of leaves V/F, whose differential geometry, i.e. the transverse geometry of the foliation, will then be fully used. We refer to Section 7 of Chapter III for the general index formula.

II.10. The Analytic Assembly Map and Lie Groups

In this section we shall construct for general smooth groupoids G a group $K^*_{\mathrm{top}}(G)$ which is computable from the standard tools of equivariant K-theory, K_H, where H is a compact group, and a map

$$\mu : K^*_{\mathrm{top}}(G) \to K(C^*(G))$$

which will extend to the general case the constructions of the analytic assembly map of Sections 4, 5, 7, and 8 above.

The new ingredient needed to reach the general case is to understand the case when G has torsion, i.e. when the groups $G^x_x = \{\gamma \in G; r(\gamma) = s(\gamma) = x\}$ contain elements of finite order. We have carefully avoided this case in the examples of Sections 4, 7, and 8 by requiring in Sections 4 and 7 that the discrete group Γ be torsion-free, and in Section 8 that the holonomy groups be torsion-free.

After formulating the analytic assembly map μ in general, we shall then deal in detail with the case of Lie groups.

10.α Geometric cycles for smooth groupoids

In the examples of Sections 4, 5, 7, and 8 we constructed an additive map

$$\mu : K_{*,\tau}(BG) \to K(C^*(G))$$

where BG is the classifying space of the topological groupoid G and $K_{*,\tau}$ is K-homology, twisted by a suitable real vector bundle on BG. As we said above, this is only appropriate in the case when G is torsion-free, and obviously μ is not a surjection onto $K(C^*(G))$ when G is, say, a finite group. Elements of $K_{*,\tau}(BG)$ were described as geometric cycles (cf. Proposition 7.6 and Proposition 8.4)

$$W \xrightarrow{f} BG$$

or equivalently, since BG classifies G-principal bundles, as free and proper G-spaces.

But since G was assumed to be torsion-free, the hypothesis of freeness did follow automatically from the properness of the action. We shall now give the general definition of a geometric cycle, and then explain how it specializes to the previous one.

Definition 1. *Let G be a smooth groupoid. Then a G-manifold is given by:*

a) *A manifold P with a submersion $\alpha : P \to G^{(0)}$.*

b) *A right action of G on P : $P \times_\alpha G \to P$, $(p, \gamma) \to p \circ \gamma$ such that $(p \circ \gamma_1) \circ \gamma_2 = p \circ (\gamma_1 \gamma_2)$ $\forall (\gamma_1, \gamma_2) \in G^{(2)}$, where $P \times_\alpha G = \{(p, \gamma) \in P \times G\,;\, \alpha(p) = r(\gamma)\}$.*

Such a G-manifold provides us with a contravariant functor from the small category G to the category of smooth manifolds, because, since α is a submersion, its fibers $\alpha^{-1}\{x\}$, $x \in G^{(0)}$, are manifolds P_x, while $p \to p \circ \gamma$ is, for $\gamma \in G$ with $\gamma : x \to y$, a diffeomorphism from P_y to P_x.

We shall say that a G-manifold P is *proper* when the following map is proper:

$$P \times_\alpha G \to P \times P\ ,\ (p, \gamma) \to (p, p \circ \gamma).$$

Let us put a groupoid structure on $P \times_\alpha G$, by letting $(P \times_\alpha G)^{(0)} = P$ with range, source maps and composition given by

$$r(p, \gamma) = p\ ,\ s(p, \gamma) = p \circ \gamma$$

$$(p, \gamma) \circ (p', \gamma') = (p, \gamma \circ \gamma').$$

We thus get a smooth groupoid, which if the G-manifold P is proper satisfies the following definition:

Definition 2. *A smooth groupoid G is* proper *if the map $G \to G^{(0)} \times G^{(0)}$ given by $\gamma \to (r(\gamma), s(\gamma))$ is a proper map.*

The computation of $K(C^*(G))$ for smooth *proper* groupoids is perfectly doable with the usual tools of equivariant K-theory, K_H of spaces, where H is a compact group. Indeed, if we look at the space of objects $G^{(0)}$ of G, modulo isomorphism, i.e. at the quotient of $G^{(0)}$ by the equivalence relation $\mathcal{R} = (r, s)(G)$, we get a locally compact space $X = G^{(0)}/\mathcal{R}$ because of the *properness* hypothesis. Next, given $x \in G^{(0)}$, its automorphism group $G_x^x = \{\gamma \in G; r(\gamma) = s(\gamma) = x\}$ is compact, so that we are essentially dealing with a family $(H_z)_{z \in X}$ of compact groups indexed by X. For a compact group H the K-theory of the C^*-algebra $C^*(H)$ is naturally isomorphic to the representation ring $R(H)$ which is the free abelian group generated by classes of irreducible representations of H. The general case of $K(C^*(G))$ for G a smooth proper groupoid, though more complicated, is part of standard geometry.

We now complete the definition of geometric cycles for smooth groupoids. Given a G-manifold P we get another G-manifold $T_G P$ by replacing each P_x, $x \in G^{(0)}$, with its tangent bundle TP_x. Thus the total space $P' = T_G P$ is the space of vectors tangent to P which belong to the kernel of the map α_* tangent to α. Note that $T_G(P)$ is a *proper* G-manifold whenever P is a proper G-manifold.

Definition 3. *Let G be a smooth groupoid. Then a geometric cycle for G is given by a* proper *G-manifold P and an element $y \in K_*(C^*(T_G P \rtimes G))$.*

Note that $T_G P$ is a proper G-manifold, so that, as we explained above, the K-theory of $C^*(T_G P \rtimes G)$ is computable by standard geometric tools; thus the qualifier "geometric" in Definition 3 is appropriate. Note also that the properness of the action of G on P yields the existence of a G-invariant Euclidean metric on the real vector bundle $T_G(P)$ over P, so that there is no real distinction between T_G and T_G^*. Now, let P_1 and P_2 be two proper G-manifolds and let

$$f : P_1 \to P_2$$

be a smooth G-equivariant map. This means, with obvious notation, that we have $\alpha_2(f(p_1)) = \alpha_1(p_1)\ \forall p_1 \in P_1$, and that

$$f(p_1 \circ \gamma) = f(p_1) \circ \gamma \quad \forall p_1 \in P_1\ ,\ \gamma \in G\ ,\ \alpha(p_1) = r(\gamma).$$

It implies that f defines a morphism from the functor $x \to P_{1,x}$ to the functor $x \to P_{2,x}$ and, because of the *properness* of both P_j's, we can put together the deformations

$$(Tf_x)! \in E(TP_{1,x}, TP_{2,x})$$

defined in Section 6. Note that the tangent maps Tf_x are G-equivariantly K-oriented. This thus yields

$$(Tf)! \in E(C^*(T_G P_1 \rtimes G),\ C^*(T_G P_2 \rtimes G))$$

and hence a corresponding map of K-theories

$$K(C^*(T_G P_1 \rtimes G)) \to K(C^*(T_G P_2 \rtimes G)).$$

Proposition 4. *Let G be a smooth groupoid. The following relation is an equivalence relation among geometric cycles: $(P_1, y_1) \sim (P_2, y_2)$ iff there exists a proper G-manifold P and G-equivariant smooth maps $f_j : P_j \to P$ such that*

$$(Tf_1)!(y_1) = (Tf_2)!(y_2).$$

In order to show the transitivity of the above relation, one shows that given the diagram of proper G-manifolds and G-equivariant smooth maps

$$\begin{array}{ccc} & & P \\ & \overset{h}{\nearrow} & \\ P'' & & \\ & \underset{h'}{\searrow} & \\ & & P' \end{array}$$

there are a proper G-manifold P''' and G-equivariant smooth maps $f : P \to P'''$ and $f' : P' \to P'''$ such that $f \circ h$ is G-equivariantly homotopic to $f' \circ h'$.

Under the above equivalence relation any finite number of geometric cycles (P_j, y_j) are equivalent to geometric cycles (P, z_j) with the same P, using say $P = \bigcup P_j$ and the obvious maps $f_j : P_j \to P$. Moreover, the group law on equivalence classes given by

$$(P, z_1) + (P, z_2) = (P, z_1 + z_2)$$

is independent of the choice of P and coincides with the disjoint sum. We shall denote by $K^*_{\text{top}}(G)$ the additive group of equivalence classes of geometric cycles and call it the group of topological G-indices.

Proposition 5. *In the examples of Sections 4, 5, 7, and 8 one has*

$$K^*_{\text{top}}(G) = K_{*,\tau}(BG).$$

Proof. Let us treat the case (Section 7) of a torsion-free discrete group Γ acting on a manifold V. Thus $G = V \rtimes \Gamma$ is the semidirect product of V by Γ.

Let (P, y) be a geometric cycle for the smooth groupoid G. Then the action of G on P yields an action of Γ on P, which is proper by hypothesis, and is thus *free* since Γ has no torsion. The map $\alpha : P \to V = G^{(0)}$ is a smooth submersion and is Γ-equivariant since

$$zg = z \circ (\alpha(z), g) \quad \forall z \in P\,,\ g \in \Gamma$$

so that $\alpha(zg) = s(\alpha(z), g) = \alpha(z)g$.

We thus get, according to Definition 7.2, a smooth map from the manifold $W = P/\Gamma$ to V/Γ, and a corresponding map

$$\alpha' : W \to V \times_\Gamma E\Gamma.$$

Next, since the action of Γ on P is free and proper, the crossed product C^*-algebra $C_0(T_G P) \rtimes \Gamma$ is strongly Morita equivalent, by Proposition 7.1, to the C^*-algebra $C_0(T_G P/\Gamma) = C_0(E)$, where E is the total space of a real vector bundle E over $W = P/\Gamma$ such that

$$E \oplus \alpha'^*(\tau) = TW.$$

One has $y \in K(C_0(T_G P) \rtimes \Gamma) = K(C_0(E)) = K^*(E) = K_{*,\alpha'^*(\tau)}(W)$ by Poincaré duality on W.

Thus $\alpha'(y) \in K_{*,\tau}(V \times_\Gamma E\Gamma)$ is well-defined. One thus obtains a natural map $j : K^*_{\text{top}}(G) \to K_{*,\tau}(BG)$. It passes to equivalence classes as in Proposition 7.6. To check that this map j is surjective one uses Proposition 7.6 a), and remarks that in that statement we may assume that the map $\tilde{g} : \tilde{N} \to V$ is a submersion. Indeed, one can factor $\tilde{g}$ as $\text{pr}_2 \circ \tilde{h}$ where $\text{pr}_2 : \tilde{N} \times V \to V$ is the second projection while $\tilde{h} : \tilde{N} \to \tilde{N} \times V$ is given by

$$\tilde{h}(z, v) = (z, \tilde{g}(v)) \quad \forall z \in \tilde{N}\,,\ v \in V.$$

The injectivity of j follows from the definition of the equivalence of geometric cycles.

This proves the proposition for the examples of Sections 4 and 7. The case of foliations follows from [Hi-S$_1$]. The case of the tangent groupoid of a manifold follows from the following proposition, which allows one to compute $K^*_{\text{top}}(G)$ in many cases.

Proposition 6. *Let G be a smooth groupoid, and let C_G be the category of proper G-manifolds and homotopy classes of smooth G-equivariant maps. Then if P is a final object for C_G one has*

$$K^*_{\text{top}}(G) = K_*(C^*(T_G P \underset{\alpha}{\rtimes} G)).$$

This is easy to check.

This proposition will apply in particular to compute $K^*_{\text{top}}(G)$ when G is a Lie group (cf. β) below).

We can now construct the analytic assembly map in general. Let (P, y) be a geometric cycle for the smooth groupoid G. Then, for each $x \in G^{(0)}$, let $P_x = \alpha^{-1}\{x\}$ be the corresponding manifold, and G_{P_x} be its tangent groupoid. Let G' be the smooth groupoid obtained by putting together the G_{P_x}, $x \in G^{(0)}$, and let $G'' = G' \rtimes G$ be the semidirect product of G' by the natural action of G (which follows from the naturality of the construction of the tangent groupoid). Then $C^*(G'')$ gives us a natural deformation of $C^*(T_G^* P \rtimes G)$ to a C^*-algebra which is strongly Morita equivalent to $C^*(G)$. We shall let $\mu(P, y) \in K_*(C^*(G))$ be the image of $y \in K_*(C^*(T_G P^* \rtimes G))$ by this deformation.

Theorem 7. *The above construction yields an additive map $\mu : K^*_{\text{top}}(G) \to K(C^*(G))$ which extends to the general case the previously constructed analytic assembly maps of Section 4, 5, 7, and 8.*

Composing μ with the canonical surjection $r : C^*(G) \to C^*_r(G)$ one obtains the map $\mu_r : K^*_{\text{top}}(G) \to K(C^*_r(G))$.

We shall now investigate in more detail the case of Lie groups, where torsion plays an important role.

10.β Lie groups and deformations

Let G be a connected Lie group, and let us compute the geometric group $K^*_{\text{top}}(G)$ and the analytic assembly map μ in this special case.

Proposition 8. *Let H be a maximal compact subgroup of G and $V = T_e(H\backslash G)$ the tangent space at the origin e to the homogeneous space $H\backslash G$. Then one has a canonical isomorphism*

$$K_H(V) \simeq K^*_{\text{top}}(G).$$

Here K_H is equivariant K-theory, and the vector space V is an H-space for the isotropy representation of H.

Proof. First, by a result of H. Abels and A. Borel ([Ab]), the category C_G of proper G-manifolds and homotopy classes of smooth G-equivariant maps admits the homogeneous space $P = H\backslash G$ as a final object. Thus (Proposition 6) one has

$$K^*_{\mathrm{top}}(G) = K_*(C^*(TP \rtimes G)).$$

The smooth groupoid $TP \rtimes G$ is equivalent to $V \rtimes H$, so that $C^*(TP \rtimes G)$ is strongly Morita equivalent to $C^*(V \rtimes H) = C_0(V) \rtimes H$. Now by results in [Ju] the K-theory of $C_0(V) \rtimes H$ is the same as the H-equivariant K-theory of $V, K_H(V)$.

Note that the latter group is easy to compute. If the dimension of V is even and the isotropy representation of H in V lifts to the covering $\mathrm{Spin}(V)$ of $SO(V)$, one has a natural isomorphism $\tau : R(H) \to K_H(V)$, the Thom isomorphism of [At-Si$_1$], where $R(H)$ is the representation ring of H. We shall come back later to the general case (Proposition 14).

Let us now compute the analytic assembly map μ. We let G_0 be the Lie group $V \rtimes H$, semi-direct product of the vector group V by the compact group H. As pointed out quite early in the theoretical physics literature ([Mu$_1$][Mu$_2$] [M$_1$] [W-I] [Se$_5$]), there is a close resemblance between the representation theories of G and of G_0, which is tied up with a natural deformation from G_0 to G which we now describe. To that end we just need to describe the relevant topology on the groupoid $G_1 = G_0 \cup (G \times\,]0,1])$. Here G_0 is a closed subset, and a sequence (g_n, ε_n), $g_n \in G$, $\varepsilon_n \to 0$, converges to an element $g = (\xi, k) \in G_0 = V \rtimes H$ of G_0 iff there holds

$$g_n \to k \in H\,, \quad \frac{k - g_n}{\varepsilon_n} \to \xi \in T_e(H\backslash G) = V.$$

We thus obtain a deformation $\delta \in E(C^*(G_0), C^*(G))$, and since $C^*(G_0) = C^*(V \rtimes H) \simeq C_0(V) \rtimes H$, we get a natural map from $K_H(V) = K_*(C_0(V) \rtimes H)$ to $K(C^*(G))$.

Proposition 9. *The above map is exactly the same as the analytic assembly map* $\mu : K^*_{\mathrm{top}}(G) \to K(C^*(G))$.

Proof. By definition (cf. Theorem 7), the map μ is obtained from the semi-direct product of the tangent groupoid G_P by the Lie group G. Thus, let $G_2 = G_P \rtimes G$. Then $P \times [0,1]$ is the space $G_2^{(0)}$ of units of G_2, and G_2 is equivalent to the smooth groupoid $G_{2,Z} = \{\gamma \in G_2 \,;\, r(\gamma) \in Z\,,\, s(\gamma) \in Z\}$ where $Z = e \times [0,1] \subset G_2^{(0)}$. One then checks that $G_{2,Z}$ is the smooth groupoid G_1 of the deformation from G_0 to G discussed above.

Thus the deformation from G_0 to G gives, in all cases where μ_r is shown to be an isomorphism (cf. Theorem 20 below), an isomorphism of the K-theories of the group C^*-algebras $C^*_r(G)$, which gives a precise meaning to the observations of [M$_1$].

10.γ The G-equivariant index of elliptic operators on homogeneous spaces of Lie groups

As above, let G be a connected Lie group and H a compact subgroup, with $P = H\backslash G$ the corresponding homogeneous space. Every G-equivariant (Hermitian) complex vector bundle E on P is obtained from a corresponding (unitary) complex representation ε of the isotropy group H in the fiber E_e. We shall still denote by ε_E the natural extension of this representation to the convolution algebra $C^{-\infty}(H)$ of distributions on H. Since distributions push forward, the latter algebra $C^{-\infty}(H)$ is, in a natural manner, a subalgebra of the convolution algebra $C_c^{-\infty}(G)$ of distributions with compact support on G. When E is finite-dimensional the module ε_E over $C^{-\infty}(H)$ is finite and projective, and, using the inclusion $C^{-\infty}(H) \subset C_c^{-\infty}(G)$, we can consider the corresponding induced module $\tilde{\varepsilon}_E$ which is finite and projective over $C_c^{-\infty}(G)$. It is given through the action by convolution of $C_c^{-\infty}(G)$ in the space $C_c^{-\infty}(P, E)$ of distributional sections of E with compact support on P.

Proposition 10. 1) *The convolution algebra $J = C_c^\infty(G)$ is a two-sided ideal in the algebra $\mathcal{A} = C_c^{-\infty}(G)$.*

2) *Let D be a G-invariant elliptic differential operator on P from a G-equivariant bundle E_1 to a G-equivariant bundle E_2. Then it follows that the triple $(C_c^{-\infty}(P, E_i), D)$ defines a quasi-isomorphism for the pair $J, \mathcal{A}$.*

Proof. 1) is standard ([Co-M$_4$]). 2) follows from the existence of a quasi-inverse for D modulo $C_c^\infty(G)$, as proven in [Co-M$_4$] Proposition 1.3.

Definition 11. *We let* $\mathrm{Ind}(D) \in K_0(C_c^\infty(G))$ *be the index of the quasi-isomorphism of Proposition 10 2) (cf. Section 9 α)).*

We shall now show in a simple example that this index is not in general invariant under homotopy of elliptic operators, unlike the C^*-algebra index $\mathrm{Ind}_a(D)$ obtained from the natural map

$$K_0(C_c^\infty(G)) \xrightarrow{j} K_0(C^*(G))$$

coming from the inclusion $C_c^\infty(G) \subset C^*(G)$. This will, in particular, show that the map j is far from injective in general.

We take $G = \mathbb{R}$ and specialize to scalar operators so that the two bundles E_j over $P = \mathbb{R}$ are trivial.

Proposition 12. *The map $D \to \mathrm{Ind}(D) \in K_0(C_c^\infty(\mathbb{R}))$ is an injection of the projective space of non-zero polynomials $D = P\left(\frac{\partial}{\partial x}\right)$ into $K_0(C_c^\infty(\mathbb{R}))$.*

Proof. Consider the exact sequence of algebras

$$0 \to C_c^\infty(\mathbb{R}) \to \mathcal{A}_0 \xrightarrow{\sigma} K \to 0$$

where $\mathcal{A}_0$ is the convolution algebra of distributions on $\mathbb{R}$ with compact support and singular support $\{0\}$. Then the quotient $K = \mathcal{A}_0/C_c^\infty(\mathbb{R})$ is a commutative field. Any invariant elliptic differential operator D on $\mathbb{R}$ defines an element of $\mathcal{A}_0$ and $\sigma(D) \in K$, $\sigma(D) \neq 0$. By construction, the index, $\mathrm{Ind}(D) \in K_0(C_c^\infty(\mathbb{R}))$, is the image of $\sigma(D)$ by the connecting map $\partial : K_1(K) \to K_0(C_c^\infty(\mathbb{R}))$ of algebraic K-theory ([Mi$_1$]). Moreover, since $\mathcal{A}_0$ is a commutative ring, one has a natural map

$$\det : K_1(\mathcal{A}_0) \to \mathcal{A}_0^*$$

given by the determinant. It follows that for any $\alpha \in K$, $\alpha \neq 0$, $\partial(\alpha) = 0$, there exists an element β of $\mathcal{A}_0^*$ such that $\sigma(\beta) = \alpha$. Thus to show that ∂ is injective on $K^*/\mathbb{C}^*$ it is enough to show that given any $\alpha \in K$, $\alpha \neq 0$, it is impossible to find β invertible in $\mathcal{A}_0$ with $\sigma(\beta) = \alpha$. This follows from the Paley-Wiener theorem, which shows that scalar multiples $\lambda\delta_0$, $\lambda \in \mathbb{C}$, of the Dirac mass at 0 are the only invertible elements of $\mathcal{A}_0$. (If $\alpha \in \mathcal{A}_0^*$, then its Fourier transform $\hat{\alpha}$ must vanish somewhere in $\mathbb{C}$ or be constant.)

The C^*-algebra index, namely $\mathrm{Ind}_a(D) = j_* \,\mathrm{Ind}(D)$, where

$$j : C_c^\infty(G) \to C^*(G),$$

is homotopy invariant and ranges over a countable abelian group. It thus depends only upon the K-theory class of the principal symbol σ_D of D, and is given in general by the following analogue of Lemma 5.6:

Proposition 13. *Let G be a connected Lie group, $P = H\backslash G$ a proper G-homogeneous space, and D a G-invariant elliptic differential operator on P. Then $\mathrm{Ind}_a(D) = j_*(\mathrm{Ind}\, D) \in K(C^*(G))$ is given by*

$$\mathrm{Ind}_a(D) = \mu([\sigma_D])$$

where the principal symbol σ_D of D defines a K-theory class

$$[\sigma_D] \in K_H(T_e(P)) \simeq K(C^*(TP \rtimes G)).$$

We have used the strong Morita equivalence of the algebras $C^*(TP \rtimes G)$ and $C^*(T_eP \rtimes H)$ and the isomorphism $K(C^*(T_eP \rtimes H)) \simeq K_H(T_eP)$. A convenient description of the H-equivariant K-theory of the vector space $T_eP = V$ which fits with symbols of differential operators is the following: The basic objects to start with are smooth H-equivariant maps $\alpha : S \to \mathrm{Iso}(E_1, E_2)$, where S is the unit sphere in V with respect to an H-invariant metric, and E_1 and E_2 are finite-dimensional unitary H-modules. Two such maps

$$\alpha_0 \in (C^\infty(S, \mathrm{Iso}(E_0, F_0))^H \ , \ \alpha_1 \in (C^\infty(S, \mathrm{Iso}(E_1, F_1))^H$$

are called isomorphic, and we shall write $\alpha_0 \simeq \alpha_1$, if there exist $\varphi \in \mathrm{iso}_H(E_0, E_1)$ and $\psi \in \mathrm{Iso}_H(F_0, F_1)$, where Iso_H denotes the H-equivariant isomorphisms, such that

$$\psi\alpha_0(\xi) = \alpha_1(\xi)\varphi, \text{ for any } \xi \in S.$$

Further, α_0 and α_1 are said to be homotopic if there can be found an $\alpha \in (C^\infty(S \times I, \mathrm{Iso}(E,F)))^H$, where $I = [0,1]$ with the trivial action of H, such that $\alpha|S \times \{0\} \simeq \alpha_0$ and $\alpha|S \times \{1\} \simeq \alpha_1$. The set of all homotopy classes of such maps will be denoted by C. This is an abelian semigroup, under the obvious direct sum operation. Let C_0 denote the subsemigroup of all classes which can be represented by a constant map α, $\alpha(\xi) = \varphi$, $\xi \in S$, with $\varphi \in \mathrm{Iso}_H(E_1, E_2)$. Then C/C_0 is not only a semigroup but actually a group, which is isomorphic to $K_H(V)$.

The elementary properties of the K-theory of C^*-algebras and of the G-invariant pseudo-differential calculus ([Co-M$_4$]) show a priori that the index $\mathrm{Ind}_a(D) \in K(C^*(G))$ only depends upon the K-theory class $[\sigma_D] \in K_H(T_eP)$ of the principal symbol of D (see [Co-M$_4$] for more details). The proof of Proposition 13 then follows from the asymptotic pseudo-differential calculus ([Wi$_2$]).

Let us assume that the isotropy group H is connected. Then, if the homogeneous space $P = H\backslash G$ is odd-dimensional, the Bott periodicity theorem implies that the equivariant K-theory $K_H^0(V)$ is zero, so that Proposition 13 has no content in this case. Let us thus assume that $P = H\backslash G$ is *even*-dimensional and relax the condition of connectedness of H by requiring only that H preserves the orientation of $V = T_e(P)$. We shall then determine $K_H(V)$ and show that all its elements are obtained as symbols $[\sigma_D]$, where D is a G-invariant order-one elliptic differential operator on P.

Let $\pi : H \to SO(V)$ be the representation of H in V, and $\rho : \mathrm{Spin}(V) \to SO(V)$ be the Spin covering. In general π does not lift to $\mathrm{Spin}(V)$, and so we form the double covering of H given by:

$$\tilde{H} = \{(h,s) \in H \times \mathrm{Spin}(V) \; ; \; \pi(h) = \rho(s)\}.$$

Let $u \in \mathrm{Spin}(V)$, $u^2 = 1$, be the generator of the kernel of ρ, so that, viewed as the element $(1,u) \in \tilde{H}$, it is central in $\tilde{H}$ and yields the central extension

$$0 \to \mathbb{Z}/2 \to \tilde{H} \to H \to 1.$$

In particular, the representation ring $R(\tilde{H})$ inherits from u a $\mathbb{Z}/2$ grading and

$$R(\tilde{H}) = R(\tilde{H})_+ \oplus R(\tilde{H})_-$$

where $R(\tilde{H})_\pm$ is generated by the equivalence classes of irreducible representations ε of $\tilde{H}$ such that $\varepsilon(u) = \pm 1$. Clearly one has

$$R(\tilde{H})_+ \simeq R(H)$$

and $R(\tilde{H})_-$ is by construction an $R(H)$-module.

Let $\varepsilon \in R(\tilde{H})_-$. We shall now construct a Dirac-type operator $\partial\!\!\!/_\varepsilon$ on P. Let $S^\pm$ be the half-spin representation of $\mathrm{Spin}(V)$. Since $u \in \tilde{H}$ acts by -1 on both ε and $S^\pm$ it follows that $\varepsilon^\pm = \varepsilon \otimes S^\pm$ are unitary H-modules. We let $E^\pm$ be the associated G-equivariant Hermitian vector bundles over P. Let

$$\nabla_\varepsilon^\pm : C^\infty(P, E^\pm) \to C^\infty(P, T_{\mathbb{C}}^*P \otimes E^\pm)$$

be G-invariant connections, compatible with the Hermitian structure of $E^{\pm}$. We define the $(\pm)$ Dirac operators with coefficients in ε as being the compositions

$$\partial\!\!\!/_{\varepsilon}^{\pm} : C_c^{\infty}(P, E^{\pm}) \xrightarrow{\nabla_{\varepsilon}^{\pm}} C_c^{\infty}(P, T_{\mathbb{C}}^{*}P \otimes E^{\pm}) \xrightarrow{\gamma} C_c^{\infty}(P, E^{\mp})$$

where γ is the bundle homomorphism induced by the Clifford multiplication. Clearly $\partial\!\!\!/_{\varepsilon}^{\pm}$ are G-invariant first-order elliptic differential operators.

Proposition 14. *The following map from $R(\widetilde{H})_{-}$ to $K_H(V)$, $V = T_e(P)$, is an isomorphism: $\varepsilon \in R(\widetilde{H})_{-} \to \sigma(\partial\!\!\!/_{\varepsilon}^{+}) =$ symbol of $\partial\!\!\!/_{\varepsilon}^{+}$.*

We refer to [Co-M$_4$] for details. It thus follows that, when the dimension of P is even, for $H \subset G$ a maximal compact subgroup (with $H \backslash G$ equivariantly oriented) the analytic assembly map $\mu : K_{\text{top}}^{*}(G) \to K(C^{*}(G))$ is the same as the *Dirac induction* map

$$\varepsilon \in R(\widetilde{H})_{-} \to \operatorname{Ind}_a(\partial\!\!\!/_{\varepsilon}^{+}) \in K(C^{*}(G)).$$

Dirac induction was introduced by Schmid and Parthasarathy in [Sch] [Par] in order to solve the problem of geometric realization of the discrete series of representations of semi-simple Lie groups. They proved the following result:

Theorem 15. [Sch] [Par] *Let G be a semisimple Lie group with finite center, H a maximal compact subgroup. For each discrete-series representation π of G there is a unique H-module ε such that*

$$\pi = \text{Kernel } \partial\!\!\!/_{\varepsilon}^{+}, \text{ Kernel } \partial\!\!\!/_{\varepsilon}^{-} = \{0\}.$$

The relation between their use of Dirac induction and the map

$$\varepsilon \in R(\widetilde{H})_{-} \to \operatorname{Ind}_a(\partial\!\!\!/_{\varepsilon}^{+}) \in K(C^{*}(G))$$

is provided by the following lemma:

Lemma 16. *Let D be a G-invariant elliptic differential operator on a proper G-homogeneous space P. Assume that the corresponding Hilbert space operator $L^2(P, E_1) \xrightarrow{D} L^2(P, E_2)$ is* surjective. *Then* $\ker D$, *viewed as a $C_r^{*}(G)$-module, is finite and projective and in $K_0(C_r^{*}(G))$,*

$$[\ker D] = \operatorname{Ind}_a(D).$$

Of course, in general one does not have vanishing theorems for the twisted Dirac operators $\partial\!\!\!/_{\varepsilon}^{+}$; moreover, there are examples of operators D with non-zero analytical indices $\operatorname{Ind}_a(D)$ in situations where both $\ker D$ and $\ker D^{*}$ are reduced to $\{0\}$.

In order to prove, in the context of semi-simple Lie groups G, that the L^2-kernel of a Dirac-type operator was not $\{0\}$, Atiyah and Schmid relied on the index theorem for covering spaces [At$_5$] [Sin$_2$], using the existence of discrete cocompact subgroups $\Gamma \subset G$. This theorem was then extended in [Co-M$_4$] to cover the general case of homogeneous spaces of *unimodular* Lie groups. We shall now explain this result in detail ([Co-M$_4$]). We assume, as above, that $P = H\backslash G$ is even-dimensional and oriented. Let $D : C_c^\infty(P, E_1) \to C_c^\infty(P, E_2)$ be a G-invariant elliptic differential operator. There is no ambiguity in extending D as a Hilbert space operator. One specifies an H-invariant volume form $\omega \in \wedge^m V$, $V = T_e(P)$, $m = \dim V$, which fixes the orientation of P as well as a G-invariant 1-density on P. The Hilbert spaces $L^2(P, E_j)$ of L^2-sections of the G-equivariant bundles E_j are then well-defined, and ([Co-M$_4$] Lemma 3.1) the closure (in the Hilbert space sense) of the densely defined operator

$$D : C_c^\infty(P, E_1) \to C_c^\infty(P, E_2)$$

is equal to the distributional extension of D whose domain is

$$\{\xi \in L^2(P, E_1)\ ;\ D\xi \in L^2(P, E_2) \text{ in the distributional sense}\}.$$

Next, since G is *unimodular*, a choice of Haar measure dg on G uniquely specifies the Plancherel trace tr_G on the von Neumann algebra of the (left) regular representation λ of G in $L^2(G)$. It is a semifinite faithful normal trace such that for any $f \in L^2(G)$ with $\lambda(f)$ bounded one has

$$\mathrm{tr}_G(\lambda(f)^*\lambda(f)) = \int |f(g)|^2\, dg.$$

The restriction of this trace to $C_r^*(G)$ is semifinite and semicontinuous. This trace tr_G gives a unique dimension, $\dim_G(\pi)$ (with positive real values), to any representation π of G which is quasi-contained in the regular representation of G. The existence of a quasi-inverse for D immediately yields

Lemma 17. *Let D be a G-invariant elliptic differential operator on P, and let* $\ker D$ *be its L^2-kernel. Then*

$$\dim_G(\ker D) < \infty.$$

The index theorem for homogeneous spaces permits one to compute

$$\mathrm{Ind}_G(D) = \dim_G(\ker D) - \dim_G(\ker D^*)$$

in terms of the K-theory class of the principal symbol of D

$$[\sigma_D] \in K_H(V)\ ,\ V = T_e(P).$$

By Proposition 14 we just need to treat the case of a Dirac-type operator ∂_ε^+, where $\varepsilon \in R(\tilde{H})$.

The first ingredient we need is the Chern character

$$\mathrm{ch} : R(\tilde{H}) \to H^*(\mathfrak{g}, H, \mathbb{R}).$$

Here $H^*(\mathfrak{g}, H, \mathbb{R})$ denotes the relative Lie algebra cohomology with trivial coefficients, i.e. the cohomology of the complex $(C(\mathfrak{g}, H, \mathbb{R}), d)$, where

$$C^q(\mathfrak{g}, H, \mathbb{R}) = \{\alpha \in \Lambda^q g^* \; ; \; \iota_X \alpha = 0 \text{ for } X \in \mathfrak{h} \, , \; \mathrm{Ad}^*(h)\alpha = \alpha \text{ for } h \in H\}$$

and $d : C^q(\mathfrak{g}, H, \mathbb{R}) \to C^{q+1}(\mathfrak{g}, H, \mathbb{R})$ is given by

$$d\alpha(x_1, \ldots, x_{q+1}) = \frac{1}{q+1} \sum_{1 \le i < j \le q+1} (-1)^{i+j+1} \alpha([x_i, x_j], x_1, \ldots, \hat{x}_i, \ldots, \hat{x}_j, \ldots, x_{q+1}).$$

Let us fix, for the moment, an $\mathrm{Ad}(H)$-invariant splitting of $\mathfrak{g}$, $\mathfrak{g} = \mathfrak{h} \oplus \mathfrak{m}$. This specifies a G-invariant connection on the principal bundle $H \to G \to M$, whose connection form is given by the projection $\theta : \mathfrak{g} \to \mathfrak{h}$ parallel to $\mathfrak{m}$, and whose curvature form is prescribed by

$$\Theta(X, Y) = \frac{1}{2} \, \theta([X, Y]) \, , \quad X, Y \in \mathfrak{m}.$$

Now let ε be a unitary representation of $\widetilde{H}$ on E. We denote by the same letter its differential, $\varepsilon : \mathfrak{h} \to g\ell(E)$, and we then define $\Theta_\varepsilon \in \Lambda^2 \mathfrak{m}^* \otimes g\ell_{\mathbb{C}}(E)$ by the formula

$$\Theta_\varepsilon(X, Y) = \frac{1}{2i\pi} \, \varepsilon(\Theta(X, Y)) \, , \quad X, Y \in \mathfrak{m}.$$

Further, let us consider the form $\operatorname{tr}\exp\Theta_\varepsilon \in \Lambda \mathfrak{m}^* \otimes \mathbb{C}$. Actually, because ε is unitary, $\operatorname{tr}\exp\Theta_\varepsilon \in \Lambda m^*$. In addition, since for $\widetilde{h} \in \widetilde{H}$ with image $h \in H$ one has

$$\Theta_\varepsilon(\mathrm{Ad}(h)X, \mathrm{Ad}(h)Y) = \varepsilon(\widetilde{h})\Theta_\varepsilon(X, Y)\varepsilon(\widetilde{h})^{-1} \, , \quad X, Y \in \mathfrak{m},$$

one can see that $\operatorname{tr}\exp\Theta_\varepsilon$ is H-invariant. This shows that its pull-back to $\Lambda \mathfrak{g}$, via the projection $I - \theta : \mathfrak{g} \to \mathfrak{m}$, defines a cochain in $\sum_q^{\oplus} C^q(\mathfrak{g}, H, \mathbb{R})$, which we will continue to denote by the same symbol. Standard arguments in the Chern-Weil approach to characteristic classes imply first that $\operatorname{tr}\, \exp\Theta_\varepsilon$ is closed, and next that the cohomology class in $H^*(\mathfrak{g}, H, \mathbb{R})$ it defines, and which will be denoted $\operatorname{ch}\varepsilon$, does not depend on the choice of the $\mathrm{Ad}(H)$-invariant splitting of $\mathfrak{g}$. Finally, it is clear that $\operatorname{ch}\varepsilon_1 = \operatorname{ch}\varepsilon_2$ if ε_1 and ε_2 are equivalent unitary representations, so that we can define unambiguously $\operatorname{ch} : R(\widetilde{H}) \to H^*(\mathfrak{g}, H, \mathbb{R})$.

The last ingredient we need is the analogue of the $\hat{A}$-polynomial of Hirzebruch. To define it, we start from the (real) H-module V (which can also be viewed as an $\widetilde{H}$-module) and form, as above, $\Theta_V \in \Lambda^2 \mathfrak{m}^* \otimes g\ell_{\mathbb{C}}(V)$. Then we construct the element in $\Lambda \mathfrak{m}^*$

$$(\det)^{1/2} \frac{\Theta_V}{\exp\left(\frac{1}{2}\Theta_V\right) - \exp\left(-\frac{1}{2}\Theta_V\right)},$$

pull it back to an element in $\Lambda \mathfrak{g}^*$, and, after noting that it is in fact a cocycle, we define $\widehat{\mathcal{A}}(\mathfrak{g}, H) \in H^*(\mathfrak{g}, H, \mathbb{R})$ as being its cohomology class.

Remark now that $\dim H^m(\mathfrak{g}, H, \mathbb{R}) = 1$, since G and H are unimodular; in fact, $H^m(\mathfrak{g}, H, \mathbb{R}) = C^m(\mathfrak{g}, H, \mathbb{R}) \simeq \Lambda^m V$. If $\Omega = \sum \Omega^{(q)} \in H^*(\mathfrak{g}, H, \mathbb{R})$, we define the scalar $\Omega[V]$ by the relation $\Omega^{(m)} = (\Omega[V])\omega$.

We can now define a natural map $\mathrm{Ind}_t : K_H(V) \to \mathbb{R}$ uniquely, using Proposition 14, by

$$\mathrm{Ind}_t(\sigma(\partial\!\!\!/_\varepsilon^+)) = (\mathrm{ch}(\varepsilon)\ \widehat{\mathcal{A}}(\mathfrak{g}, H))[V]$$

for any element ε of $R(\widetilde{H})_-$.

The index theorem for homogeneous spaces is then ([Co-M4]):

Theorem 18. *Let D be a G-invariant elliptic differential operator on the proper oriented homogeneous space $P = H\backslash G$, and let $[\sigma_D] \in K_H(V)$, $V = T_e(P)$, be the K-theory class of its principal symbol. Then* $\dim_G(\ker D)$ *and* $\dim_G(\ker D^*)$ *are finite, and*

$$\dim_G(\ker D) - \dim_G(\ker D^*) = \mathrm{Ind}_t([\sigma_D]).$$

We refer to [Co-M4] for the detailed proof of this theorem. It is important to note that for large classes of unimodular Lie groups the L^2-kernels $\ker D$ of G-invariant elliptic operators on proper homogeneous spaces are automatically finite direct sums of irreducible discrete-series representations. Thus, for instance ([Co-M4]):

Theorem 19. *Let G be a connected semisimple Lie group with finite center, and let $P = H\backslash G$ be a proper homogeneous space over G. Then for any G-invariant (pseudo) differential elliptic operator D on P the unitary representation of G on the space of L^2-solutions of the equation $Du = 0$ is a finite sum of irreducible discrete-series representations.*

10.δ The K-theory $K(C_r^*(G))$ for Lie groups

Let G be a connected Lie group, and H a maximal compact subgroup. If $P = H\backslash G$ is even-dimensional then, by Propositions 13 and 14, the Dirac induction map

$$\varepsilon \in R(\widetilde{H})_- \to \mathrm{Ind}_a(\partial\!\!\!/_\varepsilon^+) \in K(C^*(G))$$

coincides with the analytic assembly map

$$\mu : K^0_{\mathrm{top}}(G) \to K_0(C^*(G)).$$

When P is odd-dimensional the relevant part of the analytic assembly map

$$\mu : K^*_{\mathrm{top}}(G) \to K_*(C^*(G))$$

concerns K^1_{top} instead of K^0_{top}. Replacing G by $G \times \mathbb{R}$ and using Bott periodicity, one gets a similar interpretation of μ in operator-theoretic terms (cf. [Co-M4]).

When G is a connected solvable Lie group, then ([Co-M4] [Kas]) the Thom isomorphism (Appendix C) shows that the analytic assembly map is an isomorphism

$$K^*_{\mathrm{top}}(G) \xrightarrow{\mu} K_*(C^*(G)).$$

For semisimple Lie groups the results of [Sch], [Par], [At-S_1] and [At-Sc_2] on the geometric realisation of all discrete-series representations by Dirac induction, together with [Kas] and [Co-M_4] Section 7.5, suggested that μ_r should be an isomorphism

$$\mu_r : K_{\text{top}}^*(G) \to K_*(C_r^*(G)).$$

It is crucial here to have composed μ with the natural morphism

$$C^*(G) \xrightarrow{r} C_r^*(G)$$

of restriction to the reduced C^*-algebra, i.e. to the support of the Plancherel measure. The point is that μ itself fails to be surjective (Proposition 21) as soon as G satisfies Kazhdan's property T.

After several important partial results ([V_1] [Kas_4] [Pen-Pl]), the bijectivity of μ_r was proven for general connected linear reductive groups by A. Wassermann [Wa_1].

Theorem 20. [Wa_1] *Let G be a connected linear reductive group. Then the analytic assembly map is an isomorphism*

$$\mu_r : K_{\text{top}}^*(G) \to K_*(C_r^*(G)).$$

The proof ([Wa_1]) relies on fundamental results of Arthur and Harish-Chandra on the Plancherel theorem for the Schwartz space of G, which yield a complete description of the reduced C^*-algebra $C_r^*(G)$ as a finite direct sum of C^*-algebras associated to each generalized principal series. The latter C^*-algebras, though not strongly Morita equivalent to commutative ones, are of the same nature as the C^*-algebra of Example 2 β).

It is of course desirable to obtain a direct proof of Theorem 20 not relying on the explicit knowledge of $C_r^*(G)$ provided by the work of Harish-Chandra. Such a direct proof for the injectivity of μ follows from Kasparov's proof of the Novikov conjecture for discrete subgroups of Lie groups. It implies, in particular:

Theorem 21. [Kas_5] *Let G be a connected Lie group. Then the analytic assembly map $\mu_r : K_{\text{top}}^*(G) \to K_*(C_r^*(G))$ is* injective.

Let us now explain why μ fails to be *surjective* on $K_*(C^*(G))$ as soon as G has property T of Kazhdan, which is the case for any semisimple Lie group of real rank ≥ 2.

First one has the following characterization of property T:

Proposition 22. [Ak-W] *Let G be a locally compact group. The following properties are equivalent:*

1) *G has property T of Kazhdan ([Kaz]).*

2) *There exists an orthogonal projection* $e \in C^*(G)$, $e \neq 0$, *such that* $\pi(e) = 0$ *for any unitary representation* π *of* G *disjoint from the trivial representation.*

We refer to [V_2] for the simple proof.

Let then G be a Lie group with property T and let $[e] \in K_0(C^*(G))$ be the class of the projection given by Proposition 22 2). Since $\pi_0(e) = 1$, where π_0 is the trivial representation, the class $[e]$ gives a non-torsion element of $K_0(C^*(G))$ whose image in $K_0(C_r^*(G))$ vanishes by Proposition 22 2). In particular $[e]$ does not belong to the range of μ, say, by Theorem 21.

The surjectivity of $\mu_r : K_{\text{top}}^*(G) \to K(C_r^*(G))$ is tied up with the completeness of the description of the discrete series by Dirac induction, as well as with a strong form of the Selberg principle on vanishing of non-elliptic orbital integrals of discrete series coefficients

$$g \in G \to \langle \pi(g)\xi, \eta \rangle \quad \xi, \eta \in \mathcal{H}_\pi.$$

The point is that the coefficient

$$p_\xi(g) = d_\pi \langle \xi, \pi(g)\xi \rangle$$

where ξ is a unit vector in $\mathcal{H}_\pi$, and d_π the formal degree of the square-integrable representation π, is an *idempotent*, $p^2 = p = p^*$, in the convolution algebra $C_r^*(G)$ of G. This follows immediately from the orthogonality relations ([Di_3]) for locally compact unimodular groups. The regularity of p depends on the integrability of the representation π; $p \in L^1(G)$ if π is integrable ([Di_3]). Now the orbital integral defines a trace τ on the convolution algebra of G, and the value $\tau(p)$ depends only upon the K-theory class of p in that algebra. By construction the Dirac induction, or more generally the index map for G-invariant differential operators on $P = H\backslash G$, has the following property of *localisation near* H:

Let $W \supset H$ be any open neighborhood of H. Then the K-theory class $\operatorname{Ind}(D) \in K_0(C_c^\infty(G))$ is represented by an idempotent $p \in M_k(C_c^\infty(G)^\sim)$, with $p = (p_{ij})_{i,j=1,\ldots,k}$ and

$$\text{Support } p_{ij} \subset W.$$

In particular, it follows that $\tilde{\tau}(p) = 0$, where $\tilde{\tau}$ is the trace associated to the orbital integral of a $g \in G$ whose conjugacy class does not meet W. We refer to [Ju-V_2] [B-B] [Ju-V_3] for the relations between Dirac induction, cyclic cohomology and the Selberg principle which we originally suggested.

It is, of course, desirable to find direct proofs of the surjectivity in Theorem 20. In that respect the ideas developed by Mackey in [M_1], or in the theoretical physics literature, on deformation theory should be relevant.

10.ε The general conjecture for smooth groupoids

The case of Lie groups discussed above (δ), though much simpler than the general case of smooth groupoids (which includes discrete groups, foliations etc. for which $C_r^*(G)$ fails, as a rule, to be of type I) gives a very clear indication that the analytic assembly map

$$\mu_r : K_{\text{top}}^*(G) \to K(C_r^*(G))$$

should be an *isomorphism* in general. Besides the case of discrete groups (Section 4), this general conjecture is supported by numerous explicit computations for foliations ([Co$_{11}$] [Nat] [Tor] [St] [Pen]). It has many interesting consequences; the *injectivity* of μ_r has as consequences the homotopy invariance of higher signatures (Section 4), the Gromov-Lawson-Rosenberg positive scalar curvature conjecture [Gro-L], the homotopy invariance of relative η-invariants ([We]). The *surjectivity* of μ_r has implications such as the absence of nontrivial idempotents, $e^2 = e \neq 0$ in the reduced C^*-algebra $C_r^*(\Gamma)$ of torsion-free discrete groups.

In essence the general conjecture is a form of G-equivariant Bott periodicity. Indeed, when the category C_G of Proposition 6 has a final object P which is a *contractible* proper G-manifold (i.e. each P_x is contractible for $x \in G^{(0)}$) then the bijectivity of μ asserts that, G-equivariantly, the tangent groupoid deformation (Section 5) $T^*P_x \sim$ point is a K-theory isomorphism, as in Bott's theorem.

We shall refer the reader to [Bau-C-H] for a detailed discussion of the implications of the above conjecture, and for its formulation for locally compact groups, in particular for p-adic groups ([Pl$_2$]).

One of the interests of the general formulation is to put many particular results in a common framework. Thus, for instance, the following three theorems:

1) The Atiyah [At$_5$]-Singer [Sin$_2$] index theorem for covering spaces

2) The index theorem for measured foliations (Chapter I)

3) The index theorem for homogeneous spaces (Theorem 18)

are all special cases of the same index theorem for G-invariant elliptic operators D on proper G-manifolds (Definition 1), where G is a smooth groupoid with a transverse measure Λ in the sense of [Co$_{10}$]. While it is desirable to prove such results in full generality, most of the work goes into explicit computations of relevant examples which motivated our presentation in this chapter. In Chapter III we shall construct higher invariants of K-theory based on cyclic cohomology, and use them, in particular, as a tool to control the rational injectivity of μ_r.

II. Appendix A. C^*-modules and Strong Morita Equivalence

In this appendix we expound the notions, due to M. Rieffel and W. Paschke ([Ri$_5$][Ri$_2$][Pas]) of C^*-module over a C^*-algebra and of strong Morita equivalence of C^*-algebras. Let B be a C^*-algebra; by a B-valued inner product on a right B-module $\mathcal{E}$ we mean a B-valued sesquilinear form $\langle\ ,\ \rangle$, conjugate linear in the first variable, and such that

α) $\langle \xi, \xi \rangle$ is a positive element of B for any $\xi \in \mathcal{E}$.

β) $\langle \xi, \eta \rangle^* = \langle \eta, \xi \rangle \quad \forall \xi, \eta \in \mathcal{E}$.

γ) $\langle \xi, \eta b \rangle = \langle \xi, \eta \rangle b \quad \forall b \in B\ ,\ \xi, \eta \in \mathcal{E}$.

By a *pre-C^*-module* over B we mean a right B-module $\mathcal{E}$ endowed with a B-valued inner product. A semi-norm on $\mathcal{E}$ is defined by

$$\|\xi\| = \|\langle \xi, \xi \rangle\|^{1/2} \quad \forall \xi \in \mathcal{E}$$

where $\|\langle \xi, \xi \rangle\|$ is the C^*-algebra norm of $\langle \xi, \xi \rangle \in B$.

Definition 1. *A C^*-module $\mathcal{E}$ over B is a pre-C^*-module $\mathcal{E}$ for which $\|\ \|$ is a complete norm.*

By completion, any pre-C^*-module yields an associated C^*-module. Given a C^*-module $\mathcal{E}$ over B, an *endomorphism* T of $\mathcal{E}$ is by definition a continuous endomorphism of the right B-module $\mathcal{E}$ which admits an adjoint T^*, that is, an endomorphism of the right B-module $\mathcal{E}$ such that

$$\langle \xi, T\eta \rangle = \langle T^*\xi, \eta \rangle \quad \forall \xi, \eta \in \mathcal{E}.$$

One checks that T^* is uniquely determined by T and that, endowed with this involution, the algebra $\mathrm{End}_B(\mathcal{E})$ of endomorphisms of $\mathcal{E}$ is a C^*-algebra. One has

$$\langle T\xi, T\xi \rangle \le \|T\|^2 \langle \xi, \xi \rangle \quad \forall \xi \in \mathcal{E}\ ,\ T \in \mathrm{End}_B(\mathcal{E})$$

where $\|T\|$ is the C^*-algebra norm of T.

Of particular importance are the *compact* endomorphisms obtained from the norm closure of endomorphisms of finite rank.

Proposition 2. [Ri$_5$] *Let $\mathcal{E}$ be a C^*-module over B.*

a) *For any $\xi, \eta \in \mathcal{E}$ an endomorphism $|\xi\rangle\langle\eta| \in \mathrm{End}_B(\mathcal{E})$ is defined by*

$$(|\xi\rangle\langle\eta|)\,(\alpha) = \xi\langle \eta, \alpha \rangle \quad \forall \alpha \in \mathcal{E}.$$

b) *The linear span of the above endomorphisms is a self-adjoint two-sided ideal of* $\mathrm{End}_B(\mathcal{E})$.

The usual properties of the bra-ket notation of Dirac hold in this setup, so that, for instance:

$(|\xi\rangle\langle\eta|)^* = |\eta\rangle\langle\xi| \quad \forall \xi, \eta \in \mathcal{E}$

$$(|\xi\rangle\langle\eta|)\,(|\xi'\rangle\langle\eta'|) = |\xi\langle\eta,\xi'\rangle\rangle\langle\eta'| = |\xi\rangle\langle(|\eta\rangle\langle\xi'|)\eta'| \quad \forall \xi,\xi',\eta,\eta' \in \mathcal{E}.$$

We let $\mathrm{End}_B^0(\mathcal{E})$ be the *norm closure* in $\mathrm{End}_B(\mathcal{E})$ of the above two-sided ideal (Proposition 2b). An element of $\mathrm{End}_B^0(\mathcal{E})$ is called a *compact endomorphism* of $\mathcal{E}$. Obvious corresponding notions and notations are $\mathrm{Hom}_B(\mathcal{E}_1,\mathcal{E}_2)$ and $\mathrm{Hom}_B^0(\mathcal{E}_1,\mathcal{E}_2)$ for pairs of C^*-modules over B.

To get familiar with these notions, let us consider the special case when B is a commutative C^*-algebra, so that B is the $*$-algebra $C_0(X)$ of continuous functions vanishing at ∞ on the locally compact space X. Then a complex Hermitian vector bundle E on X gives rise to a C^*-module: $\mathcal{E} = C_0(X,E)$ is the $C_0(X)$-module of continuous sections of E vanishing at ∞, and the $C_0(X)$-valued inner product is given by

$$\langle\xi,\eta\rangle(p) = \langle\xi(p),\eta(p)\rangle_{E_p} \quad \forall \xi,\eta \in \mathcal{E}\,,\ p \in X.$$

It is not true that every C^*-module over $C_0(X)$ arises in this way; they correspond to bundles of Hilbert spaces but the finite-dimensionality of the fibers and the local triviality are no longer required. Instead one requires that one has a *continuous field* of Hilbert spaces in the following sense (cf. [Di$_3$]).

Definition 3. *Let X be a topological space.* A continuous field E *of Banach spaces over X is a family $(E(t))_{t\in X}$ of Banach spaces, with a set $\Gamma \subseteq \prod_{t\in X} E(t)$ of sections such that:*

(i) *Γ is a complex linear subspace of $\prod_{t\in X} E(t)$;*

(ii) *for every $t \in X$, the set of $x(t)$ for $x \in \Gamma$ is dense in $E(t)$;*

(iii) *for every $x \in \Gamma$ the function $t \to \|x(t)\|$ is continuous;*

(iv) *let $x \in \prod_{t\in X} E(t)$ be a section; if, for every $t \in X$ and every $\varepsilon > 0$, there exists an $x' \in \Gamma$ such that $\|x(t) - x'(t)\| \le \varepsilon$ throughout some neighborhood of t, then $x \in \Gamma$.*

The elements of Γ are called the *continuous sections of E*. When each $E(t)$ is a Hilbert space (i.e. $\|\xi\| = (\langle\xi,\xi\rangle)^{1/2}$) it follows that for any $\xi,\eta \in \Gamma$ the function $t \to \langle\xi(t),\eta(t)\rangle$ is continuous. Every continuous field of Hilbert spaces E over X yields a C^*-module over $C_0(X)$, namely the space $C_0(X,E)$ of continuous sections of E which vanish at ∞, i.e. such that $\|\xi(t)\| \to 0$ when $t \to \infty$. Moreover, now every C^*-module $\mathcal{E}$ over $C_0(X)$ arises from a continuous field E of Hilbert spaces (canonically associated to $\mathcal{E}$). We shall see later (Proposition 4) how, when X is compact, the finite-dimensional Hermitian vector bundles are characterized among general continuous fields of Hilbert spaces. The latter notion implies the following *semicontinuity* of $t \to \dim(E_t)$:

$$\{t \in X; \dim(E_t) \ge n\} \text{ is an } \textit{open subset} \text{ of } X.$$

Given an open subset V of X and a continuous field E of Hilbert spaces on V there is a canonical extension $\tilde{E}$ of E to X, where the fibers of $\tilde{E}$ on the complement of V are $\{0\}$. This operation is particularly convenient for problems of excision.

Let X be a locally compact space and E a finite-dimensional Hermitian complex vector bundle over X, with $\mathcal{E} = C_0(X, E)$ the corresponding C^*-module. Then $\mathrm{End}_{C_0(X)}(\mathcal{E})$ is the algebra $C_b(X, \mathrm{End}(E))$ of bounded continuous sections of the bundle $\mathrm{End}\, E = E^* \otimes E$ of endomorphisms of E. Moreover, $\mathrm{End}^0_{C_0(X)}(\mathcal{E})$ is the subalgebra $C_0(X, \mathrm{End}(E))$ of sections vanishing at ∞.

Let us go back to the general case of noncommutative C^*-algebras and characterize *finite projective modules* among general C^*-modules.

Proposition 4. *Let B be a unital C^*-algebra.*

1) *Let $\mathcal{E}$ be a C^*-module over B such that $1_{\mathcal{E}} \in \mathrm{End}^0_B(\mathcal{E})$. Then the underlying right B-module is finite and projective.*

2) *Let $\mathcal{E}_0$ be a finite projective module over B. Then there exist B-valued inner products $\langle\ ,\ \rangle$ on $\mathcal{E}_0$ for which $\mathcal{E}_0$ is a C^*-module over B, and one has $1_{\mathcal{E}} \in \mathrm{End}^0_B(\mathcal{E})$.*

3) *Let $\langle\ ,\ \rangle$ and $\langle\ ,\ \rangle'$ be two B-valued inner products on $\mathcal{E}_0$ as in 2). Then there exists an invertible endomorphism T of $\mathcal{E}_0$ such that*

$$\langle \xi, \eta \rangle' = \langle T\xi, T\eta \rangle \quad \forall \xi, \eta \in \mathcal{E}_0.$$

For the proof see [Ri$_5$] [Mis$_3$].

Recall that, by definition, a finite projective module $\mathcal{E}_0$ over an algebra B is a direct summand of a free module $\mathcal{E}_1 = B^N$, N finite. For arbitrary C^*-modules one has the following stabilization theorem due to Kasparov ([Kas$_2$]):

Theorem 5. [Kas$_2$] *Let B be a C^*-algebra, and let $\mathcal{E}$ be a C^*-module over B with a countable subset $S \subset \mathcal{E}$ such that SB is total in $\mathcal{E}$. Let $\ell^2 \otimes B$ be the C^*-module over B which is a sum of countably many copies of B. Then*

$$\mathcal{E} \oplus (\ell^2 \otimes B) \text{ is isomorphic to } \ell^2 \otimes B.$$

We have used the notation $\ell^2 \otimes B$ as a special case of the following general notion of tensor product of C^*-modules:

Proposition 6. [Ri$_5$] *Let B and C be C^*-algebras, $\mathcal{E}'$ (resp. $\mathcal{E}''$) be a C^*-module over B (resp. C) and ρ a $*$-homomorphism $B \to \mathrm{End}_C(\mathcal{E}'')$. Then the following equality yields the structure of a pre-C^*-module over C on the algebraic tensor product $\mathcal{E} = \mathcal{E}' \otimes_B \mathcal{E}''$:*

$$\langle \xi_1 \otimes \eta_1, \xi_2 \otimes \eta_2 \rangle = \langle \rho(\langle \xi_2, \xi_1 \rangle)\eta_1, \eta_2 \rangle \in C$$

$\forall \xi_j \in \mathcal{E}', \eta_j \in \mathcal{E}''$.

We shall still denote by $\mathcal{E}' \otimes_B \mathcal{E}''$ the associated C^*-module over C. Given $T \in \mathrm{End}_B(\mathcal{E}')$, an endomorphism $T \otimes 1 \in \mathrm{End}_C(\mathcal{E}' \otimes_B \mathcal{E}'')$ is defined by

$$(T \otimes 1)(\xi \otimes \eta) = T\xi \otimes \eta \quad \forall \xi \in \mathcal{E}',\ \eta \in \mathcal{E}''.$$

By a $(B\text{-}C)$ C^*-bimodule we shall mean a C^*-module $\mathcal{E}$ over C together with a $*$-homomorphism from B to $\mathrm{End}_C(\mathcal{E})$. In particular, given a C^*-algebra B, we denote by 1_B the B-B C^* bimodule given by $\mathcal{E} = B$, the actions of B by left and right multiplications, and the B-valued inner product $\langle b_1, b_2\rangle = b_1^* b_2 \;\; \forall b_1, b_2 \in B$.

Definition 7. *Let B and C be C^*-algebras. A strong Morita equivalence $B \simeq C$ is given by a pair $(\mathcal{E}_1, \mathcal{E}_2)$ of C^*-bimodules such that*

$$\mathcal{E}_1 \otimes_C \mathcal{E}_2 = 1_B \; , \; \mathcal{E}_2 \otimes_B \mathcal{E}_1 = 1_C .$$

One can then show that the linear span in C of the set of inner products $\{\langle \xi, \eta\rangle;\ \xi, \eta \in \mathcal{E}_1\}$, is a dense two-sided ideal, and that the left action $\rho : B \to \mathrm{End}_C(\mathcal{E}_1)$ is an isomorphism of B with $\mathrm{End}_C^0(\mathcal{E}_1)$. It follows thus that $\overline{\mathcal{E}}_1$, the complex conjugate of the vector space $\mathcal{E}_1$, which is in a natural way a C-B-bimodule:

$$c \cdot \overline{\xi} \cdot b \overset{\text{def}}{=} (b^* \xi c^*)^- \quad \forall \xi \in \mathcal{E}_1$$

is also endowed with a B-valued inner product

$$\langle \overline{\xi}, \overline{\eta}\rangle = \rho^{-1}\,(|\eta\rangle\langle\xi|) \in B.$$

With this inner product, $\overline{\mathcal{E}}_1$ is a C-B C^*-bimodule. The bimodule $\mathcal{E}_1$ is then a B-C-equivalence bimodule in the sense of [Ri$_2$] and one checks that the above Definition 7 is equivalent to the existence of a B-C equivalence bimodule. One can then take $\mathcal{E}_2 = \overline{\mathcal{E}}_1$.

Let $\mathcal{K}$ be the C^*-algebra of compact operators on an infinite-dimensional separable Hilbert space.

Theorem 8. [Bro$_3$] [B-G-R] *Let B and C be two separable C^*-algebras (more generally, two C^*-algebras with countable approximate units). Then B is strongly Morita equivalent to C iff $B \otimes \mathcal{K}$ is isomorphic to $C \otimes \mathcal{K}$.*

Of particular importance is the strong Morita equivalence of a given C^*-algebra B with its full hereditary sub-C^*-algebras C. Here $C \subset B$ is

α) hereditary iff $0 \le c \le b$ and $b \in C$ implies $c \in C$.

β) full iff the two-sided ideal generated by C is dense in B.

Hereditary sub-C^*-algebras C of B correspond bijectively to closed left ideals L of B by $C = L \cap L^*$, and, for instance, any self-adjoint idempotent $e \in B$ gives rise to such a hereditary C^*-subalgebra, the *reduced* C^*-algebra

$$B_e = \{x \in B \; ; \; ex = xe = x\}.$$

The reduced subalgebra B_e is full iff the two-sided ideal generated by e is dense in B.

Let us now illustrate the general notion of strong Morita equivalence by a simple example:

Example. Let E be a real finite-dimensional Euclidean space and $\mathrm{Cliff}_{\mathbb{C}}(E)$ the complexified Clifford algebra of E, i.e. the quotient of the complex tensor algebra of E by the relation

$$\gamma(\xi)^2 = \|\xi\|^2 1 \quad \forall \xi \in E,$$

where γ is the natural linear map of E to its tensor algebra. When E is even-dimensional, the C^*-algebra $\mathrm{Cliff}_{\mathbb{C}}(E)$ is isomorphic, but not canonically, to a matrix algebra.

Let X be a compact space and E a real even-dimensional oriented Euclidean vector bundle over X. Then the algebra $A = C(X, \mathrm{Cliff}_{\mathbb{C}}(E))$ of continuous sections of the bundle of Clifford algebras $\mathrm{Cliff}_{\mathbb{C}}(E_x)$, $x \in X$, is a C^*-algebra which in general fails to be strongly Morita equivalent to a commutative C^*-algebra. Recall that a Spin^c structure on the vector bundle E is a lifting of its structure group $SO(2n)$ to its covering group $\mathrm{Spin}^c(2n)$. Any such structure gives rise, using the spin representation of $\mathrm{Spin}^c(2n)$ in $\mathbb{C}^{2^n}$, to an associated Hermitian complex vector bundle S on X, which is a module over the complexified Clifford algebra $\mathrm{Cliff}_{\mathbb{C}}(E)$. One thus obtains in this way an A-$C(X)$ C^*-bimodule which gives a strong Morita equivalence of $A = C(X, \mathrm{Cliff}_{\mathbb{C}}(E))$ with a commutative C^*-algebra (cf. [Pl$_1$]).

Proposition 9. *The above construction yields a one-to-one correspondence between* Spin^c *structures on E and strong Morita equivalences of a commutative C^*-algebra with $A = C(X, \mathrm{Cliff}_{\mathbb{C}}(E))$.*

Let B and C be C^*-algebras and $\mathcal{E}_1$ an equivalence B-C C^*-bimodule. One obtains a functor from the category of unitary representations of C to that of B by

$$\mathcal{H} \in \mathrm{Rep}\, C \to \mathcal{E}_1 \otimes_C \mathcal{H} \in \mathrm{Rep}\, B,$$

and using $\mathcal{E}_2 = \overline{\mathcal{E}}_1$ as the inverse of $\mathcal{E}_1$ one gets a natural equivalence between the two categories of representations.

It follows in particular that two strongly Morita equivalent C^*-algebras have the same space of classes of irreducible representations. In particular, if a C^*-algebra B is strongly Morita equivalent to some commutative C^*-algebra then the latter is unique, and is the C^*-algebra of continuous functions vanishing at ∞ on the space of irreducible representations of B.

Strong Morita equivalence preserves many other properties. An equivalence B-C C^*-bimodule determines an isomorphism between the lattices of two-sided ideals of B and C, and hence a homeomorphism between the primitive ideal spaces of B and C. It does also give a canonical isomorphism of the K-theory groups $K_*(B) \simeq K_*(C)$. One should, however, not conclude too hastily that nothing will change when we replace C^*-algebras by strongly Morita equivalent ones. We shall illustrate this by the following example:

Example $\pi_3(S^1)$.

By Gel'fand's theorem the category of commutative C^*-algebras and $*$-homomorphisms is dual to the category of locally compact spaces and continuous proper maps. To the notion of *homotopy* between continuous proper maps corresponds the following notion of homotopy of $*$-homomorphisms (morphisms for short) of C^*-algebras:

Definition 10. *Let A and B be C^*-algebras, and let ρ_0 and ρ_1 be two morphisms $\rho_j : A \to B$. Then a* homotopy *between ρ_0 and ρ_1 is a morphism from A to $B \otimes C[0,1]$ whose composition with evaluation at $j \in \{0,1\}$ gives ρ_j.*

In topology one usually deals with pointed spaces $(X, *)$ and base-point-preserving continuous maps. When dealing with compact spaces X, this is equivalent to locally compact spaces $X \backslash \{*\}$ and proper maps. In particular, the homotopy groups $\pi_n(X, *)$ are obtained from homotopy classes of morphisms

$$C_0(X\backslash\{*\}) \xrightarrow{\rho} C_0(\mathbb{R}^n).$$

The group law on $\pi_n(X, *)$ comes from the natural morphism

$$C_0(\mathbb{R}^n) \oplus C_0(\mathbb{R}^n) \xrightarrow{\delta} C_0(\mathbb{R}^n)$$

obtained from the $n = 1$ case and the usual map from a circle to a bouquet of two circles. All this is of course just a transposition of the usual notions of topology.

All these notions continue to make sense if we replace $C_0(\mathbb{R}^n)$ by the strongly Morita equivalent C^*-algebra $M_k(\mathbb{C}) \otimes C_0(\mathbb{R}^n) = M_k(C_0(\mathbb{R}^n))$ of $k \times k$ matrices over $C_0(\mathbb{R}^n)$. We may in particular define new homotopy groups $\pi_{n,k}(X, *)$ for a compact pointed space $(X, *)$ as the group (using δ) of homotopy classes of morphisms

$$C_0(X\backslash *) \xrightarrow{\rho} M_k(C_0(\mathbb{R}^n)).$$

Let us now show, by exhibiting a specific ρ, that $\pi_{3,2}(S^1)$ is very different from the trivial $\pi_3(S^1)$. For this we need to construct a non-trivial morphism of $C_0(S^1\backslash\{*\})$ to $M_2(C_0(\mathbb{R}^3))$, or, equivalently, a nontrivial morphism ρ of $C(S^1)$ to $M_2(C(S^3))$ whose range over the base point $*$ of S^3 is formed of scalar multiples of the identity matrix $1 \in M_2(\mathbb{C})$. To that effect, let us identify S^3 with the unit quaternions $S^3 = \{(\alpha, \beta) \in \mathbb{C}^2; |\alpha|^2 + |\beta|^2 = 1\}$ represented as 2×2 matrices, $\begin{bmatrix} \alpha & \beta \\ -\bar{\beta} & \bar{\alpha} \end{bmatrix} \in M_2(\mathbb{C})$. This yields a canonical unitary element $U \in M_2(C(S^3))$, whose value at $q = \alpha + \beta j$ is given by $U(q) = \begin{bmatrix} \alpha & \beta \\ -\bar{\beta} & \bar{\alpha} \end{bmatrix} \in M_2(\mathbb{C})$.

Let us choose the base point $*$ of S^3 to be $q = 1$, so that $U(*) = 1$ is the identity matrix. Then the morphism ρ from $C(S^1)$ to $M_2(C(S^3))$ uniquely determined by

$$\rho(e^{i\theta}) = U$$

yields an element of $\pi_{3,2}(S^1)$.

We leave it to the reader to check his understanding of K-theory by showing that the class of ρ in $\pi_{3,2}(S^1)$ is not trivial. We shall see later (Chapter 3) that this morphism is *of degree 1* in cyclic cohomology.

At an intuitive level, i.e. trying to think of the above morphism ρ at the set theoretic level as a generalized map from S^3 to S^1, one finds that to a point $q = \alpha + \beta j$, $|\alpha|^2 + |\beta|^2 = 1$, of S^3 it associates the *two* solutions $z, z' \in S^1 = \{z \in \mathbb{C}; |z| = 1\}$ of the equation

$$\det \begin{bmatrix} \alpha - z & \beta \\ -\overline{\beta} & \overline{\alpha} - z \end{bmatrix} = 0$$

or, equivalently, of $z^2 - (\alpha - \overline{\alpha})z + 1 = 0$.

II. Appendix B. *E*-theory and Deformations of Algebras

A functor F from the category of C^*-algebras and $*$-homomorphisms to the category of abelian groups is called half-exact iff for any short exact sequence of C^*-algebras

$$0 \to J \to A \to B \to 0$$

the corresponding sequence of abelian groups is exact at $F(A)$. The functor K, which assigns to A its K-theory group $K(A)$, is half-exact. G. Skandalis has shown ([Sk$_4$]) that the bivariant functor KK of Kasparov, which plays a crucial role in the construction of K-theory maps, in general, fails to be half-exact. It follows in particular that in general the connecting map of K-theory

$$K_i(B) \xrightarrow{\partial} K_{i+1}(J)$$

associated to an exact sequence of C^*-algebras, does not necessarily come from a bivariant element (belonging to $KK^1(B, J)$). The natural place for such bivariant elements is the extension theory ([Kas$_1$]) but the intersection product as constructed in [Kas$_1$] is limited to KK. In this appendix we shall expound the main features of the bivariant E-theory which solves the above difficulties. Its abstract existence as a bivariant semi-exact theory extending KK-theory was first proved by N. Higson [Hig]. In our joint work with N. Higson [Co-H], it was shown that, using the theory of deformations and the notion of asymptotic morphisms, originating in [Co-M$_1$], one can define the intersection product of extensions. The resulting theory has a number of technical advantages; in particular, the complete proofs of the main properties of the intersection product are quite short, and will be given in this appendix.

B.α Deformations of C^*-algebras and asymptotic morphisms

Let A and B be two C^*-algebras. By a strong deformation from A to B we mean a continuous field $(A(t), \Gamma)$ of C^*-algebras over $[0,1]$ (cf. [Di$_3$] Definition 10.3.1) whose fiber at 0 is $A(0) = A$, and whose restriction to the half-open interval $]0,1]$ is the constant field with fiber $A(t) = B$ for $t > 0$.

We shall now associate to such a deformation an asymptotic morphism from A to B. From the definition ([Di$_3$]) of a continuous field of C^*-algebras it follows that for any $a \in A = A(0)$ there exists a continuous section $\alpha \in \Gamma$ of the above field such that $\alpha(0) = a$. Let us choose such an $\alpha = \alpha_a$ for each $a \in A$ and set

$$\varphi_t(a) = \alpha_a\left(\frac{1}{t}\right) \in B \quad \forall t \in [1, \infty[.$$

Using the continuity of $\|\alpha(t)\|$ as a function of $t \in [0,1]$ for any continuous section $\alpha \in \Gamma$, one checks that the family φ_t of maps from A to B fulfills the following conditions:

(I) For any $a \in A$, the map $t \to \varphi_t(a)$ is norm continuous.

(II) For any $a, b \in A$, $\lambda \in \mathbb{C}$, the following norm limits vanish:

$$\lim_{t \to \infty} (\varphi_t(a) + \lambda\varphi_t(b) - \varphi_t(a + \lambda b)) = 0$$

$$\lim_{t \to \infty} (\varphi_t(ab) - \varphi_t(a)\, \varphi_t(b)) = 0$$

$$\lim_{t \to \infty} (\varphi_t(a^*) - \varphi_t(a)^*) = 0.$$

We have thus associated to the given deformation an asymptotic morphism in the following sense:

Definition 1. *Let A and B be C^*-algebras. An asymptotic morphism from A to B is given by a family $(\varphi_t)_{t \in [1,\infty[}$ of maps from A to B fulfilling conditions (I) and (II).*

Roughly speaking these conditions mean that given a finite subset F of A and an $\varepsilon > 0$, there exists a t_0 such that for $t > t_0$, φ_t behaves on F as a true homomorphism, up to the precision ε.

The notion of deformation of algebras already plays a critical role in algebraic geometry ([Ful]), quantization ([Li] [B-F-F-L-S$_2$), and the construction of quantum groups ([Dr] [Fa-R-T]). We have given in this chapter many concrete examples of deformations of noncommutative spaces (Sections 5, 6, 7, 8, and 10). To get some feeling for asymptotic morphisms versus ordinary morphisms one can consider the following very simple example.

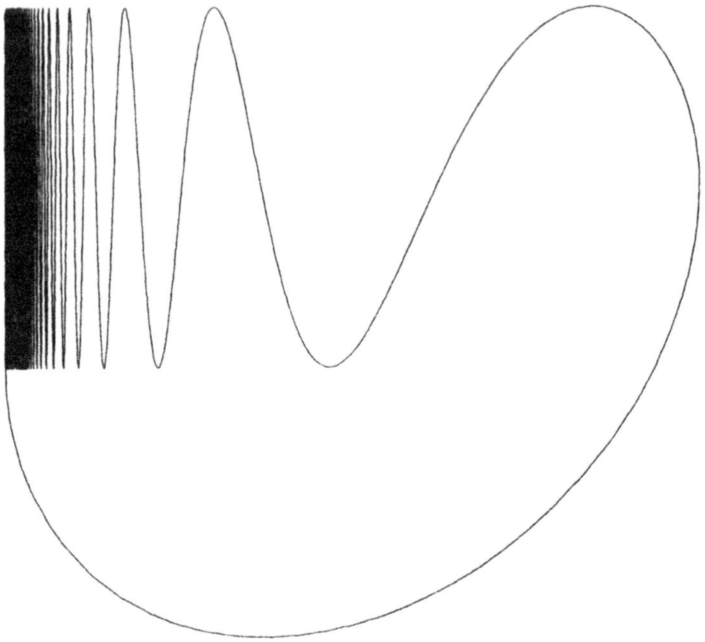

Figure II.9. A compact space with $\pi_1(X) = \{0\}$

We let X be the compact space shown in Figure 9. One can easily show that $\pi_1(X) = \{0\}$, i.e. there are no interesting continuous maps $f : S^1 \to X$, or equivalently no interesting morphisms of $C(X)$ to $C(S^1)$. However, for each $\varepsilon > 0$, the ε-neighborhood of X,

$$X_\varepsilon = \{z \in \mathbb{C} \ ; \ \mathrm{dist}(z, X) \leq \varepsilon\}$$

obviously has $\pi_1(X_\varepsilon) = \mathbb{Z}$. With a little work, one can get a family f_ε of degree-one continuous maps $f_\varepsilon : S^1 \to X_\varepsilon$, $\varepsilon > 0$ such that $f_\varepsilon(x)$ is for fixed $x \in S^1$ a continuous function of $\varepsilon > 0$. Of course, for $\varepsilon \to 0$ these maps f_ε do not converge to a continuous map of S^1 to X. One obtains a corresponding asymptotic morphism $\varphi_t : C(X) \to C(S^1)$ by setting:

$$\varphi_t(a) = \tilde{a} \circ f_{1/t} \quad \forall a \in C(X) \ , \ t \in [1, \infty[$$

where for each $a \in C(X)$, $\tilde{a}$ is a continuous extension of a to $\mathbb{C}$. This easy example shows that the notion of asymptotic morphism is relevant even in the topology of ordinary compact spaces. Thus, in extending the generalized homology theory associated to a spectrum Σ

$$H_n(X, \Sigma) = \pi_n(X \wedge \Sigma)$$

beyond spaces X which are simplicial complexes, so as to include arbitrary compact (metrisable) X, [Ka-Kam-S], one should replace the continuous maps

$S^n \to Y$ involved in the definition of $\pi_n(Y)$ by asymptotic maps (i.e. the transposes of asymptotic morphisms).

Coming back to the general discussion, we shall say that two asymptotic morphisms (φ_t) and (φ'_t) from A to B are *equivalent* iff

$$\lim_{t\to\infty} (\varphi'_t(a) - \varphi_t(a)) = 0 \quad \forall a \in A.$$

In other words, if we let B_∞ be the quotient C^*-algebra

$$B_\infty = C_b\left([1,\infty[,B\right)/C_0\left([1,\infty[,B\right)$$

where C_b means bounded continuous functions, then the equivalence classes of asymptotic morphisms from A to B correspond exactly to the morphisms $\tilde{\varphi} : A \to B_\infty$ by the formula

$$\tilde{\varphi}(a)_t = \varphi_t(a) \quad \forall a \in A\,,\ t \in [1,\infty[.$$

Since $\tilde{\varphi}$ is a morphism of C^*-algebras it is always a contraction.

Giving a strong deformation from A to B is equivalent to giving the associated class of asymptotic morphism $\varphi_t : A \to B$ constructed above. The convergence $\|\varphi_t(a)\| \to \|a\|$ $\forall a \in A$ characterizes the classes of asymptotic morphisms thus obtained. Any injective morphism $\psi : A \to B_\infty$ gives rise canonically to a *deformation* from A to B in the following sense (cf. [Lor]): one has a C^*-subalgebra C of $C_b(]0,1],B)$ containing $B(]0,1]) = C_0(]0,1],B)$ and an isomorphism $A \simeq C/B(]0,1])$. Given ψ, one just lets $C = \{(\psi_{1/t}(a) + b(t))_{t\in]0,1]};\ a \in A, b \in B(]0,1])\}$.

We can now define homotopy between asymptotic morphisms.

Definition 2. *Two asymptotic morphisms* $(\varphi^i_t) : A \to B$, $i = 0,1$, *are homotopic iff there exists an asymptotic morphism* (φ_t) *from* A *to* $B[0,1] = C([0,1]) \otimes B$ *whose evaluation at* $i = 0,1$ *gives* (φ^i_t).

This is an equivalence relation between asymptotic morphisms. Given two C^*-algebras A and B, we shall let $[[A,B]]$ denote the set of homotopy classes of asymptotic morphisms from A to B. A change of parameter $r(t) \to \infty$, i.e. the replacement of (φ_t) by $(\varphi_{r(t)}) = (\psi_t)$ does not affect the homotopy class of φ.

B.β Composition of asymptotic morphisms

In this section we shall describe the composition of asymptotic morphisms, i.e. the analogue in E-theory of the main tool of the KK-theory, which is the intersection product. The idea for defining the composition $\psi\circ\varphi$ of asymptotic morphisms $(\varphi_t) : A \to B$ and $(\psi_t) : B \to C$ is quite simple. In general the plain composition $\psi_t \circ \varphi_t$ *does not* satisfy the conditions of Definition 1 because of lack of uniformity of the hypothesis on ψ_t; but one may easily compensate by a change of parameter, i.e. the replacement of ψ_t by $\psi_{r(t)}$, where $r(t)$ is determined by a diagonal process. The resulting composition is then well defined at the level of *homotopy classes* of asymptotic morphisms. Let us now give the details.

Let $(\varphi_t)_{t\in[1,\infty[}$, $\varphi_t : A \to B$, be an asymptotic morphism and let $K \subset A$ be a subset of A. We shall say that (φ_t) is *uniform* on K iff

α) $(t,a) \to \varphi_t(a)$ is a continuous map from $[1,\infty[\times K$ to B

β) For any $\varepsilon > 0$, $\exists T < \infty$ such that for any $t \geq T$ one has:

$\|\varphi_t(a) + \lambda\varphi_t(a') - \varphi_t(a+\lambda a')\| < \varepsilon$, $\forall a, a' \in K$, $\forall \lambda, |\lambda| \leq 1$

$\|\varphi_t(a)\varphi_t(a') - \varphi_t(aa')\| < \varepsilon$, $\forall a, a' \in K$

$\|\varphi_t(a)^* - \varphi_t(a^*)\| < \varepsilon$, $\forall a \in K$

$\|\varphi_t(a)\| < \|a\| + \varepsilon$, $\forall a \in K$.

The Bartle-Graves selection theorem applied to the quotient map

$$C_b([1,\infty[, B) \to B_\infty$$

shows that each asymptotic morphism (φ_t) is equivalent to an asymptotic morphism (φ'_t) which is uniform on every (norm) compact subset K of A.

Let $\mathcal{A} \subset A$ be a dense involutive subalgebra of A, and let us assume that $\mathcal{A}$ is a countable union $\mathcal{A} = \bigcup K_n$ of compact subsets K_n of A. One can choose the K_n's so that $K_n + K_n \subset K_{n+1}$, $K_nK_n \subset K_{n+1}$, and $\lambda K_n \subset K_n$ $\forall \lambda \in \mathbb{C}$, $|\lambda| \leq 1$. It follows that if K is a compact convex subset of A and $K \subset \mathcal{A}$ then one has $K \subset K_n$ for some n.

Any asymptotic morphism (φ_t) from $\mathcal{A}$ to B which is uniform on compact convex subsets of $\mathcal{A}$ defines a homomorphism from A to B_∞, and hence an equivalence class of asymptotic morphisms from A to B. The composition is given by the following lemma:

Lemma 3. *Let $(\varphi_t) : A \to B$, and $(\psi_t) : B \to C$ be asymptotic morphisms of C^*-algebras. Let $\mathcal{A}$ be a dense σ-compact involutive subalgebra of A. Assume that (φ_t) is uniform on every compact convex subset of $\mathcal{A}$, while (ψ_t) is uniform on every compact subset of B. Then there exists a continuous increasing function $r : [1,\infty[\to [1,\infty[$ such that for any increasing continuous function $s(t) \geq r(t)$, the composition $\theta_t = \psi_{s(t)} \circ \varphi_t$ is an asymptotic morphism from $\mathcal{A}$ to C, uniform on compact convex subsets.*

The proof is quite simple; with $\mathcal{A} = \bigcup K_n$ as above, let $t_n \in [1,\infty[$ be such that φ_t satisfies conditions β) above on K_n for $\varepsilon = \frac{1}{n}$, $t \geq t_n$. Then let $K'_n \subset B$ be the compact subset

$$K'_n = \{\varphi_t(a)\,;\, a \in K_{n+3}\,,\, t \leq t_{n+1}\},$$

and let $r_n \in [1,\infty[$ be such that ψ_t satisfies β) on K'_n for $\varepsilon = \frac{1}{n}$ and $t \geq r_n$. Then one checks that any continuous increasing function $r : [1,\infty[\to [1,\infty[$ such that $r(t_n) \geq r_n$ does the job.

Let $(\varphi_t) : A \to B$, $(\psi_t) : B \to C$ be asymptotic morphisms of C^*-algebras which are uniform on compact subsets, and let $\mathcal{A} \subset A$ be a dense σ-compact involutive subalgebra of A. For $s : [1,\infty[\to [1,\infty[$ as in Lemma 3, we let $\overline{\theta}_t : A \to C$ be the extension to A of the composition $\psi_{s(t)} \circ \varphi_t = \theta_t$. It is an asymptotic morphism (well-defined up to equivalence) from A to C.

Proposition 4. 1) *The homotopy class $[\overline{\theta}] \in [[A,C]]$ is independent of the choices of $\mathcal{A}$ and of $s(t)$, and only depends upon the homotopy classes $[\varphi] \in [[A,B]]$ and $[\psi] \in [[B,C]]$.*

2) *The composition $[\psi] \circ [\varphi]$ of homotopy classes is associative.*

To prove the first statement note that the involutive subalgebra of A generated by two σ-compact subalgebras is still σ-compact, whence the independence in the choice of $\mathcal{A}$. To prove the second statement, use the involutive subalgebra $\mathcal{B} \subset B$ of B generated by the $\varphi_t(\mathcal{A})$ where (φ_t) is uniform on compact convex subsets of $\mathcal{A}$. Since $\mathcal{B}$ is still σ-compact the conclusion follows.

One can define the external tensor product $\varphi \otimes \psi$ of asymptotic morphisms $(\varphi_t) : A \to C$ and $(\psi_t) : B \to D$ as an asymptotic morphism, unique up to equivalence

$$(\varphi \otimes \psi) : A \otimes_{\max} B \to C \otimes_{\max} D$$

using the maximal tensor product of C^*-algebras ([T_1]). This follows from the following useful lemma whose proof is immediate using the C^*-algebra C_∞.

Lemma 5. *Let A, B and C be C^*-algebras and let $(\varphi_t) : A \to C$, and $(\psi_t) : B \to C$ be asymptotic morphisms such that, for any $a \in A$ and $b \in B$, the commutator $[\varphi_t(a), \psi_t(b)]$ converges to 0 in norm when $t \to \infty$. Then there exists an asymptotic morphism $(\theta_t) : A \otimes_{\max} B \to C$, unique up to equivalence, such that*

$$\theta_t(a \otimes b) - \varphi_t(a)\, \psi_t(b) \to 0 \quad \forall a \in A\,,\ b \in B.$$

The external tensor product $\varphi \otimes \psi$ of asymptotic morphisms passes to homotopy classes.

B.γ Asymptotic morphisms and exact sequences of C^*-algebras

Let $0 \to J \to A \overset{p}{\to} B \to 0$ be an exact sequence of separable C^*-algebras. By [Vo_2] there exists, in the ideal J, a quasi-central continuous approximate unit $u_t \in J, 0 \le u_t \le 1,\ t \in [1,\infty[$. This means that the following holds:

a) $\|xu_t - x\| \to 0$, $\|u_t x - x\| \to 0$ when $t \to \infty$, $\forall x \in J$

b) $\|[u_t, y]\| \to 0$ when $t \to \infty$, $\forall y \in A$

c) $t \to u_t$ is norm continuous.

Let us use the shorthand notation, for any C^*-algebra A

$$SA = C_0(]0,1[) \otimes A$$

which is consistent with the topological notation for the suspension of a space.

Lemma 6. *Let* $0 \to J \to A \xrightarrow{p} B \to 0$ *be an exact sequence of separable* C^*-*algebras.*

1) *For any continuous quasi-central approximate unit* (u_t), $0 \leq u_t \leq 1$, *and any section* $b \in B \to b' \in A$ *of* p, *the following equality defines an asymptotic morphism* $(\varphi_t) : SB \to J$

$$\varphi_t(f \otimes b) = f(u_t)b' \quad \forall f \in C_0(]0,1[) \ , \ b \in B.$$

2) *The homotopy class* $[\varphi] = \varepsilon_p$ *of* (φ_t) *depends only upon the morphism* $p : A \to B$.

The first statement is a direct application of Lemma 5. The second follows from the convexity of the set of quasi-central approximate units.

This Lemma 6 is a key result which allows one to associate to any extension of C^*-algebras a class of asymptotic morphisms, while to get a KK-class some further hypothesis, such as the existence of a completely positive lifting of p, is required.

It is obvious that an asymptotic morphism of C^*-algebras yields a corresponding map of K-theories, since in a C^*-algebra the equations $e = e^* = e^2$ are stable: if such an equation is fulfilled by $x \in A$, up to ε in norm, then x is close to a solution (for ε small enough). Thus if $(\varphi_t) : A \to B$ is an asymptotic morphism and $e \in \mathrm{Proj}\,(A)$ is a projection ($e = e^* = e^2$), then $\varphi_t(e)$ is close, for t large, to a projection $f \in \mathrm{Proj}\,(B)$ whose K-theory class is well defined. We urge the reader to check directly at this point that the K-theory map thus associated to ε_p

$$(\varepsilon_p)_* : K(SB) \to K(J)$$

does coincide with the connecting map of the six-term exact sequence of K-groups associated to an exact sequence of C^*-algebras.

To end this section, let us note that the above construction of ε_p is coherent with the construction of the asymptotic morphism $(\varphi_t) : A \to B$ associated to a deformation of C^*-algebras. Indeed, given such a deformation one gets an exact sequence:

$$0 \to SB \to C \xrightarrow{\pi} A \to 0$$

where C is the C^*-algebra of the restriction of the continuous field $(A(t), \Gamma)$ to the half-open interval $[0, 1[$ and where π is the evaluation at 0. One then checks the following equality:

$$\varepsilon_\pi = 1 \otimes \varphi \ \text{ in } \ [[SA, SB]].$$

B.δ The cone of a map and half-exactness

Before giving the precise definition of E-theory we shall, in this section, give the main ingredient of the proof of its half-exactness. The point is that we shall then get a byproduct of this proof even for ordinary compact spaces.

Let $0 \to J \xrightarrow{j} A \xrightarrow{p} B \to 0$ be an exact sequence of C^*-algebras. The *cone* C_p of the morphism $p : A \to B$ is, by definition, the C^*-algebra fibered product of $CB = C_0(]0,1]) \otimes B$ and A using the following morphisms to B:

$$\rho : CB \to B\ ,\ \rho(b) = b(1)\ \ \forall b = (b(s))_{s\in]0,1]} \in CB$$
$$p : A \to B.$$

In other words, an element x of C_p is a pair $((b(s))_{s\in]0,1]}, a)$ with $b(1) = p(a) \in B$. One has a natural morphism $i : J \to C_p$ given by

$$i(y) = (0, y)\ \ \forall y \in J$$

and its composition with the evaluation map, $\mathrm{ev} : C_p \to A$,

$$\mathrm{ev}((b_s)_{s\in]0,1]}, a) = a \in A$$

is the morphism j

$$\mathrm{ev} \circ i = j : J \to A.$$

The main point in proving half-exactness is to invert the morphism i. That this is likely to be possible comes from the contractibility of the third term (the cone CB of B) in the exact sequence

$$0 \to J \xrightarrow{i} C_p \to CB \to 0.$$

The inverse of i will be given by the asymptotic morphism ε_σ associated to the following exact sequence of C^*-algebras:

$$0 \to SJ \to CA \xrightarrow{\sigma} C_p \to 0$$

where $CA = C_0(]0,1]) \otimes A$ is the cone of A and σ is given by $\sigma((a(s))_{s\in]0,1]}) = ((p(a(s)))_{s\in]0,1]}, a(1)) \in C_p$. It is clear that the kernel of σ is $C_0(]0,1[) \otimes J = SJ$. Let (Lemma 6) $\varepsilon_\sigma \in [[SC_p, SJ]]$ be the corresponding asymptotic morphism.

Lemma 7. *Let $Sj = \mathrm{id} \otimes j : SJ \to SA$. Then the composition $Sj \circ \varepsilon_\sigma : SC_p \to SA$ is homotopic to $S\,\mathrm{ev} = \mathrm{id} \otimes \mathrm{ev}$, where $\mathrm{ev} : C_p \to A$ is the evaluation morphism $\mathrm{ev}((b_s)_{s\in]0,1]}, a) = a \in A$.*

For the proof, let $(u_t)_{t\in[1,\infty[}$ (resp. $(h_t)_{t\in[1,\infty[}$) be a quasi-central approximate unit for the ideal J of A (resp. for $C_0(]0,1[) \subset C([0,1])$). Thus $0 \le h_t(s) \le 1\ \ \forall s \in [0,1]$, $h_t(0) = h_t(1) = 0$, and $h_t(s) \to 1$ when $t \to \infty$, for any $s \in]0,1[$. By construction, the asymptotic morphism $\varepsilon_\sigma : SC_p \to SJ$ is given by

$$\varphi_t(f \otimes x) = f(h_t \otimes u_t)\tilde{x} \qquad \forall f \in C_0(]0,1[)\ ,\ x \in C_p$$

where $\tilde{x} \in CA$ is such that $\sigma(\tilde{x}) = x$. Here $\tilde{x} = (\tilde{x}_s)_{s\in]0,1]}$ and $f(h_t \otimes u_t)\tilde{x} \in SJ$ is given by the function

$$s \in]0,1[\to f(h_t(s)u_t)\tilde{x}_s \in J.$$

The composition $Sj \circ \varepsilon_\sigma$ is given by the same formula, but now $f(h_t(s)u_t)\tilde{x}_s$ is viewed as an element of A. It is thus clear that it is homotopic through asymptotic morphisms to the following asymptotic morphism: $(\psi_t)_{t\in[1,\infty[}$, from SC_p to SA given by

$$\psi_t(f \otimes x) = (f(h_t(s))\tilde{x}_s)_{s\in]0,1[} \qquad \forall f \in C_0(]0,1[)\ ,\ x \in C_p.$$

As above, the class of (ψ_t) is independent of the choices $x \to \tilde{x}$, and one checks directly that (ψ_t) is homotopic to the morphism ρ

$$\rho(f \otimes x) = (f(1-s)\,\tilde{x}_1)_{s\in]0,1[}.$$

An easy corollary of this lemma and of the technique of the Puppe exact sequence ([Cu-S]) is the following:

Proposition 8. *Let $0 \to J \xrightarrow{j} A \xrightarrow{p} B \to 0$ be an exact sequence of separable C^*-algebras and D a C^*-algebra.*

1) Let $h \in [[A,D]]$ be such that $h \circ j$ is homotopic to 0. Then there exists $k \in [[S^2B, S^2D]]$ such that S^2h is homotopic to $k \circ S^2p \in [[S^2A, S^2D]]$.

2) Let $h \in [[D,A]]$ be such that $p \circ h$ is homotopic to 0. Then there exists $k \in [[SD, SJ]]$ such that $Sh = Sj \circ k$.

Let us briefly sketch the proof of 1). With the notation of Lemma 7 one has $S\,\mathrm{ev} = Sj \circ \varepsilon_\sigma$; thus the hypothesis: $h \circ j \sim 0$ implies that $Sh \circ S\,\mathrm{ev} = (Sh \circ Sj) \circ \varepsilon_\sigma = S(h \circ j) \circ \varepsilon_\sigma \sim 0$. Thus, provided one applies S once, one may assume that $h \circ \mathrm{ev} \sim 0$. But such a homotopy yields by restriction to $SB \subset C_p$ an asymptotic morphism, k, from SB to SD with $k \circ Sp \sim Sh$, thus the conclusion.

Similarly, for 2) a homotopy $p \circ h \sim 0$ determines precisely an element k_1 of $[[D, C_p]]$ such that $h = \mathrm{ev} \circ k_1$. Then by Lemma 7 one gets $Sh = S\,\mathrm{ev} \circ Sk_1 = Sj \circ (\varepsilon_\sigma \circ Sk_1)$, so that $k = \varepsilon_\sigma \circ Sk_1$ gives the desired factorisation.

The proposition implies the half-exactness of the E-theory to be defined below, but it also applies to the topology of ordinary compact spaces. Let us first recall a few definitions from topology. We work in the category of pointed topological spaces and use the standard notations: $X \vee Y = (X \times *) \cup (* \times Y)$ is called the wedge of X and Y; $X \wedge Y$ is the space obtained from $X \times Y$ by smashing the subspace $X \vee Y$ to a point and is called the reduced join, or smash product, of X and Y. In particular $X \wedge S^1 = SX$ is called the suspension of X.

Definition 9. [Ada] *A spectrum Σ is a sequence of spaces Σ_n and maps $\sigma_n : S\Sigma_n \to \Sigma_{n+1}$ such that each Σ_n is a CW-complex and the maps σ_n are embeddings of CW-complexes.*

The generalized homology theory h_* associated to a spectrum Σ is defined by the equality

$$h_n(X,\Sigma) = \varinjlim \pi_{n+k}(X \wedge \Sigma_k). \qquad (*)$$

This definition works well for spaces X such as CW-complexes, and Kahn, Kaminker and Schochet [Ka-Kam-S] have shown, using duality, how to extend such a theory to the category of pointed (metrisable) compact spaces so that the axioms of *Steenrod generalized homology* are satisfied:

Definition 10. *A Steenrod homology theory h_* on the category of pointed (metrisable) compact spaces is a sequence of covariant, homotopy invariant functors h_n from this category to that of abelian groups which fulfills the following axioms for all n and X:*

Exactness. *If A is a closed subset of X then the sequence*

$$h_n(A) \to h_n(X) \to h_n(X/A)$$

is exact.

Suspension. *There is a natural equivalence $h_n(X) \overset{\sigma}{\simeq} h_{n+1}(SX)$.*

Strong Wedge. *Suppose X_j is a compact pointed metrisable space, $j = 1,2,\ldots,$. Then the natural map*

$$h_n\left(\varprojlim_k (X_1 \vee \cdots \vee X_k)\right) \to \prod_j h_n(X_j)$$

is an isomorphism.

We shall now show that equality $(*)$ gives a Steenrod homology theory provided one modifies the usual definition of homotopy groups $\pi_n(X)$ of a pointed topological space X to take care of spaces such as the compact space of Figure 9. The new homotopy groups $\underline{\pi}_n(X)$ coincide with the usual ones for simplicial complexes but do not agree with the usual ones for arbitrary compact spaces.

Let first K be a compact space with a base point $* \in K$. The usual homotopy group $\pi_n(K)$ involves homotopy classes of (base point preserving) maps of the n-sphere to K, or equivalently of morphisms

$$C_0(K\backslash\{*\}) \to C_0(\mathbb{R}^n).$$

It is straightforward to see that the same construction can be done if we use homotopy classes of *asymptotic morphisms*, i.e.

$$\underline{\pi}_n(K) \overset{\text{def}}{=} [[C_0(K\backslash\{*\}), C_0(\mathbb{R}^n)]].$$

In particular the group structure of $\pi_n(K)$ comes from the natural morphism

$$C_0(\mathbb{R}) \oplus C_0(\mathbb{R}) \overset{\delta}{\to} C_0(\mathbb{R})$$

corresponding to the usual map: $S^1 \to S^1 \vee S^1$.

When K is a finite simplicial complex, one can embed it in Euclidean space as a deformation retract of an open neighborhood, and one checks in this way that the new definition of π_n agrees with the old one for such spaces. However, for the compact space of Figure 9 one has

$$\pi_1(K) = \{0\} \ , \ \underline{\pi}_1(K) = \mathbb{Z}.$$

For arbitrary pointed topological spaces $(X, *)$ we extend the above definition as follows:

$$\underline{\pi}_n(X, *) = \varinjlim \underline{\pi}_n(K, *)$$

where K varies over compact subsets of X containing $*$.

Then Proposition 8 yields easily:

Theorem 11. *Let Σ be a spectrum. The following equality defines a Steenrod homology theory $h_*(\cdot, \Sigma)$ on the category of pointed metrisable compact spaces, which agrees with the homology theory defined by $(*)$ on simplicial complexes*

$$h_n(X, \Sigma) = \varinjlim \underline{\pi}_{n+k}(X \wedge \Sigma_k).$$

Let us now return to noncommutative C^*-algebras.

B.ε ***E*-theory**

Let $\mathcal{K}$ be the elementary C^*-algebra of all compact operators on a separable ∞-dimensional Hilbert space. Since such a Hilbert space $\mathcal{H}$ is isomorphic to $\mathcal{H} \oplus \mathcal{H}$, one has a natural isomorphism

$$\rho : M_2(\mathcal{K}) \simeq \mathcal{K}$$

where $M_2(\mathcal{K}) = \mathcal{K} \otimes M_2(\mathbb{C})$ is the C^*-algebra of 2×2 matrices over $\mathcal{K}$. We let E be the category whose objects are separable C^*-algebras, while the set $E(A, B)$ of morphisms from A to B is

$$E(A, B) = [[SA \otimes \mathcal{K}, SB \otimes \mathcal{K}]]$$

i.e. the set of homotopy classes of asymptotic morphisms from $SA \otimes \mathcal{K} = A \otimes C_0(\mathbb{R}) \otimes \mathcal{K}$ to $SB \otimes \mathcal{K}$. The composition $E(A, B) \times E(B, C) \to E(A, C)$ is given by the composition of homotopy classes of asymptotic morphisms.

Using the above isomorphism $\rho : M_2(\mathcal{K}) \simeq \mathcal{K}$ one defines the sum $\varphi + \psi$ of elements φ and ψ of $E(A, B)$ as the asymptotic morphism $SA \otimes \mathcal{K} \to SB \otimes M_2(\mathcal{K}) = M_2(SB \otimes \mathcal{K})$ given by

$$\theta_t(a) = \begin{bmatrix} \varphi_t(a) & 0 \\ 0 & \psi_t(a) \end{bmatrix} \quad \forall a \in SA \otimes \mathcal{K}$$

with obvious notation.

One checks that with this operation E becomes an additive category. The opposite $-[\varphi]$ of a given element of $E(A, B)$ is obtained using the reflexion $s \to -s$ as an automorphism of $C_0(\mathbb{R})$. Let C^*-Alg be the category of separable C^*-algebras and $*$-homomorphisms. Then, let $j : C^*\text{-Alg} \to E$ be the functor which associates to $\varphi : A \to B$ the asymptotic morphism $\varphi_t = \varphi \otimes \text{id}$ from $SA \otimes \mathcal{K}$ to $SB \otimes \mathcal{K}$.

Theorem 12. [Co-H] (I) *The bifunctor $E(A, B)$ from the category C^*-Alg to the category of abelian groups is half-exact in each of its arguments.*

(II) *Any functor F from the category C^*-Alg to that of abelian groups which is unchanged by $A \to A \otimes \mathcal{K}$, homotopy invariant, and half-exact, factorises through the category E.*

The proof of (I) follows from Proposition 8. See [Co-H] for the proof of (II). As the functor j verifies the hypothesis of the next corollary, the latter gives a characterization of the category E.

Corollary 13. *Let $F : C^*\text{-Alg} \to Z$ be a functor to an additive category Z which is unchanged by $A \to A \otimes \mathcal{K}$, homotopy invariant, and half exact, as a bifunctor to abelian groups. Then F factorises uniquely through the category E.*

The E-theory is thus a concrete realisation of the category whose existence was proven by N. Higson in [Hig].

To get other corollaries of Theorem 12 one specialises to the functors $E(A, \cdot)$ and $E(\cdot, B)$ the following general properties of functors F from C^*-Alg to abelian groups, due to G. Kasparov [Kas_1] and J. Cuntz [Cu_3].

Lemma 14. [Kas_1] *Let F be a covariant, homotopy invariant, half-exact functor from the category C^*-Alg to abelian groups. Then there corresponds to any exact sequence: $0 \to J \to A \xrightarrow{p} B \to 0$ of separable C^*-algebras a long exact sequence of abelian groups*

$$\hookrightarrow F_{-n}(J) \to F_{-n}(A) \to F_{-n}(B) \ldots$$

$$\ldots \hookrightarrow F(J) \to F(A) \to F(B)$$

where $F_{-n}(A) = F(S^n A)$.

There is a similar dual statement for contravariant functors.

Of course, this lemma applies, in particular, to the functor $E(\cdot, D)$ and $E(D, \cdot)$ for a fixed C^*-algebra D and yields corresponding long exact sequences involving the functors $E(S^n \cdot, D)$ and $E(D, S^n \cdot)$. But in fact the situation is very much simplified by the built-in stability of E under the replacement of a C^*-algebra by its tensor product with $\mathcal{K}$. Indeed, one has natural isomorphisms in the E-category between $C_0(\mathbb{R}^2)$ and $\mathbb{C}$. Moreover, the suspension map S :

$E(A,B) \to E(SA,SB)$ is an isomorphism of abelian groups for any C^*-algebras A and B. Thus $E(S^k A, S^\ell B) \xrightarrow{\sim} E(A,B)$ for $k+\ell$ even and is equal to $E(A,SB) \simeq E(SA,B)$ for $k+\ell$ odd. We shall denote the latter group by $E^1(A,B)$. All these are variants of Bott periodicity, about which the cleanest statement is given by the following result of J. Cuntz, whose direct proof is simple ([Cu$_3$]).

Lemma 15. [Cu$_3$] *Let F be a functor from the category C^*-Alg to the category of abelian groups which is homotopy invariant, half-exact and unchanged by $A \to A \otimes \mathcal{K}$. Then there is a natural equivalence between $F(\cdot)$ and the double suspension $F(S^2 \cdot)$.*

The proof uses the Toeplitz C^*-algebra, i.e. the C^*-algebra τ generated by an isometry U, with $U^*U = 1$ and $UU^* \neq 1$. All such C^*-algebras are canonically isomorphic, and one has an exact sequence

$$0 \to \mathcal{K} \to \tau \to C(S^1) \to 0$$

where $C(S^1)$ is obtained as a quotient of τ by adding the relation $UU^* = 1$. Taking the kernel of the evaluation $C(S^1) \to \mathbb{C}$ at some point $* \in S^1$ yields an exact sequence

$$0 \to \mathcal{K} \to \tau_0 \xrightarrow{p} C_0(\mathbb{R}) \to 0.$$

Cuntz's proof consists in showing that $F(\tau_0) = \{0\}$. In fact the corresponding asymptotic morphism $\varepsilon_p \colon C_0(\mathbb{R}^2) \to \mathcal{K}$ is the Heisenberg deformation, and the proof shows that ε_p yields an E-theory isomorphism

$$C_0(\mathbb{R}^2) \xrightarrow{\sim} \mathbb{C}.$$

Finally, we note that for any C^*-algebra A one has a natural map from $K^1(A)$ to $E(C_0(\mathbb{R}), A \otimes \mathcal{K})$, which to a unitary $U \in (A \otimes \mathcal{K})^\sim$, with $\varepsilon(U) = 1$, associates the corresponding ordinary morphism from $C_0(\mathbb{R})$ to $A \otimes \mathcal{K}$. One then easily checks

Proposition 16. *The above map is a canonical isomorphism*

$$E^1(\mathbb{C}, A) \simeq K_1(A)\,,\ E^0(\mathbb{C}, A) \simeq K_0(A).$$

In particular, any E-theory class $y \in E(A,B)$ yields by composition a corresponding K-theory map

$$x \in K(A) = E(\mathbb{C}, A) \to y \circ x \in E(\mathbb{C}, B) = K(B).$$

For further developments of E-theory see [Dad$_{1,2}$], [D-L] and [Lor$_{1,2}$].

II. Appendix C. Crossed Products of C^*-algebras and the Thom Isomorphism

Let A be a C^*-algebra, G a locally compact group, and $\alpha : G \to \text{Aut}(A)$ a continuous action of G on A. Thus, for each $g \in G$, $\alpha_g \in \text{Aut}(A)$ is a $*$-automorphism of A, and for each $x \in A$ the map $g \mapsto \alpha_g(x)$ is norm continuous.

Definition 1. *A covariant representation π of (A, α) is given by unitary representations π_A of A, and π_G of G on a Hilbert space $\mathcal{H}$ such that*

$$\pi_G(g)\pi_A(x)\pi_G(g)^{-1} = \pi_A(\alpha_g(x))$$

$\forall g \in G, x \in A$.

Of course, the unitary representation of G is assumed to be strongly continuous ([Pe_2]). The above definition would continue to make sense if G were just a topological group, but in the locally compact case there exists a natural C^*-algebra, the crossed product $B = A \rtimes_\alpha G$, whose unitary representations correspond exactly to the covariant representations of (A, α). Indeed, let dg be a left Haar measure on G and let us endow the linear space $C_c(G, A)$ of continuous compactly supported maps from G to A with the following involutive algebra structure:

$$(f_1 * f_2)(g) = \int f_1(g_1)\, \alpha_{g_1}(f_2(g_1^{-1}g))\mathrm{d}g_1 \quad \forall f_j \in C_c(G, A), g \in G$$

$$(f^*)(g) = \delta(g)^{-1}\alpha_g(f(g^{-1})^*) \quad \forall f \in C_c(G, A), g \in G$$

where $\delta : G \to \mathbb{R}_+^*$ is the modular function of the not necessarily unimodular group G, i.e. the homomorphism from G to $\mathbb{R}_+^*$ defined by the equality

$$\mathrm{d}(g^{-1}) = \delta(g)^{-1}\mathrm{d}g.$$

These algebraic operations on $C_c(G, A)$ are uniquely prescribed in order to get an involutive representation $\tilde{\pi}$ of $C_c(G, A)$ from a covariant representation π of (A, α) by the following formula

$$\tilde{\pi}(f) = \int \pi_A(f(g))\pi_G(g)\mathrm{d}g \quad \forall f \in C_c(G, A).$$

It is not difficult to check then ([Pe_2]) that the completion $B = A \rtimes_\alpha G$ of $C_c(G, A)$ for the following norm is a C^*-algebra whose unitary representations correspond exactly to the covariant representations of (A, α):

$$\|f\| = \sup\{\|\tilde{\pi}(f)\|; \pi \text{ a covariant representation of } (A, \alpha)\}. \qquad (*)$$

Proposition 2. a) *The map $\pi \to \tilde{\pi}$ is a natural equivalence between the category of covariant representations of (A, α) and the category of representations of the C^*-algebra crossed product $A \rtimes_\alpha G$.*

b) Let $0 \to J \to A \to B \to 0$ be a G-equivariant exact sequence of C^*-algebras; then the crossed product sequence is exact:

$$0 \to J \rtimes G \to A \rtimes G \to B \rtimes G \to 0.$$

To prove b) one just notices that the covariant representations of A vanishing on $J \rtimes G$ are exactly the covariant representations of B.

In order to define the *reduced* crossed product, $A \rtimes_{\alpha,r} G$, which differs from the above only when G fails to be amenable (in fact when the *action* of G fails to be amenable), let us consider the (right) C^*-module over A given by

$$\mathcal{E} = L^2(G, dg) \otimes A.$$

We can view $\mathcal{E}$ as the completion, for the norm $\|\xi\| = \|\langle \xi, \xi \rangle\|^{1/2}$ of $C_c(G, A)$ with A-valued inner product given by

$$\langle \xi, \eta \rangle = \int \xi(g)^* \eta(g) dg \in A \quad \forall \xi, \eta \in C_c(G, A)$$

while the right action of A is given by

$$(\xi a)(g) = \xi(g) a \quad \forall \xi \in C_c(G, A), a \in A, g \in G.$$

We then define the following left-regular representation of (A, α) as endomorphisms of $\mathcal{E}$

$$(\pi(a)\xi)(g) = \alpha_{g^{-1}}(a)\xi(g) \quad \forall \xi \in C_c(G, A), a \in A, g \in G$$

$$(\pi(g)\xi)(k) = \xi(g^{-1}k) \quad \forall \xi \in C_c(G, A), \ g, k \in G.$$

One checks that these formulae define elements of $\mathrm{End}_A(\mathcal{E})$, and one thus gets a natural representation

$$\lambda : A \rtimes_\alpha G \to \mathrm{End}_A(\mathcal{E}).$$

The image of λ is a C^*-algebra, called the *reduced crossed product* of A by α and denoted

$$A \rtimes_{\alpha,r} G.$$

When $A = \mathbb{C}$ the reduced crossed product $\mathbb{C} \rtimes_r G$ is, of course, the reduced C^*-algebra of G, $C_r^*(G)$, i.e. the C^*-algebra generated in the left regular representation of G by the left action of the convolution algebra $C_c(G)$ or equivalently by $L^1(G, dg)$. Similarly, $\mathbb{C} \rtimes G$ is the C^*-algebra $C^*(G)$ of G. The nuance between these two C^*-algebras exists only in the non-amenable case. For instance, when G is a semisimple Lie group the unitary representations of G which come from representations of $C_r^*(G)$ form the support of the Plancherel measure and are called tempered representations.

Proposition 3. [Pe$_2$] *When G is amenable the representation λ of $A \rtimes_\alpha G$ in $\mathrm{End}_A(\mathcal{E})$ is injective.*

We shall now specialise to the case when G is abelian and describe the Takesaki-Takai duality theorem for crossed products [T$_0$]. Since abelian groups are amenable, the distinction between the two crossed products disappears. Let $\hat{G}$ be the Pontryagin dual of G, and let $\langle g, g' \rangle$, for $g \in G$ and $g' \in \hat{G}$, be the canonical pairing with values in $\mathbb{T} = \{z \in \mathbb{C}; |z| = 1\}$.

Proposition 4. [T$_0$] *Let G be an abelian locally compact group, α an action of G on a C^*-algebra A. The following equality defines a canonical action $\hat{\alpha}$ of the Pontryagin dual $\hat{G}$ of G on the crossed product $A \rtimes_\alpha G$:*

$$(\hat{\alpha}_{g'}(f))(g) = \langle g, g' \rangle f(g) \quad \forall f \in C_c(G, A) \subset A \rtimes_\alpha G \;,\; g \in G, g' \in \hat{G}.$$

Indeed, one easily checks that the norm $(*)$ is preserved by the $\hat{\alpha}_{g'}$, which are obviously automorphisms of the involutive algebra $C_c(G, A)$. Also one can implement the automorphisms $\hat{\alpha}_{g'}$ in the representation λ of $A \rtimes_\alpha G$ as endomorphisms of $\mathcal{E} = L^2(G, dg) \otimes A$. One defines a representation ρ of $\hat{G}$ as automorphisms of $\mathcal{E}$ by the equality

$$(\rho(g')\xi)(g) = \langle g, g' \rangle \xi(g) \quad \forall \xi \in C_c(G, A) \;;\; g \in G, g' \in \hat{G}.$$

One then checks that the pair (λ, ρ) is a covariant representation of $(A \rtimes_\alpha G, \hat{\alpha})$ as endomorphisms of the C^*-module $\mathcal{E}$. We thus obtain a homomorphism $\hat{\lambda} : ((A \rtimes_\alpha G) \rtimes_{\hat{\alpha}} \hat{G} \to \mathrm{End}_A(\mathcal{E})$, and one easily checks that the range of $\hat{\lambda}$ consists of *compact* endomorphisms (cf. Appendix A). The main content of the Takesaki-Takai duality theorem is that $\hat{\lambda}$ is actually an *isomorphism*

$$\hat{\lambda} : (A \rtimes_\alpha G) \rtimes_{\hat{\alpha}} \hat{G} \simeq \mathrm{End}_A^0(\mathcal{E}) = A \otimes \mathcal{K}$$

where $\mathcal{K}$ is the elementary C^*-algebra of compact operators in $L^2(G, dg)$. To state this duality with the required precision we need the following definition (compare with Chapter V).

Definition 5. *Let A be a C^*-algebra, G a locally compact group, and α, α' two actions of G on A. Then α and α' are outer equivalent iff there exists a map $g \to u_g$ from G to the group $\mathcal{U}$ of unitary automorphisms of the C^*-module A over A such that:*

1) *$g \to u_g \xi$ is norm continuous for any $\xi \in A$.*
2) *$u_{g_1 g_2} = u_{g_1} \alpha_{g_1}(u_{g_2}) \quad \forall g_1, g_2 \in G$.*
3) *$\alpha'_g(a) = u_g \alpha_g(a) u_g^* \quad \forall a \in A, g \in G$.*

The endomorphisms $\mathrm{End}_A(A)$ of the C^*-module A over A are called the multipliers of A (cf. [Pe$_2$]) and denoted $M(A)$. The continuity invoked in 1) is called strong continuity. An automorphism $\beta \in \mathrm{Aut}(A)$ of the form $\beta(a) = uau^*$, where u is a unitary element of $M(A)$, is called an *inner* automorphism. We thus see that if α and α' are outer equivalent then $\alpha'_g \alpha_g^{-1}$ is inner for any $g \in G$, but the converse does not hold in general. The crossed product $A \rtimes_\alpha G$

of the C^*-algebra A by the action α of G is unaffected if one replaces α by an outer equivalent action α'. The canonical isomorphism

$$A \rtimes_\alpha G \overset{\theta}{\simeq} A \rtimes_{\alpha'} G$$

is given by the formula

$$(\theta(f))(g) = u_g f(g) \quad \forall f \in C_c(G, A).$$

Theorem 6. [T_0] *Let G be an abelian locally compact group, and let α be an action of G on a C^*-algebra A. There is a canonical isomorphism $\hat{\lambda}$ of the double crossed product $(A \rtimes_\alpha G) \rtimes_{\hat{\alpha}} \hat{G}$ with $A \otimes \mathcal{K}$ which transforms the double dual action $\hat{\hat{\alpha}}$ of G into an action outer equivalent to the action $\alpha \otimes 1$ of G on $A \otimes \mathcal{K}$.*

Note that this theorem could not even be formulated had we decided to only consider commutative C^*-algebras. Indeed, even for $A = \mathbb{C}$ it expresses the fact that the C^*-algebra $C^*(G) \rtimes \hat{G}$ is isomorphic to the noncommutative elementary C^*-algebra of compact operators. We shall see in particular how it immediately implies a far-reaching analogue of the Bott periodicity theorem [Co_{20}] (motivated by [Pi-V_1]). We specialise to the case $G = \mathbb{R}$, the additive group of real numbers, and our first task is to construct a natural map

$$\phi_\alpha : K_i(A) \to K_{i+1}(A \rtimes_\alpha \mathbb{R})$$

for any C^*-algebra A with one-parameter automorphism group α. When the action α of $\mathbb{R}$ on A is trivial, $\alpha_s = \mathrm{id}\ \forall s \in \mathbb{R}$, the crossed product $A \rtimes_\alpha \mathbb{R}$ is canonically isomorphic to $SA = A \otimes C_0(\hat{\mathbb{R}})$ and the map ϕ_α is the natural suspension isomorphism

$$K_i(A) \simeq K_{i+1}(SA).$$

To define ϕ_α one reduces the general situation to the particular case of the trivial action using the following lemma, whose geometric meaning is the absence of curvature in the *one-dimensional* situation.

Lemma 7. [Co_{20}] *Let A be a C^*-algebra, α a one-parameter group of automorphisms of A, $e = e^* = e^2$ a self-adjoint projection, $e \in A$, of A. There exists an equivalent projection $f \sim e$, $f \in A$, and an outer equivalent action α' of $\mathbb{R}$ on A such that $\alpha'_t(f) = f\ \forall t \in \mathbb{R}$.*

One first replaces e by an equivalent projection f such that the map $t \to \alpha_t(f) \in A$ is of class C^∞, and then one replaces the derivation $\delta = \left(\frac{d}{dt}\alpha_t\right)_{t=0}$ which generates α by the new derivation $\delta' = \delta + \mathrm{ad}(h)$ where $\mathrm{ad}(h)x = hx - xh\ \forall x \in A$, and $h = f\delta(f) - \delta(f)f$.

From Lemma 7 and the canonical isomorphism $A \rtimes_\alpha \mathbb{R} \simeq A \rtimes_{\alpha'} \mathbb{R}$ for outer equivalent actions one gets the construction of

$$\phi_\alpha^0 : K_0(A) \to K_1(A \rtimes_\alpha \mathbb{R}).$$

Replacing A by SA, one gets similarly

$$\phi_\alpha^1 : K_1(A) \to K_0(A \rtimes_\alpha \mathbb{R}).$$

Theorem 8. [Co$_{20}$] *Let A be a C^*-algebra, and let α be a one-parameter group of automorphisms of A. Then $\phi_\alpha : K_i(A) \to K_{i+1}(A \rtimes_\alpha \mathbb{R})$ is an isomorphism of abelian groups for $i = 0, 1$. The composition $\phi_{\hat{\alpha}} \circ \phi_\alpha$ is the canonical isomorphism of $K_i(A)$ with $K_i(A \otimes \mathcal{K})$, where the double crossed product is identified with $A \otimes \mathcal{K}$ by Theorem 6.*

The proof is simple since it is clearly enough to prove the second statement, and to prove it only for $i = 0$. One then uses Lemma 7 to reduce to the case $A = \mathbb{C}$, where it is easy to check ([Co$_{20}$]).

The above theorem immediately extends to arbitrary simply connected solvable Lie groups H in place of $\mathbb{R}$.

II. Appendix D. Penrose Tilings

In this section, I review for the reader's convenience the classical results ([Gru-S]) on R. Robinson and R. Penrose's quasiperiodic tilings of the plane.

One first considers two types, L_A and S_A, of triangular tiles as represented in Figure 10. The vertices are colored white or black, and the edge between two vertices of the same color is oriented. The tile L_A has two black vertices and one white, whereas the tile S_A has two whites and a black.

A tiling of type A of the plane is defined to be a triangulation of the plane by triangles isometric to L_A or S_A, in such a way that the colors of common vertices, and the orientations of common edges, are the same.

The reader is referred to [Gru-S] for the proof of the existence of such tilings of the plane (see Figure 3 in Section 3 of this chapter). We shall say that two tilings T and T' are identical if they can be obtained from each other by an isometry of the Euclidean plane.

Let X be the set of all type A tilings (up to isometry) of the plane. The essential result we shall use is the possibility of parametrizing X by the set K of infinite sequences $(a_n)_{n\in\mathbb{N}}$, $a_n \in \{0, 1\}$, of zeros and ones, satisfying

$$a_n = 1 \Longrightarrow a_{n+1} = 0$$

in such a way that every tiling T of type A arises from a sequence, $T = T(a)$, and two sequences a and b yield the same tiling if and only if there exists an index N such that $a_n = b_n$ for all $n \geq N$. As this is an important point, I shall describe the correspondence $a \mapsto T(a)$ explicitly ([Gru-S], p. 568).

If T is a tiling of type A, and if we delete from the triangulation T all the short edges that join two vertices of different colors and separate an L_A-tile

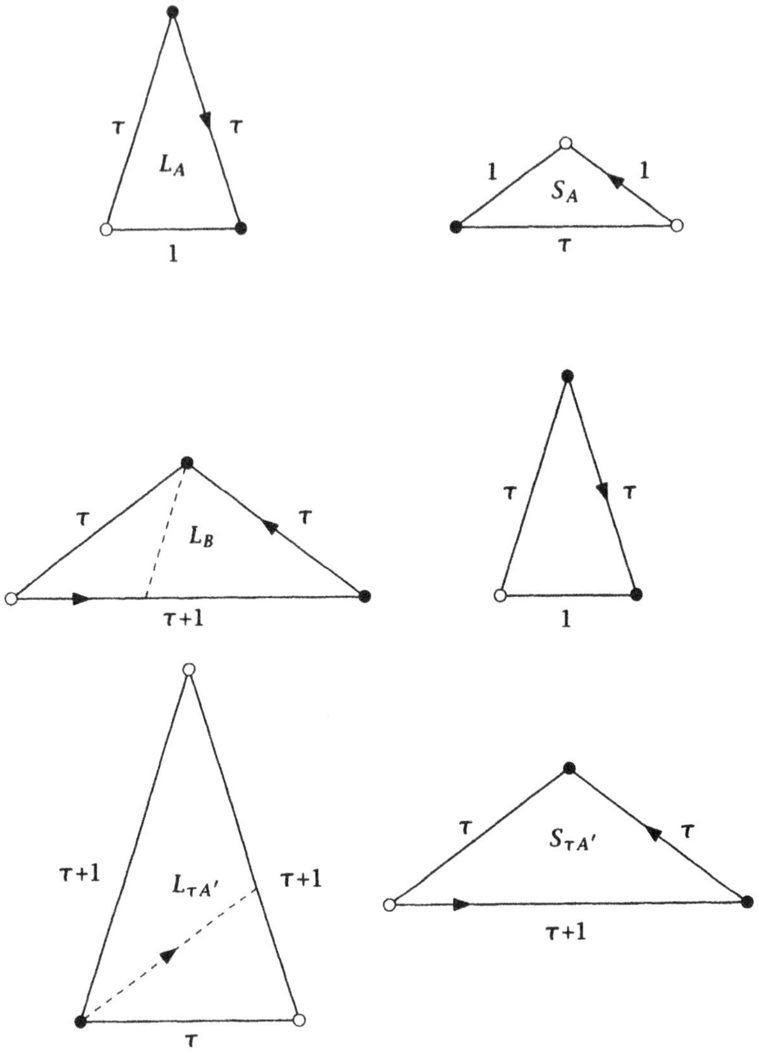

Figure II.10. Elementary tiles (see [Gru-S])

from an S_A-tile (Figure 11), we obtain a new triangulation T_1 of the plane whose triangles are isometric to one of the two triangles L_B, S_B of Figure 10. We obtain in this way a tiling of the plane of type B, that is, a triangulation of the plane by triangles isometric to L_B or S_B, such that the colors of common vertices, and the orientations of common edges, are the same. If, in the triangulation T_1, we delete all the edges that join two vertices of the same color and separate a L_B-tile from a S_B-tile, we obtain a new triangulation T_2 whose triangles are isometric to one of the triangles $L_{\tau A'}$, $S_{\tau A'}$ of Figure 10. Iterating the procedure, we obtain in this way a sequence T_n of triangulations of the plane. To each of them there correspond triangles L_n and S_n.

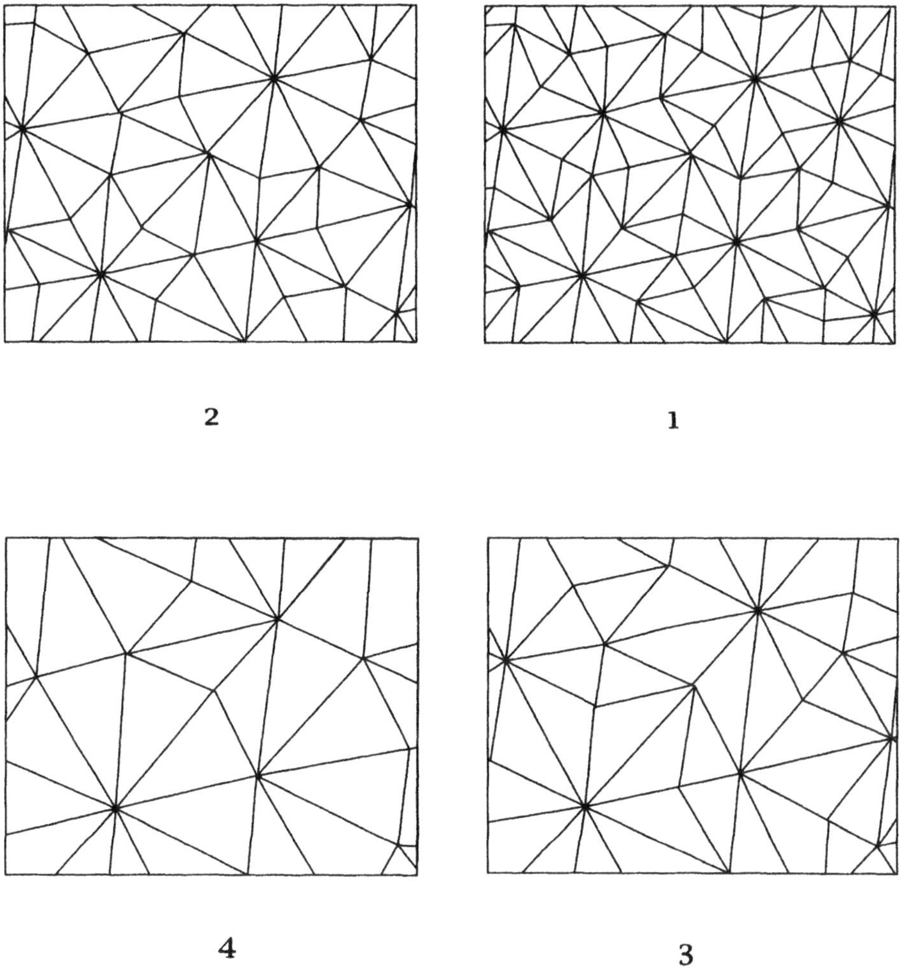

Figure II.11. A sequence of triangulation patches (see [Gru-S])

Given a tiling T of type A and a triangle α of the triangulation T (Figure 11), we associate with the latter the sequence $(a_n)_{n\in\mathbb{N}}$, where a_n is 0 or 1 according as the triangle of T_n that contains α is large (i.e., isometric to L_n) or small. We denote by $i(T,\alpha)$ the sequence obtained in this way.

We then denote by K the set of sequences $(a_n)_{n\in\mathbb{N}}$, $a_n \in \{0,1\}$, such that $a_n = 1 \Longrightarrow a_{n+1} = 0$. It is clear that every sequence of the form $i(T,\alpha)$ belongs to K, because the triangle S_n is not used for constructing $S_{n+1} = L_n$. Conversely ([Gru-S], p. 568), one shows that every element a of K is of the form

$i(T, \alpha)$ for a suitable tiling of type A and a suitable α.

Two triangles α, β of a tiling T occur, for n sufficiently large, in the same triangle of T_n. It follows that the sequences $a = i(T, \alpha)$ and $b = i(T, \beta)$ satisfy the condition

$$a_m = b_m \quad \forall m \geq n. \tag{$*$}$$

Conversely ([Gru-S], p.568), one shows that if $a = i(T, \alpha)$ and $b \in K$ satisfy $(*)$, then there exists a triangle β of T such that $b = i(T, \beta)$.

One obtains in this way a bijection between the set X of Penrose tilings and the quotient $K/\mathcal{R}$ of the set K by the equivalence relation $\mathcal{R}$ defined by the relation $(*)$ ([Gru-S], p.568).

III

CYCLIC COHOMOLOGY AND DIFFERENTIAL GEOMETRY

In this chapter we shall extend de Rham homology beyond its usual commutative framework to obtain numerical invariants of K-theory classes on noncommutative spaces.

In the commutative case, $A = C(X)$ for X a compact space, we have at our disposal in K-theory a tool of capital importance, the Chern character

$$\mathrm{ch} : K^*(X) \to H^*(X, \mathbb{Q})$$

which relates the K-theory of X to the cohomology of X. When X is a smooth manifold the Chern character may be calculated explicitly by the differential calculus of forms, currents, connections and curvature ([Mi_4]). More precisely, we have first the isomorphism

$$K_0(\mathcal{A}) \simeq K_0(A)$$

where $\mathcal{A} = C^\infty(X)$ is the algebra of smooth functions on X, a dense subalgebra of the C^*-algebra $C(X) = A$. Then, given a smooth vector bundle E over X, or equivalently the finite projective module, $\mathcal{E} = C^\infty(X, E)$ over $\mathcal{A} = C^\infty(X)$ of smooth sections of E, the Chern character of E

$$\mathrm{ch}(E) \in H^*(X, \mathbb{R})$$

is represented by a closed differential form:

$$\mathrm{ch}(E) = \text{ trace } (\exp(\nabla^2/2\pi i))$$

for any connection ∇ on the vector bundle E. Any closed de Rham current C on the manifold X determines a map φ_C from $K^*(X)$ to $\mathbb{C}$ by the equality

$$\varphi_C(E) = \langle C, \mathrm{ch}(E) \rangle$$

where the pairing between currents and differential forms is the usual one.

One obtains in this way numerical invariants of K-theory classes whose knowledge for arbitrary closed currents C is equivalent to that of $\mathrm{ch}(E)$.

In this chapter we shall adapt the above classical construction to the noncommutative case. This requires defining the analogue of the de Rham homology, which was introduced in [Co$_{17}$] under the name of cyclic cohomology, for arbitrary noncommutative algebras. This first step is *purely algebraic*: One starts with an algebra $\mathcal{A}$ over $\mathbb{C}$, which plays the role that $C^\infty(X)$ had in the commutative case, and one develops the analogue of the de Rham homology of X, the pairing with the algebraic K-groups $K_0(\mathcal{A})$ and $K_1(\mathcal{A})$, and tools, such as connections and curvature, to perform the computations.

The resulting theory is a *contravariant* functor HC^* from noncommutative algebras to graded modules over the polynomial ring $\mathbb{C}[\sigma]$ with one generator σ of degree 2. Being contravariant for algebras means that the above functor is covariant for the corresponding "space", whence its *homological* nature. In the definition of this functor the finite cyclic groups play a crucial role and this is why $HC^*(\mathcal{A})$ is called the cyclic cohomology of the algebra $\mathcal{A}$.

We shall expound the theory below in a way suitable for the applications we have in mind, but we refer the reader to several books [Kast$_1$] [Kar$_3$] [Lo] for more information.

The second step, crucial in order to apply the above construction to the K-theory of C^*-algebras of noncommutative spaces, involves *analysis*.

Thus, the noncommutative algebra $\mathcal{A}$ is now a dense subalgebra of a C^*-algebra A and the problem is, given a cyclic cocycle C on $\mathcal{A}$ as above, satisfying a suitable condition relative to A, to extend the K-theory invariant

$$\varphi_C : K_0(\mathcal{A}) \to \mathbb{C}$$

to a map from $K_0(A)$ to $\mathbb{C}$.

In the simplest situation the algebra $\mathcal{A} \subset A$ is stable under holomorphic functional calculus (cf. Appendix C; this means, essentially, that if $a \in \mathcal{A}$ is invertible in A its inverse is in $\mathcal{A}$). The inclusion $\mathcal{A} \subset A$ is then an isomorphism in K-theory and the above problem of extension of φ_C is trivially solved.

This method still applies when $\mathcal{A}$ is not stable under holomorphic functional calculus but when the cocycle C extends to the closure $\overline{\mathcal{A}}$ of $\mathcal{A}$ under this calculus. This will apply to the cyclic cocycles associated to bounded group cocycles on Gromov's word hyperbolic groups.

However, a new step is required to treat the transverse fundamental class for the leaf space $X = V/F$ of a transversely oriented foliation, or for the orbit space $X = W/\Gamma$, where Γ is a discrete group acting by orientation-preserving diffeomorphisms of a manifold W.

The main difficulty to be overcome is that, in general, the transverse bundle of a foliation admits no holonomy-invariant metric. Similarly, an action by diffeomorphisms on a manifold does not preserve any Riemannian metric (if so it would then factorize through the Lie group of isometries).

It follows then that the cyclic cocycle, which is the transverse fundamental class, is very hard to control with respect to the C^*-algebra norm.

However, we have already met a similar situation in Chapter I, in the measure theory discussion, where the type III situation was precisely characterized by the absence of a holonomy-invariant volume element on the transverse bundle. We saw, moreover, that one could reduce type III to type II by replacing the noncommutative space X with the total space Y of a principal $\mathbb{R}_+^*$-bundle over X, whose fiber Y_L over each point $L \in X$ is the space of transverse volume elements at this point.

We adapted this method to overcome the above difficulty, through the following steps:

α) To the problem of finding a holonomy-invariant metric on the transverse bundle τ ($\dim \tau = q$) corresponds a principal $GL(q,\mathbb{R})$-bundle over the noncommutative space X. We let Y be the total space of the bundle over X associated to the action of $GL(q,\mathbb{R})$ on $GL(q,\mathbb{R})/O(q,\mathbb{R})$. Then the transverse structure group of Y reduces to a group of *triangular matrices with orthogonal diagonal blocks.*

β) Using the above reduction of the transverse structure group of Y we control the cyclic cocycle associated to the transverse fundamental class of Y.

γ) We use the fact that the symmetric space $H = GL(q,\mathbb{R})/\,O(q,\mathbb{R})$, which is the fiber of the projection

$$p : Y \longrightarrow X$$

has non-positive curvature and the Kasparov bivariant theory to construct a map $K(C^*(X)) \longrightarrow K(C^*(Y))$ of K-theories, whose composition with the map

$$\varphi_C : K(C^*(Y)) \longrightarrow \mathbb{C}$$

given by the fundamental class of Y will yield that of X.

In the context of foliations we shall thus construct K-theory invariants

$$K(C_r^*(V,F)) \xrightarrow{\varphi_C} \mathbb{C}$$

making the following diagram commutative:

$$\begin{array}{ccc} K_{*,\tau}(BG) & \xrightarrow{\mu} & K(C_r^*(V,F)) \\ \downarrow \mathrm{ch}_* & & \downarrow \varphi_{C(\omega)} \\ H_*(BG,\mathbb{R}) & \xrightarrow{\omega} & \mathbb{C} \end{array}$$

where μ is the analytic assembly map of Chapter II, and $\omega \in H^*(BG,\mathbb{R})$ is any cohomology class on the classifying space BG of the holonomy groupoid G of

(V, F) belonging to the subring $\mathcal{R} \subset H^*(BG, \mathbb{R})$ generated by the Pontryagin classes of the transverse bundle, the Chern classes of any holonomy equivariant vector bundle on V and the pull-back of the Gel'fand-Fuchs characteristic classes $y \in H^*(WO_q)$.

As applications of the above construction we shall get geometric corollaries which no longer involve C^*-algebras in their formulation, namely:

1) *The Novikov conjecture is true for Gromov's word hyperbolic groups* ([Co-M$_1$]).

2) *The Gel'fand-Fuchs cocycles on* Diff(M) *satisfy the Novikov conjecture* ([Co$_{14}$] [Co-G-M]).

3) *Let M be a compact oriented manifold and assume that the rational number $\hat{A}(M)$ is non-zero (since M is not assumed to be a spin manifold, $\hat{A}(M)$ need not be an integer). Let F be an integrable spin subbundle of TM. There exists no metric on F for which the scalar curvature (of the leaves) is strictly positive ($\geq \varepsilon > 0$) on M.*

Moreover, we shall see, in Section 6 of this chapter, how the Godbillon-Vey class appears naturally in our setup from the lack of invariance of the transverse fundamental class $[V/F]$ of a codimension-1 foliation under the modular automorphism group σ_t of the von Neumann algebra of V/F (Chapter I). In particular, we shall obtain the following corollary showing, in the noncommutative case, the depth of the interplay between characteristic classes and measure theory:

4) *Let M be the von Neumann algebra of the foliation (V, F) of codimension one. If the Godbillon-Vey class $GV \in H^3(V, \mathbb{R})$ is nonzero, then the flow of weights* mod(M) *admits a finite invariant measure.*

This implies in particular the remarkable result of S. Hurder [Hu-K$_1$] that, under the hypothesis of the theorem, the von Neumann algebra M is *not semifinite* and is of type III if the foliation is ergodic.

III.1. Cyclic Cohomology

Given any (possibly noncommutative) algebra $\mathcal{A}$ over $\mathbb{C}$, $HC^*(\mathcal{A})$ is the cohomology of the complex (C^n_λ, b), where C^n_λ is the space of $(n+1)$-linear functionals φ on $\mathcal{A}$ such that

$$\varphi(a^1, \ldots, a^n, a^0) = (-1)^n \varphi(a^0, \ldots, a^n) \qquad \forall a^i \in \mathcal{A}$$

and where b is the Hochschild coboundary map given by

$$(b\varphi)(a^0, \ldots, a^{n+1}) = \sum_{j=0}^{n} (-1)^j \varphi(a^0, \ldots, a^j a^{j+1}, \ldots, a^{n+1}) + (-1)^{n+1} \varphi(a^{n+1} a^0, \ldots, a^n).$$

We shall develop the main properties of this cohomology theory for algebras in the following sections.

1.α Characters of cycles and the cup product in HC^*

In order to clarify the geometric significance of the basic operations, such as B and the cup product $\varphi \# \psi$ of cyclic cohomology, we shall base our discussion on the following notion:

Definition 1. a) *A cycle of dimension n is a triple $(\Omega, d, \int)$ where $\Omega = \bigoplus_{j=0}^{n} \Omega^j$ is a graded algebra over $\mathbb{C}$, d is a graded derivation of degree 1 such that $d^2 = 0$, and $\int : \Omega^n \to \mathbb{C}$ is a closed graded trace on Ω.*

b) *Let $\mathcal{A}$ be an algebra over $\mathbb{C}$. Then a cycle over $\mathcal{A}$ is given by a cycle $(\Omega, d, \int)$ and a homomorphism $\rho : \mathcal{A} \to \Omega^0$.*

As we shall see below, a cycle of dimension n over $\mathcal{A}$ is, essentially, determined by its *character*, the $(n+1)$-linear function τ,

$$\tau(a^0, \ldots, a^n) = \int \rho(a^0)\, d(\rho(a^1))\, d(\rho(a^2)) \cdots d(\rho(a^n)) \qquad \forall a^j \in \mathcal{A}$$

and the functionals thus obtained are exactly the elements of $\ker b \cap C^n_\lambda$.

Given two cycles Ω and Ω' of dimension n, their sum $\Omega \oplus \Omega'$ is defined as the direct sum of the two differential graded algebras, with $\int(\omega, \omega') = \int \omega + \int \omega'$. Given cycles Ω and Ω' of dimensions n and n', their tensor product $\Omega'' = \Omega \otimes \Omega'$ is the cycle of dimension $n + n'$, which, as a differential graded algebra, is the tensor product of (Ω, d) by (Ω', d'), and where

$$\int (\omega \otimes \omega') = (-1)^{nn'} \int \omega \int \omega' \qquad \forall \omega \in \Omega\,,\ \omega' \in \Omega'.$$

One defines in the corresponding ways the notions of direct sum and tensor product of cycles over $\mathcal{A}$.

Example 2. a) Let V be a smooth compact manifold, and let C be a closed de Rham current of dimension q ($\leq \dim V$) on V. Let Ω^i, $i \in \{0, \ldots, q\}$, be the space $C^\infty(V, \wedge^i T^*V)$ of smooth differential forms of degree i. With the usual product structure and differentiation, $\Omega = \bigoplus_{i=0}^{q} \Omega^i$ is a differential algebra, on which the equality $\int \omega = \langle C, \omega \rangle$, for $\omega \in \Omega^q$, defines a closed graded trace.

b) Let V be a smooth oriented manifold, and $(x, g) \in V \times \Gamma \to xg \in V$ an action, by orientation preserving diffeomorphisms, of the discrete group Γ on V. Let $A^*(V)$ be the graded differential algebra $C_c^\infty(V, \wedge^* T_{\mathbb{C}}^*)$ of smooth differential forms on V with compact support. The group Γ acts by automorphisms of this differential graded algebra. For any $g \in \Gamma$, let ψ_g be the associated diffeomorphism $\psi_g(x) = xg$ $\forall x \in V$; then

$$g\omega = \psi_g^* \omega \quad \forall \omega \in A^*(V).$$

It follows that the algebraic crossed product

$$\Omega^* = A^*(V) \rtimes \Gamma$$

is also a graded differential algebra. Let us describe it in more detail. As a linear space, Ω^p is the space $C_c^\infty(V \times \Gamma, \wedge^p T_{\mathbb{C}}^*)$ of smooth forms with compact support on the (disconnected) manifold $V \times \Gamma$. Algebraically, it is convenient to write such a form as a finite sum

$$\omega = \sum_{g \in \Gamma} \omega_g U_g , \quad \omega_g \in A^*(V)$$

where the U_g's are symbols. The algebraic rules are then:

$$\left(\sum \omega_g U_g\right)\left(\sum \omega'_k U_k\right) = \sum \omega_g \wedge g(\omega'_k) U_{gk}$$

$$d\left(\sum \omega_g U_g\right) = \sum (d\omega_g) U_g$$

$$\int \sum \omega_g U_g = \int_V \omega_e \quad (e \text{ the unit of } \Gamma).$$

The invariance under diffeomorphisms (preserving the orientation) of the integral of top-dimensional forms then shows that the triple $(\Omega, d, \int)$ defines a cycle of dimension $n = \dim V$ over the algebra $C_c^\infty(V) \rtimes \Gamma$.

c) Let Γ be a discrete group, $\mathcal{A} = \mathbb{C}\Gamma$ the group ring of Γ over $\mathbb{C}$. Let $\Omega^*(\Gamma)$ be the graded differential algebra of finite linear combinations of symbols

$$g_0 dg_1 dg_2 \cdots dg_n , \quad g_i \in \Gamma$$

with product and differential given by

$$\begin{aligned}(g_0 dg_1 \cdots dg_n)&(g_{n+1} dg_{n+2} \cdots dg_m) \\ &= \sum_{j=1}^{n} (-1)^{n-j} g_0 dg_1 \cdots d(g_j g_{j+1}) \cdots dg_n\, dg_{n+1} \cdots dg_m \\ &\qquad + (-1)^n g_0 g_1 dg_2 \cdots dg_m , \\ d(g_0 dg_1 \cdots dg_n) &= dg_0 dg_1 \cdots dg_n .\end{aligned}$$

Then any *normalized* group cocycle $c \in Z^k(\Gamma, \mathbb{C})$ determines a k-dimensional cycle through the following closed graded trace on $\Omega^*(\Gamma)$:

$$\int g_0 dg_1 \cdots dg_n = 0 \quad \text{unless } n = k \text{ and } g_0 g_1 \cdots g_n = 1,$$

$$\int g_0 dg_1 \cdots dg_k = c(g_1, \ldots, g_k) \quad \text{if } g_0 \cdots g_k = 1.$$

We recall that group cohomology $H^*(\Gamma, \mathbb{C})$ is by definition the cohomology of the classifying space $B\Gamma$, or equivalently of the complex (C^*, b), where C^p is the space of all functions $\gamma : \Gamma^{p+1} \to \mathbb{C}$ such that

$$\gamma(gg_0, \ldots, gg_p) = \gamma(g_0, \ldots, g_p) \ \forall g, g_j \in \Gamma$$

$$(b\gamma)(g_0, \ldots, g_{p+1}) = \sum_{i=0}^{p+1} (-1)^i \gamma(g_0, \ldots, g_{i-1}, g_{i+1}, \ldots, g_{p+1}).$$

The group cocycle associated to $\gamma \in C^k$, $b\gamma = 0$, is given by

$$c(g_1, \ldots, g_k) = \gamma(1, g_1, g_1 g_2, \ldots, g_1 g_2 \cdots g_k).$$

The *normalization* required above is the following:

$$c = 0 \text{ if any } g_i = 1 \text{ or if } g_1 \cdots g_k = 1.$$

Any group cocycle can be normalized, without changing its cohomology class, because the above complex can be replaced, without altering its cohomology, by the subcomplex of skew-symmetric cochains for which, for all $g_i \in \Gamma$ and $\sigma \in S_{p+1}$,

$$\gamma^\sigma(g_0, \ldots, g_n) = \gamma(g_{\sigma(0)}, \ldots, g_{\sigma(p)}) = \text{Sign}(\sigma)\, \gamma(g_0, \ldots, g_p).$$

In this last example c) the differential algebra $\Omega^*(\Gamma)$ is independent of the choice of the cocycle c. The construction of $\Omega^*(\Gamma)$ from the group ring $\mathcal{A} = \mathbb{C}\Gamma$ is a special case of the *universal differential algebra* $\Omega^*(\mathcal{A})$ associated to an algebra $\mathcal{A}$ ([Arv$_3$] [Kar$_3$]), which we briefly recall.

Proposition 3. *Let $\mathcal{A}$ be a not necessarily unital algebra over $\mathbb{C}$.*

1) *Let $\Omega^1(\mathcal{A})$ be the linear space $\tilde{\mathcal{A}} \otimes_{\mathbb{C}} \mathcal{A}$, where $\tilde{\mathcal{A}} = \mathcal{A} \oplus \mathbb{C}1$ is the algebra obtained by adjoining a unit to $\mathcal{A}$. Then the following equalities define the structure of an $\mathcal{A}$*-bimodule *on $\Omega^1(\mathcal{A})$ and a derivation $d : \mathcal{A} \to \Omega^1(\mathcal{A})$: for any $a, b, x, y \in \mathcal{A}$, $\lambda \in \mathbb{C}$*

$$x((a + \lambda 1) \otimes b)y = (xa + \lambda x) \otimes by - (xab + \lambda xb) \otimes y$$

$$da = 1 \otimes a \in \Omega^1(\mathcal{A}) \quad \forall a \in \mathcal{A}.$$

2) *Let $\mathcal{E}$ be an $\mathcal{A}$-bimodule and $\delta : \mathcal{A} \to \mathcal{E}$ a derivation; then there exists a bimodule morphism $\rho : \Omega^1(\mathcal{A}) \to \mathcal{E}$ such that $\delta = \rho \circ d$.*

Thus, $(\Omega^1(\mathcal{A}), d)$ is the universal derivation of $\mathcal{A}$ in an $\mathcal{A}$-bimodule.

The universal graded differential algebra $\Omega^*(\mathcal{A})$ is then obtained by letting $\Omega^n(\mathcal{A}) = \Omega^1(\mathcal{A}) \otimes_{\mathcal{A}} \Omega^1(\mathcal{A}) \cdots \otimes_{\mathcal{A}} \Omega^1(\mathcal{A})$ be the n-fold tensor product of the bimodule $\Omega^1(\mathcal{A})$, while the differential $d : \mathcal{A} \to \Omega^1(\mathcal{A})$ extends uniquely to a square-zero graded derivation of that tensor algebra. Remark that one has a natural isomorphism of *linear spaces*

$$j : \tilde{\mathcal{A}} \otimes \mathcal{A}^{\otimes n} \to \Omega^n(\mathcal{A})$$

with $j((a^0 + \lambda 1) \otimes a^1 \otimes \cdots \otimes a^n) = a^0 da^1 \cdots da^n + \lambda da^1 da^2 \cdots da^n \ \forall a^j \in \mathcal{A}$, $\lambda \in \mathbb{C}$.

(Note that the cohomology of the complex $(\Omega^*(\mathcal{A}), d)$ is 0 in all dimensions, including 0 if we set $\Omega^0(\mathcal{A}) = \mathcal{A}$.)

The product in $\Omega^*(A)$ is given in a way analogous to that in $\Omega^*(\Gamma)$ above; thus

$$\begin{aligned}(a^0da^1\cdots da^n)(a^{n+1}da^{n+2}\cdots da^m)\\ =\sum_{j=1}^{n}(-1)^{n-j}a^0da^1\cdots d(a^ja^{j+1})\cdots da^nda^{n+1}\cdots da^m\\ +(-1)^na^0a^1da^2\cdots da^m.\end{aligned}$$

This product can be seen as ensuing from the requirements that $\Omega^*(\mathcal{A})$ be a right $\mathcal{A}$-module and the derivation property.

We can now characterize cyclic cocycles as the characters of cycles over an algebra $\mathcal{A}$.

Proposition 4. *Let τ be an $(n+1)$-linear functional on $\mathcal{A}$. Then the following conditions are equivalent:*

1) *There is an n-dimensional cycle $(\Omega, d, \int)$ and a homomorphism $\rho : \mathcal{A} \to \Omega^0$ such that*

$$\tau(a^0,\ldots,a^n)=\int\rho(a^0)\,d(\rho(a^1))\cdots d(\rho(a^n))\quad \forall a^0,\ldots,a^n\in\mathcal{A}.$$

2) *There exists a closed graded trace $\hat{\tau}$ of dimension n on $\Omega^*(\mathcal{A})$ such that*

$$\tau(a^0,\ldots,a^n)=\hat{\tau}(a^0da^1\cdots da^n)\qquad \forall a^0,\ldots,a^n\in\mathcal{A}.$$

3) *One has $\tau(a^1,\ldots,a^n,a^0)=(-1)^n\,\tau(a^0,\ldots,a^n)$ and*

$$\sum_{i=0}^{n}(-1)^i\,\tau(a^0,\ldots,a^ia^{i+1},\ldots,a^{n+1})+(-1)^{n+1}\,\tau(a^{n+1}a^0,\ldots,a^n)=0$$

for any $a^0,\ldots,a^{n+1}\in\mathcal{A}$.

Proof. The universality of $\Omega^*(\mathcal{A})$ shows that 1) and 2) are equivalent. Let us show that 3) $\Longrightarrow$ 2). Given any $(n+1)$-linear functional φ on $\mathcal{A}$, define $\hat{\varphi}$ as a linear functional on $\Omega^n(\mathcal{A})$ by

$$\hat{\varphi}\circ j((a^0+\lambda^0 1)\otimes a^1\otimes\cdots\otimes a^n)=\varphi(a^0,a^1,\ldots,a^n).$$

By construction, one has $\hat{\varphi}(d\omega)=0$ for all $\omega\in\Omega^{n-1}(\mathcal{A})$. Now, with τ satisfying 3) let us show that $\hat{\tau}$ is a graded trace. We have to show that

$$\begin{aligned}&\hat{\tau}((a^0da^1\cdots da^k)(a^{k+1}da^{k+2}\cdots da^{n+1}))\\ &=(-1)^{k(n-k)}\,\hat{\tau}((a^{k+1}da^{k+2}\cdots da^{n+1})(a^0da^1\cdots da^k)).\end{aligned}$$

Using the definition of the product in $\Omega^*(\mathcal{A})$ the left-hand side gives

$$\sum_{j=0}^{k}(-1)^{k-j}\,\tau(a^0,\ldots,a^ja^{j+1},\ldots,a^{n+1}),$$

and the right-hand side gives

$$\sum_{j=0}^{n-k} (-1)^{k(n-k)+n-k-j}\, \tau(a^{k+1},\ldots,a^{k+1+j}a^{k+1+j+1},\ldots,a^0,a^1,\ldots,a^k),$$

where we let $a^{n+2} = a^0$. The cyclic permutation λ such that $\lambda(\ell) = k+1+\ell$ has a signature $\varepsilon(\lambda)$ equal to $(-1)^{n(k+1)}$ so that, as $\tau^\lambda = \varepsilon(\lambda)\tau$ by hypothesis, the right-hand side gives

$$-\sum_{j=k+1}^{n} (-1)^{k-j}\, \tau(a^0,\ldots,a^j a^{j+1},\ldots,a^{n+1}) + (-1)^{k-n}\tau(a^{n+1}a^0,a^1,\ldots,a^n).$$

Hence the equality follows from the second hypothesis on τ.

Let us show that 1) $\Longrightarrow$ 3). We can assume that $\mathcal{A} = \Omega^0$. One has

$$\begin{aligned}\tau(a^0,a^1,\ldots,a^n) &= \int (a^0da^1)(da^2\cdots da^n) = (-1)^{n-1}\int (da^2\cdots da^n)(a^0da^1)\\ &= (-1)^n\int (da^2\cdots da^n da^0)a^1 = (-1)^n\,\tau(a^1,\ldots,a^n,a^0).\end{aligned}$$

To prove the second property we shall only use the equality

$$\int a\omega = \int \omega a \quad \text{for } \omega\in\Omega^n,\ a\in\mathcal{A}.$$

From the equality $d(ab) = (da)b + adb$ it follows that

$$\begin{aligned}(da^1\cdots da^n)a^{n+1} &= \sum_{j=1}^{n}(-1)^{n-j}\, da^1\cdots d(a^j a^{j+1})\cdots da^{n+1}\\ &\quad + (-1)^n\, a^1 da^2\cdots da^{n+1},\end{aligned}$$

thus the second property follows from

$$\int a^{n+1}\,(a^0da^1\cdots da^n) = \int (a^0da^1\cdots da^n)a^{n+1}.$$

Let us now recall the definition of the Hochschild cohomology groups $H^n(\mathcal{A},\mathcal{M})$ of $\mathcal{A}$ with coefficients in a bimodule $\mathcal{M}$ ([Car-E]). Let $\mathcal{A}^e = \mathcal{A}\otimes\mathcal{A}^o$ be the tensor product of $\mathcal{A}$ by the opposite algebra. Then any bimodule $\mathcal{M}$ over $\mathcal{A}$ becomes a left $\mathcal{A}^e$-module and, by definition, $H^n(\mathcal{A},\mathcal{M}) = \mathrm{Ext}^n_{\mathcal{A}^e}(\mathcal{A},\mathcal{M})$, where $\mathcal{A}$ is viewed as a bimodule over $\mathcal{A}$ via $a(b)c = abc$, $\forall a,b,c\in\mathcal{A}$. As in [Car-E], one can reformulate the definition of $H^n(\mathcal{A},\mathcal{M})$ using the standard resolution of the bimodule $\mathcal{A}$. One forms the complex $(C^n(\mathcal{A},\mathcal{M}),b)$, where

a) $C^n(\mathcal{A},\mathcal{M})$ is the space of n-linear maps from $\mathcal{A}$ to $\mathcal{M}$;

b) for $T\in C^n(\mathcal{A},\mathcal{M})$, bT is given by

$$\begin{aligned}(bT)(a^1,\ldots,a^{n+1}) &= a^1T(a^2,\ldots,a^{n+1})\\ &\quad + \sum_{i=1}^{n}(-1)^i\, T(a^1,\ldots,a^ia^{i+1},\ldots,a^{n+1})\\ &\quad +(-1)^{n+1}\, T(a^1,\ldots,a^n)a^{n+1}.\end{aligned}$$

Definition 5. *The Hochschild cohomology of $\mathcal{A}$ with coefficients in $\mathcal{M}$ is the cohomology $H^n(\mathcal{A},\mathcal{M})$ of the complex $(C^n(\mathcal{A},\mathcal{M}),b)$.*

The space $\mathcal{A}^*$ of all linear functionals on $\mathcal{A}$ is a bimodule over $\mathcal{A}$ by the equality $(a\varphi b)(c) = \varphi(bca)$, for $a,b,c \in \mathcal{A}$. We consider any $T \in C^n(\mathcal{A},\mathcal{A}^*)$ as an $(n+1)$-linear functional τ on $\mathcal{A}$ by the equality

$$\tau(a^0,a^1,\ldots,a^n) = T(a^1,\ldots,a^n)(a^0) \quad \forall a^i \in \mathcal{A}.$$

To the boundary bT corresponds the $(n+2)$-linear functional $b\tau$, given by

$$\begin{aligned}(b\tau)(a^0,\ldots,a^{n+1}) &= \tau(a^0a^1,a^2,\ldots,a^{n+1})\\ &+ \sum_{i=1}^{n}(-1)^i\,\tau(a^0,\ldots,a^ia^{i+1},\ldots,a^{n+1})\\ &+(-1)^{n+1}\,\tau(a^{n+1}a^0,\ldots,a^n).\end{aligned}$$

Thus, with this notation, the condition 3) of Proposition 4 becomes

a) $\tau^\gamma = \varepsilon(\gamma)\tau$ for any *cyclic* permutation γ of $\{0,1,\ldots,n\}$;

b) $b\tau = 0$.

Now, though the Hochschild coboundary b does not commute with cyclic permutations, it maps cochains satisfying a) to cochains satisfying a). More precisely, let A be the linear map of $C^n(\mathcal{A},\mathcal{A}^*)$ to $C^n(\mathcal{A},\mathcal{A}^*)$ defined by

$$(A\varphi) = \sum_{\gamma\in\Gamma}\varepsilon(\gamma)\,\varphi^\gamma,$$

where Γ is the group of cyclic permutations of $\{0,1,\ldots n\}$. Obviously the range of A is the subspace $C^n_\lambda(\mathcal{A})$ of $C^n(\mathcal{A},\mathcal{A}^*)$ of cochains which satisfy a). One has

Lemma 6. *$b\circ A = A\circ b'$ where $b' : C^n(\mathcal{A},\mathcal{A}^*) \to C^{n+1}(\mathcal{A},\mathcal{A}^*)$ is defined by the equality*

$$(b'\varphi)(x^0,\ldots,x^{n+1}) = \sum_{j=0}^{n}(-1)^j\,\varphi(x^0,\ldots,x^jx^{j+1},\ldots,x^{n+1}).$$

Proof. One has, dropping the composition sign $\circ$ for notational simplicity,

$$((Ab')\varphi)(x^0,\ldots,x^{n+1}) = \sum_{k=0}^{n+1}\sum_{i=0}^{n}(-1)^{i+(n+1)k}\,\varphi(x^k,\ldots,x^{k+i}x^{k+i+1},\ldots,x^{k-1})$$

where we adopt the convention that the indices cycle back to 0 after $n+1$. Then also

$$\begin{aligned}&((bA)\varphi)(x^0,\ldots,x^{n+1})\\ &= \sum_{j=0}^{n}(-1)^j\,(A\varphi)(x^0,\ldots,x^jx^{j+1},\ldots,x^{n+1}) + (-1)^{n+1}\,(A\varphi)(x^{n+1}x^0,\ldots,x^n).\end{aligned}$$

For $j \in \{0, \dots n\}$ one has

$$(A\varphi)(x^0, \dots, x^j x^{j+1}, \dots, x^{n+1}) = \sum_{k=0}^{j} (-1)^{nk} \varphi(x^k, \dots, x^j x^{j+1}, \dots, x^{k-1})$$
$$+ \sum_{k=j+2}^{n+1} (-1)^{n(k-1)} \varphi(x^k, \dots, x^{n+1}, x^0, \dots, x^j x^{j+1}, \dots, x^{k-1}).$$

Also,

$$(A\varphi)(x^{n+1}x^0, \dots, x^n) = \varphi(x^{n+1}x^0, \dots, x^n) + \sum_{j=1}^{n} (-1)^{jn} \varphi(x^j, \dots, x^n, x^{n+1}x^0, \dots, x^{j-1}).$$

In all these terms, the x^j's remain in cyclic order, with only two consecutive x^j's replaced by their product. There are $(n+1)(n+2)$ such terms, which all appear in both $bA\varphi$ and $Ab'\varphi$. Thus, we just have to check the signs in front of $T_{k,j}$ $(k \neq j+1)$, where $T_{k,j} = \varphi(x^k, \dots, x^j x^{j+1}, \dots, x^{k-1})$. For Ab' we get $(-1)^{i+(n+1)k}$ where $i \equiv j - k \pmod{n+2}$ and $0 \le i \le n$. For bA we get $(-1)^{j+nk}$ if $j \ge k$ and $(-1)^{j+n(k-1)}$ if $j < k$. When $j \ge k$ one has $i = j - k$; thus the two signs agree. When $j < k$ one has $i = n + 2 - k + j$. Then as

$$n + 2 - k + j + (n+1)k \equiv j + n(k-1) \text{ modulo } 2$$

the two signs still agree.

Corollary 7. *$(C_\lambda^n(\mathcal{A}), b)$ is a subcomplex of the Hochschild complex.*

We let $HC^n(\mathcal{A})$ be the n-th cohomology group of the complex (C_λ^n, b) and we call it the cyclic cohomology of the algebra $\mathcal{A}$. For $n = 0$, $HC^0(\mathcal{A}) = Z_\lambda^0(\mathcal{A})$ is exactly the linear space of traces on $\mathcal{A}$.

For $\mathcal{A} = \mathbb{C}$, one has $HC^n = 0$ for n odd but $HC^n = \mathbb{C}$ for any even n. This example shows that the subcomplex C_λ^n is not a retraction of the complex C^n, which for $\mathcal{A} = \mathbb{C}$ has a trivial cohomology for all $n > 0$.

To each homomorphism $\rho : \mathcal{A} \to \mathcal{B}$ corresponds a morphism of complexes $\rho^* : C_\lambda^n(\mathcal{B}) \to C_\lambda^n(\mathcal{A})$ defined by

$$(\rho^*\varphi)(a^0, \dots, a^n) = \varphi(\rho(a^0), \dots, \rho(a^n))$$

and hence an induced map $\rho^* : HC^n(\mathcal{B}) \to HC^n(\mathcal{A})$.

The map ρ^* only depends upon the class of ρ modulo inner automorphisms, as follows from:

Proposition 8. *Let u be an invertible element of $\mathcal{A}$ and θ, defined by $\theta(x) = uxu^{-1}$ for $x \in \mathcal{A}$, the corresponding inner automorphism. Then the induced map $\theta^* : HC^*(\mathcal{A}) \to HC^*(\mathcal{A})$ is the identity.*

Proof. Let $a \in \mathcal{A}$ and let δ be the corresponding inner derivation of $\mathcal{A}$ given by $\delta(x) = ax - xa$. Given $\varphi \in Z_\lambda^n(\mathcal{A})$ let us check that ψ, given by

$$\psi(a^0, \dots, a^n) = \sum_{i=0}^{n} \varphi(a^0, \dots, \delta(a^i), \dots, a^n),$$

is a coboundary, i.e. that $\psi \in B_\lambda^n(\mathcal{A})$. Let $\psi_0(a^0, \dots, a^{n-1}) = \varphi(a^0, \dots, a^{n-1}, a)$ with a as above. Let us compute $bA\psi_0 = Ab'\psi_0$. One has

$$\begin{aligned}(b'\psi_0)(a^0, \dots, a^n) &= \sum_{i=0}^{n-1} (-1)^i \varphi(a^0, \dots, a^i a^{i+1}, \dots, a^n, a) \\ &= (b\varphi)(a^0, \dots, a^n, a) - (-1)^n \varphi(a^0, \dots, a^{n-1}, a^n a) \\ &\quad + (-1)^n \varphi(aa^0, \dots, a^{n-1}, a^n).\end{aligned}$$

Since $b\varphi = 0$ by hypothesis, only the last two terms remain and one gets $Ab'\psi_0 = (-1)^n \psi$. Thus, $\psi = (-1)^n Ab'\psi_0 = b((-1)^n A\psi_0) \in B_\lambda^n(\mathcal{A})$.

Now let u be an invertible element of $\mathcal{A}$, let $\varphi \in Z_\lambda^n(\mathcal{A})$ and define $\theta(x) = uxu^{-1}$ for $x \in \mathcal{A}$. To prove that φ and $\varphi \circ \theta$ are in the same cohomology class, one can replace $\mathcal{A}$ by $M_2(\mathcal{A})$, u by $v = \begin{bmatrix} u & 0 \\ 0 & u^{-1} \end{bmatrix}$, and φ by φ_2 where, for $a^i \in \mathcal{A}$ and $b^i \in M_2(\mathbb{C})$,

$$\varphi_2(a^0 \otimes b^0, a^1 \otimes b^1, \dots, a^n \otimes b^n) = \varphi(a^0, \dots, a^n) \operatorname{Trace}(b^0 \cdots b^n).$$

Now $v = v_1 v_2$ with $v_1 = \begin{bmatrix} u & 0 \\ 0 & 1 \end{bmatrix} \begin{bmatrix} 0 & -1 \\ 1 & 0 \end{bmatrix} \begin{bmatrix} u^{-1} & 0 \\ 0 & 1 \end{bmatrix}$, $v_2 = \begin{bmatrix} 0 & 1 \\ -1 & 0 \end{bmatrix}$. One has $v_i = \exp a_i$, $a_i = \frac{\pi}{2} v_i$, thus the result follows from the above discussion (cf. [Lo] for a purely algebraic proof).

We shall now characterize the *coboundaries* as the cyclic cocycles which extend to cyclic cocycles on arbitrary algebras containing $\mathcal{A}$. In fact, extendibility to a certain tensor product $C \otimes_{\mathbb{C}} \mathcal{A}$ algebra will suffice.

Following Karoubi [Kar$_4$] [Kar$_5$], let C be the algebra of infinite matrices $(a_{ij})_{i,j \in \mathbb{N}}$ with $a_{ij} \in \mathbb{C}$, such that

α) the set of complex number entries $\{a_{ij}\}$ is finite,

β) the number of nonzero a_{ij}'s per line or column is bounded.

Then for any algebra $\mathcal{A}$ the algebra $C\mathcal{A} = C \otimes_{\mathbb{C}} \mathcal{A}$ is algebraically contractible, in that it verifies the hypothesis of the following lemma, and hence has trivial cyclic cohomology.

Lemma 9. *Let $\mathcal{A}$ be a unital algebra. Assume that there exists a homomorphism $\rho : \mathcal{A} \to \mathcal{A}$ and an invertible element X of $M_2(\mathcal{A})$ such that $X \begin{bmatrix} a & 0 \\ 0 & \rho(a) \end{bmatrix} X^{-1} = \begin{bmatrix} 0 & 0 \\ 0 & \rho(a) \end{bmatrix}$ for $a \in \mathcal{A}$. Then $HC^n(\mathcal{A}) = 0$ for all n.*

Proof. Let $\varphi \in Z_\lambda^n(\mathcal{A})$ and φ_2 be the cocycle on $M_2(\mathcal{A})$ defined in the proof of Proposition 8. For $a \in \mathcal{A}$, let $\alpha(a) = \begin{bmatrix} a & 0 \\ 0 & \rho(a) \end{bmatrix}$ and $\beta(a) = \begin{bmatrix} 0 & 0 \\ 0 & \rho(a) \end{bmatrix}$. By hypothesis α and β are homomorphisms of $\mathcal{A}$ into $M_2(\mathcal{A})$ and, by Proposition 8, $\varphi_2 \circ \alpha$ and $\varphi_2 \circ \beta$ are in the same cohomology class. From the definition of φ_2 one has

$$\varphi_2(\alpha(a^0),\ldots,\alpha(a^n)) = \varphi(a^0,\ldots,a^n) + \varphi(\rho(a^0),\ldots,\rho(a^n)),$$
$$\varphi_2(\beta(a^0),\ldots,\beta(a^n)) = \varphi(\rho(a^0),\ldots,\rho(a^n)).$$

Definition 10. *We shall say that a cycle vanishes when the algebra Ω^0 satisfies the condition of Lemma* 9.

Given an n-dimensional cycle $(\Omega, d, \int)$ and a homomorphism $\rho : \mathcal{A} \to \Omega^0$, recall that its *character* is given by

$$\tau(a^0,\ldots,a^n) = \int \rho(a^0)\, d(\rho(a^1)) \cdots d(\rho(a^n)).$$

Proposition 11. *Let τ be an $(n+1)$-linear functional on $\mathcal{A}$; then*
α) $\tau \in Z_\lambda^n(\mathcal{A})$ *if and only if τ is the character of a cycle;*
β) $\tau \in B_\lambda^n(\mathcal{A})$ *if and only if τ is the character of a vanishing cycle.*

Proof. α) is just a restatement of Proposition 4.

β) For $(\Omega, d, \int)$ a vanishing cycle, one has $HC^n(\Omega^0) = 0$, thus the character is a coboundary. Conversely, if $\tau \in B_\lambda^n(\mathcal{A})$, $\tau = b\psi$ for some $\psi \in C_\lambda^{n-1}(\mathcal{A})$, one can extend ψ to $C\mathcal{A} = C \otimes \mathcal{A}$ as an n-linear functional $\tilde{\psi}$ such that

$$\tilde{\psi}(1 \otimes a^0,\ldots,1 \otimes a^{n-1}) = \psi(a^0,\ldots,a^{n-1}) \text{ for all } a^i \in \mathcal{A},$$

and such that $\tilde{\psi}^\lambda = \varepsilon(\lambda)\tilde{\psi}$ for any cyclic permutation λ of $\{0,\ldots,n-1\}$. Let $\rho : \mathcal{A} \to C\mathcal{A}$ be the obvious homomorphism $\rho(a) = 1 \otimes a$. Then $\tau' = b\tilde{\psi}$ is an n-cocycle on $C\mathcal{A}$ and $\tau = \rho^*\tau'$, so that by the implication 3) $\Longrightarrow$ 2) of Proposition 4, τ is the character of a cycle Ω with $\Omega^0 = C\mathcal{A}$.

Let us now pass to the definition of the cup product

$$HC^*(\mathcal{A}) \otimes HC^*(\mathcal{B}) \to HC^*(\mathcal{A} \otimes \mathcal{B}).$$

In general one does not have $\Omega^*(\mathcal{A} \otimes \mathcal{B}) = \Omega^*(\mathcal{A}) \otimes \Omega^*(\mathcal{B})$ (where the right-hand side is the graded tensor product of differential graded algebras) but, from the universal property of $\Omega^*(\mathcal{A} \otimes \mathcal{B})$, we get a natural homomorphism $\pi : \Omega^*(\mathcal{A} \otimes \mathcal{B}) \to \Omega^*(\mathcal{A}) \otimes \Omega^*(\mathcal{B})$.

Thus, for arbitrary cochains $\varphi \in C^n(\mathcal{A}, \mathcal{A}^*)$ and $\psi \in C^m(\mathcal{B}, \mathcal{B}^*)$, one can define the cup product $\varphi \# \psi$ by the equality

$$(\varphi \# \psi)\hat{} = (\hat{\varphi} \otimes \hat{\psi}) \circ \pi.$$

Theorem 12. 1) *The cup product* $\varphi \otimes \psi \mapsto \varphi \# \psi$ *defines a homomorphism*

$$HC^n(\mathcal{A}) \otimes HC^m(\mathcal{B}) \longrightarrow HC^{n+m}(\mathcal{A} \otimes \mathcal{B}).$$

2) *The character of the tensor product of two cycles is the cup product of their characters.*

Proof. First, let $\varphi \in Z_\lambda^n(\mathcal{A})$ and $\psi \in Z_\lambda^m(\mathcal{B})$; then $\hat{\varphi}$ is a closed graded trace on $\Omega^*(\mathcal{A})$ and similarly for $\hat{\psi}$ on $\Omega^*(\mathcal{B})$, and thus $\hat{\varphi} \otimes \hat{\psi}$ is a closed graded trace on $\Omega^*(\mathcal{A}) \otimes \Omega^*(\mathcal{B})$ and $\varphi \# \psi \in Z_\lambda^{n+m}(\mathcal{A} \otimes \mathcal{B})$ by Proposition 4.

Next, given cycles Ω and Ω' and homomorphisms $\rho : \mathcal{A} \to \Omega$ and $\rho' : \mathcal{B} \to \Omega'$, one has a commutative triangle

$$\begin{array}{ccc} & & \Omega^*(\mathcal{A}) \otimes \Omega^*(\mathcal{B}) \\ & \overset{\pi}{\nearrow} & \Big\downarrow {\scriptstyle \tilde{\rho} \otimes \tilde{\rho}'} \\ \Omega^*(\mathcal{A} \otimes \mathcal{B}) & & \\ & \underset{\widetilde{(\rho \otimes \rho')}}{\searrow} & \\ & & \Omega \otimes \Omega' \end{array}$$

Thus, 2) follows.

It remains to show that if $\varphi \in B_\lambda^n(\mathcal{A})$ then $\varphi \# \psi$ is a coboundary:

$$\varphi \# \psi \in B_\lambda^{n+m}(\mathcal{A} \otimes \mathcal{B}).$$

This follows from Proposition 11 and the trivial fact that the tensor product of any cycle with a vanishing cycle is vanishing.

Corollary 13. 1) *$HC^*(\mathbb{C})$ is a polynomial ring with one generator σ of degree 2.*

2) *Each $HC^*(\mathcal{A})$ is a module over the ring $HC^*(\mathbb{C})$.*

Proof. 1) It is obvious that $HC^n(\mathbb{C}) = 0$ for n odd and $HC^n(\mathbb{C}) = \mathbb{C}$ for n even. Let e be the unit of $\mathbb{C}$; then any $\varphi \in Z_\lambda^n(\mathbb{C})$ is characterized by $\varphi(e, \ldots, e)$. Let us compute $\varphi \# \psi$, where $\varphi \in Z_\lambda^{2m}(\mathbb{C})$ and $\psi \in Z_\lambda^{2m'}(\mathbb{C})$. Since e is an idempotent, one has in $\Omega^*(\mathbb{C})$ the equalities

$$de = ede + (de)e\,,\; e(de)e = 0\,,\; e(de)^2 = (de)^2 e.$$

Similar identities hold for $e \otimes e$ and $\pi(e \otimes e) \in \Omega^*(\mathbb{C}) \otimes \Omega^*(\mathbb{C})$ and one has

$$\pi((e \otimes e)d(e \otimes e)d(e \otimes e)) = edede \otimes e + e \otimes edede.$$

Thus, one gets $(\varphi \# \psi)(e, \ldots, e) = \frac{(m+m')!}{m!m'!}\, \varphi(e, \ldots, e)\, \psi(e, \ldots, e)$.

2) Let $\varphi \in Z_\lambda^n(\mathcal{A})$. Let us choose as a generator of $HC^2(\mathbb{C})$ the 2-cocycle $\sigma(e, e, e) = 1$. Let us then check that $\sigma \# \varphi = \varphi \# \sigma$, and at the same time write an explicit formula for the corresponding map $S : HC^n(\mathcal{A}) \to HC^{n+2}(\mathcal{A})$.

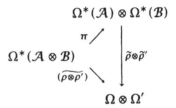

With the notation of 1) one has

$$\begin{aligned}(\varphi\#\sigma)(a^0,\dots,a^{n+2}) &= (\hat\varphi\otimes\hat\sigma)((a^0\otimes e)d(a^1\otimes e)\cdots d(a^{n+2}\otimes e))\\ &= \hat\varphi(a^0a^1a^2da^3\cdots da^{n+2})\\ &+ \hat\varphi(a^0da^1(a^2a^3)da^4\cdots da^{n+2})+\cdots\\ &+ \hat\varphi(a^0da^1\cdots da^{i-1}(a^ia^{i+1})da^{i+2}\cdots da^{n+2})+\cdots\\ &+ \hat\varphi(a^0da^1\cdots da^n(a^{n+1}a^{n+2})).\end{aligned}$$

The computation of $\sigma\#\varphi$ gives the same result.

For $\varphi\in Z^n_\lambda(\mathcal{A})$, let $S\varphi=\sigma\#\varphi=\varphi\#\sigma\in Z^{n+2}_\lambda(\mathcal{A})$. By Theorem 12 we know that $SB^n_\lambda(\mathcal{A})\subset B^{n+2}_\lambda(\mathcal{A})$ but we do not have a definition of S as a morphism of cochain complexes. We shall now explicitly construct such a morphism.

Recall that $\varphi\#\psi$ is already defined at the cochain level by

$$(\varphi\#\psi)\hat{}=(\hat\varphi\otimes\hat\psi)\circ\pi.$$

Lemma 14. *For any cochain $\varphi\in C^n_\lambda(\mathcal{A})$ let $S\varphi\in C^{n+2}_\lambda(\mathcal{A})$ be defined by $S\varphi=\frac{1}{n+3}A(\sigma\#\varphi)$; then*

a) $\frac{1}{n+3}A(\sigma\#\varphi)=\sigma\#\varphi$ *for $\varphi\in Z^n_\lambda(\mathcal{A})$, so S extends the previously defined map.*

b) $bS\varphi=\frac{n+1}{n+3}Sb\varphi$ *for $\varphi\in C^n_\lambda(\mathcal{A})$.*

Proof. a) If $\varphi\in Z^n_\lambda(\mathcal{A})$ then $(\sigma\#\varphi)^\lambda=\varepsilon(\lambda)\sigma\#\varphi$ for any cyclic permutation λ of $\{0,1,\dots,n+2\}$.

b) We shall leave to the reader the tedious check in the special case $\psi=\sigma$ of the equality $(bA\varphi)\#\psi=bA(\varphi\#\psi)$ for $\varphi\in C^m(\mathcal{A},\mathcal{A}^*)$. It is based on the following explicit formula for $A(\varphi\#\sigma)$. For any subset with two elements $s=\{i,j\}$, $i<j$, of $\{0,1,\dots,n+2\}=\mathbb{Z}/(n+3)$ one defines

$$\alpha(s)=\varphi(a^0,\dots,a^{i-1},a^ia^{i+1},\dots,a^ja^{j+1},\dots,a^{n+2}).$$

in the special case $j=n+2$ one takes

$$\begin{aligned}\alpha(s)&=\varphi(a^{n+2}a^0,\dots,a^ia^{i+1},\dots,a^{n+1}) &&\text{if } i<n+1,\\ \alpha(s)&=\varphi(a^{n+1}a^{n+2}a^0,\dots,a^n) &&\text{if } i=n+1.\end{aligned}$$

Then one gets $A(\sigma\#\varphi)=\sum_{i=1}^{1+[n/2]}(-1)^{i+1}\,(n+3-2i)\psi_i$ where, for n even, one has

$$\psi_i=\alpha(\{0,i\})+\alpha(\{1,i+1\})+\cdots+\alpha(\{n+2,i-1\}),$$

and for n odd

$$\psi_i=\alpha(\{0,i\})+\cdots+\alpha(\{n+2-i,n+2\})-\alpha(\{n+2-i+1,0\})\cdots-\alpha\{n+2,i-1\}.$$

We shall end this section with the following proposition. One can show in general that, if $\varphi\in Z^n(\mathcal{A},\mathcal{A}^*)$ and $\psi\in Z^m(\mathcal{B},\mathcal{B}^*)$ are Hochschild cocycles, then $\varphi \# \psi$ is still a Hochschild cocycle $\varphi \# \psi \in Z^{n+m}(\mathcal{A} \otimes \mathcal{B}, \mathcal{A}^* \otimes \mathcal{B}^*)$ and that the corresponding product of cohomology classes is related to the product $\vee$ of [Car-E], p.216, by $[\varphi \# \psi] = \frac{(n+m)!}{n!m!} [\varphi] \vee [\psi]$. Since $\sigma \in Z^2(\mathbb{C}, \mathbb{C})$ is a Hochschild coboundary one has:

Proposition 15. *For any cocycle $\varphi \in Z_\lambda^n(\mathcal{A})$, $S\varphi$ is a Hochschild coboundary: $S\varphi = b\psi$, where*

$$\psi(a^0, \ldots, a^{n+1}) = \sum_{j=1}^{n+1} (-1)^{j-1} \hat{\varphi}(a^0(da^1 \cdots da^{j-1})\, a^j (da^{j+1} \cdots da^{n+1})).$$

Proof. One checks that the coboundary of the j-th term in the sum defining ψ gives

$$\hat{\varphi}(a^0(da^1 \cdots da^{j-1})\, a^j a^{j+1} (da^{j+2} \cdots da^{n+2})).$$

1.β Cobordisms of cycles and the operator B

By a *chain* of dimension $n+1$ we shall mean a quadruple $(\Omega, \partial\Omega, d, \int)$ where Ω and $\partial\Omega$ are differential graded algebras of dimensions $n+1$ and n with a given surjective morphism $r : \Omega \to \partial\Omega$ of degree 0, and where $\int : \Omega^{n+1} \to \mathbb{C}$ is a graded trace such that

$$\int d\omega = 0, \ \forall \omega \in \Omega^n \ \text{ such that } \ r(\omega) = 0.$$

By the *boundary* of such a chain we mean the cycle $(\partial\Omega, d, \int')$ where for $\omega' \in (\partial\Omega)^n$ one takes $\int' \omega' = \int d\omega$ for any $\omega \in \Omega^n$ with $r(\omega) = \omega'$. One easily checks, using the surjectivity of r, that $\int'$ is a graded trace on $\partial\Omega$ which is closed by construction.

Definition 16. *Let $\mathcal{A}$ be an algebra, and let $\mathcal{A} \xrightarrow{\rho} \Omega$ and $\mathcal{A} \xrightarrow{\rho'} \Omega'$ be two cycles over $\mathcal{A}$. We shall say that these cycles are cobordant over $\mathcal{A}$ if there exists a chain Ω'' with boundary $\Omega \oplus \widetilde{\Omega'}$ (where $\widetilde{\Omega'}$ is obtained from Ω' by changing the sign of $\int$) and a homomorphism $\rho'' : \mathcal{A} \to \Omega''$ such that $r \circ \rho'' = (\rho, \rho')$.*

Using a fiber product of algebras one checks that the relation of cobordism is transitive. It is obviously symmetric. Let us check that any cycle over $\mathcal{A}$ is cobordant to itself. Let $\Omega^0 = C^\infty([0,1])$, Ω^1 be the space of C^∞ 1-forms on $[0,1]$, and d be the usual differential. Set $\partial\Omega = \mathbb{C} \oplus \mathbb{C}$ and take $\int$ to be the usual integral. Then taking for r the restriction of functions to the boundary, one gets a chain of dimension 1 with boundary $(\mathbb{C} \oplus \mathbb{C}, d, \varphi)$, by defining $\varphi(a,b) = a - b$. Tensoring a given cycle over $\mathcal{A}$ with the above chain gives the desired cobordism. Equivalently, one could replace smooth functions in the above chain by polynomials.

Thus, cobordism is an equivalence relation. The main result of this section is a precise description of its meaning for the characters of the cycles. We shall assume throughout that the algebra $\mathcal{A}$ is unital.

Lemma 17. *Let τ_1, τ_2 be the characters of two cobordant cycles over $\mathcal{A}$. Then there exists a Hochschild cocycle $\varphi \in Z^{n+1}(\mathcal{A}, \mathcal{A}^*)$ such that $\tau_1 - \tau_2 = B_0\varphi$, where*

$$(B_0\varphi)(a^0, \dots, a^n) = \varphi(1, a^0, \dots, a^n) - (-1)^{n+1}\, \varphi(a^0, \dots, a^n, 1).$$

Proof. With the notation of Definition 16, let

$$\varphi(a^0, \dots, a^{n+1}) = \int \rho''(a^0)\, d\rho''(a^1) \cdots d\rho''(a^{n+1}), \ \forall a^i \in \mathcal{A}.$$

Let

$$\omega = \rho''(a^0)\, d\rho''(a^1) \cdots d\rho''(a^n) \in (\Omega'')^n.$$

Then by hypothesis one has

$$(\tau_1 - \tau_2)(a^0, a^1, \dots, a^n) = \int d\omega.$$

Since $\rho''(a^0) = \rho''(1)\, \rho''(a^0)$ one has

$$d\omega = (d\rho''(1))\, \rho''(a^0)\, d\rho''(a^1) \cdots d\rho''(a^n) + \rho''(1)\, d\rho''(a^0) \cdots d\rho''(a^n).$$

Using the tracial property of $\int$ one gets

$$\int d\omega = (-1)^n\, \varphi(a^0, a^1, \dots, a^n, 1) + \varphi(1, a^0, \dots, a^n).$$

Using again the tracial property of $\int$ one checks that φ is a Hochschild cocycle.

Lemma 18. *Let $\tau_1, \tau_2 \in Z_\lambda^n(\mathcal{A})$ and assume that $\tau_1 - \tau_2 = B_0\varphi$ for some $\varphi \in Z^{n+1}(\mathcal{A}, \mathcal{A}^*)$. Then any two cycles over $\mathcal{A}$ with characters τ_1 and τ_2 are cobordant.*

Proof. Let $\mathcal{A} \xrightarrow{\rho} \Omega$ be a cycle over $\mathcal{A}$ with character τ. Let us first show that it is cobordant with $(\Omega^*(\mathcal{A}), \hat{\tau})$. In the above cobordism of Ω with itself, with restriction maps r_0 and r_1, we can consider the subalgebra $\{\omega; r_1(\omega) \in \Omega'\}$, where Ω' is the graded differential subalgebra of Ω generated by $\rho(\mathcal{A})$. This defines a cobordism of Ω with Ω'. Now the homomorphism $\tilde{\rho} : \Omega^*(\mathcal{A}) \to \Omega'$ is surjective, and satisfies $\tilde{\rho}^* \int = \hat{\tau}$. Thus, one can modify the restriction map in the canonical cobordism of $(\Omega^*(\mathcal{A}), \hat{\tau})$ with itself to get a cobordism of $(\Omega^*(\mathcal{A}), \hat{\tau})$ with Ω'.

Let us show that $(\Omega^*(\mathcal{A}), \hat{\tau}_1)$ and $(\Omega^*(\mathcal{A}), \hat{\tau}_2)$ are cobordant. Let μ be the linear functional on $\Omega^{n+1}(\mathcal{A})$ defined by

1) $\mu(a^0 da^1 \cdots da^{n+1}) = \varphi(a^0, \dots, a^{n+1})$,

2) $\mu(da^1 \cdots da^{n+1}) = (B_0\varphi)(a^1, \dots, a^{n+1})$.

Let us check that μ is a graded trace on $\Omega^*(\mathcal{A})$. We already know by the Hochschild cocycle property of φ that

$$\mu(a(b\omega)) = \mu((b\omega)a)\,,\ \forall a,b\in\mathcal{A}\,,\ \omega\in\Omega^{n+1}.$$

Let us check that $\mu(a\omega) = \mu(\omega a)$ for $\omega = da^1\cdots da^{n+1}$. The right side, $\mu(\omega a)$, gives

$$\begin{aligned}
&\mu(\sum_{j=1}^{n}(-1)^{n+1-j}\, da^1\cdots d(a^j a^{j+1})\cdots da^{n+1}\, da \\
&\quad + da^1 da^2\cdots da^n d(a^{n+1}a)) + (-1)^{n+1}\,\mu(a^1 da^2\cdots da^{n+1}da) \\
&= \sum_{j=1}^{n}(-1)^{n+1-j}\,(B_0\varphi)(a^1,\dots,a^j a^{j+1},\dots,a^{n+1},a) \\
&\quad + (B_0\varphi)(a^1,a^2,\dots,a^{n+1}a) + (-1)^{n+1}\,\varphi(a^1,a^2,\dots,a^{n+1},a) \\
&= (-1)^n\,((b'B_0\varphi)-\varphi)(a^1,a^2,\dots,a^{n+1},a).
\end{aligned}$$

Now one checks that for an arbitrary cochain $\varphi\in C^{n+1}(\mathcal{A},\mathcal{A}^*)$ one has

$$B_0 b\varphi + b'B_0\varphi = \varphi - (-1)^{n+1}\,\varphi^\lambda,$$

where λ is the cyclic permutation $\lambda(i) = i-1$. Here φ is a cocycle, $b\varphi = 0$ and $b'B_0\varphi - \varphi = (-1)^n\,\varphi^\lambda$ so that $\mu(\omega a) = \varphi(a,a^1,\dots,a^{n+1}) = \mu(a\omega)$.

It remains to check that for any $a\in\mathcal{A}$ and $\omega\in\Omega^n$ one has

$$\mu((da)\omega) = (-1)^n\,\mu(\omega da).$$

For $\omega\in d\Omega^{n-1}$ this follows from the fact that $B_0\varphi\in C^n_\lambda$ (recall that $B_0\varphi = \tau_1-\tau_2$). For $\omega = a^0 da^1\cdots da^n$ it is a consequence of the cocycle property of $B_0\varphi$. Indeed, one has $bB_0\varphi = 0$, hence $b'B_0\varphi(a^0,a^1,\dots,a^n,a) = (-1)^n\,B_0\varphi(aa^0,a^1,\dots,a^n)$ and since $b'B_0\varphi = \varphi - (-1)^{n+1}\,\varphi^\lambda$ we get

$$\varphi(a^0,\dots,a^n,a) - (-1)^{n+1}\,\varphi(a,a^0,\dots,a^n) = (-1)^n\,(B_0\varphi)(aa^0,a^1,\dots,a^n),$$

i.e. that $\mu((da)a^0da^1\cdots da^n) = (-1)^n\,\mu(a^0da^1\cdots da^n da)$.

To end the proof of Lemma 18 one modifies the natural cobordism between $(\Omega^*(\mathcal{A}),\hat{\tau}_1)$ and itself, given by the tensor product of $\Omega^*(\mathcal{A})$ with the algebra of differential forms on $[0,1]$, by adding to the integral the term $\mu\circ r_1$, where r_1 is the restriction map to $\{1\}\subset[0,1]$.

Putting together Lemmas 17 and 18 we see that two cocycles $\tau_1,\tau_2\in Z^n_\lambda(\mathcal{A})$ correspond to cobordant cycles if and only if $\tau_1-\tau_2$ belongs to the subspace $Z^n_\lambda(\mathcal{A})\cap B_0(Z^{n+1}(\mathcal{A},\mathcal{A}^*))$.

We shall now work out a better description of this subspace. Since $A\tau = (n+1)\tau$ for any $\tau\in C^n_\lambda(\mathcal{A})$, where A is the operator of cyclic antisymmetrisation, the above subspace is clearly contained in $B(Z^{n+1}(\mathcal{A},\mathcal{A}^*))$, where $B = AB_0 : C^{n+1}\to C^n$.

Lemma 19. a) *One has* $bB = -Bb$.

b) *One has* $Z_\lambda^n(\mathcal{A}) \cap B_0(Z^{n+1}(\mathcal{A},\mathcal{A}^*)) = BZ^{n+1}(\mathcal{A},\mathcal{A}^*)$.

Proof. a) For any cochain $\varphi \in C^{n+1}(\mathcal{A},\mathcal{A}^*)$, one has

$$B_0 b\varphi + b' B_0 \varphi = \varphi - (-1)^{n+1}\,\varphi^\lambda,$$

where λ is the cyclic permutation, $\lambda(i) = i-1$ modulo $n+2$. Applying A to both sides gives $AB_0 b\varphi + Ab'B_0\varphi = 0$. Thus, the result follows from Lemma 6.

b) By a) one has $BZ^{n+1}(\mathcal{A},\mathcal{A}^*) \subset Z_\lambda^n(\mathcal{A})$. Let us show that

$$BZ^{n+1}(\mathcal{A},\mathcal{A}^*) \subset B_0 Z^{n+1}(\mathcal{A},\mathcal{A}^*).$$

Let $\beta \in BZ^{n+1}(\mathcal{A},\mathcal{A}^*)$, so that $\beta = B\varphi$, $\varphi \in Z^{n+1}(\mathcal{A},\mathcal{A}^*)$.

We shall construct in a canonical way a cochain $\psi \in C^n(\mathcal{A},\mathcal{A}^*)$ such that $\frac{1}{n+1}\,\beta = B_0(\varphi - b\psi)$. Let $\theta = B_0\varphi - \frac{1}{n+1}\,\beta$; this implies $A\theta = 0$. Thus, there exists a canonical ψ such that $\psi - \varepsilon(\lambda)\psi^\lambda = \theta$, where λ is the generator of the group of cyclic permutations of $\{0,1,\dots,n\}$, $\lambda(i) = i-1$. We just have to check the equality

$$B_0 b\psi = \theta.$$

Using the equality $B_0 b\psi + b'B_0\psi = \psi - \varepsilon(\lambda)\psi^\lambda$, we just have to show that $b'B_0\psi = 0$. One has

$$\begin{aligned}
B_0\psi(a^0,\dots,a^{n-1}) &= \psi(1,a^0,\dots,a^{n-1}) - (-1)^n\,\psi(a^0,\dots,a^{n-1},1)\\
&= (-1)^{n-1}\,(\psi - \varepsilon(\lambda)\psi^\lambda)(a^0,\dots,a^{n-1},1) = (-1)^{n-1}\,\theta(a^0,\dots,a^{n-1},1)\\
&= (-1)^{n-1}\,(\varphi(1,a^0,\dots,a^{n-1},1) - (-1)^{n+1}\,\varphi(a^0,\dots,a^{n-1},1,1))\\
&\quad + \tfrac{1}{n+1}\,(-1)^n\,\beta(a^0,\dots,a^{n-1},1).
\end{aligned}$$

The contribution of the first two terms to $b'B_0\psi(a^0,\dots,a^n)$ is, since $b\varphi = 0$,

$$\begin{aligned}
&(-1)^{n-1}\sum_{j=0}^{n-1}(-1)^j\,(\varphi(1,a^0,\dots,a^ja^{j+1},\dots,a^n,1)\\
&\qquad + (-1)^n\,\varphi(a^0,\dots,a^ja^{j+1},\dots,a^n,1,1))\\
&\quad = (-1)^n\,(b\varphi(1,a^0,\dots,a^n,1) - \varphi(a^0,\dots,a^n,1))\\
&\qquad - (b\varphi(a^0,\dots,a^n,1,1) - (-1)^n\,\varphi(a^0,\dots,a^n,1)) = 0.
\end{aligned}$$

The contribution of the second term is proportional to

$$\sum_{j=0}^{n-1}(-1)^j\,\beta(a^0,\dots,a^ja^{j+1},\dots,a^n,1) = b\beta(a^0,\dots,a^n,1) = 0.$$

Corollary 20. 1) *The image of* $B: C^{n+1} \to C^n$ *is exactly* C_λ^n.

2) $B_\lambda^n(\mathcal{A}) \subset B_0 Z^{n+1}(\mathcal{A},\mathcal{A}^*)$.

Proof. 1) $\Rightarrow$ 2) since, assuming 1), any $b\varphi$, $\varphi \in C_\lambda^{n+1}$, is of the form $bB\psi = -Bb\psi$ and hence belongs to $BZ^{n+1}(\mathcal{A},\mathcal{A}^*)$, so that the conclusion follows from Lemma 19b). To prove 1) let $\varphi \in C_\lambda^n$. Choose a linear functional φ_0 on $\mathcal{A}$ with $\varphi_0(1) = 1$, and then let

$$\psi(a^0,\ldots,a^{n+1}) = \varphi_0(a^0)\,\varphi(a^1,\ldots,a^{n+1}) + (-1)^n\,\varphi((a^0 - \varphi_0(a^0)1),a^1,\ldots,a^n)\,\varphi_0(a^{n+1}).$$

One has $\psi(1,a^0,\ldots,a^n) = \varphi(a^0,\ldots,a^n)$ and

$$\psi(a^0,\ldots,a^n,1) = \varphi_0(a^0)\,\varphi(a^1,\ldots,a^n,1) + (-1)^n\,\varphi(a^0,a^1,\ldots,a^n) + (-1)^{n+1}\,\varphi_0(a^0)\,\varphi(1,a^1,\ldots,a^n) = (-1)^n\,\varphi(a^0,\ldots,a^n).$$

Thus, $B_0\psi = 2\varphi$ and $\varphi \in \operatorname{im} B$.

We are now ready to state the main result of this section. By Lemma 19 a) one has a well-defined map B from the Hochschild cohomology group $H^{n+1}(\mathcal{A},\mathcal{A}^*)$ to $HC^n(\mathcal{A})$.

Theorem 21. *Two cycles over $\mathcal{A}$ are cobordant if and only if their characters $\tau_1,\tau_2 \in HC^n(\mathcal{A})$ differ by an element of the image of B, where*

$$B : H^{n+1}(\mathcal{A},\mathcal{A}^*) \to HC^n(\mathcal{A}).$$

It is clear that the direct sum of two cycles over $\mathcal{A}$ is still a cycle over $\mathcal{A}$ and that cobordism classes of cycles over $\mathcal{A}$ form a group $M^*(\mathcal{A})$. The tensor product of cycles gives a natural map: $M^*(\mathcal{A}) \times M^*(\mathcal{B}) \to M^*(\mathcal{A} \otimes \mathcal{B})$. Since $M^*(\mathbb{C})$ is equal to $HC^*(\mathbb{C}) = \mathbb{C}[\sigma]$ as a ring, each of the groups $M^*(\mathcal{A})$ is a $\mathbb{C}[\sigma]$-module and, in particular, a vector space. By Theorem 21 this vector space is $HC^*(\mathcal{A})/\operatorname{im} B$.

The same group $M^*(\mathcal{A})$ has a closely related interpretation in terms of graded traces on the differential algebra $\Omega^*(\mathcal{A})$ defined following Proposition 3. Recall that, by Proposition 4, the map $\tau \mapsto \hat{\tau}$ is an isomorphism of $Z_\lambda^n(\mathcal{A})$ with the space of *closed* graded traces of degree n on $\Omega^*(\mathcal{A})$.

Theorem 22. *The map $\tau \mapsto \hat{\tau}$ gives an isomorphism of $HC^n(\mathcal{A})/\operatorname{im} B$ with the quotient of the space of closed graded traces of degree n on $\Omega^*(\mathcal{A})$ by those of the form $d^t\mu$, μ a graded trace on $\Omega^*(\mathcal{A})$ of degree $n+1$.*
Here d^t denotes the natural differential induced on the graded traces.

Proof. We have to show that, given $\tau \in Z_\lambda^n(\mathcal{A})$, one has $\hat{\tau} = d^t\mu$ for some graded trace μ if and only if $\tau \in \operatorname{im} B \supset B_\lambda^n$. Assume first that $\hat{\tau} = d^t\mu$. Then as in Lemma 17, one gets $\tau = B_0\varphi$ where $\varphi \in Z^{n+1}(\mathcal{A},\mathcal{A}^*)$ is the Hochschild cocycle

$$\varphi(a^0,a^1,\ldots,a^{n+1}) = \mu(a^0da^1\cdots da^{n+1}),\ \forall a^i \in \mathcal{A}.$$

Thus, $\tau = \frac{1}{n+1}AB_0\varphi \in \operatorname{im} B$.

Conversely, if $\tau \in \operatorname{im} B$, then by Lemma 19 b) one has $\tau = B_0\varphi$ for some $\varphi \in Z^{n+1}(\mathcal{A}, \mathcal{A}^*)$. Defining the linear functional μ on $\Omega^{n+1}(\mathcal{A})$ as in Lemma 18 we get a graded trace such that

$$\mu(da^0 da^1 \cdots da^n) = \tau(a^0, \dots, a^n) \quad \forall a^i \in \mathcal{A}$$

i.e.

$$\mu(d\omega) = \hat{\tau}(\omega), \quad \forall \omega \in \Omega^n(\mathcal{A}).$$

Thus, $M^*(\mathcal{A})$ is the homology of the complex of graded traces on $\Omega^*(\mathcal{A})$ with the differential d^t. This theory is dual to the theory obtained as the cohomology of the quotient of the complex $(\Omega^*(\mathcal{A}), d)$ by the subcomplex of commutators. The latter appears independently in the work of M. Karoubi [Kar$_1$] as a natural range for the higher Chern character defined on all the Quillen algebraic K-theory groups $K_i(\mathcal{A})$. Thus, Theorem 22 (and the analogous dual statement) allows one

1) to apply Karoubi's results [Kar$_1$] to extend the pairing of Section 3 (below) to all $K_i(\mathcal{A})$;

2) to apply the results of section γ) (below) to compute the cohomology of the complex $(\Omega^*(\mathcal{A})/[\ ,\ ], d)$.

1.γ The exact couple relating $HC^*(\mathcal{A})$ to Hochschild cohomology

By construction, the complex $(C^n_\lambda(\mathcal{A}), b)$ is a subcomplex of the Hochschild complex $(C^n(\mathcal{A}, \mathcal{A}^*), b)$, i.e. the identity map I is a morphism of complexes and gives an exact sequence

$$0 \to C^n_\lambda \xrightarrow{I} C^n \to C^n/C^n_\lambda \to 0.$$

To this there corresponds a long exact sequence of cohomology groups.

We shall prove in this section that the cohomology of the complex C/C_λ is $H^n(C/C_\lambda) = H^{n-1}(C_\lambda)$.

Thus, the long exact sequence of the above triple will take the form

$$\begin{aligned} 0 \to HC^0(\mathcal{A}) &\xrightarrow{I} H^0(\mathcal{A}, \mathcal{A}^*) \to HC^{-1}(\mathcal{A}) \to HC^1(\mathcal{A}) \xrightarrow{I} H^1(\mathcal{A}, \mathcal{A}^*) \\ &\to HC^0(\mathcal{A}) \to HC^2(\mathcal{A}) \xrightarrow{I} \cdots \\ &\to HC^n(\mathcal{A}) \xrightarrow{I} H^n(\mathcal{A}, \mathcal{A}^*) \to HC^{n-1}(\mathcal{A}) \\ &\to HC^{n+1}(\mathcal{A}) \xrightarrow{I} H^{n+1}(\mathcal{A}, \mathcal{A}^*) \to \cdots . \end{aligned}$$

On the other hand we have already constructed morphisms of cochain complexes S and B which have precisely the right degrees:

$$S : HC^{n-1}(\mathcal{A}) \to HC^{n+1}(\mathcal{A}),$$

$$B : H^n(\mathcal{A}, \mathcal{A}^*) \to HC^{n-1}(\mathcal{A}).$$

We shall prove that these are exactly the maps involved in the above long exact sequence, which now takes the form

$$\cdots HC^n(\mathcal{A}) \xrightarrow{I} H^n(\mathcal{A},\mathcal{A}^*) \xrightarrow{B} HC^{n-1}(\mathcal{A}) \xrightarrow{S} HC^{n+1}(\mathcal{A}) \xrightarrow{I} \cdots.$$

Finally, to the pair b, B corresponds a double complex as follows: $C^{n,m} = C^{n-m}(\mathcal{A},\mathcal{A}^*)$ (i.e. $C^{n,m}$ is 0 above the main diagonal) where the first differential $C^{n,m} \to C^{n+1,m}$ is given by the Hochschild coboundary b and the second differential $C^{n,m} \to C^{n,m+1}$ is given by the operator B.

By Lemma 19 one has the graded commutation of b and B. In addition, one checks that $B^2 = 0$. By construction, the cohomology of this double complex depends only upon the parity of n, and we shall prove that the sum of the even and odd groups is canonically isomorphic to

$$HC^*(\mathcal{A}) \underset{HC^*(\mathbb{C})}{\otimes} \mathbb{C} = H^*(\mathcal{A})$$

where $HC^*(\mathbb{C})$ acts on $\mathbb{C}$ by evaluation at $\sigma = 1$.

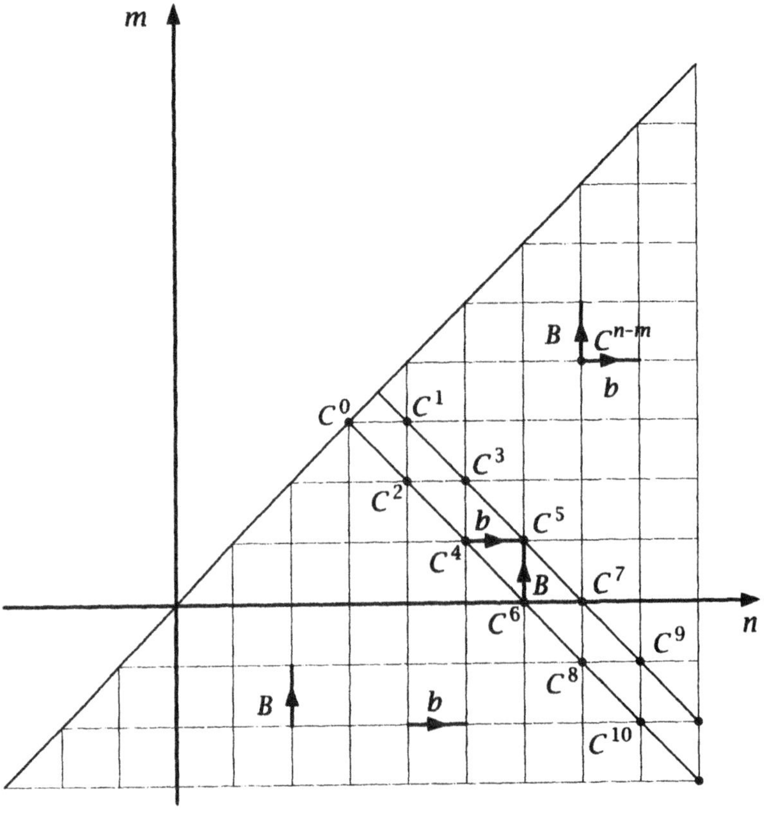

Figure III.1. The (b, B) bicomplex

The second filtration of this double complex $\left(F^q = \sum_{m \geq q} C^{n,m}\right)$ yields the same filtration of $H^*(\mathcal{A})$ as the filtration by dimensions of cycles. The associated spectral sequence is convergent and coincides with the spectral sequence coming from the above exact couple. All these results are based on the next two lemmas.

Lemma 23. *Let $\psi \in C^n(\mathcal{A}, \mathcal{A}^*)$ be such that $b\psi \in C_\lambda^{n+1}(\mathcal{A})$. Then $B\psi \in Z_\lambda^{n-1}(\mathcal{A})$ and $SB\psi = n(n+1)b\psi$ in $HC^*(\mathcal{A})$.*

Proof. One has $B\psi \in C_\lambda^{n-1}$ by construction, and $bB\psi = -Bb\psi = 0$ since $b\psi \in C_\lambda^{n+1}$. Thus, $B\psi \in Z_\lambda^{n-1}$. In the same way $b\psi \in Z_\lambda^{n+1}$.

Let $\varphi = B\psi$; by Proposition 15 one has $S\varphi = b\psi'$ where

$$\psi'(a^0, \ldots, a^n) = \sum_{j=1}^{n} (-1)^{j-1}\, \hat{\varphi}(a^0(da^1 \cdots da^{j-1})\, a^j (da^{j+1} \cdots da^n)).$$

It remains to find ψ'' such that $\psi'' - \psi \in B^n$ and such that

$$\psi' - \varepsilon(\lambda)\, \psi'^{\lambda} = n(n+1)(\psi'' - \varepsilon(\lambda)\, \psi''^{\lambda})$$

where $\lambda(i) = i - 1$ for $i \in \{1, \ldots, n\}$ and $\lambda(0) = n$. Let us first check that

$$(\psi' - \varepsilon(\lambda)\, \psi'^{\lambda})(a^0, \ldots, a^n) = (-1)^{n-1}\, (n+1)\, \varphi(a^n a^0, a^1, \ldots, a^{n-1}).$$

One has

$$\psi'^{\lambda}(a^0, \ldots, a^n) = \sum_{j=0}^{n-1} (-1)^j\, \hat{\varphi}((da^0 \cdots da^{j-1})\, a^j (da^{j+1} \cdots da^{n-1}) a^n).$$

Let

$$\omega_j = a^0(da^1 \cdots da^{j-1})\, a^j (da^{j+1} \cdots da^{n-1}) a^n.$$

Then

$$\begin{aligned} d\omega_j &= (da^0 \cdots da^{j-1})\, a^j (da^{j+1} \cdots da^{n-1}) a^n \\ &+ (-1)^{j-1}\, a^0 (da^1 \cdots da^j \cdots da^{n-1}) a^n \\ &+ (-1)^n\, a^0 (da^1 \cdots da^{j-1})\, a^j (da^{j+1} \cdots da^n). \end{aligned}$$

Thus, for $j \in \{1, \ldots, n-1\}$ one has

$$\begin{aligned} &(-1)^{j-1}\, \hat{\varphi}(a^0(da^1 \cdots da^{j-1})\, a^j (da^{j+1} \cdots da^n)) - \\ &\quad \varepsilon(\lambda)(-1)^j\, \hat{\varphi}(da^0 \cdots da^{j-1})\, a^j (da^{j+1} \cdots da^{n-1}) a^n) \\ &\quad = (-1)^{n-1}\, \varphi(a^n a^0, a^1, \ldots, a^{n-1}). \end{aligned}$$

Taking into account the cases $j = 0$ and $j = n$ gives the desired result.

Let us now determine ψ'', $\psi'' - \psi \in B^n(\mathcal{A}, \mathcal{A}^*)$ such that

$$(\psi'' - \varepsilon(\lambda)\, \psi''^{\lambda})(a^0, \ldots, a^n) = \tfrac{(-1)^{n-1}}{n}\, \varphi(a^n a^0, \ldots, a^{n-1}).$$

Let $\theta = B_0\psi$ and write $\theta = \theta_1 + \theta_2$ with $A\theta_1 = 0$, $\theta_2 \in C_\lambda^{n-1}(\mathcal{A})$ so that $\theta_2 = \frac{1}{n}\varphi$. Since $A\theta_1 = 0$ there exists $\psi_1 \in C^{n-1}$ such that $\theta_1 = D\psi_1$ where $D\psi_1 = \psi_1 - \varepsilon(\lambda)\,\psi_1^\lambda$.

Parallel to Lemma 6 one checks that $D \circ b = b' \circ D$ and hence $D(b\psi_1) = b'\theta_1$. Let $\psi'' = \psi - b\psi_1$. As $D = B_0 b + b' B_0$ we get $D\psi = b' B_0 \psi = b'\theta_1 + b'\theta_2$ hence $D\psi'' = b'\theta_2 = \frac{1}{n} b'\varphi$. Finally since $b\varphi = 0$ one has

$$b'\varphi = (-1)^{n-1}\,\varphi(a^n a^0, a^1, \dots, a^{n-1}).$$

As an immediate application of this lemma we get:

Corollary 24. *The image of $S : HC^{n-1}(\mathcal{A}) \to HC^{n+1}(\mathcal{A})$ is the kernel of the map $I : HC^{n+1}(\mathcal{A}) \to H^{n+1}(\mathcal{A}, \mathcal{A}^*)$.*

This is a really useful criterion for deciding when a given cocycle is a cup product by $HC^2(\mathbb{C})$, a question which arises naturally in determining the dimension of a given class in $H^*(\mathcal{A})$. In particular, it shows that if V is a compact manifold of dimension m, and if we take $\mathcal{A} = C^\infty(V)$, any cocycle τ in $HC^n(\mathcal{A})$ (satisfying the obvious continuity requirements (cf. Section 2)) is in the image of S for $n > m = \dim V$.

Let us now prove the second important lemma:

Lemma 25. *The natural map*

$$(\operatorname{im} B \cap \ker b)/b(\operatorname{im} B) \to (\ker B \cap \ker b)/b(\ker B)$$

is bijective.

Proof. Let us show the injectivity. Let $\varphi \in \operatorname{im} B \cap \ker b$, say $\varphi \in Z_\lambda^{n+1}(\mathcal{A})$, and assume $\varphi \in b(\ker B)$. Then Lemma 23 shows that φ and $S(0) = 0$ are in the same class in $HC^{n+1}(\mathcal{A})$, and hence $\varphi \in b(\operatorname{im} B)$.

Let us show the surjectivity. Let $\varphi \in Z^{n+1}(\mathcal{A}, \mathcal{A}^*)$, $B\varphi = 0$ and $\psi \in C^n(\mathcal{A}, \mathcal{A}^*)$, with $\psi - \varepsilon(\lambda)\psi^\lambda = B_0\varphi$. As above one gets $B_0 b\psi = B_0\varphi$. This shows that $\varphi' = \varphi - b\psi \in Z_\lambda^{n+1}(\mathcal{A})$ since $D\varphi' = B_0 b\varphi' + b' B_0 \varphi' = 0$. Let us show that $B\psi \in bC_\lambda^{n-2}$. Since $\psi - \varepsilon(\lambda)\,\psi^\lambda = B_0 b\psi$ one has $b' B_0 \psi = 0$. One checks easily that $b'^2 = 0$ and that the b' cohomology on $C^n(\mathcal{A}, \mathcal{A}^*)$ is trivial (if $b'\varphi_1 = 0$ one has $b'\varphi_1(a^0, \dots, a^n, 1) = 0$, i.e. $\varphi_1 = b'\varphi_2$, where

$$\varphi_2(a^0, \dots, a^{n-1}) = (-1)^{n-1}\,\varphi_1(a^0, \dots, a^{n-1}, 1)).$$

Thus, $B_0\psi = b'\theta$ for some $\theta \in C^{n-2}$, and $B\psi = Ab'\theta = bA\theta \in bC_\lambda^{n-2}$.

Thus, since $C_\lambda^{n-2} = \operatorname{im} B$ one has $B\psi = bB\theta_1$ for some $\theta_1 \in C^{n-1}$, i.e. $\psi + b\theta_1 \in \ker B$ and $b\psi \in b(\ker B)$. As $\varphi - b\psi \in Z_\lambda^{n+1}$ this ends the proof of the surjectivity.

Putting together the above Lemmas 23 and 25 we arrive at an expression for $S : HC^{n-1}(\mathcal{A}) \to HC^{n+1}(\mathcal{A})$ involving b and B:

$$S = n(n+1)\,bB^{-1}.$$

More explicitly, given $\varphi \in Z_\lambda^{n-1}(\mathcal{A})$ one has $\varphi \in \operatorname{im} B$, thus $\varphi = B\psi$ for some ψ, and this uniquely determines $b\psi \in (\ker b \cap \ker B)/b(\ker B) = HC^{n+1}(\mathcal{A})$. To check that $b\psi$ is equal to $\frac{1}{n(n+1)} S\varphi$ one chooses ψ as in Proposition 15

$$\psi(a^0,\ldots,a^n) = \frac{1}{n(n+1)} \sum_{j=1}^{n} (-1)^{j-1}\, \hat{\varphi}(a^0(da^1\cdots da^{j-1})\, a^j(da^{j+1}\cdots da^n)).$$

As an immediate corollary we get

Theorem 26. *The following triangle is exact:*

$$\begin{array}{ccc} & H^*(\mathcal{A},\mathcal{A}^*) & \\ {}^{B}\nearrow & & \searrow{}^{I} \\ HC^*(\mathcal{A}) & \xrightarrow{\ S\ } & HC^*(\mathcal{A}) \end{array}$$

Proof. We have already seen that $\operatorname{im} S = \ker I$. By the above description of S one has $\ker S = \operatorname{im} B$. Next $B \circ I = 0$ since B is equal to 0 on C_λ. Finally, if $\varphi \in Z^n(\mathcal{A},\mathcal{A}^*)$ and $B\varphi \in B_\lambda^{n-1}$, $B\varphi = bB\theta$ for some $\theta \in C^{n-1}$, so that

$$\varphi + b\theta \in \ker B \cap \ker b \subset \operatorname{im} I + b(\ker B)$$

by Lemma 25. Thus, $\ker B = \operatorname{im} I$.

We shall now identify the long exact sequence given by Theorem 26 with the one derived from the exact sequence of complexes

$$0 \to C_\lambda \to C \to C/C_\lambda \to 0.$$

Corollary 27. *The morphism of complexes $B : C/C_\lambda \to C$ induces an isomorphism of $H^n(C/C_\lambda)$ with $HC^{n-1}(\mathcal{A})$ and identifies the above triangle with the long exact sequence derived from the exact sequence of complexes $0 \to C_\lambda \to C \to C/C_\lambda \to 0$.*

Proof. This follows from the five lemma applied to

$$\begin{array}{ccccccccc} H^n(C_\lambda) & \to & H^n(C) & \to & H^n(C/C_\lambda) & \to & H^{n+1}(C_\lambda) & \to & H^{n+1}(C) \\ \| & & \| & & \downarrow B & & \| & & \| \\ HC^n(\mathcal{A}) & \xrightarrow{I} & H^n(\mathcal{A},\mathcal{A}^*) & \xrightarrow{B} & HC^{n-1}(\mathcal{A}) & \xrightarrow{S} & HC^{n+1}(\mathcal{A}) & \to & H^{n+1}(\mathcal{A},\mathcal{A}^*). \end{array}$$

Together with Theorem 21 we get:

Corollary 28. a) *Two cycles with characters τ_1 and τ_2 are cobordant if and only if $S\tau_1 = S\tau_2$ in $HC^*(\mathcal{A})$.*

b) *One has a canonical isomorphism*

$$M^*(\mathcal{A}) \underset{M^*(\mathbb{C})}{\otimes} \mathbb{C} = H^*(\mathcal{A}).$$

c) *Under that isomorphism the canonical filtration $F^n H^*(\mathcal{A})$ corresponds to the filtration of the left side by the dimension of the cycles.*

Proof of 28 b). Both sides are identical with the inductive limit of the system $(HC^n(\mathcal{A}), S)$.

Let us now carefully normalize the two differentials d_1 and d_2 of the double complex C^* so that the map S is given simply by $d_1 d_2^{-1}$. This is done as follows:

a) $C^{n,m} = C^{n-m}(\mathcal{A}, \mathcal{A}^*)$, $\forall n, m \in \mathbb{Z}$;

b) for $\varphi \in C^{n,m}$, $d_1\varphi = (n - m + 1)\, b\varphi \in C^{n+1,m}$;

c) for $\varphi \in C^{n,m}$, $d_2\varphi = \frac{1}{n-m} B\varphi \in C^{n,m+1}$ (if $n = m$, the latter is 0).

Note that $d_1 d_2 = -d_2 d_1$ follows from $Bb = -bB$.

Theorem 29. a) *The initial term E_2 of the spectral sequence associated to the first filtration $F_p C = \sum_{n \geq p} C^{n,m}$ is equal to 0.*

b) *Let $F^q C = \sum_{m \geq q} C^{n,m}$ be the second filtration; then $H^p(F^q C) = HC^n(\mathcal{A})$ for $n = p - 2q$.*

c) *The cohomology of the double complex C is given by*

$$H^n(C) = H^{\mathrm{ev}}(\mathcal{A}) \ if\ n\ is\ even$$

and

$$H^n(C) = H^{\mathrm{odd}}(\mathcal{A}) \ if\ n\ is\ odd.$$

d) *The spectral sequence associated to the second filtration is convergent: it converges to the associated graded $\sum F^q H^*(\mathcal{A}) / F^{q+1} H^*(\mathcal{A})$ and it coincides with the spectral sequence associated with the exact couple. In particular, its initial term E_2 is*

$$\ker(I \circ B) / \operatorname{im}(I \circ B).$$

Proof. a) Let us consider the exact sequence of complexes of cochains $0 \to \operatorname{im} B \to \ker B \to \ker B / \operatorname{im} B \to 0$, where the coboundary is b. By Lemma 25 the first map, $\operatorname{im} B \to \ker B$, becomes an isomorphism in cohomology, thus the b cohomology of the complex $\ker B / \operatorname{im} B$ is 0.

b) Let $\varphi \in (F^q C)^p = \sum_{m \geq q, n+m=p} C^{n,m}$, satisfy $d\varphi = 0$, where $d = d_1 + d_2$. By a) it is cohomologous in $F^q C$ to an element ψ of $C^{p-q,q}$. Then $d\psi = 0$

means $\psi \in \ker b \cap \ker B$, and $\psi \in \operatorname{im} d$ means $\psi \in b(\ker B)$. Thus, using the isomorphism from Lemma 25,

$$(\ker b \cap \ker B)/b(\ker B) = HC^{p-2q}(\mathcal{A})$$

one gets the result.

c) By the computation in Lemma 23 of S as $d_1 d_2^{-1}$ we see that the map from $H^p(F^q C)$ to $H^p(F^{q-1}C)$ is the map $(-S)$ from $HC^{p-2q}(\mathcal{A})$ to $HC^{p-2q+2}(\mathcal{A})$; thus the result is immediate.

d) The convergence of the spectral sequence is obvious, since $C^{n,m} = 0$ for $m > n$. Since the filtration of $H^n(C)$ given by $H^n(F^q C)$ coincides with the natural filtration of $H^*(\mathcal{A})$ (cf. the proof of c)), the limit of the spectral sequence is the associated graded

$$\sum_q F^q\, H^{\mathrm{ev}}(\mathcal{A})/F^{q+1}\, H^{\mathrm{ev}}(\mathcal{A}) \quad \text{for } n \text{ even},$$

$$\sum_q F^q\, H^{\mathrm{odd}}(\mathcal{A})/F^{q+1}\, H^{\mathrm{odd}}(\mathcal{A}) \quad \text{for } n \text{ odd}.$$

It is clear that the initial term E_2 is $\ker I \circ B / \operatorname{im} I \circ B$. One then checks that it coincides with the spectral sequence of the exact couple.

Remark 30. a) We shall throughout this book identify the cyclic cohomology $HC^n(\mathcal{A})$ with $H^p(F^q C)$, $n = p - 2q$, in the (d_1, d_2) bicomplex using the following sign convention dictated by the $-$ sign appearing in the proof of Theorem 29 c)

$$\varphi \in Z_\lambda^n(\mathcal{A}) \to (-1)^{[n/2]}\, \varphi \in C^{p-q,q}.$$

This sign is important in comparing the various expressions for the pairing of cyclic cohomology with K-theory.

b) *Action of $H^*(\mathcal{A}, \mathcal{A})$.* Using the product $\vee$ of [Car-E]

$$H^n(\mathcal{A}, \mathcal{M}_1) \otimes H^m(\mathcal{A}, \mathcal{M}_2) \to H^{n+m}(\mathcal{A}, \mathcal{M}_1 \otimes_{\mathcal{A}} \mathcal{M}_2)$$

one sees that $H^*(\mathcal{A}, \mathcal{A})$ becomes a graded commutative algebra (making use of the equality $\mathcal{A} \otimes_{\mathcal{A}} \mathcal{A} = \mathcal{A}$, as $\mathcal{A}$-bimodules) which acts on $H^*(\mathcal{A}, \mathcal{A}^*)$ (since $\mathcal{A} \otimes_{\mathcal{A}} \mathcal{A}^* = \mathcal{A}^*$). In particular any derivation δ of $\mathcal{A}$ defines an element $[\delta]$ of $H^1(\mathcal{A}, \mathcal{A})$. The explicit formula of [Car-E] for the product $\vee$ would give, at the cochain level

$$(\varphi \vee \delta)(a^0, a^1, \ldots, a^{n+1}) = \varphi(\delta(a^{n+1})a^0, a^1, \ldots, a^n), \quad \forall \varphi \in Z^n(\mathcal{A}, \mathcal{A}^*).$$

One checks that at the level of cohomology classes it coincides with

$$(\delta \# \varphi)(a^0, a^1, \ldots, a^{n+1})$$
$$= \frac{1}{n+1} \sum_{j=1}^{n+1} (-1)^j\, \hat{\varphi}(a^0(da^1 \cdots da^{j-1})\, \delta(a^j)(da^{j+1} \cdots da^{n+1}))\,.$$

With the latter formula one checks the equality

$$\delta^*\varphi = (I \circ B)(\delta\#\varphi) + \delta\#((I \circ B)\varphi) \text{ in } H^{n+1}(\mathcal{A}, \mathcal{A}^*)$$

where $\delta^*\varphi(a^0,\ldots,a^n) = \sum_{i=1}^n \varphi(a^0,\ldots,\delta(a^i),\ldots,a^n)$ for all $a^i \in \mathcal{A}$. This is the natural extension of the basic formula of differential geometry $\partial_X = di_X + i_X d$, expressing the Lie derivative with respect to a vector field X on a manifold.

c) *Homotopy invariance of* $H^*(\mathcal{A})$. Let $\mathcal{A}$ be an algebra (with unit), $\mathcal{B}$ a locally convex topological algebra and $\varphi \in Z_\lambda^n(\mathcal{B})$ a continuous cocycle (cf. Appendix B). Let ρ_t, $t \in [0,1]$, be a family of homomorphisms $\rho_t : \mathcal{A} \to \mathcal{B}$ such that

$$\text{for all } a \in \mathcal{A}\,,\ \text{the map } t \in [0,1] \to \rho_t(a) \in \mathcal{B} \text{ is of class } C^1.$$

Then the images by S of the cocycles $\rho_0^*\varphi$ and $\rho_1^*\varphi$ coincide. To prove this one extends the Hochschild cocycle $\varphi\#\psi$ on $\mathcal{B} \otimes C^1([0,1])$ giving the cobordism of φ with itself (i.e. $\psi(f^0,f^1) = \int_0^1 f^0 df^1$, $\forall f^i \in C^1([0,1])$) to a Hochschild cocycle on the algebra $C^1([0,1],\mathcal{B})$ of C^1-maps from $[0,1]$ to $\mathcal{B}$. Then the map $\rho : \mathcal{A} \to C^1([0,1],\mathcal{B})$, $(\rho(a))_t = \rho_t(a)$, defines a chain over $\mathcal{A}$ and is a cobordism of $\rho_0^*\varphi$ with $\rho_1^*\varphi$. This shows that if one restricts to continuous cocycles, one has

$$\rho_0^* = \rho_1^* : H^*(\mathcal{B}) \to H^*(\mathcal{A}).$$

d) *Excision.* Since the cohomology theory $HC^*(\mathcal{A})$ is defined from the cohomology of a complex (C_λ^n, b), a relative theory $HC^*(\mathcal{A},\mathcal{B})$ can be developed for pairs $\mathcal{A} \overset{\pi}{\to} \mathcal{B}$ of algebras, where π is a surjective homomorphism. To the exact sequence of complexes

$$0 \to C_\lambda^n(\mathcal{B}) \overset{\pi^*}{\to} C_\lambda^n(\mathcal{A}) \to C_\lambda^n(\mathcal{A},\mathcal{B}) = C_\lambda^n(\mathcal{A})/C_\lambda^n(\mathcal{B}) \to 0$$

corresponds a long exact sequence of cohomology groups. Using the five lemma, the results of this section on the absolute groups extend easily to the relative groups, provided that one also extends Hochschild theory $H^*(\mathcal{A},\mathcal{A}^*)$ to the relative case.

M. Wodzicki ([Wo$_1$]) has characterized the non-unital algebras which satisfy excision in Hochschild and cyclic homology by the very simple property of H-unitality: An algebra $\mathcal{A}$ (over $\mathbb{C}$) is H-unital iff the b' complex is acyclic.

In [C-Q$_4$] J. Cuntz and D. Quillen have shown that excision holds in full generality in periodic cyclic cohomology.

III.2. Examples

The results of Section 1.γ), such as Theorem 26, provide very powerful tools to compute the cyclic cohomology of a given algebra $\mathcal{A}$ by tying it up with the standard tools of homological algebra available to compute Hochschild cohomology from projective resolutions of $\mathcal{A}$ viewed as an $\mathcal{A}$-bimodule. We shall illustrate this by two examples α) and β). In example γ) we describe the result of D. Burghelea for the cyclic cohomology of the group ring $\mathcal{A} = \mathbb{C}\Gamma$ of a discrete group Γ.

2.α $\mathcal{A} = C^\infty(V)$, V a compact smooth manifold

We endow this algebra $\mathcal{A} = C^\infty(V)$ with its natural Fréchet space topology and only consider *continuous* multilinear forms on $\mathcal{A}$ (cf. Appendix B). Thus, the topology of $\mathcal{A}$ is given by the seminorms $p_n(f) = \sup_{|\alpha|\le n} |\partial^\alpha f|$, using local charts in V. As a locally convex vector space, $C^\infty(V)$ is then nuclear, so that topological tensor products are uniquely defined, and for any integer n the n-fold topological tensor product

$$C^\infty(V)\hat{\otimes}\cdots\hat{\otimes}C^\infty(V) = C^\infty(V)^{\hat{\otimes}n}$$

is canonically isomorphic to $C^\infty(V^n)$.

In particular the algebra $\mathcal{B} = \mathcal{A}\hat{\otimes}\mathcal{A}^0$ is canonically isomorphic to the algebra $C^\infty(V\times V)$. Thus, $\mathcal{A}$, viewed as an $\mathcal{A}$-bimodule, corresponds to the $\mathcal{B}$-module given by the diagonal inclusion

$$\Delta : V \to V\times V\,,\ \Delta(p) = (p,p)\ \ \forall p\in V.$$

Using a projective resolution of the diagonal in $V\times V$ one gets

Proposition 1. [Co$_{17}$] *Let V be a smooth compact manifold and $\mathcal{A}$ the locally convex topological algebra $C^\infty(V)$. Then*

a) *The following map $\varphi \to C_\varphi$ is a canonical isomorphism of the continuous Hochschild cohomology group $H^k(\mathcal{A},\mathcal{A}^*)$ with the space $\mathcal{D}_k$ of k-dimensional de Rham currents on V:*

$$\langle C_\varphi, f^0 df^1\wedge\cdots\wedge df^k\rangle = \frac{1}{k!}\sum_{\sigma\in S_k}\varepsilon(\sigma)\,\varphi(f^0, f^{\sigma(1)},\dots f^{\sigma(k)})$$

$\forall f^0,\dots,f^k \in C^\infty(V)$.

b) *Under the isomorphism C the operator $I\circ B : H^k(\mathcal{A},\mathcal{A}^*) \to H^{k-1}(\mathcal{A},\mathcal{A}^*)$ is (k times) the de Rham boundary d^t for currents.*

We refer to [Co$_{17}$] for the proof. The analogous statement for the algebra of polynomials on an affine variety is due to Hochschild-Kostant-Rosenberg [Ho-K-R].

In particular, we recover from this proposition the de Rham complex of the manifold V without any appeal to the commutativity of the algebra $\mathcal{A}$, which is used in a crucial manner in the construction of the exterior algebra. Thus, for instance, if we replace the algebra $\mathcal{A}$ in Proposition 1 by the algebra $M_q(\mathcal{A}) = M_q(\mathbb{C}) \otimes \mathcal{A}$ of matrices over $\mathcal{A}$ we lose the commutativity and the exterior algebra but the cohomology groups $H^k(\mathcal{A}, \mathcal{A}^*)$, the operators $I, B \ldots$ still make sense and yield, using the Morita invariance of Hochschild cohomology ([Car-E]) the same result as for $k = 1$. The formula (a) is no longer valid and the Hochschild cocycle (class) associated to a current $C \in \mathcal{D}_k$ is now given by the formula:

$$\varphi_C(f^0 \otimes \mu^0, f^1 \otimes \mu^1, \ldots, f^k \otimes \mu^k) = \langle C, f^0 df^1 \wedge \cdots \wedge df^k \rangle \operatorname{Trace}(\mu^0 \cdots \mu^k)$$
$$\forall f^j \in C^\infty(V)\,,\ \mu^j \in M_q(\mathbb{C}).$$

Let us now compute the cyclic cohomology $HC^*(C^\infty(V))$.

Theorem 2. [Co$_{17}$] *Let $\mathcal{A}$ be the locally convex topological algebra $C^\infty(V)$. Then*

1) *For each k, $HC^k(\mathcal{A})$ is canonically isomorphic to the direct sum*

$$\ker b (\subset \mathcal{D}_k) \oplus H_{k-2}(V, \mathbb{C}) \oplus H_{k-4}(V, \mathbb{C}) \oplus \cdots$$

where $H_q(V, \mathbb{C})$ is the usual de Rham homology of V and b the de Rham boundary.

2) *$H^*(\mathcal{A})$ is canonically isomorphic to the de Rham homology $H_*(V, \mathbb{C})$, with filtration by dimension.*

Proof. 1) Let us explicitly describe the isomorphism. Let $\varphi \in HC^k(\mathcal{A})$. Then the current C associated to $I(\varphi)$ is given by

$$\langle C, f^0 df^1 \wedge \cdots \wedge df^k \rangle = \frac{1}{k!} \sum_{\sigma \in S_k} \varphi(f^0, f^{\sigma(1)}, \ldots, f^{\sigma(k)}).$$

It is closed since $B(I(\varphi)) = 0$, so that the cochain

$$\overline{\varphi}(f^0, f^1, \ldots, f^k) = \langle C, f^0 df^1 \wedge \cdots \wedge df^k \rangle$$

belongs to $Z^k_\lambda(\mathcal{A})$. The class of $\varphi - \overline{\varphi}$ in $HC^k(\mathcal{A})$ is well determined, and is by construction in the kernel of I. Thus, by Theorem 26 there exists $\psi \in HC^{k-2}(\mathcal{A})$ with $S\psi = \varphi - \overline{\varphi}$, and ψ is unique modulo the image of B. Thus, the homology class of the closed current $C(I(\psi))$ is well determined. Moreover, the class of $\psi - \overline{\psi}$ in $HC^{k-2}(\mathcal{A})$ is well determined. Repeating this process one gets the desired sequence of homology classes $\omega_j \in H_{k-2j}(V, \mathbb{C})$. By construction, φ is in the same class in $HC^k(\mathcal{A})$ as $\widetilde{C} + \sum_{j=1}^{\infty} S^j \widetilde{\omega}_j$, where for any closed current ω_j in the class one takes

$$\widetilde{\omega}_j(f^0, f^1, \ldots, f^{k-2j}) = \langle \omega_j, f^0 df^1 \wedge \cdots \wedge df^{k-2j} \rangle.$$

This shows that the map that we just constructed is an injection of $HC^k(\mathcal{A})$ to $\ker b \oplus H_{k-2}(V,\mathbb{C}) \oplus \cdots \oplus H_{k-2i}(V,\mathbb{C}) \oplus \cdots$.

The surjectivity is obvious.

2) In 1) we see, by the construction of the isomorphism, that $S : HC^k(\mathcal{A}) \to HC^{k+2}(\mathcal{A})$ is the map which associates to each $C \in \ker b$ its homology class. The inclusion follows. Note that in this example the spectral sequence of Theorem 26 is degenerate, so that its E_2 term is already the de Rham homology of V.

Theorem 2 shows, in particular, that the periodic cyclic cohomology of $C^\infty(V)$, with its natural filtration, is the de Rham homology of the manifold V. One should not however conclude too hastily that cyclic cohomology gives nothing new in this case of smooth manifolds. We shall now see indeed from the example of the fundamental class $[\mathbb{S}^1]$ of the circle $\mathbb{S}^1$ that its image under the periodicity operator

$$S^k[\mathbb{S}^1] \in HC^{2k+1}(C^\infty(\mathbb{S}^1))$$

extends to the much larger algebra $C^\alpha(\mathbb{S}^1)$ of Hölder continuous functions of exponent $\alpha > \frac{1}{2k+1}$, i.e. of functions f such that

$$|f(x) - f(y)| \le C\, d(x,y)^\alpha$$

where d is the usual metric on $\mathbb{S}^1$.

To get a nice formula for a cyclic cocycle $\tau_k \sim S^k[\mathbb{S}^1]$ which extends to C^α we shall write it as a cocycle on $C_c^\alpha(\mathbb{R})$, but a similar formula works for $\mathbb{S}^1$.

Proposition 3. a) *The following defines a cyclic cocycle $\tau_k \in HC^{2k+1}(C_c^\alpha(\mathbb{R}))$:*

$$\tau_k(f^0,f^1,\ldots,f^{2k+1}) = \int f^0(x^0)\frac{f^1(x^1)-f^1(x^0)}{x^1-x^0}\,\frac{f^2(x^2)-f^2(x^1)}{x^2-x^1}\cdots$$
$$\cdots\frac{f^{2k}(x^{2k})-f^{2k}(x^{2k-1})}{x^{2k}-x^{2k-1}}\,\frac{f^{2k+1}(x^0)-f^{2k+1}(x^{2k})}{x^0-x^{2k}}\prod_{i=0}^{2k} dx^i.$$

b) *The restriction of τ_k to $C_c^\infty(\mathbb{R})$ is equal in HC^{2k+1} to $c_k S^k \tau_0$, where τ_0 is the homology fundamental class of $\mathbb{R}$*

$$\tau_0(f^0,f^1) = \int f^0 df^1$$

and

$$c_k = \frac{(2\pi i)^{2k}}{2^k(2k+1)(2k-1)\cdots 3.1.}$$

The multiple integral makes sense since each of the terms

$$\frac{f^j(x^j)-f^j(x^{j-1})}{x^j-x^{j-1}}$$

is $O(|x^j - x^{j-1}|^{\alpha-1})$. One then checks that τ_k is a cyclic cocycle and that b) holds.

Thus, using a similar formula for $S^k[\mathbb{S}^1] \in HC^{2k+1}(C^\alpha(\mathbb{S}^1))$ we see that we get an explicit formula for the push-forward $\psi_*[\mathbb{S}^1]$ without perturbing the map $\psi : \mathbb{S}^1 \to V$, Hölder continuous of exponent $\alpha > \frac{1}{2k+1}$

$$\psi_*[\mathbb{S}^1](f^0, \dots, f^{2k+1}) = \tau_k(\psi^* f^0, \psi^* f^1, \dots, \psi^* f^{2k+1})$$

$\forall f^0, \dots, f^{2k+1} \in C^\infty(V)$, with $\psi^* f = f \circ \psi$.

This gives a cyclic cocycle $\psi_*[\mathbb{S}^1] \in HC^{2k+1}(C^\infty(V))$ whose formula does not use any smoothing of the Hölder continuous map ψ.

The naive formula

$$\int_{\mathbb{S}^1} \psi^* f^0 d(\psi^* f^1)$$

does not make sense unless the functions $g^j = \psi^* f^j$ on $\mathbb{S}^1$ belong to the Sobolev space $W^{\frac{1}{2}}$ governed by the finiteness of

$$\sum n |\hat{g}(n)|^2$$

where $\hat{g}$ is the Fourier transform of g. Thus, it does not make sense for $\psi^* f$, $f \in C^\infty(V)$, when f is only Hölder of exponent $\alpha < \frac{1}{2}$.

Of course in the periodic cyclic cohomology of $C^\infty(V)$, the above cocycle coincides with $S^k \psi'_*[\mathbb{S}^1]$ where ψ' is smooth and homotopic to ψ.

Remark 4. a) Let, as above, $\mathcal{A} = C^\infty(V)$ and assume, to simplify, that the Euler characteristic of V vanishes. Then let X be a nonvanishing section of the vector bundle on $V \times V$ which is the pull-back by the second projection $\mathrm{pr}_2 : V \times V \to V$ of the complexified tangent bundle $T_{\mathbb{C}}(V)$, and which satisfies:

For $(a, b) \in V \times V$ close enough to the diagonal, $X(a, b)$ coincides with the real tangent vector $\exp_b^{-1}(a)$, where $\exp_b : T_b(V) \to V$ is the exponential map associated to a given Riemannian metric on V.

Let E_k be the complex vector bundle on $V \times V$ which is the pull-back by the second projection pr_2 of the exterior power $\wedge^k T^*_{\mathbb{C}}(V)$. The contraction i_X by the section X gives a well-defined complex of $C^\infty(V \times V)$ modules, i.e. of $C^\infty(V)$ bimodules

$$C^\infty(V) \xleftarrow{\Delta^*} C^\infty(V \times V) \xleftarrow{i_X} C^\infty(V \times V, E_1) \leftarrow \cdots \leftarrow C^\infty(V \times V, E_n) \leftarrow 0$$

where $n = \dim V$ and Δ is the diagonal $V \to V \times V$. This gives an explicit projective resolution $\mathcal{M}'$ of the $\mathcal{A}$-bimodule $\mathcal{A}$, and a proof of Proposition 1 ([Co$_{17}$]).

Now let $\mathcal{M}_k = (\mathcal{A} \hat{\otimes} \mathcal{A}^o) \hat{\otimes} \mathcal{A}^{\hat{\otimes} k}$ be the standard resolution of the bimodule $\mathcal{A}$, with the boundary

$$b_k : \mathcal{M}_k \to \mathcal{M}_{k-1}$$

given by the equality

$$\begin{aligned} b_k(1 \otimes a_1 \otimes \cdots \otimes a_k) &= (a_1 \otimes 1) \otimes (a_2 \otimes \cdots \otimes a_k) \\ &+ \sum_{j=1}^{k-1} (-1)^j \, (1 \otimes 1) \otimes a_1 \otimes \cdots \otimes a_j a_{j+1} \otimes \cdots \otimes a_k \\ &+ (-1)^k \, (1 \otimes a_k^0) \otimes (a_1 \otimes \cdots \otimes a_{k-1}). \end{aligned}$$

Then an explicit homotopy of the resolutions is given by

$$F : \mathcal{M}' \to \mathcal{M}$$

$$(F\omega)(a,b,x^1,\ldots,x^k) = \langle X(x^1,b) \wedge \cdots \wedge X(x^k,b)\ ,\ \omega(a,b)\rangle$$

$$\forall \omega \in \mathcal{M}'_k = C^\infty(V^2, E_k)\ ,\ \forall a,b,x^1,\ldots,x^k \in V.$$

Working out the homotopy formulas explicitly yields for any given cyclic cocycle $\varphi \in HC^q(C^\infty(V))$ explicit closed currents ω_j of dimension $q-2j$ such that

$$\varphi = \omega_0 + \sum S^j \, \omega_j \ \text{ in } \ HC^q(C^\infty(V)).$$

b) Let $W \subset V$ be a submanifold of V, $i^* : C^\infty(V) \to C^\infty(W)$ the restriction map, and $0 \to \ker i^* \to C^\infty(V) \to C^\infty(W) \to 0$ the corresponding exact sequence of algebras. For the ordinary homology groups one has a long exact sequence

$$\cdots \to H_q(W) \to H_q(V) \to H_q(V,W) \to H_{q-1}(W) \to \cdots$$

where the connecting map is of degree -1.

Since HC^n is defined as a *cohomology* theory, i.e. from a cochain complex, the long exact sequence

$$\begin{aligned} \cdots \to HC^q(C^\infty(W)) &\to HC^q(C^\infty(V)) \\ &\to HC^q(C^\infty(V), C^\infty(W)) \xrightarrow{\partial} HC^{q+1}(C^\infty(W)) \to \cdots \end{aligned}$$

has a connecting map of degree $+1$. So one may wonder how this is compatible with Theorem 2. The point is that the connecting map for the long exact sequence of Hochschild cohomology groups is 0 (any current on W whose image in V is zero does itself vanish), thus $\mathrm{im}(\partial) \subset HC^{q-1}(C^\infty(W))$.

c) Only very trivial cyclic cocycles on $C^\infty(V)$ do extend continuously to the C^*-algebra $C(V)$ of continuous functions on a compact manifold. In fact, for any compact space X the continuous Hochschild cohomology of $\mathcal{A} = C(X)$ with coefficients in the bimodule $\mathcal{A}^*$ is trivial in dimension $n \geq 1$. Thus, by Theorem 26 the cyclic cohomology of $\mathcal{A}$ is given by $HC^{2n}(\mathcal{A}) = HC^0(\mathcal{A})$ and $HC^{2n+1}(\mathcal{A}) = 0$. This remark extends to arbitrary nuclear C^*-algebras.

2.β The cyclic cohomology of the noncommutative torus $\mathcal{A} = \mathcal{A}_\theta$, $\theta \in \mathbb{R}/\mathbb{Z}$ Let $\lambda = \exp 2\pi i\theta$. Denote by $S(\mathbb{Z}^2)$ the space of sequences $(a_{n,m})_{n,m\in\mathbb{Z}^2}$ of rapid decay (i.e. $(|n| + |m|)^q |a_{n,m}|$ is bounded for any $q \in \mathbb{N}$).

Let $\mathcal{A}_\theta$ be the algebra of which the generic element is a formal sum, $\sum a_{n,m} U_1^n U_2^m$, where $(a_{n,m}) \in S(\mathbb{Z}^2)$ and the product is specified by the equality $U_2 U_1 = \lambda U_1 U_2$.

We let τ be the canonical trace on $\mathcal{A}_\theta$ given by

$$\tau(\sum a_\nu U^\nu) = a_{(0,0)}$$

where we let $U^\nu = U_1^{n_1} U_2^{n_2}$ for $\nu = (n_1, n_2) \in \mathbb{Z}^2$. For $\theta \in \mathbb{Q}$ this algebra is Morita equivalent to the commutative algebra of smooth functions on the 2-torus. Thus in the case $\theta \in \mathbb{Q}$, the computation of $H^*(\mathcal{A}_\theta)$ follows from Theorem 2.

We shall now do the computation for arbitrary θ. The first step is to compute the Hochschild cohomology $H(\mathcal{A}_\theta, \mathcal{A}_\theta^*)$, where, of course, $\mathcal{A}_\theta$ is considered as a locally convex topological algebra (using the seminorms $p_q(a) = \sup(1 + |n| + |m|)^q |a_{n,m}|$). We say that θ *satisfies a Diophantine condition* if the sequence $|1 - \lambda^n|^{-1}$ is $O(n^k)$ for some k.

Proposition 5. [Co$_{17}$] a) *Let $\theta \notin \mathbb{Q}$. One has $H^0(\mathcal{A}_\theta, \mathcal{A}_\theta^*) = \mathbb{C}$.*

b) *If $\theta \notin \mathbb{Q}$ satisfies a Diophantine condition, then $H^j(\mathcal{A}_\theta, \mathcal{A}_\theta^*)$ is of dimension 2 for $j = 1$, and of dimension 1 for $j = 2$.*

c) *If $\theta \notin \mathbb{Q}$ does not satisfy a Diophantine condition, then H^1 and H^2 are infinite-dimensional non-Hausdorff spaces.*

Recall that by Theorem 2, $H^j(\mathcal{A}_\theta, \mathcal{A}_\theta^*)$ is infinite-dimensional for $j \leq 2$ when $\theta \in \mathbb{Q}$.

At this point, it might seem hopeless to compute the periodic cyclic cohomology $H^*(\mathcal{A}_\theta)$ when θ is an irrational number not satisfying a Diophantine condition, since the Hochschild cohomology is already quite complicated. We shall see however, that even in that case, where $H^*(\mathcal{A}_\theta, \mathcal{A}_\theta^*)$ is infinite-dimensional non-Hausdorff, the homology of the complex $(H^n(\mathcal{A}_\theta, \mathcal{A}_\theta^*), I \circ B)$ is *still finite-dimensional.*

Proposition 6. [Co$_{17}$] *For $\theta \notin \mathbb{Q}$ one has $HC^0(\mathcal{A}_\theta) = \mathbb{C}$ and the map*

$$I : HC^1(\mathcal{A}_\theta) \to H^1(\mathcal{A}_\theta, \mathcal{A}_\theta^*)$$

is an isomorphism.

Thus, in particular, any 1-dimensional current is closed.

Theorem 7. [Co$_{17}$] a) *For all values of θ, $H^{\text{ev}}(\mathcal{A}_\theta) \simeq \mathbb{C}^2$ and $H^{\text{odd}}(\mathcal{A}_\theta) \simeq \mathbb{C}^2$.*

b) *Let τ be the canonical trace on $\mathcal{A}_\theta$. A basis of*

$$H^{\text{odd}}(\mathcal{A}_\theta) = H^1(\mathcal{A}_\theta, \mathcal{A}_\theta^*)/\operatorname{im}(I \circ B)$$

is provided by the cyclic cocycles φ_1 *and* φ_2 *with* $\varphi_j(x^0,x^1) = \tau(x^0\delta_j(x^1))$ $\forall x^i \in \mathcal{A}_\theta$.

c) *One has* $H^{\mathrm{ev}}(\mathcal{A}_\theta) = H^2(\mathcal{A}_\theta)$; *it is a vector space of dimension 2 with basis* $S\tau$ (τ *the canonical trace) and the functional* φ *given by*

$$\varphi(x^0,x^1,x^2) = (2\pi i)^{-1}\,\tau(x^0(\delta_1(x^1)\delta_2(x^2) - \delta_2(x^1)\delta_1(x^2))) \qquad \forall x^i \in \mathcal{A}_\theta.$$

In these formulas, δ_1, δ_2 are the basic derivations of $\mathcal{A}_\theta : \delta_1(U^\nu) = 2\pi i n_1 U^\nu$, $\delta_2(U^\nu) = 2\pi i n_2 U^\nu$.

2.γ The cyclic cohomology of the group ring $\mathbb{C}\Gamma$ for Γ a discrete group

First let G be a compact group and X a topological space with a continuous action of G on it. Then the equivariant cohomology H_G^* of X is defined as the cohomology of the homotopy quotient $X_G = X \times_G EG$, where EG is the total space of the universal principal G-bundle over the classifying space BG. In particular $H_G^*(X)$ is, in a natural manner, a module over $H_G^*(\mathrm{pt}) = H^*(BG)$.

For $G = \mathbb{S}^1$, the one-dimensional torus, $B\mathbb{S}^1 = P_\infty(\mathbb{C})$ and $H_{\mathbb{S}^1}^*(\mathrm{pt})$ is a polynomial ring in one generator of degree 2.

The formal analogy between cyclic cohomology and $\mathbb{S}^1$-equivariant cohomology is well understood from the equality ([Co$_{26}$])

$$B\Lambda = B\mathbb{S}^1$$

where Λ is the small category introduced in [Co$_{26}$] which governs cyclic cohomology (cf. Appendix A) through the equality

$$HC^*(\mathcal{A}) = \mathrm{Ext}_\Lambda^*(\mathcal{A}^\natural, \mathbb{C}^\natural)$$

where the functor $\mathcal{A} \to \mathcal{A}^\natural$ from algebras to Λ-modules (Appendix A) gives the appropriate linearization of the nonabelian category of algebras. For group rings, i.e. for algebras over $\mathbb{C}$ of the form

$$\mathcal{A} = \mathbb{C}\Gamma$$

where Γ is a discrete (countable) group, the following theorem of D. Burghelea ([Bu]) yields a natural $\mathbb{S}^1$-space whose $\mathbb{S}^1$-equivariant cohomology (with complex coefficients) is the cyclic cohomology of $\mathcal{A}$. Recall that the *free loop space* $Y^{\mathbb{S}^1}$ of a given topological space Y is the space $\mathrm{Map}(\mathbb{S}^1, Y)$ of continuous maps from $\mathbb{S}^1$ to Y with the compact-open topology. This space is naturally an $\mathbb{S}^1$-space, using the action of $\mathbb{S}^1$ on $\mathbb{S}^1$ by rotations.

Theorem 8. [Bu] *Let* Γ *be a discrete group,* $\mathcal{A} = \mathbb{C}\Gamma$ *its group ring.*

a) *The Hochschild cohomology* $H^*(\mathcal{A}, \mathcal{A}^*)$ *is canonically isomorphic to the cohomology* $H^*((B\Gamma)^{\mathbb{S}^1}, \mathbb{C})$ *of the free loop space of the classifying space of* Γ.

b) *The cyclic cohomology* $HC^*(\mathcal{A})$ *is canonically isomorphic to the* $\mathbb{S}^1$-*equivariant cohomology* $H_{\mathbb{S}^1}^*((B\Gamma)^{\mathbb{S}^1}, \mathbb{C})$.

Moreover, the isomorphism b) is compatible with the module structure over $H^*(B\Lambda) = H^*(B\mathbb{S}^1)$, and under the isomorphisms a) and b) the long exact sequence of Theorem 26 becomes the Gysin exact sequence relating H^* to $H^*_{\mathbb{S}^1}$.

As a corollary of this theorem we get the computation of the cyclic cohomology of $\mathbb{C}\Gamma$ in terms of the cohomology of the subgroups of Γ defined as follows:

1) For any $g \in \Gamma$ let $C_g = \{h \in \Gamma; gh = hg\}$ be the *centralizer* of g.

2) For any $g \in \Gamma$ let $N_g = C_g/g^{\mathbb{Z}}$ be the quotient of C_g by the (central) subgroup generated by g.

Then let $\langle\Gamma\rangle$ be the set of conjugacy classes of Γ, let $\langle\Gamma\rangle' \subset \langle\Gamma\rangle$ be the subset of classes of elements $g \in \Gamma$ of *finite order* and $\langle\Gamma\rangle''$ its complement.

By construction, the groups C_g and N_g only depend upon the class $\hat{g} \in \langle\Gamma\rangle$ of g.

Corollary 9. [Bu]

1) $H^*(\mathbb{C}\Gamma, (\mathbb{C}\Gamma)^*) = \prod_{\hat{g}\in\langle\Gamma\rangle} H^*(C_g, \mathbb{C})$

2) $HC^*(\mathbb{C}\Gamma) = \prod_{\hat{g}\in\langle\Gamma\rangle'} (H^*(N_g, \mathbb{C}) \otimes HC^*(\mathbb{C})) \times \prod_{\hat{g}\in\langle\Gamma\rangle''} H^*(N_g, \mathbb{C})$.

Moreover, the structure of an $HC^*(\mathbb{C})$-module on $HC^*(\mathbb{C}\Gamma)$, i.e. the operator S, decomposes over the conjugacy classes and is the obvious one for the finite classes $\hat{g} \in \langle\Gamma\rangle'$. For the infinite classes the operator S is the product by the 2-cocycle $\omega_g \in H^2(N_g, \mathbb{C})$ of the central extension:

$$0 \to \mathbb{Z} \to C_g \to N_g \to 1$$

where we used the inclusion $\mathbb{Z} \subset \mathbb{C}$, and the map $H^2(N_g, \mathbb{Z}) \to H^2(N_g, \mathbb{C})$. In particular, the infinite conjugacy classes $\hat{g} \in \langle\Gamma\rangle''$ may contribute nontrivially to the periodic cyclic cohomology $H^*(\mathbb{C}\Gamma)$ (cf. [Bu]). We shall now give the details of the proof of Theorem 8 (cf. [Bu]). First, with $\mathcal{A} = \mathbb{C}\Gamma$, the associated cyclic vector space $C(\mathcal{A})$ (cf. Appendix A) is the linear span of the following cyclic set (Y_n, d_n^i, s_n^j, t_n):

$$Y_n = \Gamma^{n+1} = \{(g_0, g_1, \dots, g_n); g_i \in \Gamma\}$$

$$d_n^i(g_0, g_1, \dots, g_n) = (g_0, \dots, g_i g_{i+1}, \dots, g_n) \in Y_{n-1}, \text{ for } 0 \le i \le n-1$$

$$d_n^n(g_0, g_1, \dots, g_n) = (g_n g_0, g_1, \dots, g_{n-1}) \in Y_{n-1},$$

$$s_n^i(g_0, \dots, g_n) = (g_0, \dots, g_i, 1, g_{i+1}, \dots, g_n) \in Y_{n+1}$$

$$t_n(g_0, \dots, g_n) = (g_n, g_0, \dots, g_{n-1}) \in Y_n.$$

Thus, it follows that the Hochschild (resp. cyclic) cohomology $H^*(\mathcal{A}, \mathcal{A}^*)$ (resp. $HC^*(\mathcal{A})$) is the cohomology (resp. $\mathbb{S}^1$-equivariant cohomology) of the geometric realization $|Y|$ of Y

$$H^*(\mathcal{A}, \mathcal{A}^*) = H^*(|Y|, \mathbb{C})$$

$$HC^*(\mathcal{A}) = H^*_{\mathbb{S}^1}(|Y|)$$

where the *cyclic* structure of Y endows $|Y|$ with a canonical action of $\mathbb{S}^1$ (cf. Appendix A).

As we shall now show, the $\mathbb{S}^1$-space $|Y|$ turns out to be $\mathbb{S}^1$-equivariantly homeomorphic to the space of Γ-valued configurations of the oriented circle. By definition a Γ-valued configuration on $\mathbb{S}^1$ is a map $\alpha : \mathbb{S}^1 \to \Gamma$ such that $\alpha(\theta) = 1_\Gamma$ except on a finite subset of $\mathbb{S}^1$, called the support: $\mathrm{supp}(\alpha)$. The topology of the configuration space $C_{\mathbb{S}^1}(\Gamma)$ is generated by the open sets

$$\mathcal{U}(I_1,\dots,I_k,g_1,\dots,g_k) = \left\{\alpha;\ \mathrm{Supp}(\alpha) \subset \cup I_j\ ,\ \textstyle\prod_{I_j}\alpha(\theta) = g_j\right\}$$

where the I_j are open intervals in $\mathbb{S}^1$, the g_j belong to Γ, and the product $\prod_{I_j} \alpha(\theta)$ is the time-ordered product $\alpha(\theta_1)\cdots\alpha(\theta_k)$ of the values of α at the times $\theta_1 < \theta_2 < \cdots < \theta_k$ where $\mathrm{Supp}(\alpha) \cap I_j = \{\theta_1,\dots,\theta_k\}$. The natural homeomorphism $h : |Y| \to C_{\mathbb{S}^1}(\Gamma)$ is defined as follows. At the set theoretic level one has the decomposition

$$|Y| = \bigcup_{n=0}^{\infty} (Y_n \backslash \mathrm{Deg}\ Y_n) \times \mathrm{Int}(\Delta_n)$$

where the degeneracy $\mathrm{Deg}\ Y_n$ is the union of the images of the maps $s^i_{n-1} : Y_{n-1} \to Y_n$ and $\mathrm{Int}(\Delta_n)$ is the interior of the n-simplex $\Delta_n = \{(\lambda_i)_{i=0,\dots,n}\ ;\ \lambda_i \geq 0\ ,\ \sum \lambda_i = 1\}$.

Then the image under h of $((g_i),(\lambda_i)) \in (Y_n \backslash \mathrm{Deg}\ Y_n) \times \mathrm{Int}\,\Delta_n$ is the configuration with support contained in the set $\{0, \lambda_0, \lambda_0 + \lambda_1, \dots, \lambda_0 + \cdots + \lambda_{n-1}\}$ and such that

$$\alpha(0) = g_0\ ,\ \alpha(\lambda_0) = g_1, \dots, \alpha\left(\sum_{k=0}^{i-1} \lambda_k\right) = g_i\ ,\ \text{for } i \leq n.$$

Note that since $(g_i) \in Y_n \backslash \mathrm{Deg}\ Y_n$ one has $g_i \neq 1_\Gamma$ for $i \neq 0$, but one may have $g_0 = 1_\Gamma$, in which case the cardinality of the support of the configuration α is equal to n.

It is then not difficult to check that h is an $\mathbb{S}^1$-equivariant homeomorphism

$$h : |Y| \to C_{\mathbb{S}^1}(\Gamma).$$

Thus, Theorem 8 follows from the following proposition due to J. Milnor and G. Segal:

Proposition 10. *The configuration space $C_{\mathbb{S}^1}(\Gamma)$ is $\mathbb{S}^1$-equivariantly weakly homotopy equivalent to the free loop space $(B\Gamma)^{\mathbb{S}^1}$.*

More explicitly, let $I \subset \mathbb{S}^1$ be the open interval $\mathbb{S}^1 \backslash \{0\}$, so that then $C_I(\Gamma)$ is canonically homeomorphic to $B\Gamma$, and the above weak homotopy equivalence associates to a configuration $\alpha \in C_{\mathbb{S}^1}(\Gamma)$ the loop $\beta = j(\alpha)$,

$$\beta(\theta) = \text{ restriction to } I \text{ of } \alpha \circ R_\theta \in C_I(\Gamma) \simeq B\Gamma\ ,\ \forall \theta \in \mathbb{S}^1$$

where R_θ is the rotation $R_\theta(t) = t + \theta \quad \forall t \in \mathbb{S}^1$.

The classifying space $B\Gamma$ is the geometric realization $|X|$ of the simplicial set X given by

$$\begin{aligned} X^n &= \Gamma^n = \{(g_1,\ldots,g_n); g_i \in \Gamma\} \\ d_n^0(g_1,\ldots,g_n) &= (g_2,\ldots,g_n) \\ d_n^i(g_1,\ldots,g_n) &= (g_1,\ldots,g_i g_{i+1},\ldots,g_n) \\ d_n^n(g_1,\ldots,g_n) &= (g_1,\ldots,g_{n-1}) \\ s_n^i(g_1,\ldots,g_n) &= (g_1,\ldots,g_i,1_\Gamma,g_{i+1},\ldots,g_n). \end{aligned}$$

The natural homeomorphism $C_I(\Gamma) \simeq |X|$ associates to a configuration α on $I =]0,1[$, with supp $\alpha = \{t_1,t_2,\ldots,t_n\}$, $t_i < t_{i+1}$, $\alpha(t_i) = g_i$, the following element of $(X_n \backslash \mathrm{Deg}\, X_n) \times \mathrm{Int}\,\Delta_n$:

$$(g_1,\ldots,g_n) \times (t_1, t_2 - t_1, t_3 - t_2, \ldots, t_n - t_{n-1}, 1 - t_n).$$

To prove Proposition 10 one compares the two fibrations in

$$\begin{array}{ccccc} \Gamma & \longrightarrow & C_{\mathbb{S}_1}(\Gamma) & \xrightarrow{\text{res}} & C_I(\Gamma) \\ \downarrow & & \downarrow & & \downarrow \\ \Omega B\Gamma & \longrightarrow & (B\Gamma)^{\mathbb{S}^1} & \xrightarrow{\text{ev}} & B\Gamma \end{array}$$

where for $\alpha \in C_{\mathbb{S}^1}(\Gamma)$ its restriction res(α) to I determines α up to the value $\alpha(0) \in \Gamma$. In the second fibration $\Omega B\Gamma$ is the loop space of $B\Gamma$. The vertical arrows $C_I(\Gamma) \to B\Gamma$ and $\Gamma \to \Omega B\Gamma$ are weak homotopy equivalences and thus so is

$$j : C_{\mathbb{S}^1}(\Gamma) \to (B\Gamma)^{\mathbb{S}^1}.$$

2.δ Cyclic cohomology of $C_c^\infty(V \rtimes \Gamma)$

Let V be a smooth manifold and Γ a discrete group acting on V by diffeomorphisms. In this section we shall describe the cyclic cohomology of the crossed product algebra $\mathcal{A} = C_c^\infty(V) \rtimes \Gamma$. The analogues of the above results of D. Burghelea are due to Feigin and Tsygan [F-T], V. Nistor [Nis], and J.L. Brylinski [Bry$_2$] [Bry-N]. We shall describe an earlier result [Co$_{14}$], namely that the periodic cyclic cohomology $H^*(\mathcal{A})$ contains as a direct factor the twisted cohomology groups $H_\tau^*(V_\Gamma, \mathbb{C})$. We adopt here the notation of Section II.7. Thus, V_Γ is the homotopy quotient $V \times_\Gamma E\Gamma$ and τ is the real vector bundle on V_Γ associated to the (Γ-equivariant) tangent bundle TV of V. The algebra $\mathcal{A}$ is the convolution algebra of smooth functions with compact support on $V \times \Gamma$ and the convolution product is given by

$$(f_1 f_2)(x,g) = \sum_{g_1 g_2 = g} f_1(x,g_1)\, f_2(x g_1, g_2). \tag{1}$$

We let U_g be the element of $\mathcal{A}$ given, when V is compact, by $U_g(x,k) = 0$ if $g \neq k$, $U_g(x,k) = 1$ if $g = k$. When V is not compact this does not define an element of $\mathcal{A}$ but a multiplier of $\mathcal{A}$. In both cases, any element f of $\mathcal{A}$ can be uniquely written as a finite sum

$$f = \sum f_g U_g \quad , \quad f_g \in C_c^\infty(V) \tag{2}$$

and one has the algebraic rule

$$(U_g\, h\, U_g^{-1})(x) = h(xg) \qquad \forall x \in V\,,\ g \in \Gamma\,,\ h \in C_c^\infty(V). \tag{3}$$

Since the homotopy quotient V_Γ is the geometric realization of a simplicial manifold (cf. Appendix A) we can describe the twisted cohomology $H_\tau^*(V_\Gamma)$ as the cohomology of a double complex ([Bot$_1$] Theorem 4.5). More explicitly, we can view the space $E\Gamma$ as the geometric realization of the simplicial set $e\Gamma$ where $(e\Gamma)_n = \Gamma^{n+1}$ and

$$\begin{aligned} d_i(g_0,\dots,g_n) &= (g_0,\dots,\check{g}_i,\dots,g_n) \qquad \forall i = 0,1,\dots,n\,;\ g_j \in \Gamma, \\ s_j(g_0,\dots,g_n) &= (g_0,\dots,g_j,g_j,\dots,g_n) \qquad \forall j = 0,1,\dots,n\,;\ g_i \in \Gamma, \end{aligned} \tag{4}$$

where the superscript $^\vee$ denotes omission. The group Γ acts on the right on both V and $e\Gamma$ and by [Bot$_1$] Theorem 4.5 the τ-twisted cohomology $H_\tau^*(V_\Gamma, \mathbb{C})$ is the cohomology of the bicomplex of Γ-invariant simplicial τ-twisted forms on $V \times e\Gamma$. Now a twisted form on V is the same thing as a smooth de Rham current: to the twisted form ω one associates the smooth current C with values

$$C(\alpha) = \int_V \alpha \wedge \omega \tag{5}$$

for any differential form α with compact support such that deg (α)+deg (ω) = dim V.

It will be more convenient to describe the double complex (C^*, d_1, d_2) used in computing $H_\tau^*(V_\Gamma, \mathbb{C})$ in terms of currents on V rather than twisted forms.

We let $C^{n,m} = \{0\}$ unless $n \geq 0$ and $-\dim V \leq m \leq 0$. Otherwise we let $C^{n,m}$ be the space of totally antisymmetric maps $\gamma : \Gamma^{n+1} \to \Omega_{-m}(V)$, from Γ^{n+1} to the space $\Omega_{-m}(V)$ of de Rham currents of dimension $-m$ on V, which satisfy

$$\gamma(g_0 g, g_1 g, \dots, g_n g) = \gamma(g_0,\dots,g_n)g \qquad \forall g_i \in \Gamma\,,\ g \in \Gamma. \tag{6}$$

We have used the right action of Γ on V to act on currents. The coboundary $d_1 : C^{n,m} \to C^{n+1,m}$ is given by

$$(d_1\gamma)(g_0,\dots,g_{n+1}) = (-1)^m \sum_{j=0}^{n+1} (-1)^j\, \gamma(g_0,\dots,\check{g}_j,\dots,g_{n+1}) \qquad \forall g_i \in \Gamma. \tag{7}$$

The coboundary $d_2 : C^{n,m} \to C^{n,m+1}$ is the de Rham boundary

$$(d_2\gamma)(g_0,\ldots,g_n) = d^t(\gamma(g_0,\ldots,g_n)). \tag{8}$$

We can summarize the above discussion by:

Proposition 11. *The twisted cohomology $H^*_\tau(V_\Gamma,\mathbb{C})$ is naturally isomorphic, with a shift in dimension of* $\dim V$, *to the cohomology of the above bicomplex* (C^*,d_1,d_2).

This holds without regard to the regularity imposed on the currents and we shall perform our constructions with arbitrary currents. It is important, though, to note that the filtration of the cohomology of the bicomplex, given by the maximal value of $n-m$ on the support of cocycles $(\gamma_{n,m})$, does depend on the regularity. Let us illustrate this by an example. We let $V = \mathbb{S}^1$, $\Gamma = \mathbb{Z}$ and the action of Γ on V be given by a diffeomorphism $\varphi \in \mathrm{Diff}^+(\mathbb{S}^1)$. We choose φ with a Liouville rotation number, and not C^∞-conjugate to a rotation (cf. [Her]). The homotopy quotient V_Γ is the mapping torus, the quotient of $V \times \mathbb{R}$ by the diffeomorphism $\tilde{\varphi}$ given by

$$\tilde{\varphi}(x,s) = (\varphi(x), s+1) \qquad \forall x \in V = \mathbb{S}^1\ ,\ \forall s \in \mathbb{R}. \tag{9}$$

It is, by construction, a 2-torus which fibers over $B\Gamma = \mathbb{R}/\mathbb{Z}$. Its cohomotopy group $\pi^1(V_\Gamma)$ has another generator given by a continuous map $V_\Gamma \to V = \mathbb{S}^1$. We want to compute the corresponding cocycle in the bicomplex (C^*,d_1,d_2). (Here the group Γ preserves the orientation so that we can ignore the twisting by τ. Thus, one has a canonical map $\pi^1(V_\Gamma) \to H^1_\tau(V_\Gamma,\mathbb{C})$). If we use arbitrary currents the corresponding cocycle is easy to describe; it is given by the following element of $C^{0,0}$:

$$\gamma_0 \in \Omega_0 \text{ is the unique } \Gamma\text{-invariant probability measure on } V = \mathbb{S}^1. \tag{10}$$

Of course this measure is a 0-dimensional current, and one has $d_1\gamma = d_2\gamma = 0$.

Since φ is not C^∞-conjugate to a rotation the current γ_0 is not smooth, and we now have to describe a smooth cocycle γ' in the above bicomplex belonging to the same cohomology class. For this we let $\gamma'_0 \in \Omega_0$ be any *smooth* 0-dimensional current such that $\langle 1,\gamma'_0\rangle = 1$. It is not φ-invariant, but the equation $\varphi\gamma'_0 - \gamma'_0 = d^t\gamma'_1$ can easily be solved with γ'_1 a smooth 1-dimensional current on $V = \mathbb{S}^1$. This yields the desired smooth cocycle γ'. As we shall see, while γ_0 gives rise to a *trace* on the crossed product algebra $\mathcal{A} = C^\infty_c(V) \rtimes \Gamma$, the cocycle γ' gives rise to a cyclic 2-cocycle on $\mathcal{A}$, with the same class in periodic cyclic cohomology.

We shall now describe in full generality a morphism Φ of bicomplexes from (C^*,d_1,d_2) to the (b,B) bicomplex of the algebra $\mathcal{A}$. This will give the desired map from $H^*_\tau(V_\Gamma,\mathbb{C})$ to the periodic cyclic cohomology of $\mathcal{A}$ and will make full use of the (b,B) description of the latter (Section 1). Our construction of the morphism Φ works for smooth groupoids for which the maps r and s

are étale but the case $G = V \rtimes \Gamma$ has interesting special features due to the *total antisymmetry* of the cochains $\gamma(g_0, \dots, g_n)$, $g_i \in \Gamma$. To exploit this, let us introduce an auxiliary graded differential algebra C. As an *algebra*, C is the crossed product $C = \mathcal{B} \rtimes_\alpha \Gamma$, where $\mathcal{B}$ is the graded tensor product:

$$\mathcal{B} = A^*(V) \otimes \wedge^*(\mathbb{C}\Gamma') \tag{11}$$

of the graded algebra $A^*(V)$ of smooth compactly supported differential forms on V by the exterior algebra $\wedge^*(\mathbb{C}\Gamma')$ of the linear space $\mathbb{C}\Gamma'$ with basis the δ_g, $g \in \Gamma$, with $\delta_e = 0$, where e is the unit of Γ. The action α of Γ on $\mathcal{B}$ by automorphisms is defined as the tensor product, $\alpha_g = \alpha_{1,g} \otimes \alpha_{2,g} \quad \forall g \in \Gamma$. Here α_1 is the natural action of Γ on $A^*(V)$ commuting with the differential and satisfying

$$\alpha_{1,g}(f)(x) = f(xg) \qquad \forall f \in C_c^\infty(V)\,,\ x \in V\,,\ g \in \Gamma. \tag{12}$$

The action α_2 of Γ on $\wedge^*(\mathbb{C}\Gamma')$ preserves the subspace $\mathbb{C}\Gamma' = \wedge^1 \mathbb{C}\Gamma'$ and is given by the equality

$$\alpha_{2,g}\, \delta_k = \delta_{kg^{-1}} - \delta_{g^{-1}} \qquad \forall g, k \in \Gamma. \tag{13}$$

Since the action α of Γ on $\mathcal{B}$ preserves the (bi)grading of $\mathcal{B}$ the crossed product $C = \mathcal{B} \rtimes \Gamma$ has a canonical bigrading. We shall write the generic element of C as a finite sum

$$c = \sum_\Gamma b_g\, U_g\,,\ b_g \in \mathcal{B}.$$

We endow the algebra $\mathcal{B}$ with the differential

$$d(\omega \otimes \varepsilon) = d\omega \otimes \varepsilon \qquad \forall \omega \in A^*(V)\,,\ \varepsilon \in \wedge^* \tag{14}$$

where $d\omega$ is the usual differential of forms. Thus, we have endowed $\wedge^*$ with the 0-differential.

Finally we define the differential d in the algebra C by

$$d(b\, U_g) = (db)\, U_g + (-1)^{\partial b}\, b\, U_g\, \delta_g \qquad \forall b \in \mathcal{B}\,,\ g \in \Gamma. \tag{15}$$

Lemma 12. *(C, d) is a graded differential algebra.*

Proof. By construction, C is a bigraded algebra; thus it is enough to show that the two components d' and d'' of d given by

$$d'(b\, U_g) = (-1)^{\partial b}\, b\, U_g\, \delta_g\,,\ d''(b\, U_g) = (db)\, U_g \qquad \forall b \in \mathcal{B}\,,\ g \in \Gamma \tag{16}$$

are derivations of C such that $d'^2 = d''^2 = d'd'' + d''d' = 0$.

It is clear that d'' is a derivation and that $d''^2 = 0$. To check that d' is a derivation one has to show that for $g_1, g_2 \in \Gamma$,

$$d'(U_{g_1 g_2}) = (d'\, U_{g_1})U_{g_2} + U_{g_1}(d' U_{g_2}).$$

This follows from $U_{g_1}\,\delta_{g_1}\,U_{g_2} = U_{g_1g_2}\,\alpha_{g_2^{-1}}(\delta_{g_1}) = U_{g_1g_2}(\delta_{g_1g_2} - \delta_{g_2})$ by (13). Since $\delta_g^2 = 0$ one has $d'^2 = 0$. Finally $d'd'' = -d''d'$ from (16).

Now let $\gamma \in C^{n,m}$ be a cochain in the bicomplex (C^*, d_1, d_2). We associate to γ a linear form $\tilde{\gamma}$ on C defined by

$$\tilde{\gamma}(\omega \otimes \delta_{g_1} \ldots \delta_{g_n}) = \langle \omega, \gamma(1, g_1, \ldots, g_n)\rangle \qquad \forall g_i \in \Gamma\,,\ \omega \in A^{-m}(V) \tag{17}$$

$$\tilde{\gamma}(b\,U_g) = 0 \text{ unless } g = 1_\Gamma \text{ and } b \in A^{-m}(V) \otimes \wedge^n\,. \tag{18}$$

The relation between the coboundaries d_1 and d_2 of C^* (see (7) (8)) and the derivations d' and d'' of C is given by:

Lemma 13. *Let* $\gamma \in C^{n,m}$.

a) $\tilde{\gamma}(a_1a_2 - (-1)^{\partial a_1 \partial a_2}\, a_2a_1) = (-1)^{\partial a_1}\, (\widetilde{d_1\gamma})(a_1\, d'a_2) \quad \forall a_j \in C$

b) $\tilde{\gamma}(da) = \tilde{\gamma}(d''a) = (\widetilde{d_2\gamma})(a) \quad \forall a \in C.$

Proof. a) We can assume that $a_j = b_j\,U_{g_j}$ with $b_j \in \mathcal{B}$, and that $g_1g_2 = 1$. Then $a_1a_2 = b_1\,\alpha_{g_1}(b_2)$ and $(-1)^{\partial a_1 \partial a_2}\, a_2a_1 = (-1)^{\partial a_1 \partial a_2}\, b_2\,\alpha_{g_2}(b_1) = \alpha_{g_2}(b_1\,\alpha_{g_1}(b_2))$ using the graded commutativity of $\mathcal{B}$. In addition, we have $a_1\,d'a_2 = (-1)^{\partial a_2}\, b_1\,\alpha_{g_1}(b_2)\delta_{g_2}$. Thus, (with $b = b_1\,\alpha_{g_1}(b_2)$) it is enough to show that for any $g \in \Gamma$,

$$\tilde{\gamma}(b - \alpha_g(b)) = (-1)^{n+m}\,(\widetilde{d_1\gamma})(b\,\delta_g) \qquad \forall b \in \mathcal{B}. \tag{19}$$

One can assume that $b = \omega \otimes \delta_{g_1} \cdots \delta_{g_n}$ with $g_i \in \Gamma$, $\omega \in A^{-m}(V)$. One has $\alpha_g(b) = \alpha_{1,g}(\omega) \otimes \alpha_{2,g}(\delta_{g_1} \cdots \delta_{g_n})$ and

$$\begin{aligned}\alpha_{2,g}(\delta_{g_1} \cdots \delta_{g_n}) &= (\delta_{g_1g^{-1}} - \delta_{g^{-1}}) \cdots (\delta_{g_ng^{-1}} - \delta_{g^{-1}})\\ &= \delta_{g_1g^{-1}}\,\delta_{g_2g^{-1}} \cdots \delta_{g_ng^{-1}} - \delta_{g^{-1}}\,\delta_{g_2g^{-1}} \cdots \delta_{g_ng^{-1}}\\ &\quad - \delta_{g_1g^{-1}}\delta_{g^{-1}} \cdots \delta_{g_ng^{-1}} - \cdots - \delta_{g_1g^{-1}}\delta_{g_2g^{-1}} \cdots \delta_{g_{n-1}g^{-1}}\delta_{g^{-1}}.\end{aligned}$$

Thus, using (6) and the definition of d_1 (7) we get (19).

b) follows from (17) and the definition of d_2 (8).

Note that since $d = d''$ on $\mathcal{B}$ the equality b) still holds for d instead of d''. We can now give a general construction of cyclic cocycles on $\mathcal{A} = C_c^\infty(V) \rtimes \Gamma$ ([Co$_{14}$]):

Theorem 14. a) *The following map* Φ *is a morphism of the bicomplex* (C^*, d_1, d_2) *to the* (b, B) *bicomplex of* $\mathcal{A}$*: For* $\gamma \in C^{n,m}$, $\Phi(\gamma)$ *is the* $(n - m + 1)$*-linear form on* $\mathcal{A}$ *given, with* $\ell = n - m + 1$, *by*

$$\Phi(\gamma)(x^0, \ldots, x^\ell) = \lambda_{n,m} \sum_{j=0}^{\ell} (-1)^{j(\ell-j)}\, \tilde{\gamma}(dx^{j+1} \cdots dx^\ell\, x^0\, dx^1 \cdots dx^j)$$

$\forall x^j \in \mathcal{A}$, *where* $\lambda_{n,m} = \frac{n!}{(\ell+1)!}$.

b) *The corresponding map of cohomology groups gives a canonical inclusion* $\Phi_* : H_\tau^*(V_\Gamma, \mathbb{C}) \to H^*(\mathcal{A})$ *of* $H_\tau^*(V_\Gamma, \mathbb{C})$ *as a direct factor of the periodic cyclic cohomology of* $\mathcal{A} = C_c^\infty(V) \rtimes \Gamma$.

Proof. a) Let us first compute $b(\Phi(\gamma))(x^0, \dots x^{\ell+1})$. The Hochschild coboundary of the functional $(x^i) \to \tilde{\gamma}(dx^{j+1} \cdots dx^\ell \, x^0 \, dx^1 \cdots dx^j)$ gives the result $(-1)^j \, \tilde{\gamma}(a_j \, x^{j+1} - x^{j+1} \, a_j)$, where $a_j = dx^{j+2} \cdots dx^{\ell+1} \, x^0 \, dx^1 \cdots dx^j$. Thus, by Lemma 13 a) we get

$$b \, \Phi(\gamma)(x^0, \dots, x^{\ell+1}) = (-1)^\ell \, \lambda_{n,m} \sum_{j=0}^{\ell} (-1)^{j(\ell-j+1)} \, \widetilde{(d_1\gamma)} \, (a_j \, d' \, x^{j+1}). \quad (20)$$

Since $d_1^2 = 0$, $\widetilde{(d_1\gamma)}$ is a graded trace on C by Lemma 13, and we can rewrite (20) as

$$b \, \Phi(\gamma)(x^0, \dots, x^{\ell+1}) = \lambda_{n,m} \sum_{j=0}^{\ell} \widetilde{(d_1\gamma)}(x^0 \, dx^1 \cdots dx^j \, d'x^{j+1} \, dx^{j+2} \cdots dx^{\ell+1}). \quad (21)$$

Using $dx^k = d'x^k + d''x^k$ and considering the terms in which d' appears $n+1$ times in the product

$$x^0(d'x^1 + d''x^1) \cdots (d'x^{\ell+1} + d''x^{\ell+1})$$

we get

$$b \, \Phi(\gamma)(x^0, \dots, x^{\ell+1}) = (n+1) \, \lambda_{n,m} \, \widetilde{(d_1\gamma)} \, (x^0 \, dx^1 \cdots dx^{\ell+1}). \quad (22)$$

Since $\widetilde{(d_1\gamma)}$ is a graded trace on C one has

$$\Phi(d_1\gamma)(x^0, \dots, x^{\ell+1}) = \lambda_{n+1,m} \, (\ell + 2) \, \widetilde{(d_1\gamma)} \, (x^0 \, dx^1 \cdots dx^{\ell+1})$$

so that using (22) we get $\Phi(d_1\gamma) = b \, \Phi(\gamma)$.

Let us now compute $B \, \Phi(\gamma)$. One has

$$B_0 \, \Phi(\gamma) \, (x^0, \dots, x^{\ell-1}) = \lambda_{n,m} \sum_{j=0}^{\ell} (-1)^{j(\ell-j)} \, \tilde{\gamma}(dx^j \cdots dx^{\ell-1} \, dx^0 \cdots dx^{j-1}).$$

(For instance, for $\ell = 2$ one gets three terms: $\tilde{\gamma}(dx^0 \, dx^1) - \tilde{\gamma}(dx^1 \, dx^0) + \tilde{\gamma}(dx^0 \, dx^1)$.) Thus one has

$$B\Phi(\gamma)(x^0, \dots, x^{\ell-1}) = (\ell+1) \, \lambda_{n,m} \sum_{j=0}^{\ell-1} (-1)^{j(\ell-1)} \tilde{\gamma}(dx^j \cdots dx^{\ell-1} \, dx^0 \cdots dx^{j-1}).$$

By Lemma 13 b) we have

(23)

$$B \, \Phi(\gamma) \, (x^0, \dots, x^{\ell-1}) = (\ell + 1) \, \lambda_{n,m} \sum_{j=0}^{\ell-1} (-1)^{(j-1)(\ell-j)} \, \widetilde{(d_2\gamma)} \, (dx^j \cdots dx^{\ell-1} \, x^0 \, dx^1 \cdots dx^{j-1}).$$

Thus, $B\Phi(\gamma) = \Phi(d_2\gamma)$.

b) We shall describe in Section 7 γ), in the context of foliations, a natural retraction $\lambda : H^*(\mathcal{A}) \to H^*_\tau(V_\Gamma, \mathbb{C})$ using localization. The conclusion follows from the equality $\lambda \circ \Phi_* = \mathrm{id}$.

Remark 15. a) In the special case $V = \{\mathrm{pt}\}$ the above construction gives a cycle $(C, d, \tilde{\gamma})$ on the algebra $\mathcal{A} = \mathbb{C}\Gamma$, for any group cocycle $\gamma \in Z^n(\Gamma, \mathbb{C})$ represented by a totally antisymmetric right invariant cochain $\gamma : \Gamma^{n+1} \to \mathbb{C}$ with $d_1\gamma = 0$. The algebra $C = (\wedge^* \mathbb{C}\Gamma') \rtimes \Gamma$ is a nontrivial quotient of the universal differential algebra $\Omega\mathbb{C}\Gamma$ used in Section 1.

b) Let us explain the analogue of the above construction (Theorem 14) in the general case of smooth groupoids G such that r and s are *étale* maps. The bicomplex (C^*, d_1, d_2) of Proposition 11 is now the bicomplex of twisted differential forms on the simplicial manifold $\mathrm{Mr}(G)$ (Appendix A) which is the nerve of the small category G. We describe it in terms of currents, so that $C^{n,m}$ is the space of de Rham currents of dimension $-m$ on the manifold

$$G^{(n)} = \{(\gamma_1, \ldots, \gamma_n) \in G^n \; ; \; s(\gamma_i) = r(\gamma_{i+1}) \quad \forall i = 1, \ldots, n-1\}. \tag{24}$$

The first coboundary d_1 is the simplicial one,

$$d_1 = (-1)^m \sum (-1)^j \, d_j^* \tag{25}$$

where $d_0(\gamma_1, \ldots, \gamma_n) = (\gamma_2, \ldots, \gamma_n)$, $d_j(\gamma_1, \ldots, \gamma_n) = (\gamma_1, \ldots, \gamma_j\gamma_{j+1}, \ldots, \gamma_n)$, $d_n(\gamma_1, \ldots, \gamma_n) = (\gamma_1, \ldots, \gamma_{n-1})$.

Note that we *pull back* currents in Formula 25, which is possible since we are only using étale maps.

The second coboundary d_2 is the de Rham d^t, as in formula (8). We restrict to the *normalized* subcomplex of currents which vanish if any γ_j is a unit or if $\gamma_1 \cdots \gamma_n$ is a unit.

Let us describe the analogue of the bigraded differential algebra (C, d', d'') of Lemma 12.

As a *linear space*, the space $C^{n,m}$ of elements of C of bidegree (n, m) is the quotient of the space of smooth compactly supported differential forms of degree m on $G^{(n+1)}$ by the subspace of forms with support in $\{(\gamma_0, \ldots, \gamma_n)$; γ_j is a unit for some $j \neq 0\}$. The differential d'' is the ordinary differential of forms. The product and the differential d' are given as follows:

(26)

$$\begin{aligned}
(\omega_1\omega_2)(\gamma_0, \ldots, \gamma_{n_1}, \ldots, \gamma_{n_1+n_2})
&= \sum_{\gamma\gamma' = \gamma_{n_1}} \omega_1(\gamma_0, \ldots, \gamma_{n_1-1}, \gamma) \wedge \omega_2(\gamma', \gamma_{n_1+1}, \ldots, \gamma_{n_1+n_2}) \\
&+ \sum_{j=0}^{n_1-2} (-1)^{n_1-j-1} \sum_{\gamma\gamma' = \gamma_j} \omega_1(\gamma_0, \ldots, \gamma_{j-1}, \gamma, \gamma', \ldots, \gamma_{n_1-1}) \\
&\qquad \wedge \omega_2(\gamma_{n_1}, \ldots, \gamma_{n_1+n_2})
\end{aligned}$$

where we have used the étale maps $r, s : G \to G^{(0)}$ to identify the corresponding cotangent spaces and perform the wedge product

$$(27) \qquad (d'\omega)(\gamma_0, \ldots, \gamma_{n+1}) = 0 \text{ unless } \gamma_0 \text{ is a unit,}$$
$$\text{and}$$
$$(d'\omega)(\gamma_0, \ldots, \gamma_{n+1}) = \omega(\gamma_1, \ldots, \gamma_{n+1}) \text{ if } \gamma_0 \text{ is a unit.}$$

Let $c \in C^{n,m}$ be a cochain in the bicomplex (C^*, d_1, d_2). We associate to c the linear map $\tilde{c}$ on $C^{n,m}$ obtained from the push-forward of the current c by the map

$$(\gamma_1, \ldots, \gamma_n) \in G^{(n)} \to ((\gamma_1 \cdots \gamma_n)^{-1}, \gamma_1, \ldots, \gamma_n) \in G^{(n+1)}.$$

Then Lemma 13 holds, but with the important difference that part a) only holds for $a_2 \in C^{(0,0)} = C_c^\infty(G)$. This is the price to pay for the loss of the total antisymmetry of cochains. The map Φ is defined as in Theorem 14, and this theorem still holds because its proof only used the above weaker form of Lemma 13. We shall see in Section 6 many concrete examples of cyclic cocycles on $C_c^\infty(G)$ constructed using Φ.

To conclude this section we refer the reader to [F-T] [Nis] [Bry-N] and [Ge-J$_2$] for the complete description of the cyclic cohomology of crossed products.

III.3. Pairing of Cyclic Cohomology with K-Theory

Let $\mathcal{A}$ be a unital noncommutative algebra and $K_0(\mathcal{A})$ and $K_1(\mathcal{A})$ its algebraic K-theory groups (cf. [Mi$_1$]). By definition, $K_0(\mathcal{A})$ is the group associated to the semigroup of stable isomorphism classes of finite projective modules over $\mathcal{A}$. Furthermore $K_1(\mathcal{A})$ is the quotient of the group $GL_\infty(\mathcal{A})$ by its commutator subgroup, where $GL_\infty(\mathcal{A})$ is the inductive limit of the groups $GL_n(\mathcal{A})$ of invertible elements of $M_n(\mathcal{A})$, with the embedding maps $x \to \begin{bmatrix} x & 0 \\ 0 & 1 \end{bmatrix}$.

In this section we shall define, by straightforward formulae, a pairing between $HC^{\text{ev}}(\mathcal{A})$ and $K_0(\mathcal{A})$ and between $HC^{\text{odd}}(\mathcal{A})$ and $K_1(\mathcal{A})$. For an equivalent approach see [Kar$_3$].

The pairing satisfies $\langle S\varphi, e\rangle = \langle \varphi, e\rangle$, for $\varphi \in HC^*(\mathcal{A})$, $e \in K(\mathcal{A})$, and hence is in fact defined on $H^*(\mathcal{A}) = HC^*(\mathcal{A}) \otimes_{HC^*(\mathbb{C})} \mathbb{C}$. As a computational device we shall also formulate the pairing in terms of connections and curvature as one does for the usual Chern character for smooth manifolds.

This will show the Morita invariance of $HC^*(\mathcal{A})$, and will give in the case $\mathcal{A}$ abelian an action of the ring $K_0(\mathcal{A})$ on $HC^*(\mathcal{A})$.

Lemma 1. *Assume $\varphi \in Z_\lambda^n(\mathcal{A})$ and let $p, q \in \operatorname{Proj} M_k(\mathcal{A})$ be two idempotents of the form $p = uv$, $q = vu$ for some $u, v \in M_k(\mathcal{A})$. Then the following cocycles* *on* $\mathcal{B} = \{x \in M_k(\mathcal{A}); xp = px = x\}$ *differ by a coboundary:*

$$\psi_1(a^0,\dots,a^n) = (\varphi \# \mathrm{Tr})(a^0,\dots,a^n),$$
$$\psi_2(a^0,\dots,a^n) = (\varphi \# \mathrm{Tr})(va^0u,\dots,va^nu).$$

Proof. First, replacing $\mathcal{A}$ by $M_k(\mathcal{A})$, one may assume that $k = 1$. Then one can replace p, q, u, v by $\begin{bmatrix} p & 0 \\ 0 & 0 \end{bmatrix}, \begin{bmatrix} 0 & 0 \\ 0 & q \end{bmatrix}, \begin{bmatrix} 0 & u \\ 0 & 0 \end{bmatrix}, \begin{bmatrix} 0 & 0 \\ v & 0 \end{bmatrix}$ and hence assume the existence of an invertible element w such that $wpw^{-1} = q$, $u = pw^{-1} = w^{-1}q$, $v = qw = wp$; it suffices to take $w = \begin{bmatrix} 1-p & u \\ v & 1-q \end{bmatrix}$. Then the result follows from Proposition 1.8.

Recall that an equivalent description of $K_0(\mathcal{A})$ is as the abelian group associated to the semigroup of stable equivalence classes of idempotents $e \in \mathrm{Proj}\, M_k(\mathcal{A})$.

Proposition 2. a) *The following equality defines a bilinear pairing between* $K_0(\mathcal{A})$ *and* $HC^{\mathrm{ev}}(\mathcal{A})$: $\langle [e], [\varphi] \rangle = (m!)^{-1}(\varphi \# \mathrm{Tr})(e,\dots,e)$ *for* $e \in \mathrm{Proj}\, M_k(\mathcal{A})$ *and* $\varphi \in Z_\lambda^{2m}(\mathcal{A})$.

b) *One has* $\langle [e], [S\varphi] \rangle = \langle [e], [\varphi] \rangle$.

c) *Let* φ *(resp.* ψ*) be an even cyclic cocycle on an algebra* $\mathcal{A}$ *(resp.* $\mathcal{B}$*); then for any* $e \in K_0(\mathcal{A})$ *and* $f \in K_0(\mathcal{B})$ *one has*

$$\langle e \otimes f \,,\, \varphi \# \psi \rangle = \langle e, \varphi \rangle \, \langle f, \psi \rangle.$$

Proof. First, if $\varphi \in B_\lambda^{2m}(\mathcal{A})$, $\varphi \# \mathrm{Tr}$ is also a coboundary, $\varphi \# \mathrm{Tr} = b\psi$ and hence $(\varphi \# \mathrm{Tr})(e,\dots,e) = b\psi(e,\dots,e) = \sum_{i=0}^{2m} (-1)^i \, \psi(e,\dots,e) = \psi(e,\dots,e) = 0$, since $\psi^\lambda = -\psi$. This, together with Lemma 1, shows that $(\varphi \# \mathrm{Tr})(e,\dots,e)$ only depends on the equivalence class of e. Since replacing e by $\begin{bmatrix} e & 0 \\ 0 & 0 \end{bmatrix}$ does not change the result, one gets the additivity and hence a).

b) One has $S\varphi(e,\dots,e) = \sum_{j=1}^{2m} \hat{\varphi}(e(de)^{j-1} \, e(de)^{n-j+1})$ and, since $e^2 = e$, one has $ede\,e = 0$, $e(de)^2 = (de)^2 e$, so that

$$S\varphi(e,\dots,e) = (m+1)\varphi(e,\dots,e).$$

c) Follows from b).

We shall now describe the odd case.

Proposition 3. a) *The following equality defines a bilinear pairing between* $K_1(\mathcal{A})$ *and* $HC^{\mathrm{odd}}(\mathcal{A})$

$$\langle [u], [\varphi] \rangle = \frac{1}{\sqrt{2i}} \, 2^{-n} \, \Gamma\left(\frac{n}{2} + 1\right)^{-1} \, (\varphi \# \mathrm{Tr})(u^{-1} - 1, u - 1, u^{-1} - 1, \dots, u - 1)$$

where $\varphi \in Z_\lambda^n(\mathcal{A})$ and $u \in GL_k(\mathcal{A})$.

b) *One has $\langle [u], [S\varphi] \rangle = \langle [u], [\varphi] \rangle$.*

c) *Let φ (resp. ψ) be an even (resp. odd) cyclic cocycle on an algebra $\mathcal{A}$ (resp. $\mathcal{B}$); then for any $e \in K_0(\mathcal{A})$, $u \in K_1(\mathcal{B})$ one has*

$$\langle [e \otimes u + (1-e) \otimes 1] \, , \, \varphi \# \psi \rangle = \langle e, \varphi \rangle \, \langle u, \psi \rangle .$$

Proof. a) Let $\widetilde{\mathcal{A}}$ be the algebra obtained from $\mathcal{A}$ by adjoining a unit. Since $\mathcal{A}$ is already unital, $\widetilde{\mathcal{A}}$ is isomorphic to the product of $\mathcal{A}$ by $\mathbb{C}$, by means of the homomorphism $\rho : (a, \lambda) \to (a + \lambda 1, \lambda)$ of $\widetilde{\mathcal{A}}$ to $\mathcal{A} \times \mathbb{C}$. Let $\tilde{\varphi} \in Z_\lambda^n(\widetilde{\mathcal{A}})$ be defined by the equality

$$\tilde{\varphi}((a^0, \lambda^0), \ldots, (a^n, \lambda^n)) = \varphi(a^0, \ldots, a^n) \quad \forall (a^i, \lambda^i) \in \widetilde{\mathcal{A}}.$$

Let us check that $b\tilde{\varphi} = 0$. For $(a^0, \lambda^0), \ldots, (a^{n+1}, \lambda^{n+1}) \in \widetilde{\mathcal{A}}$ one has

$$\begin{aligned} \tilde{\varphi}((a^0, \lambda^0), \ldots, (a^i, \lambda^i)(a^{i+1}, \lambda^{i+1}), \ldots, (a^{n+1}, \lambda^{n+1})) \\ = \varphi(a^0, \ldots, a^i a^{i+1}, \ldots, a^{n+1}) \\ + \lambda^i \varphi(a^0, \ldots, a^{i-1}, a^{i+1}, \ldots, a^{n+1}) \\ + \lambda^{i+1} \varphi(a^0, \ldots, a^i, a^{i+2}, \ldots, a^{n+1}). \end{aligned}$$

Thus

$$\begin{aligned} b\tilde{\varphi}((a^0, \lambda^0), \ldots, (a^{n+1}, \lambda^{n+1})) \\ = \lambda^0 \varphi(a^1, \ldots, a^{n+1}) + (-1)^{n-1} \lambda^0 \varphi(a^{n+1}, a^1, \ldots, a^n) = 0. \end{aligned}$$

Now for $u \in GL_1(\mathcal{A})$ one has

$$\varphi(u^{-1} - 1, u - 1, \ldots, u^{-1} - 1, u - 1) = (\tilde{\varphi} \circ \rho^{-1})(\overline{u}^{-1}, \overline{u}, \ldots, \overline{u}^{-1}, \overline{u})$$

where $\overline{u} = (u, 1) \in \mathcal{A} \times \mathbb{C}$. Thus, to show that this function $\chi(u)$ satisfies

$$\chi(uv) = \chi(u) + \chi(v) \quad \text{for } u, v \in GL_1(\mathcal{A}),$$

one can assume that $\varphi(1, a^0, \ldots, a^{n-1}) = 0$ for $a^i \in \mathcal{A}$, and replace χ by

$$\chi(u) = \varphi(u^{-1}, u, \ldots, u^{-1}, u).$$

Now one has with $U = \begin{bmatrix} uv & 0 \\ 0 & 1 \end{bmatrix}$ and $V = \begin{bmatrix} u & 0 \\ 0 & v \end{bmatrix}$

$$\chi(uv) = (\varphi \# \mathrm{Tr})(U^{-1}, U, \ldots, U^{-1}, U),$$

$$\chi(u) + \chi(v) = (\varphi \# \mathrm{Tr})(V^{-1}, V, \ldots, V^{-1}, V).$$

Since U is connected to V by the smooth path

$$U_t = \begin{bmatrix} u & 0 \\ 0 & 1 \end{bmatrix} \begin{bmatrix} \sin t & -\cos t \\ \cos t & \sin t \end{bmatrix} \begin{bmatrix} 1 & 0 \\ 0 & v \end{bmatrix} \begin{bmatrix} \sin t & \cos t \\ -\cos t & \sin t \end{bmatrix}$$

it is enough to check that

$$\frac{d}{dt}(\varphi \# \mathrm{Tr})(U_t^{-1}, U_t, \ldots, U_t) = 0.$$

Using $(U_t^{-1})' = -U_t^{-1}U_t'U_t^{-1}$ the desired equality follows easily. We have shown that the right-hand side of 3 a) defines a homomorphism of $GL_k(\mathcal{A})$ to $\mathbb{C}$. The compatibility with the inclusion $GL_k \subset GL_{k'}$ is obvious.

In order to show that the result is 0 if φ is a coboundary, one may assume that $k = 1$, and, using the above argument, that $\varphi = b\psi$ where $\psi \in C_\lambda^{n-1}$ and $\psi(1, a^0, \ldots, a^{n-2}) = 0$ for $a^i \in \mathcal{A}$. (One has $b\tilde{\varphi} = (b\psi)^\sim$ for $\psi \in C_\lambda^{n-1}$.) Then one gets $b\psi(u^{-1}, \ldots, u^{-1}, u) = 0$.

b) The proof is left to the reader.

Remark. The normalization 2 a) of the pairing between K_0 and HC^{ev} is uniquely specified by conditions 2 b) c). The normalization of the pairing between K_1 and HC^{odd} is only specified up to an overall multiplicative constant λ, independent of n, by conditions 3 b) c). Our choice (3 a)) is, up to the choice of the square root of $2i$, the only one for which the following formula holds:

$$\langle u \wedge v \, , \, \varphi\#\psi \rangle = \langle u, \varphi \rangle \, \langle v, \psi \rangle$$

where the product $\wedge : K_1(\mathcal{A}) \times K_1(\mathcal{B}) \to K_0(\mathcal{A}\hat{\otimes}\mathcal{B})$ is defined in the context of pre-C^*-algebras (cf. Chapter IV Section 9). To check the above formula one just needs to know that if $e \in C^\infty(\mathbb{T}^2, P_1(\mathbb{C}))$ is a degree-1 map of the 2-torus to $P_1(\mathbb{C}) \subset M_2(\mathbb{C})$ then $\tau(e,e,e) = 2\pi i$, where τ is the cyclic 2-cocycle on $C^\infty(\mathbb{T}^2)$ given by $\tau(f^0, f^1, f^2) = \int_{\mathbb{T}^2} f^0 \, df^1 \wedge df^2$, $\forall f^j \in C^\infty(\mathbb{T}^2)$. Note that the Bott generator b of the K-theory of $P_1(\mathbb{C})$ corresponds to the class of $[1-e]-1$ (cf. [At$_3$] p.61) while the definition of the product $\varphi\#\psi$ (Section 1 α)) introduces a $-$ sign in the odd case.

Let $H^*(\mathcal{A}) = HC^*(\mathcal{A}) \otimes_{HC^*(\mathbb{C})} \mathbb{C}$ be the periodic cyclic cohomology of $\mathcal{A}$. Here $HC^*(\mathbb{C})$, which by Corollary 1.13 is identified with a polynomial ring $\mathbb{C}[\sigma]$, acts on $\mathbb{C}$ by $P(\sigma) \mapsto P(1)$. This homomorphism of $HC^*(\mathbb{C})$ to $\mathbb{C}$ is the pairing given by Proposition 2 with the generator of $K_0(\mathbb{C}) = \mathbb{Z}$.

By construction, $H^*(\mathcal{A})$ is the inductive limit of the groups $HC^n(\mathcal{A})$ under the map $S : HC^n(\mathcal{A}) \to HC^{n+2}(\mathcal{A})$, or equivalently the quotient of $HC^*(\mathcal{A})$ by the equivalence relation $\varphi \sim S\varphi$. As such, it inherits a natural $\mathbb{Z}/2$ grading and a filtration

$$F^n H^*(\mathcal{A}) = \mathrm{Im}\, HC^n(\mathcal{A}).$$

Propositions 2 and 3 define a canonical pairing $\langle \, , \, \rangle$ between $H^{\mathrm{ev}}(\mathcal{A})$ and $K_0(\mathcal{A})$, and between $H^{\mathrm{odd}}(\mathcal{A})$ and $K_1(\mathcal{A})$.

Corollary 4. *Let $\mathcal{A}$ be a locally convex algebra and τ a continuous even (resp. odd) cyclic cocycle on $\mathcal{A}$ (Appendix B); then the pairing of τ with $e \in K_0(\mathcal{A})$ (resp. $u \in K_1(\mathcal{A})$) only depends upon the homotopy class of e (resp. u).*

This follows from Remark 1.30 c).

The following notion will be important both in explicit computations of the above pairing (this is already clear in the case $\mathcal{A} = C^\infty(V)$, V a smooth manifold) and in the discussion of Morita equivalences.

Definition 5. *Let $\mathcal{A} \xrightarrow{\rho} \Omega$ be a cycle over $\mathcal{A}$, and $\mathcal{E}$ a finite projective module over $\mathcal{A}$. Then a connection ∇ on $\mathcal{E}$ is a linear map $\nabla : \mathcal{E} \to \mathcal{E} \otimes_{\mathcal{A}} \Omega^1$ such that*

$$\nabla(\xi x) = (\nabla \xi)x + \xi \otimes d\rho(x)\,,\ \forall \xi \in \mathcal{E}\,,\ x \in \mathcal{A}.$$

Here $\mathcal{E}$ is a *right* module over $\mathcal{A}$ and Ω^1 is considered as a bimodule over $\mathcal{A}$ using the homomorphism $\rho : \mathcal{A} \to \Omega^0$ and the ring structure of Ω^*. Let us list a number of obvious properties:

Proposition 6. a) *Let $e \in \operatorname{End}_{\mathcal{A}}(\mathcal{E})$ be an idempotent and ∇ a connection on $\mathcal{E}$; then $\xi \mapsto (e \otimes 1)\nabla\xi$ is a connection on $e\mathcal{E}$.*

b) *Any finite projective module $\mathcal{E}$ admits a connection.*

c) *The space of connections is an affine space over the vector space*

$$\operatorname{Hom}_{\mathcal{A}}(\mathcal{E}, \mathcal{E} \otimes_{\mathcal{A}} \Omega^1).$$

d) *Any connection ∇ extends uniquely to a linear map of $\widetilde{\mathcal{E}} = \mathcal{E} \otimes_{\mathcal{A}} \Omega$ into itself such that*

$$\nabla(\xi \otimes \omega) = (\nabla\xi)\omega + \xi \otimes d\omega\,,\ \forall \xi \in \mathcal{E}\,,\ \omega \in \Omega.$$

Proof. a) One multiplies the equality of Definition 5 by $e \otimes 1$ (on the left).

b) By a) one can assume that $\mathcal{E} = \mathbb{C}^k \otimes \mathcal{A}$ for some k. Then, with $(\xi_i)_{i=1,\dots,k}$ the canonical basis of $\mathcal{E}$, put

$$\nabla\left(\sum \xi_i a_i\right) = \sum \xi_i \otimes d\rho(a_i) \in \mathcal{E} \otimes_{\mathcal{A}} \Omega^1.$$

Note that, if $k = 1$, for instance, then $\mathcal{A} \otimes_{\mathcal{A}} \Omega^1 = \rho(1)\Omega^1$ and $\nabla a = \rho(1)d\rho(a)$ for any $a \in \mathcal{A}$ since $\mathcal{A}$ is unital. This differs in general from d, even when $\rho(1)$ is the unit of Ω^0.

c) Immediate.

d) By construction, $\widetilde{\mathcal{E}}$ is the projective module over Ω induced by the homomorphism ρ. The uniqueness statement is obvious since $\nabla\xi$ is already defined for $\xi \in \mathcal{E}$. The existence follows from the equality

$$\nabla(\xi a)\omega + \xi a \otimes d\omega = (\nabla\xi)a\omega + \xi \otimes d(a\omega)$$

for any $\xi \in \mathcal{E}$, $a \in \mathcal{A}$ and $\omega \in \Omega$.

We shall now construct a cycle over $\operatorname{End}_{\mathcal{A}}(\mathcal{E})$. We start with the graded algebra $\operatorname{End}_\Omega(\widetilde{\mathcal{E}})$, where T is of degree k if $T\widetilde{\mathcal{E}^j} \subset \widetilde{\mathcal{E}^{j+k}}$ for all j. For any

$T \in \mathrm{End}_\Omega(\tilde{\mathcal{E}})$ of degree k we let $\delta(T) = \nabla T - (-1)^k T\nabla$. By the equality d) one gets

$$\nabla(\xi\omega) = (\nabla\xi)\omega + (-1)^{\deg\xi}\,\xi d\omega \quad \text{for } \xi \in \tilde{\mathcal{E}},\ \omega \in \Omega,$$

and hence that $\delta(T) \in \mathrm{End}_\Omega(\tilde{\mathcal{E}})$, and is of degree $k+1$. By construction, δ is a graded derivation of $\mathrm{End}_\Omega(\tilde{\mathcal{E}})$. Next, since $\tilde{\mathcal{E}}$ is a finite projective module, the graded trace $\int : \Omega^n \to \mathbb{C}$ defines a trace, which we shall still denote by $\int$, on the graded algebra $\mathrm{End}_\Omega(\tilde{\mathcal{E}})$.

Lemma 7. *One has* $\int \delta(T) = 0$ *for any* $T \in \mathrm{End}_\Omega(\tilde{\mathcal{E}})$ *of degree* $n-1$.

Proof. First, replacing the connection ∇ by $\nabla' = \nabla + \Gamma$, in which we have $\Gamma \in \mathrm{Hom}_{\mathcal{A}}(\mathcal{E}, \mathcal{E} \otimes_{\mathcal{A}} \Omega^1)$, the extension corresponding to $\tilde{\mathcal{E}}$ is $\nabla' = \nabla + \tilde{\Gamma}$, where $\tilde{\Gamma} \in \mathrm{End}_\Omega(\tilde{\mathcal{E}})$ and $\tilde{\Gamma}$ is of degree 1. Thus, it is enough to prove the lemma for some connection on $\mathcal{E}$. Hence we can assume that $\mathcal{E} = e\mathcal{A}^k$ for some $e \in \mathrm{Proj}\, M_k(\mathcal{A})$ and that ∇ is given by 6 a) from a connection ∇_0 on $\mathcal{A}^k$. Then using the equality $\delta(T) = e\delta_0(T)e$ for $T \in \mathrm{End}\,\tilde{\mathcal{E}} \subset \mathrm{End}\,\tilde{\mathcal{E}}_0$ $(\mathcal{E}_0 = \mathcal{A}^k)$, as well as

$$\delta_0(T) = \delta_0(eTe) = \delta_0(e)T + \delta(T) + (-1)^{\deg T}\, T\delta_0(e),$$

one is reduced to the case $\mathcal{E} = \mathcal{A}^k$, with ∇ given by 6 b). Let us end the computation, say with $k = 1$. Let $e = \rho(1)$. One has $\tilde{\mathcal{E}} = e\Omega$, $\mathrm{End}_\Omega(\tilde{\mathcal{E}}) = e\Omega e$ and $\delta(a) = e(da)e$. Thus, $\int \delta(a) = \int (d(eae) - (de)a - (-1)^{\partial a}\, ade) = 0$.

Now we do not yet have a cycle over $\mathrm{End}_{\mathcal{A}}(\mathcal{E})$ by taking the obvious homomorphism of $\mathrm{End}_{\mathcal{A}}(\mathcal{E})$ in $\mathrm{End}_\Omega(\tilde{\mathcal{E}})$, the differential δ and the integral $\int$. In fact the crucial property $\delta^2 = 0$ is not satisfied:

Proposition 8. a) *The map* $\theta = \nabla^2$ *of* $\tilde{\mathcal{E}}$ *to* $\tilde{\mathcal{E}}$ *is an endomorphism:* $\theta \in \mathrm{End}_\Omega(\tilde{\mathcal{E}})$ *and* $\delta^2(T) = \theta T - T\theta$ *for all* $T \in \mathrm{End}_\Omega(\tilde{\mathcal{E}})$.

b) *One has* $\langle [\mathcal{E}], [\tau] \rangle = \frac{1}{m!} \int \theta^m$ *when* n *is even,* $n = 2m$, *where* $[\mathcal{E}] \in K_0(\mathcal{A})$ *is the class of* $\mathcal{E}$, *and* τ *is the character of* Ω.

Proof. a) One uses the rules $\nabla(\eta\omega) = (\nabla\eta)\omega + (-1)^{\deg\eta}\,\eta d\omega$ and $d^2 = 0$ to check that $\nabla^2(\eta\omega) = \nabla^2(\eta)\omega$.

b) Let us show that $\int \theta^m$ is independent of the connection ∇. The result is then easily checked by taking on $\mathcal{E} = e\mathcal{A}^k$ the connection of Proposition 6. Thus, let $\nabla' = \nabla + \Gamma$ where Γ is an endomorphism of degree 1 of $\tilde{\mathcal{E}}$. It is enough to check that the derivative of $\int \theta_t^m$ is 0, where θ_t corresponds to $\nabla_t = \nabla + t\Gamma$. In fact it suffices to do it for $t = 0$; we obtain

$$\frac{d}{dt}\int \theta_t^m = \sum_{k=0}^{m-1} \int \theta_t^k \left(\frac{d}{dt}\,\theta_t\right)\, \theta_t^{m-k-1}.$$

As $\left(\frac{d}{dt}\,\theta_t\right)_{t=0} = \Gamma\nabla + \nabla\Gamma = \delta(\Gamma)$ one has

$$\left(\frac{d}{dt}\int \theta_t^m\right)_{t=0} = m \int \delta(\theta^{m-1}\,\Gamma) = 0.$$

By a), while $\delta^2 \neq 0$, there exists $\theta \in \Omega' = \mathrm{End}_\Omega(\tilde{\mathcal{E}})$ such that

$$\delta^2(T) = \theta T - T\theta \quad \forall T \in \Omega'.$$

We shall now construct a cycle from the quadruple $(\Omega', \delta, \theta, \int)$.

Lemma 9. *Let $(\Omega', \delta, \theta, \int)$ be a quadruple such that Ω' is a graded algebra, δ a graded derivation of degree 1 of Ω' and $\theta \in \Omega'^2$ satisfies*

$$\delta(\theta) = 0 \text{ and } \delta^2(\omega) = \theta\omega - \omega\theta \text{ for } \omega \in \Omega'.$$

Then one constructs canonically a cycle by adjoining to Ω' an element X of degree 1 with $dX = 0$, such that $X^2 = \theta$, $\omega_1 X \omega_2 = 0$, $\forall \omega_i \in \Omega'$.

Proof. Let Ω'' be the graded algebra obtained by adjoining X. Any element of Ω'' has the form $\omega = \omega_{11} + \omega_{12}X + X\omega_{21} + X\omega_{22}X$, $\omega_{ij} \in \Omega'$. Thus, as a vector space, Ω'' coincides with $M_2(\Omega')$; the product is such that

$$\omega\omega' = \begin{bmatrix} \omega_{11} & \omega_{12} \\ \omega_{21} & \omega_{22} \end{bmatrix} \begin{bmatrix} 1 & 0 \\ 0 & \theta \end{bmatrix} \begin{bmatrix} \omega'_{11} & \omega'_{12} \\ \omega'_{21} & \omega'_{22} \end{bmatrix}$$

and the grading is obtained by considering X as an element of degree 1; thus $[\omega_{ij}]$ is of degree k when ω_{11} is of degree k, ω_{12} and ω_{21} of degree $k-1$ and ω_{22} of degree $k-2$. One checks easily that Ω'' is a graded algebra containing Ω'. The differential d is given by the conditions $d\omega = \delta(\omega) + X\omega + (-1)^{\deg \omega}\omega X$ for $\omega \in \Omega' \subset \Omega''$, and $dX = 0$. One gets

$$d\begin{bmatrix} \omega_{11} & \omega_{12} \\ \omega_{21} & \omega_{22} \end{bmatrix} = \begin{bmatrix} \delta(\omega_{11}) & \delta(\omega_{12}) \\ -\delta(\omega_{21}) & -\delta(\omega_{22}) \end{bmatrix} + \begin{bmatrix} 0 & -\theta \\ 1 & 0 \end{bmatrix} \begin{bmatrix} \omega_{11} & \omega_{12} \\ \omega_{21} & \omega_{22} \end{bmatrix} + (-1)^{\deg \omega} \begin{bmatrix} \omega_{11} & \omega_{12} \\ \omega_{21} & \omega_{22} \end{bmatrix} \begin{bmatrix} 0 & 1 \\ -\theta & 0 \end{bmatrix}$$

and checks that the two terms on the right define graded derivations of Ω'' and that $d^2 = 0$. Finally one checks that the equality

$$\int (\omega_{11} + \omega_{12}X + X\omega_{21} + X\omega_{22}X) = \int \omega_{11} - (-1)^{\deg \omega} \int \omega_{22}\theta$$

defines a closed graded trace.

Putting together Proposition 8 a) and Lemma 9 we have

Corollary 10. *Let $\mathcal{A} \xrightarrow{\rho} \Omega$ be a cycle over $\mathcal{A}$, $\mathcal{E}$ a finite projective module over $\mathcal{A}$ and $\mathcal{A}' = \mathrm{End}_{\mathcal{A}}(\mathcal{E})$. To each connection ∇ on $\mathcal{E}$ corresponds canonically a cycle $\mathcal{A}' \xrightarrow{\rho'} \Omega'$ over $\mathcal{A}'$.*

One can show that the character $\tau' \in Z_\lambda^n(\mathcal{A}')$ of this new cycle has a class $[\tau'] \in HC^n(\mathcal{A}')$ independent of the choice of the connection ∇, which coincides with the class given by Lemma 1. One can easily check a reciprocity formula which takes care of the Morita equivalence.

Corollary 11. *Let $\mathcal{A}$ and $\mathcal{B}$ be unital algebras and $\mathcal{E}$ an $(\mathcal{A},\mathcal{B})$-bimodule, finite projective on both sides, with $\mathcal{A} = \mathrm{End}_{\mathcal{B}}(\mathcal{E})$ and $\mathcal{B} = \mathrm{End}_{\mathcal{A}}(\mathcal{E})$. Then $HC^*(\mathcal{A})$ is canonically isomorphic to $HC^*(\mathcal{B})$.*

Finally, when $\mathcal{A}$ is abelian, and one is given a finite projective module $\mathcal{E}$ over $\mathcal{A}$, then one has an obvious homomorphism of $\mathcal{A}$ to $\mathcal{A}' = \mathrm{End}_{\mathcal{A}}(\mathcal{E})$. Thus, in this case, by restriction to $\mathcal{A}$ of the cycle of Corollary 10 one obtains

Corollary 12. *When $\mathcal{A}$ is abelian, $HC^*(\mathcal{A})$ is in a natural manner a module over the ring $K_0(\mathcal{A})$.*

To give some meaning to this statement we shall compute an example. We let V be a compact oriented smooth manifold. Let $\mathcal{A} = C^\infty(V)$ and Ω be the cycle over $\mathcal{A}$ given by the ordinary de Rham complex and integration of forms of degree n. Let E be a complex vector bundle over V and $\mathcal{E} = C^\infty(V,E)$ the corresponding finite projective module over $\mathcal{A} = C^\infty(V)$. Then the notion of connection given by Definition 5 coincides with the usual notion. Thus, Corollary 12 yields a new cocycle $\tau \in Z^n_\lambda(\mathcal{A})$ canonically associated to ∇. We shall leave as an exercise the following proposition.

Proposition 13. *Let ω_k be the differential form of degree $2k$ on V which gives the component of degree $2k$ of the Chern character of the bundle E with connection $\nabla : \omega_k = (1/k!)\,\mathrm{Tr}(\theta^k)$, where θ is the curvature form. Then one has the equality*

$$\tau = \sum S^k\, \widetilde{\omega}_k,$$

where $\widetilde{\omega}_k \in Z^{n-2k}_\lambda(\mathcal{A})$ is given by

$$\widetilde{\omega}_k(f^0,\dots,f^{n-2k}) = \int f^0 df^1 \wedge \cdots \wedge df^{n-2k} \wedge \omega_k\,, \quad \forall f^i \in \mathcal{A} = C^\infty(V),$$

and where τ is the restriction to $\mathcal{A} = C^\infty(V)$ of the character of the cycle associated to the bundle E, the connection ∇, and the de Rham cycle of $\mathcal{A}$ by Corollary 10.

Of course in this example of $\mathcal{A} = C^\infty(V)$ the pairing between cyclic cohomology and $K_0(\mathcal{A})$ gives back the ordinary Chern character of vector bundles.

As a more sophisticated example let us consider Example 2, β), i.e. the irrational rotation algebra. The smooth subalgebra $\mathcal{A}_\theta \subset A_\theta$ is stable under holomorphic functional calculus (Appendix C) and thus by a result of Pimsner and Voiculescu ([Pi-V$_2$]) one has $K_0(\mathcal{A}_\theta) = \mathbb{Z}^2$. An explicit description of finite projective modules over $\mathcal{A}_\theta$ was given in [Co$_{13}$] and shown by M. Rieffel [Ri$_4$] to classify up to isomorphism all the finite projective modules over $\mathcal{A}_\theta$. We have already seen it in a slightly different guise in Section II.8β).

Let $(p,q) \in \mathbb{Z}^2$, $q > 0$, be a pair of relatively prime integers ($p = 0$ being allowed). Let us construct a finite projective module $\mathcal{E} = \mathcal{E}_{p,q}$ over $\mathcal{A}_\theta$ as follows.

We let $S(\mathbb{R})$ be the usual Schwartz space of complex-valued functions on the real line and let two operators V_1 and V_2 on $S(\mathbb{R})$ be defined by

$$(V_1\xi)(s) = \xi(s - \varepsilon)\ ,\ (V_2\xi)(s) = e(s)\xi(s)\ ,\ s \in \mathbb{R}\ ,\ \xi \in S(\mathbb{R})$$

where $\varepsilon = \frac{p}{q} - \theta$ and $e(s) = \exp(2\pi i s)\ \forall s \in \mathbb{R}$.

One has, of course,

$$V_2V_1 = e(\varepsilon)V_1V_2.$$

Next, let K be a finite-dimensional Hilbert space and w_1 and w_2 be unitary operators on K such that

$$w_2w_1 = \bar{e}(p/q)w_1w_2\ ,\ w_1^q = w_2^q = 1.$$

In other words, K is a (finite-dimensional) representation of the Heisenberg commutation relations for the finite cyclic group $\mathbb{Z}/q\mathbb{Z}$, and thus it is a direct sum of d equivalent irreducible representations.

It is clear now that in the tensor product $\mathcal{E} = S(\mathbb{R}) \otimes K$ one has

$$(V_2 \otimes w_2)(V_1 \otimes w_1) = \bar{\lambda}(V_1 \otimes w_1)(V_2 \otimes w_2),\ \lambda = e(\theta).$$

We turn $\mathcal{E}$ into a right $\mathcal{A}_\theta$-module as follows:

$$\xi U_1 = (V_1 \otimes w_1)\xi$$
$$\xi U_2 = (V_2 \otimes w_2)\xi \qquad \forall \xi \in S(\mathbb{R}) \otimes K = \mathcal{E}.$$

The above equality shows that this is compatible with the presentation of $\mathcal{A}_\theta$. It is easy to check that for any $a \in \mathcal{A}_\theta$, the element ξa still belongs to $\mathcal{E}$ using, for instance, the general properties of nuclear spaces.

By results in [Co_{13}] the right module over $\mathcal{A}_\theta$ obtained is *finite and projective*. In fact:

Theorem 14. [Ri_4] *Let $\mathcal{E}$ be a finite projective module over $\mathcal{A}_\theta$; then either $\mathcal{E}$ is free, $\mathcal{E} = \mathcal{A}^p$ for some $p > 0$, or $\mathcal{E}$ is isomorphic to $S(\mathbb{R}) \otimes K$ with the above module structure.*

Let us now recall (Theorem 7 Section 2) that $H^{\mathrm{ev}}(\mathcal{A}_\theta)$ is a vector space of dimension 2 with basis $S\tau$ and φ, where τ is the canonical trace of $\mathcal{A}_\theta$ and where the cyclic 2-cocycle φ is

$$\varphi(x^0, x^1, x^2) = (2\pi i)^{-1}\ \tau(x^0(\delta_1(x^1)\delta_2(x^2) - \delta_2(x^1)\delta_1(x^2)))\ \ \forall x^i \in \mathcal{A}_\theta$$

where δ_1, δ_2 are the natural commuting derivations of $\mathcal{A}_\theta$. The pairing of K_0 with $S\tau$ is the same as with the trace τ, and given by the Murray and von Neumann dimension

$$\langle \tau, \mathcal{E}_{p,q} \rangle = p - \theta q.$$

To compute the pairing of K_0 with φ let us use the following 2-dimensional cycle with character φ.

As a graded algebra Ω^* is the tensor product of $\mathcal{A}_\theta$ by the exterior algebra $\wedge^* \mathbb{C}^2$ of the two-dimensional vector space $\mathbb{C}^2 = \mathbb{C}e_1 + \mathbb{C}e_2$. The differential d is uniquely specified by the equality

$$d(a \otimes \alpha) = \delta_1(a) \otimes (e_1 \wedge \alpha) + \delta_2(a) \otimes (e_2 \wedge \alpha)$$

for any $a \in \mathcal{A}_\theta$, $\alpha \in \wedge^* \mathbb{C}^2$.

Finally the graded trace, $\Omega^2 \to \mathbb{C}$, is

$$a \otimes (e_1 \wedge e_2) \to (2\pi i)^{-1} \tau(a) \in \mathbb{C}.$$

A connection (Definition 5) on a finite projective module $\mathcal{E}$ over $\mathcal{A}_\theta$ is thus given by a pair ∇_1, ∇_2 of covariant differentials, satisfying

$$\nabla_j(\xi a) = (\nabla_j \xi) a + \xi \delta_j(a) \qquad \forall \xi \in \mathcal{E}\,,\ a \in \mathcal{A}_\theta$$

and the curvature Θ is easily identified as

$$(\nabla_1 \nabla_2 - \nabla_2 \nabla_1) \otimes (e_1 \wedge e_2).$$

Proposition 15. [Co$_{13}$] *Let $\mathcal{E}_{p,q} = S(\mathbb{R}, K)$ be as above; then the following formulae define a connection ∇ on $\mathcal{E}$:*

$$(\nabla_1 \xi)(s) = 2\pi i \frac{s}{\varepsilon}\, \xi(s)\,,\ (\nabla_2 \xi)(s) = \frac{d\xi}{ds}(s)$$

where $\varepsilon = \frac{p}{q} - \theta$.

The curvature of this connection is constant, equal to $-\frac{2\pi i}{\varepsilon} \otimes (e_1 \wedge e_2)$ and by Proposition 8 b) the value of the pairing is

$$\langle \mathcal{E}_{p,q}, \varphi \rangle = \frac{1}{2\pi i} \left(\frac{-2\pi i}{\varepsilon} \right) \tau(\mathrm{id}_\varepsilon) = -\frac{1}{\varepsilon}(p - \theta q) = q.$$

Corollary 16. *Let $\varphi \in HC^2(\mathcal{A}_\theta)$ be the above cyclic 2-cocycle; then*

$$\langle K_0(\mathcal{A}_\theta), \varphi \rangle \subset \mathbb{Z}.$$

This integrality result will be fully understood and exploited in Chapter IV Section 6.

Finally we remark that (for $\theta \notin \mathbb{Q}$) the filtration of $H^{\mathrm{ev}}(\mathcal{A}_\theta)$ by dimensions is not compatible with the lattice dual to $K_0(\mathcal{A}_\theta)$, since the 0-dimensional class of τ does not pair integrally with K_0. In the next section we shall compute the pairing of K-theory with cyclic cohomology in the case of group rings.

III.4. The Higher Index Theorem for Covering Spaces

Let Γ be a discrete group acting properly and freely on a smooth manifold $\widetilde{M}$ with compact quotient $M = \widetilde{M}/\Gamma$. Any Γ-invariant elliptic differential operator D on $\widetilde{M}$ yields by the parametrix construction detailed below a K-theory class

$$\mathrm{Ind}(D) = K_0(\mathcal{R}\Gamma)$$

where $\mathcal{R}\Gamma$ is the group ring of Γ with coefficients in the algebra $\mathcal{R}$ of matrices $(a_{ij})_{i,j\in\mathbb{N}}$ with rapid decay

$$\sup_{i,j\in\mathbb{N}} i^k\, j^\ell\, |a_{ij}| < \infty \quad \forall k,\ell \in \mathbb{N}.$$

In this section we shall extend the Atiyah-Singer index theorem for covering spaces [At$_5$] [Sin$_2$], which computes the pairing of $\mathrm{Ind}(D)$ with the natural trace of $\mathcal{R}\Gamma$, to the case of arbitrary group cocycles ([Co-M$_1$]). More specifically, given a normalized group cocycle $c \in Z^k(\Gamma,\mathbb{C})$ we have seen (Section 1 α)) that there is a corresponding cyclic cocycle τ_c on the group ring $\mathbb{C}\Gamma$ given by the equality

$$\tau_c(g_0,\ldots,g_k) = 0 \ \text{ if } g_0 g_1 \cdots g_k \neq 1$$

$$\tau_c(g_0,\ldots,g_k) = c(g_1,\ldots,g_k) \ \text{ if } g_0 g_1 \cdots g_k = 1.$$

Since we have as a trace on $\mathcal{R}$

$$\mathrm{Trace}(a) = \sum a_{jj}$$

the cup product $\tau_c\#\mathrm{Trace}$ is a k-dimensional cyclic cocycle on $\mathcal{R}\Gamma$ and, for k even, it makes sense to pair it with $\mathrm{Ind}_\Gamma D$. Let us carefully define the latter index.

4.α The smooth groupoid of a covering space

As above, let Γ act freely and properly on the smooth manifold $\widetilde{M}$. Let us give the manifold G, the quotient of $\widetilde{M}\times\widetilde{M}$ by the diagonal action of Γ, the groupoid structure

$$G^{(0)} = \widetilde{M}/\Gamma = M$$

$$r(\tilde{x},\tilde{y}) = x \in M\,,\ s(\tilde{x},\tilde{y}) = y \in M \qquad \forall \tilde{x},\tilde{y} \in \widetilde{M}$$

$$(\tilde{x},\tilde{y})\circ(\tilde{y},\tilde{z}) = (\tilde{x},\tilde{z}) \ \ \forall \tilde{x},\tilde{y},\tilde{z} \in \widetilde{M}.$$

Since $(\tilde{x},\tilde{y}) = (\tilde{x}g,\tilde{y}g)$ $\forall \tilde{x},\tilde{y} \in \widetilde{M}$, $g \in \Gamma$, the last equality above is sufficient to define the composition of composable elements. One checks that G is a smooth groupoid. When $\Gamma = \pi_1(M)$ acts on the universal cover $\widetilde{M}$ of M then G is the *fundamental groupoid* ([Sp]) of M, i.e. the groupoid of homotopy classes of paths in M. Note that, while $G^{(0)} = M$ is compact, the groupoid G is not compact if Γ is infinite.

Let $J = C_c^\infty(G)$ be the convolution algebra of the smooth groupoid G (cf. Chapter II) and let us, by fixing the volume form on M associated to a fixed Riemannian metric, ignore the $\frac{1}{2}$-density bundle.

Let $\mathcal{A}$ be the convolution algebra of distributions $T \in C_c^{-\infty}(G)$ which are multipliers of J, i.e. satisfy

$$T * f \in C_c^\infty(G)\ ,\ f * T \in C_c^\infty(G) \qquad \forall f \in C_c^\infty(G).$$

Any such $T \in \mathcal{A}$ is characterized by the corresponding Γ-invariant operator on $C_c^\infty(\widetilde{M})$, with the distribution T as Schwartz kernel. In particular, we have a canonical inclusion

$$C^\infty(M) \subset \mathcal{A}$$

of the algebra of smooth functions on M as Γ-invariant multiplication operators on $C_c^\infty(\widetilde{M})$, and thus as a subalgebra of $\mathcal{A}$. Now let D be a Γ-invariant elliptic differential operator

$$D : C_c^\infty(\widetilde{M}, \widetilde{E}_1) \to C_c^\infty(\widetilde{M}, \widetilde{E}_2)$$

where the $\widetilde{E}_j$ are the Γ-equivariant vector bundles on $\widetilde{M}$ corresponding to the vector bundles E_j on M. The above inclusion, $C^\infty(M) \subset \mathcal{A}$, allows their induction to finite projective modules $\mathcal{E}_j$ on $\mathcal{A}$. Moreover, the existence of a parametrix for D modulo the ideal $J = C_c^\infty(G)$ yields (cf. Section II.9):

Proposition 1. *The triple $(\mathcal{E}_1, \mathcal{E}_2, D)$ defines a quasi-isomorphism over the algebras $J = C_c^\infty(G)$ and $\mathcal{A} \supset J$.*

We refer to [Co-M$_1$] for the detailed proof.

We shall then let $\mathrm{Ind}(D) \in K_0(C_c^\infty(G))$ be the index associated to the quasi-isomorphism $(\mathcal{E}_1, \mathcal{E}_2, D)$ by Proposition II.9.2.

The groupoid G has two important properties. The first is that as a small category it is *equivalent* to the discrete group Γ. This will allow us in β) to relate $C_c^\infty(G)$ to the group ring $\mathcal{R}\Gamma$. The second is that, as a smooth groupoid, it is *locally isomorphic* to the trivial groupoid $M \times M$ (cf. Chapter II); this allows us to compute the pairing of $\mathrm{Ind}(D)$ with cyclic cocycles (γ).

4.β The group ring $\mathcal{R}\Gamma$

Let $\mathcal{R}$ be, as above, the ring of infinite matrices with rapid decay. Let us construct a canonical map

$$K_0(C_c^\infty(G)) \to K_0(\mathcal{R}\Gamma)$$

playing at this algebraic level the same role as the K-theory isomorphism

$$K(C^*(G)) \simeq K(C^*(\Gamma))$$

coming from the strong Morita equivalence $C^*(G) \simeq C^*(\Gamma)$ associated to the equivalence of groupoids $G \simeq \Gamma$.

The following lemma is the algebraic counterpart of Proposition II.4.1.

Lemma 2. *Let the algebra* $C^\infty(M) \otimes \mathbb{C}\Gamma$ *act on the right on the vector space* $C_c^\infty(\widetilde{M})$ *by*

$$(\xi(fU_g))(\tilde{x}) = f(x)\,\xi(g\tilde{x}) \qquad \forall \xi \in C_c^\infty(\widetilde{M}),\ f \in C^\infty(M),\ g \in \Gamma,\ \tilde{x} \in \widetilde{M}$$

with $x \in M$ *the class of* $\tilde{x}$. *Then* $C_c^\infty(\widetilde{M})$ *is a finite projective module over* $C^\infty(M) \otimes \mathbb{C}\Gamma$.

In fact, it is useful to describe explicitly a corresponding idempotent $E \in M_n(C^\infty(M) \otimes \mathbb{C}\Gamma)$ and an isomorphism of right modules $U \in \mathrm{Hom}_{\mathcal{B}}(\mathcal{E}, E\mathcal{B}^n)$, where we let $\mathcal{B} = C^\infty(M) \otimes \mathbb{C}\Gamma$ and $\mathcal{E} = C_c^\infty(\widetilde{M})$ be the above module. Thus, let $(V_i)_{i=1,\dots,n}$ be a finite open cover of M, $\beta_i : V_i \to \widetilde{M}$ be local continuous sections of the projection

$$\pi : \widetilde{M} \to M$$

and $(\chi_i)_{i=1,\dots,n}$ be a smooth partition of unity in M subordinate to the covering $(V_i)_{i=1,\dots,n}$. We may assume that the $\chi_j^{1/2}$ are smooth functions as well.

Then one defines the module homomorphisms $U \in \mathrm{Hom}_{\mathcal{B}}(\mathcal{E}, \mathcal{B}^n)$ and $U^* \in \mathrm{Hom}_{\mathcal{B}}(\mathcal{B}^n, \mathcal{E})$, given by

$$(U\xi)(x,g,j) = \xi(g^{-1}\,\beta_j(x))\,\chi_j^{1/2}(x) \quad \forall x \in M, g \in \Gamma, j \in \{1,\dots,n\}, \xi \in \mathcal{E}$$

and

$$(U^*\eta)(\tilde{x}) = \sum_{j=1}^{n} \chi_j^{1/2}(x)\,\eta(x, g(\beta_j(x),\tilde{x}), j) \quad \forall \tilde{x} \in \widetilde{M},\ \pi(\tilde{x}) = x,\ \eta \in \mathcal{B}^n,$$

and where $g(\beta_j(x),\tilde{x})$ is the unique $g \in \Gamma$ such that

$$g\tilde{x} = \beta_j(x)\ ,\ x = \pi(\tilde{x}).$$

One checks directly that both U and U^* are $\mathcal{B}$-module morphisms and that $U^*U = \mathrm{id}_{\mathcal{E}}$.

Thus, the product

$$E = UU^* \in M_n(\mathcal{B})$$

is an idempotent and U gives an isomorphism

$$\mathcal{E} \simeq E\mathcal{B}^n.$$

The components E_{jk} of the matrix $E \in M_n(\mathcal{B})$ are easily computed and are given by

$$E_{jk} = \chi_j^{1/2}\,\chi_k^{1/2} \otimes \beta_j\,\beta_k^{-1}$$

where $\beta_j\,\beta_k^{-1}$ is a short-hand notation for the 1-cocycle with values in Γ associated to the covering V_j and sections β_j.

The above isomorphism, $U : \mathcal{E} \simeq E\mathcal{B}^n$, is an isomorphism of $\mathcal{B}$-modules and thus, *a fortiori*, of $\mathbb{C}\Gamma$-modules, where $\mathbb{C}\Gamma$ is the subalgebra $1 \otimes \mathbb{C}\Gamma$ of $C^\infty(M) \otimes \mathbb{C}\Gamma$.

Thus, the homomorphism θ, given by $\theta(T) = UTU^*$, can transform Γ-invariant linear operators on $C_c^\infty(\widetilde{M})$ into $\mathbb{C}\Gamma$-endomorphisms of $E\mathcal{B}^n$. Since we have that $\mathcal{B} = C^\infty(M) \otimes \mathbb{C}\Gamma$, an arbitrary $\mathbb{C}\Gamma$-endomorphism S of $\mathcal{B}^n$ is given by a matrix $(S_{ij})_{i,j=1,\dots,n}$, where $S_{ij} \in \mathrm{End}_{\mathbb{C}}(C^\infty(M)) \otimes \mathbb{C}\Gamma$.

Now let $\mathcal{R}_M$ be the algebra of smoothing operators on $C^\infty(M)$.

Proposition 3. *The homomorphism θ maps the algebra $C_c^\infty(G)$ to $M_n(\mathcal{R}_M \otimes \mathbb{C}\Gamma)$.*

Indeed, the map is given explicitly by the equality

$$\theta(k)_{ij}(x,y,g) = \chi_i(x)^{1/2}\, \chi_j(y)^{1/2}\, k(\beta_j(x), g\beta_j(y))$$
$$\forall k \in C_c^\infty(\widetilde{M} \times_\Gamma \widetilde{M}), x, y \in M, g \in \Gamma$$

which shows that each $\theta(k)_{ij}(g)$ is a smoothing operator.

Proposition 4. *The induced map*

$$K_0(C_c^\infty(G)) \to K_0(\mathcal{R}_M \otimes \mathbb{C}\Gamma)$$

is independent of the choice of $(U_j, \beta_j, \chi_j)_{j=1,\dots,n}$.

Indeed, though the choice of the $\mathcal{B}$-module morphism $U \in \mathrm{Hom}_{\mathcal{B}}(\mathcal{E}, \mathcal{B}^n)$ involved in the construction of θ, is not unique, we know that two such choices U and U' are equivalent. In fact, at the expense of replacing n by $2n$ we may even assume that there exists $W \in M_n(\mathcal{B})$, invertible, such that

$$U' = W \circ U.$$

Then $\theta'(T) = W\theta(T)W^{-1} \quad \forall T \in C_c^\infty(G)$, where W is a multiplier of the algebra $M_n(\mathcal{R}_M \otimes \mathbb{C}\Gamma)$; thus the equality of θ_* and $\theta'_* : K_0(C_c^\infty(G)) \to K_0(\mathcal{R}_M \otimes \mathbb{C}\Gamma)$. Finally the manifold M has disappeared since one has:

Lemma 5. *Let M be a compact manifold,* $\dim M > 0$. *Then there is a canonical class of isomorphisms*

$$\rho : \mathcal{R}_M \to \mathcal{R}$$

unique modulo inner automorphisms of $\mathcal{R}$.

Indeed, the algebra $\mathcal{R}_M$ depends functorially on the pair, $L^2(M) \supset C^\infty(M)$, of a Hilbert space and a dense (nuclear) subspace. Moreover, it is easy to check, using , for instance, an orthonormal basis of $L^2(M)$ of eigenvectors for a Laplacian, that the above pair is isomorphic to the pair $\ell^2(\mathbb{N}) \supset S(\mathbb{N})$, where $S(\mathbb{N})$ is the subspace of ℓ^2 of sequences of rapid decay

$$\mathrm{Sup}\, n^k |a_n| < \infty \quad \forall k \in \mathbb{N}.$$

Putting together Proposition 4 and Lemma 5 we get a well defined K-theory map

$$j : K_0(C_c^\infty(G)) \to K_0(\mathcal{R}\Gamma).$$

Definition 6. *The* equivariant index *of a Γ-invariant elliptic differential operator, $D : C_c^\infty(\widetilde{M}, E_1) \to C_c^\infty(\widetilde{M}, E_2)$, is defined as* $\mathrm{Ind}_\Gamma(D) = j(\mathrm{Ind}(D)) \in K_0(\mathcal{R}\Gamma)$.

Note that it is *not possible* to define $\mathrm{Ind}_\Gamma(D)$ as an element of $K_0(\mathbb{C}\Gamma)$ since the latter group is too small in general, and in particular for torsion-free groups.

4.y The index theorem

Let $c \in Z^q(\Gamma, \mathbb{C})$ be a group cocycle, with $q = 2k$ an even integer. The cup product $\tau_c \# \text{Trace} = \varphi_c$ is a well defined cyclic cocycle on the algebra $\mathcal{R}\Gamma$, and thus we can pair it with the group $K_0(\mathcal{R}\Gamma)$ as described in detail in Section 3. The following result computes the pairing $\varphi_c \operatorname{Ind}(D)$ in terms of the principal symbol σ_D of D, the classifying map

$$\psi : M \to B\Gamma$$

of the principal Γ-bundle $\widetilde{M}$ over M and the usual ingredients of the Atiyah-Singer index theorem, namely the Todd genus $\operatorname{Td}(T_{\mathbb{C}}M)$ of the complexified tangent bundle of M.

Theorem 7. [Co-M$_1$] *Let Γ be a (countable) discrete group acting properly and freely on a manifold $\widetilde{M}$ and D a Γ-invariant elliptic differential operator on $\widetilde{M}$. For any group cocycle $c \in Z^{2k}(\Gamma, \mathbb{C})$ one has*

$$\langle \tau_c \# \text{Trace}, \operatorname{Ind}(D) \rangle = \frac{(-1)^{\dim M}}{(2\pi i)^k} \frac{1}{(2k)!} \langle \operatorname{ch}(\sigma_D) \operatorname{Td}(T_{\mathbb{C}}M) \psi^*(c) \, , \, [T^*M] \rangle$$

where $M = \widetilde{M}/\Gamma$, $\psi : M \to B\Gamma$ is the classifying map, and $\psi^*(c)$ is the pull-back of the class of c in $H^{2k}(B\Gamma, \mathbb{C})$.

We refer to [Co-M$_1$] for the proof, which relies on the local isomorphism of the groupoid G of the covering space with the groupoid $M \times M$, thus reducing the statement to an index theorem for germs of cocycles near the diagonal on the groupoid $M \times M$, i.e. to Alexander-Spanier cohomology. Note that we use the pairing as defined in Proposition 3.2 which eliminates the term $k!$ of [Co-M$_1$]. The sign $(-1)^{\dim M}$ can be eliminated by a suitable choice of orientation of $[T^*M]$.

As a special case of Theorem 7 one gets that the equivariant index of the *signature operator* D_{sign} on $\widetilde{M}$, paired with $\tau_c \# \text{Trace}$, gives (up to a numerical factor) the Novikov higher signature

$$\operatorname{Sign}_c(M, \psi) = \langle L(M) \cup \psi^*(c), [M] \rangle.$$

Thus, if we knew that the equivariant index

$$\operatorname{Ind}_\Gamma(D_{\text{sign}}) \in K_0(\mathcal{R}\Gamma)$$

is a *homotopy invariant* of the pair (M, ψ), then the Novikov conjecture would follow.

But the only thing we know, thanks to the index theorem of Mishchenko and Kasparov (cf. Section II.4), is that:

Lemma 8. *The image of $\operatorname{Ind}_\Gamma(D_{\text{sign}})$ in $K_0(C^*\Gamma)$ under the morphism $\mathcal{R}\Gamma \to \mathcal{K} \otimes C^*(\Gamma)$ is the Mishchenko-Kasparov $C^*\Gamma$-signature, and is homotopy invariant.*

We refer to [Co-M$_1$] for the details of the proof. We are thus confronted with the crucial problem of extending the K-theory invariant

$$\varphi : K_0(\mathcal{R}\Gamma) \to \mathbb{C}$$

given by $\varphi_c(x) = \langle \tau_c \# \mathrm{Trace}, x \rangle$, to the larger C^*-algebra $\mathcal{K} \otimes C^*(\Gamma)$.

We can formulate the following corollary of Theorem 7 and Lemma 8:

Corollary 9. *Let Γ be a discrete group and $c \in Z^{2k}(\Gamma, \mathbb{C})$ a group cocycle. Then if the pairing φ_c with the cyclic cocycle τ_c extends from $K_0(\mathcal{R}\Gamma)$ to $K_0(\mathcal{K} \otimes C^*(\Gamma))$, the pair (Γ, c) satisfies the Novikov conjecture.*

As we mentioned already in the introduction this extension problem will occupy the second part of this chapter. We shall first describe what can be done directly for discrete groups.

III.5. The Novikov Conjecture for Hyperbolic Groups

We shall see in this section that the extension problem for K-theory invariants given by group cocycles (Corollary 4.9) can be solved for Gromov's word hyperbolic groups, thus proving the Novikov conjecture for this large class of groups ([Co-M$_1$]).

5.α Word hyperbolic groups

There are (cf. [Gro$_3$]) several equivalent definitions of word hyperbolic groups Γ. We shall use the definition in terms of the hyperbolicity of the (discrete) metric space (Γ, d), where d is the left invariant distance $d(g_1, g_2) = \ell(g_1^{-1} g_2)$, $\forall g_1, g_2 \in \Gamma$ and ℓ is the word length. This is relative to a finite system $F \subset \Gamma$ of generators of Γ, with $F = F^{-1}$, and is defined as the length of the shortest word in the generators representing g

$$\ell(g) = \mathrm{Inf}\{n ; g \in F^n\}.$$

Definition 1. [Gro$_3$] *Let (X, d) be a metric space. It is* hyperbolic *if for some $x_0 \in X$, there exists $\delta_0 > 0$ such that the inequality*

$$\delta(x, z) \geq \mathrm{Inf}(\delta(x, y), \delta(y, z)) - \delta_0 \quad \forall x, y, z \in X$$

where $\delta(x, y) = \frac{1}{2}(d(x, x_0) + d(y, x_0) - d(x, y))$, is satisfied.

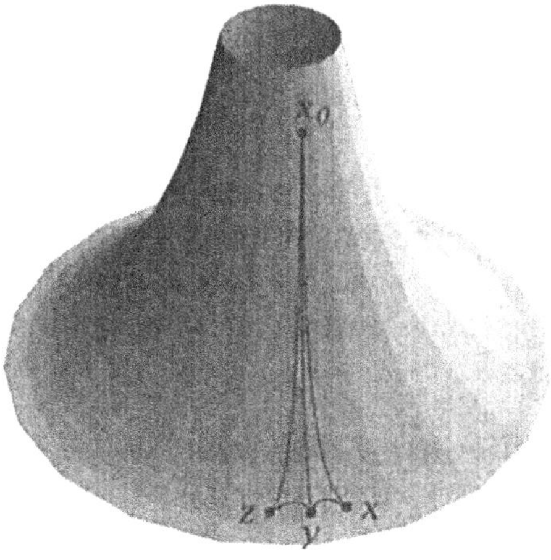

Figure III.2. Hyperbolic metric space

It then follows that the same holds for any $x_0 \in X$ (with the constant $2\delta_0$) so that the choice of x_0 is irrelevant. The simplest example of hyperbolic metric spaces, besides the line and trees, is given by complete simply connected Riemannian manifolds with sectional curvature everywhere bounded above by $-\varepsilon < 0$.

Definition 2. [Gro$_3$] *Let* Γ *be a finitely generated discrete group. Then* Γ *is* hyperbolic *if the metric space* (Γ, d) *(associated to a finite set of generators* F *of* Γ*) is hyperbolic.*

This definition does not depend upon the set of generators, at least in the case we are interested in, namely that of finitely presented groups. Any *finitely presented* group Γ is the fundamental group $\pi_1(P)$ of a 2-dimensional cell complex associated naturally to the presentation of Γ. Then Γ is hyperbolic if and only if the following isoperimetric inequality is satisfied by a smooth connected neighborhood V of $P \subset \mathbb{R}^n$, $n \geq 5$.

There exists $C < \infty$ *such that every smooth simple curve* S *in* V *which is contractible in* V *bounds a smooth embedded disk* $D \subset V$ *such that*

$$\text{Area } D \leq C \text{ Length } S.$$

Finally, the third equivalent formulation of hyperbolicity is that the metric space (Γ, d) looks treelike when seen from very far ([Gro$_3$]). Nevertheless one should not conclude too hastily that hyperbolic groups are one-dimensional objects, and for instance:

Proposition 3. [Gro$_3$] *For every finite polyhedron V_0 one can find an aspherical polyhedron V with word hyperbolic fundamental group $\Gamma = \pi_1(V)$ such that $\dim V = n = \dim V_0$ and V admits a continuous map $f : V \to V_0$ which is injective on the cohomology $H^*(V_0)$. Furthermore, if V_0 is a manifold, then one also can choose V to be a manifold, such that $f^* : H^*(V_0, \mathbb{Q}) \to H^*(V, \mathbb{Q})$ sends the Pontryagin classes of V_0 to those of V.*

In order to get familiar with Definition 2 of hyperbolic groups let us use it to estimate the number of decompositions $g = g_1 g_2$ of a given $g \in \Gamma$ such that $\ell(g_1) + \ell(g_2) \leq \ell(g) + \text{const}$. We fix a system of generators F of Γ; let ℓ be the corresponding length function on Γ and δ_0 the constant, provided by Definition 1, such that

$$\delta(g_1, g_3) \geq \operatorname{Inf}(\delta(g_1, g_2), \delta(g_2, g_3)) - \delta_0 \qquad \forall g_1, g_2, g_3 \in \Gamma$$

where $\delta(g_1, g_2) = \frac{1}{2}(\ell(g_1) + \ell(g_2) - \ell(g_1^{-1} g_2))$ $\forall g_i \in \Gamma$.

Lemma 4. a) *Let $g_1, g_2 \in \Gamma$, $n_i = \ell(g_i)$, and let ℓ be such that $\ell(g_1 g_2) = \ell(g_1) + \ell(g_2) - 2\ell$. Then there exist $g'_1, g'_2 \in \Gamma$ such that $g'_1 g'_2 = g_1 g_2$, $\ell(g'_i) = \ell(g_i) - \ell$, and $g_1 = g'_1 k$ with*

$$\ell(g'_1) + \ell(k) \leq \ell(g_1) + 2\delta_0.$$

b) *Let $\delta < \infty$. There exists $C < \infty$ such that the number of decompositions $g = g_1 g_2$ of any $g \in \Gamma$, with $\ell(g_1)$ fixed and $\ell(g_1) + \ell(g_2) \leq \ell(g) + \delta$, is bounded above by C.*

Proof. a) Note first that $\ell \leq \inf(n_1, n_2)$ with the above notation, so that $n_i - \ell \geq 0$. Writing $g = g_1 g_2$ as a product of $\ell(g)$ elements of F, we can define g'_1 as the product of the first $n_1 - \ell$ elements and g'_2 as the product of the last $n_2 - \ell$ elements. One has

$$2\delta(g'_1, g) = \ell(g'_1) + \ell(g) - \ell(g'_2) = 2n_1 - 2\ell$$

$$2\delta(g_1, g) = \ell(g_1) + \ell(g) - \ell(g_2) = 2n_1 - 2\ell.$$

Thus, the hyperbolicity of Γ yields

$$\delta(g_1, g'_1) \geq (n_1 - \ell) - \delta_0.$$

Hence $\ell(g_1) + \ell(g'_1) - d(g_1, g'_1) \geq 2(n_1 - \ell) - 2\delta_0$ and $d(g_1, g'_1) \leq \ell + 2\delta_0$. Thus, $k = (g'_1)^{-1} g_1$ has length less than $\ell + 2\delta_0$.

b) Let $g = g_1 g_2$ with $\ell(g_1) = n_1$ and $\ell(g) = \ell(g_1) + \ell(g_2) - 2\ell$ as above. By hypothesis, $2\ell \leq \delta$; thus with g'_1 and g'_2 as in a) one has $d(g_1, g'_1) \leq \delta + 2\delta_0$, and the result follows with C the cardinality of the ball of radius $\delta + 2\delta_0$.

5.β The Haagerup inequality

In work on the approximation property for C^*-algebras ([H$_6$]), U. Haagerup proved an important estimate for the norm of the convolution operator $\lambda(x)$, $x \in \mathbb{C}\Gamma$, acting in the Hilbert space $\ell^2(\Gamma)$, where Γ is a free group. This estimate was then extended to word hyperbolic groups in [Jol] [Harp]. We shall give its proof below using Lemma 4 above.

Theorem 5. [Jol] [Harp] *Let Γ be a word hyperbolic group and $g \to \ell(g)$ the word length. Then for some $k < \infty$ and $C < \infty$ one has, for any $x \in \mathbb{C}\Gamma$*

$$\|\lambda(x)\| = \|x\|_{C_r^*} \leq C\, \nu_k(x)$$

where $\nu_k(x) = (\sum_{g\in\Gamma}(1 + \ell(g))^{2k}\, |x_g|^2)^{1/2}$, $x = \sum x_g g$.

Here $\lambda(x)$ is the operator on $\ell^2(\Gamma)$ of left convolution by x

$$(\lambda(x)\xi)(g) = \sum_{g_1 g_2 = g} x_{g_1}\, \xi(g_2).$$

Writing $x = \sum x^{(n)}$ where $x_g^{(n)} = 0$ unless $\ell(g) = n$ we just need to find an inequality of the form

$$\|\lambda(x^{(n)})\| \leq P(n)\, \|x^{(n)}\|_2 \tag{$*$}$$

where P is a polynomial in n and $\|y\|_2$ is the ℓ^2 norm. One then has $\|\lambda(x)\| \leq \sum \|\lambda(x^{(n)})\| \leq \sum P(n)\|x^{(n)}\|_2 \leq \text{const}\left(\sum (nP(n))^2\, \|x^{(n)}\|_2^2\right)^{1/2}$.

We can thus fix $n \in \mathbb{N}$ and assume that $x_g = 0\ \ \forall g \in \Gamma$, $\ell(g) \neq n$. Let us then consider the orthogonal decomposition

$$\ell^2(\Gamma) = \mathcal{H} = \bigoplus_m \mathcal{H}_m$$

where, with ε_g, $g \in \Gamma$, the canonical orthonormal basis of $\ell^2(\Gamma)$, the subspace $\mathcal{H}_m$ is the closed linear span of the ε_g with $\ell(g) = m$. Let $\lambda(x) = (T_{m_1,m_2})_{m_j\in\mathbb{N}}$ be the matrix of operators

$$T_{m_1,m_2} : \mathcal{H}_{m_2} \to \mathcal{H}_{m_1}$$

corresponding to $\lambda(x)$. The triangle inequality shows that $T_{m_1,m_2} = 0$ unless $|m_1 - m_2| \leq n$, since $kh_2 = h_1$, $\ell(k) = n$, $\ell(h_i) = m_i$ implies $|m_1 - m_2| \leq n$. Thus, in order to prove $(*)$ it is enough to find a polynomial $P(n)$ such that

$$\|T_{m_1,m_2}\| \leq P(n)\|x\|_2 \quad \forall m_1, m_2.$$

Let us fix $m, q \geq 0$. Let ξ be such that $\xi(g) = 0$ unless $\ell(g) = m$. One has

$$\|T_{m+n-2q,m}\xi\|^2 = \sum_{\ell(g)=m+n-2q} |(x * \xi)(g)|^2$$

$$(x * \xi)(g) = \sum_{\ell(g_1)=n,\ell(g_2)=m,g_1g_2=g} x(g_1)\, \xi(g_2).$$

For $g \in \Gamma$, $\ell(g) = m+n-2q$, let $g_1', g_2' \in \Gamma$ be such that $g_1' g_2' = g$, $\ell(g_1') = n-q$, $\ell(g_2') = m-q$. Then by Lemma 4 we can, for any decomposition $g = g_1 g_2$, with $\ell(g_1) = n$ and $\ell(g_2) = m$, find $k \in \Gamma$ with $g_1 = g_1' k$ and $\ell(k) \le q + 2\delta_0$. Then $g_2 = k^{-1} g_2'$. Thus, the Schwarz inequality gives

$$|(x * \xi)(g)|^2 \le \left(\sum_{\substack{\ell(g_1' k_1) = n \\ \ell(k_1) \le q + 2\delta_0}} |x(g_1' k_1)|^2 \right) \left(\sum_{\substack{\ell(k_2^{-1} g_2') = m \\ \ell(k_2) \le q + 2\delta_0}} |\xi(k_2^{-1} g_2')|^2 \right).$$

Adding up these inequalities we get

$$\|(T_{m+n-2q,m})\xi\|^2 \le C^2 \left(\sum |x(g_1)|^2\right) \left(\sum |\xi(g_2)|^2\right)$$

where C is the constant of Lemma 4 b) for $\delta = 2\delta_0$.

Remark 6. The above estimate of the convolution norm $\|\lambda(x)\|$, $x \in \mathbb{C}\Gamma$, in terms of the word length norms $\nu_k(x)$ fails in general, and in particular for any solvable group with exponential growth. Indeed, for any amenable group the trivial representation is weakly contained in the regular representation and it follows that for $x \in \mathbb{C}\Gamma$, with $x_g \ge 0\ \forall g \in \Gamma$, one has

$$\sum x_g \le \|\lambda(x)\|.$$

But the inequality $\sum x_g \le C\, \nu_k(x)$ is easily violated for any C and k when Γ has exponential growth. We refer to [Jol] for a general study of groups for which Theorem 5 is valid.

5.γ Extension to $C_r^*(\Gamma)$ of K-theory invariants

Let Γ be a word hyperbolic group and $[c] \in H^p(\Gamma, \mathbb{C})$ a group cohomology class (assume p even > 0). By [Gro$_3$] we may let $c \in Z^p(\Gamma, \mathbb{C})$ be a *bounded* group cocycle in that class, i.e. $|c(g_1, \dots, g_k)| \le C\ \ \forall g_i \in \Gamma$. Let $\tau_c \in Z_\lambda^p(\mathbb{C}\Gamma \otimes \mathcal{R})$ be the corresponding cyclic cocycle on the group ring $\mathbb{C}\Gamma \otimes \mathcal{R} = \mathcal{R}\Gamma$ of Γ with coefficients in the ring $\mathcal{R}$ of matrices of rapid decay (cf. Section 4).

Theorem 7. [Co-M$_1$] *The K-theory invariant $\tau_c : K_0(\mathcal{R}\Gamma) \to \mathbb{C}$ extends to the K-theory of the C^*-algebra $C_r^*(\Gamma) \otimes \mathcal{K} \supset \mathcal{R}\Gamma$.*

To prove this theorem we shall show that the cyclic cocycle τ_c extends by continuity to a cyclic cocycle on the closure C of $\mathcal{R}\Gamma$ under holomorphic functional calculus in $C_r^*(\Gamma) \otimes \mathcal{K}$. Any element x of $C_r^*(\Gamma) \otimes \mathcal{K}$ yields a matrix (x_{ij}) of elements of $C_r^*(\Gamma)$ and we can thus define

$$N_k(x) = \left(\sum_{i,j} (\nu_k(x_{ij}))^2 \right)^{1/2}.$$

For $x \in C_r^*(\Gamma)$ the components $x_g \in \mathbb{C}$ are well defined, $x_g = \langle x \varepsilon_g, \varepsilon_g \rangle$, so that $\nu_k(x) = \left(\sum (1 + \ell(g)^{2k}) |x_g|^2\right)^{1/2}$ is unambiguous.

Let us show that for any element x of the holomorphic closure C of $\mathcal{R}\Gamma$ in $C_r^*(\Gamma) \otimes \mathcal{K}$ one has

$$N_k(x) < \infty \quad \forall k \in \mathbb{N}.$$

We shall construct a subalgebra $\mathcal{B}$ of $C_r^*(\Gamma) \otimes \mathcal{K}$ which contains $\mathbb{C}\Gamma \otimes \mathcal{R}$, is closed under holomorphic functional calculus (Appendix C) and such that $N_k(x) < \infty$ for any $x \in \mathcal{B}$ and $k \in \mathbb{N}$.

To this end, we let Δ (resp. D) be the (unbounded) operator on $\ell^2(\mathbb{N})$ (resp. $\ell^2(\Gamma)$) defined by

$$\Delta \delta_j = j \delta_j \quad (j \in \mathbb{N}) \quad (\text{resp. } D\delta_g = |g| \delta_g \quad (g \in \Gamma)).$$

Next, we consider the unbounded derivations $\partial = \operatorname{ad} D$ of $\mathcal{L}(\ell^2(\Gamma))$ and $\tilde{\partial} = \operatorname{ad}(D \otimes 1)$ of $\mathcal{L}(\ell^2(\Gamma) \otimes \ell^2(\mathbb{N}))$, and set

$$\mathcal{B} = \{x \in C_r^*(\Gamma) \otimes \mathcal{K};\ \ \tilde{\partial}^k(x) \circ (1 \otimes \Delta) \text{ is bounded } \ \forall k \in \mathbb{N}\}.$$

First we claim that $\mathcal{B}$ contains $\mathbb{C}\Gamma \otimes \mathcal{R}$. Indeed, if $x = \lambda(g) \otimes S$, with $g \in \Gamma$, λ the left regular representation and $S \in \mathcal{R}$, then $\tilde{\partial}^k(x) \circ (1 \otimes \Delta) = \partial^k(\lambda(g)) \otimes S\Delta$; but both $\partial^k(\lambda(g))$ and $S\Delta$ are bounded.

Secondly, since $\{x \in C_r^*(\Gamma) \otimes \mathcal{K};\ \ x \circ (1 \otimes \Delta) \text{ bounded}\}$ is evidently a left ideal, $\mathcal{B}$ is a subalgebra, in fact a left ideal in

$$\mathcal{B}_\infty = \bigcap_{k \geq 0} \operatorname{Domain} \tilde{\partial}^k.$$

Now $\mathcal{B}_\infty$ is stable under holomorphic functional calculus and, therefore, so is $\mathcal{B}$.

Finally, let us check that $N_k(x) < \infty$, $\forall x \in \mathcal{B}$ and $k \in \mathbb{N}$. As $D\delta_e = 0$, one has

$$\begin{aligned}(\tilde{\partial}^k(x) \circ 1 \otimes \Delta)(\delta_e \otimes \delta_j) &= j \sum_{i \in \mathbb{N}} D^k\,(x_{ij}\,\delta_e) \otimes \delta_i \\ &= j \sum_{(g,i) \in \Gamma \times \mathbb{N}} |g|^k\, x_{ij}(g)\, \delta_g \otimes \delta_i.\end{aligned}$$

Thus, there exists $C < \infty$ such that

$$\sum_{(g,i) \in \Gamma \times \mathbb{N}} |g|^{2k}\, |x_{ij}(g)|^2 < C j^{-2};$$

therefore,

$$\sum_{i,j \in \mathbb{N}} \sum_{g \in \Gamma} |g|^{2k}\, |x_{ij}(g)|^2 < \infty\,, \quad \forall k \in \mathbb{N},$$

which implies $N_k(x) < \infty$, $\forall k \in \mathbb{N}$.

To conclude the proof of Theorem 7 we just need to show that the cyclic cocycle $\tau_c \#\text{Trace}$ on $\mathbb{C}\Gamma \otimes \mathcal{R}$ is continuous for the norm N_k, k large enough, and thus extends by continuity to C. It is clear from the definition of N_k that $\tau_c \#\text{Trace}$ is continuous with respect to N_k if τ_c is continuous with respect to ν_k. The continuity of τ_c with respect to ν_k, for k large enough, follows from

Theorem 5 and the boundedness of the group cocycle c (its polynomial growth is sufficient) ([Jol]). Indeed, one has

$$\tau_c(x^0, x^1, \ldots, x^p) = \sum_{g_0 \cdots g_k = 1} x^0(g_0) \cdots x^p(g_p)\, c(g_1, \ldots, g_p)$$

$$|\tau_c(x^0, x^1, \ldots, x^p)| \le \|c\| \sum_{g_0 \cdots g_k = 1} y^0(g_0) \cdots y^p(g_p)$$

where $y^j(g) = |x^j(g)| \in \mathbb{C}\ \ \forall g \in \Gamma$. Thus

$$|\tau_c(x^0, x^1, \ldots, x^p)| \le \|c\| (y^0 * y^1 * \cdots * y^p)(1)$$

$$\le \|c\| C^p \left(\prod_1^p \nu_k(y^j) \right) \nu_0(y^0) = \|c\| C^p \left(\prod_1^p \nu_k(x^j) \right) \nu_0(x^0).$$

As a corollary of Theorem 7 and Section 4 Corollary 9, this results in

Theorem 8. [Co-M$_1$] *Let Γ be a word hyperbolic group. Then Γ satisfies the Novikov conjecture.*

Remark 9. 1) For another subsequent proof of this result cf. [Co-G-M].

2) The proof of the above theorem has been extended by C. Ogle to show that even in the presence of torsion for Γ the analytic assembly map (Chapter II Section 10) is rationally injective ([O]).

III.6. Factors of Type III, Cyclic Cohomology and the Godbillon-Vey Invariant

Let (V, F) be a transversely oriented codimension-1 foliation. The Godbillon-Vey class $GV \in H^3(V, \mathbb{R})$ is a 3-dimensional cohomology class, the simplest nontrivial example of a Gel'fand-Fuchs cohomology class. It is defined as follows: Since (V, F) is transversely oriented there exists a nowhere vanishing smooth 1-form ω on V which defines the foliation by the equality

$$F_x = \{X \in T_x(V)\ ;\ \omega_x(X) = 0\} \quad \forall x \in V.$$

The condition of integrability of the bundle $F \subset TV$ is equivalent to the existence of a 1-form α on V such that

$$d\omega = \alpha \wedge \omega.$$

One then checks the following:

Lemma 1. [Go-V] *The 3-form $\alpha \wedge d\alpha$ is closed and its cohomology class $GV \in H^3(V, \mathbb{R})$ is independent of the choices of ω and α once (V, F) is given.*

Indeed, fixing ω first and changing α to α', the equality $(\alpha' - \alpha) \wedge \omega = 0$ shows that $\alpha' = \alpha + \rho\omega$ for some $\rho \in C^\infty(V)$; it results that

$$\alpha' \wedge d\alpha' = \alpha \wedge d\alpha + d(\rho\omega \wedge \alpha).$$

This shows that the class of $\alpha \wedge d\alpha$ is independent of the choice of α. If one replaces ω by $\omega' = f\omega$ with $f \in C^\infty(V, \mathbb{R}^*)$ then one has

$$d\omega = \alpha \wedge \omega \Longrightarrow d\omega' = (f^{-1}df + \alpha) \wedge \omega'.$$

Thus, the class of $\alpha \wedge d\alpha$ is independent of the choice of ω, and is hence closed.

The cohomology class $GV \in H^3(V, \mathbb{R})$ is not rational in general; in fact when V is a compact oriented 3-manifold, the Godbillon-Vey invariant

$$\langle GV, [V] \rangle \in \mathbb{R}$$

can assume any real value ([Thu$_2$]).

With a little more work one can define GV as a 3-dimensional cohomology class

$$GV \in H^3(BG, \mathbb{R})$$

where BG is the classifying space of the holonomy groupoid G of (V, F) (cf. [Ha$_{1,2,3}$] and γ) below).

In this section we shall prove that the Godbillon-Vey class yields K-theory invariants

$$\varphi_z : K_0(C_r^*(V, F)) \to \mathbb{R}$$

for the C^*-algebra of the foliation, parametrized by elements $z \in \mathrm{mod}(M)$ of the flow of weights of the von Neumann algebra M of (V, F). These invariants enjoy two properties:

1) $\varphi_{\theta_\lambda(z)} = \varphi_z \quad \forall \lambda \in \mathbb{R}_+^*$, $z \in \mathrm{mod}(M)$, where $\theta_\lambda \in \mathrm{Aut}\,\mathrm{mod}(M)$ is the flow of weights (Chapter V Section 8).

2) φ_1 is related to $GV \in H^3(BG, \mathbb{R})$ by the equality

$$\varphi_1(\mu(x)) = \langle \mathrm{ch}_*(x), GV \rangle \quad \forall x \in K_{*,\tau}(BG)$$

where μ is the analytic assembly map (Chapter II Section 8).

As an immediate corollary of the existence of such K-theory invariants it follows that if $GV \neq 0$ as an element of $H^3(BG, \mathbb{R})$, then the flow of weights $\mathrm{mod}(M)$ admits a normal invariant probability measure (Theorem 21) [Co$_{14}$] and the von Neumann algebra M has a nontrivial type III component ([Hu-K$_1$]).

The existence of φ_z follows from the interplay between the transverse fundamental class of (V, F), which is a certain canonical cyclic cohomology class $[V/F] \in HC^1(C_c^\infty(G))$ encoding the 1-dimensional oriented transverse structure, and the canonical dynamics $\delta(t) \in \mathrm{Out}\,M = \mathrm{Aut}\,M/\mathrm{Inn}\,M$ of the von Neumann algebra M of the foliation, given by the modular automorphism group of M (cf. Chapter I Section γ).

The point is that, in general, $[V/F]$ fails to be invariant under the evolution $\delta(t)$, but its *second* time derivative $\frac{d^2}{dt^2}[V/F]$ does vanish and the first time derivative $\frac{d}{dt}[V/F]$ yields by contraction $i_\delta \frac{d}{dt}[V/F]$, a cyclic 2-cohomology class

$$i_\delta \frac{d}{dt}[V/F] \in HC^2(C_c^\infty(G))$$

whose associated K-theory invariant, once extended to $C_r^*(V,F)$, yields φ_1.

The relation, for any $x \in K_{*,\tau}(BG)$,

$$\langle i_\delta \frac{d}{dt}[V/F], \mu(x)\rangle = \langle GV, \mathrm{ch}_*(x)\rangle$$

expresses the deep interplay between noncommutative measure theory and Gel'fand-Fuchs cohomology.

Let us now enter into more detail, beginning with a technique to extend K-theory invariants

$$\varphi : K(\mathcal{A}) \to \mathbb{C}$$

from a subalgebra $\mathcal{A} \subset A$ of a C^*-algebra to

$$\tilde{\varphi} : K(A) \to \mathbb{C}.$$

6.α Extension of densely defined cyclic cocycles on Banach algebras

Let A be a unital C^*-algebra; then the traces τ on A are the only interesting continuous cyclic cocycles on A ([Co$_{17}$]). As a rule the relevant cocycles are only defined on a dense subalgebra

$$\mathcal{A} \subset A$$

and are not continuous for the C^*-algebra norm of A.

In this section we shall give a *relative* continuity condition on cyclic cocycles on $\mathcal{A}$ which will ensure that the K-theory invariant

$$\varphi : K(\mathcal{A}) \to \mathbb{C}$$

associated to the cyclic cocycle, extends to $K(A)$.

Let us begin by giving a simple condition implying that two (continuous) traces τ_0, τ_1 yield the same map

$$\tau_j : K_0(A) \to \mathbb{C}.$$

If we were to consider τ_0 and τ_1 as elements of $HC^0(A)$ then a sufficient condition would be $\tau_0 - \tau_1 \in B(\ker b)$ (cf. Section 3), i.e. the existence of a Hochschild 1-cocycle $\psi \in H^1(A, A^*)$ such that $B\psi = \tau_0 - \tau_1$, or, in other words, with $\tau = \tau_0 - \tau_1$,

$$\psi(1,a) - \psi(a,1) = \tau(a) \quad \forall a \in A.$$

Since the Hochschild cohomology $H^1(A, A^*)$ always vanishes ([H$_3$]) the above condition is of no use and only yields trivial homologies between traces.

As we shall see this is no longer the case if we allow ψ to be unbounded. A Hochschild 1-cocycle $\psi \in Z^1(A, A^*)$ is a *derivation* from the algebra A to the bimodule A^* of continuous linear forms on A and the notion of *unbounded derivation* from A to A^* makes perfect sense. Let us recall the standard terminology:

Given two Banach spaces B_1, B_2 an *unbounded operator* T from B_1 to B_2 is given by a linear subspace $\mathrm{Dom}\, T \subset B_1$ and a linear map $T : \mathrm{Dom}\, T \to B_2$.

One says that T is *densely defined* when $\mathrm{Dom}\, T$ is dense in B_1, and that it is *closable* when the closure of its graph

$$\mathrm{graph}(T) = \{(\xi, T\xi) \in B_1 \times B_2 \,;\, \xi \in \mathrm{Dom}\, T\}$$

is the graph of an unbounded operator $\overline{T}$, the closure of T. The adjoint T^* of a densely defined operator T is the unbounded operator from B_2^*, the dual Banach space of B_2, to B_1^* defined by

$$\mathrm{Dom}\, T^* = \{\eta \in B_2^* \,;\, \exists c < \infty \text{ with } |\langle T\xi, \eta\rangle| \leq c\|\xi\| \text{ for any } \xi \in \mathrm{Dom}\, T\}$$

$$\langle T^*\eta, \xi\rangle = \langle \eta, T\xi\rangle \quad \forall \eta \in \mathrm{Dom}\, T^* \,,\ \xi \in \mathrm{Dom}\, T.$$

The adjoint T^* is densely defined provided T is closable.

Let B_1 be a Banach algebra and B_2 a Banach B_1-bimodule such that

$$\|b_1 \xi b_1'\| \leq \|b_1\|\, \|\xi\|\, \|b_1'\| \quad \forall b_1, b_1' \in B_1 \,,\ \xi \in B_2$$

Then an unbounded operator from B_1 to B_2 is called a *derivation* if $\mathrm{Dom}\, \delta$ is a *subalgebra* of B_1 and

$$\delta(ab) = \delta(a)b + a\delta(b) \quad \forall a, b \in \mathrm{Dom}\, \delta.$$

One then has the following standard result:

Lemma 2. *Let δ be a densely defined and closable derivation from B_1 to B_2; then its closure $\overline{\delta}$ is still a derivation, and its domain $\mathrm{Dom}\, \overline{\delta}$ is a subalgebra of B_1 stable under holomorphic functional calculus.*

Proof. One has $a \in \mathrm{Dom}\, \overline{\delta}$ iff there exists $a_n \in B_1$, with $a_n \to a$, $\delta(a_n)$ convergent. Then $\overline{\delta}(a) = \lim \delta(a_n)$. It follows easily that $\mathrm{Dom}\, \overline{\delta}$ is an algebra and $\overline{\delta}$ a derivation. To prove the stability under holomorphic functional calculus of $\mathrm{Dom}\, \overline{\delta}$, let us first show that if $a \in \mathrm{Dom}\, \overline{\delta}$ is invertible in B_1 then $a^{-1} \in \mathrm{Dom}\, \overline{\delta}$.

As $\mathrm{Dom}\, \overline{\delta} = \mathcal{A}$ is dense in B_1, there exists $b \in \mathcal{A}$ with $\|1 - ab\| < 1$, $\|1 - ba\| < 1$, hence it is enough to show that if $a \in \mathcal{A}$, $\|a\| < 1$ then $(1-a)^{-1} \in \mathcal{A}$. This is clear since $\sum_0^n a^k \to (1-a)^{-1}$, and $\overline{\delta}(\sum_{k=0}^n a^k)$ is norm convergent, because $\|\overline{\delta}(a^k)\| \leq k\|a\|^{k-1}\, \|\overline{\delta}(a)\|$ for any $k \in \mathbb{N}$.

Applying this to the resolvent $(a - \lambda)^{-1}$, $\lambda \notin \mathrm{Spectrum}_{B_1}(a)$ one sees that $\mathrm{Dom}\, \overline{\delta}$ is stable under holomorphic functional calculus.

Let us now return to homology between traces and prove the following ([Co14]).

Proposition 3. *Let B be a unital Banach algebra, δ a densely defined derivation of B with values in the dual space B^* (viewed as a bimodule over B with $\langle a\varphi b, x\rangle = \langle \varphi, bxa\rangle$ $\forall a,x,b \in B$, $\varphi \in B^*$) and assume that the unit 1_B belongs to the domain of the adjoint δ^* of δ. Then*

a) $\tau = \delta^*(1)$ *is a trace on B.*

b) *The map of $K_0(B)$ to $\mathbb{C}$ given by τ is equal to 0.*

Proof. a) One has $\tau(xy) = \langle xy, \delta^*(1)\rangle = \langle \delta(xy), 1\rangle = \langle \delta(x), y\rangle + \langle \delta(y), x\rangle = \tau(yx)$ $\forall x,y \in \operatorname{Dom}\delta$.

b) The equality $\langle \delta(x), y\rangle + \langle \delta(y), x\rangle = \tau(xy)$ for $x,y \in \operatorname{Dom}\delta$ shows that $\operatorname{Dom}\delta \subset \operatorname{Dom}\delta^*$, with $\delta^*(x) = -\delta(x) + x\tau$ for any $x \in \operatorname{Dom}\delta$. This shows that δ is a closable operator from the Banach space B to the dual Banach space B^*. Let $\overline{\delta}$ be the closure of δ, so it follows that the domain of $\overline{\delta}$, $\mathcal{A} = \operatorname{Dom}\overline{\delta}$, is a subalgebra of B and $\overline{\delta}$ is a derivation from $\mathcal{A}$ to B^*. Moreover, by Lemma 2, the inclusion $\mathcal{A} \subset B$ gives an isomorphism

$$K_0(\mathcal{A}) \to K_0(B)$$

(cf. Appendix C). Thus, it is enough to show that the map from $K_0(\mathcal{A})$ to $\mathbb{C}$ given by τ is equal to 0, which follows from the equality $\tau = B\psi$, where $\psi \in Z^1(\mathcal{A}, \mathcal{A}^*)$ is given by

$$\psi(a^0, a^1) = \langle a^0, \overline{\delta}(a^1)\rangle \quad \forall a^0, a^1 \in \mathcal{A}.$$

Let us illustrate Proposition 3 by a very simple example.

Assume μ to be a Radon measure of 0 total mass on a connected compact manifold V. Let us express the homology between the current μ and the current 0 in C^*-algebra terms:

Example 4. *If $\mu(V) = 0$ there exists a densely defined derivation δ from the C^*-algebra $A = C(V)$ to its dual A^* such that 1_A belongs to the domain of the adjoint δ^* and $\delta^*(1) = \mu$.*

Proof. Choose a Riemannian metric on V and assign in a Borel manner a geodesic path $\pi_{p,q} : [0,1] \to V$ to each pair (p,q) of elements of V. Assuming for simplicity that μ is real, let $\mu = \mu^+ - \mu^-$ be its Jordan decomposition, and put, for $f,g \in C^\infty(V)$,

$$\tau(f,g) = \int \left(\int_0^1 \pi_{p,q}^*(f\, dg) \right) d\mu^+(p)\, d\mu^-(q).$$

Then the equality $\langle \delta(g), f\rangle = \tau(f,g)$ gives a densely defined derivation δ from A to A^* (considered as a bimodule over A) and $\delta^*(1) = \mu^+(V)\mu$ since

$$\tau(1,g) = \int (g(p) - g(q))\, d\mu^+(p)\, d\mu^-(q) = \mu^+(V) \int g\, d\mu.$$

Let B be a Banach algebra. We shall now show how to construct maps from $K_1(B)$ to $\mathbb{C}$ using instead of a trace a homology (in the sense of Proposition 3) between the trace $\tau = 0$ and itself.

Definition 4. *Let B be a Banach algebra. By a 1-trace on B we mean a densely defined derivation δ from B to B^* such that*

$$\langle \delta(x), y \rangle = -\langle \delta(y), x \rangle \quad \forall x, y \in \text{Dom}\, \delta.$$

Proposition 5. *Let δ be a 1-trace on B; then:*

a) *δ is closable.*

b) *There exists a unique map of $K_1(B)$ to $\mathbb{C}$ such that, for $u \in GL_n(\text{Dom}\, \overline{\delta})$,*

$$\varphi(u) = \langle u^{-1}, \overline{\delta}(u) \rangle.$$

Proof. We can assume that B is unital with $\delta(1) = 0$.

a) Any skew-symmetric operator from B to B^* is closable.

b) We can assume that δ is closed; let $\mathcal{A}$ be its domain. By Lemma 2, any element of $\mathcal{A}$ which is invertible in B is invertible in $\mathcal{A}$, and the same holds for $M_n(\mathcal{A}) \subset M_n(B)$. Thus, since the open set $M_n(B)^{-1}$ of invertible elements in $M_n(B)$ is locally convex, two elements u and v of $GL_n(\mathcal{A})$ which are in the same connected component of $GL_n(B)$ are connected by a piecewise affine path in $GL_n(\mathcal{A})$.

On such a path $t \to u_t$ the function $f(t) = \langle u_t^{-1}, \delta(u_t) \rangle$ is constant, since its derivative is $-\langle u_t^{-1}\, \dot{u}_t\, u_t^{-1}, \delta(u_t) \rangle + \langle u_t^{-1}, \delta(\dot{u}_t) \rangle\rangle = 0$ since $\langle u_t^{-1}, \delta(\dot{u}_t) \rangle = -\langle \delta(u_t^{-1}), \dot{u}_t \rangle = \langle u_t^{-1}\, \delta(u_t) u_t^{-1}, \dot{u}_t \rangle$. The result then follows.

We shall now give non-trivial examples of 1-traces on C^*-algebras. The simplest example is to take a one-parameter group α_t of automorphisms of A, and an α-invariant trace τ on A. Let D be the generator of (α_t), i.e. the closed derivation

$$D(x) = \lim_{t \to 0} \frac{1}{t} \, (\alpha_t(x) - x).$$

The equality $x \in \text{Dom}\, D \to \delta(x) = D(x)\tau \in A^*$ defines a 1-trace on A, and the map of $K_1(A)$ to $\mathbb{C}$ given by this 1-trace coincides with the one defined in [Co_{20}]. We shall now give other examples of 1-traces, not of the above form (the algebra A will have no nonzero traces in some examples) and use them to prove the non-triviality of $K_1(A)$ for crossed products $A = C(\mathbb{S}^1) \rtimes \Gamma$ of $\mathbb{S}^1$ by a group of *orientation-preserving* homeomorphisms. Note that we shall use the *reduced crossed product* so that the result we obtain is stronger than for the maximal crossed product.

Theorem 6. *Let Γ be a countable group of orientation-preserving homeomorphisms of $\mathbb{S}^1$ and $A = C(\mathbb{S}^1) \rtimes_r \Gamma$ be the reduced crossed product C^*-algebra. Then the canonical homomorphism $i : C(\mathbb{S}^1) \to A$ is an injection of $K_1(C(\mathbb{S}^1)) = \mathbb{Z}$ in $K_1(A)$.*

Proof. On $B = C(\mathbb{S}^1)$, we have a natural 1-trace obtained as the weak closure of the derivation δ, with domain $C^\infty(\mathbb{S}^1)$, which assigns to each $f \in C^\infty(\mathbb{S}^1)$ the differential df viewed as an element of B^*. (This uses the orientation.) This weak closure of δ is easily identified, using distribution theory, as the derivation (also denoted δ) with domain $BV(\mathbb{S}^1)$, the space of functions $f \in C(\mathbb{S}^1)$ of bounded variation, i.e. such that df is a measure, $\delta(f) = df$. A function f is of bounded variation iff the sums $\sum |f(x_{i+1}) - f(x_i)|$ are bounded when the finite subset $(x_i)_{i=0,\dots,n}$ of $\mathbb{S}^1$ varies with x_i in the same order on $\mathbb{S}^1$ as i, (i.e. x_{j+1} between x_j and x_{j+2} for $j = 0, \cdots, n-2$ and x_n between x_{n-1} and x_0). It is then clear that if φ is any orientation preserving homeomorphism of $\mathbb{S}^1$, $\varphi^* : C(\mathbb{S}^1) \to C(\mathbb{S}^1)$ leaves $\mathrm{Dom}\,\delta = BV(\mathbb{S}^1)$ invariant, and that $\delta(f \circ \varphi) = \varphi(\delta(f))$. Using this Γ-equivariant 1-trace on $C(\mathbb{S}^1)$ we shall now construct a 1-trace on A. Let $\mathcal{A} = \{a \in A\,;\ a = \sum_\Gamma a_g U_g$ with $a_g \neq 0$ only for finitely many $g \in \Gamma$ and $a_g \in BV(\mathbb{S}^1)\ \ \forall g \in \Gamma\}$. For any $a \in A$, let $\delta(a) \in A^*$ be the linear functional given by

$$\langle x, \delta(a)\rangle = \int \sum_\Gamma x_g\, g(da_{g^{-1}}) \qquad \forall x \in A.$$

Here, for any $g \in \Gamma$, $da_{g^{-1}}$ is a measure on $\mathbb{S}^1$ and $g(da_{g^{-1}})$ is its image under the action of $g \in \Gamma$ on $\mathbb{S}^1$. For any $x \in A$, $x = \sum x_g U_g$, x_g is (in a faithful representation) a matrix element of x so that one has $\|x_g\| \leq \|x\|$ where the latter is the norm in the reduced crossed product. Thus $\delta(a) \in A^*$. Let us check that δ is a 1-trace. For $a, b \in \mathcal{A}$ one has

$$\begin{aligned}\langle b, \delta(a)\rangle &= \int \sum b_g\, g(da_{g^{-1}}) \\ &= -\int \sum db_{g^{-1}}\, g^*\, a_g \\ &= -\int \sum a_g\, g(db_{g^{-1}}) = -\langle a, \delta(b)\rangle.\end{aligned}$$

For $a, b, c \in A$ one has

$$\begin{aligned}&\langle ab, \delta(c)\rangle - \langle a, \delta(bc)\rangle + \langle ca, \delta(b)\rangle \\ &\quad = \int \sum_{g_0 g_1 g_2 = 1} a_{g_0}((g_0^{-1})^* b_{g_1})(g_0 g_1) dc_{g_2} \\ &\qquad - \int \sum_{g_0 g_1 g_2 = 1} a_{g_0} g_0(d(b_{g_1}(g_1^{-1})^* c_{g_2})) \\ &\qquad + \int \sum_{g_0 g_1 g_2 = 1} c_{g_0}(g_0^{-1})^* a_{g_1}(g_0 g_1) db_{g_2} \\ &\quad = -\sum_{g_0 g_1 g_2 = 1} \int a_{g_0} g_0(db_{g_1})(g_0 g_1)^{-1*} c_{g_2} \\ &\qquad + \sum_{h_0 h_1 h_2 = 1} \int c_{h_0}(h_0^{-1})^* a_{h_1}(h_0 h_1) db_{h_2} = 0.\end{aligned}$$

Thus, δ is a 1-trace on A, and for any unitary $u \in BV(\mathbb{S}^1)$ with winding number equal to 1 one has

$$\langle u^{-1}, \delta(u)\rangle = \int u^{-1}\, du = 2i\pi.$$

This shows that u^n is a nontrivial element of $K_1(A)$ for any $n \in \mathbb{Z}$, $n \neq 0$.

In a similar manner, we shall now construct a 1-trace τ on the C^*-algebra $C_r^*(V,F)$ of a transversely oriented codimension-1 foliation and detect nontrivial elements of $K_1(C_r^*(V,F))$.

Let (V,F) be such a foliation and G its holonomy groupoid. Let $T \subset V$ be a submanifold of V of dimension one which is everywhere transverse to F and intersects every leaf of (V,F). We do not assume that T is compact or connected, so that such a T always exists. We shall work with the reduced groupoid

$$G_T = \{\gamma \in G\,;\, r(\gamma) \in T\,,\; s(\gamma) \in T\}$$

which is equivalent to G and is a one-dimensional manifold (we assume that it is Hausdorff and refer to [Co$_{11}$] for the general case). One has by construction $G_T^{(0)} = T$ and the maps r, s from G_T to $G_T^{(0)}$ are étale maps. We shall now construct a 1-trace on the C^*-algebra $A = C_r^*(G_T)$, which is strongly Morita equivalent to $C_r^*(V,F) = C_r^*(G)$ and hence has the same K-theory.

Proposition 7. *Let G be a 1-dimensional smooth groupoid such that $G^{(0)}$ is one-dimensional and that G and $G^{(0)}$ are oriented with r and s both orientation-preserving. Then the following equality defines a cyclic 1-cocycle on $C_c^\infty(G)$ which is a 1-trace on $C_r^*(G)$:*

$$\tau(f^0, f^1) = \int_G f^0(\gamma^{-1})\,(df^1)\,(\gamma).$$

Proof. Let us first check that τ is a cyclic cocycle on $C_c^\infty(G)$. The map $\gamma \to \gamma^{-1}$ is an orientation-preserving diffeomorphism of G, so

$$\tau(f^0, f^1) = -\tau(f^1, f^0) \quad \forall f^j \in C_c^\infty(G).$$

The product in $C_c^\infty(G)$ is given by

$$(f * f')(\gamma) = \sum_{\gamma_1 \circ \gamma_2 = \gamma} f(\gamma_1)\, f'(\gamma_2).$$

Since the source and range maps are étale we can use them to identify the bundles T^*G, $s^*(T^*(T))$, and $r^*(T^*(T))$ over G. We then have

$$d(f_1 * f_2) = (df_1) * f_2 + f_1 * df_2$$

in the space of 1-forms on G. To prove this one may assume that the supports of f_j are so small that both r and s are diffeomorphisms of neighborhoods of Support f_j with open sets in T. The required equality is then just the Leibniz rule for d. The equality $b\tau = 0$ follows from the above and the equivalence, for $\gamma_1, \gamma_2, \gamma_3 \in G$, between $\gamma_1\gamma_2\gamma_3 \in G^{(0)}$ and $\gamma_2\gamma_3\gamma_1 \in G^{(0)}$. Finally, to show that τ is a 1-trace on $C_r^*(G)$ one just needs to show that for any $f_1 \in C_c^\infty(G)$ the linear functional

$$f \in C_c^\infty(G) \to \int_G f(\gamma)\, df_1(\gamma^{-1}) = L(f)$$

is continuous in the C^*-algebra norm of $C_r^*(G)$, which is straightforward using the left regular representation of $C_r^*(G)$ in $L^2(G)$.

As an immediate corollary we get a way to detect the fundamental class of V/F in K-theory. The latter class makes sense for any transversely oriented foliation, and yields $[V/F]^* \in K_*(C^*(V,F))$ with the same parity as the codimension q of F. To define $[V/F]^*$ choose an orientation-preserving transversal embedding $\psi : D^q \to V$ where D^q is the q-dimensional open disk, and consider $\psi!(\beta_q) \in K_*(C^*(V,F))$, where $\beta_q \in K^*(D_q)$ is the canonical Bott generator of $K^*(D_q) = \mathbb{Z}$. One then checks (Chapter II Section 8):

Lemma 8. *The class $\psi!(\beta_q) \in K_*(C^*(V,F))$ is independent of the choice of ψ.*

We shall denote this class by $[V/F]^* \in K_*(C^*(V,F))$. Using Propositions 5 and 7 we get:

Theorem 9. *Let (V,F) be a transversely oriented foliation of codimension 1; then $[V/F]^*$ is a non-torsion element of $K_1(C_r^*(V,F))$.*

Proof. Let, as above, $T \subset V$ be a transverse one-dimensional manifold such that the holonomy groupoid G of (V,F) is equivalent to G_T. Then $C_r^*(G)$ is strongly Morita equivalent to $C_r^*(G_T)$, and under this equivalence the element $[V/F]^* \in K_1(C_r^*(V,F))$ becomes $j_*(\alpha)$ where $j : C_0(T) \subset C_r^*(G_T)$ is the canonical inclusion of $C_0(T)$ in $C_r^*(G_T)$ and α is any element of $K_1(C_0(T))$ associated to the Bott generator of one of the oriented connected components of T. As the restriction of the 1-trace τ of Proposition 7 to $C_0(T)$ is the fundamental class of this oriented manifold, the result follows from Proposition 5.

Note that we used $C_r^*(V,F)$ instead of $C^*(V,F)$ in the statement of Theorem 9 and hence prove a stronger result.

Remark 10. 1) We shall see in Section 7 that Theorem 9 is true in any codimension but that the transverse fundamental class in cyclic cohomology, i.e. τ, is much easier to analyze in codimension 1.

2) If B is a non-unital Banach algebra, Proposition 3 still holds if one replaces the equality $\tau = \delta^*(1)$ by

$$\tau(xy) = \langle \delta(x), y\rangle + \langle \delta(y), x\rangle \quad \forall x, y \in \operatorname{Dom}\delta.$$

3) Let A be a non-unital C^*-algebra, and τ a densely defined semi-continuous weight which is a trace: $\tau(x^*x) = \tau(xx^*)\ \forall x \in A$. Then, using 2) above, one can show that if δ is a densely defined derivation from A to A^* such that

a) $\mathrm{Dom}\,\delta \cap \mathrm{Dom}_{1/2}(\tau)$ is dense in A, where $\mathrm{Dom}_{1/2}(\tau) = \{x \in A; \tau(x^*x) < \infty\}$,

b) $\langle \delta(x), y\rangle + \langle \delta(y), x\rangle = \tau(xy) \quad \forall x, y \in \mathrm{Dom}\,\delta \cap \mathrm{Dom}_{1/2}(\tau)$,
then τ defines the 0-map from $K_0(A)$ to $\mathbb{C}$.

This happens if there exists a one-parameter group of automorphisms θ_t of A such that $\tau \circ \theta_t = e^t \tau \; \forall t \in \mathbb{R}$. Thus, if $\varphi \in A^*$ is a KMS state for a one-parameter group θ_t then the associated trace τ on the crossed product $\hat{A} = A \rtimes_\alpha \mathbb{R}$ (cf. Chapter V) is always homologous to 0.

4) The conclusion of Theorem 6 does not hold when Γ fails to preserve the orientation of $\mathbb{S}^1$; in fact if $[u] \in K^1(D^1)$ is the generator, then $2i_*[u] = 0$ in $K_1(A)$.

We shall now extend Proposition 5 to higher dimensional densely defined cyclic cocycles on Banach algebras. Let B be a, not necessarily unital, Banach algebra. Recall (Section 1) that to any $(n+1)$-linear functional τ on an algebra $\mathcal{A}$ we associate the linear form $\hat{\tau}$ on $\Omega^n(\mathcal{A})$ given by

$$\hat{\tau}(a^0\, da^1\, da^2 \cdots da^n) = \tau(a^0, a^1, \dots, a^n) \qquad \forall a^i \in \mathcal{A}.$$

This gives a meaning to $\hat{\tau}((x^1\, da^1)(x^2\, da^2)(x^3\, da^3)\cdots(x^n\, da^n))$ for a^i, $x^i \in \mathcal{A}$, but of course one could also define it directly by a formula. Thus, for $n = 2$, $\hat{\tau}((x^1\, da^1)(x^2\, da^2)) = \tau(x^1, a^1x^2, a^2) - \tau(x^1a^1, x^2, a^2)$. The crucial definition of this section is the following:

Definition 11. *Let B be a Banach algebra. By an n-trace on B we mean an $(n+1)$-linear functional τ on a dense subalgebra $\mathcal{A}$ of B such that*

a) *τ is a cyclic cocycle on $\mathcal{A}$.*

b) *For any $a^i \in \mathcal{A}$, $i = 1, \dots, n$, there exists $C = C_{a^1,\dots,a^n} < \infty$ such that*

$$|\hat{\tau}((x^1\, da^1)(x^2\, da^2)\cdots(x^n\, da^n))| \le C\|x^1\|\cdots\|x^n\| \qquad \forall x^i \in \tilde{\mathcal{A}}.$$

Note that the conditions a) and b) are still satisfied if we replace $\mathcal{A}$ by any subalgebra which is still dense in B.

Example 12. a) Let V be a smooth manifold (not necessarily compact). Let $A = C_0(V)$ be the C^*-algebra of continuous functions vanishing at ∞. Recall that a de Rham current C on V of dimension p is a linear functional on the space $C_c^\infty(V, \wedge^p T_{\mathbb{C}}^* V)$ of differential forms of degree p on V, which is continuous in the following sense: for any compact $K \subset V$ and family $\omega_\alpha \in C_c^\infty(V, \wedge^p T_{\mathbb{C}}^*(V))$, with Support $\omega_\alpha \subset K$ for all α, converging to 0 in the C^k-topology on the coefficients of the forms ω_α, one has $C(\omega_\alpha) \to 0$. In other words, C is *of order k* when for any $\omega \in C_c^\infty(V, \wedge^p T_{\mathbb{C}}^*(V))$ the linear functional $f \in C^\infty(V) \to C(f\omega)$ is continuous in the C^k-topology.

Assume C to be a *closed* current of dimension p and *order* 0 on V, and put

$$\tau(f^0, \dots, f^p) = C(f^0\, df^1 \wedge \cdots \wedge df^p) \quad \forall f^0, \dots, f^p \in C_c^\infty(V).$$

Let us check that this defines a p-trace on the C^*-algebra $C_0(V)$. Its domain $C_c^\infty(V)$ is a dense subalgebra of $C_0(V)$, and one easily checks the cyclic cocycle property of τ using the closedness of C. One has

$$\begin{aligned}\hat{\tau}(x^1 da^1 \cdots x^p da^p) \\ = C((x^1 da^1) \wedge \cdots \wedge (x^p da^p)) = C(x^1 \cdots x^p da^1 \wedge \cdots \wedge da^p).\end{aligned}$$

Thus, since $da^1 \wedge \cdots \wedge da^p = \omega$ belongs to $C_c^\infty(V, \wedge^p T_{\mathbb{C}}^*(V))$ there exists, as C is of order 0, a $C_{a_1 \cdots a_p} < \infty$ such that

$$|\hat{\tau}(x^1 da^1 \cdots x^p da^p)| \leq C_{a_1 \cdots a_p} \prod_{j=1}^{p} \|x^j\|.$$

One checks that the map φ of Theorem 13 below from $K_i(C_0(V)) = K^i(V)$ to $\mathbb{C}$ is given, up to normalization, by

$$\varphi([e]) = \langle \operatorname{ch} e, [C] \rangle,$$

where $\operatorname{ch} : K_c^*(V) \to H_c^*(V, \mathbb{R})$ is the usual Chern character and $[C] \in H_*(V, \mathbb{C})$ is the homology class of the closed current C. Since there are always enough closed currents of order 0 to yield all of $H_*(V, \mathbb{C})$ we have not lost any information on the Chern character $\operatorname{ch} e \in H_c^*(V, \mathbb{R})$ in this presentation.

b) Let Δ be a locally finite simplicial complex, and $X = |\Delta|$ the associated locally compact space. Let us construct enough p-traces on the C^*-algebra $A = C_0(|\Delta|)$ to recover the usual Chern character, as in a). Let $\gamma = \sum \lambda_i s_i$ be a locally finite cycle of dimension p (i.e., the s_i are all oriented p-simplices of Δ and the λ_i's are complex numbers, with $b\gamma = \sum \lambda_i bs_i = 0$). Put

$$\tau(f^0, \ldots, f^p) = \sum \lambda_i \int_{s_i} f^0 df^1 \wedge \cdots \wedge df^p$$

where the $f^j \in C_c(X)$ have the following property (cf. [Su$_6$]): On each simplex s of Δ the restriction of f is equal to the restriction of a C^∞ function on the affine space of s. This space $C_c^\infty(\Delta) \subset C_0(X)$ is a dense subalgebra and one checks as in a) that τ is a p-trace on $C_0(X)$. Again the map φ of Theorem 13 below from $K_c^i(V)$ to $\mathbb{C}$ is given up to normalization by

$$\varphi(x) = \langle \operatorname{ch} x, [\gamma] \rangle$$

where ch is the usual Chern character and $[\gamma]$ is the homology class of γ in the homology of locally finite chains on X, dual to the cohomology with compact supports $H_c^*(X)$.

c) Let (A, G, α) be a C^*-dynamical system, i.e. A is a C^*-algebra on which the locally compact group G acts by automorphisms $\alpha_g \in \operatorname{Aut} A \;\; \forall g \in G$. Assume that G is a Lie group and let τ be an α-invariant trace on A. Let Ω be the graded differential algebra of right-invariant differential forms (with complex coefficients) on G. By construction Ω is, as a graded algebra, identical with $\wedge_{\mathbb{C}} T_e^*(G)$, the exterior algebra on the dual of the Lie algebra of G. Let

$t \in H_p(\Omega^*)$ be a p-homology class in the dual chain complex. Considering t as a closed linear form on Ω^p, put

$$\sigma(x^0,\ldots,x^p) = (\tau \otimes t)(x^0 dx^1 \cdots dx^p) \quad x^j \in A^\infty.$$

Here, A^∞ is the dense subalgebra of A formed of elements $x \in A$ for which $g \to \alpha_g(x)$ is a smooth function from G to the Banach space A. The differentials dx belong to the tensor product algebra, $A^\infty \otimes \Omega$, and are defined as follows. Let $X_i \in \mathrm{Lie}\, G$ be a basis of the Lie algebra of G, $\omega^i \in (\mathrm{Lie}\, G)^*$ the dual basis, and $\delta_i \in \mathrm{Der}\,(A)$ the unbounded derivations of A given by $(\partial_{X_i} \alpha_g)_{g=1}$. One takes

$$dx = \sum \delta_i(x) \otimes \omega^i \in A^\infty \otimes \Omega^1 \quad \forall x \in A^\infty.$$

One checks that $\tau \otimes t$ is a closed graded trace on the differential algebra $A^\infty \otimes \Omega$ and it follows that σ is a cyclic cocycle on the algebra A^∞. Let us show that it is a p-trace. For fixed $a^1,\ldots,a^p \in A^\infty$, the expression $\hat{\sigma}(x^1 da^1 \cdots x^p da^p)$ is a finite linear combination of terms of the form

$$\tau(x^1 y^1 x^2 y^2 \cdots x^p y^p)$$

where $y^j \in A^\infty$ is of the form $\delta_k(a^j)$. Since, by hypothesis, $\tau \in A^*$ one has $|\tau(x^1 y^1 \cdots x^p y^p| \le C\|x^1 y^1 \cdots x^p y^p\| \le C'\|x^1\| \cdots \|x^p\|$; thus the conclusion.

Applying Theorem 13 below one recovers the Chern character introduced in [Co_{13}], from $K(A)$ to $H^*(\Omega)$ (which is dual to $H_*(\Omega)$).

The above examples are easy instances of the following general construction of K-theory invariants which will be used extensively in β) and in Section 7.

Theorem 13. *Let τ be an n-trace on a Banach algebra B. Then there exists a canonically associated map φ of $K_i(B)$, $i \equiv n (\mathrm{mod}\, 2)$, to $\mathbb{C}$ such that:*

a) *If n is even and $e \in \mathrm{Proj}\, M_q$ (Dom τ) then*

$$\varphi([e]) = (\tau \# \mathrm{Tr})(e,\ldots,e).$$

b) *If n is odd and $u \in GL_q$ (Dom τ) then*

$$\varphi([u]) = (\tau \# \mathrm{Tr})(u^{-1}, u, u^{-1}u, \ldots, u^{-1}, u).$$

We refer to [Co_{14}] for the proof. Note that by [Co_{14}], τ extends uniquely to an n-trace with domain the closure of Dom τ under holomorphic functional calculus in B. In particular, if τ and τ' agree on a common dense domain they define the same K-theory invariants: $\varphi = \varphi'$.

Remark 14. Combining the construction of n-traces in Example b) with the known results on the Chern character, we see that if X is a locally compact space coming from a locally finite simplicial complex all maps from $K(C_0(X))$ to $\mathbb{C}$ which are additive come from n-traces on $A = C_0(X)$.

At this point it would be tempting to define, for arbitrary n, a homology relation between n-traces on an arbitrary Banach algebra extending that given by Proposition 3 and to work out an analogue of homology in this context. We shall see, however, that this would be premature, because it would overlook a purely noncommutative phenomenon which prevents some very natural densely defined n-cocycles on C^*-algebras from being n-traces, for $n \geq 2$.

As a simple example take $A = C_r^*(\Gamma)$, the reduced C^*-algebra of the following solvable discrete group Γ. We let $\mathbb{Z}$ act by automorphisms on the group $\mathbb{Z}^2$ using the unimodular matrix $\alpha = \begin{bmatrix} 1 & 1 \\ 1 & 2 \end{bmatrix}$, and let $\Gamma = \mathbb{Z}^2 \rtimes_\alpha \mathbb{Z}$ be the semi-direct product. Now consider the group cocycle $c \in Z^2(\Gamma, \mathbb{R})$ given by the equality

$$c(g_1, g_2) = v_1 \wedge \alpha^{n_1}(v_2) \quad \forall g_j = (v_j, n_j) \in \Gamma.$$

We represent an element of Γ as a pair (v, n) where $v \in \mathbb{Z}^2 \subset \mathbb{R}^2$ and $n \in \mathbb{Z}$, the $\wedge$ is the usual exterior product, and the cocycle takes values in $\wedge^2 \mathbb{R}^2 = \mathbb{R}$.

Given any discrete group and group cocycle c which is normalized so that $c(g_1, \ldots, g_n) = 0$ if any $g_i = 1$ or $g_1 \cdots g_n = 1$ one gets (Section 1) a cyclic n-cocycle on the group ring $\mathcal{A} = \mathbb{C}\Gamma$ by the equality

$$\tau(f^0, \ldots, f^n) = \sum_{g_0 g_1 \cdots g_n = 1} f^0(g_0) \cdots f^n(g_n)\, c(g_1, \ldots, g_n).$$

Thus, here we have a cyclic 2-cocycle on $\mathcal{A} = \mathbb{C}\Gamma$, and thus a densely defined cyclic cocycle on $A = C_r^*(\Gamma)$. One has

$$\begin{aligned} \hat{\tau}(x^1 da^1 x^2 da^2) &= \tau(x^1, a^1 x^2, a^2) - \tau(x^1 a^1, x^2, a^2) \\ &= \sum_{g_0 g_1 g_2 g_3 = 1} x^1(g_0)\, a^1(g_1)\, x^2(g_2)\, a^2(g_3)\, (c(g_1 g_2, g_3) - c(g_2, g_3)). \end{aligned}$$

Let $a^1, a^2 \in \mathbb{C}\Gamma$. If, for fixed g_1 and g_3, the function of g_2 defined by $g_2 \mapsto c(g_1 g_2, g_3) - c(g_2, g_3)$ were bounded one would easily get the desired estimate

$$|\hat{\tau}(x^1 da^1 x^2 da^2)| \leq C_{a^1, a^2}\, \|x^1\|\, \|x^2\|.$$

However, this fails precisely in our example, since with $g_1 = (v_1, 0)$ and $g_3 = (v_2, 0)$ we get, for $g_2 = (0, n)$,

$$c(g_1 g_2, g_3) - c(g_2, g_3) = v_1 \wedge \alpha^n(v_2).$$

Moreover, fixing such a choice of v_1 and v_2 and identifying Γ with a subgroup of the unitary group of $\mathbb{C}\Gamma \subset C_r^*(\Gamma)$ we get

$$\hat{\tau}(x^1 da^1 x^2 da^2) = v_1 \wedge \alpha^n(v_2)$$

for $x^1 = g_2^{-1} h^{-n} g_1^{-1}$, $a^1 = (v_1, 0)$, $x^2 = h^n$, $a^2 = (v_2, 0)$ with $h = (0, 1) \in \Gamma$. This shows that τ is not a 2-trace.

Since, as one easily checks, there is for each pair a^1, a^2 in $\mathbb{C}\Gamma$ a constant C_{a_1, a_2} such that

$$|\hat{\tau}(x\, da^1 da^2)| \leq C_{a_1, a_2}\, \|x\| \quad \forall x \in \mathbb{C}(\Gamma)$$

the obstruction to τ being a 2-trace comes from the noncommutativity $da^1 x^2 \neq x^2 da^1$. The analysis of this lack of commutativity will occupy us in Section 7 in the special case of the cyclic cocycle on $C_0(V) \rtimes \Gamma$ coming from the Γ-equivariant fundamental class of a manifold V on which the discrete group Γ acts by orientation-preserving diffeomorphisms. The above 2-cocycle on $\mathbb{C}(\mathbb{Z}^2 \rtimes_\alpha \mathbb{Z})$ is a special case of this more general problem, since it is exactly the equivariant fundamental class of the 2-torus dual to $\mathbb{Z}^2$ on which $\mathbb{Z}$ acts by α.

6.β The Bott-Thurston cocycle and the equality $GV = i_\delta \frac{d}{dt}[V/F]$

Let, as above, (V, F) be a transversely oriented codimension-one foliation and G_T the reduction of its holonomy groupoid by a transversal T. Since G_T is equivalent to G we shall start with an arbitrary one-dimensional smooth groupoid G as in Proposition 7 above, and leave the translation into foliation language to the reader.

To a choice of a smooth nowhere vanishing 1-density ρ on the one-dimensional manifold $G^{(0)}$ corresponds a faithful normal weight on the von Neumann algebra of the regular representation of G; i.e. in foliation language, on the von Neumann algebra of the foliation (Chapter I). This weight φ_ρ is given on $C_c^\infty(G)$ by

$$\varphi_\rho(f) = \int_{G^{(0)}} f\rho.$$

The modular automorphism group σ_t of φ_ρ leaves the subalgebra $C_c^\infty(G)$ globally invariant and is given by the following formula which one can check, for instance, by verifying the KMS_1 condition for the pair φ_ρ, σ_{-t} (cf. Chapter V)

$$\sigma_t(f)(\gamma) = \delta(\gamma)^{it} f(\gamma) \quad \forall \gamma \in G, \ f \in C_c^\infty(G)$$

where the modular homomorphism $\delta : G \to \mathbb{R}_+^*$ is given by the ratio of the 1-densities on G

$$\delta = r^*\rho / s^*\rho.$$

We shall let $\ell = \log \delta$ be the corresponding homomorphism from G to $\mathbb{R}$.

Now let τ be the cyclic 1-cocycle on $C_c^\infty(G)$, corresponding to the transverse fundamental class $[V/F]$ in the foliation case (Proposition 7).

Proposition 15. a) *The following equality defines a 1-trace on $C_r^*(G)$ with domain $C_c^\infty(G)$:*

$$\dot{\tau}(f^0, f^1) = \lim_{t \to 0} \frac{1}{t} \left(\tau(\sigma_t(f^0), \sigma_t(f^1)) - \tau(f^0, f^1) \right)$$

b) *The 1-trace $\dot{\tau}$ is invariant under the automorphisms σ_t of $C_r^*(G)$.*

Proof. a) Let D be the derivation of $C_c^\infty(G)$ given by $\lim_{t\to 0} \frac{1}{t}(\sigma_t - 1)$. Then

$$\begin{aligned}\dot\tau(f^0,f^1) &= \tau(Df^0,f^1) + \tau(f^0,Df^1)\\ &= \int_G \ell(\gamma^{-1})\, f^0(\gamma^{-1})\, df^1(\gamma) + \int_G f^0(\gamma^{-1})\, d(\ell(\gamma)f^1(\gamma))\\ &= \int_G f^0(\gamma^{-1})\, f^1(\gamma)\, d\ell(\gamma)\end{aligned}$$

where $d\ell$ is the 1-form on G which is the differential of the real-valued function ℓ. One checks that for fixed $f^1 \in C_c^\infty(G)$, the linear form L, $L(f^0) = \dot\tau(f^0,f^1)$, is continuous on $C_r^*(G)$, so that a) follows.

b) Since $\delta(\gamma) = \delta(\gamma^{-1})^{-1}$ the above formula

$$\dot\tau(f^0,f^1) = \int_G f^0(\gamma^{-1})\, f^1(\gamma)\, d\ell(\gamma)$$

is obviously invariant under σ_t.

We shall now define in general the contraction of a cyclic n-cocycle by the generator D of a one-parameter group of automorphisms fixing the cocycle and check that the n-trace property is preserved.

Lemma 16. *Let φ be an n-trace on a C^*-algebra A, invariant under the one-parameter group of automorphisms $(\sigma_t)_{t\in\mathbb{R}}$ with generator D. Assume that* $\mathrm{Dom}\,\varphi \cap \mathrm{Dom}\,D$ *is dense in A. Then the following formula defines an* $(n+1)$-*trace on A, $\psi = i_D\varphi$,*

$$\psi(x^0,\dots,x^{n+1}) = \sum_{j=1}^{n+1} (-1)^j \hat\varphi\left(x^0(dx^1\cdots dx^{j-1})D(x^j)(dx^{j+1}\cdots dx^{n+1})\right).$$

We already saw this formula in Section 1 Remark 30 b). It clearly gives a Hochschild cocycle. The invariance of φ under σ_t implies that ψ is cyclic. The $(n+1)$-trace property is straightforward using Dom $\varphi \cap$ Dom D as a domain (cf. [Co_{14}] for details).

When φ is a 1-cocycle the formula for $\psi = i_D\varphi$ can be written as

$$\psi(x^0,x^1,x^2) = \varphi(D(x^2)x^0,x^1) - \varphi(x^0D(x^1),x^2).$$

Now let $\varphi = \dot\tau$ be the time derivative of the transverse fundamental class (Proposition 15), and let us compute $i_D\varphi$. One has

$$\varphi(f^0,f^1) = \int_G f^0(\gamma^{-1})\, f^1(\gamma)\, d\ell(\gamma)$$

and $(Df)(\gamma) = \ell(\gamma)f(\gamma)$. Thus, the computation is straightforward and gives

$$\psi(f^0,f^1,f^2) = \int_{\gamma_0\gamma_1\gamma_2\in G^{(0)}} f^0(\gamma_0)\, f^1(\gamma_1)\, f^2(\gamma_2)\, c(\gamma_1,\gamma_2)$$

where $c(\gamma_1,\gamma_2) = \ell(\gamma_2)\, d\ell(\gamma_1) - \ell(\gamma_1)\, d\ell(\gamma_2)$ is the Bott-Thurston 2-cocycle, a 2-cocycle on local diffeomorphisms of a 1-manifold with values in 1-forms on that manifold. The latter cocycle is the Godbillon-Vey 3-dimensional cohomology class in another guise and this is reflected in the following:

Theorem 17. *Let $\mu_r : K_{*,\tau}(BG) \to K(C_r^*(G))$ be the analytic assembly map, where $G = G_T$ is the reduced holonomy groupoid of the transversely oriented codimension-1 foliation (V,F). Let $\psi = i_D\dot{\tau}$ be the 2-trace on $C_r^*(G)$ given by Proposition 15 and Lemma 16. Then for any $x \in K_{*,\tau}(BG)$ one has*

$$\langle \mu_r(x), i_D\dot{\tau}\rangle = \langle \mathrm{ch}_*(x), GV\rangle.$$

Here GV is viewed as a cohomology class $GV \in H^3(BG,\mathbb{R})$ and the twisting $K_{*,\tau}$ in the K-homology introduces a shift of parity, so that it is $K^0_{*,\tau}(BG)$ which is mapped to $H_*^{\mathrm{odd}}(BG)$ by the Chern character ch_*. The proof of this result is a special case of the general index theorem of Section 7 below. We refer to [Co$_{14}$] for a direct proof. We shall now localise the above K-theory invariant to the flow of weights $\mathrm{mod}(M)$ of the von Neumann algebra M of the foliation.

6.γ Invariant measures on the flow of weights

As above, let (V,F) be a codimension-one transversely oriented foliation and let $G_T = G$ be the reduced holonomy groupoid. The von Neumann algebra of the foliation (Chapter I) is the same, up to tensoring by a factor of type I_∞, as the von Neumann algebra M of the left regular representation of $C_c^\infty(G)$ in $L^2(G)$. The flow of weights of the von Neumann algebra of the foliation (cf. Chapter V) is the same as the flow of weights $\mathrm{mod}(M)$. It is thus given by the dual action θ of the Pontryagin dual $\hat{\mathbb{R}} = \mathbb{R}_+^*$ restricted to the center $Z(M \rtimes_\sigma \mathbb{R})$ of the von Neumann algebra crossed product of M by the modular automorphism group $\sigma_t = \sigma_t^{\varphi_\rho}$ of Subsection β). Since the latter group σ_t leaves the subalgebra $C_c^\infty(G) \subset M$ (and its norm closure $C_r^*(G)$) globally invariant, we can use the C^*-algebra crossed product

$$B = C_r^*(G) \rtimes_\sigma \mathbb{R}$$

in order to investigate the center of $M \rtimes_\sigma \mathbb{R}$. By construction, the C^*-algebra B is weakly dense in $M \rtimes_\sigma \mathbb{R} = N$, and the center $Z(N)$ will appear in the following definition.

Definition 18. *Let B be a C^*-algebra, τ a 1-trace on B and $\delta : B \to B^*$ the corresponding unbounded derivation, so that $\tau(x^0,x^1) = \langle x^0, \delta(x^1)\rangle \quad \forall x^j \in$* Dom τ. *Then τ* is anabelian *if the domain of the adjoint $\delta^* : B^{**} \to B^*$ contains the* center $Z(B^{**})$ *and $\delta^*(z) = 0 \quad \forall z \in Z(B^{**})$.*

Recall that the *bidual* B^{**} of a C^*-algebra is a von Neumann algebra so that its center makes good sense. The term *anabelian* is motivated by the following trivial point:

If B is commutative, any anabelian 1-trace is 0.

In other words, such 1-traces can exist only in the noncommutative context.

Proposition 19. 1) *Let τ be an anabelian 1-trace on the C^*-algebra B; then for any $z \in Z(B^{**})$ the following equality defines a 1-trace on B:*

$$\tau_z(x^0,x^1) = \langle x^0, z\delta(x^1)\rangle \qquad \forall x^j \in \operatorname{Dom}\tau.$$

2) *For any $u \in K^1(B)$, the map $z \in Z(B^{**}) \to \langle \tau_z, u\rangle$ is a normal linear form on $Z(B^{**})$.*

Proof. 1) As B^{**} acts by multiplication on its predual B^* the element $z\delta(x^1) \in B^*$ makes sense and $z\delta(x^1) = \delta(x^1)z$ since $z \in Z(B^{**})$. It is clear that τ_z is a Hochschild cocycle; let us check that $\tau_z(x^0,x^1) = -\tau_z(x^1,x^0)$ $\forall x^j \in \operatorname{Dom}\tau$. One has

$$\tau_z(x^1,x^0) = \langle x^1, z\delta(x^0)\rangle = \langle x^1 z, \delta(x^0)\rangle = \langle z, \delta(x^0)x^1\rangle = -\langle z, x^0\delta(x^1)\rangle$$

where the last equality follows from the hypothesis on τ

$$\langle z, \delta(y)\rangle = 0 \qquad \forall y \in \operatorname{Dom}\tau.$$

Finally $\langle z, x^0\delta(x^1)\rangle = \langle zx^0, \delta(x^1)\rangle = \langle x^0, z\delta(x^1)\rangle = \tau_z(x^0,x^1)$.

2) By construction, $\langle u^{-1}, z\delta(u)\rangle$ is a normal linear form in the variable $z \in B^{**}$.

Next, let us take $B = C_r^*(G) \rtimes_\sigma \mathbb{R}$ and let

$$\phi_\sigma : K_0(C_r^*(G)) \to K_1(B)$$

be the Thom isomorphism in K-theory (Chapter II Appendix C). Let ψ be the 2-trace on $C_r^*(G)$ related to the Godbillon-Vey class by Theorem 17.

Theorem 20. *There exists an* anabelian *1-trace τ on $B = C_r^*(G) \rtimes_\sigma \mathbb{R}$ such that:*

1) *τ is invariant under the dual action $(\theta_\lambda)_{\lambda \in \mathbb{R}_+^*}$*

2) *τ corresponds to ψ by the Thom isomorphism, i.e.*

$$\langle \tau, \phi_\sigma(y)\rangle = \langle \psi, y\rangle \qquad \forall y \in K_0(C_r^*(G)).$$

Before we explain the proof of this theorem, let us deduce from it the following corollary.

Theorem 21. [Co$_{14}$] *Let (V,F) be a codimension-1 transversely oriented foliation, M its von Neumann algebra, $(W(M),\theta_\lambda)$ its flow of weights.*

a) *If the Godbillon-Vey class $GV \in H^3(BG,\mathbb{R})$ is not zero (which holds if $GV \in H^3(V,\mathbb{R})$ is nonzero) there exists a θ_λ-invariant probability measure in the normal measure class on $W(M)$.*

b) *The conclusion of a) holds if the Bott-Thurston 2-cocycle ψ of Theorem 17 pairs nontrivially with $K_0(C^*(V,F))$.*

Corollary 22. [Hu-K$_1$] *If $GV \neq 0$ then the von Neumann algebra M has a nontrivial type III component.*

The corollary is immediate since if M is semifinite its flow of weights is given, up to multiplicity, by the action of $\mathbb{R}_+^*$ by translations on $L^\infty(\mathbb{R}_+^*, ds/s)$ which admits no invariant probability measure in the Lebesgue measure class.

Assuming Theorem 20, the proof of Theorem 21 is the following. First it is enough to prove b) since, by Theorem 17, if $GV \neq 0$ in $H^3(BG,\mathbb{R})$ then there exists an element $x \in K_0(C^*(V,F))$, in the image of the analytic assembly map, such that $\langle x, \psi\rangle \neq 0$. Next, by Theorem 20 2) and the Thom isomorphism (Chapter II Appendix C) one gets an element $u \in K_1(B)$ such that

$$\langle \tau, u\rangle \neq 0.$$

Let us then consider the normal linear functional L on the center $Z(M \rtimes_\sigma \mathbb{R})$ given by

$$L(z) = \langle \tau_z, u\rangle \qquad \forall z \in Z(M \rtimes_\sigma \mathbb{R}).$$

It makes sense because $M \rtimes_\sigma \mathbb{R}$ is the von Neumann algebra B_E^{**}, that is B^{**} reduced by a suitable central projection, so that $Z(M \rtimes_\sigma \mathbb{R}) \subset Z(B^{**})$. It is a *normal* linear functional on $Z(M \rtimes_\sigma \mathbb{R})$ by Proposition 19 2), and it is not zero since $L(1) = \langle \tau, u\rangle \neq 0$. We shall show that it is invariant under the dual action $(\theta_\lambda)_{\lambda\in\mathbb{R}_+^*}$. By Theorem 20 1), the 1-trace τ is θ_λ invariant so that

$$\langle \tau_{\theta_\lambda(z)}, u\rangle = \langle \tau_z, \theta_\lambda(u)\rangle.$$

But since the action θ_λ is pointwise norm-continuous in the parameter λ, it follows that it acts trivially on $K_1(B)$ and we see that $\langle \tau_{\theta_\lambda(z)}, u\rangle = \langle \tau_z, u\rangle$, i.e. that

$$L \circ \theta_\lambda = L \qquad \forall \lambda \in \mathbb{R}_+^*.$$

It now remains to construct a 1-trace on B fulfilling the conditions 1) and 2) of Theorem 20. By Chapter II Proposition 5.7, one has $B = C_r^*(G_1)$ where the smooth groupoid G_1 is associated to G and to the homomorphism $\delta : G \to \mathbb{R}_+^*$ as follows:

$$G_1 = G \times \mathbb{R}_+^*\,,\ G_1^{(0)} = G^{(0)} \times \mathbb{R}_+^*$$
$$r(\gamma,\lambda) = (r(\gamma),\lambda)\,,\ s(\gamma,\lambda) = (s(\gamma),\lambda\delta(\gamma))\ ;\ \forall \gamma \in G\,,\ \lambda \in \mathbb{R}_+^*$$
$$(\gamma_1,\lambda_1)(\gamma_2,\lambda_2) = (\gamma_1 \circ \gamma_2, \lambda_1)\ \text{ for any composable pair.}$$

In the foliation context, this means replacing the foliation (V,F) by a foliation of codimension 2 on the total space of the principal $\mathbb{R}_+^*$-bundle over V of transverse densities (Chapter I Section 4). The dual action $(\theta_\lambda)_{\lambda\in\mathbb{R}_+^*}$, $\theta_\lambda \in \operatorname{Aut} C_r^*(G_1)$, is given by the obvious automorphisms of G_1

$$\theta_\lambda(\gamma, y) = (\gamma, \lambda y) \qquad \forall \lambda \in \mathbb{R}_+^*\,,\ (\gamma, y) \in G_1.$$

The *anabelian* 1-trace τ on $B = C_r^*(G_1)$ is given by the formula

$$\tau(f^0, f^1) = \int_{G_1} f^0(\gamma^{-1})\, f^1(\gamma)\, \omega(\gamma) \qquad \forall f^j \in C_c^\infty(G_1)$$

where ω is the 2-form on the 2-manifold G_1 given by

$$\omega = d\ell \wedge \frac{d\lambda}{\lambda}$$

where $\ell = \log \delta$ as above and λ is the variable $\lambda \in \mathbb{R}_+^*$. Since ω is invariant under θ_λ one checks Theorem 20 1). Since $\ell(\gamma^{-1}) = -\ell(\gamma)$ one checks that $\tau(f^1, f^0) = -\tau(f^0, f^1)$ $\forall f^j \in C_c^\infty(G_1)$. The cocycle property $\ell(\gamma_1 \circ \gamma_2) = \ell(\gamma_1) + \ell(\gamma_2)$ implies the cocycle property for ω, so that τ is a Hochschild 1-cocycle. To show that the 1-trace τ is *anabelian* one proves that any element z of the center of the von Neumann algebra generated by B in $L^2(G_1)$ is given by a function $f(\gamma)$, which is in $L^\infty(G_1)$ and such that

$$f\omega \text{ is zero almost everywhere.}$$

We refer to [Co$_{14}$] for the detailed proof, which can be taken as an exercise. Finally, Statement 2 of Theorem 20 is a special case of the following general fact (applied to τ and θ_λ).

Proposition 23. a) *Let (B, θ) be a C^*-dynamical system and τ a θ-invariant 1-trace on B. The following equality defines a 1-trace $\hat{\tau}$ on $\hat{B} = B \rtimes_\theta \mathbb{R}$, invariant under the dual action*

$$\hat{\tau}(f^0, f^1) = \int \tau\left(f^0(t), \theta_t(f^1(-t))\right) dt$$

for $f^j \in C_c^\infty(\mathbb{R}, \mathrm{Dom}\,\tau) \subset B \rtimes_\theta \mathbb{R}$.

b) *Let $\varphi_\theta : K_1(B) \to K_0(B \rtimes_\theta \mathbb{R})$ be the Thom isomorphism, and $D = \frac{d}{dt}\sigma_t$ be the generator of the dual action. Then for any $y \in K_1(B)$ one has*

$$\langle \tau, y \rangle = \langle i_D \hat{\tau}, \varphi_\theta(y) \rangle$$

where $i_D\hat{\tau}$ is the contraction of $\hat{\tau}$ by D (Lemma 16).

We refer to [Co$_{14}$] for the simple proof.

III.7. The Transverse Fundamental Class for Foliations and Geometric Corollaries

7.α The transverse fundamental class

Let (V,F) be a foliated manifold and $\tau = TV/F$ the transverse bundle of the foliation. The holonomy groupoid G of (V,F) acts in a natural way on τ by the differential of the holonomy. Thus, every $\gamma : x \to y$, $\gamma \in G$, determines a linear map

$$h(\gamma) : \tau_x \to \tau_y.$$

It is not in general possible to find a Euclidean metric on τ which is *invariant* under the above action of G. In fact, we have already seen in Chapter I that in the type III situation there does not even exist a measurable volume element,

$$v_x \in \wedge^q \tau_x \ , \quad q = \text{codim}\, F$$

which is invariant under the action of G. The nonexistence of a holonomy invariant measurable volume element is equivalent to the outer property for some $t \in \mathbb{R}$ of the modular automorphism group of the von Neumann algebra of the foliation.

Here we have to take account of the full information on $h(\gamma)$ (not only its determinant and not only in the measurable category). To that end we shall construct the analogue of Tomita's modular operator Δ and modular morphism $\Delta \cdot \Delta^{-1}$ of conjugation by the modular operator. Let g be an arbitrary smooth Euclidean metric on the real vector bundle τ. Thus, for $\xi \in \tau_x$ we let $\|\xi\|_g = (\langle \xi, \xi \rangle_g)^{1/2}$ be the corresponding norms and inner products and drop the subscript g if no ambiguity can arise.

Using g we define a C^*-module $\mathcal{E} = \mathcal{E}_g$ on the C^*-algebra $C_r^*(V,F)$ of the foliation. Recall that $C_r^*(V,F)$ is the completion of the convolution algebra $C_c^\infty(G,\Omega^{1/2})$ defined in Section II.8. The algebra $C_c^\infty(G,\Omega^{1/2})$ acts by right convolution on the linear space $C_c^\infty(G,\Omega^{1/2} \otimes r^*(\tau_{\mathbb{C}}))$

$$(\xi f)(\gamma) = \int_{G^y} \xi(\gamma_1)\, f(\gamma_1^{-1}\gamma) \qquad \text{where } y = r(\gamma). \tag{1}$$

Endowing the complexified bundle $\tau_{\mathbb{C}}$ with the inner product associated to g and antilinear in the first variable, the following formula defines a $C_c^\infty(G,\Omega^{1/2})$-valued inner product

$$\langle \xi, \eta \rangle(\gamma) = \int_{G^y} \langle \xi(\gamma_1^{-1}), \eta(\gamma_1^{-1}\gamma) \rangle \tag{2}$$

for any $\xi, \eta \in C_c^\infty(G,\Omega^{1/2} \otimes r^*(\tau_{\mathbb{C}}))$.

One then checks that the completion $\mathcal{E} = \mathcal{E}_g$ of the space $C_c^\infty(G,\Omega^{1/2} \otimes r^*(\tau_{\mathbb{C}}))$ for the norm

$$\|\xi\| = \left(\|\langle \xi, \xi \rangle\|_{C_r^*(V,F)}\right)^{1/2}$$

becomes, using formulas (1) and (2), a C^*-module over $C_r^*(V,F)$. The above construction *did not* involve the action of G on τ. We shall now use this action h to define a left action of $C_c^\infty(G,\Omega^{1/2})$ on $\mathcal{E}$ by the equality

$$(f\xi)(\gamma) = \int_{G^y} f(\gamma_1)\, h(\gamma_1)\, \xi(\gamma_1^{-1}\gamma) \quad \forall f \in C_c^\infty(G,\Omega^{1/2})\,,\ \xi \in \mathcal{E}. \tag{3}$$

The analogue of the modular operator Δ then appears in the following:

Proposition 1. *For any $f \in C_c^\infty(G,\Omega^{1/2})$ the formula (3) defines an endomorphism $\lambda(f)$ of the C^*-module $\mathcal{E}$ whose adjoint $\lambda(f)^*$ is given by the equality*

$$(\lambda(f)^*\xi)(\gamma) = \int_{G^y} f^\#(\gamma_1)\, h(\gamma_1)\, \xi(\gamma_1^{-1}\gamma)$$

where $f^\#(\gamma) = \bar{f}(\gamma^{-1})\,\Delta(\gamma)$, and $\Delta(\gamma) \in \mathrm{End}(\tau_{\mathbb{C}}(r(\gamma)))$,

$$\Delta(\gamma) = (h(\gamma)^{-1})^t\, h(\gamma)^{-1}.$$

The proof is the same as [Co_{14}] Lemma 3.1.

This proposition shows that, unless the metric on τ is G-invariant, the representation λ is not a $*$-representation, the nuance between $\lambda(f)^*$ and $\lambda(f^*)$ being measured by Δ. In particular λ is not in general bounded for the C^*-algebra norms on both $\mathrm{End}_{C_r^*(V,F)}\mathcal{E}$ and $C_r^*(V,F) \supset C_c^\infty(G,\Omega^{1/2})$. However:

Lemma 2. *The densely defined homomorphism λ is a* closable *homomorphism of C^*-algebras.*

(cf. [Co_{14}]) In other words the closure of the graph of λ is the graph of a densely defined homomorphism. Then with the graph norm $\|x\|_\lambda = \|x\| + \|\lambda(x)\|$ the domain B of the closure $\bar{\lambda}$ of λ is a Banach algebra, which is dense in the C^*-algebra $A = C_r^*(V,F)$. The C^*-module $\mathcal{E}$ is then a B-A-bimodule.

The information contained in this structure is easy to formulate when the bundle τ is trivial over V, with, say, a basis $(\xi_i)_{i=1,\dots,q}$ of orthonormal sections. Then $\mathcal{E}$ is the C^*-module A^q over A, and the closable homomorphism $\lambda : A \to M_q(A) = \mathrm{End}_A(\mathcal{E})$ corresponds to the coaction of $GL(q,\mathbb{R})$ on A associated to the homomorphism

$$h : G \to GL(q,\mathbb{R}).$$

It is only for $q = 1$ that such a coaction can be simply described as an action by automorphisms of the Pontryagin dual group.

To complete the description of the transverse structure of the foliation we shall define *transverse differentiation.* First note that the above construction of $\mathcal{E}$ as a bimodule applies equally well to any G-equivariant vector bundle E over V endowed with a (not invariant) Euclidean structure. The domain B of the left action of A on $\mathcal{E}$ depends, of course, upon the choice of E. At the formal level the tensor calculus is still available, but one has to be careful about domain problems. Thus, if we let Ω^j be the bimodule associated to transverse differential forms of degree j, it is the C^*-module completion of

$$\Omega^j_{c,\infty} = C_c^\infty(G, \Omega^{1/2} \otimes r^*(\wedge^j \tau^*))$$

and one has a densely defined product, compatible with the bimodule structure

$$\Omega^j \times \Omega^k \xrightarrow{\wedge} \Omega^{j+k}$$

$$(\omega \wedge \omega')(\gamma) = \int_{G^\gamma} \omega(\gamma_1) \wedge \gamma_1 \omega'(\gamma_1^{-1}\gamma) \quad \forall \omega, \omega' \in \Omega_{c,\infty}.$$

One still has $\omega f \wedge \omega' = \omega \wedge f\omega'$ $\forall f \in C_c^\infty(G, \Omega^{1/2})$, but it is *no longer true* that $\omega \wedge \omega' = \pm\omega' \wedge \omega$ since this does not even hold for $j = k = 0$, the algebra $C_c^\infty(G, \Omega^{1/2})$ being noncommutative.

Transverse differentiation is not quite canonical and requires the harmless choice of a *horizontal distribution* H, i.e. of a vector bundle $H \subset TV$ such that for each $x \in V$, H_x is a complement of the integrable bundle F_x. Of course we do not require the integrability of H and one can, for instance, take $H = F^\perp$ for a fixed Riemannian metric on V. We refrain from using the word connection to describe H (even though it will be used as a connection on the bundle $L^2(\tilde{L})$ over V/F) in order to avoid confusion with connections on the bundle τ over V/F.

Elements of the algebra $C_c^\infty(G, \Omega^{1/2})$ are $\frac{1}{2}$-densities, i.e. sections over G of the bundle $r^*(\Omega_F^{1/2}) \otimes s^*(\Omega_F^{1/2})$ where $\Omega_F^{1/2}$ is the line bundle over V whose fiber at x is the space of maps

$$\rho : \wedge^k F_x \to \mathbb{C}\,, \; \rho(\lambda v) = |\lambda|^{1/2} \rho(v) \quad \forall \lambda \in \mathbb{R}$$

with $k = \dim F$.

Thus, to differentiate elements of $C_c^\infty(G, \Omega^{1/2})$ we first need to know how to differentiate a $\frac{1}{2}$-density $\rho \in C^\infty(V, \Omega_F^{1/2})$. Let us start by defining transverse differentiation for sections of the bundle $\wedge^r F^* \otimes \wedge^s \tau^*$ over V, so that

$$d_H : C^\infty(V, \wedge^r F^* \otimes \wedge^s \tau^*) \to C^\infty(V, \wedge^r F^* \otimes \wedge^{s+1} \tau^*).$$

Using the decomposition $T = F + H$ we get an isomorphism j_x of the exterior algebras:

$$j_x : \wedge F_x^* \otimes \wedge \tau_x^* \to \wedge T_x^* \quad \forall x \in V.$$

Letting $j : C^\infty(V, \wedge F^* \otimes \wedge \tau^*) \xrightarrow{\simeq} C^\infty(V, \wedge T^*)$ be the corresponding isomorphism, we define $d_H \omega$ for $\omega \in C^\infty(V, \wedge^r F^* \otimes \wedge^s \tau^*)$ as the component of type $(r, s+1)$ of $j^{-1} d(j(\omega))$, where $d(j(\omega))$ is the differential of $j(\omega) \in C^\infty(V, \wedge^{(r+s)} T^*)$ on V. Note that unlike $d_H \omega$, which depends upon the choice of H, the component of type $(r+1, s)$ of $j^{-1} d(j(\omega))$, i.e. the longitudinal differential of ω, is canonical, once (V, F) is given.

Next, as any $\frac{1}{2}$-density $\rho \in C^\infty(V, \Omega_F^{1/2})$ can be written, at least locally, in the form $\rho = f|\omega|^{1/2}$, where $f \in C^\infty(V)$ and $\omega \in C^\infty(V, (\wedge^k F)^*)$, $k = \dim F$, we can define its transverse differential $d_H\rho$ unambiguously by the formula

$$d_H(f|\omega|^{1/2}) = (d_H f)|\omega|^{1/2} + f|\omega|^{1/2}\,\frac{1}{2}(d_H\omega/\omega). \qquad (\alpha)$$

More explicitly, if we work in local coordinates around $x \in V$, we can take a domain of foliation chart of the form $\mathbb{R}^k \times \mathbb{R}^q$. Then the $\frac{1}{2}$-density ρ is of the form

$$\rho(t,u) = r(t,u)\ |du^1 \wedge \cdots \wedge du^k|^{1/2}.$$

Given a vector field $X(t) = \sum_i X^i(t)\ \frac{\partial}{\partial t^i}$ on the transversal $\mathbb{R}^q$, let then $\widetilde{X}$ be its unique horizontal lift

$$\widetilde{X}(t,u) = \sum_i X^i(t)\ \frac{\partial}{\partial t^i} + \sum_j X^j(t,u)\ \frac{\partial}{\partial u^j}.$$

Then $(d_H\rho)/\rho$ evaluated at (t,u) on the vector $X(t) \in \tau_x$ is

$$\widetilde{X}(t,u)r(t,u) + \left(\sum_j \frac{\partial}{\partial u^j}\ X^j(t,u)\right) r(t,u).)$$

Let us now consider the smooth groupoid G. The map $\pi = (r,s) : G \xrightarrow{\pi} V \times V$ is an immersion (with, in general, a fairly complicated range: the equivalence relation that x and y belong to the same leaf). Given $\gamma \in G$, $\gamma : x \to y$ and $X \in \tau_y = r^*(\tau)_\gamma$, let $\widetilde{X} \in H_y$ be the horizontal lift of X and let $\widetilde{X}' \in H_x$ be the horizontal lift of $h(\gamma)^{-1}X \in \tau_x$, i.e. the transverse vector at x corresponding to X by holonomy. Then the vector $(\widetilde{X}, \widetilde{X}') \in T_{\pi(\gamma)}(V \times V)$ belongs to $\pi_*(T_\gamma(G))$ and is the image by π_* of a unique tangent vector $Y \in T_\gamma(G)$. We can thus define $d_H f \in C_c^\infty(G, r^*(\tau))$ for $f \in C_c^\infty(G)$ by

$$(d_H f)(X) = df(Y). \qquad (\beta)$$

Let us now state the main properties of transverse differentiation:

Proposition 3. *Let (V,F) be a foliated manifold, H a horizontal distribution.*

1) *The following equality uniquely defines a* derivation d_H *from* $C_c^\infty(G, \Omega^{1/2})$ *to the bimodule* $C_c^\infty(G, \Omega^{1/2} \otimes r^*(\tau^*))$

$$d_H(r^*(\rho)fs^*(\rho)) = r^*(d_H\rho)fs^*(\rho) + r^*(\rho)(d_H f)s^*(\rho) + r^*(\rho)fs^*(d_H\rho)$$

for $\rho \in C^\infty(V, \Omega_F^{1/2})$, $f \in C_c^\infty(G)$.

2) *The derivation d_H extends uniquely to a derivation of the graded algebra* $C_c^\infty(G, \Omega^{1/2} \otimes r^*(\wedge\tau^*))$ *such that for any* $\omega \in C^\infty(V, \wedge\tau^*)$ *and* $f \in C_c^\infty(G, \Omega^{1/2})$ *one has*

$$d_H(f\ r^*(\omega)) = (d_H\ f)\ r^*(\omega) + f\ r^*(d_H\ \omega).$$

In 1) and 2) the algebra structures are of course given by convolution on the groupoid G and the exterior product in $\wedge\tau^*$.

Due to the lack of integrability of the subbundle H of TV, it is not true in general that $d_H^2 = 0$. We shall however, easily overcome this difficulty using Lemma 9 of Section 3 and that d_H^2 is an inner derivation of $C_c^\infty(G, \Omega^{1/2} \otimes r^*(\wedge \tau^*))$. Let us first consider d_H^2 acting on differential forms on V

$$d_H^2 : C^\infty(V, \wedge^r F^* \otimes \wedge^s \tau^*) \to C^\infty(V, \wedge^r F^* \otimes \wedge^{s+2} \tau^*).$$

The full differential $d\omega$ of a form of type $(1,0)$, $\omega \in C^\infty(V, F^*)$, has 3 components: the longitudinal differential $d_L\omega \in C^\infty(V, \wedge^2 F^*)$, the transverse differential $d_H\omega \in C^\infty(V, F^* \otimes \tau^*)$ and also a component of type $(0,2)$. It is given by the contraction $\theta\omega$ of ω with the section $\theta \in C^\infty(V, F \otimes \wedge^2 \tau^*)$:

$$\theta(p_\tau(X), p_\tau(Y)) = p_F([X,Y])$$

for any pair of horizontal vector fields $X, Y \in C^\infty(V, H)$, where (p_F, p_τ) is the isomorphism $T \to F \oplus \tau$ given by H. The following equality then holds for arbitrary forms:

$$d\omega = d_L\omega + d_H\omega + \theta\omega$$

where the last term stands for contraction with θ. Then, computing the component of type $(r, s+2)$ of $d^2\omega$ for ω of type (r,s), one gets

$$d_H^2\omega = (d_L\theta + \theta d_L)\omega \quad \forall \omega.$$

The operator $d_L\theta + \theta d_L$ just involves longitudinal Lie derivatives, and thus there exists a (unique) element

$$\Theta \in C_c^{-\infty}(G, \Omega^{1/2} \otimes r^*(\wedge^2 \tau^*))$$

with support contained in $G^{(0)}$ and whose action on $C^\infty(V, \Omega_F^{1/2})$ is the Lie derivative $d_L\theta + \theta d_L$.

Lemma 4. *For any $\omega \in C_c^\infty(G, \Omega^{1/2} \otimes r^*(\wedge \tau^*))$ one has $d_H^2\omega = \Theta \wedge \omega - \omega \wedge \Theta$. Moreover, $d_H\Theta = 0$.*

Indeed, one knows a priori that d_H^2 is a derivation and the equality is easy to check using Proposition 3.

By Lemma 9 of Section 3 we thus get a differential graded algebra $\tilde{\Omega}_{c,\infty}$ whose elements are two-by-two matrices with entries

$$\omega_{ij} \in C_c^\infty(G, \Omega^{1/2} \otimes r^*(\wedge \tau^*)).$$

The product is given by

$$\omega \cdot \omega' = \begin{bmatrix} \omega_{11} & \omega_{12} \\ \omega_{21} & \omega_{22} \end{bmatrix} \begin{bmatrix} 1 & 0 \\ 0 & \Theta \end{bmatrix} \begin{bmatrix} \omega'_{11} & \omega'_{12} \\ \omega'_{21} & \omega'_{22} \end{bmatrix},$$

and the differential d by

$$d\omega = \begin{bmatrix} d_H\omega_{11} & d_H\omega_{12} \\ -d_H\omega_{21} & -d_H\omega_{22} \end{bmatrix} + \begin{bmatrix} 0 & -\Theta \\ 1 & 0 \end{bmatrix} \omega + (-1)^{\partial\omega}\, \omega \begin{bmatrix} 0 & 1 \\ -\Theta & 0 \end{bmatrix}.$$

We have thus achieved the equality $d^2 = 0$ while retaining the previous properties of C_c^∞ and d_H.

Let us now define, with $q = \operatorname{codim} F$, the integral of forms

$$\omega \in C_c^\infty(G, \Omega^{1/2} \otimes \wedge^q \tau^*)$$

as an immediate adaptation of the measure theoretic Proposition I.4.6.

Lemma 5. *Let (V, F) be a transversely oriented foliated manifold. The following equality defines a closed graded trace on $(C_c^\infty(G, \Omega^{1/2} \otimes r^*(\wedge \tau^*)), d_H)$:*

$$\tau(\omega) = \int_{G^{(0)}} \omega \qquad (\tau(\omega) = 0 \text{ if } \deg \omega \neq q).$$

Here the restriction of $\omega \in C_c^\infty(G, \Omega^{1/2} \otimes r^*(\wedge^q \tau^*))$ to the subspace $G^{(0)} \subset G$ yields a section of $\Omega_F \otimes \wedge^q \tau^*$ on V and has a canonical integral over V using the transverse orientation. Thus, one has $\tau(\omega_2 \wedge \omega_1) = (-1)^{\partial_1 \partial_2} \tau(\omega_1 \wedge \omega_2)$ and $\tau(d_H \omega) = 0$.

Let $\tilde{\tau}$ be the extension of τ to $\tilde{\Omega}_{c,\infty}$ given by

$$\tilde{\tau} \begin{bmatrix} \omega_{11} & \omega_{12} \\ \omega_{21} & \omega_{22} \end{bmatrix} = \tau(\omega_{11}) - (-1)^q \tau(\omega_{22} \Theta).$$

Proposition 6. *The triple $(\tilde{\Omega}, d, \tilde{\tau})$ is a* cycle *over the algebra $C_c^\infty(G, \Omega^{1/2})$.*

We shall call this cycle the *fundamental cycle* of the transversely oriented foliation (V, F). In particular it yields a cyclic cocycle φ_H of dimension q on $C_c^\infty(G, \Omega^{1/2})$, which depends on the auxiliary choice of the horizontal distribution H but whose cohomology class is independent of this choice.

Definition 7. *Let (V, F) be a transversely oriented foliation of codimension q. Then its transverse fundamental class is the cyclic cohomology class $[V/F] \in HC^q\left(C_c^\infty(G, \Omega^{1/2})\right)$ of the above cycle.*

7.β Geometric corollaries

The above construction of the transverse fundamental class $[V/F]$ of a transversely oriented foliation yields a cyclic cohomology class on the dense subalgebra $\mathcal{A} = C_c^\infty(G, \Omega^{1/2})$ of the C^*-algebra $C_r^*(V, F)$ of the foliation and hence a corresponding map of K-theory

$$K(\mathcal{A}) \xrightarrow{\varphi} \mathbb{C}.$$

We have solved in [Co$_{14}$] the problem of topological invariance of this map, i.e. of extension of φ to a map

$$K(C_r^*(V, F)) \xrightarrow{\tilde{\varphi}} \mathbb{C}.$$

Let us now state several corollaries of the solution of this problem ([Co_{14}]).

Theorem 8. *Let (V,F) be a not necessarily compact foliated manifold which is transversely oriented. Let $G = \mathrm{Graph}(V,F)$ be its holonomy groupoid and let $\pi : BG \to B\Gamma_q$ be the map classifying the natural Haefliger structure ($q = \operatorname{codim} V$),and let τ the bundle over BG given by the transverse bundle of (V,F). Let $\mathcal{R} \subset H^*(BG,\mathbb{C})$ be the ring generated by the Pontryagin classes of τ, the Chern classes of holonomy equivariant bundles on V and $\pi^*\left(H^*(WO_q)\right)$.*

For any $P \in \mathcal{R}$ there exists an additive map φ of $K_*\,(C_r^*(V,F))$ to $\mathbb{C}$ such that:

$$\varphi(\mu_r(x)) = \langle \phi \circ \mathrm{ch}(x), P\rangle \qquad \forall x \in K_{*,\tau}(BG).$$

Here $K_{*,\tau}(BG)$ is the geometric group as defined in Chapter II Proposition 8-4, ch is the Chern character: $K_{*,\tau}(BG) \to H_*(B\tau, S\tau)$, where τ is the bundle on BG corresponding to the transverse bundle of (V,F), and $\phi : H_*(B\tau,S\tau) \to H_*(BG)$ is the Thom isomorphism.

An important step in the proof of this theorem is the longitudinal index theorem for foliations (Chapter II, Theorem 9-6).

Let (V,F) be a transversely oriented foliation of codimension q of a connected manifold V. Let W be the 0-dimensional manifold consisting of one point, and $f : W \to V/F$ the map from W to V/F associated to a leaf and K-oriented by the transverse orientation of (V,F). Then the triple (W,y,f), where y is the class of the trivial bundle on W, is a geometric cycle for (V,F) (cf. Chapter II), and the corresponding K-theory class $f!(y) \in K_q(C_r^*(V,F))$ is independent of the choice of the leaf. We shall denote this class by $[V/F]^*$ and call it the orientation class in K-theory.

Corollary 9. *Let (V,F) be a transversely oriented foliation of codimension q; then the class $[V/F]^*$ is a non-torsion element of $K_q(C_r^*(V,F))$.*

Note that V need not be compact. When (V,F) is not transversely orientable with V connected, $[V/F]^*$ is a 2-torsion element (see [Tor] for relevant computations). We now pass to two other corollaries which are purely geometric, i.e. the C^*-algebra $C_r^*(V,F)$ does not appear in the statement. Its role is to allow one to integrate in K-theory *in two steps:* 1) along the leaves of the foliation, which provides under a suitable K-orientation hypothesis a map from $K^*(V)$ to $K_*(C_r^*(V,F))$; 2) over the space of leaves, which provides a map from $K_*(C_r^*(V,F))$ to $\mathbb{C}$.

That the composition of these two steps is the same as integration in K-theory over V is a corollary of Theorem 8.

It follows from [Li] and [Ros_1] that if the bundle F is a Spin bundle which can be endowed with a Euclidean metric with strictly positive (bounded below by $\varepsilon > 0$) scalar curvature (this makes sense since F is integrable) then the longitudinal integral of the trivial bundle does vanish, i.e. the K-theory index of the longitudinal Dirac operator is equal to 0. Thus, the integral over V of the trivial bundle vanishes, or, in other words, $\hat{A}(V) = 0$. Note that here only F is assumed to be Spin so that $\hat{A}(V)$ is not a priori an integer. More precisely:

Corollary 10. *Let V be a compact foliated oriented manifold. Assume that the integrable bundle $F \subset TV$ is a Spin bundle and is endowed with a metric of strictly positive ($\geq \varepsilon > 0$) scalar curvature. Let $\mathcal{R}$ be the subring of $H^*(V,\mathbb{C})$ generated by the Pontryagin classes of $\tau = TV/F$, the Chern classes of holonomy equivariant bundles and the range of the natural map $H^*(WO_q) \to H^*(V,\mathbb{C})$. Then $\langle \hat{A}(F)\omega, [V]\rangle = 0 \ \forall \omega \in \mathcal{R}$, where $\hat{A}(F)$ is the $\hat{A}$-genus of this Spin bundle.*

When $F = TV$, i.e. when the foliation has just one leaf, this is exactly the content of the well-known vanishing theorem of A. Lichnerowicz ([Li]).

As an immediate application we see that no spin foliation of a compact manifold V, with non-zero $\hat{A}$-genus, $\hat{A}(V) \neq 0$, admits a metric of strictly positive scalar curvature.

Proof. The projection $V \to V/F$ is K-oriented by the Spin structure on F and hence defines a geometric cycle $x \in K_{*,\tau}(BG)$. The argument of [Ros$_1$] shows that the analytical index of the Dirac operator along the leaves of (V,F) is equal to 0 in $K_*(C_r^*(V,F))$, so that one has $\mu_r(x) = 0$ with μ_r the analytic assembly map (Chapter II Section 8).

Let $f : V \to BG$ be the map associated to the projection $V \to V/F$. There exists a polynomial P in the Pontryagin classes of τ, with leading coefficient 1, such that

$$\phi \circ \mathrm{ch}(x) = f_*(\hat{A}(F) \cap [V]) \cap P \in H_*(BG).$$

Thus, since $\mu_r(x) = 0$, the result follows from Theorem 8.

Corollary 11. [Bau-C] *Let (V,F) and (V',F') be oriented and transversely oriented compact foliated manifolds. Let $f : V \to V'$ be a smooth, orientation preserving, leafwise homotopy equivalence. Then for any element P of the ring $\mathcal{R} \subset H^*(V,\mathbb{C})$ of Corollary 10 one has*

$$\langle (f^*L(V') - L(V)) \,,\, P \cap [V]\rangle = 0$$

where $L(V)$ (resp. (V')) is the L-class of V (resp. V').

7.γ Index formula for longitudinal elliptic operators

The main difficulty in the proof of Theorem 8 is to show the topological invariance of the cyclic cohomology map $\varphi : K(\mathcal{A}) \to \mathbb{C}$, $\mathcal{A} = C_c^\infty(G, \Omega^{1/2})$. We refer to [Co$_{14}$] for the proof. Here we shall explain how to compute the pairing

$$\langle \Phi(c), \mathrm{Ind}(D) \rangle \in \mathbb{C} \tag{1}$$

of the cyclic cohomology of $\mathcal{A}$ with the index $\mathrm{Ind}(D) \in K_0(\mathcal{A})$ of an arbitrary longitudinal elliptic operator D, as defined in Section II.9 α). The result is stated quite generally in the theorem on p.888 of [Co$_{23}$], but we shall make it more specific, using the natural map

$$\Phi_* : H_\tau^*(BG) \to H^*(\mathcal{A}) \tag{2}$$

constructed in Section 2 δ) Theorem 14 and Remark b).

It is worthwhile to formulate the general result in terms of invariant elliptic operators on G-manifolds, where G is a smooth groupoid (cf. Section II.10 α)). We shall assume that G is *étale*, i.e. that the maps r and s are étale. This suffices to cover the case of foliations since the holonomy groupoid G of a foliation is equivalent to the reduced groupoid G_T, for T a suitable transversal, and the latter is étale. We let Φ be the morphism of bicomplexes constructed in Section 2 δ) from the bicomplex of twisted simplicial forms on the nerve of G to the (b, B) bicomplex of the algebra $\mathcal{A} = C_c^\infty(G)$.

When formulated in terms of the (b, B) bicomplex, the pairing between cyclic cohomology and K-theory of Proposition 3.2 is given by the formula

$$\langle (\varphi_{2n}), [e] \rangle = \sum_{n=0}^{\infty} (-1)^n \frac{(2n)!}{n!} \varphi_{2n}\left(e - \tfrac{1}{2}, e, \dots, e\right) \tag{3}$$

where (φ_{2n}) is a (b, B)-cocycle, i.e. $b\varphi_{2n} + B\varphi_{2n+2} = 0 \quad \forall n$, and e is an idempotent.

Equivalently, if (ψ_{2n}) is a cocycle in the (d_1, d_2) bicomplex of Theorem I.29 the pairing reads

$$\langle (\psi_{2n}), [e] \rangle = \sum_{n=0}^{\infty} (-1)^n \frac{1}{n!} \psi_{2n}\left(e - \tfrac{1}{2}, e, \dots, e\right). \tag{4}$$

We shall come back to this point in great detail in Section IV.7; the replacement of e by $e - \frac{1}{2}$, due to [Ge-S], improves the original formula. We refer to Remark I-30 a) for the sign $(-1)^n$.

Now, let P be a proper G-manifold (Definition 1 of Section II.10). Thus, to P corresponds a contravariant functor from the small category G to the category of manifolds and diffeomorphisms. To each $x \in V = G^{(0)}$ it assigns the fiber $\alpha^{-1}(\{x\})$ of the submersion $\alpha : P \to G^{(0)}$. The right action $p \to p \circ \gamma$ gives a diffeomorphism from P_y to P_x for $\gamma \in G$, $r(\gamma) = y$, $s(\gamma) = x$. To keep things simple we shall assume that the discrete groups $G_x^x = \{\gamma \in G ; r(\gamma) = s(\gamma)\}$ are torsion-free for any $x \in G^{(0)}$. Then the properness of P implies that G_x^x acts freely on P_x.

Let D be a G-invariant elliptic differential operator on the G-manifold P, i.e. a family D_x of elliptic differential operators on P_x which is smooth on the total space P and is invariant under the action of G.

When the quotient manifold P/G of P by the action of G, is *compact*, the constructions of Section II.9 α), or of Section 4 α) of this chapter, yield a well defined K-theory class

$$\mathrm{Ind}(D) \in K_0(C_c^\infty(G) \otimes \mathcal{R}). \tag{5}$$

Finally, the K-theory class of the principal symbol σ_D of D yields, as in Section II.10 α), a well defined element $[\sigma_D]$ of the geometric group $K_{\mathrm{top}}^*(G)$. Since G is étale and torsion-free the latter group is equal to the τ-twisted K-homology group $K_{*,\tau}(BG)$. We can now state:

Theorem 12. *Let G be a smooth étale groupoid without torsion. Let D be a G-invariant elliptic differential operator on a proper G-manifold P with P/G compact. Let c be a cocycle of total degree $2q$ in the bicomplex of simplicial currents on the nerve of G and $\Phi(c)$ the associated cyclic cohomology class in the (b,B) bicomplex of $\mathcal{A} = C_c^\infty(G) \otimes \mathcal{R}$. Then*

$$\langle \Phi(c), \mathrm{Ind}(D)\rangle = (2\pi i)^{-q} \langle c, \mathrm{ch}_\tau([\sigma_D])\rangle.$$

Here $\mathrm{ch}_\tau([\sigma_D]) \in H_{*,\tau}(BG)$ is the twisted Chern character which was defined in Section II.7 Remark 12 by the equality $\mathrm{ch}_\tau(x) = \mathrm{Td}(\tau_{\mathbb{C}})^{-1}\, \mathrm{ch}(x)$.

Note that $c \in H_\tau^*(BG)$ pairs with $\mathrm{ch}_\tau([\sigma_D]) \in H_{*,\tau}(BG)$.

We recall from Section 2 γ) that in the bicomplex (C^*, d_1, d_2) of simplicial currents one had $C^{n,m} = 0$ unless $n \geq 0$, $-\dim G^{(0)} \leq m \leq 0$. In particular the total degree $n+m$ can vary from $-\dim G^{(0)}$ to $+\infty$.

Let us explain briefly why Theorem 12 reduces to Theorem 4.7 in the case of discrete groups $G = \Gamma$. Let c be a group cocycle, $c \in Z^{2q}(\Gamma, \mathbb{C})$. Then by Section 2 γ) one has

$$\Phi(c) = \tau_c. \tag{6}$$

But the cyclic cocycle τ_c of Theorem 4.7 is here viewed as a cocycle in the (b,B) bicomplex. The point then is that the formula (3) for the pairing of (b,B) cocycles with K-theory accounts for the strange numerical factor $1/(2q)!$ of Theorem 4.7 (cf. [Co-M$_1$] for a discussion of this numerical factor).

Let us now specialize Theorem 12 to the case of foliations. We let (V,F) be a foliated compact manifold. In Section α) we have carefully described the cyclic cohomology class on the algebra $\mathcal{A} = C_c^\infty(G, \Omega^{1/2})$ corresponding under the natural Morita equivalence to the fundamental class of reduced groupoids, G_T, with T a transversal. The same procedure applies to map the periodic cyclic cohomology $H^*(C_c^\infty(G_T))$ to $H^*(\mathcal{A})$. Combined with the construction of the morphism Φ of Section 2 α), we thus obtain a canonical map

$$\Phi_* : H_\tau^*(BG) \to H^*(\mathcal{A}), \tag{7}$$

where we have used the equivalence of smooth groupoids $G \simeq G_T$ to replace BG_T by BG.

Applying Theorem 12 to the G_T manifold $P = \{\gamma \in G\,;\, s(\gamma) \in T\}$ one obtains:

Corollary 13. [Co23] *Let (V,F) be a compact foliated manifold, D a longitudinal elliptic operator, and* $\mathrm{Ind}(D) \in K_0(\mathcal{A})$, $\mathcal{A} = C_c^\infty(G,\Omega^{1/2})$ *its analytical index (cf. II.9 α)). Then for any cohomology class $\omega \in H_\tau^{2q+\dim\tau}(BG)$, one has*

$$\langle \Phi_*(\omega), \mathrm{Ind}(D)\rangle = (2\pi i)^{-q}\,\langle \omega, \mathrm{ch}_\tau(\sigma_D)\rangle.$$

To end this section we shall explain, in this context of foliations, how to construct a left inverse λ of the map Φ_* ([Co29])

$$\lambda : H^*(\mathcal{A}) \to H_\tau^*(BG). \tag{8}$$

In fact, we shall only describe the composition of this map with the pull-back $H_\tau^*(BG) \to H_\tau^*(V)$. When one varies V without varying V/F one gets the desired information. The idea of the construction of

$$\lambda_V : H^*(\mathcal{A}) \to H_\tau^*(V) \tag{9}$$

is to exploit the local triviality of the foliation as follows: For each open set $U \subset V$ one lets $\mathcal{A}(U)$ be the algebra associated by the functor $C_c^\infty(\cdot,\Omega^{1/2})$ to the restriction of the foliation to U. If $U_1 \subset U_2$ one has an obvious inclusion

$$\mathcal{A}(U_1) \subset \mathcal{A}(U_2). \tag{10}$$

Thus, one obtains for each n a presheaf Γ^n on V by setting

$$\Gamma^n(U) = C^n(\mathcal{A}(U)^\sim, \mathcal{A}(U)^{\sim *}), \tag{11}$$

the space of continuous $(n+1)$-linear forms on $\mathcal{A}(U)^\sim$, the algebra obtained by adjoining a unit to $\mathcal{A}(U)$.

Because the construction of the (b,B) bicomplex is functorial it yields a presheaf of bicomplexes: $(\Gamma^{(n,m)}, b, B)$. Choosing a covering $\mathcal{U} = (U_\alpha)$ of V sufficiently fine so that the multiple intersections are domains of foliation charts, we get a triple complex

$$\Big(\Gamma^{n,m,p} = \bigoplus_{U_i \in \mathcal{U}} \Gamma^{(n,m)}(U_0 \cap U_1 \cap \ldots \cap U_p)\,,\ b, B, \delta\Big) \tag{12}$$

where δ is the Čech coboundary.

Moreover, there is an obvious forgetful map ϕ from the periodic cyclic cohomology $H^*(\mathcal{A})$, $\mathcal{A} = \mathcal{A}(V)$, to the cohomology of the triple complex (12). So far we have just localized the periodic cyclic cohomology of $\mathcal{A}$ in a straightforward manner. The interesting point is that the cohomology of the triple complex (12) is easy to compute, and is directly related to the resolution of the orientation sheaf of the transverse bundle τ by transverse currents. More precisely, one defines another triple complex (Γ', d_1, d_2, d_3) as follows: For each $k \in 0, 1, \ldots, \operatorname{codim} F$, let Ω_k be the sheaf of holonomy invariant transverse currents. Thus, for any open set U, $\Omega_k(U)$ is the kernel of the longitudinal differential in the space $C^{-\infty}(U, \wedge^{\operatorname{codim} F - k}\, \tau^*)$ of generalized sections of the exterior power of the dual of the transverse bundle. If U is a domain of foliation chart we are just dealing with currents on the space of plaques U/F. We let

$$\Gamma'^{n,m,p} = \bigoplus_{U_i \in \mathcal{U}} \Omega_{n-m} \, (U_0 \cap \ldots \cap U_p) \tag{13}$$

and the coboundaries are $d_1 = 0$, $d_2 =$ de Rham coboundary, and $d_3 =$ Čech coboundary.

The construction of subsection α), involving the additional choice of the subbundle $H \subset TV$ transverse to F, applies with minor modifications to yield a morphism of triple complexes

$$(\Gamma', d_1, d_2, d_3) \xrightarrow{\theta} (\Gamma, b, B, \delta) \tag{14}$$

which defines an isomorphism in cohomology independent of the choice of H. One thus obtains the desired construction of λ_V as

$$\lambda_V = \theta_*^{-1} \circ \phi, \tag{15}$$

since the cohomology of (Γ', d_1, d_2, d_3) is, almost by construction, equal to $H_\tau^*(V)$. Indeed, the sheaves Ω_k provide a resolution of the orientation sheaf of τ. One can in particular reformulate the index formula (Corollary 13) in terms of λ_V ([Co$_{23}$]). One can also compare the natural filtration of $H^*(\mathcal{A})$ by the dimensions of cyclic cocycles with the filtration of $H_\tau^*(V)$ given by the above resolution of the orientation sheaf of τ.

III. Appendix A. The Cyclic Category Λ

The definition (Section 1) of the cyclic cohomology functors HC^n from the category of algebras to vector spaces is simple and direct. It is however, important from a conceptual point of view to obtain these functors as derived functors from the simplest of them, HC^0, which to an algebra $\mathcal{A}$ assigns the linear space of traces on $\mathcal{A}$, i.e. of linear forms τ such that

$$\tau(xy) = \tau(yx) \quad \forall x, y \in \mathcal{A}.$$

This is done in this Appendix (cf. [Co$_{26}$] [Lo]) using a natural functor from the nonabelian category of algebras to the abelian category of Λ-modules described below. In some sense this appendix analyzes the machinery underlying the algebraic manipulations of Section 1.

A.α The simplicial category Δ

We first review some standard notions ([Lo]).

Let Δ be the small category whose objects are the totally ordered finite sets

$$[n] = \{0 < 1 < 2 < \ldots < n\} \quad , \quad n \in \mathbb{N},$$

and whose morphisms are the increasing maps (where increasing means not decreasing).

Definition 1. *Let C be a category. A simplicial object in C is a contravariant functor from Δ to C.*

Such a functor X is uniquely specified by the objects X_n corresponding to $[n]$ and by the morphisms $d_i : X_n \to X_{n-1}$, $0 \le i \le n$, and $s_j : X_n \to X_{n+1}$, $0 \le j \le n$ which correspond to the following morphisms δ_i and σ_j of Δ. For each n and $i \in \{0, 1, \ldots, n\}$, $\delta_i^n : [n-1] \to [n]$ is the injection which misses i while $\sigma_j^n : [n+1] \to [n]$ is the surjection such that $\sigma_j(j) = \sigma_j(j+1) = j$.

The conditions fulfilled by the morphisms d_i and s_j of C in a simplicial object are obtained by transposing the following standard presentation of Δ.

Proposition 2. *The morphisms σ_i^n and δ_j^n generate Δ, which admits the following presentation:*

$$\delta_j \delta_i = \delta_i \delta_{j-1} \quad \text{for} \quad i < j$$

$$\sigma_j \sigma_i = \sigma_i \sigma_{j+1} \quad \text{for} \quad i \le j$$

$$\sigma_j \delta_i = \begin{cases} \delta_i \sigma_{j-1} & \text{if} \quad i < j, \\ 1_n & \text{if} \quad i = j \text{ or } i = j+1, \\ \delta_{i-1} \sigma_j & \text{for} \quad i > j+1. \end{cases}$$

Let us now describe the geometric realization $|X|$ of a simplicial set X. One lets Δ be the following functor from the small category Δ to the category of topological spaces. For each n, $\Delta(n)$ is the standard n-simplex: $\{(t_i) \in \mathbb{R}^{n+1} \, ; \, 0 \le t_i \le 1 \, , \, \Sigma \, t_i = 1\}$. To δ_i corresponds the *face map*

$$\delta_i((t_0, \ldots, t_{n-1})) = (t_0, \ldots, t_{i-1}, 0, t_i, \ldots, t_{n-1})$$

and to σ_j the *degeneracy map*

$$\sigma_j((t_0, \ldots, t_{n+1})) = (t_0, \ldots, t_{j-1}, t_j + t_{j+1}, t_{j+2}, \ldots, t_{n+1}).$$

The geometric realization $|X|$ of a simplicial set X is then the quotient of the topological space $X \times_\Delta \Delta = \bigcup_{n \ge 0} X_n \times \Delta^n$ by the equivalence relation which identifies $(x, \alpha_*(y))$ with $(\alpha^*(x), y)$ for any morphism α of Δ. This continues to make sense when X is a simplicial topological space (cf. [Lo]).

With these notations we can give a formal definition of the classifying space BC of a small category C. By definition the *nerve* of C is a simplicial set $\mathrm{Mr}(C)$. The elements of $\mathrm{Mr}_n(C)$ are the composable n-uples of morphisms belonging to C, while the faces d_i (resp. degeneracies s_j) are obtained using composition of adjacent morphisms (resp. the identity morphism). More precisely, with $f_1, \ldots, f_n$ composable morphisms one lets

$$\begin{aligned} d_0(f_1, \ldots, f_n) &= (f_2, \ldots, f_n) \\ d_i(f_1, \ldots, f_n) &= (f_1, \ldots, f_i f_{i+1}, \ldots, f_n) \text{ for } 1 \le i \le n-1 \\ d_n(f_1, \ldots, f_n) &= (f_1, \ldots, f_{n-1}) \end{aligned}$$

and

$$s_i(f_1, \ldots, f_n) = (f_1, \ldots, f_i, 1, f_{i+1}, \ldots, f_n).$$

Definition 3. *The classifying space BC of a small category C is the geometric realization of the simplicial set* $\mathrm{Mr}(C)$.

This applies, for instance, to discrete groups Γ viewed as small categories with a single object. Taking account of topological categories (cf. [Lo]) it also applies to smooth groupoids as described in Chapter II.

A.β The cyclic category Λ

For $n \in \mathbb{N}$ we let $\mathbb{Z}/n+1$ be the cyclic group, the quotient of $\mathbb{Z}$ by the subgroup $(n+1)\mathbb{Z}$. We identify $\mathbb{Z}/n+1$ with the group of $(n+1)$-st roots of unity by the homomorphism

$$e_n : \mathbb{Z}/n+1 \to \mathbb{S}^1 = \{\lambda \in \mathbb{C} \,;\, |\lambda| = 1\},$$

$$e_n(k) = \exp(2\pi i k/(n+1)).$$

We endow $\mathbb{S}^1$ with its usual trigonometric orientation and for $\lambda, \mu \in \mathbb{S}^1$ let $[\lambda, \mu]$ be the closed interval from λ to μ. We say that a map $\psi : \mathbb{Z}/n+1 \to \mathbb{Z}/m+1$ is *increasing* if for any $\lambda, \mu \in \mathbb{Z}/n+1$ the image by ψ of the interval

$$[\lambda, \mu] \cap \mathbb{Z}/n+1 = \{\lambda, \lambda+1, \lambda+2, \ldots, \mu\} \subset \mathbb{Z}/n+1$$

is contained in the interval $[\psi(\lambda), \psi(\mu)] \cap \mathbb{Z}/m+1$.

The cyclic category Λ is the small category with one object Λ_n for each $n \in \mathbb{N}$ and with as morphisms $f \in \mathrm{Hom}(\Lambda_n, \Lambda_m)$ the homotopy classes of continuous increasing maps from $\mathbb{S}^1$ to $\mathbb{S}^1$, of degree 1 and such that

$$\varphi(\mathbb{Z}/n+1) \subset \mathbb{Z}/m+1.$$

For $\lambda \in \mathbb{Z}/n+1$ the value of $\varphi(\lambda) \in \mathbb{Z}/m+1$ is independent of the choice of φ in the homotopy class f and gives a well defined increasing map $\tilde{f} : \mathbb{Z}/n+1 \to \mathbb{Z}/m+1$. One thus obtains a functor $f \to \tilde{f}$ from Λ to the category of sets.

Proposition 4. a) *Let $n, m \in \mathbb{N}$ and $\psi : \mathbb{Z}/n+1 \to \mathbb{Z}/m+1$ be a non-constant increasing map. Then there exists a unique $f \in \mathrm{Hom}(\Lambda_n, \Lambda_m)$ such that $\tilde{f} = \psi$.*

b) *Let $\psi_k \in \mathrm{Hom}(\Lambda_n, \Lambda_m)$ be the constant map with range $\{k\}$. There are $n+1$ elements $f \in \mathrm{Hom}(\Lambda_n, \Lambda_m)$ such that $\tilde{f} = \psi_k$.*

Proof. Composing ψ with a rotation we may assume (in additive notation) that $\psi(0) = 0$. Then for any $j_1 < j_2 \in \{0, 1, \ldots, n\}$ one has $\psi(j_1) \in [0, \psi(j_2)] \subset \{0, 1, \ldots, m\}$ so that ψ is a non-decreasing map of $\{0 < 1 < \ldots < n\}$ to $\{0 < 1 < \ldots < m\}$. When ψ is not equal to 0 it determines uniquely the homotopy class of φ, $\tilde{\varphi} = \psi$. When ψ is constant, equal to 0, one has the choice of the interval $[e_n(j), e_n(j+1)]$ mapped onto $\mathbb{S}^1$ by φ, $\tilde{\varphi} = \psi$.

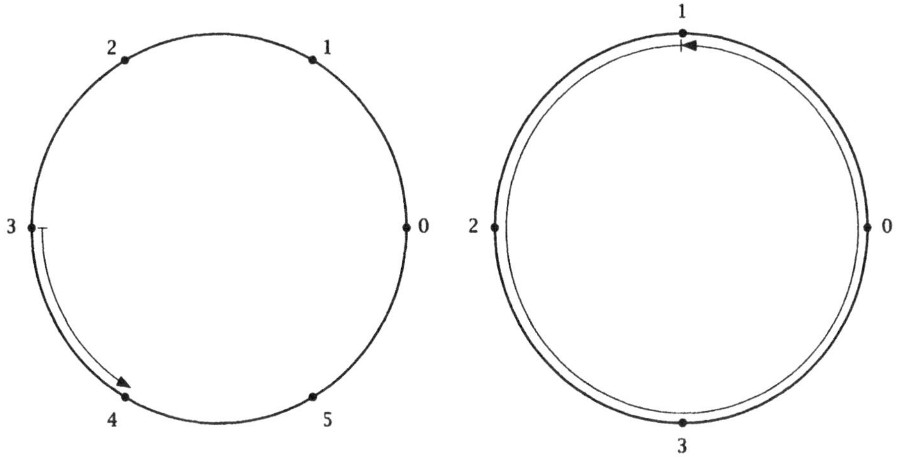

Figure III.3. $f \in \mathrm{Hom}(\Lambda_5, \Lambda_3)$, $\tilde{f}(j) = 1 \quad \forall j \in \mathbb{Z}/6\mathbb{Z}$, f is constant on $[4, 3]$ and equal to 1; it makes a complete turn on $[3, 4]$

We see in particular that Λ_0 is not a final object in the small category Λ and $\mathrm{Hom}(\Lambda_n, \Lambda_0)$ has cardinality $n + 1$. Any element $f \in \mathrm{Hom}(\Lambda_n, \Lambda_m)$ is characterized by the (possibly empty) following intervals of $\mathbb{S}^1$:

$$f^{-1}(j) = \bigcap_{\varphi \in f} \varphi^{-1}\{j\} \ , \ j \in \mathbb{Z}/m+1.$$

This follows from Proposition 4 a) if $\tilde{f}$ is not constant, and from 4 b) if $\tilde{f}$ is constant.

A collection $(I_j)_{j \in \mathbb{Z}/m+1}$ of (possibly empty) intervals comes from an $f \in \mathrm{Hom}(\Lambda_n, \Lambda_m)$ iff they satisfy:

1) If $I_j \neq \emptyset$ then $I_j = [\lambda, \mu]$ for some $\lambda, \mu \in \mathbb{Z}/n+1$.

2) The $I_j \cap \mathbb{Z}/n+1$ form a partition of $\mathbb{Z}/n+1$.

3) If $I_j = [\lambda, \mu] \neq \emptyset$, $I_{j+1} = \ldots = I_{\ell-1} = \emptyset$, $I_\ell = [\lambda', \mu'] \neq \emptyset$ then $\lambda' = \mu + 1$ in $\mathbb{Z}/n+1$.

This follows directly from Proposition 4. One has

$$I_j \cap \mathbb{Z}/n+1 = \tilde{f}^{-1}\{j\} \quad \forall j \in \mathbb{Z}/m+1,$$

and if $\tilde{f}$ is not constant, $\tilde{f}^{-1}\{j\}$ determines $f^{-1}(j)$. We shall identify the category Δ of Subsection α) with a subcategory of Λ as follows. To each $h \in$

$\mathrm{Hom}_\Delta([n],[m])$ we associate the corresponding increasing map $j \to h(j)$ from $\mathbb{Z}/n+1 = \{0,1,\ldots,n\}$ to $\mathbb{Z}/m+1 = \{0,1,\ldots,m\}$. This specifies uniquely an element $h_* \in \mathrm{Hom}(\Lambda_n,\Lambda_m)$ such that $\widetilde{h}_* = h$ provided that h is not constant. To handle the constant map, $h(i) = k\ \forall\ i$, we just need to specify the interval $h_*^{-1}(k) = [0,n]$.

Proposition 5. [Co$_{26}$] a) *The functor* $* : \Delta \to \Lambda$ *identifies* Δ *with a subcategory of* Λ.

b) *Any* $f \in \mathrm{Hom}(\Lambda_n,\Lambda_m)$ *can be uniquely written as the product* $h_* k$ *where* $h \in \Delta$ *and* k *is an automorphism.*

In other words $\Lambda = \Delta C$, where C is the subcategory of Λ with the same objects and with as morphisms the invertible morphisms of Λ. By Proposition 4, C is a groupoid which is the union of the cyclic groups $\mathrm{Aut}(\Lambda_n) \simeq \mathbb{Z}/n+1$.

Proof. a) For each n let $\theta_n \in]n/(n+1),1[$ and $\alpha_n = \exp(2\pi i\theta_n)$. By imposing the condition

$$\varphi(\alpha_n) = \alpha_m$$

on the continuous increasing degree-1 maps involved in the above definition of $\mathrm{Hom}(\Lambda_n,\Lambda_m)$, we obtain a subcategory of Λ. One easily verifies that this subcategory is the image of Δ by $*$ and that the functor $f \to \widetilde{f}$ composed with $*$ is the identity on Δ.

b) Let $f \in \mathrm{Hom}(\Lambda_n,\Lambda_m)$ and φ be a representative of f. There exists a unique interval $I = [e_n(j), e_n(j+1)]$ of $\mathbb{S}^1$ such that $\alpha_m \in \varphi(I)$. Moreover, I does not depend on the choice of $\varphi \in f$. Thus, there exists a unique $k \in \mathrm{Aut}(\Lambda_n)$ such that the interval associated to $f \circ k^{-1}$ is $[e_n(n),1]$. By a) one then has $f \circ k^{-1} \in \Delta$. This proves the existence and uniqueness of the decomposition.

For each n let $\tau_n \in \mathrm{Aut}(\Lambda_n)$ be the generator of the cyclic group such that $\widetilde{\tau}_n(j) = j-1\quad \forall j$. With a slight abuse of notation, we let δ_j, σ_i (instead of δ_{j*}, σ_{i*}) be the canonical generators of Δ (cf. Proposition 2) viewed as elements of Λ. By Propositions 2 and 5 the δ_j^n, σ_i^n and τ_n generate Λ.

Proposition 6. [Lo] *The relations of Proposition* 2 *together with the following relations give a presentation of* Λ:

$$\tau_n\,\delta_i = \delta_{i-1}\,\tau_{n-1} \text{ for } 1 \le i \le n\,,\ \tau_n\,\delta_0 = \delta_n$$

$$\tau_n\,\sigma_i = \sigma_{i-1}\,\tau_{n+1} \text{ for } 1 \le i \le n\,,\ \tau_n\,\sigma_0 = \sigma_n\,\tau_{n+1}^2$$

$$\tau_n^{n+1} = 1_n.$$

We refer to [Lo] for the proof. In [Lo] J.-L. Loday takes the above presentation as the definition of Λ. This leads him to many interesting generalizations of the cyclic category involving dihedral, symmetric and braid groups ([Fied-L]). Transposing the relations of Proposition 6 one obtains a description of *cyclic objects* in a category in terms of the morphisms d_n^i, s_n^j, and t_n corresponding to the above generators of Λ.

Definition 7. *Let C be a category. A cyclic object in C is a* contravariant functor *X from Λ to C.*

This definition differs only superficially from the one given in [Co$_{26}$] which used covariant functors. We shall indeed describe a canonical contravariant functor $f \to f^*$ from Λ to Λ which gives an isomorphism $\Lambda^{\mathrm{op}} \simeq \Lambda$. The composition of any covariant functor X with $*$ is then a cyclic object. Given $f \in \mathrm{Hom}(\Lambda_n, \Lambda_m)$ we define $f^* \in \mathrm{Hom}(\Lambda_m, \Lambda_n)$ by the intervals $J_k = (f^*)^{-1}(k)$, $k \in \mathbb{Z}/n+1$ constructed as follows: If $\tilde{f}(k-1) \neq \tilde{f}(k)$, we let $J_k = [\tilde{f}(k-1)+1, \tilde{f}(k)]$ (i.e. the interval $[e_m(\tilde{f}(k-1)+1), e_m(\tilde{f}(k))]$ in $\mathbb{S}^1$). If $\tilde{f}(k-1) = \tilde{f}(k)$ we let $J_k = \varnothing$ if the representatives φ of f are constant in the interval $[k-1,k]$ and $J_k = [\tilde{f}(k-1)+1, \tilde{f}(k)]$ otherwise (cf. Figure 4).

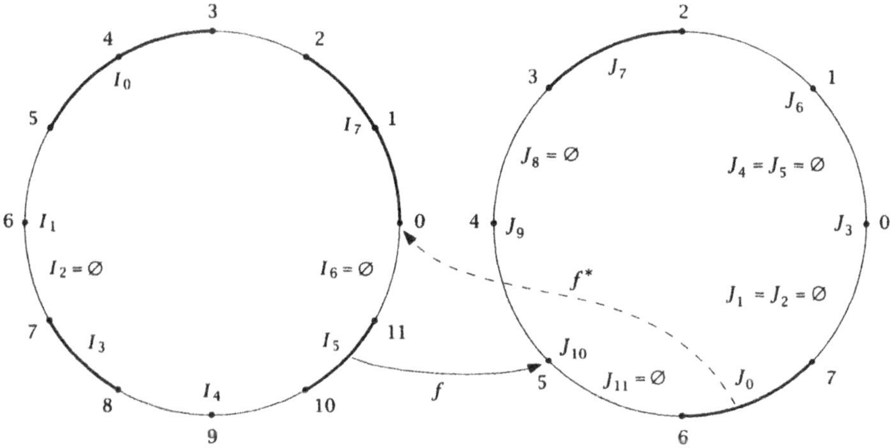

Figure III.4. An $f \in \mathrm{Hom}(\Lambda_{11}, \Lambda_7)$ with f is constant and equal to j on the interval I_j. The adjoint $f^* \in \mathrm{Hom}(\Lambda_7, \Lambda_{11})$ is constant and equal to ℓ on the interval J_ℓ.

Proposition 8. a) *The intervals* $(J_k)_{k\in\mathbb{Z}/n+1}$ *satisfy conditions* 1), 2), *and* 3) *and uniquely define the element* $f^* \in \mathrm{Hom}(\Lambda_m, \Lambda_n)$ *such that* $f^{*-1}(k) = J_k$.

b) *The map* $f \to f^*$ *is an isomorphism of the opposite category* Λ^{op} *with the category* Λ.

The proof of a) is straightforward. To understand b) without any tedious checking one compares, for an arbitrary monoid Γ, the following covariant and contravariant functors from Λ to the category of sets:

Covariant

$$X_n = \Gamma^{n+1} \quad \forall n \in \mathbb{N} \quad , \quad X(f) : \Gamma^{n+1} \to \Gamma^{m+1},$$

$$X(f)(a_0, \ldots, a_n) = (b_0, \ldots, b_m) \quad \text{with} \quad b_j = \prod_{f^{-1}(j)} a_\ell.$$

When $f^{-1}(j) = \varnothing$ we let $b_j = 1 \in \Gamma$, otherwise $f^{-1}(j)$ is a well specified interval, $f^{-1}(j) \cap \mathbb{Z}/m+1 = \{r, r+1, \ldots, s\}$, and the product $a_r a_{r+1} \ldots a_s$ is well defined. (Note that when $f^{-1}(j)$ contains $\mathbb{Z}/m+1$ we need more than $f^{-1}(j) \cap \mathbb{Z}/m+1$ to specify the endpoints.)

Contravariant

$$Y_n = \Gamma^{n+1} \quad \forall n \in \mathbb{N} \quad , \quad Y(f) : \Gamma^{m+1} \to \Gamma^{n+1},$$

$$Y(f)(a_0, \ldots, a_m) = (b_0, \ldots, b_n) \quad \text{with} \quad b_j = \prod_{f(j-1) < \ell \leq f(j)} a_\ell.$$

Again there is a small ambiguity in the notation and we mean $b_j = \prod_{J_j} a_\ell$, where the J_j are the intervals used in Proposition 8.

The covariance and contravariance of the above functors shows that $f \to f^*$ is a contravariant functor and that $Y(f) = X(f^*) \quad \forall f$. That $*$ is bijective is obvious since we described two equivalent ways of labeling all collections of intervals satisfying conditions 1), 2), and 3) above.

A.γ The Λ-module $\mathcal{A}^\natural$ associated to an algebra $\mathcal{A}$

Let $\mathcal{A}$ be a unital algebra over a field k. Let us define a covariant functor from Λ to k-vector spaces as follows. For each $n \in \mathbb{N}$ we let

$$\mathcal{A}_n^\natural = \mathcal{A}^{\otimes n+1} = \mathcal{A} \otimes \mathcal{A} \otimes \cdots \otimes \mathcal{A} \quad (n+1 \text{ terms}).$$

For each $f \in \mathrm{Hom}(\Lambda_n, \Lambda_m)$ the action $f_\natural$ of f is given by

$$f_\natural(x^0 \otimes \cdots \otimes x^n) = y^0 \otimes \cdots \otimes y^m \quad \forall x^j \in \mathcal{A},$$

where $y^j = \prod_{f^{-1}(j)} x^\ell$. (We use the same conventions as in β).)

This covariant functor $\mathcal{A}^\natural$ defines a $k(\Lambda)$-module whose underlying vector space is $\bigoplus_{n=0}^{\infty} \mathcal{A}^{\otimes(n+1)}$, and which we still denote by $\mathcal{A}^\natural$.

By composition with the contravariant functor $* : \Lambda \to \Lambda$ of Proposition 8, one obtains a cyclic vector space $C(\mathcal{A})$, whose faces, degeneracies, and permutations are given by

$$\begin{aligned}
d^i(x_0 \otimes x_1 \otimes \cdots \otimes x_n) &= (x_0 \otimes \cdots \otimes x_i x_{i+1} \otimes \cdots \otimes x_n) \quad 0 \leq i \leq n-1 \\
d^n(x_0 \otimes x_1 \otimes \cdots \otimes x_n) &= (x_n x_0 \otimes x_1 \otimes \cdots \otimes x_{n-1}) \\
s^j(x_0 \otimes \cdots \otimes x_n) &= (x_0 \otimes \cdots \otimes x_i \otimes 1 \otimes x_{i+1} \otimes \cdots \otimes x_n) \\
t_n(x_0 \otimes \cdots \otimes x_n) &= (x_n \otimes x_0 \otimes \cdots \otimes x_{n-1}) \quad \forall\, x_j \in \mathcal{A}.
\end{aligned}$$

Depending on the context it may be more convenient to use the $k(\Lambda)$-module $\mathcal{A}^\natural$ or the cyclic vector space $C(\mathcal{A})$.

We let $k^\natural$ be the trivial $k(\Lambda)$-module.

Proposition 9. 1) *Let τ be a trace on $\mathcal{A}$; then the equality*

$$\tau'(x^0 \otimes \cdots \otimes x^n) = \tau(x^0 x^1 \cdots x^n)$$

defines an element τ' of $\mathrm{Hom}_{k(\Lambda)}(\mathcal{A}^\natural, k^\natural)$.

2) *The map $\tau \to \tau'$ is an isomorphism of the linear space of traces on $\mathcal{A}$ with* $\mathrm{Hom}_{k(\Lambda)}(\mathcal{A}^\natural, k^\natural)$.

The proof is straightforward.

The above discussion of k-algebras works with minor changes for rings and so we let $\mathcal{A}^\natural$ be the $\mathbb{Z}(\Lambda)$-module associated to a ring $\mathcal{A}$. The categories of $k(\Lambda)$-modules (or $\mathbb{Z}(\Lambda)$-modules) are abelian categories, and we use the standard notation Ext^n for the derived functors of the functor Hom (cf. [Car-E]). By Proposition 9 the functor

$$\text{Algebra}\, \mathcal{A} \to HC^0(\mathcal{A}) = \{\text{traces on } \mathcal{A}\}$$

is the composition of $\mathrm{Hom}_{k(\Lambda)}(\cdot, k^\natural) = \mathrm{Ext}^0_{k(\Lambda)}(\cdot, k^\natural)$ with $\natural$.

The higher cyclic cohomology groups HC^n are just the derived functors Ext^n as follows:

Theorem 10. [Co$_{26}$] *For any k-algebra $\mathcal{A}$ the cyclic cohomology group $HC^n(\mathcal{A})$ is canonically isomorphic to*

$$\mathrm{Ext}^n_{k(\Lambda)}(\mathcal{A}^\natural, k^\natural).$$

This theorem yields, in particular, obvious definitions of the cyclic cohomology functors for rings, $\mathrm{Ext}^n_{\mathbb{Z}(\Lambda)}(\mathcal{A}^\natural, \mathbb{Z}^\natural)$, and of the bivariant theory for algebras, as $\mathrm{Ext}^n_{k(\Lambda)}(\mathcal{A}^\natural, \mathcal{B}^\natural)$. A simple nontrivial example of an element of $\mathrm{End}_{k(\Lambda)}(\mathcal{A}^\natural, \mathcal{A}^\natural)$ is provided, for $\mathcal{A} = \mathbb{C}\Gamma$ a group ring, by the projection on a given conjugacy class (cf. [Lo]).

For the proof of Theorem 10 (cf. [Co$_{26}$]) one constructs a projective resolution of the trivial $\mathbb{Z}(\Lambda)$-module $\mathbb{Z}^\natural$.

Theorem 11. [Co$_{26}$] a) *The ring* $\mathrm{Ext}^*_\Lambda(\mathbb{Z}^\natural, \mathbb{Z}^\natural)$ *is the polynomial ring $\mathbb{Z}[\sigma]$, where the generator σ is of degree* 2.

b) *The classifying space $B\Lambda$ of the small category Λ is $B\mathbb{S}^1 = P_\infty(\mathbb{C})$.*

The Yoneda product with the generator $\sigma \in \mathrm{Ext}^2(\mathbb{Z}^\natural, \mathbb{Z}^\natural)$ defines the S operation of cyclic cohomology (Section 1) in full generality and justifies the normalization of S in Section 1.

The equality $B\Lambda = B\mathbb{S}^1$ brings out the analogy between cyclic cohomology and $\mathbb{S}^1$-equivariant cohomology.

A.δ Cyclic spaces and $\mathbb{S}^1$ spaces

The inclusion of Δ as a subcategory of Λ (Proposition 5) shows that any cyclic object in a category C is, in particular, a simplicial object in C. Let us apply this to cyclic sets X (or more generally to cyclic topological spaces, called cyclic spaces for short) and denote by $|X|$ the geometric realization of the underlying simplicial set (resp. space). One has

Proposition 12. [Bur-F] [Good] [Lo] *Let X be a cyclic space. Then $|X|$, the geometric realization of the underlying simplicial space, admits a canonical action of $\mathbb{S}^1$. One obtains in this way a functor from the category of cyclic spaces to the category of $\mathbb{S}^1$-spaces.*

We refer to [Lo] for a detailed proof. We shall briefly describe the $\mathbb{S}^1$ action. First, the forgetful functor, which to a cyclic space assigns the underlying simplicial space, has a left adjoint: the induction functor F from simplicial spaces to cyclic spaces. For any simplicial space Y (viewed as a right Δ-space) the cyclic space $F(Y)$ (viewed as a right Λ-space) is

$$F(Y) = Y \times_\Delta \Lambda$$

where Λ is viewed as a left Δ-space and a right Λ-space. Using the equality $\Lambda = \Delta C$ of Proposition 5 one describes more concretely the cyclic space $F(Y)$ as follows:

$$F_n(Y) = Y_n \times C_n \quad \forall n \in \mathbb{N},$$

the right action of $f \in \mathrm{Hom}(\Lambda_n, \Lambda_m)$ being given by

$$(y,c)f = (yh,c') \quad \forall y \in Y_m\,,\ c \in C_m$$

where $h \in \Delta$, $c' \in C_n$ and $hc' = cf$ (using Proposition 5). The geometric realization $|F(Y)|$ of the underlying simplicial space is homeomorphic to $|Y| \times \mathbb{S}^1$ using the canonical maps $p : |F(Y)| \to |Y|$ and $q : |F(Y)| \to \mathbb{S}^1$. Recall (cf. α)) that $|F(Y)|$ is a quotient of $\bigcup_{n\geq 0} F_n(Y) \times \Delta^n$ and use the obvious action of C_n on Δ^n to map $F_n(Y) \times \Delta^n = Y_n \times C_n \times \Delta^n$ to $Y_n \times \Delta^n$. One then checks (cf. [Lo]) that this map p is compatible with the equivalence relations yielding $p : |F(Y)| \to |Y|$. Next, q is just obtained by functoriality of F using the constant map from Y to the trivial simplicial set $\{\mathrm{pt}\}$ and the equality $|F(\mathrm{pt})| = \mathbb{S}^1$.

All of this holds for any simplicial space Y and we shall simply write $|F(Y)| \simeq |Y| \times \mathbb{S}^1$.

When X is a cyclic space its cyclic structure yields a morphism ρ of cyclic spaces from $F(X)$ to X given by the right action of Λ on X

$$\rho : X \times_\Delta \Lambda \to X\,,\ (y,f) \to yf.$$

With the above identifications this yields the desired action of $\mathbb{S}^1$ on $|X|$

$$|\rho| : |F(X)| = |X| \times \mathbb{S}^1 \to |X|.$$

We refer to [Lo] for a clear and detailed proof.

Let X be a simplicial set and $\mathbb{Z}(X)$ the corresponding simplicial abelian group. It is then straightforward that, letting $\mathbb{Z} = \mathbb{Z}(\mathrm{pt})$ be the trivial simplicial abelian group, one has $\mathrm{Ext}^n_{\mathbb{Z}(\Delta)}(\mathbb{Z}(X),\mathbb{Z}) \simeq H^n(|X|,\mathbb{Z})$ (resp. $\mathrm{Tor}_n^{\mathbb{Z}(\Delta)}(\mathbb{Z},\mathbb{Z}(X)) \simeq H_n(|X|,\mathbb{Z})$) where $|X|$ is the geometric realization of X (cf., for instance, [Lo] Theorem 6.2.2). Let us now assume that X is a cyclic set and let us endow $|X|$ with the $\mathbb{S}^1$-action given by Proposition 12. The following theorem is a natural generalization of Theorem 11:

Theorem 13. [Bur-F] *One has canonical isomorphisms:*

$$\mathrm{Ext}^n_{\mathbb{Z}(\Lambda)}(\mathbb{Z}(X),\mathbb{Z}) \simeq H^n_{\mathbb{S}^1}(|X|,\mathbb{Z})$$

$$\mathrm{Tor}_n^{\mathbb{Z}(\Lambda)}(\mathbb{Z},\mathbb{Z}(X)) \simeq H_n^{\mathbb{S}^1}(|X|,\mathbb{Z})$$

where $\mathbb{Z}(X)$ is the cyclic abelian group associated to the cyclic space X.

A similar statement holds for an arbitrary ring of coefficients, and the Gysin sequence relating the equivariant $\mathbb{S}^1$-cohomology $H^*_{\mathbb{S}^1}$ corresponds to the Gysin sequence relating the Ext functors for $\mathbb{Z}(\Lambda)$- and $\mathbb{Z}(\Delta)$-modules, of which Theorem 1.26 is a special case ([Co_{26}]).

Problem 14. While the resolution used in the proof of Theorem 10 gives a concrete way of formulating $\mathrm{Ext}^n_{\mathbb{Z}(\Lambda)}(\cdot,\mathbb{Z})$ for arbitrary cyclic abelian groups, one is missing a similar description for the bivariant theory and, in particular, for $\mathrm{Ext}^n_{k(\Lambda)}(\mathcal{A}^\natural,\mathcal{B}^\natural)$ where $\mathcal{A}$ and $\mathcal{B}$ are k-algebras, for k a field.

III. Appendix B. Locally Convex Algebras

We shall briefly indicate how Section 1 adapts to a topological situation. Thus, we shall assume now that the algebra $\mathcal{A}$ is endowed with a locally convex topology, for which the product $\mathcal{A} \times \mathcal{A} \to \mathcal{A}$ is continuous. In other words, for any continuous seminorm p on $\mathcal{A}$ there exists a continuous seminorm p' such that $p(ab) \le p'(a)\,p'(b)$, $\forall a,b \in \mathcal{A}$. Then we replace the algebraic dual $\mathcal{A}^*$ of $\mathcal{A}$ by the topological dual, and the space $C^n(\mathcal{A},\mathcal{A}^*)$ of $(n+1)$-linear functionals on $\mathcal{A}$ by the space of continuous $(n+1)$-linear functionals: $\varphi \in C^n$ if and only if for some continuous seminorm p on $\mathcal{A}$ one has

$$|\varphi(a^0,\dots,a^n)| \le p(a^0)\cdots p(a^n)\,, \quad \forall a^i \in \mathcal{A}.$$

Since the product is continuous one has $b\varphi \in C^{n+1}$, $\forall \varphi \in C^n$. Since the formulae for the cup product of cochains involve only the product in $\mathcal{A}$ they still make sense for continuous multilinear functions and all the results of Section 1 apply with no change.

There is however, an important point which we wish to discuss: the use of resolutions in the computation of the Hochschild cohomology. Note first that we may as well assume that $\mathcal{A}$ is complete, since C^n is unaffected if one replaces $\mathcal{A}$ by its completion, which is still a locally convex topological algebra.

Let $\mathcal{B}$ be a complete locally convex topological algebra. By a *topological module* over $\mathcal{B}$ we mean a locally convex vector space $\mathcal{M}$, which is a $\mathcal{B}$-module, and is such that the map $(b,\xi) \mapsto b\xi$ is continuous from $\mathcal{B} \times \mathcal{M}$ to $\mathcal{M}$. We say that $\mathcal{M}$ is *topologically projective* if it is a direct summand of a topological module of the form $\mathcal{M}' = \mathcal{B}\hat{\otimes}_\pi E$, where E is a *complete* locally convex vector space and $\hat{\otimes}_\pi$ means the projective tensor product ([Grot$_1$]). In particular, $\mathcal{M}$ is complete, as a closed subspace of the complete locally convex vector space $\mathcal{M}'$.

It is then clear that if $\mathcal{M}_1$ and $\mathcal{M}_2$ are topological $\mathcal{B}$-modules which are complete (as locally convex vector spaces) and $p : \mathcal{M}_1 \to \mathcal{M}_2$ is a continuous $\mathcal{B}$-linear map with a continuous $\mathbb{C}$-linear cross-section s, one can complete the triangle of continuous $\mathcal{B}$-linear maps

$$\begin{array}{ccc} & & \mathcal{M}_1 \\ & \tilde{f}\nearrow & \downarrow p \\ \mathcal{M} & \xrightarrow{f} & \mathcal{M}_2 \end{array}$$

for any continuous $\mathcal{B}$-linear map $f : \mathcal{M} \to \mathcal{M}_2$.

Definition 1. *Let $\mathcal{M}$ be a topological $\mathcal{B}$-module. By a (topological) projective resolution of $\mathcal{M}$ we mean an exact sequence of projective $\mathcal{B}$-modules and continuous $\mathcal{B}$-linear maps*

$$\mathcal{M} \xleftarrow{\varepsilon} \mathcal{M}_0 \xleftarrow{b_1} \mathcal{M}_1 \xleftarrow{b_2} \mathcal{M}_2 \leftarrow \cdots$$

which admits a $\mathbb{C}$-linear continuous homotopy $s_i : \mathcal{M}_i \to \mathcal{M}_{i+1}$ such that

$$b_{i+1}\, s_i + s_{i-1}\, b_i = \mathrm{id}\,, \quad \forall i.$$

As in [Car-E] the module $\mathcal{A}$ over $\mathcal{B} = \mathcal{A}\hat{\otimes}_\pi\mathcal{A}^o$ (tensor product of the algebra $\mathcal{A}$ by the opposite algebra $\mathcal{A}^o$) given by

$$(a \otimes b^o)\, c = acb\,, \quad a,b,c \in \mathcal{A}$$

admits the following canonical projective resolution:

1) $\mathcal{M}_n = \mathcal{B}\hat{\otimes}_\pi E_n$ (as a $\mathcal{B}$-module), with $E_n = \mathcal{A}\hat{\otimes}_\pi \cdots \hat{\otimes}_\pi \mathcal{A}$ (n factors);

2) $\varepsilon : \mathcal{M}_0 \to \mathcal{A}$ is given by $\varepsilon(a \otimes b^o) = ab$, $a,b \in \mathcal{A}$;

3) $b_n(1\otimes a_1 \otimes \cdots \otimes a_n) = (a_1 \otimes 1) \otimes a_2 \otimes \cdots \otimes a_n + \sum_{j=1}^{n-1} (-1)^j (1\otimes 1) \otimes a_1 \otimes \cdots \otimes a_j a_{j+1} \otimes \cdots \otimes a_n + (-1)^n\, (1 \otimes a_n^o) \otimes (a_1 \otimes \cdots \otimes a_{n-1})$.

The usual section is obviously continuous:

$$s_n((a \otimes b^0) \otimes (a_1 \otimes \cdots \otimes a_n)) = (1 \otimes b^0) \otimes (a \otimes a_1 \otimes \cdots \otimes a_n)).$$

Comparing this resolution with an arbitrary topological projective resolution of the module $\mathcal{A}$ over $\mathcal{B}$ yields:

Lemma 2. *For every topological projective resolution* $(\mathcal{M}^n, b_n)$ *of the module* $\mathcal{A}$ *over* $\mathcal{B} = \mathcal{A}\hat{\otimes}_\pi \mathcal{A}^o$*, the Hochschild cohomology* $H^n(\mathcal{A}, \mathcal{A}^*)$ *coincides with the cohomology of the complex*

$$\mathrm{Hom}_{\mathcal{B}}(\mathcal{M}^0, \mathcal{A}^*) \xrightarrow{b_1^*} \mathrm{Hom}_{\mathcal{B}}(\mathcal{M}^1, \mathcal{A}^*) \to \cdots$$

(where $\mathrm{Hom}_{\mathcal{B}}$ *means continuous* $\mathcal{B}$*-linear maps).*

Of course, this lemma extends to any complete topological bimodule over $\mathcal{A}$.

We refer to [Helem] for more refined tools in this topological context.

III. Appendix C. Stability under Holomorphic Functional Calculus

Let A be a Banach algebra over $\mathbb{C}$ and $\mathcal{A}$ a subalgebra of A, and let $\tilde{A}$ and $\tilde{\mathcal{A}}$ be obtained by adjoining a unit.

Definition 1. *$\mathcal{A}$ is stable under holomorphic functional calculus if for any* $a \in \tilde{\mathcal{A}}$ *and any function* f *holomorphic in a neighborhood of the spectrum of* a *in* $\tilde{A}$ *one has* $f(a) \in \tilde{\mathcal{A}}$.

The first important result is that if $\mathcal{A}$ is dense in A the above property is inherited by the subalgebra $M_n(\mathcal{A})$ of $M_n(A)$ for any n.

Proposition 2. *If the dense subalgebra* $\mathcal{A}$ *of the Banach algebra* A *is stable under holomorphic functional calculus then so is* $M_n(\mathcal{A})$ *in* $M_n(A)$ *for any* $n \in \mathbb{N}$.

We refer to [Sch] for a detailed proof. One can use (cf. [Bos$_1$]) the following identity in $M_2(A)$ as a substitute for the determinant of matrices

$$\begin{bmatrix} a_{11} & a_{12} \\ a_{21} & a_{22} \end{bmatrix}^{-1} = \begin{pmatrix} a_{11}^{-1} & 0 \\ 0 & (a_{22} - a_{21}a_{11}^{-1}a_{12})^{-1} \end{pmatrix} \begin{pmatrix} 1 & -a_{12}(a_{22} - a_{21}a_{11}^{-1}a_{12})^{-1} \\ 0 & 1 \end{pmatrix} \begin{pmatrix} 1 & 0 \\ -a_{21}a_{11}^{-1} & 1 \end{pmatrix}, \tag{1}$$

in a suitable neighborhood of the identity $\begin{bmatrix} 1 & 0 \\ 0 & 1 \end{bmatrix}$ in $M_2(A)$.

Let $\mathcal{A} \subset A$ be stable under holomorphic functional calculus. In particular, one has $GL_n(\tilde{\mathcal{A}}) = GL_n(\tilde{A}) \cap M_n(\tilde{\mathcal{A}})$; hence if we endow $GL_n(\tilde{\mathcal{A}})$ with the induced topology we get a topological group which is locally contractible as a topological space. We recall the density theorem (cf. [At$_3$], [Kar$_2$]).

Proposition 3. *Let* $\mathcal{A}$ *be a dense subalgebra of* A*, stable under holomorphic functional calculus.*

a) *The inclusion* $i : \mathcal{A} \to A$ *is an isomorphism of* K_0*-groups*

$$i_* : K_0(\mathcal{A}) \to K_0(A).$$

b) *Let* $GL_\infty(\widetilde{\mathcal{A}})$ *be the inductive limit of the topological groups* $GL_n(\widetilde{\mathcal{A}})$. *Then* i_* *yields an isomorphism,*

$$\pi_k(GL_\infty(\widetilde{\mathcal{A}})) \to \pi_k(GL_\infty(\widetilde{A})) = K_{k+1}(A).$$

IV

QUANTIZED CALCULUS

The basic idea of this chapter, and of noncommutative differential geometry ([Co$_{17}$]), is to *quantize* the differential calculus using the following operator theoretic notion for the differential

$$df = [F, f].$$

Here f is an element of an involutive algebra $\mathcal{A}$ of operators in a Hilbert space $\mathcal{H}$, while F is a selfadjoint operator of square one ($F^2 = 1$) in $\mathcal{H}$. At first one should think of f as a function on a manifold, i.e. of $\mathcal{A}$ as an algebra of functions, but one virtue of our construction is that it will apply in the noncommutative case as well.

Since the word *quantization* is often overused we feel the need to justify its use in our context.

First, in the case of manifolds the above formula replaces the differential df by an operator theoretic expression involving a *commutator,* which is similar to the replacement of the Poisson brackets of classical mechanics by commutators (cf. Chapter I Section 1).

Second, the *integrality* aspect of quantization (such as the integrality of the energy levels of the harmonic oscillators ([Dir] and Chapter V Section 11)) will have as a counterpart the integrality of the index of a Fredholm operator, which will play a crucial role in our context (Sections 5, 6 and Proposition 2 below).

We shall also see (Appendix D) how the *deformation* aspect of quantization fits with our context.

Before we begin to develop a calculus based on the above formula, we need to specify the required properties of the triple $(\mathcal{A}, \mathcal{H}, F)$, or equivalently of the representation of $\mathcal{A}$ in the pair $(\mathcal{H}, F)$. The following notion of *Fredholm representation*, or equivalently of *Fredholm module*, is due to Atiyah [At$_2$], Mishchenko [Mis$_1$], Brown, Douglas and Fillmore [BDF], and Kasparov [Kas$_7$].

Definition 1. *Let $\mathcal{A}$ be an involutive algebra (over $\mathbb{C}$). Then a Fredholm module over $\mathcal{A}$ is given by:*

1) *an involutive representation π of $\mathcal{A}$ in a Hilbert space $\mathcal{H}$;*

2) *an operator $F = F^*$, $F^2 = 1$, on $\mathcal{H}$ such that*

$$[F, \pi(a)] \text{ is a compact operator for any } a \in \mathcal{A}.$$

Such a Fredholm module will be called *odd*. An *even* Fredholm module is given by an odd Fredholm module $(\mathcal{H}, F)$ as above together with a $\mathbb{Z}/2$ grading γ, $\gamma = \gamma^*$, $\gamma^2 = 1$ of the Hilbert space $\mathcal{H}$ such that:

a) $\gamma\pi(a) = \pi(a)\gamma \ \forall a \in \mathcal{A}$

b) $\gamma F = -F\gamma$.

(In the context of $\mathbb{Z}/2$-graded algebras a) becomes $\gamma\pi(a) = (-1)^{\deg(a)}\pi(a)\gamma$ (cf. [Kas$_1$]).)

When no confusion can arise we systematically omit the representation π and write $a\xi$ instead of $\pi(a)\xi$, for $a \in \mathcal{A}$, $\xi \in \mathcal{H}$.

The above Definition 1 is, up to trivial changes, the same as Atiyah's definition of abstract elliptic operators, and the same as Kasparov's definition for the cycles in K-homology, $KK(A, \mathbb{C})$, when A is a C^*-algebra.

These trivial changes address the conditions $F = F^*$ and $F^2 = 1$, which can be replaced by $\pi(a)(F - F^*) \in \mathcal{K}$ and $\pi(a)(F^2 - 1) \in \mathcal{K}$, where $\mathcal{K}$ is the ideal of compact operators (cf. Appendix A). But the condition $F^2 = 1$ is important in our calculus.

Atiyah's motivation for the definition of abstract elliptic operators comes from the following example of an even Fredholm module over the C^*-algebra $C(V)$ of continuous functions on a smooth compact manifold V ([At$_2$]). Let $E^\pm$ be smooth Hermitian complex vector bundles over V and $P : C^\infty(V, E^+) \to C^\infty(V, E^-)$ an *elliptic* pseudo-differential operator of order 0. Then, being of order 0, it extends to a bounded operator

$$P : L^2(V, E^+) \to L^2(V, E^-),$$

and the existence of a parametrix Q for P (such that both $PQ - 1$ and $QP - 1$ are compact) shows that P is a *Fredholm* operator which *almost intertwines* the natural representations of $C(V)$ by multiplication operators in $L^2(V, E^\pm)$

$$\pi^\pm(f)\xi = f\xi \quad \forall \xi \in L^2(V, E^\pm), \ f \in C(V).$$

Indeed, one has ([At$_2$])

$$P\pi^+(f) - \pi^-(f)P \in \mathcal{K} \quad \forall f \in C(V).$$

Let us then consider the Hilbert space $\mathcal{H} = L^2(V,E^+) \oplus L^2(V,E^-) = \mathcal{H}^+ \oplus \mathcal{H}^-$ with its natural $\mathbb{Z}/2$-grading $\gamma = \begin{bmatrix} 1 & 0 \\ 0 & -1 \end{bmatrix}$. Let π be the representation of $C(V)$ in $\mathcal{H}$ given by

$$\pi(f) = \begin{bmatrix} \pi^+(f) & 0 \\ 0 & \pi^-(f) \end{bmatrix} \quad \forall f \in C(V)$$

and let $F = \begin{bmatrix} 0 & Q \\ P & 0 \end{bmatrix}$.

One has $[F,\pi(f)] \in \mathcal{K}$, $\forall f \in C(V)$, and $F^2 - 1 \in \mathcal{K}$, so that, up to an easy modification (Appendix A), we get an even Fredholm module over $C(V)$. The role of such modules in index theory is provided by the following proposition, which in the manifold context yields the index of P with coefficients in an auxiliary vector bundle (cf. [At$_2$]).

Proposition 2. [At$_2$][Kas$_7$] *Let $\mathcal{A}$ be an involutive algebra, $(\mathcal{H},F)$ a Fredholm module over $\mathcal{A}$, and for $q \in \mathbb{N}$ let $(\mathcal{H}_q,F_q)$ be the Fredholm module over $M_q(\mathcal{A}) = \mathcal{A} \otimes M_q(\mathbb{C})$ given by*

$$\mathcal{H}_q = \mathcal{H} \otimes \mathbb{C}^q\,,\ F_q = F \otimes 1\,,\ \pi_q = \pi \otimes \mathrm{id}\,.$$

We extend the action of $\mathcal{A}$ on $\mathcal{H}$ to a unital action of $\widetilde{\mathcal{A}}$.

a) *Let $(\mathcal{H},F)$ be even, with $\mathbb{Z}/2$ grading γ, and let $e \in \mathrm{Proj}(M_q(\widetilde{\mathcal{A}}))$. Then the operator $\pi_q^-(e)F_q\pi_q^+(e)$ from $\pi_q^+(e)\mathcal{H}_q^+$ to $\pi_q^-(e)\mathcal{H}_q^-$ is a Fredholm operator. An additive map φ of $K_0(\mathcal{A})$ to $\mathbb{Z}$ is determined by*

$$\varphi([e]) = \mathrm{Index}\left(\pi_q^-(e)F_q\pi_q^+(e)\right).$$

b) *Let $(\mathcal{H},F)$ be odd and let $E = \left(\frac{1+F}{2}\right)$. Let $u \in GL_q(\widetilde{\mathcal{A}})$. Then the operator $E_q\pi_q(u)E_q$ from $E_q\mathcal{H}_q$ to itself is a Fredholm operator. An additive map of $K_1(\mathcal{A})$ to $\mathbb{Z}$ is determined by*

$$\varphi([u]) - \mathrm{Index}(E_q\pi_q(u)E_q).$$

When $\mathcal{A} = A$ is a C^*-algebra, the group $K_1(\mathcal{A})$ of b) can be replaced by $K_1^{\mathrm{top}}(A)$, and in both even and odd cases the index map φ only depends upon the *K-homology* class

$$[(\mathcal{H},F)] \in KK(A,\mathbb{C})$$

of the Fredholm module, in the Kasparov KK group. This is an easy special instance of the Kasparov intersection product ([Kas$_1$]). Given a C^*-algebra A, its K-homology $K^*(A) = KK(A,\mathbb{C})$, as defined by Kasparov, is the abelian group of stable homotopy classes of Fredholm modules over A ([Kas$_7$]) (cf. Appendix A).

When $A = C(X)$ is a commutative C^*-algebra (say with unit), then the K-homology $K^*(A) = K^*(C(X))$ coincides with the Steenrod K-homology of the compact space $X = \mathrm{Spec}(A)$ (cf. [B-D-F] [Kas$_7$] [Kam-S]), as gradually emerged from the work of Brown, Douglas and Fillmore on extension theory for the odd case. The Chern character in K-homology (cf. [Bau-D]), dual to the usual Chern character, then gives a map

$$\mathrm{ch}_* : K^*(C(X)) \to H_*(X, \mathbb{Q})$$

which allows one to compute the index pairing of Proposition 2 by the formula:

$$\varphi(x) = \langle \mathrm{ch}_*(\mathcal{H}, F), \mathrm{ch}^*(x) \rangle \quad \forall x \in K_*(C(X))$$

where φ is the K-theory map given by the Fredholm module (Proposition 2) $(\mathcal{H}, F)$ over $A = C(X)$ and $\mathrm{ch}_*(\mathcal{H}, F)$ is the Chern character of the corresponding Steenrod K-homology class. Note that here in order to compute $\mathrm{ch}_*(\mathcal{H}, F)$ one first needs to know the element of Steenrod K-homology of X corresponding to $(\mathcal{H}, F)$, which is not obvious.

When A is a noncommutative C^*-algebra its K-homology $K^*(A) = KK(A, \mathbb{C})$ makes good sense, and in looking for an analogue of the above Chern character

$$\mathrm{ch}_* : K^*(C(X)) \to H_*(X, \mathbb{Q})$$

we were led to invent cyclic cohomology in [Co$_{17}$]. The idea is to use, given a Fredholm module $(\mathcal{H}, F)$ over A, the following formula to evaluate quantized differential forms $\omega = \sum f^0[F, f^1] \cdots [F, f^n]$:

$$\begin{aligned} \mathrm{Tr}_s \omega &= \mathrm{Trace}\,(\omega) \quad &\text{(odd case)} \\ \mathrm{Tr}_s \omega &= \mathrm{Trace}\,(\gamma\omega) \quad &\text{(even case).} \end{aligned}$$

The functional

$$\tau(f^0, \ldots, f^n) = \mathrm{Tr}_s(f^0[F, f^1] \cdots [F, f^n])$$

is indeed, when it makes good sense, the prototype of a *cyclic cocycle*. For these formulae to make sense one needs the following refinement of Definition 1.

Definition 3. *Let $\mathcal{A}$ be an involutive algebra and $(\mathcal{H}, F)$ a Fredholm module over $\mathcal{A}$. Let $p \in [1, \infty[$. We shall say that $(\mathcal{H}, F)$ is* p-summable *if*

$$[F, a] \in \mathcal{L}^p(\mathcal{H}) \quad \forall a \in \mathcal{A}.$$

Here $\mathcal{L}^p(\mathcal{H}) = \{T \in \mathcal{K};\ \sum_{n=0}^{\infty} \mu_n(T)^p < \infty\}$ is the Schatten-von Neumann ideal of compact operators such that $\mathrm{Trace}\,(|T|^p) < \infty$. We shall use these ideals of compact operators to measure the size of the differential $[F, f]$. The Hölder inequality and its corollary

$$\mathcal{L}^{p_1} \mathcal{L}^{p_2} \cdots \mathcal{L}^{p_k} \subset \mathcal{L}^p \quad \text{for } \frac{1}{p} = \sum_{j=1}^{k} \frac{1}{p_j}$$

show that for a p-summable Fredholm module, the quantized differential forms of high enough order belong to $\mathcal{L}^1$ and hence can be integrated according to the above formulas.

The above idea is directly in line with the earlier works of Helton and Howe [Hel-H_1] [Hel-H_2], Carey and Pincus [Ca-P] and Douglas and Voiculescu [Dou-V] in the case when $\mathcal{A}$ is commutative. But even in that case it improves on earlier work since the above cyclic cocycle determines all the lower-dimensional homology classes of an extension, and not just the top-dimensional one.

In Section 1 we shall develop our calculus, define the Chern character

$$\mathrm{ch}_*(\mathcal{H}, F) \in HC^*(\mathcal{A})$$

of finitely summable Fredholm modules, exhibit the role of the periodicity operator $S : HC^n \to HC^{n+2}$ of cyclic cohomology, and give the computation of the index map of Proposition 2 using the Chern character $\mathrm{ch}_*(\mathcal{H}, F)$.

In Section 2 we shall define the Dixmier trace which plays the role of the classical integral in our calculus. We refer to the general introduction of the book for the detailed dictionary between the classical differential calculus and the quantized calculus. To avoid any confusion, we shall not, however, use the dictionary in this chapter .

The main result of Section 2, which justifies the introduction of the Dixmier trace, is Theorem 2.8. It gives the Hochschild class of the character $\mathrm{ch}_*(\mathcal{H}, F)$ in terms of the Dixmier trace.

In Sections 3, 4, 5, and 6 we shall give concrete examples of this calculus, including highly nonabelian cases (5), spaces of non-integral Hausdorff dimension (3) and the quantum Hall effect as analysed by J. Bellissard (6).

In the remaining sections we shall show that this calculus also applies to spaces of infinite dimension, such as the configuration space of constructive quantum field theory (Section 9) or the dual space of higher rank discrete groups (Section 9). The finite summability condition of Definition 3 is then relaxed to the following:

Definition 4. *Let $(\mathcal{H}, F)$ be a Fredholm module over an algebra $\mathcal{A}$. Then $(\mathcal{H}, F)$ is θ-summable if*

$$[F, a] \in J^{1/2} \qquad \forall a \in \mathcal{A}$$

where $J^{1/2}$ is the (two-sided) ideal of compact operators T on $\mathcal{H}$ such that

$$\mu_n(T) = O((\log n)^{-1/2}).$$

Exactly as the finitely summable Fredholm modules served as motivation to develop cyclic cohomology, the θ-summable ones (in their unbounded presentation as K-cycles (cf. Section 8 below)) motivate the introduction of *entire cyclic cohomology* ([Co_{24}]) which we shall discuss in Section 7.

The Chern character $\mathrm{ch}_*(\mathcal{H}, F)$ of θ-summable Fredholm modules was first constructed in [Co_{24}] and then, in a cohomologous ([Co_{30}]) but much simpler form in [J-L-O]. This will be discussed in Section 8. Finally, Section 9 will be devoted to examples.

IV.1. Quantized Differential Calculus and Cyclic Cohomology

Let $\mathcal{A}$ be an involutive algebra and $(\mathcal{H}, F)$ be a Fredholm module over $\mathcal{A}$ (Definition 1). Let n be an integer, $n \geq 0$, and assume that $(\mathcal{H}, F)$ is *even* if n is even, and is $(n+1)$-summable, that is,

$$[F, a] \in \mathcal{L}^{n+1}(\mathcal{H}) \quad \forall a \in \mathcal{A}.$$

We shall now construct an n-dimensional cycle $(\Omega, d, \int)$ over $\mathcal{A}$ in the sense of Definition 1 of Chapter III.1.

1.α The cycle associated to a Fredholm module

The graded algebra $\Omega^* = \bigoplus \Omega^k$ is obtained as follows. For $k = 0$, $\Omega^0 = \mathcal{A}$. For $k > 0$ one lets Ω^k be the linear span of the operators

$$\omega = a^0[F, a^1] \cdots [F, a^k] \quad a^j \in \mathcal{A}.$$

The Hölder inequality shows that $\Omega^k \subset \mathcal{L}^{\frac{n+1}{k}}(\mathcal{H})$. The *product* in Ω^* is the product of operators; one has

$$\omega\omega' \in \Omega^{k+k'} \text{ for any } \omega \in \Omega^k, \ \omega' \in \Omega^{k'},$$

as one checks using the equality

$$\begin{aligned}\left(a^0[F, a^1] \cdots [F, a^k]\right) a^{k+1} &= \sum_{j=1}^{k} (-1)^{k-j}\, a^0[F, a^1] \cdots [F, a^j a^{j+1}] \cdots [F, a^{k+1}] \\ &\quad + (-1)^k\, a^0 a^1 [F, a^2] \cdots [F, a^{k+1}] \qquad \forall a^j \in \mathcal{A}.\end{aligned}$$

The differential $d : \Omega^* \to \Omega^*$ is defined as follows:

$$d\omega = F\omega - (-1)^k\, \omega F \quad \forall \omega \in \Omega^k.$$

Again one checks that $d\omega$ belongs to Ω^{k+1} using the equality

$$\begin{aligned}F(a^0[F, a^1] \cdots [F, a^k]) - (-1)^k (a^0[F, a^1] \cdots [F, a^k])F \\ = [F, a^0][F, a^1] \cdots [F, a^k] \ \forall a^j \in \mathcal{A}.\end{aligned}$$

(Since $F^2 = 1$, F anticommutes with $[F, a]$ for any $a \in \mathcal{A}$ and hence

$$[F, a^1] \cdots [F, a^k]F = (-1)^k\, F[F, a^1] \cdots [F, a^k].)$$

By construction, d is a graded derivation, i.e.

$$d(\omega_1\omega_2) = (d\omega_1)\omega_2 + (-1)^{k_1}\, \omega_1\, d\omega_2 \quad \forall \omega_j \in \Omega^{k_j}.$$

Moreover, it is straightforward that $d^2 = 0$ since, for $\omega \in \Omega^k$,

$$F(F\omega - (-1)^k\, \omega F) + (-1)^k\, (F\omega - (-1)^k\, \omega F)F = 0.$$

We thus have a graded differential algebra (Ω^*, d), and it remains to define a closed graded trace of degree n

$$\mathrm{Tr}_s : \Omega^n \to \mathbb{C}.$$

For this we introduce the following notation. Given an operator T on $\mathcal{H}$ such that $FT + TF \in \mathcal{L}^1(\mathcal{H})$ we set

$$\mathrm{Tr}'(T) = \frac{1}{2}\,\mathrm{Trace}\ (F(FT + TF)).$$

Note that if $T \in \mathcal{L}^1(\mathcal{H})$ then $\mathrm{Tr}'(T) = \mathrm{Tr}(T)$.

We then define $\mathrm{Tr}_s\omega$ for $\omega \in \Omega^n$ by the formulas

$$\mathrm{Tr}_s\omega = \mathrm{Tr}'(\omega) \ \text{ if } n \text{ is } \mathit{odd}$$
$$\mathrm{Tr}_s\omega = \mathrm{Tr}'(\gamma\omega) \ \text{ if } n \text{ is } \mathit{even}.$$

In the last formula γ is the $\mathbb{Z}/2$ grading operator provided by the evenness of the Fredholm module $(\mathcal{H}, F)$ (cf. Definition 1).

These formulae do make sense; indeed, for n *odd* and $\omega \in \Omega^n$, one has $F\omega + \omega F = d\omega \in \Omega^{n+1} \subset \mathcal{L}^{(n+1)/(n+1)} = \mathcal{L}^1$, while for n *even* one has $F\gamma\omega + \gamma\omega F = \gamma d\omega \in \mathcal{L}^1$ by the same argument.

Proposition 1. [Co$_{17}$] *$(\Omega, d, \mathrm{Tr}_s)$ is a cycle of dimension n over $\mathcal{A}$.*

Proof. We just need to check that Tr_s is a closed graded trace. Since $\mathrm{Tr}_s\omega$ only involves $d\omega$ and since $d^2 = 0$, it is clearly closed

$$\mathrm{Tr}_s d\omega = 0 \qquad \forall \omega \in \Omega^n.$$

Then let $\omega \in \Omega^k$, $\omega' \in \Omega^{k'}$ with $k + k' = n$. One has, for n odd

$$\begin{aligned}\mathrm{Tr}_s\ \omega\omega' &= \frac{1}{2}\mathrm{Trace}\,(Fd(\omega\omega')) = \frac{1}{2}\mathrm{Trace}\,(F(d\omega)\omega' + (-1)^k\, F\omega\, d\omega')\\ &= \frac{1}{2}\mathrm{Trace}((-1)^{k+1}\, d\omega\, F\omega' + (-1)^k\, (F\omega)d\omega').\end{aligned}$$

As Trace $(F\omega\ d\omega')$ = Trace $(d\omega'\ F\omega)$, using the equality

$$\mathrm{Trace}\,(T_1T_2) = \mathrm{Trace}\,(T_2T_1)\ ,\ T_j \in \mathcal{L}^{p_j}\ ,\ \frac{1}{p_1} + \frac{1}{p_2} = 1,$$

we get $\mathrm{Tr}_s\ \omega\omega' = (-1)^{kk'}\mathrm{Tr}_s\ \omega'\omega$.

For n even the computation is similar.

The *character* of the above n-dimensional cycle is the cyclic cocycle τ_n

$$\tau_n(a^0, \dots, a^n) = \mathrm{Tr}'\left(a^0[F, a^1]\cdots[F, a^n]\right) \qquad \forall a^j \in \mathcal{A}$$

with γa^0 instead of a^0 when n is even.

We shall see below many examples (Sections 4, 5, and 6) of explicit computations of these cyclic cocycles, but first we shall show how to eliminate the ambiguity in the choice of the integer n. Given a Fredholm module $(\mathcal{H}, F)$ over $\mathcal{A}$, the *parity* of n is fixed but the precise value of n is only subject to a *lower bound:*

$$[F,a] \in \mathcal{L}^{n+1} \quad \forall a \in \mathcal{A}.$$

Indeed, since $\mathcal{L}^{p_1} \subset \mathcal{L}^{p_2}$ if $p_1 \le p_2$, this shows that we can always replace n by $n+2q$ for any integer $q \ge 0$. In particular, we get a sequence of cyclic cocycles of the same parity

$$\tau_{n+2q}, \quad q \in \mathbb{N}$$

(where n is the smallest integer compatible with $(n+1)$-summability).

1.β The periodicity operator S and the Chern character

It is the comparison between these cyclic cocycles which was the reason for the introduction ([Co$_{17}$]) of the periodicity operator in cyclic cohomology (Section III.1)

$$S : HC^n(\mathcal{A}) \to HC^{n+2}(\mathcal{A}).$$

Proposition 2. [Co$_{17}$] *Let $(\mathcal{H}, F)$ be an $(n+1)$-summable Fredholm module over $\mathcal{A}$ of the same parity as n. The characters τ_{n+2q} satisfy*

$$\tau_{m+2} = -\tfrac{2}{m+2} S\tau_m \text{ in } HC^{m+2}(\mathcal{A}), \; m = n+2q, q \ge 0.$$

Proof. Let n be even. By construction, τ_n is the character of the cycle $(\Omega, d, \mathrm{Tr}_s)$ associated to $(\mathcal{H}, F)$ by Proposition 1. Thus (Chapter III Lemma 1.14) $S\tau_n$ is given by

$$S\tau_n(a^0, \dots, a^{n+2}) = \sum_{j=0}^{n+1} \mathrm{Tr}_s((a^0 da^1 \cdots da^{j-1}) a^j \, a^{j+1} (da^{j+2} \cdots da^{n+2}))$$

By definition, τ_{n+2} is given by

$$\tau_{n+2}(a^0, \dots, a^{n+2}) = \mathrm{Tr}_s(a^0 da^1 \cdots da^{n+2}).$$

We just have to find $\varphi_0 \in C_\lambda^{n+1}(\mathcal{A})$ such that $b\varphi_0 = S\tau_n + \frac{n+2}{2}\tau_{n+2}$. We shall construct $\varphi \in C_\lambda^{n+1}(\mathcal{A})$ such that

$$b\varphi(a^0, \dots, a^{n+2}) = 2 \sum_{j=0}^{n+1} \mathrm{Tr}_s((a^0 da^1 \cdots da^{j-1}) a^j \, a^{j+1} (da^{j+2} \cdots da^{n+2})) + (n+2)\, \mathrm{Tr}_s(a^0 da^1 \cdots da^{n+2}).$$

We take $\varphi = \sum_{j=0}^{n+1} (-1)^{j-1} \varphi^j$, where

$$\varphi^j(a^0, \dots, a^{n+1}) = \mathrm{Tr}_s(F a^j da^{j+1} \cdots da^{n+1} da^0 \cdots da^{j-1}).$$

One has $a^j da^{j+1} \cdots da^{n+1} da^0 \cdots da^{j-1} \in \Omega^{n+1} \subset \mathcal{L}^1(\mathcal{H})$ so that the trace makes sense; moreover, by construction, one has $\varphi \in C_\lambda^{n+1}(\mathcal{A})$.

To end the proof we shall now show that

$$(-1)^{j-1} b\varphi^j(a^0,\dots,a^{n+2}) = \mathrm{Tr}_s(a^0da^1\cdots da^{n+2}) \\ + 2\,\mathrm{Tr}_s((a^0da^1\cdots da^{j-1})a^j\, a^{j+1}(da^{j+2}\cdots da^{n+2})).$$

Using the equality $d(ab) = (da)b + adb$, with $a, b \in \mathcal{A}$, we get

$$b\varphi^j(a^0,\dots,a^{n+2}) = \mathrm{Tr}_s(F(a^{j+1}da^{j+2}\cdots da^{n+2})a^0(da^1\cdots da^j)) \\ + (-1)^{j-1}\,\mathrm{Tr}_s(Fa^{j+1}(da^{j+2}\cdots da^0\cdots da^{j-1})a^j) \\ + \mathrm{Tr}_s(Fa^j(da^{j+1}\cdots da^{n+2})a^0(da^1\cdots da^{j-1})).$$

Let $\beta = (da^{j+2}\cdots da^{n+2})a^0(da^1\cdots da^{j-1}) \in \Omega^n$. Using the equality

$$\mathrm{Tr}_s(\alpha d\beta) = \mathrm{Tr}_s((F\alpha + \alpha F)\beta) \quad \forall \alpha \in \mathcal{L}(\mathcal{H}),$$

we get, with $\alpha = a^jFa^{j+1}$

$$\mathrm{Tr}_s((F\alpha + \alpha F)\beta) = (-1)^{j-1}\,\mathrm{Tr}_s(Fa^{j+1}(da^{j+2}\cdots da^0\cdots da^{j-1})a^j).$$

Thus

$$(b\varphi^j)(a^0,\dots,a^{n+2}) = -\mathrm{Tr}_s(da^jFa^{j+1}\beta) + \mathrm{Tr}_s((F\alpha+\alpha F)\beta) + \mathrm{Tr}_s(Fa^j(da^{j+1})\beta) \\ = \mathrm{Tr}_s((Fd(a^ja^{j+1}) + (F\alpha + \alpha F))\beta).$$

As $Fd(a^ja^{j+1}) + (F\alpha + \alpha F) = da^jda^{j+1} + 2a^ja^{j+1}$, we get the desired equality.
The odd case is treated similarly.

Now let, as in Chapter III.1, $H^*(\mathcal{A}) = \varinjlim(HC^n(\mathcal{A}), S)$ be the periodic cyclic cohomology of $\mathcal{A}$. Proposition 2 shows that the class in $H^*(\mathcal{A})$ of

$$(-1)^{n/2}\Gamma\left(\tfrac{n}{2}+1\right)\tau_n \in HC^n(\mathcal{A}),$$

with n large and of the same parity as the Fredholm module $(\mathcal{H}, F)$, is well defined independently of n. At this point we still have the freedom to modify these formulae by a multiplicative constant λ_+ (resp. λ_-) for the even (resp. odd) case. This might seem desirable in order to eliminate $\Gamma\left(\frac{1}{2}\right) = \sqrt{\pi}$ in the odd case. It turns out, however, that there is a unique choice of these normalization constants which is compatible with the Kasparov product

$$KK(A, \mathbb{C}) \times KK(B, \mathbb{C}) \to KK(A \otimes B, \mathbb{C})$$

so that this product becomes the product # of Chapter III.1. This choice is $\lambda_+ = 1$ and $\lambda_- = \sqrt{2i}$, and it *does not* eliminate the term $\sqrt{\pi}$ for the Chern character in the odd case.

Definition 3. *Let $(\mathcal{H}, F)$ be a finitely summable Fredholm module over the involutive algebra $\mathcal{A}$. The Chern character, $\mathrm{ch}_*(\mathcal{H}, F) \in H^*(\mathcal{A})$ is the periodic cyclic cohomology class of the following cyclic cocycles, for n large enough*

$$\lambda_n\,\mathrm{Tr}'(\gamma a^0[F,a^1]\cdots[F,a^n]) \quad \forall a^j \in \mathcal{A}\,,\ \lambda_n = (-1)^{n(n-1)/2}\,\Gamma\left(\tfrac{n}{2}+1\right)$$

Even case

$$\lambda_n \operatorname{Tr}'(a^0[F,a^1]\cdots[F,a^n]) \quad \forall a^j \in \mathcal{A}\,,\ \lambda_n = \sqrt{2i}(-1)^{n(n-1)/2}\,\Gamma\left(\tfrac{n}{2}+1\right).$$

Odd case

Note that the normalization factors above are dependent on our choice of normalization for the operation S in cyclic cohomology. If we had chosen as generator of $HC^2(\mathbb{C})$ the cocycle

$$\sigma'(1,1,1) = \lambda$$

this would introduce an additional $\lambda^{n/2}$ in Definition 3. In [Co_{17}] we took $\lambda = 2\pi i$, which has the effect of eliminating the half-integral powers of π in the odd case, in which case for $n = 2m+1$ the normalization then yields

$$(-1)^m\,(2\pi i)^m\left(m-\frac{1}{2}\right)\cdots\frac{1}{2}\operatorname{Tr}'(a^0[F,a^1]\cdots[F,a^n]) \quad \forall a^j \in \mathcal{A}.$$

However, since cyclic cohomology is important for arbitrary rings, the normalization $\sigma(1,1,1) = 1$ of Chapter III is more general. Finally, the coefficient $(-1)^{n/2}$ can be absorbed through replacing F by iF, so that $da = i[F,a]$ is now selfadjoint for a selfadjoint. With our normalization the Chern character in K-homology, $\operatorname{ch}_*(\mathcal{H},F)$, of the fundamental class of $\mathbb{R}^n$ is given by the class of the n-cocycle

$$\tau(f^0,\ldots,f^n) = (2\pi i)^{-n/2}(-1)^{n(n-1)/2}\int f^0 df^1 \wedge df^2 \wedge\cdots\wedge df^n,\ f^j \in C_c^\infty(\mathbb{R}^n),$$

which gives the correct normalization for Fourier calculus ([Gu-S] p.271).

1.γ Pairing with K-theory and index formula

Let $\mathcal{A}$ be an involutive algebra and $(\mathcal{H},F)$ a finitely summable Fredholm module over $\mathcal{A}$. We have seen (Proposition 0.2) that it determines an index map

$$\varphi : K_j(\mathcal{A}) \to \mathbb{Z}$$

where $j = 0$ or 1 according to the parity of $(\mathcal{H},F)$.

We shall now give a simple index formula for the map φ and compute it in terms of the Chern character $\operatorname{ch}_*(\mathcal{H},F) \in H^*(\mathcal{A})$ of Definition 3.

Proposition 4. [Co_{17}] *Let $(\mathcal{H},F)$ be a finitely summable Fredholm module over $\mathcal{A}$. Then for any $x \in K_j(\mathcal{A})$, one has*

$$\varphi(x) = \langle x, \operatorname{ch}_*(\mathcal{H},F)\rangle$$

where φ is the Fredholm index map of Proposition 0.2.

Proof. Let us treat the even case. Replacing $\mathcal{A}$ by $\widetilde{\mathcal{A}}$, i.e. adjoining a unit to $\mathcal{A}$, and then by $M_q(\widetilde{\mathcal{A}})$ where the class x is represented by an idempotent $e \in M_q(\widetilde{\mathcal{A}})$, we may assume that $\mathcal{A}$ is unital and that $q = 1$. The matrix of F, in the $\mathbb{Z}/2$ decomposition $\mathcal{H} = \mathcal{H}^+ \oplus \mathcal{H}^-$ such that $\gamma = \begin{bmatrix} 1 & 0 \\ 0 & -1 \end{bmatrix}$, is of the form $F = \begin{bmatrix} 0 & Q \\ P & 0 \end{bmatrix}$ with $PQ = 1$ in $\mathcal{H}^-$ and $QP = 1$ in $\mathcal{H}^+$. Let $\mathcal{H}_1 = e\mathcal{H}^+$, $\mathcal{H}_2 = e\mathcal{H}^-$ and $P' = eP|\mathcal{H}_1$, $Q' = eQ|\mathcal{H}_2$. Then P' is a Fredholm operator from $\mathcal{H}_1$ to $\mathcal{H}_2$, and Q' is a quasi-inverse of P' such that $P'Q' - 1$ and $Q'P' - 1$ are both in $\mathcal{L}^{\frac{n+1}{2}}$ if $(\mathcal{H}, F)$ is $(n+1)$-summable. Indeed, these operators are restrictions of

$$e - eFeFe = -e[F,e]^2 e \in \mathcal{L}^{\frac{n+1}{2}}.$$

It then follows that

$$\operatorname{Index} P' = \operatorname{Trace}(1 - Q'P')^k - \operatorname{Trace}(1 - P'Q')^k$$

for any integer $k \geq \frac{n+1}{2}$. Thus, we get for any $m \geq \frac{n+1}{2}$

$$\operatorname{Index} P' = \operatorname{Trace}(\gamma(e - eFeFe)^m).$$

Now the pairing of $[e] \in K_0(\mathcal{A})$ with the representative τ_{2m} of the Chern character is given by Proposition 2 of Chapter III.3: (for $m > \frac{n+1}{2}$ so that $\operatorname{Tr}' = \operatorname{Tr}$),

$$(-1)^m \operatorname{Trace}(\gamma e[F,e]^{2m})$$

which is precisely the same formula.

Corollary 5. *Any representative $\tau \in HC^*(\mathcal{A})$ of the Chern character of a finitely summable Fredholm module pairs* integrally *with K-theory, i.e.*

$$\tau(K) \subset \mathbb{Z}.$$

We refer to Sections 5 and 6 for applications of this integrality result, in particular, to the integrality of the Hall conductivity (Section 6).

We shall now investigate the extension problem (cf. Chapter III) for the cyclic cocycles τ_n associated as above to a finitely summable Fredholm module.

Lemma 6. *Let $\mathcal{A}$ be an involutive algebra, $(\mathcal{H}, F)$ an $(n+1)$-summable Fredholm module over $\mathcal{A}$ with the parity of n. Let A be the C^*-algebra norm closure of $\mathcal{A}$ (in its action on $\mathcal{H}$) and $\overline{\mathcal{A}}$ the smallest involutive subalgebra of A containing $\mathcal{A}$ and stable under holomorphic functional calculus. Then $(\mathcal{H}, F)$ is an $(n+1)$-summable Fredholm module over $\overline{\mathcal{A}}$.*

Proof. It is clearly enough to show that given $(\mathcal{H}, F)$, the algebra of operators T on $\mathcal{H}$ which satisfy

$$[F,T] \in \mathcal{L}^{n+1}(\mathcal{H})$$

is stable under holomorphic functional calculus, which is straightforward using the equality

$$[F,(T-\lambda)^{-1}] = -(T-\lambda)^{-1}\,[F,T](T-\lambda)^{-1} \qquad \forall \lambda \notin \operatorname{Spectrum}(T).$$

This lemma shows that we can restrict our attention to those involutive algebras which satisfy the following condition:

$\mathcal{A}$ is isomorphic to a $*$-subalgebra stable under holomorphic calculus in a C^*-algebra.

The latter C^*-algebra A is then unique (with $\mathcal{A}$ dense in A) since the restriction of the C^*-norm to $\mathcal{A}$ is then given by the formula

$$\|a\| = (\text{Spectral radius}(a^*a))^{1/2}$$

where the spectral radius is taken inside $\mathcal{A}$ (or $\widetilde{\mathcal{A}}$ if $\mathcal{A}$ is not unital). For such *pre-C^*-algebras* $\mathcal{A}$ (called local C^*-algebras in [Bl_1]) the following holds:

Proposition 7. *Let $\mathcal{A}$ be a pre-C^*-algebra; then:*

1) *Any Fredholm module $(\mathcal{H}, F)$ over $\mathcal{A}$ extends by continuity to a Fredholm module over the associated C^*-algebra A.*

2) *The inclusion $\mathcal{A} \subset A$ is an isomorphism in K-theory.*

Indeed, any involutive representation π of a pre-C^*-algebra is automatically norm decreasing for the norm given by

$$\|a\| = (\text{Spectral radius}(a^*a))^{1/2}.$$

The second statement follows from Chapter III Appendix C.

Now a Fredholm module over a C^*-algebra A yields a K-homology class, i.e. an element of the Kasparov group

$$KK(A, \mathbb{C}) = K^*(A).$$

Both the K-homology of A and the cyclic cohomology $HC^*(\mathcal{A})$ (or rather the periodic cyclic cohomology $H^*(\mathcal{A})$) pair with the K-theory $K(\mathcal{A}) \simeq K(A)$, and Proposition 4 thus shows that the following diagram is commutative for any *pre-C^*-algebra $\mathcal{A}$*

$$\begin{array}{ccc}
\{\text{Fredholm modules over } \mathcal{A}\} & \xrightarrow{\text{ch}_*} & H^*(\mathcal{A}) \\
\downarrow & & \downarrow \\
K^*(A) = KK(A, \mathbb{C}) & \longrightarrow & \text{Hom}(K_*(A), \mathbb{C})
\end{array}$$

where the lower horizontal arrow is the pairing we have between K-theory and K-homology given by the Fredholm index, and the right vertical arrow is the pairing of cyclic cohomology with K-theory.

If we want to deal only with C^*-algebras, then the appropriate notion of finite summability for a Fredholm module is the following:

Definition 8. *Let $(\mathcal{H}, F)$ be a Fredholm module over a C^*-algebra A. We shall say that it is* densely p-summable *if the subalgebra : $\mathcal{A} = \{a \in A; [F, a] \in \mathcal{L}^p(\mathcal{H})\}$ is dense in A.*

We refer to [Co_{17}] for the homotopy invariance of the Chern character ch_* of Fredholm modules in the above context.

IV.2. The Dixmier Trace and the Hochschild Class of the Character

In this section we shall describe the operator theoretic tools which replace, in noncommutative geometry, the differential geometric tools having to do with symbolic calculus and integration in dimension n. These replacements also work in the case of nonintegral dimension. They involve natural ideals of operators in Hilbert space, parametrized by a real number $p \in [1, \infty]$.

We shall begin with a review of the relevant properties of eigenvalues of compact operators on a Hilbert space $\mathfrak{H}$, and of the interpolation ideals.

2.α General properties of interpolation ideals $\mathcal{L}^{(p,q)}$

Let T be a compact operator on $\mathfrak{H}$ and let $|T| = (T^*T)^{1/2}$ be its absolute value. Let $\mu_0 = \mu_0(T)$, $\mu_1 = \mu_1(T)$,... be the sequence of eigenvalues of $|T|$ arranged in decreasing order and repeated as many times as their multiplicity. Thus,

$$\mu_0(T) \geq \mu_1(T) \geq \cdots \quad \text{and} \quad \mu_n(T) \to 0 \text{ as } n \to \infty. \tag{1}$$

By the minimax principle, the value of $\mu_n(T)$ is the minimum of the norms of the restrictions of T to the orthogonal complement of an n-dimensional subspace E of $\mathfrak{H}$,

$$\mu_n(T) = \min\{\|T|E^\perp\|;\ \dim E = n\}, \tag{2}$$

and this minimum is attained by taking for E the eigenspace corresponding to the first n eigenvalues $\mu_0, \ldots, \mu_{n-1}$ of $|T|$. One can also express $\mu_n(T)$ as the distance (for the metric on $\mathcal{L}(\mathfrak{H})$ given by the operator norm) from T to the subset R_n of operators of rank less than or equal to n:

$$\mu_n(T) = d(T, R_n) = \inf\{\|T - X\|;\ X \in R_n\}. \tag{3}$$

Since the distance function to any set is Lipschitz with constant 1, we have

$$|\mu_n(T_1) - \mu_n(T_2)| \leq \|T_1 - T_2\| \tag{4}$$

for any pair T_1 and T_2 of compact operators.

The inclusion $R_n + R_m \subset R_{n+m}$ also gives

$$\mu_{n+m}(T_1 + T_2) \leq \mu_n(T_1) + \mu_m(T_2) \tag{5}$$

for any integers n and m and any pair T_1, T_2 of compact operators. Similarly, we have the submultiplicative property

$$\mu_{n+m}(T_1T_2) \leq \mu_n(T_1)\mu_m(T_2); \tag{6}$$

since $\mu_0(T) = \|T\|$, it follows , in particular, that

$$\mu_n(T_1T_2) \leq \mu_n(T_1)\|T_2\|, \qquad \mu_n(T_1T_2) \leq \|T_1\|\mu_n(T_2). \tag{7}$$

The next important property of the eigenvalues μ_n involves the partial sums

$$\sigma_N(T) = \sum_{n=0}^{N-1} \mu_n(T).$$

Every operator A of finite rank is of trace class; we denote by $\|A\|_1$ its trace norm. The relevant formula for $\sigma_N(T)$ is the following:

$$\sigma_N(T) = \sup\{\|TE\|_1;\ \dim E = N\}, \tag{8}$$

where we have used the same notation for the subspace E and the corresponding orthogonal projection. Again, the supremum is attained if we take for E the eigenspace corresponding to the first N eigenvalues of $|T|$.

As an immediate corollary of (8) it follows that the σ_N are norms:

$$\sigma_N(T_1 + T_2) \leq \sigma_N(T_1) + \sigma_N(T_2) \tag{9}$$

for any N and any pair T_1, T_2 of operators on $\mathfrak{H}$.

The final two inequalities are trickier in that they make use of the exterior algebra $\bigwedge \mathfrak{H}$ (cf. [Goh-K]). If (λ_k) is the list of eigenvalues of the compact operator T with $|\lambda_k| \geq |\lambda_{k+1}|$ for all k, then

$$\sum_{k=0}^{N-1} |\lambda_k| \leq \sum_{k=0}^{N-1} \mu_k(T) \quad \text{for every } N \in \mathbb{N}. \tag{10}$$

Finally, we have the following compatibility of σ_N with the product T_1T_2 of compact operators:

$$\sum_{k=0}^{N-1} \mu_k(T_1T_2) \leq \sum_{k=0}^{N-1} \mu_k(T_1)\mu_k(T_2). \tag{11}$$

This inequality contains as a special case the inequality of decreasing rearrangement of sequences.

Let us now describe the ideals $\mathcal{L}^{(p,q)}(\mathfrak{H})$, $p, q \in [1, \infty]$, which play a central role in our analysis of the concept of dimension, based on the results of D. Voiculescu [Vo$_1$]. These are obtained canonically from the pair of ideals $\mathcal{L}^1(\mathfrak{H}) \subset \mathcal{K}$, where $\mathcal{K}$ is the ideal of compact operators and is the largest normed ideal, while the ideal $\mathcal{L}^1(\mathfrak{H})$ of trace-class operators is the smallest. They are obtained by real interpolation theory (Appendix B), which is the most useful method for proving inequalities. We shall thus start from this point of view and proceed to a more concrete description.

Given a pair B_0, B_1 of Banach spaces that are continuously embedded in a Hausdorff topological vector space, one defines the real interpolation spaces $B_{(\alpha,\beta)}$ $(0 < \alpha < 1, 0 \leq \beta \leq 1)$ as follows: for $x \in B_0 + B_1$ and $\lambda \in]0, +\infty[$, define

$$K(\lambda, x) = \inf\{\|x_0\|_{B_0} + \lambda^{-1}\|x_1\|_{B_1};\ x_0 + x_1 = x\};$$

then the norm $\|x\|_{(\alpha,\beta)}$ is the norm in $L^q(\mathbb{R}_+^*, \frac{d\lambda}{\lambda})$, $q = 1/\beta$, of the function $\lambda \to f(\lambda) = \lambda^\alpha K(\lambda, x)$:

$$\|x\|_{(\alpha,\beta)} = \left(\int_{\mathbb{R}_+^*} |f(\lambda)|^q \frac{d\lambda}{\lambda}\right)^{1/q}$$

(with the appropriate formula for $p = \infty$).

In our case, we have $B_0 = \mathcal{K}$, $B_1 = \mathcal{L}^1(\mathfrak{H})$, and the inequality $\|x\| \le \|x\|_1$ shows that they embed continuously in B_0. Let us now estimate the function $K(\lambda, x)$, given a compact operator $x \in B_0$. By construction, K is a decreasing function of λ and, since $B_1 \subset B_0$, it is bounded by $\|x\|_{B_0}$.

Let N be an integer; if $x = x_0 + x_1$, it follows from (4) that

$$\frac{1}{N}\sigma_N(x) \le \|x_0\| + \frac{1}{N}\sigma_N(x_1),$$

thus $\frac{1}{N}\sigma_N(x) \le K(N,x)$. Conversely, let E_N be the spectral projection on the first N eigenvalues of $|x|$ and write $x = x_0 + x_1$ where, with $x = u|x|$,

$$x_0 = u\mu_N\, E_N + x(1 - E_N)$$

$$x_1 = (x - u\mu_N)E_N.$$

One has $\|x_0\| = \mu_N$, $\|x_1\|_1 = \sigma_N(x) - N\mu_N$. Thus,

$$K(N,x) \le \mu_N + \frac{1}{N}(\sigma_N(x) - N\mu_N) = \frac{1}{N}\,\sigma_N(x).$$

This proves that $K(N,x) = \frac{1}{N}\,\sigma_N(x)$. With a little more work one can show that the function $\lambda K(\lambda, x)$ is affine in the intervals $[N, N+1]$ with the values $\sigma_N(x)$ and $\sigma_{N+1}(x)$ at the endpoints.

Thus, $K(\lambda, x)$ is of the same order as $\frac{1}{N}\sigma_N(x)$, where N is the least integer greater than λ.

With $\alpha = 1/p$ and $\beta = 1/q$, let $\mathcal{L}^{(p,q)}(\mathfrak{H})$ be the interpolation space obtained as above from $B_0 = \mathcal{K}$ and $B_1 = \mathcal{L}^1(\mathfrak{H})$. Thus, for $q < \infty$, a compact operator x belongs to $\mathcal{L}^{(p,q)}(\mathfrak{H})$ if and only if

$$\sum_{N=1}^{\infty} N^{(\alpha-1)q-1}\sigma_N(x)^q < \infty,$$

whereas for $q = \infty$, x belongs to $\mathcal{L}^{(p,\infty)}(\mathfrak{H})$ if and only if $N^{(\alpha-1)}\sigma_N(x)$ is a bounded sequence.

Proposition 1. *Each $\mathcal{L}^{(p,q)}(\mathfrak{H})$ $(1 < p < \infty,\ 1 \le q \le \infty)$ is a two-sided ideal in $\mathcal{L}(\mathfrak{H})$, and the inclusion*

$$\mathcal{L}^{(p_1,q_1)} \subset \mathcal{L}^{(p_2,q_2)}$$

holds if $p_1 < p_2$, or if $p_1 = p_2$ and $q_1 \le q_2$.

This follows from (7) and the general theory of interpolation. In the corresponding interpolation square (see Figure 1), we shall be concerned primarily with the diagonal $\alpha = \beta$ and the two horizontal lines $\beta = 0$ and $\beta = 1$.

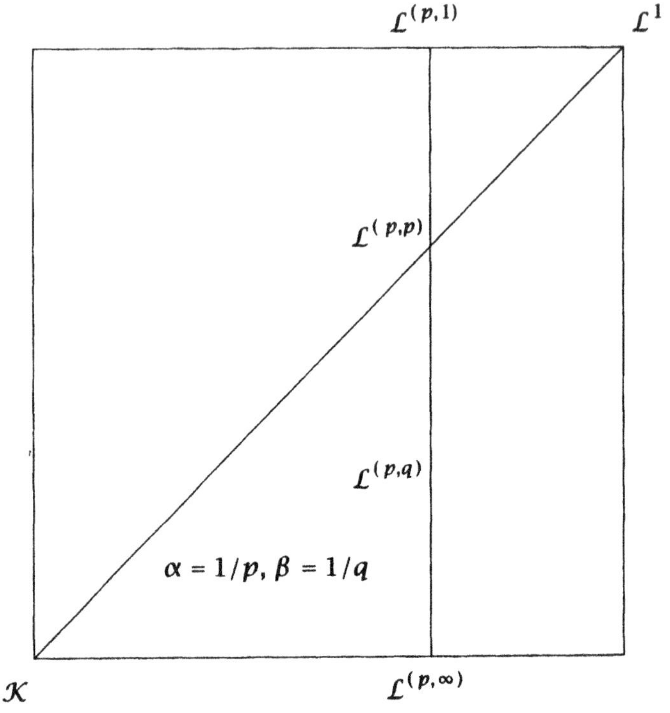

Figure IV.1. The interpolation square

As we saw above, for $1 < p < \infty$ we have

$$x \in \mathcal{L}^{(p,\infty)} \iff \sigma_N(x) = O(N^{1-\alpha}) \qquad (\alpha = 1/p),$$

which in turn is equivalent to $\mu_n(x) = O(n^{-\alpha})$, as one checks trivially. The natural norm on $\mathcal{L}^{(p,\infty)}$ is

$$\|x\|_{p,\infty} = \sup_{N\geq 1} \frac{1}{N^{1-\alpha}} \sigma_N(x).$$

The ideals $\mathcal{L}^{(p,\infty)}$ correspond to the notion of weak L^p-spaces in classical analysis, so they could be denoted $\mathcal{L}^p_{\text{weak}}$. A more compact notation is $\mathcal{L}^{p+}$.

A diagonal $\mathcal{L}^{(p,p)}(\mathfrak{H})$ is a standard Schatten ideal, and one can check, using Hardy's inequality, that the interpolation norm $\|x\|_{p,p}$ given above is equivalent (for $1 < p < \infty$) to the Schatten p-norm

$$\|x\|_p = (\text{Trace}\,(|x|^p))^{1/p} = \left(\sum_{n=0}^{\infty} \mu_n(x)^p\right)^{1/p}.$$

The ideals $\mathcal{L}^{(p,1)}(\mathfrak{H})$ play a crucial role in the work of D. Voiculescu on perturbation of the Lebesgue spectrum and quasi-central approximate units, on which we shall rely later on.

We have $x \in \mathcal{L}^{(p,1)}(\mathfrak{H})$ if and only if $\sum_{N=1}^{\infty} N^{\alpha-2}\sigma_N(x) < \infty$, or, equivalently, if and only if

$$\sum_{n=1}^{\infty} n^{\alpha-1}\mu_{n-1}(x) < \infty.$$

Moreover, it is not difficult to see that this expression defines a norm on $\mathcal{L}^{(p,1)}$ equivalent to the interpolation norm. It is sometimes convenient to replace the notation $\mathcal{L}^{(p,1)}$ by $\mathcal{L}^{p-}$. We then have

$$\mathcal{L}^{(p,1)} = \mathcal{L}^{p-} \subset \mathcal{L}^{p} \subset \mathcal{L}^{p+} = \mathcal{L}^{(p,\infty)}.$$

To conclude these generalities about the ideals $\mathcal{L}^{(p,q)}(\mathfrak{H})$, we note that the inequality (6) makes it possible to determine easily the class of a product T_1T_2 from those of T_1 and T_2. The pairing $\langle T_1, T_2\rangle = \operatorname{Trace}(T_1T_2)$ makes sense when $T_1T_2 \in \mathcal{L}^1(\mathfrak{H})$ and shows, according to general interpolation theory, that *duality* of the Banach spaces $\mathcal{L}^{(p,q)}$ is just the symmetry

$$(p,q) \to (p',q'), \quad \frac{1}{p} + \frac{1}{p'} = 1, \quad \frac{1}{q} + \frac{1}{q'} = 1$$

about the middle point $(\alpha,\beta) = (\frac{1}{2},\frac{1}{2})$ in the interpolation square, except for the following important nuance on the boundary:

$$\mathcal{L}^{(p,\infty)} \text{ is the dual of } \mathcal{L}^{(p',1)} \quad \left(1 < p < \infty, \ \frac{1}{p} + \frac{1}{p'} = 1\right).$$

However, $\mathcal{L}^{(p',1)}$ is not the dual of $\mathcal{L}^{(p,\infty)}$ but rather of the *norm-closed* ideal $\mathcal{L}_0^{(p,\infty)}$, the norm-closure of the finite rank operators in $\mathcal{L}^{(p,\infty)}$. For a compact operator x, we have

$$x \in \mathcal{L}_0^{(p,\infty)} \iff \mu_n(x) = o(n^{-\alpha}) \quad (\alpha = 1/p).$$

Thus, this situation is just a repetition of the dualities

$$(\mathcal{K})^* = \mathcal{L}^1, \quad (\mathcal{L}^1)^* = \mathcal{L}(\mathfrak{H})$$

(cf. Chapter V).

2.β The Dixmier trace

In the above interpolation square, the spaces $\mathcal{L}^{(p,q)}$ were defined only for $1 < p < \infty$ (and $1 \le q \le \infty$). Let us extend this definition to the two boundary points $p = 1$, $q = \infty$ and $p = \infty$, $q = 1$. First,

$$\mathcal{L}^{(1,\infty)}(\mathfrak{H}) = \{T \in \mathcal{K};\ \sigma_N(T) = O(\log N)\},$$

and the natural norm on $\mathcal{L}^{(1,\infty)}$ is given by

$$\|T\|_{1,\infty} = \sup_{N\ge 2} \frac{1}{\log N}\sigma_N(T);$$

as above, this is a two-sided ideal, and if $T_i \in \mathcal{L}^{(p_i,\infty)}(\mathfrak{H})$ $(i = 1,\ldots,n)$ and $\sum(1/p_i) = 1$ then $T = T_1 \cdots T_n \in \mathcal{L}^{(1,\infty)}(\mathfrak{H})$. In fact, the eigenvalues of T are of order $\mu_n = O(1/n)$ (for $p_i > 1$), by (6) and the definition of $\mathcal{L}^{(p_i,\infty)}$. $\mathcal{L}^{(1,\infty)}$ is the dual of the Macaev [Matsaev] ideal $\mathcal{L}^{(\infty,1)}(\mathfrak{H})$:

$$T \in \mathcal{L}^{(\infty,1)}(\mathfrak{H}) \iff \sum_{n\geq 1} \frac{1}{n}\mu_n(T) < \infty.$$

Also, as above, the predual of $\mathcal{L}^{(\infty,1)}$, under the pairing given by $(A,B) =$ Trace (AB), is the ideal

$$\mathcal{L}_0^{(1,\infty)}(\mathfrak{H}) = \{T \in \mathcal{K};\ \sigma_N(T) = o(\log N)\}.$$

In 1966, J. Dixmier [Di4] settled in the negative the question of the uniqueness of the trace on the type I_∞ von Neumann algebra $\mathcal{L}(\mathfrak{H})$. In doing so, he proved the existence of traces on $\mathcal{L}(\mathfrak{H})$ that vanish on the trace-class operators. By specializing his construction to the case of the sequence $a_n = \log n$, we shall get traces on the ideal $\mathcal{L}^{(1,\infty)}(\mathfrak{H})$ which will enable us to compute the Hochschild class of the character of a Fredholm module (Theorem 8 below).

Let T be a *positive* operator, $T \in \mathcal{L}^{(1,\infty)}(\mathfrak{H})$; we would like to define a functional by

$$\lim_{N\to\infty} \frac{1}{\log N} \sum_{n=0}^{N-1} \mu_n(T).$$

Since the eigenvalues $\mu_n(T)$ are unitarily invariant (i.e., $\mu_n(UTU^*) = \mu_n(T)$ for U unitary), so is the sequence $(1/\log N)\sum_{n=0}^{N-1}\mu_n(T)$. There are two problems with the above formula: its linearity and its convergence.

To handle linearity, for $T_i \geq 0$, $T_i \in \mathcal{L}^{(1,\infty)}(\mathfrak{H})$, one has to compare

$$\frac{1}{\log N} \sum_{n=0}^{N-1} \mu_n(T_1 + T_2) = \gamma_N$$

and

$$\frac{1}{\log N} \sum_{n=0}^{N-1} \mu_n(T_1) + \frac{1}{\log N} \sum_{n=0}^{N-1} \mu_n(T_2) = \alpha_N + \beta_N.$$

The inequality (9) shows that $\gamma_N \leq \alpha_N + \beta_N$. Next, by (8) we have, for $T \geq 0$,

$$\sigma_N(T) = \sup\{\text{Trace}\,(TE);\ \ \dim E = N\},$$

and it follows ([Di4]) that

$$\sigma_N(T_1) + \sigma_N(T_2) \leq \sigma_{2N}(T_1 + T_2),$$

as one sees by taking the linear span $E = E_1 \vee E_2$ of two N-dimensional subspaces E_1, E_2 of $\mathfrak{H}$. We thus have

$$\alpha_N + \beta_N \leq \left(\frac{\log 2N}{\log N}\right)\gamma_{2N}, \quad \gamma_N \leq \alpha_N + \beta_N.$$

Since $\frac{\log 2N}{\log N} \to 1$ as $N \to \infty$, we see that linearity would follow easily if we had convergence. Now, from the hypothesis $T_i \in \mathcal{L}^{(1,\infty)}(\mathfrak{H})$, the sequences α_N, β_N,and γ_N are *bounded* and thus, even without the convergence, we get a unitarily invariant positive trace on $\mathcal{L}^{(1,\infty)}(\mathfrak{H})$ for each linear form $\mathrm{Lim}_\omega = \ell$ on the space $\ell^\infty(\mathbb{N})$ of bounded sequences that satisfies the following conditions:

(α) $\mathrm{Lim}_\omega(\alpha_n) \geq 0$ if $\alpha_n \geq 0$,

(β) $\mathrm{Lim}_\omega(\alpha_n) = \lim(\alpha_n)$ if α_n is convergent,

(γ) $\mathrm{Lim}_\omega(\alpha_1, \alpha_1, \alpha_2, \alpha_2, \alpha_3, \alpha_3, \ldots) = \mathrm{Lim}_\omega(\alpha_n)$.

Condition (γ) is a crucial condition of *scale invariance*; to see how to handle it, notice first that the parameter N in sequences such as α_N is a special case of the continuous parameter $\lambda \in \mathbb{R}_+^*$ in the construction of the interpolation spaces. To any bounded sequence $(\alpha_N)_{N\in\mathbb{N}}$ we thus assign the bounded function $f_\alpha(\lambda)$ given by

$$f_\alpha(\lambda) = \alpha_N \text{ for } \lambda \in]N-1, N].$$

We can now replace f by its Cesàro mean with respect to the multiplicative group $\mathbb{R}_+^*$, with Haar measure $d\lambda/\lambda$,

$$M(f)(\lambda) = \frac{1}{\log \lambda} \int_1^\lambda f(u) \frac{du}{u}.$$

By construction, the Cesàro mean satisfies the following scale invariance, for bounded functions f,

$$|M(\theta_\mu(f))(\lambda) - M(f)(\lambda)| \to 0 \text{ as } \lambda \to \infty,$$

where $\mu > 0$ and $\theta_\mu(f)(\lambda) = f(\lambda\mu)$ $\forall \lambda \in \mathbb{R}_+^*$.

If $\tilde{\alpha} = (\alpha_1, \alpha_1, \alpha_2, \alpha_2, \ldots)$ then

$$f_{\tilde{\alpha}} = \theta_{1/2}(f_\alpha);$$

it follows that for any linear form L on $C_b(\mathbb{R}_+^*)$ satisfying α) and β), the composition

$$\alpha \to L(M(f_\alpha)) = \ell(\alpha)$$

satisfies the above conditions α), β) and γ).

We thus take a positive linear form L on the vector space of bounded continuous functions on $\mathbb{R}_+^*$, such that $L(1) = 1$, and which is zero on the subspace $C_0(\mathbb{R}_+^*)$ of functions vanishing at ∞.

Let us postpone the discussion of the choice of L, and state the properties of Tr_ω that are true for any such choice. Thus, we denote $\ell(\alpha) = L(M(f_\alpha))$ by $\mathrm{Lim}_\omega(\alpha)$, and make the following definition:

Definition 2. For $T \geq 0$, $T \in \mathcal{L}^{(1,\infty)}(\mathfrak{H})$, we set

$$\mathrm{Tr}_\omega(T) = \mathrm{Lim}_\omega \frac{1}{\log N} \sum_{n=0}^{N-1} \mu_n(T).$$

The above inequalities show that Tr_ω is additive:

$$\mathrm{Tr}_\omega(T_1+T_2)=\mathrm{Tr}_\omega(T_1)+\mathrm{Tr}_\omega(T_2)\qquad \forall T_i\geq 0,\ T_i\in\mathcal{L}^{(1,\infty)}(\mathfrak{H}).$$

Thus, Tr_ω extends uniquely by linearity to the entire ideal $\mathcal{L}^{(1,\infty)}(\mathfrak{H})$ and has the following properties:

Proposition 3. (a) *If* $T\geq 0$ *then* $\mathrm{Tr}_\omega(T)\geq 0$.

(b) *If* S *is any bounded operator and* $T\in\mathcal{L}^{(1,\infty)}(\mathfrak{H})$, *then* $\mathrm{Tr}_\omega(ST)=\mathrm{Tr}_\omega(TS)$.

(c) $\mathrm{Tr}_\omega(T)$ *is independent of the choice of the inner product on* $\mathfrak{H}$, *i.e., it depends only on the Hilbert space* $\mathfrak{H}$ *as a topological vector space.*

(d) Tr_ω *vanishes on the ideal* $\mathcal{L}_0^{(1,\infty)}(\mathfrak{H})$, *which is the closure, for the* $\|\ \|_{1,\infty}$-*norm, of the ideal of finite-rank operators.*

Property (c) follows from (b) since, for S bounded and invertible,

$$\mathrm{Tr}_\omega(STS^{-1})=\mathrm{Tr}_\omega(T)\qquad \forall T\in\mathcal{L}^{(1,\infty)}(\mathfrak{H}).$$

Property (d) has an obvious corollary:

$$\mathrm{Tr}_\omega(T)=0\qquad \forall T\in\mathcal{L}^1(\mathfrak{H}).$$

It is this vanishing property of the Dixmier trace that makes it possible to transform equalities of operators modulo the ideal $\mathcal{L}^1$ into actual numerical identities. We shall now discuss the problem of convergence of the sequence $\frac{1}{\log N}\sum_{n=0}^{N-1}\mu_n(T)$.

For a positive operator $T\in\mathcal{L}^{(1,\infty)}(\mathfrak{H})$, the complex powers T^s ($s\in\mathbb{C}$, $\mathrm{Re}(s)>1$) make sense and are of trace class, so that the equality

$$\zeta(s)=\mathrm{Trace}\,(T^s)=\sum_{n=0}^{\infty}\mu_n(T)^s$$

defines a holomorphic function in the half-plane $\mathrm{Re}(s)>1$. Now, the Tauberian theorem of Hardy and Littlewood [Har] can be stated as follows:

Proposition 4. *For* $T\geq 0$, $T\in\mathcal{L}^{(1,\infty)}(\mathfrak{H})$, *the following two conditions are equivalent:*

(1) $(s-1)\zeta(s)\to L$ *as* $s\to 1+$;

(2) $(1/\log N)\sum_{n=0}^{N-1}\mu_n\to L$ *as* $N\to\infty$.

Under these conditions, the value of $\mathrm{Tr}_\omega(T)$ is, of course, independent of ω, and if $\zeta(s)$ has a simple pole at $s=1$, this value is just the residue of ζ at $s=1$.

As an example of a rather general situation in which the above type of convergence holds, let us make the connection between the Dixmier trace Tr_ω

and the notion of residue for pseudo-differential operators, introduced by Manin [Mani$_1$], Wodzicki [Wo$_2$] and Guillemin [Gu].

Proposition 5. *Let M be an n-dimensional compact manifold and let $T \in OP^{-n}(M,E)$ be a pseudo-differential operator of order $-n$ acting on sections of a complex vector bundle E on M. Then:*

(1) *The corresponding operator T on $\mathfrak{H} = L^2(M,E)$ belongs to the ideal $\mathcal{L}^{(1,\infty)}(\mathfrak{H})$.*

(2) *The Dixmier trace $\mathrm{Tr}_\omega(T)$ is independent of ω and is equal to the residue $\mathrm{Res}(T)$.*

Let us recall that the Wodzicki residue $\mathrm{Res}(T)$ is given by a completely explicit formula involving the principal symbol $\sigma_{-n}(T) = \sigma(T)$. The latter is a homogeneous function of degree $-n$ on the cotangent bundle T^*M of M; consequently the following integral is independent of the choice (using a metric on M) of the unit-sphere bundle $S^*(M) \subset T^*(M)$ with its induced volume element:

$$\mathrm{Res}\,(T) = \frac{1}{n(2\pi)^n} \int_{S^*M} \mathrm{trace}_E(\sigma)ds.$$

The equality of Proposition 5 can be proved using Proposition 4, or using the general properties of the Dixmier trace ([Co$_{16}$]).

It is important for our later purposes that the Wodzicki residue continues to make sense for pseudo-differential operators of arbitrary order [Wo$_2$]. It is the unique trace on the algebra of pseudo-differential operators which extends the Dixmier trace on operators of order $\leq -n$. It is given by the same formula applied to the (coordinate dependent) symbol of order $-n$ of T. It is a quite remarkable result, due to Wodzicki [Wo$_2$], that the integral over S^*M of this coordinate dependent symbol is independent of any choice and defines a trace. We shall use this in Section 4.

In general the value of $\mathrm{Tr}_\omega(T)$ for $T \in \mathcal{L}^{(1,\infty)}$ does depend on the choice of the limiting procedure ω. This feature of the Dixmier trace is tied up with its non-normality (cf. [Di$_4$]). As we shall see in Section 3, this non-normality is in fact a positive feature of Tr_ω since it allows one, in the context of Julia sets, to pass, using Tr_ω, from the harmonic measure to the Hausdorff measure, which does not belong to the same measure class.

The dependence of $\mathrm{Tr}_\omega(T)$ on the choice of ω is governed by the following straightforward:

Proposition 6. a) *Let $T \geq 0$, $T \in \mathcal{L}^{(1,\infty)}$. Then $\mathrm{Tr}_\omega(T)$ is independent of ω iff the Cesàro means $M(\lambda)$ of the sequence $\frac{1}{\log N} \sum_{n=0}^{N-1} \mu_n(T)$ are convergent for $\lambda \to \infty$.*

b) *Let $\mathcal{M} = \{T \in \mathcal{L}^{(1,\infty)};\ \mathrm{Tr}_\omega(T)$ independent of $\omega\}$. Then $\mathcal{M}$ is a linear space invariant under conjugation by invertible operators on $\mathcal{H}$.*

c) *$\mathcal{M}$ contains $\mathcal{L}_0^{(1,\infty)}$ and is closed in the $(1,\infty)$-norm.*

In a) the Cesàro means are defined as above by

$$M(\lambda) = \frac{1}{\log \lambda} \int_1^\lambda f(u) \, \frac{du}{u}$$

where f is constant, equal to $\frac{1}{\log N} \sum_{n=0}^{N-1} \mu_n(T)$, on $]N-1, N]$. Of course, if a) holds then the common value of $\mathrm{Tr}_\omega(T)$ is the limit of $M(\lambda)$ when $\lambda \to \infty$.

Definition 7. *Let $T \in \mathcal{L}^{(1,\infty)}$. We shall say that T is* measurable *if $\mathrm{Tr}_\omega(T)$ is independent of ω.*

By 6b), measurability is additive, independent of the choice of inner product in Hilbert space, and invariant under conjugation by invertible elements. We shall prove in many examples below the measurability of all operators aT, $a \in A$, where A is a C^*-algebra and $T \geq 0$, $T \in \mathcal{L}^{(1,\infty)}$, by studying the positive linear functionals on A given for any ω by

$$\varphi_\omega(a) = \mathrm{Tr}_\omega(aT) \qquad \forall a \in A$$

and showing that φ_ω is independent of ω.

2.γ The residue formula for the Hochschild class of the character of Fredholm modules

The Dixmier trace Tr_ω constructed above measures, for $T \geq 0$, the logarithmic divergence of the trace of T. Other divergences such as $(\log N)^\alpha$ would yield corresponding traces. We shall see in this section that, among them, the Dixmier trace is singled out by its cohomological significance. It computes the Hochschild class $I(\tau_n) \in H^n(\mathcal{A}, \mathcal{A}^*)$ of the n-dimensional character $\tau_n \in HC^n(\mathcal{A})$ of a Fredholm module over an algebra $\mathcal{A}$ (cf. Definition 1.3). We shall need a stronger hypothesis on $(\mathcal{H}, F)$ than just its $(n+1)$-summability. Our formula will assume the existence of an unbounded selfadjoint operator D such that $FD > 0$, that $[D, a]$ is bounded for any $a \in \mathcal{A}$, and that $D^{-1} \in \mathcal{L}^{(n,\infty)}$. Choosing D is equivalent to choosing $\rho = |D|$. Such a choice is similar to the choice of a *metric* in Riemannian geometry, and will play a central role in Chapter VI. We fix an integer $n \geq 1$, and the constant λ_n as in Section 1 Definition 3.

Theorem 8. *Let $(\mathcal{H}, F)$ be a Fredholm module over an involutive algebra $\mathcal{A}$. Let D be an unbounded selfadjoint operator in $\mathcal{H}$ such that $D^{-1} \in \mathcal{L}^{(n,\infty)}$, $\mathrm{Sign}\, D = F$, and such that for any $a \in \mathcal{A}$ the operators a and $[D, a]$ are in the domain of all powers of the derivation δ, given by $\delta(x) = [|D|, x]$.*

1) *A Hochschild n-cocycle φ_ω on $\mathcal{A}$ is defined by*

$$\varphi_\omega(a^0, a^1, \ldots, a^n) = \lambda_n \, Tr_\omega\left(a^0[D, a^1] \cdots [D, a^n]|D|^{-n}\right) \qquad \forall a_j \in \mathcal{A}$$

(with γa^0 instead of a^0 in the even case).

2) *For every n-dimensional Hochschild cycle $c \in Z_n(\mathcal{A},\mathcal{A})$, $\langle \varphi_\omega, c\rangle = \langle \tau_n, c\rangle$, where $\tau_n \in HC^n(\mathcal{A})$ is the Chern character of $(\mathcal{H}, F)$ (Definition 1.3).*

By construction, the formula for φ_ω is scale invariant, i.e. it remains unchanged if we replace D by λD for $\lambda \in \mathbb{R}_+^*$. The explicit cocycle φ_ω depends, in general, on the choice of the limiting procedure ω involved in the definition of the Dixmier trace, but the equality 2) shows that its pairing with Hochschild cycles c is independent of ω. We thus obtain operators T_c of the form

$$T_c = \sum a^0[D,a^1]\cdots[D,a^n]|D|^{-n}$$

which are *measurable* in the sense of section β).

When the Hochschild cohomology is Hausdorff it can then be viewed as the dual of Hochschild homology. In that case, Condition 2) just means that φ_ω has the same Hochschild class as τ_n. Now by Proposition 1.2 the cyclic cocycle characters of $(\mathcal{H}, F)$ are related by the periodicity operator S. Thus, the following corollary of the long exact sequence of cyclic cohomology (Section III.1) shows that the Hochschild class $I(\tau_n)$ is the obstruction to a better summability of $(\mathcal{H}, F)$:

Proposition 9. *Let $\mathcal{A}$ be an algebra, $\tau \in HC^n(\mathcal{A})$ a cyclic cohomology class. Then τ belongs to the image $S(HC^{n-2}(\mathcal{A}))$ if and only if the Hochschild cohomology class $I(\tau) \in H^n(\mathcal{A},\mathcal{A}^*)$ is equal to 0.*

In particular, Theorem 8 implies nonvanishing of residues when the cohomological dimension of $\mathrm{ch}_*(\mathcal{H}, F)$ is not lower than n:

Corollary 10. *With the hypothesis of Theorem 8 and if the Hochschild class of $\mathrm{ch}_*(\mathcal{H}, F)$ pairs nontrivially with $H_n(\mathcal{A},\mathcal{A})$ one has*

$$\mathrm{Tr}_\omega(|D|^{-n}) \neq 0.$$

In other words the residue of the function $\zeta(s) = \mathrm{Trace}(|D|^{-s})$ at $s = n$ cannot vanish.

With the hypothesis of the corollary on the cohomological dimension of $\mathrm{ch}_*(\mathcal{H}, F)$ we see that it is not possible to find an operator D as in Theorem 8 and such that $D^{-1} \in \mathcal{L}_0^{(n,\infty)}$. In particular, Theorem 8 has no analogue for the smaller ideals $\mathcal{L}^{(n,q)}$, $q < \infty$.

It is important, in general, to remove the hypothesis of invertibility of the operator D. Let us recall that the resolvent $(D-\lambda)^{-1}$ is bounded for any $\lambda \notin \mathrm{Sp}(D)$, in particular $\lambda \notin \mathbb{R}$, and that for any two-sided ideal J of operators the condition

$$(D-\lambda)^{-1} \in J$$

is independent of the choice of $\lambda \notin \mathrm{Sp}\ D$.

Definition 11. *A K-cycle $(\mathcal{H}, D)$ over an involutive algebra $\mathcal{A}$ is given by a $*$-representation of $\mathcal{A}$ in a Hilbert space $\mathcal{H}$ together with an unbounded selfadjoint operator D with compact resolvent, such that $[D, a]$ is bounded for any $a \in \mathcal{A}$.*

Given a K-cycle $(\mathcal{H}, D)$, the kernel of D is finite dimensional and Theorem 8 continues to hold if we define $|D|^{-n}$ as 0 on $\ker D$. We associate, *canonically*, a Fredholm module $(\mathcal{H}', F)$ to the K-cycle $(\mathcal{H}, D)$ in the following way:

α) $\mathcal{H}' = \mathcal{H} \oplus \ker D$

β) $a(\xi, \eta) = (a\xi, 0) \qquad \forall a \in \mathcal{A}\,,\ \xi \in \mathcal{H}\,,\ \eta \in \ker D.$

γ) $F = (\operatorname{Sign} D) \oplus F_1$

where Sign D is the partial isometry of the polar decomposition of D, and where F_1 is the partial isometry which exchanges the two copies of $\ker D$. In the even case, i.e. when $\mathcal{H}$ is $\mathbb{Z}/2$-graded by γ commuting with $\mathcal{A}$ and anticommuting with D, one uses in α) the opposite $\mathbb{Z}/2$-grading on $\ker D$.

We shall refer to $(\mathcal{H}', F)$ as the Fredholm module associated to the K-cycle $(\mathcal{H}, D)$.

Theorem 8 will play a decisive role in Chapter VI. As an easy example where it applies, consider the K-cycle given by the Dirac operator on a smooth compact Spin manifold (cf. Chapter VI.1) and compute φ_ω using Proposition 2.5. A straightforward calculation gives, for all $f^j \in C^\infty$

$$\varphi_\omega(f^0, \ldots, f^n) = (2\pi i)^{-n/2}(-1)^{n(n-1)/2} \int f^0\, df^1 \wedge df^2 \wedge \cdots \wedge df^n.$$

2.δ Growth of algebras and degree of summability of K-cycles

In this section we shall show, using results of D. Voiculescu [Vo$_1$], how the existence of a finitely summable K-cycle $(\mathfrak{H}, D)$ over an algebra $\mathcal{A}$ implies, through the condition of boundedness of the commutators

$$\|[D, a]\| < \infty \qquad \forall a \in \mathcal{A},$$

restrictions on the growth of $\mathcal{A}$.

In noncommutative algebraic geometry [Mani$_2$], one does impose polynomial growth conditions, where the degree of the polynomial is related to dimension, in order to get tractable classification problems. What we show here is that the very existence of a (d, ∞)-summable K-cycle is a similar hypothesis, the d being related to dimension.

In his work on norm ideal perturbations of Hilbert space operators ([Vo$_1$]), D. Voiculescu introduced, for any normed ideal J in $\mathcal{L}(\mathfrak{H})$ with symmetric norm $\|\ \|_J$, the following quantity that measures, for a finite subset $X \subset \mathcal{L}(\mathfrak{H})$, the obstruction to finding an approximate unit quasi-central relative to X:

$$k_J(X) = \liminf_{A \in R_1^+;\, A \to 1} \|[A, X]\|_J,$$

where R_1^+ is the unit interval $0 \le A \le 1$ in the space of finite-rank operators, and

$$\|[A,X]\|_J = \sup_{T\in X} \|[A,T]\|_J.$$

Moreover, Voiculescu showed that the interpolation ideals $\mathcal{L}^{(p,1)}$, with their norms

$$\|T\|_{(p,1)} = \sum_{n=1}^{\infty} n^{(\frac{1}{p}-1)} \mu_n(T),$$

are singled out as measuring the absolutely continuous part of the spectrum, by the following remarkable result:

Theorem 12. [Vo$_1$] *Let $T_1,\ldots,T_n$ be n commuting selfadjoint operators in the Hilbert space $\mathfrak{H}$. Let $E_{ac} \subset \mathbb{R}^n$ be the absolutely continuous part of their spectral measure and let $n(x)$, $x \in E_{ac}$, be the multiplicity function. Then*

$$\int_{E_{ac}} n(x) d^n x = \alpha_n (k_J(\{T_i\}))^n,$$

where J is the normed ideal $\mathcal{L}^{(n,1)}$ and where α_n is an absolute constant, $\alpha_n \in]0,\infty[$.

This result shows that the correct ideals for dealing with the absolutely continuous spectrum are the ideals $\mathcal{L}^{(p,1)}$, but our concept of dimension discussed above involved the Dixmier trace and the *other* ideals $\mathcal{L}^{(p,\infty)}$. Even though there is a duality between $\mathcal{L}^{(p,1)}$ and $\mathcal{L}^{(q,\infty)}$, this involves p and q with $\frac{1}{p} + \frac{1}{q} = 1$ so that it is a priori unclear how to relate our hypothesis $|D|^{-1} \in \mathcal{L}^{(p,\infty)}$ with the above result of Voiculescu. This is accomplished by the next lemma.

Lemma 13. *Let $f \in C_c^\infty(\mathbb{R})$ be a smooth even function on $\mathbb{R}$ with compact support, and let D be a selfadjoint, invertible, unbounded operator on a Hilbert space $\mathfrak{H}$. Let $p \in]1,\infty[$. Then there exists a constant $C < \infty$ depending only on p and f such that, for $\epsilon > 0$,*

$$\|[f(\epsilon D),a]\|_{(p,1)} \le C\|[D,a]\| \; \|D^{-1}\|_{(p,\infty)}$$

for every bounded $a \in \mathcal{L}(\mathfrak{H})$ such that $[D,a]$ is bounded.

We shall sketch the proof of this lemma, just to see how one passes from the (p,∞)-norm to the $(p,1)$-norm.

First, it is standard that the operator norm of the commutator $[f(D),a]$ can be estimated, using the L^1-norm of the Fourier transform of f' and the equality

$$[e^{i\lambda D},a] = \int_0^1 e^{i\lambda s D}[\lambda D,a]e^{i\lambda(1-s)D} ds,$$

so that

$$\|[e^{i\lambda D},a]\| \le \lambda\|[D,a]\|, \quad \|[f(D),a]\| \le \|\hat{f}'\|_1 \|[D,a]\|.$$

Next, the $\mathcal{L}^1$-norm of $[f(\epsilon D), a]$ can also be bounded, since one controls the rank of $f(\epsilon D)$, as follows: if $\mathrm{supp}(f) \subset [-k, k]$, then the rank of $f(\epsilon D)$ is bounded above by the number of eigenvalues of $|D|^{-1}$ larger than ϵk^{-1}; but, by definition, these eigenvalues μ_n satisfy

$$\sum_{n=0}^{N} \mu_n \leq \|D^{-1}\|_{(p,\infty)} \sum_{n=1}^{N} n^{-1/p},$$

so that $\mu_N \leq \frac{1}{N} \sum_{n=0}^{N} \mu_n \leq C_p N^{-1/p}$, and $\mu_N \geq \epsilon k^{-1} \Rightarrow N \leq (C_p)^p k^p \epsilon^{-p}$. Thus, we get the inequality

$$\|[f(\epsilon D), a]\|_1 \leq 2(C_p)^p k^p \epsilon^{-p} \|[f(\epsilon D), a]\|.$$

We could conclude directly using the definition of the norm $(p, 1)$, but we prefer to use the general interpolation inequality

$$\|T\|_{(p,1)} \leq C'_p \|T\|_1^{1/p} \|T\|_\infty^{1-1/p},$$

which gives us

$$\begin{aligned} \|[f(\epsilon D), a]\|_{(p,1)} &\leq C'_p 2^{1/p} C_p k \epsilon^{-1} \|[f(\epsilon D), a]\| \\ &\leq 2^{1/p} C''_p k \|\hat{f}'\|_1 \|[D, a]\| \leq C_{f,p} \|[D, a]\|. \end{aligned}$$

As an immediate corollary of the proof and of the definition of the Dixmier trace, for each $p \in]1, \infty[$ and $f \in C_c^\infty(\mathbb{R})$ we get a constant $C < \infty$ such that

$$\liminf_{\epsilon \to 0} \|[f(\epsilon D), a]\|_{(p,1)} \leq C \|[D, a]\| (\mathrm{Tr}_\omega(|D|^{-p}))^{1/p}.$$

If $f(1) = 1$ and $0 \leq f \leq 1$, then $f(\epsilon D) \in R_1^+$ for all $\epsilon > 0$ and $f(\epsilon D) \to 1$ weakly as $\epsilon \to 0$; thus:

Proposition 14. *Let $p \in]1, \infty[$. There exists a constant $C_p < \infty$ such that for every subset X of $\mathcal{L}(\mathfrak{H})$, we have, for $J = \mathcal{L}^{(p,1)}$,*

$$k_J(X) \leq C_p \left(\sup_{T \in X} \|[D, T]\| \right) (\mathrm{Tr}_\omega(|D|^{-p}))^{1/p}.$$

If we combine this estimate with Voiculescu's theorem (Theorem 12), we get a handle on the absolutely continuous joint spectrum of elements in any abelian $*$-subalgebra of the algebra

$$\mathcal{A}_D = \{T \in \mathcal{L}(\mathfrak{H});\ [D, T] \text{ is bounded}\}.$$

Thus, for instance, we can state

Proposition 15. *Let $p \in]1, \infty[$ and let $(\mathfrak{H}, D)$ be a (p, ∞)-summable K-cycle over the $*$-algebra $\mathcal{A}$.*

(a) *The equality* $\tau(a) = \mathrm{Tr}_\omega(a|D|^{-p})$ *defines a positive trace on the algebra* $\mathcal{A}$; *moreover, this trace is nonzero if*

$$k_J(\mathcal{A}) \neq 0, \quad J = \mathcal{L}^{(p,1)}(\mathfrak{H}).$$

(b) *Let* p *be an integer and let* $a_1, \ldots, a_p \in \mathcal{A}$ *be commuting selfadjoint elements of* $\mathcal{A}$. *Then, the absolutely continuous part of their spectral measure, i.e., their Lebesgue spectrum*

$$\mu_{\mathrm{ac}}(f) = \int_{E_{\mathrm{ac}}} f(x)n(x)\mathrm{d}^p x,$$

is absolutely continuous with respect to the measure τ:

$$\tau(f) = \tau(f(a_1, \ldots, a_p)) = \mathrm{Tr}_\omega(f|D|^{-p}) \qquad \forall f \in C_c^\infty(\mathbb{R}^p).$$

The case of non-integral p and the relation between the Dixmier trace and the Hausdorff measure will be discussed now, in Section 3.

IV.3. Quantized Calculus in One Variable and Fractal Sets

The theory of distributions works well for a number of problems involving non-smooth functions, such as those generated by the variational calculus. It is, however, notoriously incompatible with *products*, i.e. products of distributions only make sense in rare cases. The reason for this is simple, since the notion of distribution on a manifold V is invariant under any continuous *linear* transformation of $C^\infty(V)$ while such linear transformations affect arbitrarily the *algebra* structure of $C^\infty(V)$. In the quantized calculus which we propose, the differential of a function f is an *operator* in Hilbert space, namely

$$df = [F, f].$$

In particular, this operator can undergo all the operations of the functional calculus such as, for instance,

$$T \to |T|^p$$

where $|T|$ is the absolute value of the operator T and p a positive real number which is not an integer.

This gives meaning to an expression such as $|df|^p$ even when f is a non-differentiable function. We shall show the power of this method by giving the formula for the Hausdorff measure Λ_p on the fractals which appear in the theory of uniformization of pairs of Riemann surfaces with the same genus:

$$\int f\, d\Lambda_p = \mathrm{Tr}_\omega(f(Z)|dZ|^p),$$

for any continuous function f on C, and with $Z : \mathbb{S}^1 \to C$ the boundary value of a conformal equivalence.

3.α Quantized calculus in one variable

Let us first quantize the calculus in one real variable, in a translation and scale invariant manner.

Our algebra $\mathcal{A}$ is the algebra of functions $f(s)$ of one real variable $s \in \mathbb{R}$; we do not specify their regularity at the moment. To quantize the calculus we need a Fredholm module over $\mathcal{A}$, i.e. a representation of $\mathcal{A}$ in a Hilbert space $\mathcal{H}$ and an operator F as in Definition 0.1. The representation of $\mathcal{A}$ is given by a measure class on $\mathbb{R}$ and a multiplicity function. Since we want the calculus to be translation invariant, the measure class is necessarily the Lebesgue class and the multiplicity is a constant. We shall take it equal to one; the more general case does not lead to anything new. Thus, so far, we have functions on $\mathbb{R}$ acting by multiplication operators on the Hilbert space $L^2(\mathbb{R})$:

$$\mathcal{H} = L^2(\mathbb{R}) \,,\; (f\xi)(s) = f(s)\,\xi(s) \quad \forall s \in \mathbb{R} \,,\; \xi \in L^2(\mathbb{R}). \tag{1}$$

Any measurable bounded function $f \in L^\infty(\mathbb{R})$ defines a bounded operator in $\mathcal{H}$ by the equality (1).

Since we want the calculus to be translation invariant, the operator F must commute with translations, and hence must be given by a convolution operator. We shall also require that it commute with dilatations, $s \to \lambda s$ with $\lambda > 0$, and it then follows easily (cf., for instance, [Stei]) that the only nontrivial choice of F, with $F^2 = 1$, is the Hilbert transform, given by

$$(F\xi)(s) = \frac{1}{\pi i} \int \frac{\xi(t)}{s-t}\, dt \tag{2}$$

where the integral is taken for $|s-t| > \varepsilon$ and then one lets $\varepsilon \to 0$.

The quantum differential $df = [F,f]$ of $f \in L^\infty(\mathbb{R})$ has a very simple expression; it is the operator on $L^2(\mathbb{R})$ associated by the equality

$$T\xi(s) = \int k(s,t)\, \xi(t)\, dt \tag{3}$$

to the following kernel $k(s,t)$ defined for $s,t \in \mathbb{R}$

$$k(s,t) = \frac{f(s) - f(t)}{s-t} \tag{4}$$

(up to the factor $\frac{1}{\pi i}$ which we ignore).

Note that the group $SL(2,\mathbb{R})$ acts by automorphisms of the Fredholm module $(\mathcal{H},F)$, generalising the above invariance by translations and homotheties. Indeed, given $g = \begin{bmatrix} a & b \\ c & d \end{bmatrix} \in SL(2,\mathbb{R})$ (so that $a,b,c,d \in \mathbb{R}, ad-bc = 1$), we let g^{-1} act on $L^2(\mathbb{R})$ as the unitary operator for which

$$(g^{-1}\,\xi)(s) = \xi\left(\frac{as+b}{cs+d}\right)(cs+d)^{-1} \quad \forall \xi \in L^2(\mathbb{R}) \,,\; s \in \mathbb{R}. \tag{5}$$

One checks that this representation of $SL(2,\mathbb{R})$ commutes with F. Its restriction to $\{\xi; F\xi = \pm\xi\}$ are the two mock discrete series. The corresponding automorphisms of the algebra of functions on $\mathbb{R}$ are given by

$$(g^{-1}\,f)(s) = f\left(\frac{as+b}{cs+d}\right) \quad \forall f \in L^\infty(\mathbb{R}) \,,\; s \in \mathbb{R}. \tag{6}$$

Using an arbitrary fractional linear transformation from the line $\mathbb{R}$ to the unit circle $\mathbb{S}^1 = \{z \in \mathbb{C}; |z| = 1\}$, such as

$$s \in \mathbb{R} \to \frac{s-i}{s+i} \in \mathbb{S}^1 \tag{7}$$

we can transport the above Fredholm module to functions on the circle $\mathbb{S}^1$. It is described as follows:

(8) $\mathcal{H} = L^2(\mathbb{S}^1, d\theta)$ with functions on $\mathbb{S}^1$ acting by multiplication (as in (1))

$$F = 2P - 1 \text{ where } P \text{ is the orthogonal projection on}$$
$$H^2(\mathbb{S}^1) = \{\xi \in L^2 \,;\, \hat{\xi}(n) = 0 \quad \forall n < 0\}$$

where $\hat{\xi}$ is the Fourier transform of ξ.

The two situations, with $\mathbb{R}$ or $\mathbb{S}^1$, are unitarily equivalent provided we take in both cases the von Neumann algebras of all measurable bounded functions. We shall keep both. Our first and easy task will be to quote a number of well known results of analysis (cf. [Stei] [Powe$_2$] [Pel]) allowing one to control the order of $df = [F, f]$ in terms of the regularity of the function $f \in L^\infty$.

The *strongest* condition we can impose on df is to belong to the smallest nontrivial ideal of operators, namely the ideal R of finite-rank operators. The necessary and sufficient condition for this to hold is a result of Kronecker (cf. [Powe$_2$]):

Proposition 1. *Let $f \in L^\infty$, then $df \in R \Longleftrightarrow f$ is a rational fraction.*

This result holds for both $\mathbb{R}$ and $\mathbb{S}^1$; in both cases the rational fraction $\frac{P(s)}{Q(s)}$ is equal a.e. to f and has no pole on $\mathbb{R}$ (resp. $\mathbb{S}^1$).

The *weakest* condition we can impose on df is to be a compact operator. In fact, we should restrict to the subalgebra of L^∞ determined by this condition if we want to comply with condition 2 of Definition 0.1.

What this means is known (cf. [Powe$_2$]) and easy to formulate for $\mathbb{S}^1$. It involves the *mean oscillation* of the function f. Let us recall that, given any interval I of $\mathbb{S}^1$, one lets $I(f)$ be the mean $\frac{1}{|I|}\int_I f\, dx$ of f on I and one defines for $a > 0$ the mean oscillation of f by

$$M_a(f) = \sup_{|I| \le a} \frac{1}{|I|} \int_I |f - I(f)|.$$

A function is said to have bounded mean oscillation (BMO) if the $M_a(f)$ are bounded independently of a. This is, of course, true if $f \in L^\infty(\mathbb{S}^1)$. A function f is said to have *vanishing* mean oscillation (VMO) if $M_a(f) \to 0$ when $a \to 0$. Let us then state a result of Fefferman and Sarason (cf. [Powe$_2$]).

Proposition 2. *If $f \in L^\infty(\mathbb{S}^1)$, then $[F, f] \in \mathcal{K} \Longleftrightarrow f \in$ VMO.*

Every continuous function $f \in C(\mathbb{S}^1)$ belongs to VMO but the algebra VMO $\cap L^\infty$ is strictly larger than $C(\mathbb{S}^1)$. Its elements are called quasi-continuous functions. For instance, the boundary values of any bounded univalent holomorphic function $f \in H^\infty(\mathbb{S}^1)$ belong to VMO but not necessarily to $C(\mathbb{S}^1)$.

The next question is how to characterize the functions $f \in L^\infty$ for which

$$[F, f] \in \mathcal{L}^p$$

for a given real number $p \in [1, \infty[$.

This question has a remarkably nice answer, due to V.V. Peller [Pel], in terms of the Besov spaces $B_p^{1/p}$ of measurable functions.

Definition 3. *Let $p \in [1, \infty[$. Then the Besov space $B_p^{1/p}$ is the space of measurable functions f on $\mathbb{S}^1$ such that*

$$\int\int |f(x+t) - 2f(x) + f(x-t)|^p \, t^{-2} \, dx \, dt < \infty.$$

For $p > 1$ this condition is equivalent to

$$\int\int |f(x+t) - f(x)|^p \, t^{-2} \, dx \, dt < \infty$$

and the corresponding norms are equivalent. For $p = 2$ one recovers the Sobolev space of Fourier series,

$$f(t) = \sum_{n \in \mathbb{Z}} a_n \exp(2\pi i n t), \quad \sum |n| \, |a_n|^2 < \infty.$$

The result of V.V. Peller is then the following:

Theorem 4. [Pel] *Let $f \in L^\infty(\mathbb{S}^1)$, $p \in [1, \infty[$; then $[F, f] \in \mathcal{L}^p \Longleftrightarrow f \in B_p^{1/p}$.*

There is a similar result for $p < 1$ due independently to Semmes and to Peller.

For $f \in L^\infty(\mathbb{S}^1)$, the operator $df = [F, f]$ anticommutes with F by construction, and is hence given by an off-diagonal 2×2 matrix in the decomposition of $L^2(\mathbb{S}^1)$ as a direct sum of eigenspaces of F. This 2×2 matrix is *lower triangular*, i.e. $(1-P)fP = 0$ with $P = \frac{1+F}{2}$, if $f \in H^\infty(\mathbb{S}^1)$, i.e. f is the boundary value of a holomorphic function in the disk.

$$\text{If } f \in L^\infty(\mathbb{S}^1), \text{ then } f \in H^\infty(\mathbb{S}^1) \Longleftrightarrow (1-P)fP = 0. \tag{9}$$

In particular, $df_1 \, df_2 = 0$ for any $f_1, f_2 \in H^\infty(\mathbb{S}^1)$.

We shall end this section by giving known reformulations of the Besov spaces $A_p^{1/p} = \{f \in B_p^{1/p}; \hat{f}(n) = 0 \text{ for } n < 0\}$. Given $f \in A_p^{1/p}$ we also denote by f the holomorphic function inside the unit disk D with f as boundary values.

Proposition 5. [Stei] a) *For $1 \le p < \infty$ one has $f \in A_p^{1/p}$ iff*

$$\int_D |f''|^p \, (1 - |z|)^{2p-2} \, dz \, d\bar{z} < \infty.$$

b) *For* $1 < p < \infty$ *one has* $f \in A_p^{1/p}$ *iff*

$$\int_D |f'|^p \, (1 - |z|)^{p-2} \, dz \, d\bar{z} < \infty.$$

One can also reformulate the condition $f \in A_p^{1/p}$ using the L^p norm of the truncations of the Fourier series of f, $\sum \hat{f}(k) \, e^{ik\theta}$, between $k = 2^n$ and $k = 2^{n+1}$. More precisely ([Powe$_2$]) one lets, for each $n \in \mathbb{N}$, w_n be the trigonometric polynomial

$$w_n = \sum_{2^{n-1}}^{2^{n+1}} c_k \, e^{ik\theta}$$

where $c_k = (k - 2^{n-1})/2^{n-1}$ for $2^{n-1} \leq k \leq 2^n$ and $c_k = (2^{n+1} - k)/2^n$ for $2^n \leq k \leq 2^{n+1}$.

Then the operator $f \mapsto f * w_n$ of convolution by w_n is the same as the multiplication of the Fourier coefficients $\hat{f}(k)$ by c_k. These operators add up to the identity, and one has:

Proposition 6. *The space* $A_p^{1/p}$ *is the space of boundary values of holomorphic functions inside D such that*

$$\sum 2^n \, \|w_n * f\|_p^p < \infty.$$

Using $w_{-n} = \overline{w}_n$ for $n < 0$ one can then check that the following conditions on $f \in L^\infty(\mathbb{S}^1)$ are equivalent, for any $p \in [1, \infty[$:

$$[F, f] \in \mathcal{L}^p \quad , \quad \sum_{n \in \mathbb{Z}} 2^{|n|} \, \|w_n * f\|_p^p < \infty.$$

3.β The class of df in $\mathcal{L}^{p,\infty}/\mathcal{L}_0^{p,\infty}$

The quantity $\int_D |f'|^p (1 - |z|)^{p-2} \, dz \, d\bar{z}$ of Proposition 5b) can easily be controlled when the function f is *univalent* in the disk, in terms of the domain $\Omega = f(\text{Disk})$. Indeed, by the Koebe $\frac{1}{4}$-theorem (cf. [Ru]) one has, for f univalent

$$\frac{1}{4}\left(1 - |z|^2\right) \, |f'(z)| \leq \operatorname{dist}(f(z), \partial\Omega) \leq \left(1 - |z|^2\right) \, |f'(z)|. \tag{10}$$

It is thus straightforward to estimate the size of the quantum differential $df = [F, f]$ of a univalent map f in terms of the geometry of the domain $f(\text{Disk}) = \Omega$.

Proposition 7. [Co-Su] *For any* $p_0 > 1$, *there exist finite constants bounding the ratio of the two quantities, which we write*

$$\operatorname{Trace}\left(|[F, f]|^p\right) \asymp \int_\Omega \operatorname{dist}(z, \partial\Omega)^{p-2} \, dz \, d\bar{z}$$

for any univalent function f *and any* $p \geq p_0$.

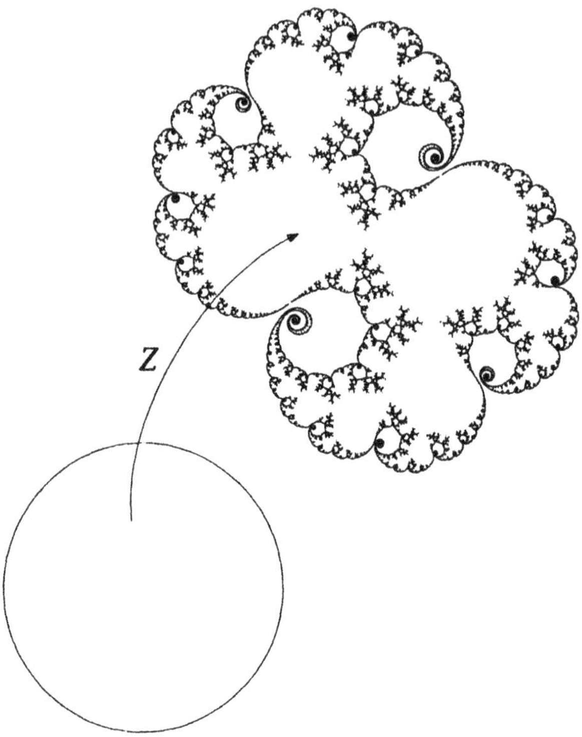

Figure IV.2. Riemann mapping

(We use the symbol $\alpha \asymp \beta$ to mean that $\frac{\alpha}{\beta}$ and $\frac{\beta}{\alpha}$ are bounded.)

The interval of p's such that the right-hand side is finite has a lower bound, known as the Minkowski dimension of the boundary $\partial\Omega$ (cf. [Fe]). It is easy to construct domains with a given Minkowski dimension $p \in]1,2[$ for $\partial\Omega$. We want to go further and relate, when $\partial\Omega$ is a fractal, the p-dimensional Hausdorff measure Λ_p with the formula

$$\mathrm{Tr}_\omega\,(f(Z)\,|dZ|^p) \qquad \forall f \in C(\partial\Omega) \tag{11}$$

where Tr_ω is the Dixmier trace and $dZ = [F, Z]$ the quantum differential. Here Z is the boundary value, $Z \in H^\infty(\mathbb{S}^1)$, of a univalent map $Z : \text{Disk} \to \Omega$. The formula (11) defines a Radon measure provided

$$dZ \in \mathcal{L}^{(p,\infty)} \tag{12}$$

which ensures that $|dZ|^p \in \mathcal{L}^{(1,\infty)}$ is in the domain of the Dixmier trace.

Of course Z is, in general, not of bounded variation, and had we taken dZ as a distribution the symbols $|dZ|$ and $|dZ|^p$ would be meaningless.

To show that, in the presence of symmetries of the domain Ω, the above Radon measure (11) has conformal weight p we shall show that, provided we work modulo the ideal $\mathcal{L}_0^{(p,\infty)} \subset \mathcal{L}^{(p,\infty)}$, the usual rules of calculus are valid.

We let $\mathcal{L}_0^{p,\infty} \subset \mathcal{L}^{p,\infty}$ be the closure in the Banach space $\mathcal{L}^{p,\infty}$ of the ideal R of finite-rank operators. One has

$$T \in \mathcal{L}_0^{p,\infty} \Longleftrightarrow \sigma_N(T) = o\left(\sum_{n=1}^{N} n^{-1/p}\right).$$

For $p > 1$ this is equivalent to $\mu_n(T) = o(n^{-1/p})$, but for $p = 1$ the condition $\mu_n(T) = o\left(\frac{1}{n}\right)$ is stronger than $T \in \mathcal{L}_0^{1,\infty}$. Then let $p > 1$ and $f \in C(\mathbb{S}^1)$ be such that its quantum differential $df = [F,f]$ belongs to $\mathcal{L}^{p,\infty}$. The main result of this section is that if we work modulo $\mathcal{L}_0^{p,\infty}$ then the following rules of calculus are valid:

a) $(df)g = g\, df \quad \forall g \in C(\mathbb{S}^1)$

b) $d(\varphi(f)) = \varphi'(f)df \quad \forall \varphi \in C^\infty(\text{Spectrum}(f))$

c) $|d(\varphi(f))|^p = |\varphi'(f)|^p\ |df|^p$.

In a) and b) the equalities mean that the following operators belong to the ideal $\mathcal{L}_0^{p,\infty}$:

$$[F,f]g - g[F,f]\ ,\ [F,\varphi(f)] - \varphi'(f)[F,f].$$

In c) the equality holds modulo $\mathcal{L}_0^{1,\infty}$:

$$|[F,\varphi(f)]|^p - |\varphi'(f)|^p\ |[F,f]|^p \in \mathcal{L}_0^{1,\infty}.$$

In fact, we shall prove that the characteristic values of the latter operator are $o\left(\frac{1}{n}\right)$, which is a stronger result.

The above rules a) b) and c) are *classical* rules of calculus, but they can now be applied to a nondifferentiable function f, for which the distributional derivative f' cannot undergo the operation $x \to |x|^p$ as the quantum differential can.

We shall thus prove, (with $p \in [1,\infty[$ for a) and b)):

Theorem 8. a) *Let $f \in L^\infty(\mathbb{S}^1)$ be such that $[F,f] \in \mathcal{L}^{p,\infty}$ and let $g \in C(\mathbb{S}^1)$. Then $[F,f]g - g[F,f] \in \mathcal{L}_0^{p,\infty}$.*

b) *Let $X_1,\dots,X_n \in C(\mathbb{S}^1)$, $X_j = X_j^*$, be such that $[F,X_j] \in \mathcal{L}^{p,\infty}$, and let $\varphi \in C^\infty(K)$ be a smooth function on the joint spectrum $K \subset \mathbb{R}^n$ of the X_j's (i.e. $K = X(\mathbb{S}^1)$) where $X = (X_1,\dots,X_n)$. Then*

$$[F,\varphi(X_1,\dots,X_n)] - \sum_{j=1}^{n} \partial_j\, \varphi(X_1,\dots,X_n)[F,X_j] \in \mathcal{L}_0^{p,\infty}.$$

c) *Let $p > 1$, $Z \in C(\mathbb{S}^1)$ be such that $[F,Z] \in \mathcal{L}^{p,\infty}$, and let φ be a holomorphic function on K = Spectrum $Z = Z(\mathbb{S}^1)$. Then*

$$|[F,\varphi(Z)]|^p - |\varphi'(Z)|^p\ |[F,Z]|^p \in \mathcal{L}_0^{p,\infty}.$$

In fact, as we already mentioned, we shall prove the stronger result that $\mu_n(T) = o\left(\frac{1}{n}\right)$ for the operator T appearing in c).

Proof of a) and b)

a) The map from $C(\mathbb{S}^1)$ to $\mathcal{L}^{p,\infty}$ given by

$$g \mapsto [F,f]g - g[F,f] = T(g)$$

is norm continuous. Thus, it is enough to show that for $g \in C^\infty(\mathbb{S}^1)$ the image $T(g)$ belongs to $\mathcal{L}_0^{p,\infty}$. In fact, one has $T(g) \in \mathcal{L}^1$ since g commutes with f while $[F,g] \in \mathcal{L}^1$.

b) The map from $C^\infty(K)$ to $\mathcal{L}^{p,\infty}$ given by

$$\varphi \mapsto [F, \varphi(X_1,\ldots,X_n)]$$

is continuous. Thus, it is enough to check that the statement is true for polynomials, which follows easily from a).

The proof of c) involves general estimates on the map $A \to |A|^p$ with respect to the norms σ_N. We first recall that by [Da] and [Ko$_3$] the map $A \to |A|$ is a Lipschitz map from $\mathcal{L}^{p,\infty}$ to itself provided that $p > 1$.

We shall need the following lemma (cf. [Bi-K-S]):

Lemma 9. *Let $\alpha \in]0,1[$. There exists $C_\alpha < \infty$ such that for any compact operators A and B on $\mathcal{H}$ one has*

$$\frac{1}{N}\,\sigma_N\left(|A|^\alpha - |B|^\alpha\right) \le C_\alpha \left(\frac{1}{N}\,\sigma_N(A-B)\right)^\alpha .$$

As an immediate corollary of this lemma we get:

Proposition 10. *Let $p > 1$.*

1) *Let A, B be bounded operators such that $A-B \in \mathcal{L}_0^{p,\infty}$. Then $|A|^\alpha - |B|^\alpha \in \mathcal{L}_0^{p/\alpha,\infty}$ for any $\alpha < 1$.*

2) *If $A, B \in \mathcal{L}^{p,\infty}$ and $A-B \in \mathcal{L}_0^{p,\infty}$, then for any $\alpha \le p$ one has $|A|^\alpha - |B|^\alpha \in \mathcal{L}_0^{p/\alpha,\infty}$.*

3) *For A, B as in 2) and $\alpha = p$ one has $\mu_N(|A|^p - |B|^p) = o(\frac{1}{N})$.*

Proof. 1) With $p_N = \frac{1}{N}\sigma_N$, one has $p_N(A-B) = o(N^{-1/p})$ by hypothesis. It then follows from Lemma 9 that $p_N\left(|A|^\alpha - |B|^\alpha\right) = o(N^{-\alpha/p})$.

2) Let $\alpha < 1$ be such that p/α is an integer k ($k > 1$). Then by 1) one has $|A|^\alpha - |B|^\alpha \in \mathcal{L}_0^{k,\infty}$, while $|A|^\alpha, |B|^\alpha \in \mathcal{L}^{k,\infty}$.

Let $S = |A|^\alpha$ and $T = |B|^\alpha$. One has $\mu_N(S) = O(N^{-1/k})$, $\mu_N(T) = O(N^{-1/k})$, and $\mu_N(S-T) = o(N^{-1/k})$. Thus, using the inequality

$$\mu_{n_1+n_2+n_3}(XYZ) \le \mu_{n_1}(X)\,\mu_{n_2}(Y)\,\mu_{n_3}(Z)$$

(cf. Section 2) as well as the equality

$$S^k - T^k = \sum S^j (S-T) T^{k-j-1}$$

we derive $\mu_N(S^k - T^k) = o(\frac{1}{N})$.

The proof of 2) is the same.

Let us apply this proposition in the proof of Theorem 8 c). First, as in b) we have

$$[F, \varphi(Z)] - \varphi'(Z)[F, Z] \in \mathcal{L}_0^{p,\infty},$$

so that by Proposition 10.2 we get

$$|[F, \varphi(Z)]|^p - |\varphi'(Z)[F, Z]|^p \in \mathcal{L}_0^{1,\infty}.$$

Thus, we just need to show that

$$|\varphi'(Z)[F, Z]|^p - |\varphi'(Z)|^p \, |[F, Z]|^p \in \mathcal{L}_0^{p,\infty}.$$

One can replace $\varphi'(Z)$ by $f = |\varphi'(Z)|$ and replace $[F, Z]$ by $T = |\,[F, Z]\,|$ since $f[F, Z] - [F, Z]f \in \mathcal{L}_0^{p,\infty}$. Thus, it is enough to use the following lemma:

Lemma 11. *Let $p > 1$, $T \in \mathcal{L}^{p,\infty}$, $T \geq 0$, and let f be bounded, $f \geq 0$, such that $fT - Tf \in \mathcal{L}_0^{p,\infty}$. Then*

$$f^{p/2} \, T^p \, f^{p/2} - \left(f^{1/2} \, T \, f^{1/2}\right)^p \in \mathcal{L}_0^{1,\infty}.$$

3.γ The Dixmier trace of $f(Z)|dZ|^p$

We shall now compare the Hausdorff measure and the Dixmier trace (Formula 11) in an example dealing with *quasi-Fuchsian circles.* We first recall how such quasi-circles are obtained from a pair of points in the Teichmüller space $\mathcal{M}_g$ of Riemann surfaces of genus $g > 1$. Let there be given Σ_+ and Σ_-, a pair of Riemann surfaces of genus g, together with an isomorphism $\pi_1(\Sigma_+) = \Gamma = \pi_1(\Sigma_-)$ of their fundamental groups corresponding to an orientation reversing homeomorphism $\Sigma_+ \to \Sigma_-$. We recall the joint uniformization theorem of L. Bers:

Theorem 12. [Ber] *With the above notation there exists an isomorphism $h : \Gamma \to PSL(2, \mathbb{C})$ of Γ with a discrete subgroup of $PSL(2, \mathbb{C})$ whose action on $P_1(\mathbb{C}) = S^2$ has a Jordan curve C as limit set and is proper with quotient $\Sigma_\pm$ on the connected components $U_\pm$ of the complement of C.*

The discrete subgroup $h(\Gamma)$ is a *quasi-Fuchsian* group and its limit set C is a *quasi-circle.* It is a Jordan curve whose Hausdorff dimension is strictly bigger than one ([Bow$_2$]). Let us choose a coordinate in $P_1(\mathbb{C}) = \mathbb{C} \cup \{\infty\}$ in such a way that $\infty \in \Sigma_-$ and use the Riemann mapping theorem to provide a conformal equivalence

$$Z : D \to \Sigma_+ \subset \mathbb{C}$$

where $D = \{z \in \mathbb{C}; |z| < 1\}$ is the unit disk. By the Carathéodory theorem the holomorphic function Z extends continuously to $\overline{D} = D \cup \mathbb{S}^1$ and yields a homeomorphism

$$Z : \mathbb{S}^1 \to C;$$

the non-differentiability of Z on $\mathbb{S}^1$ is of course a consequence of the lack of smoothness of the Jordan curve C.

Since the range of the function Z on $\mathbb{S}^1$ is equal to C, we see that the spectrum of the operator of multiplication by Z in $L^2(\mathbb{S}^1)$ is also equal to C, so that, for $p > 0$, the operator

$$f(Z)|dZ|^p$$

where f is a function on $\mathbb{C}$, and $dZ = [F, Z]$ as above, involves only the *restriction* of f to the subset C of $\mathbb{C}$, and depends, of course, linearly on f.

By construction, there is an isomorphism $g \to g_+$ of Γ with a Fuchsian subgroup Γ_+ of $PSL(2, \mathbb{R})$ such that

$$g \circ Z = Z \circ g_+ \qquad \forall g \in \Gamma, \qquad (*)$$

where we consider $PSL(2, \mathbb{R}) = PSU(1, 1)$ as the group of automorphisms of D.

Let us first use the equality $(*)$ to reexpress the condition

$$[F, Z] \in \mathcal{L}^q$$

in simpler terms.

Lemma 13. *Let $q > 1$. Then $[F, Z] \in \mathcal{L}^q$ iff the following Poincaré series is convergent for some point, and equivalently for all $z \in \Sigma_+$:*

$$\sigma(q) = \sum_{g \in \Gamma} |g'(z)|^q < \infty.$$

Moreover, there are constants c_q, C_q bounded away from 0 and ∞ for $q \geq q_0 > 1$ such that

$$c_q\, \sigma(q) \leq \operatorname{Trace} |\, [F, Z] \,|^q \leq C_q\, \sigma(q).$$

Proof. By construction, the function $Z \in C(\mathbb{S}^1)$ extends to a holomorphic function in D, so that we can apply the criterion given by Proposition 5 b), which also gives an estimate on the $\mathcal{L}^q$ norm of $[F, Z]$ in terms of

$$\int_D |Z'(z)|^q \, (1 - |z|)^{q-2} \, dz \, d\bar{z}.$$

For z in D, $1 - |z|$ and $1 - |z|^2$ are comparable, so that we may as well consider the expression

$$\int_D |Z'(z)|^q \, (1 - |z|^2)^q \, (1 - |z|^2)^{-2} \, dz \, d\bar{z}.$$

If we endow D with its canonical hyperbolic Riemannian metric of curvature -1, the last expression is equivalent to

$$\int_D \|\nabla Z\|^q \, dv$$

where ∇Z is the gradient of the function Z whose norm is evaluated with respect to the Riemannian metric, and where dv is the volume form on the Riemannian manifold D. Then let $g \in PSL(2, \mathbb{R}) = PSU(1,1)$. Since it acts as an isometry on D, one has

$$\|\nabla(Z \circ g)\|(p) = \|\nabla(Z)\|(gp) \quad \forall p \in D.$$

For $g_+ \in \Gamma_+$ one has $Z \circ g_+ = g \circ Z$, so that

$$\|\nabla(g \circ Z)\|(p) = \|\nabla(Z)\|(g_+ p) \quad \forall p \in D.$$

The left-hand side is equal to $|g'(Z(p))| \, \|\nabla Z\|(p)$, so that

$$\|\nabla(Z)\|(g_+ p) = |g'(Z(p))| \, \|\nabla Z\|(p) \quad \forall p \in D \, , \, g \in \Gamma_+ .$$

Then let $D_1 \subset D$ be a compact fundamental domain for the Fuchsian group Γ_+. We have the equality

$$\int_D \|\nabla Z\|^q \, dv = \int_{D_1} \sum_{g \in \Gamma} |g'(Z(p))|^q \; (\|\nabla Z\|(p))^q \, dv.$$

The compactness of D_1 then gives the required uniformity in $p \in D_1$, so that the conclusion follows.

Now let p be the Hausdorff dimension of the limit set C. One has $p > 1$ ([Bow$_2$]), and by [Su$_2$] it follows that the Poincaré series $\sigma(q)$ is convergent for any $q > p$, and diverges for $q = p$. Thus, we get so far:

Proposition 14. *One has $[F, Z] \in \mathcal{L}^q$ iff $q > p$ = Hausdorff dimension of C.*

But we need to know that $[F, Z] \in \mathcal{L}^{p,\infty}$ and that $\mathrm{Tr}_\omega(|[F, Z]|^p) > 0$.

Lemma 15. *One has $[F, Z] \in \mathcal{L}^{p,\infty}$.*

Proof. From real interpolation theory (Appendix B) and the above criterion, $\|\nabla Z\| \in L^q(D, dv) \Longleftrightarrow [F, Z] \in \mathcal{L}^q$, we just need to show that

$$\|\nabla Z\| \in L^{p,\infty}(D, dv)$$

where the Lorentz space $L^{p,\infty} = L^p_{\text{weak}}$ is the space of functions h on D such that for some constant $c < \infty$

$$v(\{z \in D \, ; \, |h(z)| > \alpha\}) \le c \, \alpha^{-p}.$$

Thus, the proof of Lemma 13 shows that all we need to prove is that (uniformly for $a \in D_1$) the sequences $(|g'(Z(a))|;\, g \in \Gamma)$ belong to $\ell^{p,\infty}(\Gamma)$, i.e.

$$\mathrm{Card}\{g \in \Gamma\,;\ |g'(Z(a))| > \alpha\} = O(\alpha^{-p}).$$

This follows from Corollary 10 in [Su$_2$].

Next, the pole-like behaviour of $\int_D \|\nabla Z\|^s\, dv$ for $s \to p+$, which follows from [Su$_2$], and the fact that the residue at $s = p$ is *not zero* ([Su$_2$]) imply a similar behaviour for Trace $(|[F,Z]|^s)$, so that the characteristic values

$$\mu_n = \mu_n([F,Z])$$

satisfy the following conditions:

α) $\mu_n = O(n^{-1/p})$ (by Lemma 15)

β) $(s-p)\sum_{n=0}^{\infty} \mu_n^s \geq c > 0$ for $s \in\,]p, p+\varepsilon]$.

One can then use the following Tauberian lemma:

Lemma 16. [Har] *Let μ_n be a decreasing sequence of positive real numbers satisfying α) and β). Then*

$$\liminf \frac{1}{\log N} \sum_{n=0}^{N} \mu_n^p \geq c.$$

We are now ready to prove the following theorem. In the statement we fix a Dixmier trace Tr_ω once and for all.

Theorem 17. [Co-Su] *Let $\Gamma \subset PSL(2,\mathbb{C})$ be a quasi-Fuchsian group, $C \subset P_1(\mathbb{C})$ its limit set, and $Z \in C(\mathbb{S}^1)$ the boundary values of a conformal equivalence of the disk D with the bounded component of the complement of C. Then let p be the lower bound of the set $\{q; [F,Z] \in \mathcal{L}^q\}$. One has $p =$ Hausdorff dim C and $[F,Z] \in \mathcal{L}^{p,\infty}$. Moreover, there exists a* nonzero finite *real number λ such that, with Λ_p the p-dimensional Hausdorff measure on C,*

$$\int_C f\, d\Lambda_p = \lambda\ \mathrm{Tr}_\omega(f(Z)|[F,Z]|^p) \quad \forall f \in C_0(\mathbb{C}).$$

Proof. The first part follows from Proposition 14, Lemma 15 and Lemma 16. It also follows from Lemma 16 that $\mathrm{Tr}_\omega(|[F,Z]|^p) > 0$, so that we can consider the measure μ on C determined by the equality

$$\mu(f) = \mathrm{Tr}_\omega(f(Z)\ |[F,Z]|^p) \quad \forall f \in C(C).$$

We claim that this measure has conformal weight p, i.e. that for any $g \in \Gamma$ one has the equality

$$\int f \circ g^{-1}\, d\mu = \int |g'|^p\, f\, d\mu.$$

To prove this, let $g_+ \in SL(2,\mathbb{R}) = SU(1,1)$ be the corresponding element of Γ_+. Its action on $L^2(\mathbb{S}^1)$ is given, for $g_+^{-1} = \left[\begin{smallmatrix} \alpha & \beta \\ \bar\beta & \bar\alpha \end{smallmatrix}\right]$, by

$$(g_+ \xi)(z) = \xi\left((\alpha z + \beta)(\bar\beta z + \bar\alpha)^{-1}\right)(\bar\beta z + \bar\alpha)^{-1} \quad \forall z \in \partial D.$$

This equality defines a *unitary* operator W which commutes with the Hilbert transform F, and the corresponding representations of $SL(2,\mathbb{R})$ are the mock discrete series. Moreover

$$WZW^* = Z \circ g_+ = g \circ Z.$$

Thus, we arrive at the equality

$$W[F,Z]W^* = [F, g \circ Z].$$

It implies that $W\,|[F,Z]|^p\,W^* = |[F,g\circ Z]|^p$; thus

$$W(f \circ g^{-1}(Z))\,|[F,Z]|^p W^* = f(Z)\,|[F,g\circ Z]|^p.$$

Since the Dixmier trace Tr_ω is a trace, we get

$$\mathrm{Tr}_\omega(f \circ g^{-1}(Z)\,|[F,Z]|^p) = \mathrm{Tr}_\omega(f(Z)\,|[F,g\circ Z]|^p),$$

and by Theorem 8 c) we have

$$\mathrm{Tr}_\omega(f(Z)\,|[F,g\circ Z]|^p) = \mathrm{Tr}_\omega(f(Z)|g'(Z)|^p\,|[F,Z]|^p),$$

so that

$$\int f \circ g^{-1}\,d\mu = \int f|g'|^p\,d\mu.$$

It then follows by [Su$_2$] that μ is proportional to the p-dimensional Hausdorff measure on C.

The constant λ in Theorem 17 should be independent of the choice of ω. We shall prove this in a simpler case in Section ε) where we also relate the value of $\mathrm{Tr}_\omega(|dZ|^p)$ to a normalized eigenfunction of the Laplacian in hyperbolic space (Theorem 25).

The value of the constant λ is related to the best rational approximation of the function Z using the following result.

Theorem 18. [Ad-A-K] *Let $Z \in C(\mathbb{S}^1) \cap H^\infty$. Let μ_n be the distance, in the* sup *norm on $\mathbb{S}^1$, between Z and the set R_n of rational fractions with at most n poles outside the unit disk. Then μ_n is the n-th characteristic value of the operator $[F,Z]$ in $L^2(\mathbb{S}^1)$.*

Remark 19. Let $C \subset \mathbb{C}$ be a Jordan curve whose 2-dimensional area is *positive*, $\Lambda_2(C) > 0$. (The existence of such curves is an old result of analysis.) Let Ω be a bounded simply connected domain with boundary $\partial\Omega = C$ and let $Z : \mathrm{Disk} \to \Omega$ be the Riemann mapping. Then the finiteness of the area of $\bar\Omega$ shows that

$$dZ \in \mathcal{L}^2 \quad (\text{i.e. } \mathrm{Trace}((dZ)^*\,dZ) < \infty).$$

This provides us with a very interesting example of a space C (or equivalently $\mathbb{S}^1$ with the metric $dZ\,d\bar Z$) which has nonzero 2-dimensional Lebesgue measure, but whose dimension in our sense is not 2 but rather 2−, inasmuch as $\mathrm{Trace}((dZ)^*\,dZ)$ is *finite* rather than logarithmically divergent.

3.δ The harmonic measure and non-normality of the Dixmier trace

Let $\Omega \subset \mathbb{C}$ be a Jordan domain and $X = \partial\Omega$ its boundary. Using the boundary value $Z : \mathbb{S}^1 \to X$ of a conformal equivalence with the unit disk, $D \simeq \Omega$, we can compose the Fredholm module $(L^2(\mathbb{S}^1), F)$ over $\mathbb{S}^1$ given by the Hilbert transform in $L^2(\mathbb{S}^1)$ with the isomorphism $Z^* : C(X) \to C(\mathbb{S}^1)$. We thus obtain a Fredholm module over $C(X)$ which we can describe directly with no specific reference to Z, as follows. We first need to recall that each point $z_0 \in \Omega$ defines a unique probability measure ν_{z_0} on $X = \partial\Omega$ by the equality

$$\int f \, d\nu_{z_0} = \tilde{f}(z_0) \qquad \forall f \in C(X)$$

where $\tilde{f}$ is the unique continuous harmonic function on $\overline{\Omega}$ which agrees with f on $X = \partial\Omega$. The measure ν_{z_0} is called the *harmonic measure*, and its class is independent of the choice of $z_0 \in \Omega$. We thus let $\mathcal{H}_0 = L^2(X, \nu_{z_0})$ be the corresponding Hilbert space in which $C(X)$ is represented by multiplication operators. The constant function 1 is a vector $\xi_0 \in \mathcal{H}_0$ such that

$$\langle f\xi_0, \xi_0 \rangle = \int f \, d\nu_{z_0} \qquad \forall f \in C(X).$$

Then let P_0 be the orthogonal projection of $\mathcal{H}_0$ on the subspace

$$P_0 \, \mathcal{H}_0 = \{f\xi_0 \, ; \, \tilde{f} \text{ holomorphic in } \Omega\}^-.$$

Proposition 20. *If $F_0 = 2P_0 - 1$, then $(\mathcal{H}_0, F_0)$ is a Fredholm module over $C(X)$ whose isomorphism class is independent of the choice of z_0 and equal to $Z^*(L^2(\mathbb{S}^1), F)$ where $Z : \mathbb{S}^1 \to X$ is the boundary value of a conformal equivalence with the unit disk, $D \to \Omega$.*

The proof is obvious but this presentation will adapt to Cantor subsets of $P_1(\mathbb{R})$ (cf. Section ε)).

As soon as the Hausdorff dimension p of X is > 1, the Hausdorff measure Λ_p (if it makes sense) is *singular* with respect to the harmonic measure ν (cf. [Mak]). In fact, ν is carried by a countable union of sets of finite linear measure ([Mak]) while Λ_p is really p-dimensional.

With the Fredholm module of Proposition 20, and specializing to the quasi-circles of Theorem 17, we have

$$\int f \, d\Lambda_p = \lambda \operatorname{Tr}_\omega(f(z)|dz|^p)$$

where z is the identity map $X \to \mathbb{C}$. We now claim that this formula would contradict the mutual singularity of Λ_p and ν if the Dixmier trace Tr_ω happened to be *normal* (cf. Chapter V). Indeed, for any normal weight ψ on $\mathcal{L}(\mathcal{H}_0)$, any $T \geq 0, T \in \text{Domain } \psi$, the linear form

$$f \in C(X) \to \psi(T^{1/2} \, f \, T^{1/2})$$

is a measure on $C(X)$ which is absolutely continuous with respect to the harmonic measure. It is thus a great virtue of the Dixmier trace that it allows us, by its non-normality, to go beyond the harmonic measure class. An even simpler illustration of this fact is given in Subsection ε) below.

3.ε Cantor sets, Dixmier trace and Minkowski measure

In order to understand the constant λ of Theorem 17 we shall work out in more detail the simpler but analogous Fuchsian case.

Let $K \subset \mathbb{R}$ be a compact subset of $\mathbb{R}$ and assume for simplicity that K is totally disconnected without isolated points. We shall first describe a natural Fredholm module $(\mathcal{H}, F)$ over $C(K)$ (compare with Proposition 20). Let $\Omega = K^c$ be the complement of K. We may assume for definiteness that $K \subset [0,1]$ with $0, 1 \in K$. Then except for $]-\infty, 0[\cup]1, \infty[$ the open set Ω is the disjoint union of a sequence I_j of bounded intervals. We let $D \subset K$ be the countable set of endpoints of the intervals I_j. Each such interval determines two elements b_j^+ and b_j^- of D with $b_j^- < b_j^+$, $I_j =]b_j^-, b_j^+[$. This gives a partition $D = D^- \cup D^+$ of D into two disjoint subsets. A harmonic function h on the bounded components of Ω is piecewise affine on the intervals I_j, and is hence uniquely specified by its value on D. This shows that the *harmonic measure class* on K is just the counting measure on D, an obviously 0-dimensional measure. By analogy with Proposition 20 we shall thus take

$$\mathcal{H} = \ell^2(D)\ ,\ (f\xi)(b) = f(b)\,\xi(b) \qquad \forall b \in D\ ,\ \forall f \in C(K)\ ,\ \forall \xi \in \ell^2(D).$$

The analogue of the algebra of holomorphic functions of Proposition 20 is the algebra of piecewise constant functions h on Ω. This determines the closed subspace of $\mathcal{H}$ given by

$$P\mathcal{H} = \{\xi \in \mathcal{H}\ ;\ \xi(b_j^-) = \xi(b_j^+) \qquad \forall b_j^\pm \in D\}.$$

We then let P be the orthogonal projection on this subspace, and let $F = 2P - 1$.

Proposition 21. a) *The pair $(\mathcal{H}, F)$ is a Fredholm module over $C(K)$.*

b) *The characteristic values of the operator $dx = [F, x]$ (where $x \in C(K)$ is the embedding of K in $\mathbb{R}$) are the lengths $\ell_j = |I_j|$ of the intervals I_j, each with multiplicity 2.*

Indeed, for each interval I_j the projection P_j is given by the matrix $P_j = \begin{bmatrix} 1/2 & 1/2 \\ 1/2 & 1/2 \end{bmatrix}$, so that $F_j = 2P_j - 1 = \begin{bmatrix} 0 & 1 \\ 1 & 0 \end{bmatrix}$. This shows that for any $f \in C(K)$, the operator $df = [F, f]$ is the direct sum of the

$$[F_j, f] = \left[\begin{bmatrix} 0 & 1 \\ 1 & 0 \end{bmatrix}, \begin{bmatrix} f(b_j^-) & 0 \\ 0 & f(b_j^+) \end{bmatrix}\right] = (f(b_j^+) - f(b_j^-)) \begin{bmatrix} 0 & 1 \\ -1 & 0 \end{bmatrix}.$$

One then easily gets a) and b) as well as:

Proposition 22. *Let K be Minkowski measurable and of Minkowski dimension $p \in [0,1]$. Then $|dx|^p \in \mathcal{L}^{(1,\infty)}$ and*

$$\mathrm{Tr}_\omega(|dx|^p) = 2^p(1-p)\,\mathcal{M}_p(K)$$

where $\mathcal{M}_p(K)$ is the Minkowski content of K.

Indeed, by [Lap], and with the ℓ_j in decreasing order, one has a constant L such that

$$\ell_j \sim L\, j^{-1/p} \quad \text{when} \quad j \to \infty\,;\ \mathcal{M}_p(K) = 2^{1-p}(1-p)^{-1}\, L^p.$$

Now the Minkowski measurability of K is a much stronger condition than the measurability of the operator $|dx|^p$. In particular, for the simplest self-similar Cantor sets K we shall see that, while K is not Minkowski measurable, the operator $|dx|^p$ is measurable, and the measure $f \in C(K) \mapsto \mathrm{Tr}_\omega(f(x)|dx|^p)$ is a nonzero multiple, independent of ω, of the Hausdorff measure.

Example 23. Let $q \in \mathbb{N}$, $\lambda > 0$ with $\lambda q < 1$. Let $K \subset [0,1]$ be obtained by removing q intervals of length λ so that the $q+1$ remaining intervals in $[0,1]$ have equal length, and iterating this procedure. After n steps we have $(q+1)^n$ remaining intervals of length ρ^n, $\rho = (1-\lambda q)/(q+1)$ and we remove $q(q+1)^n$ intervals of length $\lambda\rho^n$. Thus we have

$$\mathrm{Trace}(|dx|^s) = 2\sum_{n=0}^{\infty} q(q+1)^n\ (\lambda\rho^n)^s = 2q\lambda^s(1-(q+1)\rho^s)^{-1}.$$

This shows that for $p = -\frac{\log(q+1)}{\log\rho} \in\,]0,1[$ the operator $|dx|^p$ is measurable and

$$\mathrm{Tr}_\omega(|dx|^p) = \frac{1}{p}\,\mathrm{Res}_{s=p}(\mathrm{Trace}|dx|^s) = -\frac{1}{p}\,2q\lambda^p(\log\rho)^{-1} = \frac{2q\lambda^p}{\log(q+1)}.$$

One checks that K is not Minkowski measurable, that its Hausdorff dimension is p and that, with Λ_p the p-dimensional Hausdorff measure, one has

$$\mathrm{Tr}_\omega(f(x)|dx|^p) = c\int f\, d\Lambda_p \qquad \forall f \in C(K)$$

where $c = \frac{2q\lambda^p}{\log(q+1)} = \frac{2q}{\log(q+1)}(1-(q+1)^{1-1/p})^p$.

This example clearly shows that c, while independent of ω, is not just a function of p, owing to the failure of Minkowski measurability of K.

Example 24. Fuchsian groups of the second kind.

Let K be a perfect subset of the circle $\mathbb{S}^1$ (i.e. K is closed, totally disconnected and without isolated points). The above construction of the Fredholm module $(\mathcal{H}, F)$ over K works without change. We identify $\mathbb{S}^1$ with the boundary $\mathbb{S}^1 = \{z \in \mathbb{C}\,;\ |z| = 1\}$ of the open unit disk U in $\mathbb{C}$ and we endow U with the Poincaré metric of constant curvature -1. The hyperbolic distance is given by the equality

$$d(z_1,z_2) = \log\left(\frac{1+r}{1-r}\right),\ \frac{1}{1-r^2} = \frac{|1-\bar{z}_1 z_2|^2}{(1-|z_1|^2)(1-|z_2|^2)} \qquad \forall z_j \in U.$$

We let $d\sigma$ be the hyperbolic area.

Let $Y = \text{Conv}(K)$ be the geodesic convex hull of K in U. It is obtained by removing from U a half space (Figure 3) for each interval I_j component of $\mathbb{S}^1 \backslash K$.

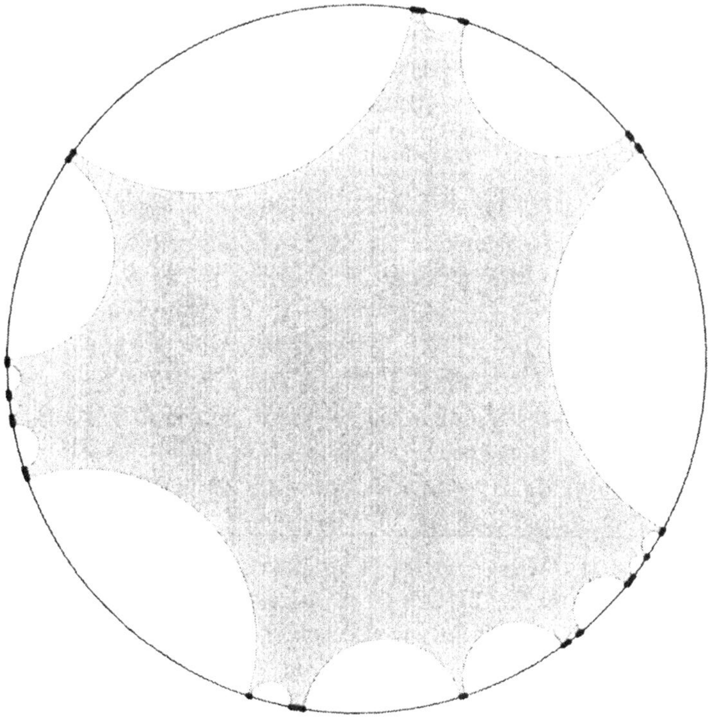

Figure IV.3. Convex hull of a perfect set

Let $G = SU(1,1)$ act by isometries on U and, for each $\alpha \in U$, let d_α be the unique Riemannian metric on $\mathbb{S}^1$ invariant under the isotropy group G_α and normalized so that $\mathbb{S}^1$ has length 2π.

For $\alpha = 0$ this is simply the round metric $|dz|^2$ on $\mathbb{S}^1$ while for arbitrary α it is $|dz_\alpha|^2$, where $z_\alpha(u) = \frac{u-\alpha}{1-\bar{\alpha}u} \quad \forall u \in \mathbb{S}^1$.

Lemma 25. *Let $\alpha \in U$, K and $Y = \text{Conv}(K)$ be as above. Assume that K is Lebesgue negligible. Let ℓ_j be the length of the component I_j of K^c in $\mathbb{S}^1$ for the metric d_α and consider the two Dirichlet integrals*

$$\zeta_1(s) = \sum \ell_j^s \,, \; \zeta_2(s) = \int_Y e^{-sd(z,\alpha)}\, d\sigma(z).$$

Then ζ_1 and ζ_2 have the same abscissa of convergence $p \in [0,1]$, the same behaviour at p, and in the divergent case one has

$$\zeta_1(p+\varepsilon) \sim c(p)\, \zeta_2(p+\varepsilon) \quad \textit{when} \quad \varepsilon \to 0+$$

where $c(p) = -4^p\,\Gamma\left(\frac{p}{2}+1\right)\,\Gamma\left(\frac{3}{2}\right)\,/\,\Gamma\left(\frac{p-1}{2}\right)$.

The proof is a simple computation of the behaviour near $\ell = 0$ of the integral $\int_0^1 (\ell - \theta(\ell, 1-\varepsilon))\,\varepsilon^{p-2}\,d\varepsilon = \varphi_p(\ell)$, where $\theta(\ell, r)$ is equal to 0 if $\frac{1+r^2}{2r}\cos\ell \geq 1$ and to $\arccos\left(\frac{1+r^2}{2r}\cos\ell\right)$ otherwise. One has (Figure 4) $\zeta_2(s) = 2\sum \psi_s(\frac{1}{2}\,\ell_j)$, where $\psi_s(\ell) = \int_0^1 (\ell - \theta(\ell, 1-\varepsilon))\,\varepsilon^{s-2}\,4(1-\varepsilon)(2-\varepsilon)^{-s-2}\,d\varepsilon$ behaves like $2^{-s}\varphi_s(\ell)$ for $\ell \to 0$. One then shows that $\varphi_s(\ell) \sim \frac{1}{2}\,4^s\,c(s)^{-1}\,\ell^s$ as $\ell \to 0$.

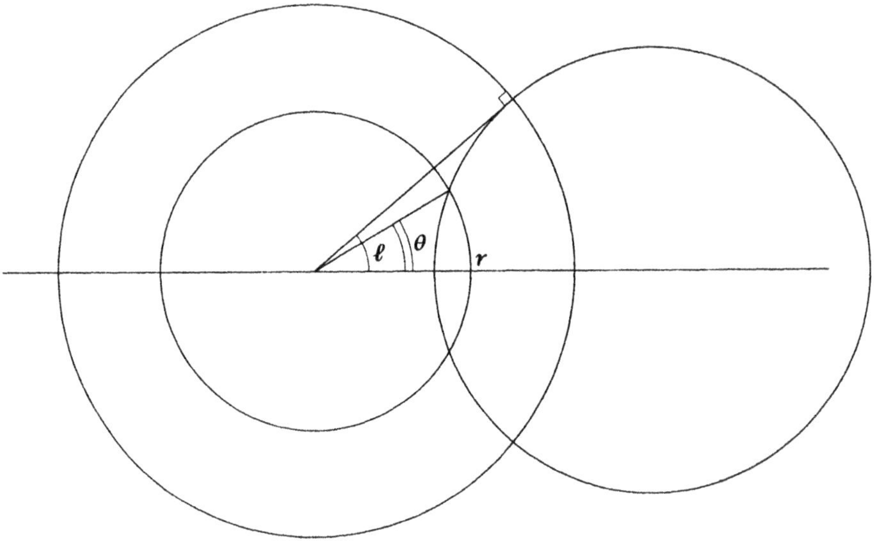

Figure IV.4. $\cos\theta = \frac{1+r^2}{2r}\cos\ell$

Now let $\Gamma \subset G$ be a Fuchsian group of the second kind. Its limit set $K \subset \mathbb{S}^1$ is then a perfect set and, by [Pat], the abscissa of convergence of the Poincaré series

$$\sum_{g\in\Gamma} e^{-sd(gz,\alpha)}$$

is equal to the Hausdorff dimension $p \in \,]0,1[$ of K.

Combining Proposition 21, Lemma 25, and the results of [Pat] and [Su] we get the following simpler but more precise analogue of Theorem 17.

Theorem 26. *Let* Γ *be a Fuchsian group of the second kind without parabolic element,* $K \subset \mathbb{S}^1$ *its limit set,* $Y = \mathrm{Conv}(K)$ *its geodesic convex hull and* $p =$

Hausdorff dim(K). *Assume $p > \frac{1}{2}$, and let h be the unique square integrable $p(1-p)$ harmonic function on U/Γ such that $\int_{U/\Gamma} h^2\, d\sigma = \int_{Y/\Gamma} h\, d\sigma$.*

Let $(\mathcal{H}, F)$ be the above Fredholm module over $C(K)$. Then for any $\alpha \in U$ the measure $f \in C(K) \to \operatorname{Tr}_\omega(f|dz_\alpha|^p)$ is independent of ω, proportional to the Hausdorff measure Λ_p for the restriction to K of the metric d_α, and its total mass is

$$\operatorname{Tr}_\omega(|dz_\alpha|^p) = (1-p)\, 2^{p-1}\, \sqrt{\pi}\, \frac{\Gamma\left(p-\frac{1}{2}\right)}{\Gamma\left(\frac{p+1}{2}\right)^2}\, h(\alpha).$$

IV.4. Conformal Manifolds

Let V be an even-dimensional oriented compact manifold endowed with a conformal structure. We shall now show how to quantize in a canonical manner the calculus on V by constructing a natural even Fredholm module $(\mathcal{H}, F, \gamma)$ over the algebra $C^\infty(V)$ of smooth functions on V ([D-S] [Co-S-T]).

4.α Quantized calculus on conformal manifolds

For the construction we shall just need the $\mathbb{Z}/2$-grading of the vector bundle $\wedge^n T^*$, $n = \frac{1}{2}\dim V$, given by the $*$ operation. Recall that given a Euclidean oriented vector space E of dimension m, the $*$ operation, $* : \wedge^* E \to \wedge^* E$, is given by the equality

$$*(e_1 \wedge \cdots \wedge e_k) = e_{k+1} \wedge \cdots \wedge e_m$$

for any orthonormal basis $e_1, \ldots, e_m$ of E compatible with the orientation. When m is even, $m = 2n$, the restriction of $*$ to $\wedge^n E$ is unaffected if one replaces the Euclidean metric of E by its scalar multiples. Moreover, one gets a $\mathbb{Z}/2$ grading on $\wedge^n_{\mathbb{C}} E$ given by the operator $\gamma = (-1)^{\frac{n(n-1)}{2}} i^n *$, which is of square 1.

Let $\mathcal{H}_0$ be the Hilbert space $L^2(V, \wedge^n_{\mathbb{C}} T^*)$ of square integrable sections of the complex vector bundle $\wedge^n_{\mathbb{C}} T^*$, with the inner product given by the complexification of the real inner product

$$\langle \omega_1, \omega_2 \rangle = \int_V \omega_1 \wedge *\omega_2 \qquad \forall \omega_1, \omega_2 \in L^2(V, \wedge^n T^*).$$

By construction, $\mathcal{H}_0$ is a module over $C^\infty(V)$ (and also $L^\infty(V)$) with

$$(f\omega)(p) = f(p)\, \omega(p) \qquad \forall f \in C^\infty(V),\ \omega \in \mathcal{H}_0,\ p \in V.$$

It is $\mathbb{Z}/2$-graded by the above operator γ of square 1,

$$(\gamma\omega)(p) = \gamma(\omega(p)) \qquad \forall p \in V,\ \omega \in \mathcal{H}_0.$$

To construct the operator F we need the following:

Lemma 1. *Let $B \subset \mathcal{H}_0$ be the closure of the image of $d : C^\infty(V, \wedge_{\mathbb{C}}^{n-1} T^*) \to C^\infty(V, \wedge_{\mathbb{C}}^{n} T^*)$. Then B is the graph of a partial isometry $S : \mathcal{H}_0^- \to \mathcal{H}_0^+$ (resp. $S^* : \mathcal{H}_0^+ \to \mathcal{H}_0^-$) and $1 - (SS^* + S^*S)$ is the orthogonal projection on the finite-dimensional space of harmonic forms.*

Proof. This follows in a straightforward manner from the Hodge decomposition ([Gi$_1$]) of $\mathcal{H}_0$ as the direct sum of the kernel of $d + d^*$, i.e. harmonic forms, and the image of $d + d^*$. Thus any $\omega \in \mathcal{H}_0^+$, orthogonal to harmonic forms, can be written as $\omega = \frac{1+\gamma}{2}\, d\alpha$, $\alpha \in L^2(V, \wedge_{\mathbb{C}}^{n-1} T^*)$, using the formula $d^* = - * d *$ for the adjoint of d. Moreover, the equality $\omega = \frac{1+\gamma}{2}\, d\alpha$ determines $d\alpha$ uniquely since

$$\left\| \tfrac{1+\gamma}{2}\, d\alpha \right\|_2 = \left\| \tfrac{1-\gamma}{2}\, d\alpha \right\|_2 \qquad \forall \alpha \in L^2(V, \wedge_{\mathbb{C}}^{n-1}\ T^*).$$

We can now define the Fredholm module $(\mathcal{H}, F, \gamma)$ over $C^\infty(V)$. We let $\mathcal{H} = \mathcal{H}_0 \oplus H^n(V, \mathbb{C})$ be the direct sum of $\mathcal{H}_0$ with the finite-dimensional Hilbert space of harmonic n-forms on V, which we identify with the n-dimensional cohomology group $H^n(V, \mathbb{C})$. We endow H^n with the opposite $\mathbb{Z}/2$-grading $-\gamma$ and with the 0-module structure over $C^\infty(V)$. The direct sum $\mathcal{H} = \mathcal{H}_0 \oplus H^n$ is thus a $\mathbb{Z}/2$-graded $C^\infty(V)$-module. We let F be the operator on $\mathcal{H}$ given as the direct sum of $\begin{bmatrix} 0 & S \\ S^* & 0 \end{bmatrix}$ acting in $\mathcal{H}_0 \ominus H^n$ and $\begin{bmatrix} 0 & 1 \\ 1 & 0 \end{bmatrix}$ acting in $H^n \oplus H^n$. Note that $\begin{bmatrix} 0 & S \\ S^* & 0 \end{bmatrix}$ acting in $\mathcal{H}_0 \ominus H^n$ is equal to $2P - 1$, where P is the orthogonal projection on $B \subset \mathcal{H}_0 \ominus H^n$.

Theorem 2. a) *The triple $(\mathcal{H}, F, \gamma)$ is a Fredholm module over $C^\infty(V)$, canonically associated to the oriented conformal structure of V. It is p-summable for any $p > 2n$.*

b) *The character $\mathrm{ch}_*(\mathcal{H}, F, \gamma)$ is 2^n times the Atiyah-Hirzebruch $\mathcal{L}$-genus multiplied by the fundamental class $[V]$.*

c) *The Fredholm module $(\mathcal{H}, F, \gamma)$ uniquely determines the oriented conformal structure of V.*

Proof. a) It is clear that $F = F^*$ and $F^2 = 1$. Thus we just have to prove that $[F, u] \in \mathcal{L}^p(\mathcal{H})$ for any $p > 2n$, $u \in C^\infty(V)$. We can assume that u is unitary and we have to show that $uFu^* - F \in \mathcal{L}^p(\mathcal{H})$. Since H^n is finite-dimensional, we just need to evaluate $uSu^* - S$. But the graph of uSu^* is given by the (closure of the) image of $udu^* = d + \alpha$, where α is the operator of exterior multiplication by the 1-form $ud(u^*)$. Since the latter is bounded, the Sobolev inequalities ([Gi$_1$]) show that $uSu^* - S$ belongs to $\mathcal{L}^p$ for any $p > 2n$ (as does any pseudo-differential operator of order -1 on V). In fact, the growth of the characteristic values $\mu_k([F, u])$ is controlled by the growth of the characteristic values of $(1 + \Delta)^{-1/2}$, where Δ is a Laplacian on V, and thus

$$\mu_k([F, u]) = O(k^{-1/2n}) \qquad \forall u \in \mathrm{Lip}(V)$$

where $\mathrm{Lip}(V)$ is the algebra of Lipschitz functions on V.

b) We shall check the equality using our knowledge of the cyclic cohomology of $C^\infty(V)$ (Chapter III, Theorem 2.2), together with the index Proposition 4 of Section 1 of this chapter, combined with the Atiyah-Singer index formula. It thus suffices, using the surjectivity of $\mathrm{ch}^* : K(V) \to H^*(V,\mathbb{C})$, to check that the index map $\varphi : K(V) \to \mathbb{Z}$ associated to $(\mathcal{H},F,\gamma)$ is the same as the index map of the signature operator. Thus, it is enough to show that the K-theory class of the symbol of F, $[\sigma(F)] \in K^0(T^*V)$, is the same as the K-theory class of the symbol of the signature operator. The latter is given by the odd endomorphism $s(x,\xi) = e_\xi + i_\xi$, $\xi \in T_x^*(V)$, of the pull-back of $\wedge^* T_{\mathbb{C}}^*$ (oriented by $\gamma = (-1)^{\frac{p(p-1)}{2}}\, i^n *$ on p-forms) to T^*V. (Here e_ξ and i_ξ are respectively the symbols of d and d^*.) The symbol $\sigma(F)$ of F is the same as the symbol of $2P-1$, where P is the orthogonal projection on the image of d. Its restriction to the unit sphere $\{\xi \in T^*V \,;\, \|\xi\| = 1\}$ is thus given by

$$\sigma(x,\xi) = e_\xi\, i_\xi - i_\xi\, e_\xi \quad \text{acting on} \quad \wedge^n T_{\mathbb{C}}^*.$$

Let $u(x,\xi) = \frac{1}{\sqrt{2}}(1 - e_\xi + i_\xi)$. Then for $\|\xi\| = 1$, it is an invertible operator in $\wedge^* T_{\mathbb{C}}^*$ with inverse $u^{-1}(x,\xi) = \frac{1}{\sqrt{2}}(1 + e_\xi - i_\xi)$. One has

$$(usu^{-1})(x,\xi) = e_\xi\, i_\xi - i_\xi\, e_\xi \quad \text{acting on} \quad \wedge^* T_{\mathbb{C}}^*.$$

By construction, u commutes with γ. This shows that the class $[s]-[\sigma]$ of K-theory is given by the symbol

$$\rho(x,\xi) = e_\xi\, i_\xi - i_\xi\, e_\xi \quad \text{acting on} \quad \wedge^* T_{\mathbb{C}}^* \ominus \wedge^n T_{\mathbb{C}}^*,$$

with the $\mathbb{Z}/2$-grading γ. But using the canonical isomorphism, for $p \neq n$,

$$\wedge^p T_{\mathbb{C}}^* \simeq \left(\wedge^p T_{\mathbb{C}}^* \oplus \wedge^{2n-p} T_{\mathbb{C}}^*\right)^{\pm} \;;\; \omega \mapsto \frac{1}{2}(\omega \pm \gamma\omega)$$

one checks that the class of ρ is equal to 0.

The precise statement of b) is then that $\mathrm{ch}_*(\mathcal{H},F,\gamma)$ is in the same cohomology class as the following cocycle (φ_{2m}) in the (d_1,d_2) bicomplex of the algebra $C^\infty(V)$

$$\varphi_{2n-4k}(f^0,\dots,f^{2n-4k})$$
$$= 2^n(2\pi i)^{-n+2k}\int_V \mathcal{L}_k(V)\wedge f^0\, df^1 \wedge\cdots\wedge df^{2n-4k} \qquad \forall f^j \in C^\infty(V)$$

where $\mathcal{L}_k = 2^{-2k} L_k$ is the component of degree $4k$ of the Atiyah-Hirzebruch $\mathcal{L}$-genus.

c) We shall use the Dixmier trace. By Proposition 5 of Section 2 we have for pseudo-differential operators T of order $-2n$

$$\mathrm{Tr}_\omega(T) = \frac{1}{2n(2\pi)^{2n}}\int_{S^*V} \mathrm{Trace}\,(\sigma(x,\xi))\, dx\, d\xi_{S^*}$$

where the trace of the principal symbol $\sigma_{-2n}(T) = \sigma(x,\xi)$ is taken in the fiber at x of the vector bundle in which T acts.

Now let $f_i, g_i \in C^\infty(V)$ and consider the quantized 1-form

$$T = \sum f_i[F, g_i].$$

The operators T, T^*T and $|T|^{2n} = (T^*T)^n$ are all pseudo-differential operators on V, acting on sections of the vector bundle $E = \wedge^n T^*_{\mathbb{C}}$, and of orders respectively -1, -2 and $-2n$. The principal symbol of T is given by

$$\sigma(T) = \sum f_i\{\sigma(F), g_i\}$$

where $\{\ ,\ \}$ is the Poisson bracket. As g_i is independent of the vector ξ, and $\sigma(F)(x,\xi) = \|\xi\|^{-2}(e_\xi\, i_\xi - i_\xi\, e_\xi)$, one gets that $\sigma(T)(x,\xi)$ depends only upon the value at x of the ordinary differential form $\alpha = \sum f_i\, dg_i$, and is given by

$$\sigma(T)(x,\xi) = -2\,\frac{\langle\alpha,\xi\rangle}{\|\xi\|^4}\,(e_\xi\, i_\xi - i_\xi\, e_\xi) + \|\xi\|^{-2}(e_\alpha\, i_\xi + e_\xi\, i_\alpha - i_\alpha\, e_\xi - i_\xi\, e_\alpha),$$

where we let e_ξ (resp. i_ξ) be exterior (resp. interior) multiplication by ξ.

Thus, for $\|\xi\| = 1$ we get $\sigma(T)(x,\xi) = 2(e_\alpha\, i_\xi + e_\xi\, i_\alpha) - 4\langle\alpha,\xi\rangle\, e_\xi\, i_\xi = 2(e_{\alpha'}\, i_\xi + e_\xi\, i_{\alpha'})$, where $\alpha' = \alpha - \langle\alpha,\xi\rangle\xi$. The computation of $|\sigma(T)|^2$ and of the trace of $\sigma(T)^{2n}\,(x,\xi)$, for $\|\xi\| = 1$, then gives

$$\operatorname{trace}(|\sigma(T)|^{2n})(x,\xi) = \|\alpha'\|^{2n}\, 2^{2n+1} \times \frac{(2(n-1))!}{((n-1)!)^2}.$$

Thus, we get the equality

$$\operatorname{Tr}_\omega\left(\left|\sum f_i[F,g_i]\right|^{2n}\right)$$
$$= c_n \int_V \left\|\sum f_i\, dg_i\right\|^{2n}, \quad c_n = \pi^{-n}\, n^{-1}((n-1)!)^{-2} \prod_{k=0}^{n-2}\left(n + \tfrac{1}{2} + k\right).$$

The right-hand side thus determines uniquely the L^{2n}-norm of 1-forms on V, and hence its conformal structure.

Let us work out in more detail the simplest example of the construction of the Fredholm module $(\mathcal{H}, F, \gamma)$ associated by Theorem 2 to an oriented conformal manifold. Thus, let $V = P_1(\mathbb{C})$ be the Riemann sphere. The Hilbert space $\mathcal{H}$ is the space of square integrable 1-forms, i.e. the direct sum $\mathcal{H} = \mathcal{H}^+ \oplus \mathcal{H}^-$ of the spaces of square integrable forms of type $(1,0)$ and $(0,1)$. Using the complex coordinate z in $P_1(\mathbb{C}) = \mathbb{C} \cup \{\infty\}$ we can write any element $\xi \in \mathcal{H}^\pm$ as $\xi(z)dz$ (resp. $\xi(z)d\bar{z}$) where ξ is a square integrable function on $\mathbb{C}$. With this notation the unitary operator S: $\mathcal{H}^- \to \mathcal{H}^+$ such that $F = \begin{bmatrix} 0 & S \\ S^* & 0 \end{bmatrix}$, is the complex Hilbert transform, given by

$$(S\xi)(z') = \frac{1}{2\pi i}\int_{\mathbb{C}} \frac{\xi(z)}{(z-z')^2}\, dz\, d\bar{z}$$

where the integral is defined as a Cauchy principal value, i.e. as the limit for $\varepsilon \to 0$ of the integral over $|z - z'| \geq \varepsilon$.

The operator S is canonically associated to the conformal structure of $P_1(\mathbb{C})$. Thus, the differential form $(z-z')^{-2}\, dz\, dz'$ on $P_1(\mathbb{C})\times P_1(\mathbb{C})$ is $SL(2,\mathbb{C})$ invariant. As an immediate corollary of Theorem 2 and of Proposition 2 of Section 1 we get:

Corollary 3. [Co-S-T] *For every integer $n \geq 1$, the following formula defines a $2n$-dimensional cyclic cocycle on the algebra $C^\alpha(P_1(\mathbb{C}))$ of Holder continuous functions of exponent $\alpha > \left(n + \frac{1}{2}\right)^{-1}$*

$$\tau_{2n}(f^0,f^1,\ldots,f^{2n}) = \int f^0(z_0)(f^1(z_1) - f^1(z_0))(f^2(z_2) - f^2(z_1))$$
$$\cdots (f^{2n}(z_0) - f^{2n}(z_{2n-1}))\ \omega(z_0,z_1,\ldots,z_{2n-1}) \quad \forall f^j \in C^\alpha(P_1(\mathbb{C}))$$

where ω is the differential form

$$(2\pi)^{-2n}\ \mathrm{im}(((z_0 - z_1)(\bar{z}_1 - \bar{z}_2)(z_2 - z_3)\cdots(\bar{z}_0 - \bar{z}_{2n-1}))^{-2})$$
$$dz_0\, d\bar{z}_0\, dz_1\, d\bar{z}_1 \cdots dz_{2n-1} d\bar{z}_{2n-1}.$$

The restriction of τ_{2n} to $C^\infty(P_1(\mathbb{C}))$ is cohomologous to $(-1)^{n-1}(n!)^{-1}\, S^{n-1}\tau_2$ where $\tau_2(f^0,f^1,f^2) = (\pi i)^{-1} \int f^0\, df^1 \wedge df^2$ for all $f^j \in C^\infty(P_1(\mathbb{C}))$.

One needs to know that $(\mathcal{H},F)$ is $(2n+1)$-summable on C^α, but this follows from Russo's theorem [Sim] applied to the kernel $\frac{f(z)-f(z')}{(z-z')^2}$ which corresponds to the commutator $[S,f]$. We refer to [R-S$_1$] for finer estimates involving Besov spaces.

Remark 4. The construction of Theorem 2 applies to arbitrary quasiconformal topological manifolds ([Co-S-T]) and yields local formulas for rational Pontryagin classes.

4.β Perturbation of Fredholm modules by the commutant von Neumann algebra

Let M be a von Neumann algebra and $M_2(M) = M_2(\mathbb{C}) \otimes M$. Let

$$G = \left\{\begin{bmatrix} a & b \\ b & a \end{bmatrix} \in M_2(M);\ \begin{bmatrix} a^* & -b^* \\ -b^* & a^* \end{bmatrix}\begin{bmatrix} a & b \\ b & a \end{bmatrix} = \begin{bmatrix} a & b \\ b & a \end{bmatrix}\begin{bmatrix} a^* & -b^* \\ -b^* & a^* \end{bmatrix} = 1\right\}.$$

In other words a and b are elements of M which fulfill the conditions

$$a^*a - b^*b = 1\ ,\ a^*b = b^*a\ ,\ aa^* - bb^* = 1\ ,\ ba^* = ab^*.$$

Proposition 5. a) *G is a subgroup of $GL_2(M)$ and is isomorphic to $GL_1(M)$.*

b) *Let $\mu = \mu^* \in M$, $\|\mu\| < 1$. Then $g(\mu) \in G$, where*

$$g(\mu) = \begin{bmatrix} a & b \\ b & a \end{bmatrix}\ ,\ a = (1-\mu^2)^{-1/2}\ ,\ b = \mu(1-\mu^2)^{-1/2}.$$

c) *Let* $\mathcal{U} = \left\{ \begin{bmatrix} u & 0 \\ 0 & u \end{bmatrix} ; u \in M , u^*u = uu^* = 1 \right\}$ *be the unitary group of M viewed as a subgroup of G. Then every element $g \in G$ is uniquely decomposable as $g = u\, g(\mu)$ for some $u \in \mathcal{U}$, $\mu \in M$, $\mu = \mu^*$, $\|\mu\| < 1$.*

Proof. a) Let $g_1, g_2 \in G$, with $g_j = \begin{bmatrix} a_j & b_j \\ b_j & a_j \end{bmatrix}$. Then one has

$$g_1 g_2 = \begin{bmatrix} a_1a_2 + b_1b_2 & a_1b_2 + b_1a_2 \\ a_1b_2 + b_1a_2 & a_1a_2 + b_1b_2 \end{bmatrix}$$

$$g_2^{-1}g_1^{-1} = \begin{bmatrix} a_2^* & -b_2^* \\ -b_2^* & a_2^* \end{bmatrix}\begin{bmatrix} a_1^* & -b_1^* \\ -b_1^* & a_1^* \end{bmatrix} = \begin{bmatrix} a_2^*a_1^* + b_2^*b_1^* & -b_2^*a_1^* - a_2^*b_1^* \\ -b_2^*a_1^* - a_2^*b_1^* & a_2^*a_1^* + b_2^*b_1^* \end{bmatrix}.$$

These equalities show that $g_1g_2 \in G$. They also show that the map $\begin{bmatrix} a & b \\ b & a \end{bmatrix} \to a + b$ is an isomorphism: $G \simeq GL_1(M)$.

b) Since $\|\mu\| < 1$, $(1 - \mu^2)^{-1/2}$ makes sense. By construction, $a = a^*$ and $b = b^*$ all commute with each other and $a^2 - b^2 = 1$. Thus, $g(\mu) \in G$.

c) Let $g = \begin{bmatrix} a & b \\ b & a \end{bmatrix} \in G$. One has $a^*a = 1 + b^*b \geq 1$, $aa^* = 1 + bb^* \geq 1$. Thus a is invertible, and we let u be the unitary of its polar decomposition: $a = u(a^*a)^{1/2}$. Replacing g by $\begin{bmatrix} u^{-1} & 0 \\ 0 & u^{-1} \end{bmatrix} g$, one can assume that a is positive. It follows then, using the equalities $b^*b = a^*a - 1 = aa^* - 1 = bb^*$, that b is normal, and $|b| = (a^2 - 1)^{1/2}$. Let $b = v|b|$ be the polar decomposition of b. Then v commutes with $|b|$, so that b commutes with a. The equality $ba = ab^*$ then shows that $b = b^*$, and it follows that $g = g(\mu)$ where $\mu = ba^{-1}$. One has $\|\mu\| < 1$ since $|b| = (a^2 - 1)^{1/2}$ and a is bounded.

Definition 6. *Let M be a von Neumann algebra. We let $\mu(M)$ be the above subgroup of $GL_2(M)$. If M is $\mathbb{Z}/2$-graded we let $\mu_{\text{ev}}(M)$ be the subgroup of $\mu(M)$ determined by the conditions:*

$$g = \begin{bmatrix} a & b \\ b & a \end{bmatrix} \in \mu_{\text{ev}} \text{ iff } a \text{ is even and } b \text{ is odd.}$$

Now let $(\mathcal{H}, F)$ be a Fredholm module over a C^*-algebra A, and let M be the commutant of A in $\mathcal{H}$. By construction, M is a von Neumann algebra, and it is $\mathbb{Z}/2$-graded when the Fredholm module is even. We shall describe a natural action of the group $\mu(M)$ (resp. $\mu_{\text{ev}}(M)$ in the even case) on the space of F's yielding a Fredholm module over A.

Proposition 7. *Let $g = \begin{bmatrix} a & b \\ b & a \end{bmatrix} \in \mu(M)$ (resp. $\mu_{\text{ev}}(M)$ in the even case). Then with $F' = g(F) = (aF + b)(bF + a)^{-1}$, the pair $(\mathcal{H}, F')$ is a Fredholm module over A. It is even if $g \in \mu_{\text{ev}}(M)$. Moreover, for any $x \in A$, the commutator $[F', x]$ belongs to the two-sided ideal generated by $[F, x]$.*

Proof. The equality $g(F) = (aF + b)F(aF + b)^{-1}$ shows that $g(F)^2 = 1$. To show that $g(F)^* = g(F)$ one has to check that

$$(aF + b)^*(bF + a) = (bF + a)^*(aF + b).$$

But this equality follows from the relations $a^*b = b^*a$ and $a^*a - b^*b = 1$. To conclude, we just need to compute $[F', x]$ in terms of $[F, x]$. One finds

$$\begin{aligned}&[(aF + b)(bF + a)^{-1}, x] \\ &\quad = a[F, x](bF + a)^{-1} - (aF + b)(bF + a)^{-1}b[F, x](bF + a)^{-1} \\ &\quad = (a - F'b)[F, x](bF + a)^{-1} = (bF + a)^{*-1}[F, x](bF + a)^{-1}.\end{aligned}$$

We have used the equality $(a - F'b)^{-1} = (bF + a)^*$.

Example 8. Let $(\mathcal{H}, F, \gamma)$ be the even Fredholm module on the C^*-algebra $C(P_1(\mathbb{C}))$ associated by Theorem 2 to the Riemann sphere, $V = P_1(\mathbb{C})$. The commutant $M = A'$ of $A = C(P_1(\mathbb{C}))$ in $\mathcal{H}$ is the von Neumann algebra of 2×2 matrices

$$a = \begin{bmatrix} f & u \\ v & g \end{bmatrix}$$

where f and g are measurable bounded functions on $V = P_1(\mathbb{C})$, and u and v are measurable bounded Beltrami differentials: $u(z, \bar{z})dz/d\bar{z}$, $v(z, \bar{z})d\bar{z}/dz$ [Ber$_1$]. In particular, an *odd* element $\mu \in M$, $\mu = \mu^*$ with $\|\mu\| < 1$, corresponds exactly to a single Beltrami differential $v(z, \bar{z})d\bar{z}/dz$, with $\|v\|_\infty < 1$ and v measurable, by the equality

$$\mu = \begin{bmatrix} 0 & v^* \\ v & 0 \end{bmatrix}.$$

Now by Proposition 5 c) all the relevant perturbations of a Fredholm module by the action of $\mu_{\text{ev}}(M)$ are obtained using the elements $g(\mu)$, μ odd, of Proposition 5 c). (The action of $\mathcal{U}$ just conjugates the Fredholm module to an equivalent one.) One checks by a direct calculation that for any $g(\mu) \in \mu_{\text{ev}}(M)$ the perturbed Fredholm module $(\mathcal{H}, g(\mu)(F))$ over $A = C(P_1(\mathbb{C}))$ is canonically isomorphic to the Fredholm module over A associated to the perturbed conformal structure on $P_1(\mathbb{C})$ associated to the measurable Beltrami differential $v(z, \bar{z})d\bar{z}/dz$.

The same interpretation of the construction of Proposition 7 holds for arbitrary Riemann surfaces. But the above case of $P_1(\mathbb{C})$ is particularly significant, since the measurable Riemann mapping theorem ([Ber]) is equivalent in that case to the *stability* of $(\mathcal{H}, F)$ under perturbations, i.e. the existence for any $g \in \mu_{\text{ev}}(M)$ of a unitary operator $U(g)$ in $\mathcal{H}$ such that:

α) $U(g)\ A\ U(g)^* = A$

β) $U(g)\ F\ U(g)^* = g(F)$.

Such a unitary is uniquely determined modulo the automorphism group $U(1) \times PSL(2, \mathbb{C})$ of the module $(\mathcal{H}, F)$.

We refer to [Ber$_1$] for a proof of the measurable Riemann mapping theorem based on the 2-dimensional Hilbert transform, i.e. on the above operator F.

4.γ The 4-dimensional analogue of the Polyakov action

In this section we shall use the quantized calculus to find the analogue in dimension 4 of the 2-dimensional Polyakov action, namely

$$I = \frac{1}{2\pi} \int_\Sigma \eta_{ij} \, dX^i \wedge * \, dX^j \tag{1}$$

for a Riemann surface Σ and a map X from Σ to a q-dimensional space $(\mathbb{R}^q, \eta)$.

Our first task will be to write the Polyakov action (1) as the Dixmier trace of the operator

$$\sum \eta_{ij} \, dX^i \, dX^j \tag{2}$$

where now $dX = [F, X]$ is the quantum differential of X taken using the canonical Fredholm module $(\mathcal{H}, F)$ of the Riemann surface Σ.

The same expression will then continue to make sense in dimension 4, i.e. with Σ replaced by a 4-dimensional conformal manifold. The action we shall get will be conformally invariant, by construction, and intimately related to the Einstein action of gravity.

In general, given an even-dimensional conformal manifold Σ, $\dim \Sigma = n = 2m$, we let $\mathcal{H} = L^2(\Sigma, \wedge_{\mathbb{C}}^m T^*)$ be, as above, the Hilbert space of square integrable forms of middle dimension, on which functions on Σ act as multiplication operators.

We let $F = 2P - 1$ be the operator on $\mathcal{H}$ obtained from the orthogonal projection P on the image of d. It is clear that both $\mathcal{H}$ and F only depend upon the conformal structure of Σ, which we assume to be *compact.*

In terms of an arbitrary Riemannian metric compatible with the conformal structure of Σ one has the formula

$$F = (dd^* - d^*d)(dd^* + d^*d)^{-1} \quad \text{on} \quad L^2(\Sigma, \wedge^m T^*) \tag{3}$$

which ignores the finite-dimensional subspace of harmonic forms, irrelevant in our later computations (cf. above for a definition of F taking this into account).

By construction, F is a pseudo-differential operator of order 0, whose principal symbol is given, as we saw above, by

$$\sigma_0(x, \xi) = (e_\xi \, i_\xi - i_\xi \, e_\xi) \, \|\xi\|^{-2}, \quad \forall (x, \xi) \in T^*(\Sigma).$$

When $n = \dim \Sigma = 2$, one has $\wedge_{\mathbb{C}}^m T^* = T_{\mathbb{C}}^*$ and σ_0 associates to any $\xi \neq 0$, $\xi \in T_x^*(\Sigma)$, the symmetry with axis ξ. For any function $f \in C^\infty(\Sigma)$, the operator $[F, f]$ is pseudo-differential of order -1. Its principal symbol is the Poisson bracket $\{\sigma_0, f\}$,

$$\{\sigma_0, f\}(x, \xi) = 2 \left(e_{df} \, i_\xi + e_\xi \, i_{df} - 2 e_\xi \, i_\xi \, \langle \xi, df \rangle \, \|\xi\|^{-2} \right) \|\xi\|^{-2}. \tag{4}$$

For $\|\xi\| = 1$, decompose df as $\langle df, \xi \rangle \xi + \eta$ where $\eta \perp \xi$. Then $\{\sigma_0, f\}(x, \xi) = 2(e_\eta \, i_\xi + e_\xi \, i_\eta)$, and its Hilbert-Schmidt norm, for $n = 2$, is given by

$$\operatorname{trace}\left(\{\sigma_0, f\}(x, \xi)^* \, \{\sigma_0, f\}(x, \xi)\right) = 8\|\eta\|^2, \quad \eta = df - \langle df, \xi \rangle \xi.$$

The Dixmier trace $\mathrm{Tr}_\omega\,(f_0[F,f_1]^*\,[F,f_2])$ is thus easy to compute for $n=2$, as the integral, on the unit sphere $S^*\Sigma$ of the cotangent bundle of Σ, of the function

$$\mathrm{trace}\,(f_0\,\{\sigma_0,f_1\}^*\,\{\sigma_0,f_2\}) = 8\,f_0(x)\,\langle df_1^\perp, df_2^\perp\rangle$$

where $df^\perp = df - \langle df,\xi\rangle\xi$ by convention. One thus gets:

Proposition 9. *Let Σ be a compact Riemann surface ($n=2$); then for any smooth map $X=(X^i)$ from Σ to $\mathbb{R}^d$ and metric $\eta_{ij}(x)$ on $\mathbb{R}^d$ one has*

$$\frac{1}{2\pi}\int_\Sigma \eta_{ij}\,dX^i\wedge *dX^j = -\tfrac{1}{2}\,\mathrm{Tr}_\omega\left(\eta_{ij}\,[F,X^i][F,X^j]\right).$$

Both sides of the equality have obvious meanings when the η_{ij} are constants. In general one just views them as functions on Σ, namely $\eta_{ij}\circ X$.

Let us now pass to the more involved 4-dimensional case. We want to compute the following action defined on smooth maps $X:\Sigma\to\mathbb{R}^d$ of a 4-dimensional compact conformal manifold Σ to $\mathbb{R}^d$, endowed with the metric $\eta_{ij}\,dx^i\,dx^j$:

$$I = \mathrm{Tr}_\omega\left(\eta_{ij}\,[F,X^i][F,X^j]\right). \tag{5}$$

Here we are beyond the natural domain of the Dixmier trace Tr_ω, but we can use the remarkable fact, due to Wodzicki, that it extends uniquely to a trace on the algebra of pseudo-differential operators (cf. [Wo$_2$]). After a lengthy computation (cf. [Co$_{31}$]) one obtains the following result:

Theorem 10. *Let Σ be a 4-dimensional conformal manifold, $X:\Sigma\to\mathbb{R}^d$ a smooth map, $\eta=\eta_{\mu\nu}\,dx^\mu\,dx^\nu$ a smooth metric on $\mathbb{R}^d$. One has*

$$\mathrm{Tr}_\omega\left(\eta_{\mu\nu}[F,X^\mu]\,[F,X^\nu]\right) = (16\pi^2)^{-1}\int_\Sigma \eta_{\mu\nu}\Big\{\frac{1}{3}\,r\,\langle dX^\mu,dX^\nu\rangle - \Delta\,\langle dX^\mu,dX^\nu\rangle + \langle\nabla\,dX^\mu,\nabla\,dX^\nu\rangle - \frac{1}{2}\,(\Delta X^\mu)(\Delta X^\nu)\Big\}\,dv$$

where r is the curvature scalar of Σ, dv its volume form, ∇ its covariant derivative, and Δ its Laplacian for an arbitrary Riemannian metric compatible with the given conformal structure.

Of course, the various terms of the formula, such as $\frac{1}{3}\,r\,\langle dX^\mu,dX^\nu\rangle$, are not themselves conformally invariant; only their sum is. It is also important to check that the right-hand side of the formula is, as the left-hand side obviously is, a Hochschild 2-cocycle. This allows one to double-check the constants in front of the various terms, except for the first one.

Theorem 10 gives a natural 4-dimensional analogue of the Polyakov action, and, in particular, in the special case when the $\eta_{\mu\nu}$ are constant, a natural conformally invariant action for scalar fields $X:\Sigma\to\mathbb{R}$,

$$I(X) = \mathrm{Tr}_\omega\left([F,X]^2\right) \tag{6}$$

which by Theorem 10, can be expressed in local terms, and defines an elliptic differential operator P of order 4 on Σ such that

$$I(X) = \int_{\Sigma} P(X)\, X\, dv. \tag{7}$$

This operator P is (up to normalization) equal to the Paneitz operator P (cf. [B-O]) already known to be the analogue of the scalar Laplacian in 4-dimensional conformal geometry.

Equation 7 uses the volume element dv so that P itself is not conformally invariant. Its principal symbol is

$$\sigma_4(P)\,(x,\xi) = \frac{1}{2}\,\|\xi\|^4 \tag{8}$$

which is *positive.*

The conformal anomaly of the functional integral

$$\int e^{-I(X)} \prod dX(x)$$

is that of $(\det P)^{-1/2}$ and can be computed (cf. [B-O]). The above discussion gives a very clear indication that the gravity theory induced from the above scalar field theory in dimension 4 should be of great interest, by analogy with the 2-dimensional case.

IV.5. Fredholm Modules and Rank-One Discrete Groups

As a rule, the construction of Fredholm modules over noncommutative spaces is a generalization of the theory of elliptic operators on a manifold. Let Γ be a discrete group and $\mathcal{A} = \mathbb{C}\Gamma$ the group ring of Γ. Then the corresponding noncommutative space is the "dual" $\hat{\Gamma}$ of Γ, and elliptic operators on $\hat{\Gamma}$ correspond to "multiplication" operators when described in the Fourier space $\ell^2(\Gamma)$. In this section we shall construct examples of Fredholm modules over $\mathbb{C}\Gamma$, or equivalently of Fredholm representations of Γ.

As an application we shall give a new proof of the beautiful result of M. Pimsner and D. Voiculescu that the reduced C^*-algebra of the free group on two generators does not contain any nontrivial idempotent [Pi-V$_3$]. This settled a long-standing conjecture of R.V. Kadison. We shall use a specific Fredholm module $(\mathcal{H},F)$ over the reduced C^*-algebra of the free group which already appears in [Pi-V$_3$] and in the simplified proof of J. Cuntz [Cu$_1$], and whose geometric meaning in terms of trees was clarified by P. Julg and A. Valette in [Ju-V].

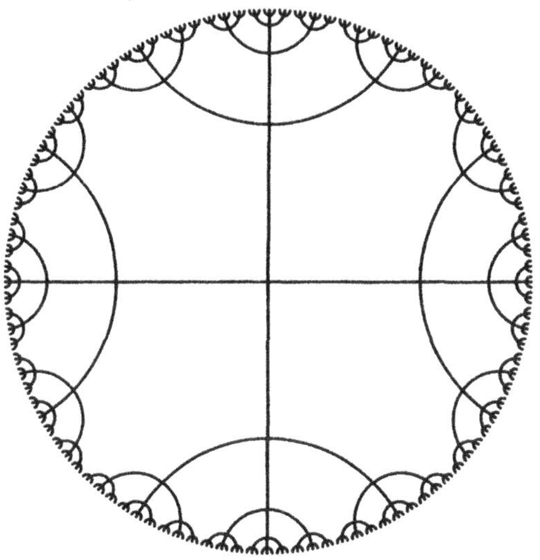

Figure IV.5. A tree

Now let Γ be an arbitrary free group, and T a tree on which Γ acts freely and transitively. By definition T is a 1-dimensional simplicial complex which is connected and simply connected. For $j = 0, 1$ let T^j be the set of j-simplices in T. Let $p \in T^0$ and $\varphi : T^0 \backslash \{p\} \to T^1$ be the bijection which associates to any $q \in T^0$, $q \neq p$, the only 1-simplex containing q and belonging to the interval $[p, q]$ (Figure 5). One readily checks that the bijection φ is *almost equivariant* in the following sense: for all $g \in \Gamma$ *one has* $\varphi(gq) = g\varphi(q)$ *except for finitely many* q's. Next, let $\mathcal{H}^+ = \ell^2(T^0)$, $\mathcal{H}^- = \ell^2(T^1) \oplus \mathbb{C}$. The action of Γ on T^0 and T^1 yields a $C_r^*(\Gamma)$-module structure on $\ell^2(T^j)$, $j = 0, 1$, and hence on $\mathcal{H}^\pm$ if we put

$$a(\xi, \lambda) = (a\xi, 0) \quad \forall \xi \in \ell^2(T^1)\,,\ \lambda \in \mathbb{C}\,,\ a \in C_r^*(\Gamma).$$

Let P be the unitary operator $P : \mathcal{H}^+ \to \mathcal{H}^-$ given by

$$P\varepsilon_p = (0, 1)\,,\ P\varepsilon_q = \varepsilon_{\varphi(q)} \quad \forall q \neq p$$

where for any set X, $(\varepsilon_x)_{x \in X}$ is the natural basis of $\ell^2(X)$. The almost equivariance of φ shows that:

Lemma 1. a) *The pair* $(\mathcal{H}, F)$, *where* $\mathcal{H} = \mathcal{H}^+ \oplus \mathcal{H}^-$, $F = \begin{bmatrix} 0 & P^{-1} \\ P & 0 \end{bmatrix}$, *is an even Fredholm module over* $A = C_r^*(\Gamma)$, *and* $\mathcal{A} = \{a; [F, a] \in \mathcal{L}^1(\mathcal{H})\}$ *is a dense subalgebra of* A.

b) *The* 0*-dimensional character of* $(\mathcal{H}, F)$ *is the canonical trace* τ *on* $C_r^*(\Gamma)$, $\tau(\sum a_g g) = a_e$, e *the unit of* Γ.

Proof. a) For any $g \in \Gamma$ the operator $gP - Pg$ is of finite rank, hence the group ring $\mathbb{C}\Gamma$ is contained in $\mathcal{A} = \{a \in C_r^*(\Gamma); [F,a] \in \mathcal{L}^1(\mathcal{H})\}$. As $\mathbb{C}\Gamma$ is dense in $C_r^*(\Gamma)$ the conclusion follows.

b) Let us compute the character of $(\mathcal{H}, F)$. Let $a \in \mathcal{A}$. Then $a - P^{-1}aP \in \mathcal{L}^1(\mathcal{H}^+)$, and

$$\frac{1}{2}\mathrm{Trace}(\gamma F[F,a]) = \mathrm{Trace}(a - P^{-1}aP).$$

Let τ be the unique positive trace on A such that $\tau(\sum a_g g) = a_e$ for any element $a = \sum a_g g$ of $\mathbb{C}\Gamma$, where $e \in \Gamma$ is the unit. Then for any $a \in A = C_r^*(\Gamma)$, $a - \tau(a)e$ belongs to the norm closure of the linear span of the elements $g \in \Gamma$, $g \neq e$.

Since the action of Γ on T^j is free, it follows that the diagonal entries in the matrix of $a - \tau(a)e$ in $\ell^2(T^j)$ are all equal to 0. This shows that for any $a \in \mathcal{A}$ one has,

$$\mathrm{Trace}(a - P^{-1}aP) = \tau(a)\ \mathrm{Trace}(e - P^{-1}eP) = \tau(a).$$

Thus, the character of $(\mathcal{H}, F)$ is the restriction of τ to $\mathcal{A}$, and since τ is faithful and positive on $C_r^*(\Gamma)$ we get:

Corollary 2. [Pi-V$_3$] *Let Γ be a free group. Then the reduced C^*-algebra $C_r^*(\Gamma)$ contains no nontrivial idempotent.*

Proof. By Corollary 1.5 one has $\tau(K_0(A)) \subset \mathbb{Z}$, which shows that if E is an idempotent, which can be assumed selfadjoint, then $\tau(E) \in \mathbb{Z} \cap [0,1] = \{0,1\}$. Thus, $E \in \{0,1\}$ since τ is faithful.

Remark 3. The proof of Lemma 1 shows that for any $a \in \mathbb{C}\Gamma$ the quantum differential da is a finite-rank operator. Then let $(\mathbb{C}\Gamma)^\sim$ be the smallest subalgebra $\mathcal{B}$ of $C_r^*(\Gamma)$ containing $\mathbb{C}\Gamma$ and having the property for any $n \in \mathbb{N}$

$$x \in M_n(\mathcal{B}) \cap M_n(C_r^*(\Gamma))^{-1} \Longrightarrow x \in M_n(\mathcal{B})^{-1}.$$

One easily checks that for any $a \in (\mathbb{C}\Gamma)^\sim$, the operator da is of finite rank. It is natural to conjecture that the converse holds. This can be proved when the free group Γ is equal to $\mathbb{Z}$. In this case the algebra $\mathcal{A}$ is the algebra of rational fractions with no pole on $\mathbb{S}^1 \subset P_1(\mathbb{C})$ and the conjecture follows from a classical result of Kronecker (Proposition 1 of Section 3).

We shall now pass to Fredholm representations due to Mishchenko [Mis$_1$] for discrete subgroups Γ of semisimple Lie groups G.

Therefore, let G be a semisimple Lie group, $H \subset G$ a maximal compact subgroup, and let $X = G/H$ be the corresponding homogeneous space over G endowed with a G-invariant Riemannian metric of non-positive curvature. Let $q = \dim G/H$ and $\rho : H \to \mathrm{Spin}^c(q)$ be a lifting of the isotropy representation to the Spin^c covering of $SO(q)$. Then let S be the associated G-equivariant Hermitian vector bundle of complex spinors on $X = G/H$. It is $\mathbb{Z}/2$-graded when q is even.

Given two points $x, y \in X$, $x \neq y$, we let $u(x,y) \in T_x(X)$ be the unit tangent vector at x tangent to the geodesic segment from x to y.

Proposition 4. [Mis$_1$] *Let Γ be a discrete subgroup of G and let $\alpha \in \Gamma\backslash X$ be an orbit of Γ in X. Let $a \in X$ with $a \notin \alpha$. Let Γ act by translations on the Hilbert space $\mathcal{H} = \ell^2(\alpha, S)$ of ℓ^2 sections of the restriction of the complex spinor bundle S to the orbit α. Then $(\mathcal{H}, F)$ is a Fredholm module over $C_r^*(\Gamma)$, where F is the following operator:*

$$(F\xi)(x) = \not{u}(x,a)\xi(x) \quad \forall x \in \alpha$$

where $\not{u}$ means Clifford xmultiplication by the vector u.

Note that $(\mathcal{H}, F)$ has the same parity as $q = \dim G/H$. One has $F^2 = 1$ since each $u(x,a)$ is a unit vector. To check that $[F,b]$ is compact for any $b \in C_r^*(\Gamma)$ it is enough to do it for elements $g \in \Gamma$, i.e. to check that $gFg^{-1} - F$ is compact. But gFg^{-1} is given by the same formula as F with the point $a \in X$ replaced by ga. As the orbit α is a discrete subset of X and as X has non-positive curvature, one has

$$\|u(x,a) - u(x,ga)\| \to 0 \text{ when } x \to \infty\,,\ x \in \alpha,$$

and the compactness of $gFg^{-1} - F$ follows.

The above proof shows that, with $\mathcal{A} = \mathbb{C}\Gamma \subset C_r^*(\Gamma)$, the above Fredholm module over $\mathcal{A}$ is p-*summable* iff there holds

$$\forall g \in \Gamma\ ,\ \sum_{x \in \alpha} \|u(x,a) - u(x,ga)\|^p < \infty.$$

When G is of real rank one, the symmetric space $X = G/H$ has strictly negative sectional curvature: $k \leq -\varepsilon < 0$. Thus, the comparison theorem shows that $\|u(x,a) - u(x,b)\|$ decreases, for fixed $a, b \in X$, like $\exp(-\varepsilon d(x,a))$ where d is the geodesic distance in X. Since the number of elements of the orbit α in the ball $B(a,R) = \{x \in X; d(x,a) \leq R\}$ is bounded by a constant times the volume of $B(a,R)$, i.e. by $C\exp(\lambda R)$, we get:

Proposition 5. *Let Γ be a discrete subgroup of a semisimple Lie group G of real rank one. Then the Fredholm module $(\mathcal{H}, F)$ of Proposition 4 is finitely summable over $\mathcal{A} = \mathbb{C}\Gamma$.*

Let us assume that Γ is torsion-free, so that it acts freely on X. Then the above construction of $(\mathcal{H}, F)$ can be reformulated using only the following map λ from Γ to the unit sphere S^{q-1} of $\mathbb{R}^q$:

$$\lambda(g) = u(x_0, g^{-1}a) \in T_{x_0}(X) = \mathbb{R}^q$$

where $x_0 \in \alpha$ is a base point in the orbit α.

Indeed, given a map λ from Γ to S^{q-1} we can let Γ act by left translations in the Hilbert space $\mathcal{H} = \ell^2(\Gamma) \otimes S(\mathbb{R}^q)$, where $S(\mathbb{R}^q)$ is the Spin representation of Spin(q), and we can define an operator F by the equality

$$(F\xi)(g) = \lambda(g)\xi(g) \quad \forall g \in \Gamma\,,\ \xi \in \mathcal{H}. \tag{$*$}$$

The only condition required for $(\mathcal{H}, F)$ to be a Fredholm module is then the following:

$$\forall g \in \Gamma \quad \|\lambda(gk) - \lambda(k)\| \to 0 \text{ when } k \to \infty \text{ in } \Gamma. \tag{$**$}$$

Proposition 6. *Let Γ be a discrete group and $\lambda : \Gamma \to S^{q-1}$ a map from Γ to the unit sphere of $\mathbb{R}^q$ satisfying $(**)$. Then let $\mathcal{H} = \ell^2(\Gamma) \otimes S(\mathbb{R}^q)$ be the tensor product of the left regular representation of Γ by the trivial representation in $S(\mathbb{R}^q)$. Let F be given by $(*)$. Then $(\mathcal{H}, F)$ is a Fredholm module over $C_r^*(\Gamma)$. It is p-summable over $\mathbb{C}\Gamma$ iff the following holds:*

$$\sum_{g\in\Gamma} \|\lambda(hg) - \lambda(g)\|^p < \infty \quad \forall h \in \Gamma.$$

For $g_0, \ldots, g_n \in \Gamma$ let us compute the operator

$$g_0[F, g_1] \cdots [F, g_n].$$

Since gFg^{-1} is given by Clifford multiplication by $\lambda(g^{-1}\cdot)$, we get the following formula:

$$(g_0[F, g_1] \cdots [F, g_n])\xi(g) = \omega(g)\xi((g_0 \cdots g_n)^{-1} g) \quad \forall g \in \Gamma$$

where ω is the map from Γ to $\mathrm{Cliff}_{\mathbb{C}}(\mathbb{R}^q)$ given

$$\omega(g) = (\lambda(k_0^{-1}g) - \lambda(k_1^{-1}g)) \cdots (\lambda(k_{n-1}^{-1}g) - \lambda(k_n^{-1}g)) \quad \forall g \in \Gamma$$

where $k_j = g_0 \cdots g_j \in \Gamma$.

We thus get:

Lemma 7. *With the notation of Proposition 6 and with n of the same parity as q, let $(\mathcal{H}, F)$ be $(n+1)$-summable. Then its n-dimensional character is given on $\mathbb{C}\Gamma$ by*

$$\tau(g_0, \ldots, g_n) = \begin{cases} 0 \text{ if } g_0 \cdots g_n \neq 1 \\ \lambda_n \sum_{g\in\Gamma} \mathrm{trace}\,(\gamma\omega(g)) \text{ if } g_0 \cdots g_n = 1 \end{cases}$$

(with λ_n as in Definition 1.3).

Here the trace is the trace on $\mathrm{Cliff}_{\mathbb{C}}(\mathbb{R}^q)$ coming from the representation in $S(\mathbb{R}^q)$ and the $\mathbb{Z}/2$-grading γ is used only in the even case. The computation of τ is particularly easy when $n = q$ with, say, q even. Let us then introduce the function from Γ^{n+1} to $\mathbb{C}$ given by

$$\sigma(k_0, k_1, \ldots, k_n) = \mathrm{trace}(\gamma(\lambda(k_0^{-1}) - \lambda(k_1^{-1})) \cdots (\lambda(k_{n-1}^{-1}) - \lambda(k_n^{-1}))).$$

Since $n = q$ one gets

$$\sigma(k_0,\ldots,k_n) = 2^{q/2}\, i^{q/2}(\lambda(k_0^{-1}) - \lambda(k_1^{-1})) \wedge \cdots \wedge (\lambda(k_{n-1}^{-1}) - \lambda(k_n^{-1}))$$

where we identify $\wedge^q \mathbb{R}^q$ with $\mathbb{R}$ using the orientation given by γ. In other words, up to a numerical factor, σ is the oriented volume in $\mathbb{R}^q$ of the simplex with vertices at the $\lambda(k_j^{-1})$, $j = 0,1,\ldots,q$. Thus, one has $b\sigma = 0$ where

$$(b\sigma)(k_0,\ldots,k_{n+1}) = \sum_{j=0}^{n+1} (-1)^j\, \sigma(k_0,\ldots,\check{k}_j,\ldots,k_{n+1}).$$

The same equality holds for every translate of σ,

$$(g\sigma)(k_0,\ldots,k_n) = \sigma(g^{-1}k_0,\ldots,g^{-1}k_n) \quad \forall k_i \in \Gamma\,,\ g \in \Gamma,$$

and hence for the sum

$$\tilde{\sigma} = \sum_{g\in\Gamma} g\sigma.$$

By construction, $\tilde{\sigma}$ is thus a cocycle in the complex (C^*, b) of Chapter III Section 1 which defines group cohomology, so that the following equality defines a group cocycle c:

$$c(g_1,\ldots,g_n) = \tilde{\sigma}(1,g_1,g_1g_2,\ldots,g_1\cdots g_n) \quad \forall g_i \in \Gamma.$$

As $\tilde{\sigma}$ is invariant under left translations, one has

$$c(g_1,\ldots,g_n) = \tilde{\sigma}(g_0,g_0g_1,\ldots,g_0g_1\cdots g_n) = \sum_{g\in\Gamma} \operatorname{trace}(\gamma\omega(g)).$$

Thus, for $g_0\cdots g_n = 1$ one gets

$$\tau(g_0,\ldots,g_n) = \lambda_n\, c(g_1,\ldots,g_n).$$

Lemma 8. *Let q be even and $(\mathcal{H},F)$ be $(q+1)$-summable. Then on $\mathbb{C}\Gamma$ the character of $(\mathcal{H},F)$ is the cyclic cocycle τ_c associated by Chapter III.1, Example 2, c) to the* group cocycle c *obtained from the sum $\tilde{\sigma} = \sum_{g\in\Gamma} g\sigma$, with*

$$\sigma(k_0,\ldots,k_q) = \text{ oriented volume of the simplex in } \mathbb{R}^q \text{ with the vertices } \lambda(k_j^{-1}).$$

We have been a bit careless in the above proof in using

$$\operatorname{Trace}(\gamma g_0[F,g_1]\cdots[F,g_n])$$

instead of $\operatorname{Tr}'(\gamma g_0[F,g_1]\cdots[F,g_n]) = \frac{1}{2}\operatorname{Trace}(\gamma F[F,g_0]\cdots[F,g_n])$. This nicety, however, does not arise in the examples to follow, where $(\mathcal{H},F)$ will be q-summable.

In general, given a proper Lipschitz map $\alpha : \Gamma \to \mathbb{R}^q$ from a finitely generated discrete group Γ with word metric to the Euclidean space $\mathbb{R}^q$, one has a natural pull-back map ([Co-G-M$_2$])

$$\alpha^* : H^*_{\text{comp}}(\mathbb{R}^q) \to H^*(\Gamma)$$

from the cohomology with compact supports in $\mathbb{R}^q$ to the group cohomology with real coefficients. Indeed, let ω be a (closed) differential form of degree q, $\omega \in C_c^\infty(\mathbb{R}^q, \wedge^q)$, with $\int_{\mathbb{R}^q} \omega = 1$, and let

$$a(g_0, \ldots, g_q) = \int_{s(\alpha(g_i))} \omega \quad \forall g_0, \ldots, g_q \in \Gamma$$

where $s(\alpha(g_i))$ is the q-simplex spanned by the $\alpha(g_i) \in \mathbb{R}^q$. Then the following sum defines a Γ-invariant element of C^q, with $b\tilde{a} = 0$ (cf. Chapter III Section 1):

$$\tilde{a}(g_0, \ldots, g_q) = \sum_{g \in \Gamma} a(g^{-1}g_0, \ldots, g^{-1}g_q) \quad \forall g_i \in \Gamma.$$

Note that the sum over Γ makes good sense since, for fixed $g_i \in \Gamma$, it involves only finitely many nonzero terms. The class of $\tilde{a}$ in $H^q(\Gamma, \mathbb{R})$ will be denoted

$$\alpha^*([\mathbb{R}^q]).$$

Next, given a discrete subgroup $\Gamma \subset G$ of a semisimple Lie group, we can apply the above procedure to any of the maps

$$\alpha_p : \Gamma \to T_p(G/K) \ , \ \alpha_p(g) = \exp_p^{-1}(g^{-1}(p))$$

where K is a maximal compact subgroup of G, $X = G/K$ is endowed with a G-invariant Riemannian metric of non-positive curvature, and $\exp_p$ is the corresponding exponential map.

The non-positivity of the curvature ensures that α_p is a contraction, and it is proper by construction. The resulting element $\alpha_p^*[\mathbb{R}^q]$, $q = \dim(G/K)$, is independent of the choice of p, and we shall call it the *volume* cocycle on Γ.

Theorem 9. *Let H^q be the q-dimensional hyperbolic space, and $\Gamma \subset SO(q,1)$ be a discrete group of isometries of H^q. Let $(\mathcal{H}, F)$ be the Fredholm module over $C_r^*(\Gamma)$ given by Proposition* 4.

a) *$(\mathcal{H}, F)$ is p-summable over $\mathbb{C}\Gamma$ iff the Poincaré series is convergent at p:*

$$\sum_{g \in \Gamma} e^{-p\, d(gx_0, a)} < \infty$$

where d is the hyperbolic distance.

b) *$(\mathcal{H}, F)$ is always q-summable over $\mathbb{C}\Gamma$, and its q-dimensional character is given by the cyclic cocycle τ_c, where c is the volume cocycle on Γ.*

Proof. a) The law of cosines for hyperbolic triangles

$$\cosh a = \cosh b \ \cosh c - \sinh b \ \sinh c \ \cos \alpha$$

shows that $\|u(gx_0, a) - u(gx_0, g_0 a)\|$ is of the order of $\exp(-d(gx_0, a))$ when g varies in Γ with $g_0 \in \Gamma$ fixed. Thus, $[F, g_0] \in \mathcal{L}^p \ \forall g_0 \in \Gamma$ iff $\sum e^{-p\, d(gx_0, a)} < \infty$, and a) follows.

b) The Poincaré series is always convergent for $p > q - 1$ (cf. [Su$_2$]) since the volume of the ball of radius R in H^q grows like $\exp((q-1)R)$. By Lemma

8 the computation of the character follows from the equality $\sigma = [\mathbb{R}^q]$, where σ is the (Alexander-Spanier) q-cocycle on $\mathbb{R}^q$ given by

$$\sigma(\xi_0, \dots, \xi_q) = q\text{-dim. volume of } s\left(\frac{\xi_i}{\|\xi_i\|}\right)$$

where $s\left(\frac{\xi_i}{\|\xi_i\|}\right)$ is the simplex spanned by the vertices $\frac{\xi_i}{\|\xi_i\|} \in S^{q-1}$. We refer to [Co$_{17}$] for a similar computation.

By Chapter III Section 5 Theorem 7, we know that for any bounded group cocycle c on a word hyperbolic group Γ, the cocycle τ_c extends by continuity to the pre-C^*-algebra $\mathcal{A}$ which is the closure of $\mathbb{C}\Gamma$ in $C_r^*(\Gamma)$ under holomorphic functional calculus. By construction, the character τ of the q-summable Fredholm module $(\mathcal{H}, F)$ is well defined on $\mathcal{A}$. It is thus natural to ask if the equality of Theorem 9 b) still holds in $HC^q(\mathcal{A})$.

For complex hyperbolic or quaternionic hyperbolic spaces the degree of summability of $(\mathcal{H}, F)$ is, in general, strictly larger than q, and so one should expect that 9 b) is replaced by the same equality in *periodic* cyclic cohomology.

The degree of summability given by Theorem 9 a) is not always an integer. The quasi-Fuchsian groups (cf. [Bow$_2$] and Section 3) are subgroups $\Gamma \subset PSL(2, \mathbb{C})$ for which the infimum of $\{p;\ (\mathcal{H}, F) \text{ is } p\text{-summable}\}$ is a real number $1 < p < 2$ which is the Hausdorff dimension of the limit set of Γ in $P_1(\mathbb{C})$.

IV.6. Elliptic Theory on the Noncommutative Torus $\mathbb{T}_\theta^2$ and the Quantum Hall Effect

As a rule, the construction of Fredholm modules over a noncommutative space is a generalization of the theory of elliptic differential operators on a manifold. In this section we shall first show that for the noncommutative space $\mathbb{T}_\theta^2$ given by the two-dimensional noncommutative torus, the elliptic theory adapts perfectly and yields, in particular, an index theorem for difference-differential operators on the real line. We shall then show, by means of an example, the role of topological invariants arising from noncommutative geometry in understanding *stable* numerical quantities, that is, quantities invariant under small variations of the natural parameters, which appear in the quantum physics of solids. The first examples of topological invariants associated with Schrödinger's equation are due independently to a mathematician, S. Novikov [No$_1$], and a physicist, D. Thouless [Tho]. The use of noncommutative geometry, which makes it possible to eliminate the rationality hypothesis of [No$_1$], is due to J. Bellissard [Bel$_2$]. We shall follow his work in sections β) and γ) below.

6.α Elliptic theory on $\mathbb{T}_\theta^2$

We have already met the noncommutative torus $\mathbb{T}_\theta^2$ in Chapter II arising from the Kronecker foliation (Section 8 β)) and in Chapter III where we computed its cyclic cohomology (Section 2 β)).

Let us recall that the topology of $\mathbb{T}_\theta^2$ is given by the C^*-algebra A_θ generated by two unitaries U_1 and U_2 such that

$$U_2 U_1 = \lambda U_1 U_2 \quad \lambda = \exp(2\pi i\theta),$$

while the smooth structure of $\mathbb{T}_\theta^2$ is given by the subalgebra $\mathcal{A}_\theta$ of A_θ;

$$\mathcal{A}_\theta = \{\sum a_{nm} U_1^n U_2^m \ ; \ a \in S(\mathbb{Z}^2)\}$$

where $S(\mathbb{Z}^2)$ is the linear space of sequences of rapid decay on $\mathbb{Z}^2$:

$$(|n|^k + |m|^k)|a_{n,m}| \text{ bounded for all } k > 0.$$

It is easy to check that $\mathcal{A}_\theta$ is a pre-C^*-algebra since it is stable under holomorphic functional calculus in A_θ.

We have seen (Theorem 14 of Section III.3) the classification of smooth vector bundles on $\mathbb{T}_\theta^2$, and in order to avoid unrevealing notational complications we shall choose one of them and expound the theory of elliptic operators acting on sections of this particular vector bundle. This space of sections is (cf. loc. cit.) the Schwartz space

$$\mathcal{E} = S(\mathbb{R})$$

of smooth functions ξ on $\mathbb{R}$ whose derivatives are of rapid decay. The right action of $\mathcal{A}_\theta$ on $\mathcal{E}$ is given (cf. loc. cit.) by

$$(\xi \cdot U_1)(s) = \xi(s+\theta) \quad \forall \xi \in S(\mathbb{R})\ , \ s \in \mathbb{R}$$

$$(\xi \cdot U_2)(s) = \exp(2\pi i s)\xi(s) \quad \forall \xi \in S(\mathbb{R})\ , \ s \in \mathbb{R}.$$

Note that while the presentation of $\mathcal{A}_\theta$ only uses $\lambda = \exp 2\pi i\theta$, i.e. the class of θ in $\mathbb{R}/\mathbb{Z}$, the action of $\mathcal{A}_\theta$ on $S(\mathbb{R})$ now involves an explicit choice for θ, which we take in the interval $]0,1]$.

By Proposition 15 of Section III.3, the vector fields on $\mathbb{T}_\theta^2$ given by the derivations δ_j, $j = 1,2$, such that

$$\delta_j(U_k) = 0 \text{ if } k \neq j\ , \ \delta_j(U_j) = 2\pi i\, U_j$$

lift to the smooth sections $\xi \in \mathcal{E}$ as the covariant derivatives

$$(\nabla_1 \xi)(s) = -2\pi i\, \frac{s}{\theta}\, \xi(s)\ , \ (\nabla_2 \xi)s = \frac{d\xi}{ds} \quad \forall \xi \in S(\mathbb{R}).$$

The exact form of these operators ∇_1 and ∇_2 is not relevant here since we shall be interested, in general, in operators D on $\mathcal{E} = S(\mathbb{R})$ of the form

$$D = \sum C_{\alpha,\beta}\, \nabla_1^\alpha\, \nabla_2^\beta$$

where the sum is a *finite* sum, while the coefficients $C_{\alpha,\beta}$ are operators of order 0 on $\mathcal{E}$, i.e. are elements of $\mathrm{End}_{\mathcal{A}_\theta}(\mathcal{E})$:

$$C_{\alpha,\beta} \in \mathrm{End}_{\mathcal{A}_\theta}(\mathcal{E}) \qquad \forall \alpha, \beta.$$

In the commutative case $\theta = 1$, say, the above operators D give exactly all the ordinary differential operators on the sections of the bundle (for $\theta = 1$ the corresponding bundle is a nontrivial line bundle). We shall now see that in the general case these operators D can still be treated in the same way as in the commutative case, and that the elliptic theory is available for them. Before we proceed let us note that an operator D as above is an *arbitrary element* of the algebra of operators on $S(\mathbb{R}) \subset L^2(\mathbb{R})$ generated by the following operators:

1) The operator of multiplication by s,

$$(T\xi)(s) = s\,\xi(s) \qquad \forall s \in \mathbb{R}$$

2) The operator d/ds of differentiation

3) The operator of multiplication by $e_\theta(s) = \exp(2\pi i\, s/\theta)$

4) The operator Δ of finite difference

$$(\Delta\xi)(s) = \xi(s+1) - \xi(s).$$

In fact, more precisely, the algebra $\mathrm{End}_{\mathcal{A}_\theta}(\mathcal{E})$ of endomorphisms of $\mathcal{E}$ is naturally isomorphic to $\mathcal{A}_{\theta'}$, $\theta' = 1/\theta$, with generators V_1 and V_2 such that

$$V_2 V_1 = \exp 2\pi i\theta' V_1 V_2$$

given by the formulae

$$(V_1\xi)(s) = \xi(s+1)$$
$$(V_2\xi)(s) = \exp\frac{-2\pi i s}{\theta}\,\xi(s).$$

Thus, one is also allowed infinite sums like $\sum a_{n,m} V_1^n V_2^m$ for $a \in S(\mathbb{Z}^2)$. They include , in particular, arbitrary smooth functions f, with period θ, acting by multiplication on $S(\mathbb{R})$.

By analogy with the commutative case we make the following definition:

Definition 1. a) *For $n \in \mathbb{N}$, we say that D is of order $\le n$ if*

$$D = \sum_{\alpha+\beta\le n} C_{\alpha,\beta}\ \nabla_1^\alpha \nabla_2^\beta\,; \quad C_{\alpha,\beta} \in \mathcal{A}_{\theta'} = \mathrm{End}_{\mathcal{A}_\theta}(\mathcal{E}).$$

b) *The symbol of order n, $\sigma_n(D)$, is defined as the map from $\mathbb{S}^1$ to $\mathcal{A}_{\theta'}$ given by $\sigma(\eta_1,\eta_2) = \sum_{\alpha+\beta=n} C_{\alpha,\beta}\eta_1^\alpha\eta_2^\beta$, for all $\eta_1,\eta_2 \in \mathbb{R}$ with $\eta_1^2+\eta_2^2 = 1$.*

c) *We say that D is* elliptic *if $\sigma(\eta_1,\eta_2)$ is an* invertible *element of $\mathcal{A}_{\theta'}$ for any $(\eta_1,\eta_2) \in \mathbb{S}^1$.*

Note that since $\mathcal{A}_{\theta'}$ is stable under holomorphic functional calculus in the C^*-algebra $A_{\theta'}$, which is its norm closure in the operators in $L^2(\mathbb{R})$, one could equivalently replace c) by the condition c′) $\sigma(\eta_1,\eta_2)$ is an invertible operator in $L^2(\mathbb{R})$ for any $(\eta_1,\eta_2)\in\mathbb{S}^1$.

We can now state the analogue in our case of the main results of the classical theory of elliptic differential operators.

Theorem 2. [Co13] *Let* $D=\sum C_{\alpha,\beta}\,\nabla_1^\alpha\nabla_2^\beta$ *be* elliptic *in the above sense. Then:*

a) *The space* $\ker D=\{\xi\in L^2(\mathbb{R});D\xi=0\}$ *is finite-dimensional.*

b) *Any* $\xi\in L^2(\mathbb{R})$ *such that* $D\xi=0$ *belongs to* $S(\mathbb{R})$.

Our next result is an index formula, that is, an explicit formula for the Fredholm index of D:

$$\operatorname{Index}D=\dim\ker D-\dim\ker D^*.$$

Here D^* is the Hilbert space adjoint of D in $L^2(\mathbb{R})$. One could equivalently consider the codimension of the image of D instead of $\dim\ker D^*$.

Our index formula will involve a specific cyclic cocycle on the algebra $C^\infty(\mathbb{S}^1,\mathcal{A}_{\theta'})$ of symbols. We have already computed in Chapter III the cyclic cohomology of $\mathcal{A}_{\theta'}$ and found as generators of HC^{even} the cyclic cocycles τ_0 and τ_2:

1)

$$\tau_0\left(\sum a_{n,m}V_1^nV_2^m\right)=a_{0,0}\quad\forall a\in S(\mathbb{Z}^2),$$

so that τ_0 is the canonical trace on $\mathcal{A}_{\theta'}$. Equivalently τ_0 is the Murray and von Neumann trace on the (hyperfinite) type II_1 factor generated by the operators V_1 and V_2 in $L^2(\mathbb{R})$.

2) τ_2 is the character of the following cycle on $\mathcal{A}_{\theta'}$ (cf. Chapter III Section 3 Proposition 15). One takes $\Omega^*=\mathcal{A}_{\theta'}\otimes\wedge^*\mathbb{C}^2$, the tensor product of $\mathcal{A}_{\theta'}$ by the graded algebra which is the exterior algebra of $\mathbb{C}^2$. The differential d is given by

$$d(a\otimes\alpha)=\delta_1'(a)e_1\wedge\alpha+\delta_2'(a)e_2\wedge\alpha\quad\forall a\in\mathcal{A}_{\theta'}\,,\ \alpha\in\wedge^*\mathbb{C}^2$$

where e_j, $j=1,2$, is the canonical basis of $\mathbb{C}^2$ and where, as for $\mathcal{A}_\theta$,

$$\delta_j'(V_k)=0\ \text{ if }\ j\neq k\,,\ \delta_j'(V_j)=2\pi iV_j\ ;\ j=1,2.$$

Finally the closed graded trace $\int:\Omega^2\to\mathbb{C}$ is given by

$$\int a\otimes(e_1\wedge e_2)=\tau_0(a)\quad\forall a\in\mathcal{A}_{\theta'}.$$

Thus, the formula for τ_2 is

$$\tau_2(a^0,a^1,a^2)=\tau_0(a^0(\delta_1'(a^1)\delta_2'(a^2)-\delta_2'(a^1)\delta_1'(a^2))).$$

By Corollary 16 of Chapter III Section 3 one has for any element E of $K_0(\mathcal{A}_{\theta'})$ that

$$\frac{1}{2\pi i}\,\tau_2(E,E,E)=q$$

where $q \in \mathbb{Z}$ is uniquely determined by the equality

$$\tau_0(E) = p + q\theta'', \quad p, q \in \mathbb{Z} \quad \text{and} \quad \theta'' = \theta' - [\theta'].$$

Let $\rho = [\mathbb{S}^1]$ be the fundamental class of $\mathbb{S}^1$, i.e. the cyclic 1-cocycle on $C^\infty(\mathbb{S}^1)$ given by

$$\rho(f^0, f^1) = \int_{\mathbb{S}^1} f^0 \, df^1,$$

or equivalently the de Rham differential algebra and integral. We then obtain two natural cyclic cocycles τ_1 and τ_3 on the algebra of symbols

$$B = C^\infty(\mathbb{S}^1, \mathcal{A}_{\theta'}) = C^\infty(\mathbb{S}^1)\hat{\otimes}\mathcal{A}_{\theta'}.$$

We thus let $\tau_1 = \rho \# \tau_0$ and $\tau_3 = \rho \# \tau_2$. They are defined on the algebraic tensor product $C^\infty(\mathbb{S}^1) \otimes \mathcal{A}_{\theta'}$, which is all we need for our symbols $\sum C_{\alpha,\beta} \xi_1^\alpha \xi_2^\beta$. They extend, however, by continuity to $C^\infty(\mathbb{S}^1, \mathcal{A}_{\theta'}) = C^\infty(\mathbb{S}^1)\hat{\otimes}\mathcal{A}_{\theta'}$. A straightforward calculation shows that:

a) $\tau_1(\sigma_0, \sigma_1) = \int_{\mathbb{S}^1} \tau_0(\sigma_0(t) \frac{d}{dt} \sigma_1(t))dt$

b) $\tau_3(\sigma_0, \sigma_1, \sigma_2, \sigma_3) = \int_{\mathbb{S}^1} \tau_0(\sigma_0 \, d\sigma_1 \wedge d\sigma_2 \wedge d\sigma_3)dt$

where we *define* $d\sigma_1 \wedge d\sigma_2 \wedge d\sigma_3$ as the element of $C^\infty(\mathbb{S}^1, \mathcal{A}_{\theta'})$ given by the map

$$t \mapsto \sum_{\pi \in S_3} \varepsilon(\pi) \, \partial_{\pi(1)} \, \sigma_1(t) \, \partial_{\pi(2)} \, \sigma_2(t) \, \partial_{\pi(3)} \, \sigma_3(t)$$

where $\varepsilon(\pi)$ is the signature of the permutation π of $\{1, 2, 3\}$, while the three derivations $\partial_1, \partial_2, \partial_3$ are given respectively by

$$(\partial_1\sigma)(t) = \delta_1'(\sigma(t)) \, , \ (\partial_2\sigma)(t) = \delta_2'(\sigma(t))$$

and

$$(\partial_3\sigma)(t) = \frac{d}{dt} \sigma(t) \, , \text{ for any } t \in \mathbb{S}^1 \, , \ \sigma \in C^\infty(\mathbb{S}^1, \mathcal{A}_{\theta'}).$$

In order to make these formulas more explicit, let us write any element a of $\mathcal{A}_{\theta'}$ as a Laurent series in V_1;

$$a = \sum_{n \in \mathbb{Z}} f_n \, V_1^n$$

where $f_n \in C^\infty(\mathbb{R}/\theta\mathbb{Z})$ is for each n a periodic function of period θ, $f_n = \sum a_{nm} V_2^m$. Of course the algebraic rule is that of the crossed product

$$ab = \sum f_n \, \alpha^n(g_n) \, V_1^{n+m}$$

with obvious notation and $\alpha(g)(s) = g(s+1) \ \ \forall s \in \mathbb{R}, \forall g \in C^\infty(\mathbb{R}/\theta\mathbb{Z})$.

With this notation, the normalized trace τ_0 is the integral of f_0 over one period, i.e.

$$\tau_0\left(\sum f_n \, V_1^n\right) = \frac{1}{\theta} \int_0^\theta f_0(s) \, ds.$$

Similarly, we can write an arbitrary element $\sigma \in C^\infty(\mathbb{S}^1, \mathcal{A}_{\theta'})$ as a Laurent series

$$\sigma = \sum_{n\in\mathbb{Z}} \sigma_n V_1^n$$

where each σ_n is a doubly periodic function $\sigma_n(t,s)$; $t \in \mathbb{S}^1$, $s \in \mathbb{R}/\theta\mathbb{Z}$. Then the derivations $\partial_1, \partial_2, \partial_3$ are given by

$$\partial_1\sigma = \sum_{n\in\mathbb{Z}} 2\pi i n \sigma_n V_1^n$$

$$(\partial_2\sigma)_n(t,s) = -\theta \frac{\partial}{\partial s} \sigma_n(t,s)$$

$$(\partial_3\sigma)_n(t,s) = \frac{\partial}{\partial t} \sigma_n(t,s).$$

One infers the formulas:

a′) $\tau_1(\sigma^0,\sigma^1) = \frac{1}{\theta}\int_0^\theta \int_{\mathbb{S}^1} (\sigma^0\partial_3\sigma^1)_0(t,s)\, dt\, ds$

b′) $\tau_3(\sigma^0,\sigma^1,\sigma^2,\sigma^3) = \frac{1}{\theta}\int_0^\theta \int_{\mathbb{S}^1} (\sigma^0 d\sigma^1 \wedge d\sigma^2 \wedge d\sigma^3)_0(t,s)\, dt\, ds.$

We can now state the analogue of the Atiyah-Singer index theorem for the elliptic operators D:

Theorem 3. [Co13] *Let* $D = \sum_{\alpha+\beta\le n} C_{\alpha,\beta}\nabla_1^\alpha\nabla_2^\beta$ *be elliptic, and let its principal symbol be* $\sigma(t) = \sum_{\alpha+\beta=n} C_{\alpha,\beta}(\cos t)^\alpha(\sin t)^\beta$. *Then*

$$\dim\ker D - \dim\ker D^* = \frac{1}{\theta}\frac{1}{(2\pi i)^2}\frac{1}{6}\tau_3(\sigma^{-1},\sigma,\sigma^{-1},\sigma) - \frac{1}{2\pi i}\tau_1(\sigma^{-1},\sigma).$$

We have thus obtained an index theorem for the highly *nonlocal* operators $D = \sum C_{\alpha\beta}\nabla_1^\alpha\nabla_2^\beta$ on the real line. As in the classical case of differential operators on manifolds, this index theorem has two important corollaries:

Corollary 4. a) *If* Index $D > 0$ *there exist nontrivial solutions* $f \in S(\mathbb{R})$ *of the equation* $Df = 0$.

b) *For any invertible* $\sigma \in C^\infty(\mathbb{S}^1, \mathcal{A}_{\theta'})$ *the following quantity is an integer:*

$$\frac{1}{\theta}\frac{1}{(2\pi i)^2}\frac{1}{6}\tau_3(\sigma^{-1},\sigma,\sigma^{-1},\sigma) - \frac{1}{2\pi i}\tau_1(\sigma^{-1},\sigma).$$

One obtains, in particular, an explanation for the integrality of $(2\pi i)^{-1}\tau_2$, meaning that $\langle\tau_2, K_0(\mathcal{A}_{\theta'})\rangle \subset 2\pi i\mathbb{Z}$. But we shall come back to this point in detail at the end of this section.

We shall now give nontrivial examples of elliptic operators D of the above form. Our first class of examples will only require *qualitative* information about the periodic functions $g_i \in C^\infty(\mathbb{R}/\theta\mathbb{Z})$ involved in the formula for it, which is

$$(Df)(s) = s\, f(s) - \sum_{k=-1}^{1} g_k(s)\, f'(s+k) \qquad \forall f \in S(\mathbb{R}).$$

The condition on the g_k is simply that for some $f_k \in C(\mathbb{R}/\theta\mathbb{Z})$, as in Figure 6, one has

$$\sum_{k=-1}^{1} |f_k(s) - g_k(s)| < 1 \qquad \forall s \in \mathbb{R}/\theta\mathbb{Z}.$$

This is easy to fulfill even with trigonometric polynomials.

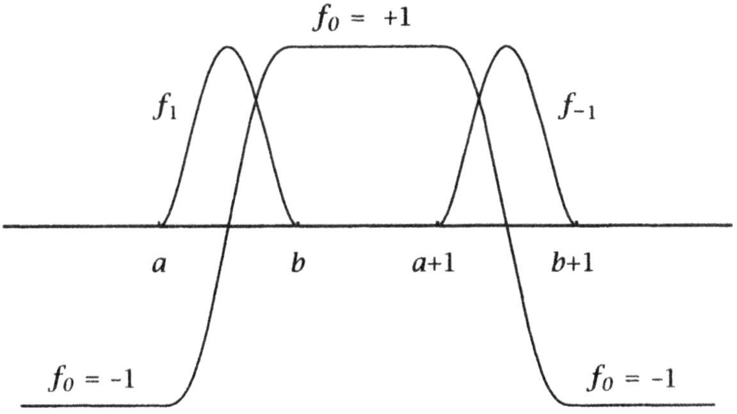

Figure IV.6. Powers-Rieffel idempotent: In this figure, we chose $I = [a, b]$ as an interval in the circle $\mathbb{R}/\theta\mathbb{Z} = \mathbb{T}$, such that $I \cap (I + 1) = \emptyset$ and let f be a continuous map from I to $[-1, 1]$ such that $f(a) = -1$, $f(b) = 1$. Then $f_0 \in C(\mathbb{T})$ is given by $f_0 = f$ on I, $f_0 = 1$ on $[b, a + 1]$, $f_0(s) = -f(s - 1)$ on $[a + 1, b + 1]$ and $f_0 = -1$ on $[b + 1, a]$. Also let $f_1 \in C(\mathbb{T})$ have support in I and satisfy $f_1^2 + f_0^2 = 1$ on I. Finally $f_{-1} \in C(\mathbb{T})$ is defined by

$$f_{-1}(s) = f_1(s - 1) \quad \forall s \in \mathbb{T}.$$

Corollary 5. *Let $g_i \in C^\infty(\mathbb{R}/\theta\mathbb{Z})$ be functions satisfying the above conditions. Then the operator D,*

$$(Df)(s) = s\, f(s) - \sum_{k=-1}^{1} g_k(s)\, f'(s + k) \qquad \forall f \in S(\mathbb{R}),$$

is elliptic in the sense of Definition 1. Its index is equal to

$$\operatorname{Index} D = 1 + 2[1/\theta]$$

where $[1/\theta]$ is the integral part of $1/\theta$. The equation $Df = 0$ admits at least $1 + 2[1/\theta]$ linearly independent real solutions $f \in S(\mathbb{R})$.

We thus get an existence theorem as well as a regularity result (Theorem 2). The function $[1/\theta]$ is discontinuous when $\theta \in \mathbb{N}^{-1} \subset]0, 1]$. The reason why this does not entail a contradiction is that when θ^{-1} gets close to an integer it becomes more and more difficult to find an interval I in $\mathbb{R}/\theta\mathbb{Z}$ such that I and

$I+1$ are disjoint, this being of course impossible for $\theta^{-1} \in \mathbb{N}$. The proof of Corollary 5 is a straightforward application of Theorem 3 or equivalently of the computation of τ_2 on the Powers-Rieffel idempotent $e_{\theta'} \in \mathcal{A}_{\theta'}$ (cf. [Ri$_1$][Co$_{13}$]).

One can construct many examples of elliptic operators with nontrivial indices simply by using the formula

$$(Df)(s) = s\, f(s) - (Tf')(s)$$

where the operator $T \in \mathcal{A}_{\theta'}$ is selfadjoint and invertible. This implies that $D = -\frac{\theta}{2\pi i} \nabla_1 - T\nabla_2$ is elliptic with principal symbol in the same class as

$$\sigma(t) = \cos t + iT \sin t.$$

The index formula of Theorem 3 reduces in that case to

$$\operatorname{Index} D = \frac{1}{2\pi i} \frac{2}{\theta} \tau_2(E,E,E) - \tau_0((2E-1))$$

where E is the spectral projection of T belonging to the interval $[0,+\infty[$,

$$E = 1_{[0,\infty[}\,(T).$$

One has $E \in \mathcal{A}_{\theta'}$ since this algebra is stable under holomorphic functional calculus.

For $q = \frac{1}{2\pi i} \tau_2(E,E,E) > 0$ one can rewrite the above formula as

$$\operatorname{Index} D = 1 + 2[q/\theta].$$

The use of the Chern character τ_2 in order to label the gaps of selfadjoint operators has been very successful in the hands of J. Bellissard, and we shall deal in detail with this in Subsection γ). Perhaps the simplest nontrivial example of a selfadjoint invertible $T \in \mathcal{A}_{\theta'}$ as above is the Peierls operator

$$(Tf)(s) = f(s+1) + f(s-1) + 2\cos\left(\frac{2\pi s}{\theta}\right) f(s) + \lambda\, f(s)$$

which, when θ is a Liouville number, is invertible for any λ outside a nowhere dense Cantor set K [B-S] on which the index changes discontinuously. We refer to [Mou] [C-E-Y] [B-S] for more information on the operator T and gap labelling. In [C-E-Y] the following q-analogue of the binomial formula is successfully used to show that the spectrum of T is a Cantor set when θ is a Liouville number:

$$(U+V)^n = \sum_{k=0}^{n} \binom{n}{k}_\lambda U^{n-k} V^k$$

whenever $VU = \lambda UV$ with λ a scalar (here $\lambda = \exp(2\pi i\theta')$), and where the Gaussian polynomial $\binom{n}{k}_\lambda$ is given by

$$\binom{n}{k}_\lambda = \frac{(1-\lambda^n)(1-\lambda^{n-1})\cdots(1-\lambda)}{(1-\lambda^k)\cdots(1-\lambda)(1-\lambda^{n-k})\cdots(1-\lambda)}.$$

As a corollary of the gap labelling of the Peierls operator and of Theorem 3 one gets the following result.

Corollary 6. *Let θ be a Liouville number, and N an integer. Then there exists $\lambda \in\ ]-2,2[$ such that the following difference-differential equation on $\mathbb{R}$ admits at least N linearly independent solutions $f \in S(\mathbb{R})$:*

$$s\, f(s) = f'(s+1) + f'(s-1) + \left(2\cos\frac{2\pi s}{\theta} + \lambda\right) f'(s).$$

We shall now pass to the application of the *integrality* of the cyclic cocycle τ_2 to the quantum Hall effect, due to J. Bellissard.

6.β The quantum Hall effect

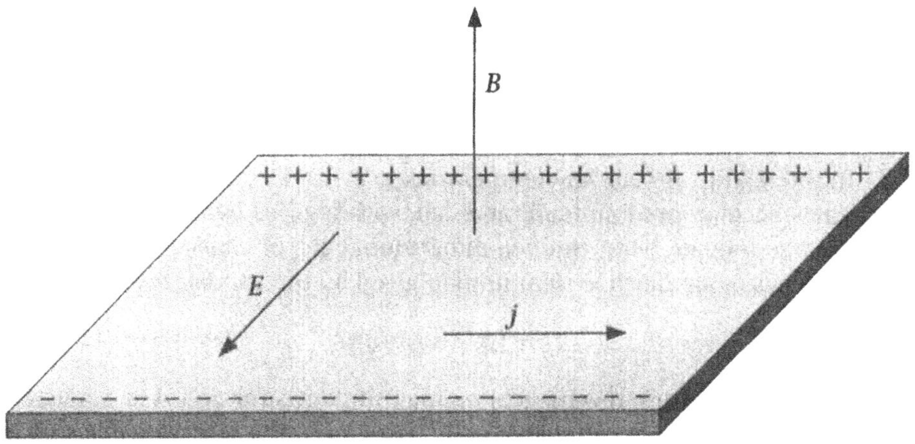

Figure IV.7. Hall current

Let us describe the experimental facts pertaining to the quantum Hall effect, starting with the classical Hall effect which goes back to 1880 ([Hal]). One considers (Figure 7) a very thin strip S of pure metal and a strong magnetic field B, uniform and perpendicular to the strip. Under these conditions elementary classical electrodynamics shows that particles of mass m and charge e in the plane S must move in circular orbits with angular frequency given by the "cyclotron frequency"

$$\omega_c = e\,|B|/m.$$

Let us view the charge carriers in S as a two-dimensional gas of classical charged particles with density N and charge e. Then if an additional electric field E is applied in the plane S (cf. Figure 7) there is a drift of the above circular orbits with velocity $|E|/|B|$ in the direction perpendicular to E. The resulting current density j perpendicular to both E and B is such that, in the stationary state the resulting force vanishes, i.e.

$$NeE + j \wedge B = 0.$$

In practice the charge carriers are scattered in a time which is short with respect to the cyclotron period, so that the observed current is mostly in the direction of the electric field E. It does, however, have a small component in the perpendicular direction, which is the *Hall current* given using the above formula by

$$j = NeB \wedge E/|B|^2,$$

which shows that $|j| = |Ne||E|/|B|$. In other words, the Hall conductivity σ_H, i.e. the ratio of the Hall current to the electric potential is equal in first approximation to a linear function of N

$$\sigma_H = Ne/|B|.$$

As early as 1880 ([Hall]) Hall observed the above drift current j and showed that the sign of the charge e may be negative or positive depending on the metal considered. This was the first evidence of what is now understood as electron or hole conduction.

In the regime of very low temperatures, $T \sim 1°K$, the effects of quantum mechanics become predominant, and can, with a gross oversimplification, be described as follows. First the two-dimensional gas of charge carriers, say of electrons, has a one-particle Hamiltonian given by the Landau formula

$$H = (p - eA)^2/2m$$

where p is the quantum mechanical momentum operator and A is a (classical) vector potential solution of $\text{curl}\, A = B$. Also, m is the effective mass of the charge carrier. It is immediate that the operators $K_j = p_j - eA_j$, $j = 1, 2$, satisfy the commutation relation

$$[K_1, K_2] = i\hbar\, e\, B$$

while $H = K^2/2m$. Thus, the energy levels of the charge carrier have discrete values which (up to a common additive constant) are all integer multiples of Planck's constant times the cyclotron frequency

$$E_n = n\hbar\, \omega_c.$$

Each of these "Landau levels" is highly degenerate, due to the translation invariance of the system, and can be filled by $\sim eB/h$ charge carriers per unit area.

A naive argument combining the filling of Landau levels with the drift velocity $|E|/|B|$ thus leads one to expect in the quantum regime a Hall current density given by

$$|j| = n(e^2/h)|E|$$

where n is the number of filled Landau levels and j is in a direction perpendicular to the electric field. In particular, this implies that the Hall conductivity σ_H is an integer multiple $\sigma_H = n\, e^2/h$ of e^2/h, provided the Fermi level is just in between two Landau levels. This argument, however, does not, in any way, account for the existence of the plateaus of conductivity which were discovered experimentally by K. von Klitzing, G. Dorda and M. Pepper ([Kl-D-P]). In their paper entitled "New Method for High-Accuracy Determination of the Fine-Structure Constant Based on Quantized Hall Resistance," the above three authors exhibited the quantization of the Hall conductivity, thus introducing the possibility of determining the fine structure constant $\alpha = \frac{e^2}{\hbar c}$ with an accuracy comparable to the best available methods.

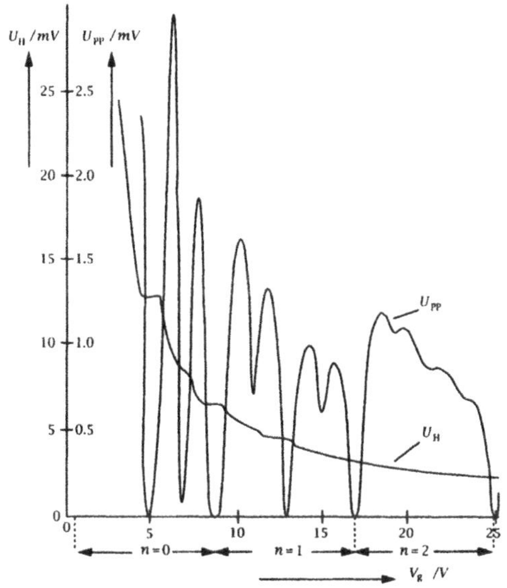

Figure IV.8. Plateaus of conductivity

We cannot end the above account of the experimental results without mentioning the fractional Hall effect found by D.C. Tsui, H. Størmer and A.C. Gossard. They observed in heterojunction devices with high electron mobility that, besides the plateaus corresponding to integral multiples of e^2/h there are other flat regions for the Hall voltage which correspond to *rational* values of $h/e^2\ \sigma_H$ with mostly odd denominators.

6.y The work of J. Bellissard on the integrality of σ_H

A first explanation for the integrality of σ_H on the plateaus of vanishing direct conductivity (cf. Figure 8) was given by Laughlin in 1981 [La] using the gauge invariance of the one-electron Hamiltonian together with a special topology of the sample. Then Avron and Seiler put the argument of [La] in rigorous mathematical form assuming that, on the plateau, the Fermi level (cf. below) belongs to a gap in the spectrum of the one-particle Hamiltonian. This approach is, however, unsatisfactory inasmuch as it uses a special sample topology and also does not account for the role of localised electron states, tied up with disorder, which imply that the plateaus cannot correspond to gaps in the spectrum of the one-particle Hamiltonian. To explain this more carefully we need to introduce a parameter other than the charge carrier density N. It is called the *Fermi level* μ, and plays the role of a chemical potential. In the approximation of a free Fermi gas, the thermal average of any observable quantity A at inverse temperature $\beta = 1/kT$ and chemical potential μ is given by

$$\langle A\rangle_{\beta,\mu} = \lim_{V\to\infty} \frac{1}{|V|} \operatorname{Trace}_V(f(H)A)$$

where Trace_V denotes the local trace of the operators in the finite volume $V \subset S$ and where f is the Fermi weight function

$$f(H) = \left(1 + e^{\beta(H-\mu)}\right)^{-1}.$$

In general, the Fermi level μ is adjusted so as to give the correct value to the charge carrier density

$$N = \lim_{V\to\infty} \frac{1}{|V|} \operatorname{Trace}_V(f(H)).$$

Note that the right-hand side $N(\beta,\mu)$ of this formula, in the limit of 0 temperature, i.e. $\beta \to +\infty$, depends only on the spectral projection E_μ of H on the interval $]-\infty,\mu]$, and is thus insensitive to the variation of μ in a spectral gap.

We shall now explain the results of J. Bellissard on the integrality of σ_H. They are directly in the line of the argument of Thouless, Kohmoto, Den Nijs and Nightingale [Tho-K-N-dN], who investigated the case of a perfectly periodic crystal with the hypothesis that the magnetic flux in units h/e is *rational*, an obviously unwarranted assumption. Their argument shows clearly that the origin of the integrality of σ_H is not the shape of the sample, but rather the topology of the so-called Brillouin zone in momentum space. When the magnetic flux is irrational this Brillouin zone becomes a noncommutative torus $\mathbb{T}^2_\theta$, and the integrality of σ_H is the integrality of the cyclic 2-cocycle τ_2 of section α) on the projection E_μ.

We shall first explain why this is true and then why, taking impurities into account, the integrality will persist. This will in particular eliminate the unwanted assumption that μ belong to a gap of the one-particle Hamiltonian, using the full force of the integrality result in Corollary 5 of Section 1 of this chapter.

The case of a periodic crystal.

Let us take as a model of metallic strip S the plane $\mathbb{R}^2$ with atoms at each vertex of a periodic lattice $\Gamma \subset \mathbb{R}^2$. The interaction of these atoms with the charge carrier, let us say the electron, thus modifies the one-particle Hamiltonian to

$$H = H_0 + V \ , \ H_0 = (p - eA)^2/2m$$

where the potential V is a Γ-periodic function on $\mathbb{R}^2$. The whole set-up is invariant under the group of plane translations belonging to Γ, so that we should get a corresponding projective representation of Γ on the one-particle quantum mechanical Hilbert space $\mathcal{H}$. We should normally write $\mathcal{H}$ as the space of L^2 sections of a complex line bundle L on $\mathbb{R}^2$ with constant curvature, but this just means that, viewing $\mathcal{H}$ as $L^2(\mathbb{R}^2)$, the correct action of the translation group is given by the unitaries, called magnetic translations

$$U(s) = \exp\ (\tfrac{i}{\hbar})(p + eA) \cdot s \qquad \forall s \in \mathbb{R}^2.$$

For $s \in \Gamma$ this unitary commutes with H, but, due to the curvature, the $U(s)$ do not commute with each other. For the generators e_1 and e_2 of Γ we have the commutation relation

$$U_2U_1 = \lambda U_1U_2 \ ; \quad \lambda = \exp\ 2\pi i\theta$$

where $U_j = U(e_j)$ and where θ is the flux of the magnetic field B through a fundamental domain for the lattice Γ, in dimensionless units. The role of the rationality of θ in the paper of Thouless et al. thus appears clearly, since we know that when θ is irrational the von Neumann algebra W of operators which have the symmetries U_ℓ, $\ell \in \Gamma$,

$$W = \{T \in \mathcal{L}(\mathcal{H})\ ;\ U_\ell T U_\ell^{-1} = T \quad \forall \ell \in \Gamma\}$$

is the hyperfinite factor of type II_∞, namely $R_{0,1}$.

In other words, if we investigate the operators which obey the natural invariance of the problem, we are not in a type I but in a type II_∞ situation. From the measure theory point of view the Brillouin zone is of type II. Moreover, the canonical trace τ on the factor W is given, using an averaging sequence V_j of compact subsets of $\mathbb{R}^2$, by

$$\tau(T) = \lim_{V \to \infty} \frac{1}{|V|}\ \mathrm{Trace}_V(T), \tag{1}$$

so that this part of the thermodynamic limit has a clear interpretation.

Since we need to understand the topology of the noncommutative Brillouin zone, we need a C^*-algebra $A \subset W$ of observables for our system. By construction, any bounded function $f(H)$ belongs to W, and in view of the formula giving the statistical average of observables, it is natural to require that A contain $f(H)$ for any $f \in C_0(\mathbb{R})$. The algebra obtained so far is commutative and is too small to allow the computation of the Hall conductivity. For that purpose we need another observable, which is the *current* j associated to the motion of the charge carrier. This current is a vector, given classically by

$$j = e\dot{X}$$

where X is the position of the charge carrier. Thus, in quantum mechanics we have

$$J = e\,\frac{i}{\hbar}\,[H,X], \tag{2}$$

where it is understood that both sides are pairs of operators, i.e. given by their components in a basis of $\mathbb{R}^2$

$$J_k = e\,\frac{i}{\hbar}\,[H,X_k]$$

with X_k the multiplication operator by the coordinate.

To understand clearly why J is invariant under the symmetries U_ℓ, $\ell \in \Gamma$, we can rewrite the formula (2) as

$$J = \frac{e}{\hbar}\,(\partial\,\alpha_s(H))_{s=0} \tag{3}$$

where the group $(\mathbb{R}^2)^\wedge$ dual of $\mathbb{R}^2$ acts by automorphisms α_s, $s \in (\mathbb{R}^2)^\wedge$, on the von Neumann algebra W by

$$\alpha_s(T) = e^{is\cdot X}\,T\,e^{-is\cdot X} \quad \forall T \in W.$$

We thus can take J as an observable (except for the trivial fact that since J is unbounded, just as for H we need to use $f(J)$, $f \in C_0(\mathbb{R}^2)$).

But in order to compute the Hall conductivity we also need to turn on an electric field E and see how our quantum statistical system reacts. This means that we replace the time evolution given by H, $\sigma_t(a) = e^{i\frac{t}{\hbar}H}\,a\,e^{-i\frac{t}{\hbar}H}$, by the time evolution associated to the perturbed Hamiltonian

$$H' = H + eE\cdot X,$$

or equivalently by the differential equation at $t = 0$

$$\frac{d}{dt}\,\sigma'_t(a) = \frac{i}{\hbar}[H,a] + \frac{e}{\hbar}\,\frac{d}{dt}\,\alpha_{tE}(a).$$

This makes it clear that the smallest C^*-algebra of observables appropriate for the computation of the Hall conductivity, besides containing $f(H)$, $f \in C_0(\mathbb{R})$ and $f(J)$, $f \in C_0(\mathbb{R}^2)$, should be invariant under the automorphism group α_s of W. In fact (cf. [Bel]), it is not difficult to see that the C^*-algebra A generated by the $\alpha_s(f(H))$ does contain the functions of the current, and is thus the natural algebra of observables for our problem. On $A \subset W$ we have the semifinite semicontinuous trace τ coming from the von Neumann algebra W (formula (1)) and the automorphism group (α_s) with as generators the derivations

$$\delta_j = (\partial_j\,\alpha_s)_{s=0}.$$

We thus get a densely defined cyclic 2-cocycle on A given by the formula

$$\tau_2(a_0,a_1,a_2) = \tau(a_0(\delta_1 a_1 \delta_2 a_2 - \delta_2 a_1 \delta_1 a_2)). \tag{5}$$

By construction, τ_2 is a 2-trace (cf. Chapter III Section 6 Definition 11) so that it pairs with $K_0(A)$. Now the crucial fact is the following formula, known as the

Kubo formula, for the Hall conductivity σ_H in the limit of 0 temperature, and assuming that the Fermi level μ *is in a gap* of the Hamiltonian H.

Lemma 7. [Bel] *If $\mu \notin$ Spec H then the Hall conductivity σ_H is given by*

$$\sigma_H = \frac{e^2}{h} \langle \tau_2, E_\mu \rangle = \frac{e^2}{h} \frac{1}{2\pi i} \tau_2(E_\mu, E_\mu, E_\mu)$$

where E_μ is the spectral projection of H corresponding to energies smaller than the Fermi level μ.

Thus, we see that the integrality of σ_H (in units of e^2/h) is implied by the integrality of the cyclic cocycle τ_2 on the C^*-algebra A. In the example at hand one can show that τ_2 is integral, as a corollary of Chapter III Section 3 Corollary 16. But we still want a conceptual reason for this integrality which survives in more difficult circumstances. This will be obtained from Corollary 5 of Section 1 by the construction of a Fredholm module over A whose Chern character is equal to τ_2.

Let us describe this even Fredholm module. The Hilbert space representation of A is just given by two copies of $\mathcal{H}$,

$$\mathcal{H}' = \mathcal{H} \otimes \mathbb{C}^2 = \mathcal{H}^+ \oplus \mathcal{H}^-$$

with the $\mathbb{Z}/2$-grading given by $\gamma = \begin{bmatrix} 1 & 0 \\ 0 & -1 \end{bmatrix}$.

The representation of A is thus $a \in A \mapsto a \otimes 1 = \begin{bmatrix} a & 0 \\ 0 & a \end{bmatrix}$. The operator F, $F = F^*$, $F^2 = 1$ in $\mathcal{H}'$ is given by

$$F = \begin{bmatrix} 0 & U \\ U^* & 0 \end{bmatrix}$$

where U is the operator in $L^2(\mathbb{R}^2)$ of multiplication by the function $u(x_1, x_2) = \frac{x_1 + ix_2}{|x_1 + ix_2|}$. It uses a Euclidean metric and orientation on $\mathbb{R}^2$.

Theorem 8. [Bel][Co$_{13}$] *The above triple $(\mathcal{H}', F, \gamma)$ is a $(2, \infty)$-summable Fredholm module over the C^*-algebra A, and its 2-dimensional character is equal to τ_2.*

The proof is the same as that of Proposition 5 and Lemma 7 of Section 5. It is interesting to relate more precisely the above construction with the construction of Section 5 and of Chapter III Section 4 (i.e. the index theorem for covering spaces). We shall briefly mention how this is done. First, with the notations of Section III.4, we take $\widetilde{M} = \mathbb{R}^2$ while the lattice Γ acts by translations on $\widetilde{M}$ with quotient the two-dimensional torus $M = \mathbb{R}^2/\Gamma$. Then let G be the fundamental groupoid of this covering (cf. III.4). Then the C^*-algebra A above is contained in (and coincides with, in the generic case) the C^*-algebra $C^*(G, \omega)$ of the smooth groupoid G twisted by the 2-cocycle ω on G associated to the line bundle L on $\mathbb{R}^2$. (One has a canonical homomorphism $\rho : G \to \mathbb{R}^2$ with $\rho(\tilde{x}, \tilde{y}) = \tilde{x} - \tilde{y}$, and $\omega = \rho^* c$, where $c \in Z^2(\mathbb{R}^2, U(1))$ corresponds to the Heisenberg central extension of $\mathbb{R}^2$ given by the magnetic field B.)

The strong Morita equivalence of $C^*(G)$ with $C^*(\Gamma)$ of Section III.4 adapts here to yield the strong Morita equivalence

$$C^*(G, \omega) \simeq A_\theta$$

where θ comes from the 2-cocycle $\omega|\Gamma \in H^2(\mathbb{Z}^2, U(1)) \simeq U(1)$. Under this equivalence and as in Section III.4, the cocycle τ_2 corresponds to the 2-cocycle φ on A_θ of Corollary 16 in Section III.3, which proves its integrality.

Next, the construction of the Fredholm module of Theorem 8 is a special case of the following variant of Proposition 3 of Section 5. Let M be a compact Riemannian manifold with non-positive curvature and β a closed 2-form on M. Then let $\Gamma = \pi_1(M)$ be its fundamental group, $\widetilde{M}$ the universal cover of M with the lifted metric, so that Γ acts by isometries on $\widetilde{M}$. Let L be a Hermitian line bundle on $\widetilde{M}$ with compatible connection ∇ and curvature equal to the lifted form $\tilde{\beta}$. This defines a 2-cocycle ω on the fundamental groupoid G of the covering $\widetilde{M}$. The value of ω on a pair $(\tilde{x}, \tilde{y})$, $(\tilde{y}, \tilde{z})$ of composable elements of G is given equivalently by the parallel transport in L around the geodesic triangle T with vertices $\tilde{x}, \tilde{y}, \tilde{z}$, or as

$$\omega(\tilde{x}, \tilde{y}, \tilde{z}) = \exp(2\pi i \int_T \tilde{\beta}).$$

By construction, the C^*-algebra $A = C^*(G, \omega)$ acts in the Hilbert space $\mathcal{H} = L^2(\widetilde{M}, L)$ of square integrable sections of the line bundle L, by the formula

$$(k\xi)(\tilde{x}) = \int k(\gamma)\, U(\gamma)\, \xi(\tilde{y})$$

where for $\gamma = (\tilde{x}, \tilde{y})/\Gamma \in G$, $U(\gamma)$ is the parallel transport in L along the geodesic from $\tilde{y}$ to $\tilde{x}$ in $\widetilde{M}$. Then let $a \in \widetilde{M}$ be a base point and S the Spin representation of $\mathrm{Spin}(T_a\widetilde{M})$. The following equality defines, as in 5.3, a Fredholm module over A:

a) $\mathcal{H}' = \mathcal{H} \otimes S$

b) For any $f \in A = C^*(G, \omega)$ one has $k(\xi \otimes s) = k\xi \otimes s$, $\forall \xi \in \mathcal{H}, s \in S$

c) F is the Clifford multiplication by the vector-valued function $u(\tilde{x}, a)$ which to $\tilde{x} \in \widetilde{M}$ associates the unit vector at a pointing towards $\tilde{x}$.

The degree of summability and the computation of the character are handled in the same way as in Section 5. We shall now close this digression and go back to the quantum Hall effect.

Real samples and localized states.

When the Fermi level μ lies in a gap I of the spectrum of H the (0-temperature limit of the) charge carrier density N as given by the formula above, is insensitive to the variation of μ. In real samples, due to the presence of a small amount of impurities, the dependence of N on μ is never of this kind, and the parameters μ and N are equivalent parameters for real samples. However, it introduces an essential new difficulty: since the spectrum of the energy H is now connected, say equal to $[0, +\infty[$, it is no longer true on the plateaus of conductivity that the spectral projection E_μ belongs to the C^*-algebra of continuous functions $f(H)$, $f \in C_0(\mathbb{R})$, since the characteristic function of $]-\infty, \mu]$ is not continuous on Spec H.

The solution of this difficulty is quite remarkable (cf. [Bel]) and deserves the attention of the reader. The qualitative idea from physics is that while the impurities eliminate the gaps in the spectrum of H, they in fact fill these gaps mostly by electron states which are localized and do not contribute to the conductivity. This is why one observes experimentally the plateaus of the conductivity as a function of the Fermi level. This qualitative idea is put on a rigorous mathematical basis by [B-G-M-S] using [F-S] and [F-M-S-S] for realistic models. Now the meaning of the *localization* of the electron states with energies in a small interval $]\mu - \varepsilon, \mu + \varepsilon[$ around the Fermi level μ is the following:

Lemma 9. *If μ lies in a gap of extended states of H then the characteristic function $E_\mu(\lambda) = (1 \text{ if } \lambda \le \mu \,;\, 0 \text{ if } \lambda > \mu)$ is* quasicontinuous *on* Spec H.

We are using here terminology from function theory which we already met in Section 3 when we developed our calculus on $\mathbb{S}^1$. Thanks to Proposition 2 of that section we can adopt the following general notion of quantum calculus.

Definition 10. *Let A be a C^*-algebra and $(\mathcal{H}, F)$ a Fredholm module over A. Let W be the von Neumann algebra weak closure of A. Then an element $f \in W$ is called* quasicontinuous *if*

$$[F, f] \in \mathcal{K}$$

(i.e. is a compact operator).

It goes without saying that elements of A are called continuous, this referring to the case when A is commutative so that $A = C_0(X)$, with $X = \operatorname{Spec} A$. In the case of Lemma 9 the C^*-algebra A is the algebra of functions $f(H)$, $f \in C_0(\mathbb{R})$, in the Hilbert space $L^2(\mathbb{R}^2)$, and the Fredholm module is the same as in Theorem 8 above. Thus, we have

$$\mathcal{H}' = L^2(\mathbb{R}^2) \otimes \mathbb{C}^2 = \mathcal{H}^+ \oplus \mathcal{H}^-.$$

The representation of $f(H)$ is by $f(H) \otimes 1 = \begin{bmatrix} f(H) & 0 \\ 0 & f(H) \end{bmatrix}$, and the operator F is $F = \begin{bmatrix} 0 & U \\ U^* & 0 \end{bmatrix}$ as above.

We shall see that, exactly as above, the Hall conductivity for μ in a gap of extended states is given by the pairing between the K-homology class of $(\mathcal{H}', F, \gamma)$ and the spectral projection E_μ, and is hence an *integer*: the Fredholm index of the operator $E_\mu U E_\mu$ (cf. Proposition 2 of the introduction to this chapter). As above this equality relies on Proposition 4 of Section 1, i.e. the computation of the character of $(\mathcal{H}', F, \gamma)$.

Let us now describe how this is done, taking into account the random parameter ω. This parameter ω belongs to a probability space (Ω, P), which is the configuration space for impurities. To the translation invariance of the system corresponds an action T of the translation group $\mathbb{R}^2$ by automorphisms of the probability space (Ω, P), such that the magnetic translations $U(X)$ satisfy

$$U(X)\, H_\omega\, U(X)^* = H_{T(X)\omega} \qquad \forall \omega \in \Omega\,,\ X \in \mathbb{R}^2.$$

There is a natural topology on Ω given by the weak topology on $\mathcal{L}(L^2(\mathbb{R}^2))$ pulled back by the map

$$\omega \mapsto V_\omega = (H_\omega + i)^{-1} \ :\Omega \to \mathcal{L}(\mathcal{H}).$$

Moreover, it is harmless to take the weak closure $\overline{V(\Omega)}$, and hence to assume that Ω is a compact space. Then the analogue of the C^*-algebra $C^*(G,\omega)$ of the periodic case is the twisted crossed product C^*-algebra

$$A = C(\Omega) \rtimes_{T,c} \mathbb{R}^2.$$

Recall (cf. Chapter II Appendix C) that any $f \in C_c(\mathbb{R}^2, C(\Omega))$ defines an element of A which we denote by

$$\int f_X\, u_X\, dX,$$

with the following algebraic rules

$$u_X\, f\, u_X^* = f \circ T_X^{-1} \qquad \forall X \in \mathbb{R}^2\,,\ \forall f \in C(\Omega)$$

$$u_X\, u_Y = c(X,Y)\, u_{X+Y} \qquad \forall X, Y \in \mathbb{R}^2$$

where $c \in Z^2(\mathbb{R}^2, U(1))$ is the Heisenberg 2-cocycle.

By construction, every $p \in \Omega$ yields a unitary representation π_p of A in $L^2(\mathbb{R}^2)$ given by

$$\pi_p\left(\int f_X\, u_X\, dX\right) = \int \pi_p(f_X)\, U(X)\, dX$$

where $U(X)$ is the magnetic translation operator (1), while $\pi_p(f)$ for $f \in C(\Omega)$ is the operator in $L^2(\mathbb{R}^2)$ of multiplication by the restriction of f to the orbit of p

$$\left(\pi_p(f)\xi\right)(Y) = f\,(T_{-Y}(p))\ \xi(Y) \qquad \forall Y \in \mathbb{R}^2\,,\ \xi \in L^2(\mathbb{R}^2).$$

The C^*-algebra norm on A is given by

$$\|f\| = \sup_{p\in\Omega}\ \|\pi_p(f)\|,$$

and $C_c(\mathbb{R}^2, C(\Omega))$ is a dense involutive subalgebra of A (cf. Chapter II Appendix C).

Let τ be the trace on A dual to the probability measure P. It is semifinite and semicontinuous, and for any element $f = \int f_X\, u_X\, dX$, $f \in C_c(\mathbb{R}^2, C(\Omega))$, one has

$$\tau(f^* f) = \int_X \int_\Omega |f(X,p)|^2\, dX\, dp.$$

If we assume ergodicity of the action T of $\mathbb{R}^2$ on (Ω, P), this trace τ coincides, using the ergodic theorem, with the trace per unit volume in almost all the representations π_p

$$\tau(f) = \lim_{V\to\infty} \frac{1}{|V|} \operatorname{Trace}_V(\pi_p(f)).$$

Next, the dual action (Chapter II Appendix C Proposition 4) of $\mathbb{R}_2 = \widehat{\mathbb{R}}^2$ on A is given by

$$\alpha_s\left(\int f_X\, u_X\, dX\right) = \int f_X\, e^{is\cdot X}\, u_X\, dX$$

$\forall f \in C_c(\mathbb{R}^2, C(\Omega))$, $\forall s \in \mathbb{R}_2$.

As in the periodic case, we get a densely defined cyclic 2-cocycle on A by the formula

$$\tau_2(f_0, f_1, f_2) = \tau(f_0(\delta_1 f_1\, \delta_2 f_2 - \delta_2 f_1\, \delta_1 f_2))$$

where δ_j is the derivation $\delta_j = (\partial_j \alpha_s)_{s=0}$.

This formula continues to make sense when the f_j are replaced by elements f (in the domains of the δ_i) of the von Neumann algebra W, the weak closure of A in the (type II) representation associated to the trace τ, such that

$$\tau(\delta_j(f)^*\, \delta_j(f)) < \infty.$$

The analogue of Lemma 7 above is now:

Lemma 11. [Bel] *If μ is in a gap of extended states, then the Hall conductivity σ_H is given by*

$$\sigma_H = \frac{e^2}{h}\frac{1}{2\pi i}\, \tau_2(E_\mu, E_\mu, E_\mu)$$

where E_μ is the spectral projection of H corresponding to energies smaller than the Fermi level.

Moreover, the integrality of τ_2 on E_μ follows exactly as above using the Fredholm modules over A given, for any $p \in \Omega$, by the representation $\pi_p \otimes 1$ of A in $\mathcal{H}' = L^2(\mathbb{R}^2) \otimes \mathbb{C}^2$, the same operator $F = \begin{bmatrix} 0 & U \\ U^* & 0 \end{bmatrix}$, and the $\mathbb{Z}/2$-grading $\gamma = \begin{bmatrix} 1 & 0 \\ 0 & -1 \end{bmatrix}$. The only new ingredient is that one computes the integral over Ω of the characters of these Fredholm modules over A to obtain τ_2. The remarkable fact is that the resulting index formula (which implies the integrality)

$$\frac{1}{2\pi i}\, \tau_2(E_\mu, E_\mu, E_\mu) = \operatorname{Index}(E_\mu U E_\mu)_p \quad \text{a.e. on } \Omega$$

continues to hold for the quasicontinuous projections involved in Lemma 11 (cf. [Bel]).

Clearly, all the above discussion of the quantum Hall effect is done in the approximation which neglects the mutual interaction of the electrons which is supposed to be responsible for the fractional Hall effect (cf. section β)). There are tentative explanations of the latter effect (cf. [Fr] and the literature there) based on conformal field theory, which in particular yield the odd denominators $2k + 1$. In order to adapt the above discussion to the interacting case it is necessary to extend the *finite-dimensional* tools used above (cyclic cohomology and finitely summable Fredholm modules) to the infinite-dimensional situation of an indefinite number of particles, i.e. of quantum field theory. This extension will be done in the next sections.

We should insist, however, on the *infrared* nature of the quantum Hall effect or of the determination of the fine structure constant, as opposed to the *ultraviolet* nature of the Fredholm modules associated to supersymmetric quantum field theories in Section 9. In the quantum Hall effect it is the differentiable structure of the Brillouin zone, i.e. of *momentum space*, which is playing a role.

Remark 12. It ought to be clear to the reader how the construction of the C^*-algebra A and of the Fredholm modules $(\mathcal{H}', F, \gamma)$ fits with the example of Penrose tilings in Chapter II or of foliations, also in Chapter II. The construction of the Fredholm modules adapts to foliations of compact manifolds with a leaf L of non-positive curvature. The Hilbert space $\mathcal{H}$ is then the tensor product of $L^2(\tilde{L})$ by the Spin representation of Spin(k), k the dimension of the leaves. The C^*-algebra of the foliation acts in $L^2(\tilde{L})$ by the unitary representation π_x of Section II.8 α), where x is a chosen origin, $x \in \tilde{L}$. The operator F in $L^2(\tilde{L}) \otimes S$ is given by Clifford multiplication

$$(F\xi)(y) = \not{u}(y,x)\ \xi(y)$$

where $u(y,x) \in T_x(\tilde{L})$ is the unit tangent vector to the geodesic from x to y in $\tilde{L}$.

IV.7. Entire Cyclic Cohomology

The results of Section 1 on the characters of Fredholm modules are limited to the *finite-dimensional case* by the hypothesis of finite summability of the modules (Definition 3 of the introduction). This hypothesis does not hold in examples coming from higher rank Lie groups (Section 5) or quantum field theory, but is replaced by the weaker θ-summability condition (Definition 4 of the introduction). Exactly as cyclic cohomology was dictated as the natural receptacle for the characters of finitely summable Fredholm modules ([Co_{17}]), the θ-summability condition (in its unbounded form, cf. Section 8) dictates, as a receptacle for the character, the *entire cyclic cohomology* ([Co24]). We shall first explain this theory and its pairing with K-theory. The construction of the character ([Co24][J-L-O]) will be done in Section 8.

7.α Entire cyclic cohomology of Banach algebras

Let A be a unital Banach algebra over $\mathbb{C}$. Let us recall the construction (Chapter III Section 1) of the fundamental (b, B) bicomplex of cyclic cohomology. For any integer $n \in \mathbb{N}$, one lets $C^n(A, A^*)$ be the space of continuous $(n+1)$-linear forms ϕ on A. For $n < 0$ one sets $C^n = \{0\}$. One defines two differentials b, B as follows:

$$1)\ b : C^n \to C^{n+1},$$
$$(b\phi)(a^0, \dots, a^{n+1}) = \sum_{j=0}^{n} (-1)^j \, \phi(a^0, \dots, a^j a^{j+1}, \dots, a^{n+1}) + (-1)^{n+1} \, \phi(a^{n+1}a^0, \dots, a^n),$$

$$2)\ B : C^n \to C^{n-1},\ B\phi = AB_0\phi, \text{ where}$$
$$(B_0\phi)(a^0, \dots, a^{n-1}) = \phi(1, a^0, \dots, a^{n-1}) - (-1)^n \, \phi(a^0, \dots, a^{n-1}, 1) \quad \forall \phi \in C^n,$$
$$(A\psi)(a^0, \dots, a^{n-1}) = \sum_{0}^{n-1} (-1)^{(n-1)j} \, \psi(a^j, a^{j+1}, \dots, a^{j-1}) \quad \forall \psi \in C^{n-1}.$$

By Lemma 19 of Section III.1 one has $b^2 = B^2 = 0$ and $bB = -Bb$, so that one obtains a bicomplex $(C^{n,m}; d_1, d_2)$, where $C^{n,m} = C^{n-m}$ for any $n, m \in \mathbb{Z}$,

$$d_1 = (n - m + 1)b : C^{n,m} \to C^{n+1,m} \ ; \ d_2 = \frac{B}{n-m} : C^{n,m} \to C^{n,m+1}.$$

The main lemma (Section III.1 Lemma 25) asserts that the b cohomology of the complex $\ker B / \operatorname{im} B$ is zero, so that the spectral sequence associated to the first filtration has the E_2 term equal to 0. Since the bicomplex $C^{n,m}$ has support in $\{(n, m); (n+m) \geq 0\}$ this spectral sequence does not converge, in general, when we take cochains with *finite* support, and by Section III.1 Theorem 29, the cohomology of the bicomplex, when taken with finite supports, is exactly the periodic cyclic cohomology $H^*(A)$. If we take cochains with arbitrary supports, without any control of their growth, then by the above Lemma 25 we get a trivial cohomology. It turns out, however, that provided we control the growth of $\|\phi_m\|$ in a cochain (ϕ_{2n}) or (ϕ_{2n+1}) of the (b, B) bicomplex, we then get the relevant cohomology to analyze infinite-dimensional spaces and cycles. Because of the periodicity $C^{n,m} \to C^{n+1,m+1}$ in the bicomplex (b, B), it is convenient just to work with

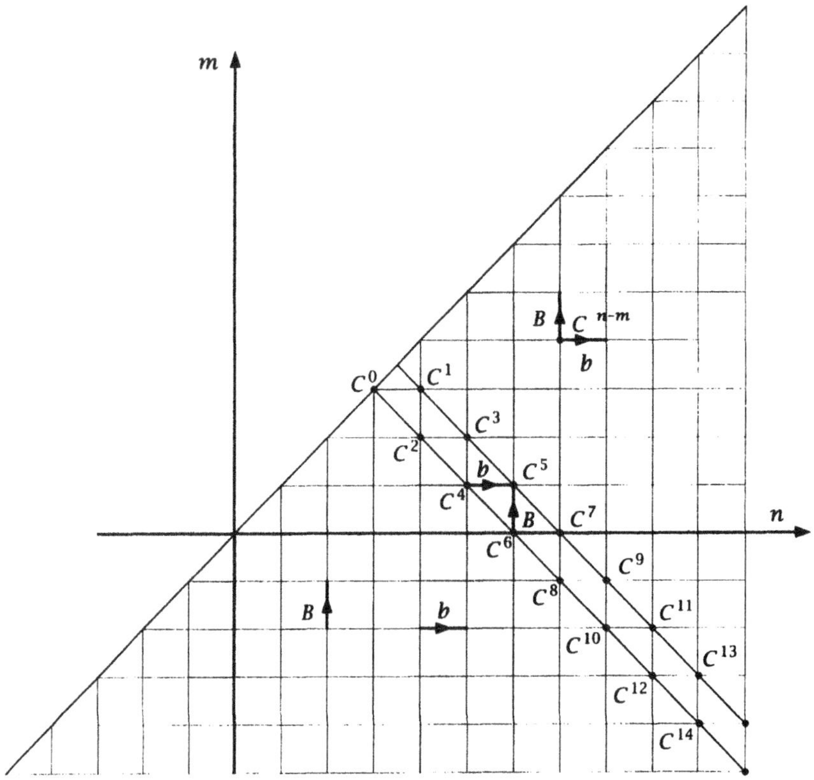

Figure IV.9. Bicomplex

$$C^{\mathrm{ev}} = \{(\phi_{2n})_{n\in\mathbb{N}}\ ;\ \phi_{2n} \in C^{2n} \quad \forall n \in \mathbb{N}\}$$

and

$$C^{\mathrm{odd}} = \{(\phi_{2n+1})_{n\in\mathbb{N}}\ ;\ \phi_{2n+1} \in C^{2n+1} \quad \forall n \in \mathbb{N}\}$$

and the boundary operator $\partial = d_1 + d_2$ which maps C^{ev} to C^{odd} and C^{odd} to C^{ev}. We shall enforce the following growth condition:

Definition 1. *An even (resp. odd) cochain $(\phi_{2n})_{n\in\mathbb{N}} \in C^{\mathrm{ev}}$ (resp. $(\phi_{2n+1})_{n\in\mathbb{N}} \in C^{\mathrm{odd}}$) is called* entire *if the radius of convergence of $\sum \|\phi_{2n}\| z^n/n!$ (resp. $\sum \|\phi_{2n+1}\|\ z^n/n!$) is infinity.*

Here, for any m and $\phi \in C^m$, the norm $\|\phi\|$ is the Banach space norm

$$\|\phi\| = \sup\{|\phi(a^0,\ldots,a^m)|\ ;\ \|a^j\| \le 1\}.$$

It follows, in particular, that any entire *even* cochain $(\phi_{2n}) \in C^{\mathrm{ev}}$ defines an entire function f_ϕ on the Banach space A by

$$f_\phi(x) = \sum_{n=0}^{\infty} (-1)^n \, \phi_{2n}(x,\dots,x)/n!.$$

Lemma 2. *If ϕ is an even (resp. odd) entire cochain, then so is $(d_1+d_2)\phi = \partial\phi$.*

Proof. For $\phi_m \in C^m$ one has $\|b\phi_m\| \le (m+2)\|\phi_m\|$ and $\|B_0\phi_m\| \le 2\|\phi_m\|$, $\|AB_0\phi_m\| \le 2m\|\phi_m\|$; thus the conclusion.

Definition 3. *Let A be a Banach algebra. Then the* entire *cyclic cohomology of A is the cohomology of the short complex*

$$C_\varepsilon^{\mathrm{ev}}(A) \underset{\partial}{\overset{\partial}{\rightleftarrows}} C_\varepsilon^{\mathrm{odd}}(A)$$

of entire cochains in A.

We thus have two groups $H_\varepsilon^{\mathrm{ev}}(A)$ and $H_\varepsilon^{\mathrm{odd}}(A)$. There is an obvious map from $H(A)$ to $H_\varepsilon(A)$, where $H(A)$ (Section III.1) is the periodic cyclic cohomology of A. We also have a natural filtration of H_ε by the dimensions of the cochains, where (ϕ_{2n}) is said to be of dimension $\le k$ if $2n > k$ implies $\phi_{2n} = 0$. However, unlike what happens for H, this filtration does not, in general, exhaust all of H_ε. In fact, it exhausts exactly the image of $H(A)$ in $H_\varepsilon(A)$.

Let us now compute H_ε for the simplest case, i.e. when $A = \mathbb{C}$ is the trivial Banach algebra. An element of $C_\varepsilon^{\mathrm{ev}}$ is given by an infinite sequence $(\lambda_{2n})_{n\in\mathbb{N}}$, $\lambda_{2n} \in \mathbb{C}$ such that $\sum |\lambda_{2n}|(z^n/n!) < \infty$ for any z and, similarly for $C_\varepsilon^{\mathrm{odd}}$. The boundary $\partial = d_1 + d_2$ of (λ_{2n}) is 0, since both b and B are 0 on even cochains. For m odd and $\phi \in C$, $\phi(a^0,\dots,a^m) = \lambda a^0 \cdots a^m$ one has

$$(b\phi)(a^0,\dots,a^{m+1}) = \lambda a^0 \cdots a^{m+1}, \quad (B\phi)(a^0,\dots,a^{m-1}) = 2m\lambda a^0 \cdots a^{m-1},$$

thus

$$(d_1\phi)(a^0,\dots,a^{m+1}) = (m+1)\lambda a^0 \cdots a^{m+1},$$
$$(d_2\phi)(a^0,\dots,a^{m-1}) = 2\lambda a^0 \cdots a^{m-1}.$$

So the boundary $\partial(\lambda)$ of an odd cochain (λ_{2n+1}) is given by $\partial(\lambda)_{2n} = 2n\lambda_{2n-1} + 2\lambda_{2n+1}$. Thus, $\partial(\lambda) = 0$ means that $\lambda_{2n+1} = (-1)^n n!\lambda_1$, and hence is possible only if $\lambda = 0$, for $\lambda \in C_\varepsilon^{\mathrm{odd}}$. Moreover, for any $(\lambda_{2n}) \in C_\varepsilon^{\mathrm{ev}}$, the series $\sigma(\lambda) = \sum_0^\infty (-1)^n(\lambda_{2n}/n!)$ is convergent and $\sigma(\lambda) = 0$ iff $\lambda \in \partial C_\varepsilon^{\mathrm{odd}}$. Thus, we have

Proposition 4. *One has $H_\varepsilon^{\mathrm{odd}}(\mathbb{C}) = \{0\}$, and $H_\varepsilon^{\mathrm{ev}}(\mathbb{C}) = \mathbb{C}$ with isomorphism given by*

$$\sigma((\phi_{2n})) = \sum_{n=0}^{\infty} \frac{(-1)^n}{n!} \phi_{2n}(1,\dots,1).$$

In order to relate entire cyclic cocycles with an infinite-dimensional version of the *cycles* of Section III.1 we need the following normalization condition:

Definition 5. *We shall say that a cocycle* (ϕ_{2n}) *(resp.* (ϕ_{2n+1})*) is* normalized *if for any* m *one has*

$$B_0\phi_m = \frac{1}{m} AB_0\phi_m. \qquad (*)$$

In other words, the cochain $B_0\phi_m$ is already cyclic: $B_0\phi_m \in C_\lambda^{m-1}$, so that $(1/m)A(B_0\phi_m) = B_0\phi_m$. Only the normalized cocycles have a natural interpretation in terms of the universal differential algebra ΩA.

Lemma 6. *For every entire cocycle there is a normalized cohomologous entire cocycle.*

We refer to [Co$_{24}$] for the original proof, and also to [C-Q$_2$].

Remark 7. a) The above definition of entire cyclic cohomology and its pairing with K-theory (cf. Subsection δ) below) adapt to arbitrary locally convex algebras $\mathcal{A}$ over $\mathbb{C}$ as follows: A cochain (ϕ_{2n}) (resp. (ϕ_{2n+1})) is called entire iff for any bounded subset $\Sigma \subset \mathcal{A}$ there exists $C = C_\Sigma < \infty$ with

$$\left|\phi_{2n}(a^0,\ldots,a^{2n})\right| \leq C_\Sigma\, n! \quad \forall a^j \in \Sigma \quad \forall n \in \mathbb{N}.$$

Replacing Σ by $\lambda^{-1}\Sigma$ for $\lambda > 0$, we see that for any bounded subset Σ of $\mathcal{A}$ and $\lambda > 0$ there exists $C = C_{\Sigma,\lambda}$ such that

$$\left|\phi(a^0,\ldots,a^{2n})\right| \leq C\,\lambda^{2n}\, n! \quad \forall a^j \in \Sigma\,,\ \forall n \in \mathbb{N}.$$

b) Let $\mathcal{A}$ be an algebra over $\mathbb{C}$. Then, with the finest locally convex topology, $\mathcal{A}$ is a locally convex algebra, so that its entire cyclic cohomology is well-defined by a). The bounded subsets of $\mathcal{A}$ are convex hulls of finite subsets Σ. Thus, a cochain ϕ is entire iff for any finite subset Σ of $\mathcal{A}$ there exists $C = C_\Sigma < \infty$ such that

$$\left|\phi(a^0,\ldots,a^{2n})\right| \leq C\, n! \quad \forall a^j \in \Sigma\,,\ \forall n \in \mathbb{N}.$$

c) We have used the (d_1, d_2) bicomplex of Section III.1 Theorem 29 so that $S = d_1 d_2^{-1}$, and the pairing of K_0 with $H_\varepsilon^{\rm ev}$ is given by the function f_ϕ whose existence fixes the growth condition. The trivial change to the (b, B) bicomplex is summarized in the following table.

(b, B) bicomplex	(d_1, d_2) bicomplex
$b : C^n \to C^{n+1}$	$d_1 = (n+1)b$
$B : C^n \to C^{n-1}$	$d_2 = \frac{1}{n} B$
$S = n(n+1) b B^{-1}$	$S = d_1 d_2^{-1}$
(ϕ_{2n})	$\psi_{2n} = (2n)!\, \phi_{2n}$
(ϕ_{2n+1})	$\psi_{2n+1} = (2n+1)!\, \phi_{2n+1}$
$\lVert\phi_{2n}\rVert \leq C_\lambda\, \lambda^{2n}/n!$	$\lVert\psi_{2n}\rVert \leq C_\lambda\, \lambda^{2n}\, n!$
$\lVert\phi_{2n+1}\rVert \leq C_\lambda\, \lambda^{2n}/n!$	$\lVert\psi_{2n+1}\rVert \leq C_\lambda\, \lambda^{2n}\, n!$
$\sum (-1)^n \frac{(2n)!}{n!}\, \phi_{2n}\,(x, \ldots, x)$	$\sum \frac{(-1)^n}{n!}\, \psi_{2n}(x, \ldots, x)$

7.β Infinite-dimensional cycles

In Chapter III we took as a starting point for cyclic cohomology the notion of a cycle of dimension n, given by a graded differential algebra (Ω, d) and a homogeneous linear form $\int$ of degree n such that

1) $\int \omega_1 \omega_2 = (-1)^{k_1 k_2} \int \omega_2 \omega_1 \quad \forall \omega_j \in \Omega^{k_j}\,,\ j = 1, 2$
2) $\int d\omega = 0 \quad \forall \omega \in \Omega^{n-1}$.

In this section we shall show that in order to handle the infinite-dimensional case one just needs to replace the homogeneous conditions 1) and 2) by the inhomogeneous condition

$$\int (\omega_1 \omega_2 - (-1)^{k_1 k_2}\, \omega_2 \omega_1) = (-1)^{k_1} \int d\omega_1\, d\omega_2 \quad \forall \omega_j \in \Omega^{k_j}.$$

We shall first work purely algebraically, and then formulate the growth condition of entire cocycles in terms of Ω.

We shall say that a linear form μ on a differential graded algebra (Ω, d) is *even* (resp. *odd*) iff $\mu(\omega) = 0\ \forall \omega \in \Omega^k$, k odd (resp. k even).

The following is the infinite-dimensional analogue of Proposition 4 in Chapter III Section 1.

Proposition 8. [Co_{24}] a) *Let $\mathcal{A}$ be an algebra over $\mathbb{C}$, (Ω, d) a graded differential algebra such that $\Omega^0 = \mathcal{A}$ and μ an even (resp. odd) linear form on Ω satisfying*

$$\mu(\omega_1\omega_2 - (-1)^{k_1k_2}\,\omega_2\omega_1) = (-1)^{k_1}\,\mu(d\omega_1\,d\omega_2) \quad \forall \omega_j \in \Omega^{k_j}, \tag{3}$$

and $\mu(da) = 0 \quad \forall a \in \mathcal{A}$.

Then the equality

$$\varphi_{2n}(a^0, \ldots, a^{2n}) = t_n\,\mu(a^0 da^1 \cdots da^{2n}) \quad \forall a_j \in \mathcal{A} \tag{4}$$

where $t_n = (-1)^n\,(2n-1)\ldots 3.1$, defines a normalized cocycle (φ_{2n}) in the (d_1, d_2) bicomplex (resp. $\varphi_{2n+1}\,(a^0, a^1, \ldots, a^{2n+1}) = t'_n\,\mu(a^0 da^1 \cdots da^{2n+1})$, with $t'_n = (-1)^n\,2n(2n-2)\cdots 4\cdot 2$).

b) *Let (φ_{2n}) (resp. (φ_{2n+1})) be a normalized cocycle in the (d_1, d_2) bicomplex. Then the following equalities define linear forms μ on the universal differential algebra $\Omega^*\mathcal{A}$ that satisfy* (3), (4), *and* (5)

$$\mu\left((a^0 + \lambda 1)da^1 \cdots da^{2n}\right) = t_n^{-1}\left(\varphi_{2n}(a^0, \ldots, a^{2n}) + \lambda(B_0\,\varphi_{2n})(a^1, \ldots, a^{2n})\right)$$

$$\mu\left((a^0 + \lambda 1)da^1 \cdots da^{2n+1}\right) = t_n'^{\,-1}\left(\varphi_{2n+1}(a^0, \ldots, a^{2n+1}) + \lambda(B_0\varphi_{2n+1})(a^1, \ldots, a^{2n+1})\right)$$

respectively, $\forall a^j \in \mathcal{A}$, $\lambda \in \mathbb{C}$.

Proof. b) Let us check the even case. We let $\psi_{2n} = t_n^{-1}\,\varphi_{2n}$ so that $B_0\,\psi_{2n}$ is cyclic (use Definition 5) and one has

$$B_0\,\psi_{2n} = b\,\psi_{2n-2} \quad \forall n.$$

We shall first show that for any $a \in \mathcal{A}$, da belongs to the centralizer of the functional μ defined by (5). The equality

$$\mu(da(da^1 \cdots da^{2n-1})) = (-1)^{2n-1}\,\mu((da^1 \cdots da^{2n-1})da)$$

follows from the cyclicity of $B_0\psi_{2n}$. One has $B_0\psi_{2n} = b\psi_{2n-2}$ so that $bB_0\psi_{2n} = 0$, and also $B_0 b\psi_{2n} = 0$, since $b\psi_{2n}$ is cyclic. Thus, the equality $B_0 b + b' B_0 = D$ (Section III.1) entails that

$$\psi_{2n}(a^0, \ldots, a^{2n-1}, a) - (-1)^{2n}\,\psi_{2n}(a, a^0, \ldots, a^{2n-1}) + (-1)^{2n}\,B_0\psi_{2n}(aa^0, a^1, \ldots, a^{2n-1}) = 0,$$

i.e. that

$$\mu(da(a^0 da^1 \cdots da^{2n-1})) = (-1)^{2n-1}\,\mu((a^0 da^1 \cdots da^{2n-1})da).$$

Thus, it follows that any $d\omega$ belongs to the centralizer of μ. Let us now show that

$$\mu(a\omega - \omega a) = \mu(da\,d\omega) \quad \forall a \in \mathcal{A}. \tag{6}$$

With $\omega = a^0 da^1 \cdots da^{2n}$ one has

$$\begin{aligned}\mu(\omega a) = \mu(a^0(da^1\cdots da^{2n})a) &= \psi_{2n}(a^0,a^1,\ldots,a^{2n-1},a^{2n}a)\\ &- \psi_{2n}(a^0,a^1,\ldots,a^{2n-1}a^{2n},a)+\cdots+\\ &+(-1)^j\,\psi_{2n}(a^0,\ldots,a^{2n-j}a^{2n-j+1},\ldots,a)+\cdots+\\ &+(-1)^{2n}\,\psi_{2n}(a^0a^1,\ldots,a).\end{aligned}$$

Thus

$$\begin{aligned}\mu(\omega a - a\omega) &= b\psi_{2n}(a^0,a^1,\ldots,a^{2n},a) = B_0\,\psi_{2n+2}(a^0,\ldots,a^{2n},a)\\ &= \mu(d\omega\, da) = -\mu(da\, d\omega).\end{aligned}$$

Finally, we just need to check that if $\omega_1 = a\,d\omega$ is of degree k_1 with $a \in \mathcal{A}$, and ω_2 is of degree k_2, one has (3). Since $d\omega$ is in the centralizer of μ we have

$$\begin{aligned}\mu(\omega_1\omega_2 - (-1)^{k_1k_2}\,\omega_2\omega_1) &= \mu(a\,d\omega\,\omega_2 - (-1)^{k_1k_2}\,\omega_2\,a\,d\omega)\\ &= \mu(a\,d\omega\,\omega_2 - d\omega(\omega_2\,a)).\end{aligned}$$

Using (6) we get

$$\mu(\omega_1\omega_2 - (-1)^{k_1k_2}\,\omega_2\omega_1) = \mu(da\,d(d\omega\omega_2)) = (-1)^{k_1}\,\mu(d\omega_1\,d\omega_2).$$

The above proof shows that, conversely, any functional μ on $\Omega^*\mathcal{A}$ which is even (resp. odd) and satisfies (3) defines an even (resp. odd) normalized cochain (ψ_{2n}) (resp. (ψ_{2n+1})) such that $b\psi_m = B_0\psi_{m+2}$ for any m, by the equality

$$\psi_m(a^0,\ldots,a^m) = \mu(a^0\,da^1\cdots da^m).$$

Thus, since $\Omega^*\mathcal{A}$ is the universal differential graded algebra over $\mathcal{A}$, we get a).

Let us now assume that A is a Banach algebra, and endow the universal differential algebra Ω^*A with the norms ([Arv$_3$])

$$\|\sum_0^\infty \omega_k\|_r = \sum_0^\infty r^k\|\omega_k\|_\pi$$

where $\|\ \|_\pi$ is the projective tensor product norm on

$$\Omega^k = \tilde{A}\otimes A\otimes\ldots\otimes A.$$

Theorem 9. *The equalities* (4) *and* (5) *of Proposition 8 establish a canonical bijection between normalized entire cocycles on A and linear forms on Ω^*A satisfying* (3) *and continuous for* all *the norms* $\|\ \|_r$.

Using Proposition 8 the proof is immediate.

The natural topology on Ω^*A provided by Theorem 9 is *not* the projective limit of the normed spaces $\|\ \|_r$, $r\to\infty$ as in [Arv$_3$]. It is the inductive limit, for $r\to 0$,

$$\varinjlim\,(\Omega^*A,\|\ \|_r).$$

For each $r > 0$ the completion of $\Omega^* A$ for the norm $\| \ \|_r$ is a Banach algebra Ω_r, and for $r' < r$ the natural homomorphism $\Omega_r \to \Omega_{r'}$, which is the identity on $\Omega^* A$, is norm decreasing. Let

$$\Omega_\varepsilon(A) = \varinjlim \Omega_r(A) \ , \ r > 0.$$

By construction, $\Omega_\varepsilon(A)$ is a locally convex algebra and the homomorphism

$$\varepsilon : \Omega_\varepsilon(A) \to \widetilde{A}$$

given on each Ω_r by the augmentation $\sum_0^\infty \omega_n \to \omega_0$ of $\Omega^* A$, is continuous.

Proposition 10. 1) *A linear form μ on $\Omega^* A$ is continuous for all the norms $\| \ \|_r$, $r > 0$ iff it is continuous on $\Omega_\varepsilon(A)$.*

2) *Let $J = \ker \varepsilon$, $\varepsilon : \Omega_\varepsilon(A) \to \widetilde{A}$. Then any element $\omega \in J$ is* quasinilpotent, *i.e. $\lambda 1 - \omega$ is invertible in $\Omega_\varepsilon(A)$ for any $\lambda \neq 0$.*

Proof. 1) is clear.

2) One has $\omega \in \Omega_r$ for some $r > 0$ and $\omega = \sum_1^\infty \omega_n$, $\omega_n \in \Omega^n A$ with $\sum r^n \|\omega_n\|_\pi < \infty$. Replacing r by $r' \ll r$, one can assume that $\|\lambda^{-1}\omega\|'_r < 1$, and the result follows since $\Omega_{r'}$ is a Banach algebra.

Finally $\Omega_\varepsilon(A)$ is, by construction, a $\mathbb{Z}/2$-graded differential algebra, since the differential d of $\Omega^* A$ is continuous for all the norms $\| \ \|_r$. The range of d is contained in the ideal J of Proposition 10.

7.γ Traces on QA and $\mathcal{E}A$

In this section we shall reformulate Theorem 8 of Subsection β) in terms of a deformed algebra structure on $\Omega^* A$ which yields the Cuntz algebra QA ([Cu$_2$]). This procedure is similar to the one used by Fedosov [Fed] in the context of manifolds. The general deformation procedure is:

Lemma 11. [Co-C] *Let (Ω, d) be a differential $\mathbb{Z}/2$-graded algebra and $\lambda \in \mathbb{C}$. The following equalities define an associative bilinear product on $\Omega = \Omega^{\rm ev} + \Omega^{\rm odd}$:*

$$\omega_1 \cdot_\lambda \omega_2 = \omega_1 \omega_2 + \lambda\, \omega_1\, d\omega_2 \ \ if \ \ \omega_1 \in \Omega^{\rm odd} \ , \ \omega_2 \in \Omega$$

$$\omega_1 \cdot_\lambda \omega_2 = \omega_1 \omega_2 \ \ if \ \ \omega_1 \in \Omega^{\rm ev} \ , \ \omega_2 \in \Omega.$$

It is easy to check directly (cf. [Co-C]). The corresponding algebra is equal to Ω for $\lambda = 0$ and is independent of λ for $\lambda \neq 0$.

The $\mathbb{Z}/2$-grading of Ω, i.e. the involutive automorphism

$$\sigma_0(\omega) = (-1)^{\deg \omega}\ \omega,$$

extends to a $\mathbb{Z}/2$-grading of the deformed algebra by the equality

$$\sigma_\lambda(\omega) = (-1)^{\deg \omega} (\omega - \lambda\, d\omega).$$

One checks directly that σ_λ is an involutive automorphism of the deformed product. We then have the following natural interpretation of condition (3) of Subsection β) above, in terms of the deformed product:

Lemma 12. [Co$_{24}$] *Let $(\Omega, d, \cdot_\lambda)$ be as in Lemma 11 and σ_λ be the $\mathbb{Z}/2$-grading of the algebra $(\Omega, \cdot_\lambda)$. For any odd linear form τ on $(\Omega, \cdot_\lambda, \sigma_\lambda)$ let $\tilde{\tau}$ be its restriction to $\Omega^{\rm odd}$ extended as 0 on $\Omega^{\rm ev}$. Then the map $\tau \to \tilde{\tau}$ is a canonical bijection between* odd traces *on $(\Omega, \cdot_\lambda, \sigma_\lambda)$ and odd linear forms on (Ω, d) such that*

$$\tilde{\tau}(\omega_1\omega_2 - (-1)^{k_1k_2}\, \omega_2\omega_1) = \frac{1}{2}\, \lambda^2\, (-1)^{k_1}\, \tilde{\tau}(d\omega_1\, d\omega_2) \quad \forall \omega_j \in \Omega^{k_j}. \quad (3')$$

Proof. First, let τ be an odd trace on $(\Omega, \cdot_\lambda)$ and $\tilde{\tau}$ the corresponding linear form on Ω. To check $(3')$ one can assume that $\omega_1 \in \Omega^{\rm odd}$, $\omega_2 \in \Omega^{\rm ev}$. Then the equalities

$$\tau(\omega_1 \cdot_\lambda \omega_2) = \tau(\omega_2 \cdot_\lambda \omega_1)$$
$$\tau(\omega) = \frac{1}{2}\, \tau(\omega - \sigma_\lambda(\omega))$$

for $\omega = \omega_1\, d\omega_2$, imply that

$$\tau(\omega_1 \cdot_\lambda \omega_2 - \lambda\omega) - \tau(\omega_2 \cdot_\lambda \omega_1) = -\frac{\lambda^2}{2}\, \tau(d\omega_1\, d\omega_2)$$

which gives the equality $(3')$.

Conversely, let μ be an odd linear form on Ω satisfying $(3')$. Then define the linear form τ on $(\Omega, \cdot_\lambda)$ by

$$\tau(\omega) = \mu(\omega) \quad \forall \omega \in \Omega^{\rm odd}, \quad \tau(\omega) = \frac{\lambda}{2}\, \mu(d\omega) \quad \forall \omega \in \Omega^{\rm ev}.$$

By construction, μ vanishes on $\Omega^{\rm ev}$ and is the restriction of τ to $\Omega^{\rm odd}$, thus $\mu = \tilde{\tau}$. Let us check that $\tau \circ \sigma_\lambda = -\tau$. One has for $\omega \in \Omega^{\rm ev}$,

$$\sigma_\lambda(\omega) = \omega - \lambda\, d\omega\,,$$
$$\tau(\sigma_\lambda(\omega)) = \tau(\omega - \lambda\, d\omega) = \frac{\lambda}{2}\, \mu(d\omega) - \lambda\, \mu(d\omega) = -\frac{\lambda}{2}\, \mu(d\omega) = -\tau(\omega).$$

For $\omega \in \Omega^{\rm odd}$ one has $\sigma_\lambda(\omega) = -\omega + \lambda\, d\omega$, $\tau(\sigma_\lambda(\omega)) = -\tau(\omega) + \lambda\, \tau(d\omega) = -\tau(\omega)$ since $d^2 = 0$.

Using $(3')$ one checks in a similar way that τ is a trace.

The above Lemma 12 has an analogue in the even case. One introduces for that purpose the algebra $\mathcal{E}_\lambda$ which is the crossed product of $(\Omega, \cdot_\lambda)$ by its $\mathbb{Z}/2$-grading automorphism σ_λ

$$\mathcal{E}_\lambda = (\Omega, \cdot_\lambda) \rtimes_{\sigma_\lambda} \mathbb{Z}/2.$$

We endow $\mathcal{E}_\lambda$ with the dual $\mathbb{Z}/2$-grading $\hat{\sigma}_\lambda$. One has $\hat{\sigma}_\lambda(F) = -F$, $\hat{\sigma}_\lambda(\omega) = \omega$ $\forall \omega \in \Omega$, where F, with $F^2 = 1$, is the element of the crossed product associated to the generator of $\mathbb{Z}/2$.

The direct analogue of Lemma 12 is then ([Co$_{24}$]):

Lemma 13. *Let $\mathcal{E}_\lambda$, $\hat{\sigma}_\lambda$, and F be as above. For any odd linear form τ on $\mathcal{E}_\lambda$ let $\tilde{\tau}$ be the restriction of $\omega \to \tau(F\omega)$ to Ω^{ev} extended by 0 on Ω^{odd}. The map $\tau \to \tilde{\tau}$ is a canonical bijection between odd traces on $\mathcal{E}_\lambda$ and even linear forms on (Ω, d) satisfying condition (3′) of Lemma 12.*

The proof is similar to that of Lemma 12.

Notice that the above lemmas have been proved in the case of arbitrary $\mathbb{Z}/2$-graded differential algebras, so that we can now apply them to the differential algebra $\Omega_\varepsilon(A)$ of Subsection β) and Theorem 9. To match conditions (3) and (3′) we shall take $\lambda = \sqrt{2}$, and for reasons which will soon become clear we shall denote by $Q_\varepsilon(A)$ the $\mathbb{Z}/2$-graded algebra obtained from $(\Omega_\varepsilon(A), d)$ by Lemma 11. Similarly, we let $\hat{Q}_\varepsilon(A)$ be obtained as above as the crossed product of $Q_\varepsilon(A)$ by its $\mathbb{Z}/2$-grading σ, $\sigma^2 = 1$.

Both Q_ε and $\hat{Q}_\varepsilon$ are locally convex algebras, and Theorem 9 together with Lemma 12 (resp. 13) yields a canonical bijection between continuous odd traces τ on Q_ε (resp. $\hat{Q}_\varepsilon$) and normalized odd (resp. even) entire cocycles on the Banach algebra A.

It remains now to identify the universal algebras $Q_\varepsilon(A)$, $\hat{Q}_\varepsilon(A)$, and their topologies.

Let us first proceed purely algebraically and denote by QA the $\mathbb{Z}/2$-graded algebra obtained from $(\Omega A, d)$ using the deformation of Lemma 11 for $\lambda = \sqrt{2}$. Also we let $\hat{Q}A$ be the crossed product of QA by its $\mathbb{Z}/2$-grading σ.

Proposition 14. [Co-C] *Let $\mathcal{A}$ be an algebra over $\mathbb{C}$.*

a) *The following pair of homomorphisms from $\tilde{\mathcal{A}}$ to $Q\mathcal{A}$ give an isomorphism of the free product $\tilde{\mathcal{A}} *_{\mathbb{C}} \tilde{\mathcal{A}}$ with $Q\mathcal{A}$*

$$\rho_1(a) = a \in \Omega^0\mathcal{A} \quad \forall a \in \tilde{\mathcal{A}}; \; \rho_2(a) = a - \sqrt{2}\, da \quad \forall a \in \tilde{\mathcal{A}}.$$

The $\mathbb{Z}/2$-grading σ of $Q\mathcal{A}$ is the automorphism which exchanges $\rho_1(a)$ with $\rho_2(a)$ for any $a \in \mathcal{A}$.

b) *The following pair of homomorphisms from $\tilde{\mathcal{A}}$ and $\mathbb{Z}/2$ to $\hat{Q}\mathcal{A}$ give an isomorphism $\rho' : \tilde{\mathcal{A}} *_{\mathbb{C}} \mathbb{C}(\mathbb{Z}/2) \to \hat{Q}\mathcal{A}$*

$$\rho_1'(a) = a \in \Omega^0\mathcal{A} \quad \forall a \in \tilde{\mathcal{A}}; \; \rho_2'(n) = F^n \quad \forall n \in \mathbb{Z}/2.$$

The $\mathbb{Z}/2$-grading σ' of $\hat{Q}\mathcal{A}$ satisfies $\sigma' \circ \rho_1' = \rho_1'$, $\sigma'(F) = -F$.

We shall first use the above isomorphisms to translate the equality (4) of Proposition 8 in terms of $Q\mathcal{A}$ and $\hat{Q}\mathcal{A}$. As in [Cu$_2$] one lets, for $a \in \tilde{\mathcal{A}}$, qa be the difference

$$qa = \rho_1(a) - \rho_2(a) \in Q\mathcal{A}.$$

We shall also omit ρ_1 when no confusion can arise, i.e. identify $\widetilde{\mathcal{A}}$ as the subalgebra $\rho_1(\widetilde{\mathcal{A}})$ of $Q\mathcal{A}$.

Using Lemmas 12 and 13 we translate Proposition 8 as follows:

Proposition 15. a) *Let τ be an odd trace on $Q\mathcal{A}$ such that $\tau(qa) = 0 \quad \forall a \in \mathcal{A}$. Then the following equality defines a normalized cocycle (φ_{2n+1}) in the (d_1, d_2) bicomplex:*

$$\varphi_{2n+1}(a^0, \ldots, a^{2n+1}) = (-1)^n \, n! \quad \tau(a^0 q a^1 \cdots q a^{2n+1}) \qquad \forall a^j \in \mathcal{A}.$$

b) *Let τ be an odd trace on $\hat{Q}\mathcal{A}$. Then the following equality defines a normalized cocycle (φ_{2n}) in the (d_1, d_2) bicomplex:*

$$\varphi_{2n}(a^0, \ldots, a^{2n}) = \Gamma\left(n + \frac{1}{2}\right) \tau\left(F\, a^0[F, a^1] \cdots [F, a^{2n}]\right) \qquad \forall a^j \in \mathcal{A}$$

Let us also mention a variant of a). The algebra $\hat{\hat{Q}}\mathcal{A} = \hat{Q}\mathcal{A} \rtimes_{\sigma'} \mathbb{Z}/2$ which is the crossed product of $\hat{Q}\mathcal{A}$ by its $\mathbb{Z}/2$-grading σ', is isomorphic to $M_2(Q\mathcal{A})$. It has a very simple presentation, as in 14b). Indeed it is generated by a copy of $\widetilde{\mathcal{A}}$ and a pair of elements F, and γ of square 1 such that

c) $\gamma a = a\gamma \quad \forall a \in \widetilde{\mathcal{A}}$, $\gamma F = -F\gamma$ and $\gamma^2 = F^2 = 1$.

These relations constitute a presentation of $\hat{\hat{Q}}\mathcal{A}$.

Then 15a) can be written as follows.

Corollary 16. *Let τ be a trace on $\hat{\hat{Q}}\mathcal{A}$ such that $\tau(\gamma a) = 0 \quad \forall a \in \mathcal{A}$. Then the following equality defines a normalized cocycle in the (d_1, d_2) bicomplex:*

$$\varphi_{2n+1}(a^0, \ldots, a^{2n+1}) = n! \, \tau\left(\gamma F\, a^0[F, a^1] \cdots [F, a^{2n+1}]\right) \qquad \forall a^j \in \mathcal{A}.$$

In [Co-C] J. Cuntz and the author give the general form of odd traces on both QA and $\hat{Q}A$, so it might appear at first sight that such traces are easy to construct and are not interesting from a cohomological point of view. It turns out, however, that the explicit construction in [Co-C] is the translation of the triviality of the first spectral sequence of the (b, B) bicomplex (Chapter III). Thus, when A is a Banach algebra, this explicit construction becomes incompatible with continuity and does not exclude the existence of nontrivial traces.

Let A be a Banach algebra. We endow the algebra QA with the locally convex topology inherited from the inductive limit topology of ΩA (Theorem 9) by the deformation isomorphism (Lemma 11). We let $Q_\varepsilon A$ correspond in the same way to $\Omega_\varepsilon A$. Thus, $QA \subset Q_\varepsilon A$ is a subalgebra of $Q_\varepsilon A$ and its topology is the restriction of the topology of $Q_\varepsilon A$. Continuous (odd) traces on both algebras correspond bijectively to each other by restriction and extension by continuity.

The crossed products $\hat{Q}A$, $\hat{\hat{Q}}A$, $\hat{Q}_\varepsilon A$, and $\hat{\hat{Q}}_\varepsilon A$ inherit their topologies in a canonical way since they are crossed products by finite groups. Theorem 9 now reads as follows:

Theorem 17. *Let A be a Banach algebra. The equalities of Proposition 15 and Corollary 16 establish a canonical bijection between continuous odd traces on $\hat{Q}_\varepsilon A$ (resp. on $\hat{\hat{Q}}_\varepsilon A$ vanishing on γA) and entire normalized even (resp. odd) cocycles on A.*

In the statement one can use either QA or $Q_\varepsilon A$. The structure of the locally convex algebra $Q_\varepsilon A$ is very similar to that of $\Omega_\varepsilon A$ as described in Proposition 10. For a start, one also has the augmentation morphism

$$\varepsilon : Q_\varepsilon A \to \tilde{A}.$$

It is given by the same formula as $\varepsilon : \Omega_\varepsilon A \to \tilde{A}$, and the fact that it is a homomorphism is insensitive to the deformation of the product (Lemma 11). In terms of the difference map q on $Q_\varepsilon A$ one has $\varepsilon \circ q = 0$.

Proposition 18. *Let $J = \ker \varepsilon$, $\varepsilon : Q_\varepsilon A \to \tilde{A}$. Any element $x \in J$ is* quasinilpotent, *i.e. $\lambda 1 - x$ is invertible in $Q_\varepsilon A$ for any $\lambda \neq 0$.*

Thus, $Q_\varepsilon A$ is a quasinilpotent extension of A. This fact plays a decisive role in the construction of the character in Section 8 below.

7.δ Pairing with $K_0(A)$

The computation (Proposition 4) of the entire cyclic cohomology of $\mathbb{C}$ yields a natural pairing of $H_\varepsilon^{\mathrm{ev}}(A)$ with $K_0(A)$ for any Banach algebra A. We developed it in [Co24] for the normalized cocycles (Definition 5) which by Lemma 6 is not a restriction on the pairing.

Lemma 19. *Let $(\phi_{2n})_{n\in\mathbb{N}}$ be a normalized entire cocycle on A. Then if $\phi \in \operatorname{im}\partial \subset C_\varepsilon^{\mathrm{ev}}$, one has*

$$\sum_0^\infty \frac{(-1)^n}{n!}\, \phi_{2n}\,(e,\dots,e) = 0$$

for any idempotent $e \in A$.

Proof. Let $(\psi_{2n+1}) \in C_\varepsilon^{\mathrm{odd}}$ be such that $\partial\psi = \phi$. Thus, for each n,

$$\phi_{2n} = 2n\; b\psi_{2n-1} + \frac{1}{2n+1}\, B\psi_{2n+1}.$$

Now, since ϕ is normalized, $B_0\phi_{2n} \in C_\lambda^{2n}$ is cyclic so that

$$B_0 b\psi_{2n-1} = \frac{1}{2n}\, B_0\phi_{2n}$$

is cyclic for any n. Let

$$\alpha_n = (B_0\psi_{2n+1})(e,\dots,e) = \frac{1}{2n+1}\, B\psi_{2n+1}(e,\dots,e).$$

One has, since $e^2 = e$, that

$$\begin{aligned}\alpha_n &= (b' B_0\psi_{2n+1})(e,\dots,e) = ((D - B_0 b)\psi_{2n+1})(e,\dots,e)\\ &= (D\psi_{2n+1})(e,\dots,e) = 2\psi_{2n+1}(e,\dots,e).\end{aligned}$$

Also

$$(b\psi_{2n+1})(e,\dots,e) = \psi_{2n+1}(e,\dots,e) = \frac{1}{2}\,\alpha_n.$$

Thus

$$\phi_{2n}(e,\dots,e) = 2n\,\frac{1}{2}\,\alpha_{n-1} + \alpha_n,$$

so that

$$\sum_0^\infty \frac{(-1)^n}{n!}\,\phi_{2n}(e,\dots,e) = \sum_0^\infty \frac{(-1)^n}{n!}\,(n\alpha_{n-1} + \alpha_n) = 0.$$

Next, let $q \in \mathbb{N}$ and $A_q = M_q \otimes A = M_q(A)$ be the Banach algebra of $q \times q$ matrices over A. For any $\phi \in C^m$, let ϕ^q be the natural extension $\phi^q = \mathrm{Tr}\#\phi$ of ϕ to $M_q(A)$ (cf. Section III.1), i.e. by definition

$$\phi^q(\mu^0 \otimes a^0,\dots,\mu^m \otimes a^m) = \mathrm{Trace}(\mu^0 \cdots \mu^m)\,\phi(a^0,\dots,a^m),$$

where $\mu^j \in M_q(\mathbb{C})$ and $a^j \in A$.

Lemma 20. 1) *For any entire even (resp. odd) cochain (ϕ_{2n}) (resp. (ϕ_{2n+1})) on A the cochain (ϕ^q_{2n}) (resp. (ϕ^q_{2n+1})) on $M_q(A)$ is also entire.*

2) *The map $\phi \to \phi^q$ is a morphism of the complexes of entire cochains.*

Proof. 1) One has an inequality of the form $\|\phi^q\| \le q^m \|\phi\|$ for $\phi \in C^m$, hence the result.

2) cf. Section III.1.

Theorem 21. *Let $\phi = (\phi_{2n})_{n\in\mathbb{N}}$ be an entire normalized cocycle on A, and f_ϕ,*

$$f_\phi(x) = \sum_{n=0}^\infty (-1)^n\,\frac{1}{n!}\,\phi^q_{2n}(x,\dots,x)$$

the corresponding entire function on $M_\infty(A)$. Then the restriction of f_ϕ to the idempotents $e = e^2$, $e \in M_\infty(A)$, defines an additive map $K_0(A) \to \mathbb{C}$. The value $\langle \phi, [e]\rangle$ of $f_\phi(e)$ only depends upon the class of ϕ in $H^{\mathrm{ev}}_\varepsilon(A)$.

Proof. Replacing A by $\tilde{A}$ and ϕ_{2n} by $\tilde{\phi}_{2n}$ leads to

$$\tilde{\phi}_{2n}(x^0 + \lambda^0 1,\dots,x^{2n} + \lambda^{2n} 1) = \phi_{2n}(x^0,\dots,x^{2n}) + \lambda^0 B_0\phi_{2n}(x^1,\dots,x^{2n}),$$

so that one can assume that each ϕ_{2n} vanishes if some x^i, $i > 0$, is equal to 1. We just need to show that the value of f_ϕ on $e \in \operatorname{Proj} M_q(A)$ only depends upon the connected component of e in $\operatorname{Proj} M_q(A)$. Since the map $\phi \to \phi^q$ is a morphism of complexes, we can assume that $q = 1$. Let $t \to e(t)$ be a C^1 map of $[0,1]$ to $\operatorname{Proj} A$. We want to show that $(d/dt)f_\phi(e(t)) = 0$. One has $(d/dt)(e(t)) = [a(t), e(t)]$, where $a(t) = (1 - 2e(t))(d/dt)e(t)$. We just need to compute $(d/dt)f_\phi(e(t))$ for $t = 0$, and therefore we let $e = e(0)$ and $a = a(0)$. We have

$$\left(\frac{d}{dt}\,\phi_{2n}\,(e(t),\ldots,e(t))\right)_{t=0} = \sum_{0}^{2n+1} \phi_{2n}(e,\ldots,[a,e],\ldots e).$$

Thus, by Lemma 19, in order to show that the above derivative vanishes, it is enough to prove that the cocycle $(\phi'_{2n})_{n\in\mathbb{N}}$ is a coboundary, with

$$\phi'_{2n}(x^0,\ldots,x^{2n}) = \sum_{j=0}^{2n+1} \phi_{2n}(x^0,\ldots,[a,x^j],\ldots,x^{2n}).$$

Let

$$\psi_{2n-1}(x^0,\ldots,x^{2n-1}) = \frac{1}{2n}\sum_{j=0}^{2n-1}(-1)^{j+1}\,\psi_{2n}(x^0,\ldots,x^j,a,x^{j+1},\ldots,x^{2n-1}).$$

Using the equality $B\psi_{2n-1} = A\theta_{2n-2}$, where

$$\theta_{2n-2}(x^0,\ldots,x^{2n-2}) = (B_0\phi_{2n})(a,x^0,\ldots,x^{2n-2}),$$

one checks that for any n one has

$$d_1\,\psi_{2n-1} + d_2\,\psi_{2n+1} = \phi'_{2n}.$$

In terms of the topological algebra $\hat{Q}_\varepsilon A$ (cf. Proposition 10), this pairing has a remarkable expression.

Theorem 22. [Co$_{24}$] *Let τ be a continuous odd trace on $\hat{Q}_\varepsilon A$. Then the map of $K_0(A)$ to $\mathbb{C}$ given by Theorem 21 and the entire even cocycle associated to τ is obtained by the formula*

$$e \in \operatorname{Proj} A \mapsto \tau\left(F\frac{e}{\sqrt{1-(qe)^2}}\right).$$

Proof. Up to an overall normalization constant, the entire cocycle ϕ associated to τ has components ϕ_{2n} given by

$$\phi_{2n}(a^0,\ldots,a^{2n}) = (-1)^n\,2^{-n}(2n-1)\cdots 3\cdot 1\;\tau(Fa^0q(a^1)\cdots q(a^{2n})).$$

Thus, the result follows from the formula giving f_ϕ.

Remark 23. 1) The normalization condition for the cocycle ϕ in Lemma 19 and Theorem 21 can be removed (cf. [Ge-S]) by the following minor modification of the formula of Lemma 19

$$\sum_{n=0}^{\infty} \frac{(-1)^n}{n!} \phi_{2n}\left(e - \frac{1}{2}, e, e, \dots, e\right).$$

2) When A is a C^*-algebra then $\mathcal{E}A$ has a natural (but not complete) C^*-algebra norm which defines a stronger topology than the one used in Theorem 14. There are, however, interesting continuous traces on $\mathcal{E}A$ for the C^*-norm, which are used in [Co-M$_1$].

3) The above pairing and Theorem 21 apply without change to the case of arbitrary algebras over $\mathbb{C}$, using Remark 7b to define entire cyclic cohomology in that generality.

7.ε Entire cyclic cohomology of $\mathbb{S}^1$

By Theorem 26 of Section III.1 the periodic cyclic cohomology $H^*(\mathcal{A})$ of an algebra $\mathcal{A}$ with finite Hochschild dimension n is exhausted by the image in $H^*(\mathcal{A})$ of the cyclic cohomology groups $HC^q(\mathcal{A})$, $q \le n$. In order to obtain a similar result for entire cyclic cohomology (either using Remark 7b or assuming that $\mathcal{A}$ is a Banach algebra), one needs to construct a homotopy σ_k, $k > n$, of the bar resolution, with good control on the size of σ_k for large k. This has been done in full generality in [Kha]. We shall do this here for the algebra $\mathcal{A}$ of Laurent polynomials, and compute its entire cyclic cohomology (as defined in Remark 7b). This will, in particular, give the analogue of Theorem 21 for the odd case. We shall then explicitly compute the map from entire cyclic cohomology of function algebras A over $\mathbb{S}^1$ to the de Rham homology of $\mathbb{S}^1$ and interpret the formulas as deformations of the algebra $\hat{Q}_\varepsilon A$. We obtained these results by direct calculations but they also follow from general results on quasifree algebras ([C-Q$_1$] [C-Q$_3$]).

Let us first recall that given an algebra $\mathcal{A}$, the standard bar resolution of the bimodule $\mathcal{A}$ over $B = \mathcal{A} \otimes \mathcal{A}^{\mathrm{o}}$ is given by the acyclic chain complex $(\mathcal{M}_k, b)$

$$\mathcal{M}_k = (\mathcal{A} \otimes \mathcal{A}^{\mathrm{o}}) \otimes \mathcal{A}^{\otimes k}$$

$$\begin{aligned} b_k((1 \otimes 1) \otimes a_1 \otimes \dots \otimes a_k) &= (a_1 \otimes 1) \otimes (a_2 \otimes \dots \otimes a_k) \\ &+ \sum_{j=1}^{k-1} (-1)^j \, (1 \otimes 1) \otimes a_1 \otimes \dots \otimes a_j a_{j+1} \otimes \dots \otimes a_k \\ &+ (-1)^k \, (1 \otimes a_k^{\mathrm{o}}) \otimes (a_1 \otimes \dots \otimes a_{k-1}) \quad \forall a_j \in \mathcal{A}, \end{aligned}$$

which suffices to define b_k as a B-module map from $\mathcal{M}_k$ to $\mathcal{M}_{k-1}$ (cf. Chapter III).

In the topological context (cf. Chapter III Appendix B) the above tensor products are π-tensor products of locally convex vector spaces.

Let us take for $\mathcal{A}$ an algebra of functions in one complex variable z. Since $\mathcal{A}$ is commutative, we can ignore, except for notational convenience, the distinction between $\mathcal{A}$ and $\mathcal{A}^o$, and consider any element $f \in \mathcal{M}_k = \mathcal{A}\otimes\mathcal{A}^o\otimes\mathcal{A}^{\otimes k}$ as a function of $k+2$ variables

$$f(z,z^0,z_1,\dots,z_k) \in \mathbb{C}.$$

We do not yet specify the domain of the complex variables nor the regularity of f, but one can have in mind the case of Laurent polynomials $\mathcal{A} = \mathbb{C}[z,z^{-1}]$.

Let us define a B-module map $\sigma_n : \mathcal{M}_n \to \mathcal{M}_{n+1}$, $n \geq 1$ by

$$(\sigma_n f)(z,z^0,z_1,\dots,z_{n+1}) = (-1)^{n+1} f(z,z^0,z_1,\dots,z_n)$$
$$+ (-1)^n \frac{z_{n+1}-z^0}{z_n - z^0}\left(f(z,z^0,z_1,\dots,z_{n-1},z_n) - f(z,z^0,z_1,\dots,z_{n-1},z^0)\right).$$

Lemma 24. *For $n > 1$ one has $b_{n+1}\,\sigma_n + \sigma_{n-1}\,b_n = \mathrm{id}_n$.*

For $g \in \mathcal{M}_n$ one has

$$(b_n\, g)(z,z^0,z_1,\dots,z_{n-1}) = g(z,z^0,z,z_1,\dots,z_{n-1})$$
$$+ \sum_{j=1}^{n-1}(-1)^j\, g(z,z^0,z_1,\dots,z_j,z_j,\dots,z_{n-1})$$
$$+ (-1)^n\, g(z,z^0,z_1,\dots,z_{n-1},z^0).$$

Using this equality one computes $(b_{n+1}\,\sigma_n + \sigma_{n-1}\,b_n)f$ and gets the desired equality.

Theorem 25. *Let $\mathcal{A} = \mathbb{C}[z,z^{-1}]$ be the algebra of Laurent polynomials. Then its (algebraic) entire cyclic cohomology is given by*

$$H_\varepsilon^{\mathrm{ev}}(\mathcal{A}) = \mathbb{C}\ ,\quad H_\varepsilon^{\mathrm{odd}}(\mathcal{A}) = \mathbb{C}$$

with generators given by the cyclic cocycles

$$\tau_0(f) = \int f(z)\,dz\ ,\quad \tau_1(f^0,f^1) = \int f^0\,df^1.$$

We shall sketch the proof and at the same time introduce useful notation for the general case. We first specialize the general homotopy σ of Lemma 24 to the bimodule $\mathcal{A}^*$ over $\mathcal{A}$, which yields linear maps

$$\alpha_n : C^{n+1}(\mathcal{A},\mathcal{A}^*) \to C^n(\mathcal{A},\mathcal{A}^*)$$

such that $\alpha_n\, b + b\,\alpha_{n-1} = \mathrm{id}\ \ \forall n > 1$.

The transposed map $\alpha_n^{\mathrm{t}} : \mathcal{A}^{\otimes n} \to \mathcal{A}^{\otimes(n+1)}$ is given by

$$(\alpha_n^{\mathrm{t}} f)(z_0,\dots,z_{n+1}) = (-1)^{n+1} f(z_0,\dots,z_n)$$
$$+ (-1)^n \frac{z_{n+1}-z_0}{z_n - z_0}\,(f(z_0,\dots,z_{n-1},z_n) - f(z_0,\dots,z_{n-1},z_0)).$$

Given a cocycle (we take it odd for definiteness) $(\varphi_{2k+1})_{k\in\mathbb{N}}$ in the (b,B) bicomplex (Chapter III Section 1) one produces a cohomologous cocycle $(\varphi'_{2k+1})_{k\in\mathbb{N}}$ with $\varphi'_{2k+1}=0\ \forall k>0$ by adding to φ the coboundary of the cochain $(\psi_{2k})_{k\in\mathbb{N}}$ whose components are given using the homotopy α by

$$\psi_2 = \alpha\,\varphi_3 + \alpha B\alpha\,\varphi_5 + \dots$$
$$\psi_{2k} = \sum_{m=0}^{\infty} \alpha(B\alpha)^m\,\varphi_{2m+2k+1}.$$

These formulae are standard homotopy formulae for cocycles with *finite support* in any bicomplex, and the only difficulty we have is to show that they continue to make sense for *entire cocycles* which have arbitrary supports. Since we are dealing with the algebraic form of entire cohomology (Remark 7b) and use the (b,B) bicomplex instead of the equivalent (d_1,d_2) bicomplex, the growth condition on our cochains is:

"For any finite subset $\Sigma\subset\mathcal{A}$ there exists $C=C_\Sigma$ such that
$$\left|\varphi_{2k+1}(a^0,\dots,a^{2k+1})\right| \le C\,(k!)^{-1}\quad \forall a^j\in\Sigma."$$

Given a finite subset Σ of the algebra $\mathcal{A}$ of Laurent polynomials, we let $d(\Sigma)$ be the maximal degree of elements of Σ, so that any $f\in\Sigma$ can be written as $\sum_{j=-d}^{d} f_j\,z^j$.

The estimate on the homotopy α which allows one to complete the proof of Theorem 25 is the following:

Lemma 26. a) *Let $f\in\mathcal{A}^{\otimes(n+1)}$ be a Laurent polynomial of degree at most d in each variable z_j; then $\alpha_n^{\mathrm{t}}\,B^{\mathrm{t}}\,f\in\mathcal{A}^{\otimes(n+1)}$ has the same property.*

b) *Let $\|\ \|_1$ be the ℓ^1 norm on $\mathcal{A}^{\otimes n}$ for any n so that*

$$\left\|\sum \lambda_{i_0\dots i_n}\, z_0^{i_0}\cdots z_n^{i_n}\right\| = \sum |\lambda_{i_0\dots i_n}|.$$

Then if the degree of f is less than d, one has

$$\|\alpha_n^{\mathrm{t}}\,B_0^{\mathrm{t}}\,f\|_1 \le (2d+2)\|f\|_1.$$

Proof. Since the cyclic permutations do not change the degree, it is enough to prove a) for $\alpha_n^{\mathrm{t}}\,B_0^{\mathrm{t}}\,f$. One has

$$(B_0^{\mathrm{t}}\,f)(z_0,\dots,z_{n+1}) = f(z_1,\dots,z_{n+1}) - (-1)^n\,f(z_0,\dots,z_n)$$

so that

$$(\alpha_{n+1}^{\mathrm{t}}\,B_0^{\mathrm{t}}\,f)(z_0,\dots,z_{n+1},z_{n+2}) = (-1)^{n+2}\,(B_0^{\mathrm{t}}\,f)(z_0,\dots,z_{n+1})$$
$$+ (-1)^{n+1}\,\frac{z_{n+2}-z_0}{z_{n+1}-z_0}\,\left(f(z_1,\dots,z_{n+1}) - f(z_1,\dots,z_n,z_0)\right).$$

The conclusion is clearly true for $(B_0^t f)(z_0,\dots,z_{n+1})$, and we just need to deal with the other term. Also we may assume that f is of the form

$$f(z_0,\dots,z_n) = h(z_0,\dots,z_{n-1})\; z_n^q \quad \text{with } |q| \le d.$$

We then only have to evaluate the Laurent polynomial

$$\frac{z_{n+2}-z_0}{z_{n+1}-z_0}\,(z_{n+1}^q - z_0^q).$$

For $q > 0$ we get

$$(z_{n+2}-z_0)\left(z_{n+1}^{q-1} + z_{n+1}^{q-2}\, z_0 + \dots + z_0^{q-1}\right).$$

For $q < 0$, $q = -p$, we get

$$(z_{n+2}-z_0)\left(z_{n+1}^{q}\, z_0^{-1} + z_{n+1}^{q+1}\, z_0^{-2} + \dots + z_{n+1}^{-1}\, z_0^{q}\right).$$

In both cases we check that the degree is less than d, which proves a). We also check that the ℓ^1 norm of the right hand side satisfies the inequality b).

We can now complete the proof of Theorem 25. First, the formula

$$\psi_{2k} = \sum_{m=0}^{\infty} \alpha(B\alpha)^m\, \varphi_{2m+2k+1}$$

is convergent. Indeed, given a finite subset Σ of $\mathcal{A}$, there exists by hypothesis $C = C_\Sigma$ such that

$$\left|\varphi_{2n+1}\,(a^0,\dots,a^{2n+1})\right| \le C\,(n!)^{-1} \quad \forall a^j \in \Sigma.$$

Thus, taking for Σ the monomials $\lambda^{-1}\, z^q$, $|q| \le d$, it follows that for any Laurent polynomials f_j of degree less than d one has

$$|\varphi_{2n+1}(f_0,\dots,f_{2n+1})| \le C_{\lambda,d}\, \frac{\lambda^{2n+2}}{n!} \prod_{j=0}^{2n+1} \|f_j\|_1.$$

Using Lemma 26 and the equality $B = AB_0$, we thus get

$$|(B\alpha)^m \varphi_{2m+2k+1}\,(f_0,\dots,f_{2k+1})|$$
$$\le C_{\lambda,d}\, \frac{\lambda^{2m+2k+2}}{(m+k)!}\,(2d+2)^m \prod_{j=1}^{m}(2k+2j) \prod_{j=0}^{2k+1} \|f_j\|_1$$

for any $f_j \in \mathcal{A}$ of degree less than d.

Taking λ small enough thus gives a constant C_d such that

$$|((B\alpha)^m\, \varphi_{2m+2k+1})\,(f_0,\dots,f_{2k+1})| \le C_d\, 2^{-m}(k!)^{-1} \prod_{j=0}^{2k+1} \|f_j\|_1$$

for any $f_j \in \mathcal{A}$ of degree less than d.

This shows the desired convergence and that (ψ_{2k}) is an entire cochain.

As an easy corollary of Theorem 25 we get for any locally convex algebra A a pairing between $K_1(A)$ and $H_\varepsilon^{\text{odd}}(A)$. Indeed, any invertible element $u \in GL_n(A)$ determines a homomorphism of the algebra $\mathcal{A}$ of Laurent polynomials

$$\rho_u : \mathcal{A} \to M_n(A),$$

and the pull-back $\rho_u^* \varphi$ of any odd entire cocycle φ on A is cohomologous to a multiple $\lambda \tau_1$ of τ_1. One sets

$$\langle u, \varphi \rangle = \lambda.$$

The explicit formula for λ follows from the above proof, but was first obtained in [C-Q$_1$] (cf. also [Ge$_3$]).

Corollary 27. *Let A be a locally convex algebra. The following equality defines a pairing between $K_1(A)$ and $H_\varepsilon^{\text{odd}}(A)$*

$$\langle u, \varphi \rangle = (2\pi i)^{-1/2} \sum_0^\infty (-1)^m \, m! \, \varphi_{2m+1}^q(u^{-1}, u, \ldots, u^{-1}, u)$$

$\forall u \in GL_q(\tilde{A}) \subset M_q(\tilde{A})$ and any normalized cocycle φ in the (b, B) bicomplex.

The notation φ^q was introduced above in Lemma 20. The above pairing agrees with the pairing defined in Proposition 3.3 of Chapter III using the natural inclusion of $HC^{\text{odd}}(A)$ in $H_\varepsilon^{\text{odd}}(A)$ (cf. Remark 30 a in Section III.1 for the signs).

We shall now go much further and compute explicitly, given an entire cocycle (φ_{2n+1}) on $\mathcal{A} = \mathbb{C}\mathbb{Z}$ as in Theorem 25, the entire cochain $\psi = (\psi_{2n})$ such that

$$\varphi_{2n+1} = b \, \psi_{2n} + B \, \psi_{2n+2} \quad \forall n \geq 1.$$

The initial formula for ψ_{2k} which we have to simplify is

$$\psi_{2k} = \alpha \, \varphi_{2k+1} + \alpha(B\alpha) \, \varphi_{2k+3} + \cdots.$$

In order to simplify the computation we shall assume that

$$\varphi_{2n+1}(f^0, \ldots, f^{2n+1}) = 0 \text{ if some } f^j = 1 \text{ for } j \neq 0. \qquad (*)$$

This is a weak normalization condition. At the level of chains, i.e. of $\mathcal{A}^{\otimes(n+1)}$, it means that we can ignore any function $f(z^0, \ldots, z^n)$ which is independent of some z^j, $j \neq 0$. This simplifies the formula for the map $\alpha_n^t B_0^t$ of Lemma 26, which now gives

$$(\alpha_n^{\mathrm{t}} B_0^{\mathrm{t}}) \, f(z_0, \ldots, z_{n+1}) = (-1)^n \, z_{n+1} \, \frac{f(z_1, \ldots, z_n) - f(z_1, \ldots, z_{n-1}, z_0)}{z_n - z_0}$$

for any $f \in \mathcal{A}^{\otimes(n+1)}$.

Thus, $\alpha_n^{\mathrm{t}} B_0^{\mathrm{t}}$ is , essentially, a divided difference, and the iterates $(\alpha^{\mathrm{t}} B^{\mathrm{t}})^n$ will invoke iterated divided differences which are known to satisfy remarkable identities.

The computation is now straightforward, and the result is best formulated in terms of the algebra $\hat{\hat{Q}}\mathcal{A}$ of Subsection γ). We recall that the latter algebra is generated by $\widetilde{\mathcal{A}}$ and two elements F, γ with $F^2 = \gamma^2 = 1$, $\gamma a = a\gamma\ \forall a \in \mathcal{A}$, and $\gamma F = -F\gamma$.

Because of the weak normalization condition $(*)$ above, the distinction between $\mathcal{A}$ and $\widetilde{\mathcal{A}}$ is unnecessary, so that the unit of $\hat{\hat{Q}}\mathcal{A}$ is now the unit of $\mathcal{A}$.

One then lets τ be a trace on $\hat{\hat{Q}}_\varepsilon\mathcal{A}$ vanishing on $\gamma\mathcal{A}$, and (φ_{2n+1}) the cocycle in the (b, B) bicomplex given by

$$\varphi_{2n+1}(a^0, \dots, a^{2n+1}) = t_n\, \tau\left(\gamma F\, a^0[F, a^1] \cdots [F, a^{2n+1}]\right) \quad \forall a^j \in \mathcal{A}$$

where $t_n^{-1} = 2^n(2n+1)(2n-1)\cdots 3 \cdot 1$. One gets, with u as the generator of $\mathcal{A}$:

Lemma 28. *With $\varphi = (\varphi_{2n+1})$ associated to the trace τ on $\hat{\hat{Q}}\mathcal{A}$ as above, the cochain $\psi = (\psi_{2n})$ such that $\varphi_{2n+1} = b\,\psi_{2n} + B\,\psi_{2n+2}\ \forall n > 1$ is given by $\psi_{2n} = \alpha\,(\varphi_{2n+1} - A\,\theta_{2n+1})$ where*

$$\theta_{2n+1}(f^0, \dots, f^{2n+1}) = \int_0^{1/2} \tau\Big(F\frac{\partial}{\partial\lambda} f^0\big(u + \lambda[F,u]\big)\left[\gamma F, f^1(u + \lambda[F,u])\right] \cdots \left[\gamma F, f^{2n+1}(u + \lambda[F,u])\right]\Big)\, d\mu_n(\lambda)$$

where $d\mu_n(\lambda) = t_n(1 - 4\lambda^2)^{n+\frac{1}{2}}\, d\lambda$.

Note that one has used the possibility of applying the Laurent polynomials $f^j \in \mathcal{A}$ to any invertible element of an algebra, and, in particular, to $u + \lambda[F, u]$ which is invertible in $\hat{\hat{Q}}_\varepsilon\mathcal{A}$.

This formula fits with our quantized calculus where the quantum differential (cf. Introduction) is given by the graded commutator with F, or equivalently γ times the commutator with γF. Thus, $f(u + \lambda[F, u])$ plays the role of $f(u + \lambda \text{ "}du\text{"})$. The following formula is the only one needed in the proof of Lemma 28:

$$\gamma\,[\gamma F, f(u + \lambda[F,u])] = (2\lambda)^{-1}\ \big(f(u + \lambda[F,u]) - f(u - \lambda[F,u])\big).$$

It relates the quantum differential of $f(u + \lambda \text{ "}du\text{"})$ to the difference quotient

$$\frac{f(u + \lambda \text{ "}du\text{"}) - f(u - \lambda \text{ "}du\text{"})}{2\lambda}.$$

The proof of this formula is straightforward for Laurent polynomials, or more generally for $f(u) = (u - z)^{-1}$, for which both sides are easy to compute.

We shall now interpret the formula of Lemma 28 using a natural deformation of the algebra $\hat{\hat{Q}}_\varepsilon\mathcal{A}$ to an exterior algebra over $\mathcal{A}$.

Let us define for $\lambda \in \left[0, \frac{1}{2}\right[$ an endomorphism σ_λ of $\hat{\hat{Q}}_\varepsilon\mathcal{A}$ by giving its value on operators as follows:

a) $\sigma_\lambda(u) = u + \lambda[F,u]$

b) $\sigma_\lambda(F) = F$

c) $\sigma_\lambda(\gamma) = (1-4\lambda^2)^{-1/2}\,\gamma(1-2\lambda F)$.

One checks easily that $\sigma_\lambda(\gamma)^2 = 1$ and that $\sigma_\lambda(\gamma)$ commutes with $\sigma_\lambda(u)$ and anticommutes with F.

Moreover, one checks that the σ_λ's restricted to the algebra generated by γ and F form a group

$$\sigma_\lambda \circ \sigma_{\lambda'} = \sigma_{\lambda''} \ \text{ with } \ 2\lambda'' = \frac{2\lambda + 2\lambda'}{1+4\lambda\lambda'}.$$

With these endomorphisms we can write, for any $f \in \mathcal{A}$,

$$[\gamma F, f(u + \lambda[F,u])] = (1-4\lambda^2)^{-1/2}\,\sigma_\lambda\,(\gamma[F,f(u)]),$$

so that the formula for θ_{2n+1} (Lemma 28) simplifies to

$$\theta_{2n+1}(f^0,\dots,f^{2n+1}) = t_n \int_0^{1/2} \tau\left(\frac{\partial}{\partial\lambda}\,\sigma_\lambda(f^0)\,\sigma_\lambda\left([F,f^1]\cdots[F,f^{2n+1}]\,\gamma F\right)\right) d\lambda.$$

Since τ is a trace one can replace the endomorphisms σ_λ by the inner conjugate ones, α_s, given, for $s \in \mathbb{R}$ by

$$\alpha_s(u) = \frac{1}{2}(u + FuF) + e^{-s}\,\frac{1}{2}(u - FuF)$$

$$\alpha_s(\gamma) = \gamma\ ,\ \ \alpha_s(F) = F.$$

One checks indeed that with $s = -\frac{1}{2}\log(1-4\lambda^2)$ one has

$$\alpha_s(x) = Z_s^{-1}\,\sigma_\lambda(x)\,Z_s \quad \forall x \in \hat{\hat{Q}}_\varepsilon \mathcal{A}$$

with $Z_s = \cosh\frac{t}{2} + F\sinh\frac{t}{2}$, $2\lambda = \tanh t$.

Thus, the above formula can be written

$$\theta_{2n+1}(f^0,\dots,f^{2n+1}) = t_n \int_0^{\infty} ds\ \tau\left(\left(\frac{\partial}{\partial s}\,\alpha_s(f_0)\right)\ \alpha_s\left([F,f^1]\cdots[F,f^{2n+1}]\,\gamma F\right)\right).$$

(One has to check that the terms in $\frac{\partial}{\partial s}\,Z_s$ do not contribute, but this is easy.)

We thus get the following reformulation of Lemma 28.

Lemma 29. *Let $(\alpha_s)_{s\in\mathbb{R}}$ be the one-parameter group of automorphisms of $\hat{\hat{Q}}_\varepsilon(\mathcal{A})$ given by $\alpha_s(u) = \frac{1}{2}(u+FuF) + e^{-s}\,\frac{1}{2}(u-FuF)$; $\alpha_s(\gamma) = \gamma$, $\alpha_s(F) = F$ $\forall s \in \mathbb{R}$. Let $\varphi = (\varphi_{2n+1})$ be the entire cocycle on $\mathcal{A}$ associated to a trace τ on $\hat{\hat{Q}}_\varepsilon\mathcal{A}$ vanishing on $\gamma\mathcal{A}$, and $\psi = (\psi_{2n})$ be the entire cochain given by Theorem 25*

$$\psi_{2n} = \sum_{m=0}^{\infty} \alpha(B\alpha)^m\,\varphi_{2m+2n+1}.$$

Then one has $\psi_{2n} = \alpha\, \psi'_{2n+1}$, where

$$\psi'_{2n+1}(f^0, \dots, f^{2n+1}) = 2t_n \int_0^\infty ds \sum_{j=1}^{2n+1} (-1)^j$$
$$(\tau \circ \alpha_s)\left(\gamma f^0 [F, f^1] \cdots [F, f^{j-1}]\, \delta(f^j)\, [F, f^{j+1}] \cdots [F, f^{2n+1}]\right)$$

where $\delta = \left(\frac{d}{ds}\, \alpha_s\right)_{s=0}$ is the derivation associated to α_s.

Given the above formula for θ_{2n+1} and the equality

$$\psi_{2n} = \alpha\,(\varphi_{2n+1} - A\theta_{2n+1})$$

of Lemma 28, the proof of Lemma 29 follows using

$$\varphi_{2n+1}(f^0, \dots, f^{2n+1}) = \sum_0^{2n+1} \int_0^\infty ds (\tau \circ \alpha_s) \left(\gamma F\, f^0 [F, f^1] \cdots \right.$$
$$\left. [F, f^{j-1}][F, \delta(f^j)]\, [F, f^{j+1}] \cdots [F, f^{2n+1}]\right) \qquad \forall f^j \in \mathcal{A},$$

which is a consequence of the vanishing of $\lim_{s \to \infty} \alpha_s\,([F, f])\ \ \forall f \in \mathcal{A}$. The next step is to compute $\alpha\, \psi'_{2n+1}$. One uses for this the identity

$$\alpha^t (f^0 \otimes f^1 \otimes \cdots \otimes f^{2k-1} \otimes f_z) = f_z\, f^0 \otimes f^1 \otimes \cdots \otimes f^{2k-1} \otimes f_z \otimes u$$

where u is the canonical generator of $\mathcal{A}$ and $f_z = (z - u)^{-1}$, where z is a formal parameter. Moreover, using the equality

$$\delta(u) = \frac{1}{2}\, [F, u] F$$

it follows that

$$[F, f_z]\, [F, u]\, f_z = 2\delta(f_z) - [F, f_z] F$$
$$-\delta(f_z)[F, u] f_z + [F, f_z]\delta(u) f_z = F\delta(f_z) + \delta(f_z) F.$$

These identities let one write the value of ψ'_{2n+1} on $\alpha^t (f^0 \otimes \cdots \otimes f^{2n-1} \otimes f_z)$ as a linear function of f_z, so that one can then replace f_z by an arbitrary function f^{2n} and obtain the formula for ψ_{2n}, with $\psi_{2n} = \psi^1_{2n} + \psi^2_{2n}$

$$\psi^1_{2n}(f^0, \dots, f^{2n}) = t_n \int_0^\infty ds (\tau \circ \alpha_s) \left(\sum_{j=1}^{2n} (-1)^j\, \gamma F\, f^0 [F, f^1] \cdots \right.$$
$$\left. [F, f^{j-1}]\, \delta(f^j)\, [F, f^{j+1}] \cdots [F, f^{2n}] - \gamma\, f^0 [F, f^1] \cdots [F, f^{2n-1}]\, F\, \delta(f^{2n}) \right)$$

$$\psi^2_{2n}(f^0, \dots, f^{2n}) = t_n \int_0^\infty ds\, \tau \circ \alpha_s \left(\sum_{j=1}^{2n-1} \gamma\, f^0 [F, f^1] \cdots \right.$$
$$\left. [F, f^{j-1}]\, \delta(f^j)\, [F, f^{j+1}] \cdots [F, f^{2n-1}]\, \delta(f^{2n}) \right).$$

We are now ready to conclude that the homotopy formula of Theorem 25 is a special case of the following general proposition now involving an *arbitrary* algebra A (which is no longer an algebra of functions in one variable).

Proposition 30. *Let A be an algebra, τ a trace on $\hat{\hat{Q}}A$ vanishing on γA, and δ a derivation of $\hat{\hat{Q}}A$ such that $\delta(\gamma) = 0$.*

a) *Let* $\varphi_{2n+1}(f^0,\dots,f^{2n+1}) = t_n\ \tau\left(\gamma F\ f^0[F,f^1]\cdots[F,f^{2n+1}]\right)\quad \forall f^j \in \mathcal{A}$ *and*

$$\psi^1_{2n}(f^0,\dots,f^{2n}) =$$

$$t_n\left(\sum_1^{2n}(-1)^j\ \tau\left(\gamma F\ f^0[F,f^1]\cdots[F,f^{j-1}]\ \delta(f^j)\ [F,f^j]\cdots[F,f^{2n}]\right)\right.$$
$$\left. -\ \tau\left(\gamma\ f^0[F,f^1]\cdots[F,f^{2n-1}]\ F\ \delta(f^{2n})\right)\right).$$

Then, for any n, one has, using the shorthand notation

$$(\delta\ \varphi_{2n+1})(f^0,\dots,f^{2n+1}) = \sum_0^{2n+1}\varphi_{2n+1}\ \left(f^0,\dots,\delta(f^j),f^{j+1},\dots,f^{2n+1}\right),$$

$$b\ \psi^1_{2n} + B\ \psi^1_{2n+2} = \frac{1}{2}\ \delta\ \varphi_{2n+1}.$$

b) *Let*

$$\psi^2_{2n}(f^0,\dots,f^{2n})$$
$$= \sum_1^{2n-1}\ (-1)^j\tau\left(\gamma f^0[F,f^1]\cdots[F,f^{j-1}]\delta(f^j)[F,f^{j+1}]\cdots[F,f^{2n-1}]\delta(f^{2n})\right)$$

$\forall f^j \in A$. *Then* $b\ \psi^2_{2n} = B\ \psi^2_{2n} = 0\ \ \forall n$.

The proof of this proposition is straightforward. We can now conclude this long computation by the following:

Theorem 31. *Let $\mathcal{A} = \mathbb{C}(\mathbb{Z})$ be the algebra of Laurent polynomials, $\varphi = (\varphi_{2n+1})$ an entire normalized cocycle on $\mathcal{A}$, and τ the corresponding trace on $\hat{\hat{Q}}\mathcal{A}$. Let (ψ_{2n}) be the entire cochain*

$$\psi_{2n} = \sum_{m=0}^{\infty}\alpha(B\alpha)^m\ \varphi_{2n+2m+1}.$$

Then

$$\psi_{2n} = \int_0^{\infty} ds\ \left(\psi^1_{2n}(s) + \psi^2_{2n}(s)\right),$$

where $\psi^j_{2n}(s)$ is given by Proposition 30 applied to the trace $\tau\circ\alpha_s$ on $\hat{\hat{Q}}\mathcal{A}$ and the derivation $\delta = \left(\frac{d}{ds}\ \alpha_s\right)_{s=0}$.

We assume that φ satisfies the weak normalization condition requiring that $\varphi_{2n+1}(f^0,\dots,f^{2n+1}) = 0$ if some $f^j = 1$ for $j \neq 0$.

IV.8. The Chern Character of θ-summable Fredholm Modules

In this section we shall extend the construction (of Section 1 of this chapter) of the Chern character in K-homology to cover the infinite-dimensional situation of θ-summable Fredholm modules. Recall that a Fredholm module $(\mathcal{H}, F)$ over an algebra $\mathcal{A}$ is θ-summable iff, for any $a \in \mathcal{A}$, one has

$$[F, a] \in J^{1/2}$$

where $J^{1/2}$ is the two-sided ideal of compact operators T in $\mathcal{H}$ such that $\mu_n(T) = O\left((\log n)^{-1/2}\right)$ for $n \to \infty$.

This is the content of Definition 4 of the introduction.

We shall first establish (Theorem 4) the existence of an unbounded self-adjoint operator D on $\mathcal{H}$ such that,

1) $\operatorname{Sign} D = F$

2) $[D, a]$ is bounded for any $a \in \mathcal{A}$

3) $\operatorname{Trace}\left(e^{-D^2}\right) < \infty$.

Condition 2) means that the pair $(\mathcal{H}, D)$ defines a K-cycle over $\mathcal{A}$ (Section 2 Definition 11).

Definition 1. *A K-cycle $(\mathcal{H}, D)$ is θ-summable if $\mathrm{Trace}(e^{-D^2}) < \infty$.*

In this section we shall construct the Chern character of θ-summable K-cycles as an entire cyclic cohomology class

$$\operatorname{ch}(\mathcal{H}, D) \in H_\varepsilon^*(\mathcal{A})$$

with the same parity as $(\mathcal{H}, D)$, i.e., even iff we are given a $\mathbb{Z}/2$-grading γ commuting with any $a \in \mathcal{A}$ and anticommuting with D.

The first construction was done in [Co24] using the algebras $Q\mathcal{A}$ and $\hat{Q}\mathcal{A}$ of Section 7 and an auxiliary quasinilpotent algebra of operator-valued distributions with support in $[0, +\infty[$. Another construction, but of the same cohomology class, was then done by [J-L-O], whose formula for the entire cyclic cocycle is simpler and easier to use than our formula ([Co30]). There are, however, important features of the construction which are apparent only in the original approach, such as normalization of the cocycles and the use of the supergroup $\mathbb{R}^{1,1}$. We shall thus expound both approaches (Subsections β and γ below). In Section 9 we shall concentrate on examples of θ-summable K-cycles coming from discrete groups and supersymmetric quantum field theory.

8.α Fredholm modules and K-cycles

Let $\mathcal{H}$ be a Hilbert space, and let

$$J^{1/2} = \left\{T \in \mathcal{K}\,;\ \mu_n(T) = O(\log n)^{-1/2} \text{ as } n \to \infty\right\}$$

$$J = \left\{T \in \mathcal{K}\,;\ \mu_n(T) = O(\log n)^{-1} \text{ as } n \to \infty\right\}.$$

By construction, one has $T \in J^{1/2} \Longleftrightarrow |T|^2 \in J$.

Lemma 2. *Both J and $J^{1/2}$ are two-sided ideals of operators, and $J^{1/2}J^{1/2} = J$.*

Proof. The inequalities $\mu_n(T_1T_2) \le \|T_1\|\,\mu_n(T_2)$, $\mu_n(T_1T_2) \le \mu_n(T_1)\,\|T_2\|$ for any $n \in \mathbb{N}$, and

$$\mu_{n+m}(T_1 + T_2) \le \mu_n(T_1) + \mu_m(T_2) \quad \forall n, m \in \mathbb{N}$$

(Section 2) show that both J and $J^{1/2}$ are two-sided ideals. The inequality

$$\mu_{n+m}(T_1T_2) \le \mu_n(T_1)\,\mu_m(T_2) \quad \forall n, m \in \mathbb{N}$$

shows that $J^{1/2}J^{1/2} \subset J$ and the conclusion follows.

The natural norm on the ideal J is given by

$$\|T\|_J = \sup_N \left(\sum_{n=2}^{N} (\log n)^{-1}\right)^{-1} \sum_{n=0}^{N} \mu_n(T).$$

The sum $\sum_{n=2}^{N} (\log n)^{-1}$ is equivalent to the logarithmic integral

$$\mathrm{Li}(x) = \int_0^x \frac{du}{\log u},$$

and it is better to define

$$\|T\|_J = \sup_N \mathrm{Li}(N)^{-1} \sum_{n=0}^{N} \mu_n(T).$$

This yields the following natural notation

$$J = \mathrm{Li}(\mathcal{H}), \qquad J^{1/2} = \mathrm{Li}^{1/2}(\mathcal{H}).$$

As for any normed ideal, one has:

Lemma 3. *Let A be a C^*-algebra and $(\mathcal{H}, F)$ a Fredholm module over A. Let $\mathcal{A} = \{a \in A\,;\ [F, a] \in \mathrm{Li}^{1/2}(\mathcal{H})\}$. Then $\mathcal{A}$ is stable under holomorphic functional calculus.*

It follows that when $(\mathcal{H}, F)$ is θ-summable over A, i.e. when $\mathcal{A}$ is norm dense in A, then $\mathcal{A}$ and A have the same K-theory.

We shall now prove the following existence theorem.

Theorem 4. *Let A be a C^*-algebra, $(\mathcal{H}, F)$ a Fredholm module over A, and $\mathcal{A} \subset A$ a countably generated subalgebra such that*

$$[F, a] \in \mathrm{Li}^{1/2}(\mathcal{H}) \quad \forall a \in \mathcal{A}.$$

Then there exists a selfadjoint unbounded operator D in $\mathcal{H}$ such that:

1) Sign $D = F$
2) $[D, a]$ *is bounded for any* $a \in \mathcal{A}$
3) Trace $\left(e^{-D^2}\right) < \infty$.

Proof. Since $B = \left\{T \in \mathcal{L}(\mathcal{H}) \; ; \; [F, T] \in \mathrm{Li}^{1/2}(\mathcal{H})\right\}$ is selfadjoint and stable under holomorphic functional calculus, we can enlarge $\mathcal{A}$ and assume that it is generated by a countable group Γ of unitaries, $u \in \Gamma$, which contains F. (One has $[F, F] = 0$ so that $F \in B$.) Using Γ we shall define an averaging operation of the form

$$\mathcal{M}(T) = \sum_{u \in \Gamma} \rho(u) \, u \, T \, u^{-1} \quad \forall T \in \mathcal{L}(\mathcal{H})$$

where the weight function ρ is such that:

a) $\rho(u) > 0 \quad \forall u \in \Gamma$
b) $\sup \left\{\rho(v)^{-1} \rho(uv) \; ; \; v \in \Gamma\right\} < \infty \quad \forall u \in \Gamma$
c) $\sum_{u \in \Gamma} \rho(u) < \infty$.

The existence of such a weight function ρ on a group Γ is inherited by any quotient group $\Gamma' = h(\Gamma)$ using

$$\rho'(u) = \sum_{h(v) = u} \rho(v) \quad \forall u \in \Gamma'.$$

It is also inherited by any subgroup $\Gamma'' \subset \Gamma$.

This existence is easy for the free group on 2 generators, taking $\rho(g) = e^{-\beta \ell(g)}$, where $\ell(g)$ is the word length of $g \in \Gamma$ and β is large enough to ensure condition c). Thus, it follows for the free group on countably many generators, which is a subgroup of the free group on 2 generators. Using quotients it then follows for arbitrary countable groups.

We shall thus fix a weight function ρ on Γ satisfying a) b) and c). The operation $\mathcal{M}$ from $\mathcal{L}(\mathcal{H})$ to $\mathcal{L}(\mathcal{H})$ is linear, positive (i.e. $\mathcal{M}(T) \geq 0$ if $T \geq 0$) *and quasi-invariant under* Γ:

$$u \, \mathcal{M}(T) \, u^{-1} \leq \lambda_u \, \mathcal{M}(T) \quad \forall T \geq 0$$

where $\lambda_u \in \mathbb{R}_+^*$ depends only upon $u \in \Gamma$. We shall normalize ρ by requiring $\sum_{u \in \Gamma} \rho(u) = 1$, so that $\mathcal{M}(1) = 1$. Also, since $F \in \Gamma$, we can assume that $\mathcal{M}(T)$ commutes with F for any T (replace $\mathcal{M}$ by $\frac{1}{2}(\mathcal{M} + F\mathcal{M}F)$ if necessary).

We shall now conjugate the linear operation $\mathcal{M}$ by the nonlinear transformation f defined for T positive and bounded by

$$f(T) = \exp(-T^{-1}).$$

(We take, of course, $f(0) = 0$.) The inverse transformation f^{-1} is defined for $0 \leq T \leq \lambda < 1$ by

$$f^{-1}(T) = -(\log T)^{-1}.$$

This nonlinear transformation is *operator monotone* (cf. [Don]), i.e.

$$0 \leq T_1 \leq T_2 \leq 1 \Longrightarrow f^{-1}(T_1) \leq f^{-1}(T_2),$$

since it is a composition of the operator monotone functions log and $-1/x$.

For any $T \in \mathcal{L}(\mathcal{H})$, $T \geq 0$, we let

$$\Theta(T) = f^{-1}(\mathcal{M}(f(T))).$$

It is, by construction, a positive bounded operator.

Next, let us consider the quantum analogue of a Riemannian metric, written $g_{\mu\nu}\, dx^\mu\, dx^\nu$, for our algebra $\mathcal{A}$ as being given by a positive operator in $\mathcal{H}$ of the form

$$G = \sum\, [F, x^\mu]^*\, g_{\mu\nu}[F, x^\nu]$$

where the x^μ, $\mu \in \mathbb{N}$, generate $\mathcal{A}$ and $g_{\mu\nu}$ is a positive matrix of elements of $\mathcal{A}$. For our purpose we shall choose the x^μ to be unitaries u^μ generating the group Γ, and take positive scalar coefficients $g_{\mu\mu} > 0$ such that

$$G = \sum\, [F, u^\mu]^*\, g_{\mu\mu}[F, u^\mu]$$

is a convergent series in $\mathrm{Li}(\mathcal{H})$, which is easy to achieve (e.g. by taking $g_{\mu\mu} = 2^{-\mu}\, \|[F, u^\mu]^*\, [F, u^\mu]\|_{\mathrm{Li}}^{-1}$).

The operator G will play the role of the Green's function, so that our choice of operator D will be specified by the equality, using the map Θ defined above,

$$D^{-2} = \Theta(G).$$

We shall now prove that $D = F(\Theta(G))^{-1/2}$ has the required properties. We have to be careful to ensure that $\ker \Theta(G) = 0$. For this let us consider the subspace of $\mathcal{H}$

$$\mathcal{H}_0 = \bigcap_{u, v \in \Gamma} \ker([F, u]v).$$

By construction, $\mathcal{H}_0$ is Γ-invariant (and hence F-invariant since $F \in \Gamma$). On $\mathcal{H}_0$ one has $[F, u] = 0\ \forall u \in \Gamma$. Thus, the Fredholm module $(\mathcal{H}, F)$ decomposes as a direct sum of the trivial part $(\mathcal{H}_0, F_0)$, with $F_0 = F|\mathcal{H}_0$ commuting with any $a \in A$, and of a Fredholm module $(\mathcal{H}', F')$ for which $\mathcal{H}_0 = \{0\}$. We can thus assume that $\mathcal{H}_0 = \{0\}$ and proceed to show that $\ker \Theta(G) = \{0\}$. By construction, $\ker f(T) = \ker T$ for any T, so that with $T = G$ one has $\ker f(G) \subset \bigcap_\mu \ker[F, u^\mu]$. Next,

$$\ker \mathcal{M}(f(G)) \subset \bigcap_{v, \mu} \ker([F, u^\mu]v) = \bigcap_{u, v \in \Gamma} \ker([F, u]v)$$

since the u^μ generate Γ. Thus, we get $\ker \Theta(G) = \ker \mathcal{M}(f(G)) = \{0\}$.

Next, we shall show that for any $u \in \Gamma$ the operator

$$u\,\Theta(G)^{-1}\,u^{-1} - \Theta(G)^{-1}$$

is bounded. One has $u\,\Theta(G)\,u^{-1} = f^{-1}(u\,\mathcal{M}(T)\,u^{-1})$. Thus

$$u\,\Theta(G)^{-1}\,u^{-1} = -\log(u\,\mathcal{M}(T)\,u^{-1}).$$

So the assertion follows from the general inequality, valid for any $T \geq 0$,

$$\lambda_{u^{-1}}^{-1}\,\mathcal{M}(T) \leq u\,\mathcal{M}(T)\,u^{-1} \leq \lambda_u\,\mathcal{M}(T)$$

and the operator monotony of the logarithm

$$\log(\mathcal{M}(T)) - \log(\lambda_{u^{-1}}) \leq \log(u\,\mathcal{M}(T)\,u^{-1}) \leq \log(\mathcal{M}(T)) + \log(\lambda_u).$$

It follows immediately that

$$u\,\Theta(G)^{-1/2}\,u^{-1} - \Theta(G)^{-1/2}$$

is bounded for any $u \in \Gamma$, and hence that, since Γ generates $\mathcal{A}$ linearly,

$$\left\|[\Theta(G)^{-1/2}, a]\right\| < \infty \quad \forall a \in \mathcal{A}.$$

Thus, $F[\Theta(G)^{-1/2}, a]$ is bounded for any $a \in \mathcal{A}$.

Since $D = F\,\Theta(G)^{-1/2}$, in order to prove 2), i.e. that $[D, a]$ is bounded for any $a \in \mathcal{A}$, it remains to show that

$$[F, x]\,\Theta(G)^{-1/2}$$

is bounded for any of the generators u^{μ} of Γ appearing in the formula for G.

For this it is enough to check that $\Theta(G) \geq \lambda G$ for some $\lambda > 0$. But

$$f^{-1}\,\mathcal{M}(f(G)) \geq f^{-1}(\rho(1)\,f(G)) = (G^{-1} - \log\rho(1))^{-1} \geq \lambda G$$

for $\lambda^{-1} = 1 - \|G\|\log\rho(1)$.

Finally we need to prove 3). Since we can replace D by an arbitrary scalar multiple of D without altering 1) and 2), it is enough to show that with the above construction one gets

$$e^{-D^2} \in \mathcal{L}^p \text{ for some finite } p.$$

Since $e^{-D^2} = \mathcal{M}(f(G))$, it is enough to show that

$$f(G) \in \mathcal{L}^p \text{ for some finite } p.$$

But $G \in \mathrm{Li}(\mathcal{H})$ so that $\mu_n(G) = O((\log n)^{-1}) \leq c(\log n)^{-1}$ for some finite c, which shows that $\mu_n(f(G)) \leq n^{-1/c}$.

On $\mathcal{H}_0$ the same construction applies and is much easier, starting with G_0, a strictly positive element of $\mathrm{Li}(\mathcal{H}_0)$.

Remark 5. a) The above proof in fact also yields the condition 4): $[D^2, a]$ is bounded for any $a \in \mathcal{A}$. This condition implies that $[|D|, a]$ is bounded for any $a \in \mathcal{A}$.

b) Let $(\mathcal{H}, D)$ be a θ-summable K-cycle over $\mathcal{A}$ so that $[|D|, a]$ is bounded for any $a \in \mathcal{A}$. Let $F = \operatorname{Sign} D$. Then $[F, a] \in \mathrm{Li}^{1/2}(\mathcal{H})$ for any $a \in \mathcal{A}$.

c) The above proof does not yield a K-cycle such that Trace $(e^{-\beta D^2}) < \infty$ for all $\beta > 0$. In fact, by Lemma 6 of Appendix C, there exists a constant $C < \infty$ such that for any unitaries u_i,

$$k_\infty\{u_i\} \leq C \sup_i \|[D^2, u_i]\|$$

where one assumes $\mathrm{Trace}(e^{-D^2}) < \infty$. Here k_∞ is the obstruction ([Vo$_1$]) to the existence of quasicentral units relative to the Macaev ideal $\mathcal{L}^{(\infty,1)}$, for which

$$T \in \mathcal{L}^{\infty,1} \text{ iff } \sum_{n=1}^{\infty} \frac{1}{n} \mu_n(T) < \infty.$$

In general, given an arbitrary normed ideal J of compact operators (cf. Appendix C), the obstruction k_J to the existence of quasicentral approximate units is defined [Vo$_1$], for any finite subset X of $\mathcal{L}(\mathcal{H})$, by

$$k_J(X) = \operatorname*{Lim\,inf}_{h \uparrow 1} \|[X, h]\|_J$$

where h runs through finite-rank operators $0 \leq h \leq 1$, while the norm is $\|[X, h]\|_J = \sup_{x \in X} \|[x, h]\|_J$.

The results of D. Voiculescu [Vo$_1$] on quasicentral approximate units relative to the Macaev ideal show that, in general, one has

$$k_\infty\{u_i\} > 0.$$

This gives a lower bound to the values of β such that Trace $(e^{-\beta D^2}) < \infty$.

8.β The supergroup $\mathbb{R}^{1,1}$ and the convolution algebra $\tilde{\mathcal{L}}$ of operator-valued distributions on $[0, +\infty[$

In this section we shall construct a convolution algebra $\tilde{\mathcal{L}}$ of operator-valued distributions and a trace τ on $\tilde{\mathcal{L}}$ which will be the natural receptacle for representations of the quasinilpotent extensions Q_ε and $\hat{Q}_\varepsilon$ of Section 7 Theorem 17.

We let $\mathcal{H}$ be a Hilbert space. By an operator-valued distribution we mean a continuous linear map T from the Schwartz space $S(\mathbb{R})$ (with its usual nuclear topology) to the Banach space $\mathcal{L}(\mathcal{H})$ of bounded operators in $\mathcal{H}$. Thus, there exists by hypothesis a continuous seminorm p on $S(\mathbb{R})$ such that $\|T(f)\| \leq p(f)\ \forall f \in S(\mathbb{R})$. We let $\mathcal{L}$ be the space of operator-valued distributions T which satisfy the following properties:

1) Support $T \subset \mathbb{R}^+ = [0, +\infty[$.

2) There exists $r > 0$ and an analytic operator-valued function $t(z)$ for $z \in C = \bigcup_{s>0} sU$, where U is the disk with center at 1 and radius r, such that

a) $t(s) = T(s)$ on $]0,+\infty[$, (i.e. $T(f) = \int f(s)\, t(s)\, ds \quad \forall f \in C_c^\infty(]0,\infty[)$)

b) the function defined for $p \in]1,\infty[$ by

$$h(p) = \sup_{z\in(1/p)U} \|t(z)\|_p$$

is bounded above by a polynomial in p as $p \to \infty$.

In b), the norm $\|t(z)\|_p$ is the Banach space norm of the Schatten ideal $\mathcal{L}^p(\mathcal{H})$. In particular, we see that $t(1) \in \mathcal{L}^1(\mathcal{H})$ is a trace class operator. The operator-valued analytic function $t(z)$ is, of course, uniquely determined by the distribution T, and we shall use the abuse of notation $T(z)$ instead of $t(z)$. Two distributions $T_1, T_2 \in \mathcal{L}$, such that $T_1(z) = T_2(z) \ \forall z \in]0,+\infty[$, differ by a distribution with support the origin, of the form $\sum a_k\, \delta_0^{(k)}$, where $a_k \in \mathcal{L}(\mathcal{H})$, and $\delta_0^{(k)}$ is the k-th derivative of the Dirac mass δ_0 at the origin.

The condition 2) essentially means that T takes its values in operators of a suitable Schatten class so that the quantity Trace $T(1)$ is well defined.

All operator-valued distributions on $\mathbb{R}$ with support $\{0\}$ belong to $\mathcal{L}$; in particular, the products $\delta_0 \times \mathrm{id}$ and $\delta_0' \times \mathrm{id}$ of the Dirac mass at 0, or of its derivative, by the identity operator in $\mathcal{L}(\mathcal{H})$, both do. To lighten the notation we shall simply write δ_0 and δ_0'.

Lemma 6. a) *Let $T \in \mathcal{L}$. Then the derivative $T' = (d/ds)T$ also belongs to $\mathcal{L}$.*

b) *Let $T \in \mathcal{L}$. There exist an integer q and $S \in \mathcal{L}$ such that $T - S^{(q)}$ has support $\{0\}$, where $S^{(q)}$ is the q-th derivative of S, and that*

$$\sup_p \sup_{z\in(1/p)U} \|S(z)\|_p < \infty,$$

where $U = \{z \in \mathbb{C}; |z-1| \le r\}$.

Proof. a) By definition, $T'(f) = -T(f') \ \forall f \in S(\mathbb{R})$, so that T' is an operator-valued distribution satisfying property 1). Let r and U be as in 2) for T, and let $r' = r/2$. Then by Cauchy's theorem the operator $T'(z)$ for $z \in (1/p)\, U'$, $U' = \{z \in \mathbb{C}; |z-1| \le r'\}$, is of the form $\int_{u\in(1/p)U} T(u)\, d\mu(u)$, where μ has total mass less than $2p/r$. Thus

$$\sup_{z\in(1/p)U'} \|T'(z)\|_p \le \frac{2p}{r} \sup_{z\in(1/p)U} \|T(z)\|_p,$$

which proves that T' satisfies property 2).

b) By hypothesis, there exist $C < \infty$ and $q \in \mathbb{N}$ such that, with the notation of 2), $h(p) \le Cp^q$. Let T_k be, for $k = 0,1,\ldots$, the operator-valued analytic function in $C = \bigcup_{s>0} sU$, defined inductively by $T_0(z) = T(z)$ and $T_{k+1}(z) = \int_1^z T_k(u)du$. For $z \in (1/p)U$ one has

$$\|T_{k+1}(z)\|_p \le 2\int_0^1 h_k\left(\left((1-t)+\frac{t}{p}\right)^{-1}\right) dt$$

where

$$h_k(p) = \sup_{z \in (1/p)U} \|T_k(z)\|_p$$

(since $\|T_k(u)\|_p \leq \|T_k(u)\|_{p'}$ for $p' \leq p$).

Thus, we see that h_k is of the order of p^{q-k} for $k < q$, and that h_q is of the order of $\log p$, while h_{q+1} is bounded. Let S be the operator-valued distribution given by

$$S(f) = \int f(s)\, T_{q+1}\,(s)\, ds \quad \forall f \in S(\mathbb{R}).$$

It is well defined, since $\|T_{q+1}(s)\|$ is bounded on $[0, 1]$ and by a polynomial for large s. By construction, the $(q+1)$-st derivative of S agrees with T outside the origin, thus the conclusion.

We can now show that $\mathcal{L}$ is an algebra under the convolution product, which at the formal level can be written

$$(T_1 * T_2)(s) = \int_0^s T_1(u)\, T_2(s-u)\, du.$$

More precisely, given $f \in S(\mathbb{R})$, one can find $a_n, b_n \in S(\mathbb{R})$ such that the restriction to $]-1, \infty[\, \times\,]-1, \infty[$ of the function $(s, u) \to f(s+u)$ is given by the convergent series $\sum a_n \otimes b_n$. Then $(T_1 * T_2)(f) = \sum T_1(a_n)\, T_2(b_n) \in \mathcal{L}(\mathcal{H})$.

Lemma 7. *If $T_1, T_2 \in \mathcal{L}$ then $T_1 * T_2 \in \mathcal{L}$.*

Proof. By Lemma 6 one can assume that T_i is given by

$$T_i(f) = \int_0^\infty f(s)\, T_i(s)\, ds,$$

where $T_i(s)$ is an analytic operator-valued function in $C = \bigcup_{t>0}(tU)$, with $U = \{z \in \mathbb{C}; |z-1| < r\}$, and where

$$C_i = \sup_p \sup_{(1/p)U} \|T_i(z)\|_p < \infty.$$

Then let $T(z) = \int_0^1 T_1(\lambda z)\, T_2((1-\lambda)z)z\, d\lambda$. It is, by construction, an analytic operator-valued function defined in C. One has, for $z \in (1/p)U$, that

$$\lambda z \in \frac{1}{p_1} U\ ,\ (1-\lambda)z \in \frac{1}{p_2} U$$

where

$$\frac{1}{p} = \frac{1}{p_1} + \frac{1}{p_2}$$

so that by Holder's inequality

$$\|T_1(\lambda z)\, T_2((1-\lambda)z)\|_p \leq \|T_1(\lambda z)\|_{p_1}\, \|T_2((1-\lambda)z)\|_{p_2} \leq C_1\, C_2.$$

Thus, we get, for any $z \in (1/p)U$

$$\|T(z)\|_p \leq \int_0^1 \|T_1(\lambda z)\, T_2((1-\lambda)z)\|_p\, |z|\, d\lambda \leq |z|\, C_1\, C_2.$$

It follows that $T(z)$ defines an element of $\mathcal{L}$, and that it coincides with the convolution product $T_1 * T_2$.

Let $\lambda = \delta_0'$ be the derivative of the Dirac mass at 0. One has $\lambda \in \mathcal{L}$, and as an operator-valued distribution λ has a natural square root (for the convolution product) given by the derivative T' of the distribution

$$T(s) = \frac{1}{\sqrt{\pi s}}.$$

But this square root does not define an element of the algebra $\mathcal{L}$, since (when $\dim \mathcal{H} = \infty$) it fails to satisfy condition 2) above, because the identity operator does not belong to any $\mathcal{L}^p$. We thus need to adjoin a formal square root $\lambda^{1/2}$ of λ to $\mathcal{L}$. For this we consider the algebra $\tilde{\mathcal{L}}$ of pairs (T_0, T_1) of elements of $\mathcal{L}$ with product given by

$$(T_0, T_1) * (S_0, S_1) = (T_0S_0 + \lambda T_1S_1, T_0S_1 + T_1S_0),$$

where we write T_0S_0 for $T_0 * S_0$, etc. Since λ belongs to the center of $\mathcal{L}$, one checks that the above product turns $\tilde{\mathcal{L}}$ into an algebra. This algebra $\tilde{\mathcal{L}}$ contains $\mathcal{L}$ (by the homomorphism $T \to (T, 0)$) and the central element $\lambda^{1/2} = (0, \delta_0)$, so that every element of $\tilde{\mathcal{L}}$ is of the form $A + B\lambda^{1/2}$ with $A, B \in \mathcal{L}$.

Lemma 8. *The equality* $\tau(T_0, T_1) = \operatorname{trace} T_1(1)$ *defines a trace* τ *on the algebra* $\tilde{\mathcal{L}}$.

Proof. By condition 2) we know that $T_1(1)$ belongs to $\mathcal{L}^1$ so that the trace is well defined. Since

$$(T_0, T_1) * (S_0, S_1) = (T_0S_0 + \lambda T_1S_1, T_0S_1 + T_1S_0),$$

it is enough to check that $T \mapsto \operatorname{trace} T(1)$ is a trace on the algebra $\mathcal{L}$. The proof of Lemma 7 shows that for $T_i \in \mathcal{L}$ of the form $T_i(f) = \int f(s)\, T_i(s)\, ds$, with

$$\sup_p \sup_{z \in (1/p)U} \|T_i(z)\|_p < \infty,$$

one has

$$\operatorname{trace}(T_1 * T_2)(s) = \operatorname{trace}(T_2 * T_1)(s) \quad \forall s \in U.$$

Thus, for any power of λ one has

$$\operatorname{trace}((\lambda^k\, T_1 * T_2)(1)) = \operatorname{trace}((\lambda^k\, T_2 * T_1)(1)).$$

Now, by Lemma 6, to show that $\operatorname{trace}(S_1 * S_2)(1) = \operatorname{trace}(S_2 * S_1)(1)$, we can assume that $S_j = \lambda^{k_j} T_j + U_j$, where T_j is as above and U_j has support the origin.

Thus, we just need to check, say, that trace$(\lambda^{k_1} T_1 U_2)(1) =$ trace $(U_2 \lambda^{k_1} T_1)(1)$, which follows from trace $(ab) =$ trace (ba), $a \in \mathcal{L}(\mathcal{H})$, $b \in \mathcal{L}^1(\mathcal{H})$.

Let us recall that the Hopf algebra H of smooth functions on the supergroup $\mathbb{R}^{1,1}$ is given as follows: as an algebra one has

$$H = C^\infty(\mathbb{R}^{1,1}) = C^\infty(\mathbb{R}) \otimes \wedge(\mathbb{R}),$$

the tensor product of the algebra of smooth functions on $\mathbb{R}$ by the exterior algebra $\wedge(\mathbb{R})$ of a one-dimensional vector space. Thus, every element of H is given by a sum $f + g\xi$, where $f, g \in C^\infty(\mathbb{R})$, $\xi^2 = 0$. The interesting structure comes from the coproduct $\Delta : H \to H \otimes H$ which corresponds to the supergroup structure; being an algebra morphism it is fully specified by its values on $C^\infty(\mathbb{R}) \subset H$ and by $\Delta(\xi) = \xi \otimes 1 + 1 \otimes \xi$; one has $(\Delta f) = \Delta_0(f) + \Delta_0(f')\, \xi \otimes \xi$, where $f' = \frac{\partial}{\partial s} f(s)$ and

$$\Delta_0 : C^\infty(\mathbb{R}) \to C^\infty(\mathbb{R}) \otimes C^\infty(\mathbb{R})$$

is the usual coproduct,

$$\Delta_0(f)(s,t) = f(s+t). \tag{4}$$

Equivalently, the (topological) dual H^* of H is endowed with a product which we can now describe. Every element of H^* is uniquely of the form (T_0, T_1), where $T_0, T_1 \in C_c^{-\infty}(\mathbb{R})$ are distributions with compact support on $\mathbb{R}$, and

$$\langle f + g\xi, (T_0, T_1) \rangle = T_0(f) + T_1(g). \tag{5}$$

The product $*$ on H^* dual to the coproduct Δ is defined by

$$\langle f + g\xi, (T_0, T_1) * (S_0, S_1) \rangle = \langle \Delta(f + g\xi), (T_0, T_1) \otimes (S_0, S_1) \rangle. \tag{6}$$

Lemma 9. *The product $*$ on H^* is given by*

$$(T_0, T_1) * (S_0, S_1) = (T_0 * S_0 + \delta_0' * T_1 * S_1, T_0 * S_1 + T_1 * S_0).$$

Using $\xi^2 = 0$, this follows from formula (4). This shows that the algebra $\widetilde{\mathcal{L}}$ is really a convolution algebra of operator valued distributions on the supergroup $\mathbb{R}^{1,1}$, thus clarifying the relations between our formulas and supersymmetry.

8.γ The Chern character of K-cycles

Let $\mathcal{A}$ be a locally convex algebra, and let $(\mathcal{H}, D)$ be a θ-summable K-cycle over $\mathcal{A}$ (cf. Definition 1 before section α)). In this section we shall construct the Chern character

$$\mathrm{ch}_*(\mathcal{H}, D) \in H_\varepsilon^*(\mathcal{A})$$

as an entire cyclic cohomology class for $\mathcal{A}$.

For this purpose we may as well replace $\mathcal{A}$ by the Banach algebra A of bounded operators

$$A = \{a \in \mathcal{L}(\mathcal{H})\ ;\ [D, a] \text{ is bounded}\}$$

which we can endow with the norm $p(a) = \|a\| + \|[D,a]\|$. By Theorem 17 of Section 7, the normalized entire even (resp. odd) cocycles on A correspond to odd traces on the algebra $\hat{Q}_\varepsilon(A)$ (resp. on $\hat{\hat{Q}}_\varepsilon$ vanishing on γA). By Proposition 18 of Section 7, $Q_\varepsilon A$ is a quasinilpotent extension of A. Thus, to construct $\mathrm{ch}_*(\mathcal{H},D)$ as a normalized cocycle requires a trace on a suitable quasinilpotent extension of the algebra A. We shall use, for that purpose, the convolution algebra $\tilde{\mathcal{L}}$ of Subsection β and the trace τ given by Lemma 8. The representation of A in $\mathcal{H}$, i.e. the inclusion $A \subset \mathcal{L}(\mathcal{H})$, yields a natural homomorphism from A to $\tilde{\mathcal{L}}$, given by

$$\rho(a) = a\,\delta_0 \quad \forall a \in A.$$

We shall now extend, using D, the homomorphism ρ to a homomorphism of $\hat{Q}_\varepsilon A$ to $\tilde{\mathcal{L}}$, when the K-cycle is *even*. In the odd case we shall get a homomorphism

$$\rho' : \hat{\hat{Q}}_\varepsilon A \to M_2(\tilde{\mathcal{L}}).$$

At the formal level, if, as above, we denote the formal square root of $\lambda = \delta_0'$ by $\lambda^{1/2}$, the formulae for ρ and ρ' are

$$\rho(F) = (D + \gamma\,\lambda^{1/2})(D^2 + \lambda)^{-1/2} \qquad \text{(even case)}$$

$$\rho'(F) = \begin{bmatrix} 0 & U \\ U^* & 0 \end{bmatrix}, \ U = (D + i\,\lambda^{1/2})(D^2 + \lambda)^{-1} \qquad \text{(odd case)}$$

$$\rho'(\gamma) = \begin{bmatrix} 1 & 0 \\ 0 & -1 \end{bmatrix}.$$

The next two lemmas are useful for both even and odd cases.

Lemma 10. *One has, with* $s = \mathrm{Re}\, z > 0$, $p \in [1,\infty[$,

$$\|e^{-zD^2}\|_p = (\mathrm{trace}(e^{-spD^2}))^{1/p}\,,$$
$$\|D\,e^{-zD^2}\|_p \le s^{-1/2}\,\|e^{-(s/2)D^2}\|_p.$$

Proof. One has

$$\|e^{-zD^2}\|_p = \|e^{-sD^2}\|_p = (\mathrm{trace}(e^{-psD^2}))^{1/p}.$$

To prove the second inequality it is enough to show that the operator norm of $\|De^{-(s/2)D^2}\|$ is bounded by $1/\sqrt{s}$. But this follows from the inequality $x\,e^{-(s/2)x^2} \le 1/\sqrt{s}$ for x real and positive.

Lemma 10 shows that we can define an element N of $\mathcal{L}$ by the equality

$$N(f) = \frac{1}{\sqrt{\pi}} \int f(s)\,\frac{1}{\sqrt{s}}\,e^{-sD^2}\,ds\,,\ f \in S(\mathbb{R}).$$

The integral makes sense since the operator norm of $\frac{1}{\sqrt{s}} e^{-sD^2}$ is integrable near the origin. We shall, however, also need to define the distribution DN, which is formally given by

$$(DN)(f) = \frac{1}{\sqrt{\pi}} \int f(s) \frac{1}{\sqrt{s}} D\, e^{-sD^2}\, ds.$$

By Lemma 10, one has an analytic operator-valued function,

$$(DN)(z) = \frac{1}{\sqrt{\pi}} \frac{1}{\sqrt{z}} D\, e^{-zD^2},$$

defined for $\mathrm{Re}\, z > 0$, and such that $\sup_{(1/p)U} \|(DN)(z)\|_p$ is of the order of p, $p \to \infty$. However, since the operator norm of $(DN)(s)$ is of the order of $1/s$, as $s \to 0$, and is not integrable, we have to be careful in the definition of the distribution DN.

Lemma 11. a) *The Laplace transform of the distribution N is given by*

$$\int_0^\infty N(s)\, e^{-s\lambda}\, ds = (D^2 + \lambda)^{-1/2}.$$

b) *There exists a unique element of $\mathcal{L}$, denoted DN, whose Laplace transform is equal to $D(D^2+\lambda)^{-1/2}$. One has $(DN)(s) = DN(s)$ for any $s > 0$.*

Proof. a) Follows from the equality

$$\int_0^\infty \frac{1}{\sqrt{\pi s}} e^{-s\alpha^2} e^{-s\lambda}\, ds = (\alpha^2 + \lambda)^{-1/2}.$$

b) The uniqueness is clear. Let us prove the existence. One has

$$\begin{aligned} &D(D^2+\lambda)^{-1/2} - D(D^2+1)^{-1/2} \\ &= \frac{1}{\pi} \int_0^\infty D((D^2+\lambda+\rho)^{-1} - (D^2+1+\rho)^{-1})\rho^{-1/2}\, d\rho \\ &= \frac{1}{\pi}(1-\lambda) \int_0^\infty D(D^2+\lambda+\rho)^{-1}\, (D^2+1+\rho)^{-1}\, \rho^{-1/2}\, d\rho. \end{aligned}$$

Now $D(D^2+1)^{-1/2}$ is the Laplace transform of the element of $\mathcal{L}$ given by

$$D(D^2+1)^{-1/2}\, \delta_0.$$

Thus, we just have to show, using Lemma 10, that

$$\int_0^\infty D(D^2+\lambda+\rho)^{-1}(D^2+1+\rho)^{-1}\, \rho^{-1/2}\, d\rho$$

is the Laplace transform of an element of $\mathcal{L}$. But

$$D(D^2+1+\rho)^{-1}\, (D^2+\lambda+\rho)^{-1}$$

is the Laplace transform of

$$D(D^2+1+\rho)^{-1}e^{-s(D^2+\rho)},$$

and it is enough to check that the operator norm of

$$T(s)=\int_0^\infty D(D^2+1+\rho)^{-1}\, e^{-s(D^2+\rho)}\, \rho^{-1/2}\, d\rho$$

is integrable near $s=0$. One has

$$\|T(s)\| \le \int_0^\infty (1+\rho)^{-1/2}\, e^{-s\rho}\, \rho^{-1/2}\, d\rho,$$

since $\|D(D^2+1+\rho)^{-1}\| \le (1+\rho)^{-1/2}$. Thus, since

$$\int_0^\infty (1+\rho)^{-1/2}\, e^{-s\rho}\, \rho^{-1/2}\, d\rho \le (3-\log s) = O(|\log s|)$$

when $s \to 0$, we see that the operator norm of $T(s)$ is integrable near 0. The same estimate works for the $\mathcal{L}^p$ norm $\|T(z)\|_p$ for $z \in (1/p)U$, and shows that $T \in \mathcal{L}$.

Let us now treat the *even* case.

Lemma 12. *The equality $F=(DN,\gamma N)$ defines an element of $\tilde{\mathcal{L}}$ of square δ_0.*

Proof. By Lemma 11, the element $DN \in \mathcal{L}$ is well defined. Since γ anticommutes with D, DN anticommutes with γN, so that the square is given by $F^2=((DN)^2+\lambda N^2, 0)$. Now the Laplace transform of $(DN)^2$ is (Lemma 11) equal to $D^2(D^2+\lambda)^{-1}$, and that of λN^2 is $\lambda(D^2+\lambda)^{-1}$. Thus, the Laplace transform of $(DN)^2+\lambda N^2$ is equal to 1 and we get $F^2=(\delta_0,0)$.

We can now state the main technical result of this section:

Theorem 13. [Co$_{24}$] *There exists a unique continuous homomorphism ρ of $\hat{Q}_\varepsilon(A)$ to $\tilde{\mathcal{L}}$ such that:*

$$\rho(a)=a\,\delta_0 \quad \forall a \in A\,, \qquad \rho(F)=(DN,\gamma N).$$

We refer to [Co$_{24}$] Section 5 for the proof of the continuity of the homomorphism ρ. The convolution algebra $\mathcal{L}$ was defined so as to make these estimates straightforward. Let τ be the trace on $\tilde{\mathcal{L}}$ given by Lemma 8.

Corollary 14. [Co$_{24}$] *Let $(\mathcal{H},D,\gamma)$ be an even θ-summable K-cycle on a locally convex algebra $\mathcal{A}$. Then the following equality defines a normalized even entire cocycle $(\varphi_{2n})_{n\in\mathbb{N}}$ on $\mathcal{A}$:*

$$\varphi_{2n}(a^0,\ldots,a^{2n})=\Gamma\left(n+\tfrac{1}{2}\right)\tau\left(\rho\left(F\,a^0[F,a^1]\cdots[F,a^{2n}]\right)\right) \quad \forall a^j \in \mathcal{A}.$$

Note that here φ is a cocycle in the (d_1, d_2) bicomplex. The corollary follows from Theorem 13 and Theorem 17 of Section 7.

The odd case is treated in the same way. One defines $\rho(F)$ as the 2×2 matrix $\begin{bmatrix} 0 & U \\ U^* & 0 \end{bmatrix} \in M_2(\widetilde{\mathcal{L}})$, where

$$U = (DN, iN)$$

with the above notation. Also $\gamma = \begin{bmatrix} 1 & 0 \\ 0 & -1 \end{bmatrix} \in M_2(\widetilde{\mathcal{L}})$.

Theorem 15. [Co$_{24}$] *There exists a unique continuous homomorphism ρ' of $\hat{\hat{Q}}_\varepsilon(A)$ to $M_2(\widetilde{\mathcal{L}})$ such that*

$$\rho'(a) = a\,\delta_0 \quad \forall a \in A\,, \qquad \rho'(F) = \begin{bmatrix} 0 & U \\ U^* & 0 \end{bmatrix}, \; \rho'(\gamma) = \begin{bmatrix} 1 & 0 \\ 0 & -1 \end{bmatrix}.$$

One has $\tau(\rho'(\gamma a)) = 0 \quad \forall a \in A$, thus, as above, one gets the following corollary.

Corollary 16. [Co$_{24}$] *Let $(\mathcal{H}, D)$ be an odd θ-summable K-cycle on a locally convex algebra $\mathcal{A}$. Then the following equality defines a normalized odd entire cocycle $(\varphi_{2n+1})_{n \in \mathbb{N}}$ on $\mathcal{A}$:*

$$\varphi_{2n+1}(a^0, \ldots, a^{2n+1}) = \frac{-1}{\sqrt{2i}}\, n!\, \tau'\, \rho'\left(\gamma\, F\, a^0 [F, a^1] \cdots [F, a^{2n+1}]\right) \quad \forall a^j \in \mathcal{A}$$

where $\tau' = \tau \otimes$ Trace on $M_2(\widetilde{\mathcal{L}}) = \widetilde{\mathcal{L}} \otimes M_2(\mathbb{C})$.

We can now adopt the following definition:

Definition 17. *Let $(\mathcal{H}, D)$ be a θ-summable K-cycle on the locally convex algebra $\mathcal{A}$. Then its* Chern character $\mathrm{ch}_*(\mathcal{H}, D)$ *is the* class *of the cocycle φ of Corollaries* 14 *and* 16 *in $H_\varepsilon^*(\mathcal{A})$.*

8.δ The index formula

Let $\mathcal{A}$ be an involutive algebra and $(\mathcal{H}, D, \gamma)$ an even K-cycle over $\mathcal{A}$. Then, as in Proposition 2 of the introduction of this chapter, the K-homology class of $(\mathcal{H}, D, \gamma)$ determines an index map

$$K_0(\mathcal{A}) \to \mathbb{Z},$$

which can be defined as follows.

Let $e \in \mathrm{Proj}\, M_q(\widetilde{\mathcal{A}})$ be an idempotent, $e \in M_q(\widetilde{\mathcal{A}})$, and let $(\mathcal{H}_q, D_q, \gamma_q)$ be the K-cycle on $M_q(\widetilde{\mathcal{A}})$ given by

$$\mathcal{H}_q = \mathcal{H} \otimes \mathbb{C}^q\,, \; D_q = D \otimes 1\,, \; \gamma_q = \gamma \otimes 1$$

with the obvious action of $\widetilde{\mathcal{A}} \otimes M_q(\mathbb{C})$ in $\mathcal{H}_q$.

Proposition 18. a) *The operator $e\, D_q^+\, e$ from $e\, \mathcal{H}_q^+$ to $e\, \mathcal{H}_q^-$ has finite-dimensional kernel and cokernel.*

b) *An additive map φ from $K_0(\mathcal{A})$ to $\mathbb{Z}$ is determined by the equality*

$$\varphi([e]) = \operatorname{Ind}(e\, D_q^+\, e).$$

The proof and the index map are the same as in Proposition 2 of the Introduction. The operator $F = \operatorname{Sign}(D)$ in $\mathcal{H}$ determines a (pre-)Fredholm module on $\mathcal{A}$ with the same index map as D.

We shall now show that, when the K-cycle $(\mathcal{H}, D, \gamma)$ is θ-summable, the index map φ is given by a formula in terms of the Chern character $\operatorname{ch}_*(\mathcal{H}, D, \gamma)$ (Definition 17).

Theorem 19. [Co24] *Let $\mathcal{A}$ be a locally convex algebra and $(\mathcal{H}, D, \gamma)$ an even θ-summable K-cycle over $\mathcal{A}$. Then for any $x \in K_0(\mathcal{A})$ one has*

$$\varphi(x) = \langle x, \operatorname{ch}_*(\mathcal{H}, D, \gamma)\rangle.$$

We use here the pairing between K-theory and entire cyclic cohomology found in Theorem 21 of Section 7.

Theorem 19 is the infinite-dimensional analogue of Proposition 4 of Section 1. To prove it one uses the invariance of the index and of the Chern character under a homotopy of the operator D. This allows one to reduce to the situation where D commutes with the idempotent e, in which case the assertion of the theorem is the McKean-Singer formula

$$\operatorname{Ind}(D^+) = \operatorname{Trace}(\gamma\, e^{-D^2}).$$

Indeed, the component φ_0 of the entire cocycle (φ_{2n}) of Corollary 14 is given by

$$\varphi_0(a) = \operatorname{Trace}(\gamma\, a\, e^{-D^2}).$$

The explicit homotopy between D and an operator D_1 which commutes with D is given by

$$D_t = D + t\delta\,,\ \delta = -e[D,e] + [D,e]e.$$

Finally, the homotopy invariance of the character follows from the next proposition applied to the algebra $B = \widetilde{\mathcal{L}}$ with trace τ of Subsection β), to the homomorphism ρ of Theorem 13, and to the elements F_t of $\widetilde{\mathcal{L}}$ corresponding to D_t (cf. [Co24] for the technical details).

Proposition 20. *Let $\rho : \mathcal{A} \to B$ be a homomorphism of algebras, with B unital, and let τ be a trace on B. Let $F_t \in B$, $t \in [0,1]$, be a C^1 family of elements of B with $F_t^2 = 1\ \forall t \in [0,1]$.*

a) *For every* $t \in [0,1]$, *a cocycle in the* (d_1, d_2) *bicomplex of* $\mathcal{A}$ *is defined for* $a^j \in \mathcal{A}$ *by*

$$\varphi_{2n}^t(a^0,\dots,a^{2n}) = \Gamma\left(n+\tfrac{1}{2}\right)\,\tau\left(F_t\,\rho(a^0)\left[F_t,\rho(a^1)\right]\cdots\left[F_t,\rho(a^{2n})\right]\right).$$

b) *For* $n \geq 0$ *and* $t \in [0,1]$, $\forall a^i \in \mathcal{A}$, *let*

$$\psi_{2n+1}^t(a^0,\dots,a^{2n+1}) = \Gamma\left(n+\tfrac{1}{2}\right)\,(2n)^{-1}\sum_{j=1}^{2n}(-1)^j\,\tau(F_t\,\rho(a^0)\left[F_t,\rho(a^1)\right]\cdots$$
$$\left[F_t,\rho(a^{j-1})\right](\frac{d}{dt}F_t)\left[F_t,\rho(a^j)\right]\cdots\left[F_t,\rho(a^{2n+1})\right]).$$

Then

$$\frac{d}{dt}\varphi^t = (d_1+d_2)\psi^t$$

for all $t \in [0,1]$.

Proof. a) follows from Proposition 15 b) of Section 7 and b) from a straightforward computation.

8.ε The JLO cocycle

We have defined above the Chern character

$$\mathrm{ch}_*(\mathcal{H},D) \in H_\varepsilon^*(\mathcal{A})$$

of K-cycles over locally convex algebras (Definition 17), and proved the index formula (Theorem 19). However, except for the component of dimension 0

$$\varphi_0(a) = \mathrm{Trace}(\gamma\,a\,e^{-D^2}),$$

the higher components of the normalized entire cocycles given by Corollaries 14 and 16 are given by complicated formulae.

Motivated by examples of θ-summable K-cycles arising in supersymmetric quantum field theory, Jaffe, Lesniewski and Osterwalter ([J-L-O]) obtained a much simpler cocycle, which was then shown ([Co$_{30}$]) to be cohomologous to the normalized cocycle of Corollaries 14 and 16.

Theorem 21. [J-L-O] *Let* $(\mathcal{H}, D)$ *be a* θ*-summable* K*-cycle over a locally convex algebra* $\mathcal{A}$. *Then the following formulae define an entire cocycle in the* (b,B) *bicomplex.*

a) *Even case:* $\forall a^j \in \mathcal{A}$

$$\varphi_{2n}(a^0,\dots,a^{2n}) = \int_{\sum s_i=1,\,s_i\geq 0} ds_0\cdots ds_{2n-1}\times$$
$$\mathrm{Trace}\,(\gamma\,a^0\,e^{-s_0D^2}\,[D,a^1]\,e^{-s_1D^2}\cdots$$
$$[D,a^{2n-1}]\,e^{-s_{2n-1}D^2}\,[D,a^{2n}]\,e^{-s_{2n}D^2}).$$

b) *Odd case*:

$$\varphi_{2n+1}(a^0,\dots,a^{2n+1}) = \sqrt{2i}\int_{\sum s_i=1,\,s_i\geq 0} ds_0\cdots ds_{2n}\times$$
$$\mathrm{Trace}\,(a^0\,e^{-s_0D^2}\,[D,a^1]\,e^{-s_1D^2}\cdots$$
$$[D,a^{2n}]\,e^{-s_{2n}D^2}\,[D,a^{2n+1}]\,e^{-s_{2n+1}D^2}).$$

To prove this we use the algebra $\mathcal{L}$ of Subsection β) and the trace τ_0 on $\mathcal{L}$ given by

$$\tau_0(T) = \text{Trace}\,(T(1)) \quad \forall T \in \mathcal{L}.$$

Let us treat the odd case. The JLO formula is, up to normalization

$$\varphi_{2n+1}(a^0,\ldots,a^{2n+1}) = \tau_0\left(a^0\,\frac{1}{D^2+\lambda}\,[D,a^1]\,\frac{1}{D^2+\lambda}\cdots[D,a^{2n+1}]\,\frac{1}{D^2+\lambda}\right).$$

In this formula λ is the element δ_0' of $\mathcal{L}$, but it is convenient to think of it as the free variable of Laplace transforms, which converts the convolution product of $\mathcal{L}$ into the ordinary pointwise product of operator-valued functions of the real positive variable λ. One cannot, however, permute the Laplace transform with the trace, since an operator like $a^0\,\frac{1}{D^2+\lambda}\,[D,a^1]\cdots[D,a^{2n+1}]\,\frac{1}{D^2+\lambda}$ is, in general, *not* of trace class for λ a scalar when D is only θ-summable.

The cocycle property of φ, $b\varphi_{2n-1} + B\varphi_{2n+1} = 0$, can be checked directly using the following straightforward equalities:

$$\begin{aligned}(b\varphi_{2n-1})&(a^0,\ldots,a^{2n})\\ &= -\tau_0\left([D,a^0]\,\frac{1}{D^2+\lambda}\,[D,a^1]\cdots\frac{1}{D^2+\lambda}\,[D,a^{2n}]\,\frac{1}{D^2+\lambda}\right),\\ (B_0\,\varphi_{2n-1})&(a^0,\ldots,a^{2n})\\ &= \tau_0\left(\frac{1}{(D^2+\lambda)^2}\,[D,a^0]\,\frac{1}{D^2+\lambda}\,[D,a^1]\cdots\frac{1}{D^2+\lambda}\,[D,a^{2n}]\right).\end{aligned}$$

One gets indeed that

$$(B\varphi_{2n-1})(a^0,\ldots,a^{2n}) = \tau_0\left(\frac{\partial}{\partial\lambda}\,T\right)\,,\quad b\varphi_{2n-1} = \tau_0(T)$$

for the element $T = -[D,a^0]\,\frac{1}{D^2+\lambda}\cdots[D,a^{2n}]\,\frac{1}{D^2+\lambda}$ of the algebra $\mathcal{L}$, so that the cocycle property follows from

$$\tau_0\left(\frac{\partial}{\partial\lambda}\,T\right) = -\tau_0(T) \quad \forall T \in \mathcal{L}.$$

Finally, it is straightforward to check that the JLO cocycle is an entire cocycle, the point being that it involves the commutators $[D,a^j]$ in a very simple manner (cf. [J-L-O]).

Theorem 22. [Co$_{30}$] *Let $(\mathcal{H},D)$ be a θ-summable K-cycle over an algebra $\mathcal{A}$. Then the JLO cocycle is cohomologous, as an entire cocycle on $\mathcal{A}$, with the Chern character of Definition* 17

$$\varphi_{\text{JLO}} \in \text{ch}_*(\mathcal{H},D) \in H_\varepsilon^*(\mathcal{A}).$$

The proof of this theorem is less straightforward than it looks, and is done in detail in [Co$_{30}$].

We have now at our disposal two equivalent cocycles for the Chern character. One is normalized and comes from a natural homomorphism ρ of $\hat{Q}_\varepsilon\mathcal{A}$

to $\widetilde{\mathcal{L}}$ (Theorem 13). The other one, i.e. the JLO cocycle, is given by a beautifully simple formula. The first formula involves the supergroup $\mathbb{R}^{1,1}$ (cf. some of Section β)). The second (JLO) formula was found in a supersymmetric context, in which the operator $H = D^2$ is the generator of time translation while D is a "supercharge" operator. In the next section we shall treat two main examples of θ-summable K cycles coming from 1) discrete subgroups of higher rank Lie groups, and 2) supersymmetric quantum field theory.

IV.9. θ-summable K-cycles, Discrete Groups and Quantum Field Theory

The examples in Sections 3, 4, 5, and 6 of Fredholm modules were limited to the finite-dimensional, i.e. finitely summable, situation. Thus, for instance, in Section 5 the rank of the Lie group G had to be equal to one, to ensure the finite summability condition. In this section we shall give two families of examples of θ-summable Fredholm modules (in the form of θ-summable K-cycles) arising from discrete subgroups of Lie groups of arbitrary rank, and from quantum field theory.

9.α Discrete subgroups of Lie groups

Before we proceed and discuss the natural θ-summable K-cycles associated to discrete subgroups of Lie groups, we prove, in order to show the need for the infinite-dimensional theory, a general result showing that the C^*-algebra $C_r^*(\Gamma)$ of a discrete *non-amenable* group Γ does not possess any *finitely summable K-cycle.*

Theorem 1. [Co_{25}] *Let Γ be a non-amenable discrete group, and $C_r^*(\Gamma)$ be the reduced C^*-algebra of Γ. Let $(\mathcal{H}, D)$ be a Hilbert space representation of $C_r^*(\Gamma)$ with an unbounded selfadjoint operator D with compact resolvent in $\mathcal{H}$ such that*

$$\{a \in C_r^*(\Gamma)\ ;\ [D,a] \ \text{bounded}\} \quad \textit{is dense in } C_r^*(\Gamma).$$

Then for any $p < \infty$ one has $\mathrm{Trace}((1 + D^2)^{-p}) = +\infty$.

Proof. Let k be an integer, and assume that $\mathrm{Trace}((1 + D^2)^{-k}) < \infty$. Then for $\varepsilon > 0$ let $T(\varepsilon) = (1 + \varepsilon\, D^2)^{-k}\ (\mathrm{Trace}(1 + \varepsilon\, D^2)^{-k})^{-1}$.

One has $T(\varepsilon) \in \mathcal{L}^1(\mathcal{H})$, $T(\varepsilon) > 0$, Trace $T(\varepsilon) = 1$. Moreover, the $\mathcal{L}^1$ norm of commutators tends to 0 with $\varepsilon \to 0$,

$$\|[T(\varepsilon), a]\|_1 \underset{\varepsilon \to 0}{\to} 0 \quad \forall a \in C_r^*(\Gamma).$$

Indeed, to prove this one can assume that $[D, a]$ is bounded, in which case the proof follows from the Hölder inequality and the simple estimate

$$\|[a, T(\varepsilon)^{1/k}]\|_k = O(\varepsilon^{1/2}) \ \text{ for } \ \varepsilon \to 0,$$

which is straightforward using the equality

$$[a,(1+\varepsilon D^2)^{-1}] = (1+\varepsilon\, D^2)^{-1}\ \varepsilon\, D[D,a]\ (1+\varepsilon\, D^2)^{-1} + (1+\varepsilon\, D^2)^{-1}\ [D,a]\ \varepsilon\, D(1+\varepsilon\, D^2)^{-1}$$

and the bound $\|\varepsilon^{1/2}\, D(1+\varepsilon\, D^2)^{-1}\| \le 1/2$.

Then let $\mathcal{L}^2$ be the Hilbert space $\mathcal{H}\hat{\otimes}\overline{\mathcal{H}}$ of Hilbert-Schmidt operators in $\mathcal{H}$, and let $\xi(\varepsilon) = T(\varepsilon)^{1/2} \in \mathcal{L}^2$ for $\varepsilon > 0$. The Powers-Størmer inequality [Pow-S] shows that as $\varepsilon \to 0$,

$$\|a\ \xi(\varepsilon) - \xi(\varepsilon)\ a\|_2 \to 0 \quad \forall a \in C_r^*(\Gamma).$$

Let π be the representation of Γ in $\mathcal{L}^2$ given by

$$\pi(g)\ \xi = g\ \xi\ g^{-1} \quad \forall g \in \Gamma\,,\ \forall \xi \in \mathcal{L}^2.$$

One has $\|\xi(\varepsilon)\| = 1 \ \ \forall \varepsilon > 0$ and $\|\pi(g)\ \xi(\varepsilon) - \xi(\varepsilon)\| \to 0$ for $\varepsilon \to 0$ and any $g \in \Gamma$. This shows that the representation π of Γ in $\mathcal{L}^2$ weakly contains the trivial representation of Γ, which contradicts the non-amenability of Γ since π, as a representation of the reduced C^*-algebra of Γ, is weakly contained in the regular representation of Γ.

Let us now pass to a general construction of θ-summable K-cycles.

Proposition 2. *Let $\Gamma \subset G$ be a countable subgroup of a semisimple Lie group G, and let $X = G/K$ be the quotient of G by a maximal compact subgroup K, endowed with its canonical G-invariant Riemannian metric. Let $\mathcal{H}$ be the Hilbert space of L^2 differential forms on X, on which Γ acts by left translations. Then the pair $(\mathcal{H}, D)$ is a θ-summable K-cycle over $\mathcal{A} = \mathbb{C}\Gamma$, where $D = d_\tau + (d_\tau)^*$, $d_\tau = e^{-\tau\varphi}\ d\ e^{\tau\varphi}$, and 2φ is the Morse function on X given by the square of the geodesic distance to the point $K \in X$. (One assumes $\tau \neq 0$.)*

The proof is simple and works in general ([Co$_{25}$]), but in order to get a very explicit formula for the operator $H = D^2$, making clear that $e^{-\beta H}$ is of trace class for any $\beta > 0$ and allowing further computations, we shall make the further hypothesis that G is an analytic subgroup of $SL(n,\mathbb{R})$. One can then choose the faithful matrix representation of G so that $G^{\mathrm{t}} = G$ and $K = G \cap O(n,\mathbb{R})$. Then X imbeds as a totally geodesic subspace of the Riemannian space $P(n,\mathbb{R})$ of positive invertible matrices, by the map μ

$$\mu(gK) = g\ g^{\mathrm{t}} \quad \forall gK \in X = G/K.$$

The metric on $P = P(n,\mathbb{R})$ is given by

$$\|dp/dt\|^2 = \operatorname{Trace}((p^{-1}\dot{p})^2)$$

for any differentiable path $p(t)$ in P.

Given $p \in X \subset P$, the geodesic distance $d(p,1)$ of p from $1 \in X$ is the same in X or in P, and is equal to the Hilbert-Schmidt norm of $\log p$:

$$2\varphi(p) = d(p,1)^2 = \operatorname{Trace}\ ((\log p)^2).$$

The geodesic from 1 to p, parametrized by arc length, is $t \mapsto p^{\alpha t}$, where $\alpha = d(p,1)^{-1}$. The gradient of φ is the tangent vector at $t = 0$ to the path $t \mapsto p^{(1+t)}$ or equivalently $t \mapsto p + t\, p \log p \in P$.

For any differential form ω on X, we let $a^*(\omega)$ be the operator (in $\mathcal{H}$) of exterior multiplication by ω

$$a^*(\omega)\, \xi(p) = \omega(p) \wedge \xi(p) \quad \forall p \in X\,,\ \forall \xi \in \mathcal{H}.$$

Its adjoint $a(\omega)$ is given by contraction i_ω by ω.

One has, obviously, with $\omega = d\varphi$

$$d_\tau = d + \tau\, a^*(\omega)$$

$$D = d_\tau + d_\tau^* = (d + d^*) + \tau(a(\omega) + a^*(\omega)).$$

Let us first compute the commutator of D with a group element g acting on X as an isometry, and check that it is bounded. One has

$$[D, g] = g((g^{-1}\, Dg) - D),$$

$$g^{-1}\, Dg - D = \tau(a(\alpha_g) + a^*(\alpha_g))$$

where the differential form α_g is equal to $g^*\omega - \omega$.

Thus $\|[D,g]\| = \sup_{p \in X} \|\alpha_g(p)\|$, and we just have to check that this is finite. Now $g^*\omega = d(g^*\varphi)$ and

$$g^*\varphi(p) = \varphi(gp) = \frac{1}{2}\, d(gp,1)^2 = \frac{1}{2}\, d(p, g^{-1}(1))^2.$$

Hence the tangent vector at $p \in X$ corresponding to $\alpha_g(p) \in T_p^*(X)$ is

$$\exp_p^{-1}(1) - \exp_p^{-1}(g^{-1}(1))$$

where $\exp_p : T_p(X) \to X$ denotes the exponential map of the Riemannian manifold X.

Since $\exp_p^{-1}$ is a contraction one has

$$\|\alpha_g(p)\| \le d(1, g^{-1}(1)) \quad \forall p \in X.$$

It follows that if Γ is any subgroup of G, then $[D, a]$ is bounded for any element a of $\mathbb{C}\Gamma$.

We shall now compute $H = D^2$. By [Wit] it is the sum of three terms

$$D^2 = (d + d^*)^2 + \tau\, T + \tau^2\, \|d\varphi\|^2$$

where $(d + d^*)^2$ is the Laplacian Δ on X, $\|d\varphi\|$ is the operator of multiplication by the scalar function $\|(d\varphi)(p)\|^2 = 2\varphi(p) = \operatorname{trace}((\log p)^2)$, and where the operator T is the following operator of order 0

$$T = \sum_{\mu\nu} (\nabla d\varphi)_{\mu\nu}\, [a_\mu^*, a_\nu]$$

where $\nabla d\varphi$ is the covariant differential of $d\varphi$, viewed as an endomorphism of $T_p(X)$ and a_μ^* (resp. a_ν) is exterior (resp. interior) multiplication by elements of a basis.

Let us compute $\nabla d\varphi$ and check that its norm is of the order of $d(p,1)$ when p tends to ∞ in X. Since X is a totally geodesic submanifold of $P = P(n,\mathbb{R})$, we just need to do the computation in P. Given $p \in P$ we can identify $T_p(P)$ with the linear space $S(n,\mathbb{R})$ of symmetric matrices, using the isometry $p^{-1/2} \cdot p^{-1/2}$. Then a straightforward computation gives

$$(\nabla d\varphi)_p = \frac{\delta}{\tanh(\delta)} \ , \ \delta = \frac{1}{2}\,\mathrm{ad}(\log p).$$

The eigenvalues of δ in $S(n,\mathbb{R})$ are the differences $\frac{1}{2}(\lambda_i - \lambda_j)$, $j = 1,\ldots,n$, where the λ_i are the eigenvalues of $\log p$. Thus, the eigenvalues of $(\nabla d\varphi)_p$ are the $h(\frac{\lambda_i-\lambda_j}{2})$, $i,j = 1,\ldots,n$, where $h(x) = \frac{x}{\tanh x}$. Since $\tanh(x) \to 1$ as $x \to \infty$ there is an inequality of the form $h(x) \leq |x| + c$ and hence

$$\|(\nabla d\varphi)_p\| = O(d(p,1)) \text{ as } p \to \infty \text{ in } P.$$

The operators in $\mathcal{H}$ given by $\Delta = (d + d^*)^2$ and $\tau\, T + \tau^2\, \|d\varphi\|^2 = \tau\, T + 2\tau^2\, \varphi$ are bounded below since $\varphi(p) = d(p,1)^2$. Thus, the Golden-Thompson inequality of ([Ar$_3$]) applies and gives for any $\beta > 0$

$$\mathrm{Trace}(e^{-\beta D^2}) \leq \mathrm{Trace}\left(e^{-\beta\Delta}\, e^{-\beta(\tau T + 2\tau^2 \varphi)}\right) < \infty.$$

The operator $e^{-\beta(\tau T + 2\tau^2\varphi)}$ $(\tau \neq 0)$ is of order 0, and its norm behaves like $e^{-\lambda\varphi(p)}$ as $p \to \infty$ for some $\lambda > 0$.

By construction, the K-cycle $(\mathcal{H}, D)$ is even, with the $\mathbb{Z}/2$-grading given by the degree of differential forms

$$\gamma\omega = (-1)^k\, \omega \quad \forall \omega \in L^2(X, \wedge^k\, T^*).$$

Let us now assume that Γ is a *discrete* subgroup of G. Then $\mathcal{H}$ is in a natural manner an *$\mathcal{A}$-$\mathcal{B}$-bimodule*, where, $\mathcal{A} = \mathbb{C}\Gamma$, as above, is the group ring of Γ while the algebra $\mathcal{B}$ is the algebra of Γ-periodic Lipschitz bounded functions on X

$$\mathcal{B} = \{f \in C_b(\Gamma\backslash X)\ ;\ \|\nabla f\| \text{ bounded}\}.$$

It acts by multiplication operators in $\mathcal{H} = L^2(X, \wedge^*)$

$$(f\omega)(p) = f(p)\, \omega(p) \quad \forall f \in \mathcal{B}\,,\ \omega \in \mathcal{H}\,,\ p \in X$$

and this action commutes, by construction, with the action of Γ.

Proposition 3. *Let Γ be a discrete subgroup of G.*

a) *$(\mathcal{H}, D, \gamma)$ is a θ-summable K-cycle over the closure C under holomorphic functional calculus of $\mathcal{A} \otimes \mathcal{B}$ in the C^*-algebra $C_b(\Gamma\backslash X, C_r^*(\Gamma))$, where $C_r^*(\Gamma)$ is the reduced C^*-algebra of Γ.*

b) *For any* $a \in \mathcal{A}$ *and* $b \in \mathcal{B}$ *one has*

$$[[D, a], b] = 0.$$

Proof. For any $f \in \mathcal{B}$ the commutator $[d_\tau, f]$ is equal to $[d, f]$, and is hence bounded since f is a Lipschitz function. Thus, $[D, f]$ is bounded for any $f \in \mathcal{B}$. Moreover, both $[d_\tau, f]$ and $[d_\tau^*, f]$ are Γ-invariant operators, so that $[D, f]$ commutes with Γ for any $f \in \mathcal{B}$.

As the algebras $\mathcal{A}$ and $\mathcal{B}$ commute with each other this proves b).

Next, since Γ is discrete in G, its representation in $L^2(X, \wedge^*)$ is a direct integral of representations which are quasi-equivalent to the regular representation of Γ (they are equivalent when Γ is torsion-free). Thus, the representation of $\mathcal{A} \otimes \mathcal{B}$ in $\mathcal{H}$ extends to a representation of the C^*-algebra

$$C_b(\Gamma \backslash X, C_r^*(\Gamma)).$$

Since the algebra $\{T \in \mathcal{L}(\mathcal{H}); [D, T] \text{ bounded}\}$ is stable under the holomorphic functional calculus in $\mathcal{L}(\mathcal{H})$ we get a).

By construction, C is a pre-C^*-algebra. It contains the closure under the holomorphic functional calculus of $\mathbb{C}\Gamma$ in $C_r^*(\Gamma)$. Similarly, it contains $\mathcal{B}$, which is, by construction, closed under the holomorphic functional calculus (but not dense) in $C_b(\Gamma \backslash X)$.

The index map of the K-cycle $(\mathcal{H}, D, \gamma)$ (Proposition 8.18) thus yields a bilinear map

$$K(C_r^*(\Gamma)) \times K(\mathcal{B}) \xrightarrow{\text{Ind}D} \mathbb{Z}.$$

This map is easy to compute for elements of $K(C_r^*(\Gamma))$ which are in the range of the analytic assembly map (Chapter II)

$$\mu_r : K_*(B\Gamma) \to K(C_r^*(\Gamma)).$$

Theorem 4. *For any* $x \in K_*(B\Gamma)$ *and* $y \in K(\mathcal{B})$ *the index of* (D, y) *evaluated on* $\mu_r(x) \otimes y \in K(C)$ *is given by*

$$\text{Ind}D(\mu_r(x) \otimes y) = \langle \text{ch}_*(x), \text{ch}^*(y) \rangle$$

where $\text{ch}_*(x) \in H_*(B\Gamma) \simeq H_*(\Gamma \backslash X)$ *is the Chern character of the* K*-homology class* x*, and* $\text{ch}^*(y) \in H^*(\Gamma \backslash X)$ *is the Chern character of the* K*-theory class* y.

Note that $\text{ch}_*(x)$ is a homology class with compact support in $\Gamma \backslash X$, so that it pairs with the cohomology class (with no support restriction) $\text{ch}^*(y)$.

The proof of Theorem 4 follows in a straightforward manner from the results of Kasparov [Kas$_5$] in his proof of the Novikov conjecture for discrete subgroups of Lie groups (cf. Appendix A Theorem 19). It follows from Proposition 3 that every Γ-equivariant Hermitian vector bundle E on X with Γ-invariant compatible connection ∇ determines canonically a K-cycle $(\mathcal{H}_E, D_E, \gamma_E)$ over $C_r^*(\Gamma)$ which can be concretely described as follows:

$$\mathcal{H}_E = L^2(X, \wedge^* \otimes E)$$

is the Hilbert space of L^2-forms on X with coefficients in E, in which Γ acts by left translations

$$D_E = \nabla_\tau + (\nabla_\tau)^*$$

is given by the same formula as $D = d_\tau + d_\tau^*$, but with the covariant differentiation ∇ of sections of E instead of the differential d,

$$\gamma_E\, \omega = (-1)^k\, \omega \quad \forall \omega \in L^2(X, \wedge^k \otimes E).$$

We can now state the following corollaries of Propositions 2 and 3, and Theorem 4.

Corollary 5. *Let $\Gamma \subset G$ be a discrete subgroup of a semisimple Lie group, and A the pre-C^*-algebra closure under the holomorphic functional calculus of $\mathbb{C}\Gamma$ in $C_r^*(\Gamma)$.*

a) *The K-cycle $(\mathcal{H}_E, D_E, \gamma_E)$ over A is θ-summable.*

b) *Let $\varphi_E \in H_\varepsilon^{\mathrm{ev}}(A)$ be the Chern character of $(\mathcal{H}_E, D_E, \gamma_E)$. Then for any $x \in K_*(B\Gamma)$ one has*

$$\langle \varphi_E, \mu_r(x) \rangle = \langle \mathrm{ch}_*(x), \mathrm{ch}^*(E) \rangle.$$

The proof of Proposition 2 is easily modified to give a).

By Theorem 4 the index map $\mathrm{Ind} D_E$ determined by the K-cycle $(\mathcal{H}_E, D_E, \gamma_E)$ satisfies

$$\mathrm{Ind} D_E(\mu_r(x)) = \langle \mathrm{ch}_*(x), \mathrm{ch}^*(E) \rangle \quad \forall x \in K_*(B\Gamma).$$

Thus, the answer follows from Theorem 19 of Section 8.

Let us now combine Corollary 5 with the higher index theorem of Chapter III Section 4.

Proposition 6. *Let G, Γ and $(\mathcal{H}_E, D_E, \gamma_E)$ be as above. There exists a unique element $\beta \in H^{\mathrm{ev}}(B\Gamma, \mathbb{C})$ such that the associated cyclic cohomology class $\tau_\beta \in HC^{\mathrm{ev}}(\mathbb{C}\Gamma)$, has the same pairing with $\mu(K_0(B\Gamma)) \subset K_0(\mathcal{R}\Gamma)$ as the entire cyclic cocycle φ_E. One has moreover*

$$\beta = \mathrm{ch}\ E.$$

In particular, if we take for E the trivial one-dimensional vector bundle, i.e. for $D_E = D$, we get that, *on the image of μ_r*, the entire cyclic cocycle $\varphi = \mathrm{ch}_*(\mathcal{H}, D, \gamma)$ pairs in the same way as the normalized trace τ

$$\tau : C_r^*(\Gamma) \to \mathbb{C}$$

$$\tau\left(\sum_{g\in\Gamma} a_g\, g\right) = a_e \quad \forall a = \sum a_g\, g \in \mathbb{C}\Gamma.$$

The equality of τ with φ on $\mu_r(K_*(B\Gamma))$ is not sufficient to conclude that $\tau(K_0(C_r^*(\Gamma))) \subset \mathbb{Z}$ for Γ torsion-free, since we do not know that μ_r is surjective.

The conclusion $\tau(K_0(C_r^*(\Gamma))) \subset \mathbb{Z}$ would follow if we could show that the entire cyclic cocycles φ and τ on A are cohomologous. We shall take a first step in this direction by proving that φ is cohomologous to a simpler entire cyclic cocycle ψ whose component of degree 0, ψ_0, is equal to τ when Γ is torsion-free. In fact, this simplification will be obtained for any of the entire cocycles φ_E, which will all be in the range of a natural map Φ from closed differential forms on $M = \Gamma\backslash X$ to entire cyclic cocycles on A.

We shall assume, to simplify the discussion, that Γ is cocompact and torsion-free, so that M is a compact manifold. The general case is not more difficult but requires heavier notation.

Let S be the spinor bundle on X with its natural G-equivariant structure, and let ∇^s be the spin connection. Let Σ be the Hilbert bundle over M whose fiber at each $p \in M$ is the Hilbert space

$$\Sigma_p = \ell^2(\pi^{-1}(p), S)$$

of ℓ^2 sections of the restriction of the bundle S to the Γ-orbit $\pi^{-1}(p) \subset X$ above p. (We denote by $\pi : X \to M$ the quotient map.)

To the connection ∇^s corresponds a natural connection ∇ on the Hilbert bundle Σ over M, uniquely determined by the natural isomorphism

$$L^2(M,\Sigma) \simeq L^2(X,S).$$

More precisely, for any smooth section ξ of Σ on M, one lets ξ^s be the corresponding section of S over X, and one has

$$(\nabla_Y\, \xi)(p) = (\nabla^s_{\tilde{Y}}\, \xi^s)(q)_{\pi(q)=p} \in \Sigma_p$$

for any tangent vector $Y \in T_p(M)$ lifted as $\tilde{Y}_q \in T_q(X)$ for any $q \in X$, $\pi(q) = p$.

Next, let T be the unbounded endomorphism of the Hilbert bundle Σ over M given by

$$(T_p\, \xi)(q) = c(q,0)\, \xi$$

where 0 denotes a fixed origin in X and, for any $q \in X$, $c(q,0)$ denotes the operator in S_q of Clifford multiplication by the unique tangent vector at q whose image by the exponential map $\exp_q$ in X is this fixed origin 0.

We shall assume that X is *even-dimensional* and use the superconnection $Z = \gamma\nabla + iT$, where γ is the $\mathbb{Z}/2$-grading of the Hilbert bundle Σ (cf. [Q1] and Appendix A, Definition 16 for the notion of superconnection on a Hilbert bundle on a manifold).

The algebra $C_r^*(\Gamma)$ acts by endomorphisms of the bundle Σ since Γ acts by translations on each of the fibers $\pi^{-1}(p)$.

One then checks that the hypothesis of the following proposition is fulfilled by the action of the pre-C^*-algebra $A \subset C_r^*(\Gamma)$.

Proposition 7. [Co29] *Let M be a compact manifold, (Σ, γ) a $\mathbb{Z}/2$-graded Hilbert bundle on M equipped with a superconnection $Z = \gamma\nabla + iT$, and let A be a subalgebra of the algebra of endomorphisms of the bundle (Σ, γ) such that:*

a) *For every $a \in A$, the commutator $\delta(a) = [Z, a]$ is a bounded endomorphism of the bundle Σ.*

b) *The operator-valued differential form* $\exp(\beta\, Z^2)$ *is of trace class for any* $\beta > 0$.

Then the following formula defines, for any closed even smooth form ω on M, an entire cocycle in the (b, B) bicomplex of A:

$$\varphi_{2n}^{\omega}(a^0, \ldots, a^{2n}) = \int_M \int_{\sum s_i = 1, s_i \geq 0} \omega \wedge \operatorname{Trace}\Big(\gamma\, a^0\, e^{s_0 Z^2}\, \delta(a^1) \cdots e^{s_{2n-1} Z^2}\, \delta(a^{2n})\, e^{s_{2n} Z^2}\Big) \prod_1^{2n} ds_i .$$

Let us apply this proposition in the above context with $A \subset C_r^*(\Gamma)$ as above, and denote by Φ the corresponding map, $\Phi(\omega) = \varphi^{\omega}$, from $H^*(M)$ to $H_\varepsilon^*(A)$. The computation of the character $\mathrm{ch}_*(\mathcal{H}_E, D_E, \gamma_E) = \varphi_E$ is partially achieved by the following result.

Theorem 8. *With the above notation one has*

$$\mathrm{ch}_*(\mathcal{H}_E, D_E, \gamma_E) = \Phi(\omega)$$

where $\omega = \hat{A}(M)\, \mathrm{ch}\, E$, and $\hat{A}(M)$ is the $\hat{A}$-genus of M ([Mi4]).

Let Γ be an arbitrary countable discrete group and, as above, let A be the pre-C^*-algebra closure of the group ring $\mathbb{C}\Gamma$ under the holomorphic functional calculus in $C_r^*(\Gamma)$.

We have seen in Chapter III (Section 2.γ Theorem 8) that the cyclic cohomology of the group ring $\mathbb{C}\Gamma \subset A$ is canonically isomorphic to the $\mathbb{S}^1$-equivariant cohomology of the free loop space $(B\Gamma)^{\mathbb{S}^1}$ of $B\Gamma$. In a similar manner the entire cyclic cohomology of A, $H_\varepsilon^*(A)$, ought to have a related geometric interpretation. Theorem 7.21 shows that any element $x \in H_\varepsilon^{\mathrm{ev}}(A)$ yields a *homotopy invariant* higher signature by the formula

$$\mathrm{Sign}_x(M, \psi) = \langle \mathrm{Ind}_\Gamma(D_{\mathrm{Sign}}), x \rangle$$

with the notations of Lemma III.4.8, where M is a compact oriented manifold and ψ a continuous map from M to $B\Gamma$.

9.β Supersymmetric quantum field theory

The apparent incompatibility of quantum mechanics and special relativity (cf., for instance, [Fey]) is resolved only by the theory of quantum fields, i.e. by the quantization of a mechanical system with infinitely many degrees of freedom.

As we shall see by working out in full detail a concrete example, this infinity of the number of degrees of freedom implies, at least in supersymmetric theories, that the cohomological dimension of the algebras of local observables is infinite. One expects in general that, provided an infrared cutoff is given, i.e. space is assumed to be compact, the following holds:

(1) The supercharge operator Q in the vacuum representation Hilbert space $\mathcal{H}$, $\mathbb{Z}/2$-graded by $\Gamma = (-1)^{N_f}$ (where N_f is the fermion number operator) defines a θ-summable K-cycle over the algebras $\mathcal{U}(\mathcal{O})$ of local observables.

(2) The index map given by Q, i.e. the map from $K_0(\mathcal{U}_L(\mathcal{O}))$ to $\mathbb{Z}$, where $\mathcal{U}_L(\mathcal{O}) = \{A \in \mathcal{U}(\mathcal{O}); [Q, A] \text{ bounded}\}$, given by the index of Q with coefficients in a K-theory class, fails to be *polynomial*, in the following sense:

Definition 9. *Let $\mathcal{A}$ be an algebra (over $\mathbb{C}$), and $\psi : K_0(\mathcal{A}) \to \mathbb{C}$ an additive map. Then ψ is polynomial if there exists a cyclic cocycle $\tau \in HC^*(\mathcal{A})$ such that:*

$$\psi(e) = \langle \tau, e \rangle \quad \forall e \in K_0(\mathcal{A}).$$

We shall only prove (1) and (2) in an explicit and non-interacting example, the $N = 2$ Wess-Zumino model on a 2-dimensional cylinder space-time. The results of [J-L-W] on constructive quantum field theory prove (1) in the interacting case. They should allow one to prove (2) in the interacting case as well.

Before we embark on the description of the model, we shall work out a toy example with finitely many degrees of freedom.

This example is a slight variation of the construction of Proposition 2. We let E be a finite-dimensional Euclidean vector space over $\mathbb{R}$, and φ a nondegenerate quadratic form on E. Then let $\mathcal{H} = L^2(E, \wedge^* E_{\mathbb{C}})$ be the Hilbert space of complex, square integrable differential forms on E. We let $D = d_\varphi + (d_\varphi)^*$ where $d_\varphi = e^{-\varphi}\, d\, e^{\varphi}$, d being the exterior differentiation.

Writing $\varphi(x) = \sum_{i=1}^{n} \frac{1}{2} \lambda_i x_i^2$, $\lambda_i \neq 0$, in a suitable orthonormal basis of E, with $x_i = \langle x, e_i \rangle$, it is immediate to compute the spectrum of D and check that it is $(2n, \infty)$-summable, $n = \dim E$. Moreover, the index of D, relative to the $\mathbb{Z}/2$-grading γ of $\mathcal{H}$ given by $(-1)^k$ on $\wedge^k E_{\mathbb{C}}^*$, is equal to $(-1)^{n_-}$, where n_- is the number of negative eigenvalues $\lambda_i < 0$ of φ.

Let $k \leq n$, $\xi_1, \ldots, \xi_k \in E$, $\eta_1, \ldots, \eta_k \in E^*$, be such that

$$\langle \xi_i, \eta_j \rangle \in 2\pi\, \mathbb{Z} \quad \forall i, j = 1, \ldots, k.$$

To each ξ_i we associate the unitary U_i in $\mathcal{H}$ given by translation by $\xi_i \in E$. To each η_j we associate the unitary V_j in $\mathcal{H}$ given by multiplication by the periodic function

$$V_j(\xi) = \exp\, i\langle \xi, \eta_j \rangle \quad \forall \xi \in E.$$

The above condition shows that these $2n$ unitaries commute pairwise. As in Proposition 3, they all have bounded commutators with D, i.e. belong to the algebra $\mathcal{A}_D = \{T \in \mathcal{L}(\mathcal{H}); T\gamma = \gamma T, [D,T] \text{ bounded}\}$.

We let $[U_1 \wedge U_2 \wedge \cdots \wedge U_k \wedge V_1 \wedge \cdots \wedge V_k] \in K_0(\mathcal{A}_D)$ be the K-theory class obtained as a cup product of the above commuting unitaries. There are two equivalent ways of describing it.

The first is to consider the compact manifold $M = \mathbb{T}^{2k}$, which is a torus of dimension $2k$, and the homomorphism ρ from $C^\infty(M)$ to $\mathcal{A}_D$ given with obvious notation by

$$\rho(u_i) = U_i \ , \ \rho(v_j) = V_j$$

where $(u_1, \ldots, u_k, v_1, \ldots, v_k)$ are the natural $\mathbb{S}^1$-valued coordinates on M. Using the orientation given by this ordering of the coordinates, one gets a well defined class $\beta \in K^0(M) = K_0(C^\infty(M))$, that of the Bott element (cf. [At$_3$]).

One then defines

$$[U_1 \wedge \cdots \wedge U_k \wedge V_1 \wedge \cdots \wedge V_k] = \rho_*(\beta) \in K_0(\mathcal{A}_D).$$

The second description uses the cup product in algebraic theory, which yields an element of $K_{2k}^{\text{alg}}(\mathcal{A}_D)$ whose image by Bott periodicity is the above element. Note that $\mathcal{A}_D$, since it is stable under holomorphic functional calculus, satisfies Bott periodicity.

The computation of the index of D with coefficients in the K-theory class $[U_1 \wedge \cdots \wedge U_k \wedge V_1 \wedge \cdots \wedge V_k]$ is given by:

Lemma 10. *The index of D relative to the $\mathbb{Z}/2$-grading γ and the K-theory class $[U_1 \wedge \cdots \wedge U_k \wedge V_1 \wedge \cdots \wedge V_k]$ is equal to*

$$(-1)^{n-} \ (2\pi)^{-k} \ \det(\langle \xi_i, \eta_j \rangle) \in \mathbb{Z}.$$

The proof is a straightforward application of the Riemann-Roch theorem, but we shall give the details of the identification of D with a $\bar\partial$ operator, since it throws light on the previous example of Proposition 3.

Let us assume, to simplify the discussion, that the matrix $(\langle \xi_i, \eta_j \rangle)_{i,j}$ is not degenerate. Then the dimension of the linear span F of the ξ_i's (resp. F' of the η_j's) is equal to k, and E is the linear span of F and $(F')^\perp = \{\xi \in E; \langle \xi, \eta_j \rangle = 0 \ \ \forall j = 1, \ldots, k\}$.

Thus, there exists a lattice $\Gamma \subset E$, i.e. a discrete cocompact subgroup of E, such that

$$\xi_i \in \Gamma \ , \ \forall i = 1, \ldots, k \ ; \ \eta_j \in \Gamma^\perp \ \ \forall j = 1, \ldots, k$$

where the dual lattice $\Gamma^\perp$ is $\{\eta \in E^*; \langle \eta, \Gamma \rangle \subset 2\pi\mathbb{Z}\}$.

As in Proposition 3, we let the group ring $\mathbb{C}\Gamma$ act on $\mathcal{H}$ by translations, and $C(E/\Gamma)$ act on $\mathcal{H}$ by multiplication operators. These two commutative and commuting algebras generate a C^*-algebra A in $\mathcal{H}$, whose spectrum S is naturally the torus

$$S = \hat{\Gamma} \times E/\Gamma = E^*/\Gamma^\perp \times E/\Gamma.$$

One has $\|[D,f]\| < \infty$ for any $f \in C^\infty(S)$, and moreover the image of the map $\rho : C^\infty(M) \to \mathcal{A}_D$ is contained in $C^\infty(S)$. This shows that $e = [U_1 \wedge \cdots \wedge U_k \wedge V_1 \wedge \cdots \wedge V_k] = \rho_*(\beta)$ can be viewed as an element of $K^0(S)$, namely the pull-back of β by the projection $p : S \to M$ which is the transpose of ρ.

We shall now endow S with a complex Kähler structure and a holomorphic line bundle L, so that the K-cycle on $C^\infty(S)$ given by $(\mathcal{H}, D, \gamma)$ is isomorphic to the $\bar\partial$ operator with coefficients in L. First, given an element (X,Y) of $E \times E^*$, one gets a corresponding vector field on S, i.e. a derivation $\delta_{(X,Y)}$ of A

$$\delta_{(X,Y)}(U_g\, f) = i\langle Y, g\rangle\, U_g\, f + U_g\, \partial_X\, f$$

for any function f on E/Γ and any $g \in \Gamma$.

The Schwartz space $S(E)$ is a finite projective right module over $C^\infty(S)$ for the action of $C^\infty(S)$ given by

$$(\xi \cdot (U_g\, f))(p) = f(p)\, \xi(p-g) \quad \forall g \in \Gamma\,,\ f \in C^\infty(E/\Gamma)\,,\ \xi \in S(E)\,,\ p \in E.$$

A connection ∇ on the corresponding complex line bundle L on S (so that $S(E) \simeq C^\infty(M,L)$ as finite projective modules) is now given by

$$(\nabla_{(X,Y)}\, \xi)(p) = (\partial_X\, \xi)(p) + i\langle Y, p\rangle\, \xi(p) \quad \forall (X,Y) \in E \times E^*.$$

One checks that $\nabla_{(X,Y)}$ satisfies the Leibniz rule with respect to the derivation $\delta_{(X,Y)}$, and also that it is compatible with the metric on L given by the following $C^\infty(S)$-valued inner product on $S(E)$:

$$(\langle \xi, \eta\rangle)^\wedge\, (g, g') = \langle \xi\, U_g\, V_{g'}, \eta\rangle_{L^2(E)} \quad \forall g \in \Gamma\,,\ g' \in \Gamma^\perp.$$

The curvature of the line bundle L is given by the formula

$$\theta(\alpha_1, \alpha_2) = i\, \omega(\alpha_1, \alpha_2) \quad \forall \alpha_j = (X_j, Y_j) \in E \times E^*$$

where $\omega(\alpha_1,\alpha_2) = \langle Y_1, X_2\rangle - \langle Y_2, X_1\rangle$ is the canonical symplectic form on the tangent space $E \times E^*$ of S.

Next, let the quadratic form φ on E be expressed, as above, as $\varphi(p) = \sum \frac{1}{2}\, \lambda_j(\langle p, e_j\rangle_E)^2$ where $(e_j)_{j=1,\dots,n}$ is an orthonormal basis of the Euclidean space E. We let e^j be the dual basis of E^*, so that $\langle p, e_j\rangle_E = \langle p, e^j\rangle \ \forall p \in E$. Then the operator d_φ is given by

$$d_\varphi = \sum_{j=1}^{n} a_j^*\,(\partial_{e_j} + \lambda_j\langle \cdot, e^j\rangle)$$

where the a_j^* are the operators of exterior multiplication by e_j in $\wedge E_{\mathbb{C}}$, so that their representation is characterized by the commutation relations

$$a_j\, a_k^* + a_k^*\, a_j = \delta_{jk}.$$

Let us consider the $\mathbb{R}$-linear isomorphism of $E \times E^*$ with $\mathbb{C}^n$ given by $z_j = \frac{1}{2}(x_j + (i/\lambda_j)\, y_j)$.

Then $\frac{\partial}{\partial \bar{z}_j} = \delta_{(e_j, i\lambda_j e^j)}$ as a derivation of A, and using the connection ∇ on the complex line bundle L we get

$$d_\varphi = \sum a_j^* \, \bar{\partial}_{L,j}.$$

We endow the manifold S with the complex Kähler structure given by the above isomorphism of its real tangent bundle with the trivial bundle with fiber $\mathbb{C}^n$ and standard inner product. We then have a natural isomorphism of $\mathcal{H} = L^2(E, \wedge^* E_\mathbb{C})$ with $L^2(S, \wedge^{0,*} \otimes L)$ where $\wedge^{0,*}$ is the exterior algebra bundle over the complex manifold S. This isomorphism U is the tensor product $U_b \otimes U_f$ of the isomorphisms $U_b : L^2(E) \simeq L^2(S, L)$ and $U_f : \wedge^{0,*} \simeq \wedge^* E_\mathbb{C}$. Here U_f comes from the identification of $E_\mathbb{C}$ with $\mathbb{C}^n$ given by the basis e_j, while U_b extends the canonical isomorphism $S(E) \to C^\infty(S, L)$. The next lemma is now clear:

Lemma 11. *Under the isomorphism $U : \mathcal{H} \to L^2(S, \wedge^{0,*} \otimes L)$ the action of $A = C(S)$ becomes the action by multiplication operators, the operator d_φ becomes the $\bar{\partial}_L$ operator and the $\mathbb{Z}/2$-grading γ becomes $(-1)^q$ on $\wedge^{0,q} \otimes L$. The K-cycle $(\mathcal{H}, D, \gamma)$ is isomorphic to the K-cycle $\bar{\partial}_L + (\bar{\partial}_L)^*$ on S.*

Let us, as a first application of this lemma, compute the index of D with coefficients in $[U_1 \wedge U_2 \wedge \cdots \wedge U_k \wedge V_1 \wedge \cdots \wedge V_k] = e$, i.e. let us prove Lemma 10. Since $e = p^*(\beta)$, where $p : S \to M$ is the transpose of the homomorphism ρ defined above, and β is the Bott K-theory class, $\beta \in K^0(M)$, we can apply the Riemann-Roch theorem (cf. [Gi$_1$]) and get

$$\mathrm{Index}(D_e) = \langle \mathrm{Td}(S)\, \mathrm{ch}(L)\, \mathrm{ch}(e), [S] \rangle$$

where the Todd genus of the complex manifold S is the cohomology class $1 \in H^0(S)$, the Chern character of L is equal to $\exp(c_1)$ and its first Chern class c_1 is $\frac{1}{2\pi}\, \omega$ where ω is the canonical symplectic form. Also, the Chern character $\mathrm{ch}(e)$ is $p^* \mathrm{ch}(\beta) = p^*([M])$ where $[M] \in H^{2k}(M)$ is its fundamental class. We thus get

$$\mathrm{Index}\,(D_e) = \left\langle \exp\left(\frac{1}{2\pi}\, \omega\right)\, p^*([M]), [S] \right\rangle .$$

The class $[M]$ is represented by the differential form

$$(2\pi i)^{-2k}\, U_1^{-1}\, dU_1 \wedge \cdots \wedge U_k^{-1}\, dU_k \wedge V_1^{-1}\, dV_1 \wedge \cdots \wedge V_k^{-1}\, dV_k,$$

so that its pull-back on S is translation invariant, and the computation of the right-hand side of the above takes place in the exterior algebra $\wedge(E \oplus E^*)$. Lemma 10 follows easily.

We shall now explain in some detail the set-up of quantum field theory in a very simple case, and show that the supercharge operator Q in supersymmetric theories yields a θ-summable K-cycle. Thanks to the work of *constructive* quantum field theory (cf. [J-L-W]) these results are valid in the interacting case, but for the sake of simplicity we shall limit ourselves to the free theory. We shall show that even in the free case, the infinite number of degrees of freedom of the theory in every (non-empty) local region $\mathcal{O}$ implies that the K-theory $K(\mathcal{U}_L(\mathcal{O}))$ of the local observable algebra is infinite-dimensional.

Let us first consider the simplest field theory, namely the free massive scalar field φ on the space-time

$$X = \mathbb{S}^1 \times \mathbb{R}$$

with the Lorentzian metric.

The scalar field φ is a real-valued function on X governed by the Lagrangian

$$\mathcal{L}(\varphi) = \frac{1}{2}\left((\partial_0\varphi)^2 - (\partial_1\varphi)^2\right) - \frac{m^2}{2}\,\varphi^2$$

where $\partial_0 = \frac{\partial}{\partial t}$ is the time derivative ($t \in \mathbb{R}$) and $\partial_1 = \frac{\partial}{\partial x}$ is the spatial derivative, where the space is here compact and one-dimensional. The action functional is given by

$$I(\varphi) = \int \mathcal{L}(\varphi)\,dx\,dt = \int L(t)\,dt$$

$$L(t) = \int_{\mathbb{S}^1} \left(\frac{1}{2}\,(\dot{\varphi})^2 - \frac{1}{2}(\partial\,\varphi)^2 - \frac{m^2}{2}\,\varphi^2\right)\,dx.$$

Thus, at the classical level, one is dealing with a mechanical system with infinitely many degrees of freedom, whose configuration space C is the space of real-valued functions on $\mathbb{S}^1$.

The Hamiltonian of this classical mechanical system is the functional on the cotangent space T^*C given by

$$H(\varphi,\pi) = \frac{1}{2}\int_{\mathbb{S}^1} \left(\pi(x)^2 + (\partial\,\varphi(x))^2 + m^2\,\varphi^2(x)\right)\,dx$$

where one uses the *linear* structure of C to identify T^*C with $C \times C^*$, and where one views the field π as an element of C^*

$$\varphi \to \int_{\mathbb{S}^1} \varphi(x)\,\pi(x)\,dx \in \mathbb{C}$$

As a classical mechanical system that above is the same as a countable collection of uncoupled harmonic oscillators. Indeed, one can take as coordinates in C (resp. C^*) the Fourier components $\varphi_k = \int_{\mathbb{S}^1} \varphi(x)\,e^{-ikx}\,dx$ (resp. π_k), which are subject only to the reality condition

$$\varphi_{-k} = \overline{\varphi}_k \quad \forall k \in \mathbb{Z} \quad (\text{resp. } \pi_{-k} = \overline{\pi}_k).$$

Thus, both spaces C and T^*C are infinite products of finite-dimensional spaces and the Hamiltonian H is an infinite sum

$$H = \sum_{k \in \mathbb{Z}} \frac{1}{2}\left(\pi_k\,\overline{\pi}_k + (k^2 + m^2)\,\varphi_k\,\overline{\varphi}_k\right).$$

The quantization of a single harmonic oscillator, say a system with configuration space $\mathbb{R}$ and Hamiltonian $\frac{1}{2}(p^2+\omega^2q^2)$, *quantizes* the values of the energy in replacing $\frac{1}{2}(p^2+\omega^2q^2)$ by the operator $\frac{1}{2}\left(\left(-i\hbar\frac{\partial}{\partial q}\right)^2+\omega^2q^2\right)$ on the Hilbert space $L^2(\mathbb{R})$, whose spectrum is (up to a shift) the set $\{n\,\hbar\omega\ ;\ n\in\mathbb{N}\}$. The algebra of observables of the quantum system is generated by a single operator a^* and its adjoint a which obey the commutation relation

$$[a,a^*]=1$$

and the Hamiltonian is $H=\hbar\omega\, a^*a$. These two equations completely describe the quantum system. Its representation in $L^2(\mathbb{R})$, given by the equality

$$a=\frac{1}{\sqrt{2}}\left(\frac{\partial}{\partial q}+q\right)$$

is, up to unitary equivalence, its only irreducible representation. The (unique up to phase) normalized vector Ω such that $a\Omega=0$ is called the vacuum vector.

The reality condition $\varphi_{-k}=\overline{\varphi}_k$ shows that for $k>0$, the pair $\{-k,k\}$ yields a pair of harmonic oscillators, whose quantization yields a pair of creation operators a_k^*, a_{-k}^*. The observable algebra of the quantized field has the following presentation. It is generated by a_k^* and a_k, $k\in\mathbb{Z}$, with relations

$$[a_k,a_k^*]=1\quad\forall k\in\mathbb{Z}$$

$$[a_k,a_\ell]=0\,,\ [a_k,a_\ell^*]=0\quad\forall k\neq\ell.$$

The Hamiltonian is given by the derivation corresponding to the formal sum

$$H_b=\sum_{k\in\mathbb{Z}}\hbar\,\omega_k\,a_k^*\,a_k\ ;\ \omega_k=\sqrt{k^2+m^2}.$$

The "vacuum representation" which corresponds to the vacuum state, which is the infinite tensor product of the vacuum states, is given by the Hilbert space

$$\mathcal{H}_b=\bigotimes_{k\in\mathbb{Z}}\ (\mathcal{H}_k,\Omega_k)$$

and the tensor product representation of the algebra.

Equivalently one can describe this infinite tensor product as the Hilbert space $L^2(C,d\mu)$ where μ is a Gaussian measure on C. All this is straightforward (cf. [Gl-J]).

The spectrum of the quantum field Hamiltonian H_b acting in $\mathcal{H}_b$ is now

$$\text{Spectrum } H_b=\left\{\sum n_k\,\hbar\omega_k\ ;\ n_k\in\mathbb{N}\right\}.$$

Quantum field theory reconciles positivity of energy (i.e. H_b is a *positive* operator in $\mathcal{H}_b$) with *causality*, which means that we have commutation at space-like separation for functions of the quantum field $\varphi(x,t)=e^{it\cdot H_b}\,\varphi(x,0)\,e^{-itH_b}$, with

$$\varphi(x,0)=\sum_{k\in\mathbb{Z}}\left(a_k\,e^{ikx}+a_k^*\,e^{-ikx}\right)(2\omega_k)^{-1/2}.$$

This commutation of $\varphi(x,t)$ with $\varphi(y,s)$ for (x,t) and (y,s) space-like in the space-time $X = \mathbb{S}^1 \times \mathbb{R}$ is easy to check directly. First, the operators $\varphi(f) = \int f(x)\, \varphi(x,0)\, dx$, $f \in C^\infty$, commute with each other, using $\omega_k = \omega_{-k}$. Next, with σ_t the automorphisms given by time evolution

$$\sigma_t = e^{itH_b} \cdot e^{-itH_b},$$

one computes $[\sigma_t\, \varphi(f), \varphi(g)]$, for $f, g \in C^\infty(\mathbb{S}^1)$, and finds a scalar multiple of the identity, where the scalar is

$$\int k(x,y,t)\, f(x)\, g(y)\, dx\, dy,$$

$$k(x,y,t) = \sum_k e^{-ik(x-y)} \left(e^{-i\omega_k t} - e^{i\omega_k t}\right) \omega_k^{-1} = k(x-y,t)$$

where k satisfies the Klein-Gordon equation

$$\left(\partial_0^2 - \partial_1^2 + m^2\right) k = 0$$

and the initial conditions $k(x,0) = 0$, $\frac{\partial}{\partial t}\, k(x,0) = \delta_0$.

It then follows from elementary properties of the wave equation that $k(x,t)$ vanishes if (x,t) is space-like separated from $(0,0)$, and hence that the quantum field φ satisfies causality. All this is well known and we refer to [Bo-S], for instance, for more details. For our purposes we shall retain two easy, and well known, corollaries.

Proposition 12. [Bo-S] a) *Let $\mathcal{H}_b$ be the vacuum representation. The quantum field φ is an operator-valued distribution.*

b) *Let $\mathcal{O}$ be a local region (i.e. a bounded open set) in X and $\mathcal{U}(\mathcal{O})$ be the von Neumann algebra in $\mathcal{H}_b$ generated by the $\varphi(f)$ with Support $f \subset \mathcal{O}$. Then when $\mathcal{O}_1$ and $\mathcal{O}_2$ are space-like separated, one has*

$$\mathcal{U}(\mathcal{O}_1) \subset \mathcal{U}(\mathcal{O}_2)'.$$

The von Neumann algebras $\mathcal{U}(\mathcal{O})$ of *local* observables are an essential part of the algebraic formulation of quantum field theory ([Kast$_4$]). From our point of view in this book, the passage to a von Neumann algebra only captures the *measure theoretic* aspect, and we shall now show that in the case of supersymmetric theories we can refine the algebras $\mathcal{U}(\mathcal{O})$ to capture the finer topological and even metric aspects. We shall only treat an example but the

Figure IV.10. Space-like regions $\mathcal{O}_1$ and $\mathcal{O}_2$ in space-time cylinder

strategy works in general. The example we take is the free Wess-Zumino model in 2 dimensions, with one complex scalar field φ of mass m and one spinor field ψ, also of mass m. Thus, the Lagrangian is now

$$\mathcal{L} = \mathcal{L}_b + \mathcal{L}_f \; ; \; \mathcal{L}_b = \frac{1}{2}\left(|\partial_0 \varphi|^2 - |\partial_1 \varphi|^2 - m^2|\varphi|^2\right)$$

$$\mathcal{L}_f = i\,\overline{\psi} \sum_\mu \gamma_\mu\, \partial_\mu\, \psi - m\,\overline{\psi}\,\psi \; , \; \overline{\psi} = \psi^*\, \gamma_0$$

where the Dirac matrices γ_μ are, say, $\gamma_0 = \begin{bmatrix} 0 & 1 \\ 1 & 0 \end{bmatrix}$, $\gamma_1 = \begin{bmatrix} 0 & -1 \\ 1 & 0 \end{bmatrix}$, and the spinor field ψ is given by a column matrix

$$\psi = \begin{bmatrix} \psi_1 \\ \psi_2 \end{bmatrix} \; , \; \psi^* = [\psi_1^*, \psi_2^*] \, .$$

The bosonic part is quantized exactly as above, except that the field φ is now complex, so that the reality condition no longer holds and one needs twice as many creation operators, which are now denoted by $a_\pm^*(k)$, $k \in \mathbb{Z}$. They satisfy the commutation relations

$$[a_\pm(k), a_\pm(\ell)] = [a_\pm(k), a_\mp(\ell)] = [a_\pm(k), a_\mp^*(\ell)] = 0$$

and $[a_\pm(k), a_\pm^*(\ell)] = \delta_{k\ell}$.

The time-zero field $\varphi(x)$ is given by

$$\varphi(x) = (4\pi)^{-1/2} \sum_{k\in\mathbb{Z}} \omega(k)^{-1/2}\, (a_+^*(k) + a_-(-k))\, e^{-ikx}$$

where $\omega(k) = (k^2 + m^2)^{1/2}$. The conjugate momentum is given by

$$\pi(x) = i(4\pi)^{-1/2} \sum_{k\in\mathbb{Z}} \omega(k)^{1/2} \, (a_-^*(k) - a_+(-k)) \, e^{-ikx},$$

and one has the canonical commutation relations

$$[\varphi(x), \varphi(y)] = [\pi(x), \pi(y)] = [\pi^*(x), \varphi(y)] = 0$$

$$[\pi(x), \varphi(y)] = -i \, \delta(x - y).$$

The quantum fields φ and π are operator-valued distributions in the bosonic Hilbert space $\mathcal{H}_b$ of the vacuum representation of the creation operators $a_\pm^*(k)$, $k \in \mathbb{Z}$. In this representation the Hamiltonian H_b is given by

$$H_b = \int : |\pi(x)|^2 + |\partial_1 \varphi(x)|^2 + m^2 |\varphi(x)|^2 : dx$$

where the Wick ordering, : :, (cf. [Bo-S]) takes care of an irrelevant additive constant. Equivalently, $\mathcal{H}_b$ is the Hilbert space $L^2(C^{-\infty}(\mathbb{S}^1), d\mu)$ where $d\mu$ is the Gaussian measure, on the configuration space $C^{-\infty}(\mathbb{S}^1)$ of complex-valued distributions on $\mathbb{S}^1$, with covariance G

$$G = -(\frac{d^2}{dx^2} + m^2)^{-1/2}.$$

With this identification the field φ is represented by multiplication operators and $\pi(x)$ becomes $-i \, \delta / \delta\varphi(x)$ (cf. [Gl-J] or [J-L-W]). Next let us describe the fermionic part of the model. It is worthwhile first to describe the C^*-algebra generated by the fermionic fields together with its natural dynamics generated by a derivation, and then to find the ground state and use this vacuum state to construct the fermionic Fock space. The C^*-algebra generated by the fermionic field $\psi = \begin{bmatrix} \psi_1 \\ \psi_2 \end{bmatrix}$ is generated by its Fourier components $\psi_j(k)$, $j = 1, 2$; $k \in \mathbb{Z}$, which satisfy the following relations

$$\{\psi_i(k), \psi_j(\ell)\} = 0 \quad \forall i, j, k, \ell$$

$$\{\psi_i(k), \psi_j(\ell)^*\} = \delta_{ij} \, \delta_{k\ell} \quad \forall i, j, k, \ell$$

where $\{a, b\} = ab + ba$.

In other words, we deal with the C^*-algebra A associated to the canonical anticommutation relations in the Hilbert space of L^2 spinors on $\mathbb{S}^1$.

The quantum fields ψ_1 and ψ_2 are then given by

$$\psi_j(x) = \sum \psi_j(k) \, e^{-ikx} \quad \forall x \in \mathbb{S}^1$$

and they are A-valued distributions on $\mathbb{S}^1$.

This specifies $\psi = \begin{bmatrix} \psi_1 \\ \psi_2 \end{bmatrix}$ at time 0. Its time evolution is specified by the Hamiltonian which is given by the derivation of A

$$\delta(a) = [H_f, a]$$

where H_f is the formal expression, with $\overline{\psi} = \psi^* \gamma_0$

$$H_f = \int_{\mathbb{S}^1} (\overline{\psi}\, i\gamma_1\, \partial\psi - m\,\overline{\psi}\psi)\, dx$$
$$= \sum_k k\,(\psi_1(k)^*\, \psi_1(k) - \psi_2(k)^*\, \psi_2(k))$$
$$- m\,(\psi_1^*(k)\, \psi_2(k) + \psi_2^*(k)\, \psi_1(k)).$$

The above derivation makes perfectly good sense, and defines a one-parameter group of automorphisms σ_t of $A = \mathrm{CAR}(L^2(\mathbb{S}^1, S))$. One has $\sigma_t = \mathrm{CAR}(U_t)$ where U_t is the one-parameter group generated by the operator

$$H_1 = \begin{bmatrix} i\partial & -m \\ -m & -i\partial \end{bmatrix}.$$

It is straightforward to check that the quantum field

$$\psi(x,t) = \begin{bmatrix} \psi_1(x,t) \\ \psi_2(x,t) \end{bmatrix}$$

does satisfy *causality*, so that $\psi(x,s)$ and $\psi(y,t)$ anticommute at space-like separated points.

The only difficulty in the quantization of the spinor field is the positivity of energy, since the first-order operator H_1 does have both negative and positive eigenvalues. The answer is well known, and quite clearly stated in C^*-algebraic terms:

Lemma 13. *For any $\beta \in [0, \infty]$ there exists a unique KMS$_\beta$ state φ_β on (A, σ_t). It is a quasi-free state. For $\beta = +\infty$ the covariance of φ_∞ is given by the sign $F = H_1 |H_1|^{-1}$ of H_1.*

The representation of A associated to the ground state φ_∞ is the "Dirac sea" representation which can be described as follows. The Hilbert space $\mathcal{H}_f$ is the antisymmetric Fock space over $L^2(\mathbb{S}^1, S) = L^2(\mathbb{S}^1) \oplus L^2(\mathbb{S}^1)$. One lets $b_\pm^*(k)$ be the corresponding creation operators for $k \in \mathbb{Z}$, the label of the natural orthonormal basis of $L^2(\mathbb{S}^1, S)$. The operators $b_\pm^*(k)$ satisfy the canonical anticommutation relations, and are related to the $\psi_i(k)$ by the equalities which define the representation of A in $\mathcal{H}_f$, namely

$$\psi_1(k) = (4\pi\omega(k))^{-1/2}\, (\nu(-k)\, b_-^*(k) + \nu(k)\, b_+(-k))$$

$$\psi_2(k) = (4\pi\omega(k))^{-1/2}\, (\nu(k)\, b_-^*(k) - \nu(-k)\, b_+(-k))$$

where $\omega(k) = (k^2 + m^2)^{1/2}$ as above, and where

$$\nu(k) = (\omega(k) + k)^{1/2}.$$

In this representation the one-parameter group σ_t of automorphisms of A has a positive generator, the fermionic Hamiltonian

$$H_f = \int_{\mathbb{S}^1} :\overline{\psi}\, i\, \gamma_1\, \partial\psi - m\, \overline{\psi}\psi : dx.$$

The Hilbert space of the full model is $\mathcal{H} = \mathcal{H}_b \otimes \mathcal{H}_f$, the tensor product of the bosonic and fermionic Hilbert spaces. The quantum fields are given by the operator-valued distributions $\varphi(x) \otimes 1$ and $1 \otimes \psi_j(x)$, where φ is the (complex) bosonic field and ψ_j are the two components of the fermionic field.

As above (Proposition 12) we let $\mathcal{U}(\mathcal{O})$ be the von Neumann algebra of *bosonic* observables associated to any local region $\mathcal{O} \subset X$. The full Hamiltonian of the non-interacting theory is given by

$$H = H_b \otimes 1 + 1 \otimes H_f,$$

and since the Wick ordering constants of the bosonic and fermionic parts cancel identically, one can write it as

$$H = \int_{\mathbb{S}^1} dx \left(|\pi(x)|^2 + |\partial\varphi(x)|^2 + m^2 |\varphi(x)|^2 + \overline{\psi}(x)(i\gamma_1 \partial - m)\, \psi(x) \right).$$

It is a positive operator in $\mathcal{H}$, and admits a natural selfadjoint square root, the *supercharge* operator Q given by ([J-L-W])

$$Q = \frac{1}{\sqrt{2}} \int_{\mathbb{S}^1} \psi_1(x)\, (\pi(x) - \partial\varphi^*(x) - im\, \varphi(x)) + \psi_2(x) \times$$
$$(\pi^*(x) - \partial\varphi(x) - im\, \varphi^*(x))\; dx + \text{ h.c.}$$

where h.c. means the adjoint operator.

On $\mathcal{H} = \mathcal{H}_b \otimes \mathcal{H}_f$ we let Γ be the $\mathbb{Z}/2$-grading given by $1 \otimes (-1)^{N_f}$, where N_f is the fermion number operator.

With the above notation it is straightforward to prove:

Proposition 14. a) *The operator Q in $\mathcal{H}$ is selfadjoint, anticommutes with the $\mathbb{Z}/2$-grading Γ, and* $\mathrm{Trace}(e^{-\beta Q^2}) < \infty$ *for any $\beta > 0$.*

b) *For any local region $\mathcal{O} \subset X$ the subalgebra*

$$\mathcal{U}_L(\mathcal{O}) = \{T \in \mathcal{U}(\mathcal{O})\ ;\ [Q,T] \textit{ is bounded}\}$$

is weakly dense in the von Neumann algebra $\mathcal{U}_L(\mathcal{O})$.

We refer to [J-L-W] for the proof of a). In fact, by construction, the pair $(\mathcal{H}, Q)$ is obtained as an infinite tensor product $\bigotimes_{k \in \mathbb{N}} (\mathcal{H}_k, Q_k)$, where each $\mathcal{H}_k$ is a $\mathbb{Z}/2$-graded Hilbert space with $\mathbb{Z}/2$-grading Γ_k, and where the infinite tensor product is relative to the unique (up to a phase) normalized vector Ω_k for which $Q_k\, \Omega_k = 0$ and $\Gamma_k\, \Omega_k = \Omega_k$. It is then an easy exercise to show that $Q = \sum Q_k$ is selfadjoint in the $\mathbb{Z}/2$-graded infinite tensor product, provided each Q_k is. Here each operator Q_k, $k \in \mathbb{N}$, is of the form

$$Q_k = d_{\varphi_k} + (d_{\varphi_k})^* \text{ in } L^2(E_k, \wedge_{\mathbb{C}}^* E_k)$$

where, for $k > 0$, E_k is the real Euclidean vector space $E_k = \mathbb{C}^2$ equipped with the nondegenerate real quadratic form

$$\varphi_k(z,u) = (2\omega_k)^{-1}\; (k\, z\bar{z} - k\, u\bar{u} + m(zu + \bar{z}\,\bar{u})) \quad \forall (z,u) \in \mathbb{C}^2.$$

Also $Q^2 = H = H_b \otimes 1 + 1 \otimes H_f$, where the spectrum of H_b (resp. H_f) is the set of all finite sums

$$E = \sum n_k \, \omega_k \quad n_k \in \mathbb{N} \quad (\text{resp. } n_k \in \{0,1\})$$

so that $\exp(-\beta H)$ is of trace class for any $\beta > 0$, and the number of eigenvalues of H below a given $E > 0$ grows like $\exp(\lambda\sqrt{E})$ for $E \to \infty$, with λ a fixed constant.

b) To prove this it is enough to show that for $f \in C_c^\infty(\mathcal{O})$ the following operators in $\mathcal{H}$ belong to $\mathcal{U}_L(\mathcal{O})$:

$$\exp\, i(\varphi(f) + \varphi(f)^*)\ ,\ \exp\ i(\pi(f) + \pi(f)^*),$$

and similarly for the fermionic fields.

This follows from the boundedness of the commutators

$$[Q,\varphi(f)]\ ,\ [Q,\pi(f)].$$

One has $[Q,\varphi(f)] = \frac{-i}{\sqrt{2}} \int \psi_1(x)\, f(x)\, dx + \frac{i}{\sqrt{2}} \left(\int \psi_2(x)\, \overline{f}(x)\, dx\right)^*$, which is bounded if $f \in L^2$. Similarly, $[Q,\pi(f)]$ is bounded provided f and ∂f belong to L^2, so the result follows.

Corollary 15. *Let $\mathcal{O} \subset X$ be a local region and $A(\mathcal{O})$ be the pre-C^*-algebra in $\mathcal{U}(\mathcal{O})$ generated by the operators* $\exp i(\varphi(f)+\varphi(f)^*)$ *and* $\exp i(\pi(f)+\pi(f)^*)$ *for $f \in C_c^\infty(\mathcal{O})$. Then $(\mathcal{H},Q,\Gamma)$ is a θ-summable K-cycle over $A(\mathcal{O})$ for any local region $\mathcal{O}$.*

In fact, $A(\mathcal{O}) \subset \mathcal{U}_L(\mathcal{O})$, as follows from the proof of Proposition 14 b). We can now state the main result of this section, which shows that even though we are in a free theory, and however small a nonempty local region $\mathcal{O}$ may be, the K-theory of the local algebra $A(\mathcal{O})$ is nontrivial, pairs nontrivially with the supercharge operator Q, and attests to the presence of infinitely many bosonic degrees of freedom localized in $\mathcal{O}$.

Theorem 16. *Let $\mathcal{O} \subset X$ be a nonempty local region. Then the index map*

$$\mathrm{Ind}_Q : K_0(A(\mathcal{O})) \to \mathbb{Z}$$

is not polynomial, *and $K_0(A(\mathcal{O}))$ is of infinite rank.*

To prove this we shall use Lemma 10 and the construction of K-theory classes given in that lemma. To prove the result we can assume that $\mathcal{O}$ intersects $\mathbb{S}^1 \times \{0\} \subset X$. Then let $n \in \mathbb{N}$ and f_j, $j = 1,\ldots,n$, be real-valued smooth functions on $\mathbb{S}^1$ with support in $\mathcal{O} \cap (\mathbb{S}^1 \times \{0\})$ and such that

$$\int_{\mathbb{S}^1} f_i(x)\, f_j(x)\, dx = \delta_{ij}.$$

Let φ_r be the real quantum field $\varphi_r(x) = \varphi(x) + \varphi^*(x)$, and let π_r be the conjugate momentum. Then the following unitary operators belong to the algebra $A(\mathcal{O})$:

$$U_j = \exp 2\pi i \varphi_r(f_j)$$

$$V_k = \exp i\pi_r(f_k).$$

The canonical commutation relations

$$[\varphi_r(f_j), \pi_r(f_k)] = \delta_{jk}$$

show that the operators U_j and V_k generate a commutative subalgebra of $A(\mathcal{O})$, and we can consider the element of $K_0(A(\mathcal{O}))$ given, as in Lemma 10, by

$$x = [U_1 \wedge U_2 \wedge \cdots \wedge U_n \wedge V_1 \wedge \cdots \wedge V_n].$$

By Lemma 10, x is a nontrivial element of $K_0(A(\mathcal{O}))$, since it pairs nontrivially with the K-cycle $(\mathcal{H}, Q, \Gamma)$ and

$$\text{Index } Q_x = 1.$$

We are applying Lemma 10 in the case when E is infinite-dimensional, but the same proof works. Here $E = C$ is the configuration space of the complex scalar field φ, and the quadratic form Φ is given by

$$\Phi(\varphi) = \frac{1}{2} \operatorname{Re} \int_{S^1} \left(i(\partial\varphi)\varphi + m\, \varphi^2 \right) dx.$$

By construction, the above $x \in K_0(A(\mathcal{O}))$ pairs trivially with any cyclic cocycle on $A(\mathcal{O})$ of dimension $< 2n$, and pairs nontrivially with Q, which is enough to show that the index map is non-polynomial.

Note that the non-polynomial index map of Q is given, using Theorem 19 of Section 8, by an entire cocycle on $A(\mathcal{O})$, namely the character $\mathrm{ch}_*(\mathcal{H}, Q, \Gamma)$ of the θ-summable K-cycle $(\mathcal{H}, Q, \Gamma)$ (Corollary 15).

By the results of [J-L-W] on constructive quantum field theory, Proposition 14 still holds for the interacting Wess-Zumino model in 2 dimensions.

We shall end this section by formulating two problems.

Problem 1. Prove Theorem 16 for the interacting Wess-Zumino model.

Problem 2. Compute the entire cyclic cocycle $\mathrm{ch}_*(\mathcal{H}, Q, \Gamma)$.

IV. Appendix A. Kasparov's Bivariant Theory

In this appendix we shall give without proof the basic results of Kasparov's bivariant theory, which is an indispensable tool in constructing K-homology classes of C^*-algebras. We shall follow [Kas_1] and [Sk_1].

Let us first recall some notation from Chapter II Appendix A.

Given two C^*-algebras A and B, an A-B C^*-bimodule is defined as a C^*-module $\mathcal{E}$ over B together with a $*$-homomorphism from A to $\text{End}_B(\mathcal{E})$. We say that a C^*-algebra A is $\mathbb{Z}/2$-graded if it is endowed with a grading automorphism $\gamma \in \text{Aut}\, A$ with $\gamma^2 = \text{id}$. Any element x of A then has a unique decomposition as a sum $x = x_0 + x_1$ with $\gamma(x_0) = x_0$ and $\gamma(x_1) = -x_1$. The C^*-subalgebra $A^\gamma = \{x \in A\,;\, \gamma(x) = x\}$ is called the even part of A, and the odd part is $\{x \in A\,;\, \gamma(x) = -x\}$. The $\mathbb{Z}/2$-grading is called trivial if $A^\gamma = A$, i.e. if $\gamma = \text{id}$. Similarly, a C^*-module $\mathcal{E}$ over a $\mathbb{Z}/2$-graded C^*-algebra B is called $\mathbb{Z}/2$-graded if it is endowed with a linear grading operator $\gamma : \mathcal{E} \to \mathcal{E}$ such that

$$\langle \gamma\xi, \gamma\eta \rangle = \langle \xi, \eta \rangle \in B \quad \forall \xi, \eta \in \mathcal{E}$$

$$\gamma(\xi b) = \gamma(\xi)\, \gamma(b) \quad \forall \xi \in \mathcal{E}\,,\ b \in B.$$

When B is trivially graded one has $\gamma \in \text{End}_B(\mathcal{E})$, and γ is an involutive unitary element of that C^*-algebra. Consideration of the general case is quite useful since it allows one to treat the even and odd cases in a unified manner.

Given a $\mathbb{Z}/2$-graded C^*-module $\mathcal{E}$ over a $\mathbb{Z}/2$-graded C^*-algebra B, the C^*-algebra $\text{End}_B(\mathcal{E})$ is naturally $\mathbb{Z}/2$-graded by the rule

$$\gamma(T)\, \gamma(\xi) = \gamma(T\xi) \quad \forall T \in \text{End}_B(\mathcal{E})\,,\ \xi \in \mathcal{E}.$$

For A, B a pair of $\mathbb{Z}/2$-graded C^*-algebras, an A-B graded C^*-bimodule is given by a $\mathbb{Z}/2$-graded C^*-module $\mathcal{E}$ over B and a $\mathbb{Z}/2$-graded $*$-homomorphism from A to $\text{End}_B(\mathcal{E})$.

In the $\mathbb{Z}/2$-graded context, all commutators are graded commutators:

$$[x, y] = xy - (-1)^{\deg x \deg y}\, yx,$$

and, more generally, one uses the rule that a transposition of two elements x and y introduces the sign $(-1)^{\deg x \deg y}$ (cf. [Mi_4]). As in Appendix A of Chapter II we shall denote by $\text{End}_B^0(\mathcal{E})$ the two-sided ideal of compact endomorphisms of a C^* B-module $\mathcal{E}$. All the C^*-modules will be assumed to be countably generated (cf. Chapter II Appendix A).

Definition 1. [Kas_1] *Let A, B be $\mathbb{Z}/2$-graded C^*-algebras. A Kasparov A-B-bimodule is a pair $(\mathcal{E}, F)$, where $\mathcal{E}$ is an A-B graded C^*-bimodule, and $F \in \text{End}_B(\mathcal{E})$, $\gamma(F) = -F$, is such that for any $a \in A$ one has $[F, a] \in \text{End}_B^0(\mathcal{E})$, $a(F^2 - 1) \in \text{End}_B^0(\mathcal{E})$, and $a(F - F^*) \in \text{End}_B^0(\mathcal{E})$.*

One says that a Kasparov A-B-bimodule is degenerate if the above three operators vanish for any $a \in A$, i.e.

$$[F, a] = a(F^2 - 1) = a(F - F^*) = 0 \quad \forall a \in A.$$

The direct sum of Kasparov A-B-bimodules is defined in a straightforward manner by

$$(\mathcal{E}_1,F_1)\oplus(\mathcal{E}_2,F_2)=(\mathcal{E}_1\oplus\mathcal{E}_2\ ,\ F_1\oplus F_2).$$

Let $B[0,1]=B\otimes C[0,1]$. Then a Kasparov A-$B[0,1]$-bimodule can be equivalently described as a continuous field $(\mathcal{E}_t,F_t)$, $t\in[0,1]$, of Kasparov A-B-bimodules, and it is called a *homotopy* between $(\mathcal{E}_0,F_0)$ and $(\mathcal{E}_1,F_1)$. More specifically $(\mathcal{E}_t,F_t)$, for $t\in[0,1]$, is the Kasparov A-B-bimodule obtained from $(\mathcal{E},F)$ by composition with the evaluation morphism

$$\rho_t:B[0,1]\to B.$$

In general, given two C^*-algebras B_1 and B_2 and a $*$-homomorphism $\rho:B_1\to B_2$, one can associate to any Kasparov A-B_1-bimodule the *induced* A-B_2 Kasparov bimodule given by

$$(\mathcal{E}\otimes_{B_1}B_2\ ,\ F\otimes 1)=\rho_*(\mathcal{E},F).$$

Similarly, given a $*$-homomorphism $\sigma:A_1\to A_2$, there is an obvious notion of *restriction* of any Kasparov A_2-B-bimodule to A_1-B.

Proposition 2. [Kas$_1$] *Let A and B be two C^*-algebras. The set $KK(A,B)$ of homotopy classes of Kasparov A-B-bimodules endowed with the operation of direct sum is an* abelian group.

Any degenerate Kasparov A-B-bimodule is homotopic to the bimodule $\mathcal{E}=\{0\}$ and represents $0\in KK(A,B)$.

The bifunctor $KK(A,B)$ is, by construction, homotopy invariant in both A and B.

Kasparov proved (cf. [Kas$_1$]) that the equivalence relation of homotopy is in fact the same, modulo the degenerate bimodules, as the apparently much more restrictive *operatorial homotopy*: given a ($\mathbb{Z}/2$-graded) A-B C^*-bimodule $\mathcal{E}$, a homotopy $(\mathcal{E},F_t)_{t\in[0,1]}$ of Kasparov A-B-bimodules is called *operatorial* if the map $t\to F_t$ is norm continuous.

Theorem 3. [Kas$_1$] *Let A and B be two C^*-algebras, with A separable. Two Kasparov A-B-bimodules $(\mathcal{E}_j,F_j)$ are homotopic iff, up to addition of degenerate bimodules, they are operator homotopic.*

The key tool of KK-theory is the Kasparov product ([Kas$_1$])

$$KK(A,B)\times KK(B,C)\to KK(A,C)$$

which we shall now describe using the convenient notion of connection introduced in [Co-S].

Definition 4. *Let $\mathcal{E}_2$ be a B-C C^*-bimodule, let $\mathcal{E}_1$ be a C^*-module over B, and let $\mathcal{E}=\mathcal{E}_1\otimes_B\mathcal{E}_2$, $F_2\in\mathrm{End}_C(\mathcal{E}_2)$. An element $F\in\mathrm{End}_C(\mathcal{E})$ is called an F_2 connection on $\mathcal{E}_1$ if for any $\xi\in\mathcal{E}_1$ the following endomorphism is compact:*

$$\left[\tilde{T}_\xi,F_2\oplus F\right]\in\mathrm{End}^0_C(\mathcal{E}_2\oplus\mathcal{E})$$

where $\tilde{T}_\xi = \begin{bmatrix} 0 & T_\xi^* \\ T_\xi & 0 \end{bmatrix} \in \mathrm{End}_C(\mathcal{E}_2 \oplus \mathcal{E})$, $T_\xi \in \mathrm{Hom}_C(\mathcal{E}_2, \mathcal{E})$ *being given by* $T_\xi(\eta) = \xi \otimes \eta \in \mathcal{E}$, $\forall \eta \in \mathcal{E}_2$.

One should view F as a lift of F_2, but the terminology connection is appropriate as can be seen from examples coming from pseudodifferential operators of order 0 on smooth manifolds and involving lifts of these operators to sections of vector bundles.

The straightforward properties of connections are listed as follows [Co-S].

Proposition 5. a) *If* $[F_2, b] = 0 \;\; \forall b \in B$, *then* $1 \otimes F_2$ *makes sense and is an* F_2 *connection. Moreover,* $[T \otimes 1, 1 \otimes F_2] = 0$ *for any* $T \in \mathrm{End}_B(\mathcal{E}_1)$.

b) *If* $[F_2, b] \in \mathrm{End}^0_C(\mathcal{E}_2) \;\; \forall b \in B$, *then there exists an* F_2 *connection for any (countably generated)* C^**-module* $\mathcal{E}_1$ *over* B.

c) *The space of* F_2 *connections is an affine space with associated vector space* $V = \{T \in \mathrm{End}_C(\mathcal{E})\,;\; Tx \text{ and } xT \in \mathrm{End}^0_C(\mathcal{E}) \text{ for any } x \in \mathrm{End}^0_B(\mathcal{E}_1) \otimes 1\}$.

d) *If* F *is an* F_2 *connection then* $[F, x] \in \mathrm{End}^0_C(\mathcal{E})$ *for any* $x \in \mathrm{End}^0_B(\mathcal{E}_1) \otimes 1$.

e) *If* F *is an* F_2 *connection and* F' *is an* F_2' *connection, then* $F + F'$, FF', F^* *are respectively* $F_2 + F_2'$, F_2F_2', *and* F_2^* *connections.*

We can now give an *implicit* definition of the Kasparov product.

Definition 6. *Let* A, B *and* C *be* $\mathbb{Z}/2$*-graded* C^**-algebras,* $(\mathcal{E}_1, F_1)$ *a Kasparov* A*-*B*-bimodule, and* $(\mathcal{E}_2, F_2)$ *a Kasparov* B*-*C*-bimodule. Let* $\mathcal{E} = \mathcal{E}_1 \otimes_B \mathcal{E}_2$ *be their graded tensor product viewed as an* A*-*C C^**-bimodule. The pair* $(\mathcal{E}, F)$, $F \in \mathrm{End}_C(\mathcal{E})$, *is called a Kasparov product of* F_1 *and* F_2 *(and one writes* $F \in F_1 \# F_2$*) if:*

a) $(\mathcal{E}, F)$ *is a Kasparov* A*-*C*-bimodule (Definition 1).*

b) F *is an* F_2 *connection.*

c) *For any* $a \in A$, $a[F_1 \otimes 1, F]a^* \geq 0$ *modulo* $\mathrm{End}^0_C(\mathcal{E})$.

The motivation for a) and b) is clear. The reason for c) is the construction of the product [Kas$_1$] using the following formula

$$F = M(F_1 \otimes 1) + N(1 \otimes F_2)$$

which makes sense when $\mathcal{E}_1$ is stabilized and M and N are *positive* operators. The original construction of Kasparov ([Kas$_1$]) is slightly improved (cf. [Co-S]) as follows:

Theorem 7. *Assume that* A *is separable. Let* $(\mathcal{E}_1, F_1)$ *be a Kasparov* A*-*B*-bimodule and let* $(\mathcal{E}_2, F_2)$ *be a Kasparov* B*-*C*-bimodule.*

a) *There exists a Kasparov product* $(\mathcal{E}, F)$, $F \in F_1 \# F_2$, *and it is unique up to operatorial homotopy.*

b) *The map* $(\mathcal{E}_1, F_1) \times (\mathcal{E}_2, F_2) \to (\mathcal{E}, F)$ *passes to homotopy classes, and defines a bilinear product denoted*

$$\otimes_B : KK(A,B) \times KK(B,C) \to KK(A,C).$$

The uniqueness of F is easy. Its existence is a corollary of the technical theorem:

Theorem 8. [Kas$_1$] *Let B be a graded C^*-algebra and $\mathcal{E}$ a countably generated, graded C^*-module over B. Let E_1 and E_2 be graded subalgebras of* $\mathrm{End}_B(\mathcal{E})$, *and $f \subset \mathrm{End}_B(\mathcal{E})$ a graded vector subspace. Assume that:*

1) *E_1 has a countable approximate unit and contains* $\mathrm{End}_B^0(\mathcal{E})$.
2) *f and E_2 are separable.*
3) $E_1 \cdot E_2 \subset \mathrm{End}_B^0(\mathcal{E})$, $[f, E_1] \subset E_1$.

Then there exist $M, N \in \mathrm{End}_B(\mathcal{E})$ such that

$$M \geq 0 \,,\; N \geq 0 \,,\; M + N = 1$$

$$ME_1 \subset \mathrm{End}_B^0(\mathcal{E}) \,,\; NE_2 \subset \mathrm{End}_B^0(\mathcal{E}) \,,\; [f, M] \subset \mathrm{End}_B^0(\mathcal{E}).$$

The Kasparov product $\otimes_B$ enjoys the same general functorial properties as the composition of morphisms. Each morphism $\rho : A \to B$ of C^*-algebras defines in an obvious way an element $\rho \in KK(A,B)$ (given by the A-B C^*-module $\mathcal{E} = B$ with left action of A given by ρ, and operator $F = 0$).

There is a small notational difficulty since the composition of morphisms is written

$$\mathrm{Hom}(A,B) \times \mathrm{Hom}(B,C) \to \mathrm{Hom}(A,C)$$

$$(\rho_1, \rho_2) \mapsto \rho_2 \circ \rho_1,$$

while the natural notation for the Kasparov product is that of tensor products of bimodules, namely

$$KK(A,B) \times KK(B,C) \to KK(A,C)$$

$$(x_1, x_2) \mapsto x_1 \otimes_B x_2,$$

which is also analogous to the contraction of tensors.

In particular, (cf. [Kas$_1$]) it is convenient to extend the Kasparov product to the following relative case:

For any C^*-algebra A with countable approximate unit we denote by τ_A the operation of minimal tensor product by A, i.e.

$$\tau_A : KK(B,C) \to KK(B \otimes A, C \otimes A),$$

where the tensor products of C^*-algebras are minimal (or spatial ones) (cf. [Pe$_2$]). Given a Kasparov B-C-bimodule $(\mathcal{E}, F)$, its image under τ_A is given by $(\mathcal{E} \otimes A, F \otimes 1)$, which is a Kasparov $(B \otimes A)$-$(C \otimes A)$-bimodule.

Then let A_1, A_2, B_1, B_2 and D be C^*-algebras, with A_1 and A_2 separable, and B_1 with countable approximate unit.

Definition 9. [Kas$_1$] *Let* $x_1 \in KK(A_1, B_1 \otimes D)$ *and* $x_2 \in KK(D \otimes A_2, B_2)$. *Then* $x_1 \otimes_D x_2 \in KK(A_1 \otimes A_2, B_1 \otimes B_2)$ *is defined as*

$$x_1 \otimes_D x_2 = \tau_{A_2}(x_1) \otimes_{B_1 \otimes D \otimes A_2} \tau_{B_1}(x_2).$$

Then the associativity of the Kasparov product, i.e. the equality: ([Kas])

$$(x_1 \otimes_B x_2) \otimes_C x_3 = x_1 \otimes_B (x_2 \otimes_C x_3)$$

for $x_1 \in KK(A, B)$, $x_2 \in KK(B, C)$, $x_3 \in KK(C, D)$ implies the stronger:

Theorem 10. [Kas$_1$] *Let* A_1, A_2, A_3 *and* D_1 *be separable,* $x_1 \in KK(A_1, B_1 \otimes D_1)$, $x_2 \in KK(D_1 \otimes A_2, B_2 \otimes D_2)$, $x_3 \in KK(D_2 \otimes A_3, B_3)$; *then*

$$(x_1 \otimes_{D_1} x_2) \otimes_{D_2} x_3 = x_1 \otimes_{D_1} (x_2 \otimes_{D_2} x_3).$$

The associativity implies, in particular, that $KK(B, B)$ is naturally a ring with unit for any separable C^*-algebra B. In the form given in Theorem 10 it is particularly convenient for the formulation of Poincaré duality ([Kas$_5$] [Kas$_6$] and [Co-S]). We shall come back to this point in Chapter VI.

Before we specialize KK to cases when A or $B = \mathbb{C}$, we state two equivalent formulations of its cycles.

Definition 11. *A Kasparov A-B-bimodule* $(\mathcal{E}, F)$ *is* normalized *if* $F = F^*$ *and* $F^2 = 1$.

One then has the following simple

Proposition 12. *Let* $(\mathcal{E}, F)$ *be a Kasparov A-B bimodule. Then after addition of a degenerate bimodule* $(\mathcal{E}_0, 0)$, *it is operator homotopic to a* normalized *Kasparov A-B-bimodule.*

Moreover, since the construction is canonical, it shows that one really does obtain the same KK-theory if one restricts to normalized Kasparov bimodules and homotopies.

When $B = \mathbb{C}$ and A is unital, and acts in a unital way on $\mathcal{E} = \mathcal{H}$ which is a $\mathbb{Z}/2$-graded Hilbert space, one can normalize $(\mathcal{H}, F)$ by simply adding to $\mathcal{H}$ the finite-dimensional space $\ker F$ with opposite $\mathbb{Z}/2$-grading and 0-module structure over A. This removes the index obstruction to making F invertible.

Let $C_1 = \{\lambda + \mu\alpha\,;\, \lambda, \mu \in \mathbb{C}\}$ be the $\mathbb{Z}/2$-graded Clifford algebra over $\mathbb{C}$, with $\alpha^2 = 1$ and $\mathbb{Z}/2$-grading given by $\lambda + \mu\alpha \mapsto \lambda - \mu\alpha$.

Proposition 13. *Let A be a trivially graded* C^**-algebra.*

a) *A normalized Kasparov A-$\mathbb{C}$-bimodule is exactly an even Fredholm module over A.*

b) *Let $K = \mathbb{C}^2$ with $\mathbb{Z}/2$-grading $\gamma = \begin{bmatrix} 1 & 0 \\ 0 & -1 \end{bmatrix}$ and action of C_1 given by $\lambda + \mu\alpha \mapsto \begin{bmatrix} \lambda & \mu \\ \mu & \lambda \end{bmatrix}$ $\forall \lambda, \mu \in \mathbb{C}$. Then every normalized Kasparov $A \otimes C_1$-$\mathbb{C}$-bimodule is of the form $\left(\mathcal{H} \otimes K \,,\, F \otimes \begin{bmatrix} 0 & -i \\ i & 0 \end{bmatrix}\right)$ for a unique odd Fredholm module $(\mathcal{H}, F)$ over A.*

The proof is straightforward. The next result due to [B-J] gives a useful construction of Kasparov bimodules from unbounded operators in C^*-modules as well as a useful notion:

Definition 14. *Let B be a C^*-algebra and $\mathcal{E}$ a C^*-module over B; then an* unbounded endomorphism T *of $\mathcal{E}$ is given by a pair of closed densely defined operators T and T^* on $\mathcal{E}$ commuting with the right action of B such that*

1) $\langle T\xi, \eta\rangle = \langle \xi, T^*\eta\rangle \quad \forall \xi \in \mathrm{Dom}\, T \,,\ \eta \in \mathrm{Dom}\, T^*$

2) $1 + T^*T$ *is surjective.*

We refer to the thesis of S. Baaj [Baaj] for this notion introduced in [B-J].

Theorem 15. (cf. [B-J]) a) *Let $\mathcal{E}$ be an A-B C^*-bimodule with $\mathbb{Z}/2$-grading γ and D an unbounded selfadjoint endomorphism of $\mathcal{E}$ anticommuting with γ and such that*

α) $\{a \in A\,;\ [D, a]$ *bounded*$\}$ *is norm dense in A*

β) $a(1 + D^*D)^{-1} \in \mathrm{End}^0_B(\mathcal{E}) \quad \forall a \in A.$

Then the following equality defines a Kasparov A-B-bimodule:

$$F = D(1 + D^*D)^{-1/2}.$$

b) *Every Kasparov A-B-bimodule $(\mathcal{E}, F)$ is operator homotopic to one obtained from the construction a).*

The proof follows by using the equality

$$(1 + D^*D)^{-1/2} = \lambda \int_0^\infty \frac{1}{D^*D + 1 + \mu}\, \mu^{-1/2}\, d\mu$$

to prove that $[F, a]$ is compact for any $a \in A$.

One reason why the unbounded formulation of Kasparov bimodules is convenient is that it makes the *external* product straightforward, and given simply by the formula

$$(\mathcal{E}_1, D_1) \otimes (\mathcal{E}_2, D_2) = (\mathcal{E}_1 \otimes \mathcal{E}_2 \,,\ D_1 \otimes 1 + 1 \otimes D_2)$$

where the tensor product is a *graded* tensor product. In the odd (ungraded) case this yields

$$D = \begin{bmatrix} 0 & D_1 \otimes 1 - i \otimes D_2 \\ D_1 \otimes 1 + i \otimes D_2 & 0 \end{bmatrix}.$$

The internal Kasparov product, $KK(A,B) \times KK(B,C) \to KK(A,C)$, is more difficult to treat by this method.

At this point it is important to discuss the closely related notion of superconnection due to D. Quillen [Q_1]. Let M be a smooth manifold and $\Omega(M) = \bigoplus \Omega^p(M)$ be the algebra of smooth differential forms on M. Let E be a $\mathbb{Z}/2$-graded complex vector bundle over M and consider the *right* $\Omega(M)$-module $\Omega(M,E)$ of E-valued differential forms on M, endowed with the total $\mathbb{Z}/2$-grading.

Definition 16. [Q_1] *A superconnection on E is an operator Z of odd degree in $\Omega(M,E)$ such that*

$$Z(\xi\omega) = Z(\xi)\omega + (-1)^{\partial\xi}\, \xi\, d\omega$$

for any $\xi \in \Omega(M,E)$, $\omega \in \Omega(M)$.

(This differs slightly from [Q_1] because we use right modules instead of left modules.)

This notion plays an important role in the computation of the Chern character of K-theory classes on M given by a pair $(\mathcal{E}, D)$ as in Theorem 15 with $A = \mathbb{C}$, $B = C_0(M)$. With $\mathcal{E} = C_0(M,E)$ and ∇ an ordinary connection on the Hilbert bundle E, the equality

$$Z = \gamma\nabla + iD,$$

where γ is the $\mathbb{Z}/2$-grading of E, defines a superconnection. The Chern character of $(\mathcal{E}, D)$ is then given ([Q_1]) as trace (e^{Z^2}). All of this extends with minor changes to the context of cycles of Chapter III Section 3, and Proposition 8 continues to hold provided one deals with pre-C^*-algebras and pre-C^*-modules. It provides one with a very useful tool for computation of the pairing of cyclic cohomology with explicit K-theory classes. Finally, by adapting Proposition 9.7 to this context, one can explicitly compute the action of a bivariant KK-class on cyclic cohomology (cf. [Ni_3] [Ni_4] for a general construction of the bivariant Chern character in the finitely summable case). For our purpose it was crucial, in order to define the character of θ-summable Fredholm modules, to make them unbounded, and Theorem 4 of Section 8 achieves this goal in a quantitative manner. By Theorem 12 Appendix B of Chapter II (cf. [Co-H]) there exists a natural map

$$KK(A,B) \to E(A,B)$$

for any C^*-algebras A and B. Moreover, by [Sk_3] this map is an isomorphism if A is a nuclear (or even K-nuclear) C^*-algebra. The deformation associated to a

KK-class can be described very explicitly in terms of an unbounded representative $(\mathcal{E}, D)$ (cf. Theorem 15) of that class (cf. [Co-H] [Co-M_1]).

For general C^*-algebras the two theories do not coincide in general ([Sk_4]). Let us now briefly discuss the G-equivariant case [Kas_5].

Let G be a locally compact σ-compact group. Then by a G-C^*-algebra one means a C^*-algebra A together with a continuous action of G on A by $*$-automorphisms, i.e. a group morphism $G \to \operatorname{Aut} A$ such that for any $a \in A$ the map $g \to g(a)$ is norm continuous. When A is also $\mathbb{Z}/2$-graded the action of G is supposed to commute with the $\mathbb{Z}/2$-grading.

Definition 17. *Let G be a locally compact group, and let A and B be $\mathbb{Z}/2$-graded G-C^*-algebras. Then a G-equivariant Kasparov A-B-bimodule is a Kasparov A-B-bimodule together with an action of G on $\mathcal{E}$, $(g, \xi) \in G \times \mathcal{E} \to g\xi \in \mathcal{E}$, preserving the $\mathbb{Z}/2$-grading, and such that*

a) $g(a\xi b) = g(a)\, g(\xi)\, g(b) \quad \forall a \in A\,,\ \xi \in \mathcal{E}\,,\ b \in B\,,\ g \in G$

b) $\langle g\xi, g\eta \rangle = g(\langle \xi, \eta \rangle) \quad \forall \xi, \eta \in \mathcal{E}$

c) $a(gFg^{-1} - F) \in \operatorname{End}_B^0(\mathcal{E}) \quad \forall a \in A\,,\ g \in G.$

As above, the homotopy classes of G-equivariant Kasparov A-B-bimodules form a group, the G-equivariant KK-group

$$KK^G(A, B).$$

All the above results on the Kasparov product extend to the G-equivariant case ([Kas_5]), and the G-equivariant theory is related to the crossed product C^*-algebras $A \rtimes_r G$ and $A \rtimes G$ (reduced and maximal crossed products) by a natural map both in the reduced and maximal cases.

Theorem 18. [Kas_5] *Both crossed product constructions extend to G-equivariant Kasparov bimodules and yield natural maps, compatible with the Kasparov product:*

$$KK^G(A, B) \to KK(A \rtimes_r G, B \rtimes_r G)$$

$$KK^G(A, B) \to KK(A \rtimes G, B \rtimes G).$$

Finally we shall end this appendix with the fundamental equality of G-equivariant KK-theory when G is a semisimple Lie group ([Kas_5]). Let $H \subset G$ be a maximal compact subgroup, and assume to simplify that the isotropy representation of H in $T_e(X)$, $X = G/H$, lifts to Spin (or Spin^c) (cf. [Kas_5] for the general case). Then let S be the G-equivariant spinor bundle on X. One then has two natural G-equivariant Kasparov bimodules for the G-C^*-algebras $C_0(X)$ and $\mathbb{C}$ (with obvious G actions).

The Dirac element $\delta \in KK^G(C_0(X), \mathbb{C})$ is given by the ($\mathbb{Z}/2$-graded) Hilbert space $L^2(X, S)$ of spinors, the action of $C_0(X)$ by multiplication, and of G by left translations. In unbounded form (cf. Theorem 15) the operator is $D = \partial\!\!\!/_X$, the Dirac operator on X.

The dual Dirac element $\hat{\delta} \in KK^G(\mathbb{C}, C_0(X))$ is given by the ($\mathbb{Z}/2$-graded) C^*-module $\mathcal{E}$ over $C_0(X)$ corresponding to the Hermitian vector bundle S with the obvious G-equivariant structure. In unbounded form the operator, which depends upon the choice of a base point $a \in X$, is given by

$$(\hat{D}\xi)(p) = c(p,a)\,\xi(p) \quad \forall \xi \in \mathcal{E}\,,\ p \in X,$$

where $c(p,a)$ is the Clifford multiplication by the unique vector $(\overrightarrow{p,a}) \in T_p(X)$ whose image under the exponential map of X at p is a:

Theorem 19. [Kas$_5$] *One has* $\delta \otimes_{\mathbb{C}} \hat{\delta} = \mathrm{id} \in KK^G(C_0(X), C_0(X))$.

This result implies Theorem 4 of Section 9.

It is not true, in general, that $\hat{\delta} \otimes_{C_0(X)} \delta = \mathrm{id} \in KK^G(\mathbb{C}, \mathbb{C})$ (cf. [Sk$_3$]) but this holds for $SO(n,1)$ ([Kas$_4$]) and $SU(n,1)$ [Ju-K]).

IV. Appendix B. Real and Complex Interpolation of Banach Spaces

A very useful method of proving inequalities is given by functorial constructions of interpolation spaces $F(B_0, B_1)$ from a pair (B_0, B_1) of Banach spaces that are continuously embedded in a locally convex vector space.

The first functorial construction is *complex* interpolation [C]. For any $\theta \in [0,1]$ one defines a Banach space

$$B = [B_0, B_1]_\theta$$

as the space of values $f(\theta)$, where f is a holomorphic function in the strip $D = \{z \in \mathbb{C}; \mathrm{Re}\, z \in [0,1]\}$, with values in $B_0 + B_1$ (viewed as a Banach space with the norm $\|x\|_{B_0+B_1} = \mathrm{Inf}\{\|x_0\|_{B_0} + \|x_1\|_{B_1};\ x_0 + x_1 = x\}$. The function f is assumed continuous and bounded in the closed strip, with boundary values such that

$$f(it) \in B_0\,,\ f(1+it) \in B_1 \quad \forall t \in \mathbb{R}.$$

The norm on B is given by

$$\|x\|_\theta = \inf\{\|f\|_{0,1}\ ;\ f(\theta) = x\}$$

where

$$\|f\|_{0,1} = \sup\{\|f(it)\|_{B_0}\,,\ \|f(1+it)\|_{B_1}\ ;\ t \in \mathbb{R}\}.$$

This construction is clearly functorial in the interpolation couple (B_0, B_1), which implies:

Theorem 1. [C] *Let* (B_0, B_1) *and* (B_0', B_1') *be two interpolation couples, and let* $T : B_0 + B_1 \to B_0' + B_1'$ *be a linear operator such that* $TB_j \subset B_j'$ *and* $\|Tx\|_{B_j'} \le C_j \|x\|_{B_j}$, $\forall x \in B_j$, $j = 0,1$. *Then, for any* $\theta \in [0,1]$, T *defines a continuous linear map from* $[B_0, B_1]_\theta$ *to* $[B_0', B_1']_\theta$ *of norm less than* $C_0^{(1-\theta)} C_1^\theta$.

One also has an iteration theorem showing that, with $B_\theta = [B_0, B_1]_\theta$ and assuming that $B_0 \cap B_1$ is dense in $B_{\theta_0} \cap B_{\theta_1}$, one has

$$[B_{\theta_0}, B_{\theta_1}]_{\theta'} = [B_0, B_1]_\theta \quad \text{for} \quad \theta = (1-\theta')\theta_0 + \theta'\theta_1.$$

The simplest examples of complex interpolation spaces are:

α) If (X, μ) is a measure space, then:

$$[L^1, L^\infty]_\theta = L^p \quad \text{for} \quad 1-\theta = 1/p \; , \; p \in [1, \infty].$$

β) If $\mathcal{H}$ is a Hilbert space, then:

$$[\mathcal{L}^1(\mathcal{H}), \mathcal{K}]_\theta = \mathcal{L}^p(\mathcal{H}) \; , \; 1-\theta = 1/p \; , \; p \in [1, \infty]$$

where $\mathcal{L}^p(\mathcal{H})$ is the Schatten ideal of compact operators T in $\mathcal{H}$ such that $\mathrm{Trace}(|T|^p) < \infty$, or, equivalently, such that

$$\sum \mu_n(T)^p < \infty$$

where $\mu_n(T)$ is the n-th eigenvalue of $|T| = (T^*T)^{1/2}$.

In fact, these examples α) and β) are special cases of interpolation between $M_* = L^1(M, \tau)$ and M, where M is a semifinite von Neumann algebra with semifinite faithful normal trace τ (cf. Chapter V).

Let us now pass to *real interpolation theory* ([Lio-P]). As above, let (B_0, B_1) be an interpolation couple. For $x \in B_0 + B_1$ and $t \in \mathbb{R}_+^*$, define

$$K(t, x) = \inf\{\|x_0\|_{B_0} + t^{-1}\|x_1\|_{B_1} \; ; \; x = x_0 + x_1\}.$$

It is a continuous function of $t \in \mathbb{R}_+^*$, and by definition the real interpolation space $[B_0, B_1]_{\theta,p}$ for $\theta \in \,]0,1[\; , \; p \in [1, \infty]$, is the subspace of $X_0 + X_1$ determined by the finiteness of

$$\left(\int_0^\infty (t^\theta K(t,x))^p \frac{dt}{t}\right)^{1/p} = \|x\|_{p,\theta}.$$

For $p = \infty$ the above integral is replaced by the L^∞ norm of $t \mapsto t^\theta K(t, x)$.

Using $1/p$ in place of p, we thus get an *interpolation square.* As above it is functorial in (B_0, B_1), and one has:

Theorem 2. [Lio-P] *Let (B_0, B_1) and (B_0', B_1') be two interpolation couples, and let $T : B_0 + B_1 \to B_0' + B_1'$ be a linear operator such that $TB_j \subset B_j'$ and $\|Tx\|_{B_j'} \leq C_j \|x\|_{B_j}$, $\forall x \in B_j$, $j = 0, 1$. Then for any $\theta \in \,]0,1[\; , \; p \in [1, \infty]$, T defines a continuous linear map from $[B_0, B_1]_{\theta,p}$ to $[B_0', B_1']_{\theta,p}$ of norm less than $C_0^{(1-\theta)} C_1^\theta$.*

What is quite remarkable in real interpolation theory is that the iteration theorem applied to B_{θ_0,p_0} and B_{θ_1,p_1} gives a result independent of p_0 and p_1, thus strengthening for intermediate values of θ the inequalities proven for θ_0 and θ_1.

Let B be a Banach space such that:

$$B_0 \cap B_1 \subset B \subset B_0 + B_1.$$

Then the inclusion $[B_0, B_1]_{\theta,1} \subset B$ holds iff for some $C < \infty$,

$$\|x\|_B \leq C\|x\|_{B_0}^{1-\theta} \, \|x\|_{B_1}^{\theta} \quad \forall x \in B_0 \cap B_1.$$

The inclusion $B \subset [B_0, B_1]_{\theta,\infty}$ holds iff for some $C < \infty$,

$$t^{\theta} \, K(t,x) \leq C\|x\|_B \quad \forall x \in B.$$

If both inclusions hold, one says that B is of class $K_\theta(B_0, B_1)$.

Theorem 3. [Lio-P] *Let $\theta_0 < \theta_1$, let B_0' be of class $K_{\theta_0}(B_0, B_1)$, and let B_1' be of class $K_{\theta_1}(B_0, B_1)$. Then for any $\theta' \in]0,1[$, $p \in [1,\infty]$,*

$$[B_0', B_1']_{\theta',p} = [B_0, B_1]_{\theta,p} \, , \; \theta = (1-\theta')\theta_0 + \theta'\theta_1.$$

The combination of Theorems 2 and 3 is a remarkably powerful tool with which to prove inequalities.

Let (X, μ) be a measure space. Then the real interpolation spaces:

$$L^{(p,q)}(X,\mu) = [L^{\infty}, L^1]_{\theta,q} \, , \; p = 1/\theta \in]1,\infty[$$

are the Lorentz spaces, which are defined for any $p, q \in [1,\infty]$ as follows. For any function f on X one lets

$$\mu_f(s) = \mu\{u \in X \, ; \, |f(u)| > s\} \quad \forall s \in \mathbb{R}_+$$

$$f^*(t) = \inf\{s > 0 \, ; \, \mu_f(s) \leq t\} \quad \forall t \in \mathbb{R}_+^*.$$

Then $f \in L^{(p,q)}$ iff $\left(\int |t^{1/p} \, f^*(t)|^q \, \frac{dt}{t}\right)^{1/q} < \infty$.

For $q = \infty$ this means that $t^{1/p} \, f^*(t)$ is bounded, or in other words that for some constant $C < \infty$

$$\mu\{u \in X \, ; \, |f(u)| > s\} \leq C \, s^{-p}.$$

The Banach space norm on $L^{(p,q)}$ is obtained as

$$\|f\|_{(p,q)} = \left(\int |t^{1/p} \, f^{**}(t)|^q \, \frac{dt}{t}\right)^{1/q}$$

where $f^{**}(t) = \frac{1}{t}\int_0^t f^*(s) \, ds \quad \forall t \in \mathbb{R}_+^*$.

Similarly, let $\mathcal{H}$ be a Hilbert space and for $p \in]1,\infty[$ and $q \in [1,\infty]$ let

$$\mathcal{L}^{(p,q)} = [\mathcal{K}, \mathcal{L}^1]_{\theta,q} \, , \; p = 1/\theta.$$

These define normed ideals of compact operators in $\mathcal{H}$, and

$$T \in \mathcal{L}^{(p,q)} \text{ iff } \left(\sum_{n=1}^{\infty} (n^{1/p} \, \mu_n(T))^q \, n^{-1}\right)^{1/q} < \infty$$

where $\mu_n(T)$, for $n \in \mathbb{N}$, is the n-th eigenvalue of $|T| = (T^*T)^{1/2}$.

IV. Appendix C. Normed Ideals of Compact Operators

In this appendix we recall the standard properties of symmetrically normed ideals of compact operators ([Goh-K]), and explain the relation between the Macaev ideal and the ideal $\mathrm{Li}(\mathcal{H})$ of Section 8.

Let $\mathcal{H}$ be a Hilbert space with countable basis and $T \in \mathcal{K}$ a compact operator on $\mathcal{H}$. The characteristic values $\mu_n(T)$ are the eigenvalues of $|T|$ arranged in decreasing order and repeated according to their multiplicities. One has:

Lemma 1. (cf. [Goh-K]) a) *For any $n \in \mathbb{N}$ and $T \in \mathcal{K}$,*

$$\mu_n(T) = \inf\{\|T|E^{\perp}\| \;;\; E \text{ an } n\text{-dimensional subspace of } \mathcal{H}\}.$$

b) $\mu_n(T) = \mathrm{dist}(T, R_n)$ *where R_n is the set of operators of rank less than n and* dist *is the distance in the operator norm.*

Assertion a) is the minimax principle. For b) see [Goh-K] Theorem 2.1. This lemma has many corollaries, for instance:

$$|\mu_n(T_1) - \mu_n(T_2)| \leq \|T_1 - T_2\| \quad \forall T_1, T_2 \in \mathcal{K} \quad \forall n \in \mathbb{N}$$

$$\mu_{n+m}(T_1 + T_2) \leq \mu_n(T_1) + \mu_m(T_2) \quad \forall T_1, T_2 \in \mathcal{K} \quad \forall n, m \in \mathbb{N}$$

$$\mu_{n+m}(T_1 T_2) \leq \mu_n(T_1)\, \mu_m(T_2) \quad \forall T_1, T_2 \in \mathcal{K} \quad \forall n, m \in \mathbb{N}.$$

However, the μ_n for $n \neq 0$ are not subadditive. The natural seminorms on $\mathcal{K}$ associated to the μ_n are given by

$$\sigma_N(T) = \sum_{n=0}^{N-1} \mu_n(T) \quad \forall N \in \mathbb{N}\,,\; T \in \mathcal{K}.$$

Lemma 2. (cf. [Goh-K]) a) *For any $T \in \mathcal{K}$, $n \in \mathbb{N}$ one has*

$$\sigma_n(T) = \sup\{\|TE\|_1 \;;\; E \text{ an } n\text{-dimensional subspace of } \mathcal{H}\}$$

where $\|\;\|_1$ is the $\mathcal{L}^1$ norm: $\|S\|_1 = \mathrm{Trace}(|S|) \;\; \forall S \in \mathcal{L}^1$.

b) $\sigma_N(T_1 + T_2) \leq \sigma_N(T_1) + \sigma_N(T_2) \quad \forall T_1, T_2 \in \mathcal{K} \quad \forall N \in \mathbb{N}$.

(cf. [Goh-K] Lemma 4.2)

Definition 3. *A* symmetrically normed ideal *is a two-sided ideal J of $\mathcal{L}(\mathcal{H})$ and a norm $\|\;\|_J$ on J such that:*

a) $\|ATB\|_J \leq \|A\|\; \|T\|_J\; \|B\| \quad \forall A, B \in \mathcal{L}(\mathcal{H})\,,\; T \in J$.

b) *J is a Banach space for the norm $\|\;\|_J$.*

Note that the inclusion $J \subset \mathcal{K}$ is automatic. Also, two symmetric norms on the same ideal J are necessarily equivalent norms.

One usually requires the normalization

$$\|T\|_J = \|T\| \quad \text{for any } T \text{ of rank one.}$$

It follows from the definition that for any $T \in J$

$$\|T\|_J = \|\,|T|\,\|_J \quad \text{with } |T| = (T^*T)^{1/2}.$$

Moreover, by unitary invariance of $\| \;\|_J$, the value of $\|T\|_J$ only depends upon the list of characteristic values

$$\mu_0(T) \geq \mu_1(T) \geq \cdots \geq \mu_n(T) \geq \cdots$$

The restriction of $\| \;\|_J$ to the ideal R of finite-rank operators is thus determined by a functional Φ defined on the cone Δ of decreasing sequences $(\xi_n)_{n\in\mathbb{N}}$ of positive real numbers such that $\xi_n = 0$ for n large enough.

The corresponding functionals are called symmetrically norming functions and are characterized by the following properties: ([Goh-K])

α) $\Phi(\xi) > 0$ for $\xi \in \Delta$, $\xi \neq 0$.

β) $\Phi(\lambda\xi) = \lambda\Phi(\xi) \quad \forall \xi \in \Delta, \lambda > 0$.

γ) $\Phi(\xi + \eta) \leq \Phi(\xi) + \Phi(\eta) \quad \forall \xi, \eta \in \Delta$.

δ) Let $\xi, \eta \in \Delta$ be such that $\sum_{j=0}^{n} \xi_j \leq \sum_{j=0}^{n} \eta_j \quad \forall n \in \mathbb{N}$. Then $\Phi(\xi) \leq \Phi(\eta)$.

One can also require the normalization:

$$\Phi(1, 0, \ldots, 0) = 1.$$

Theorem 4. [Goh-K] *Let Φ be a functional on Δ satisfying α) β) γ) and δ). Then $J_\Phi = \{T \in \mathcal{K} \,;\, \sup_N \Phi(\mu^N(T)) < \infty\}$ is a symmetrically normed ideal, where $\mu^N(T) = (\mu_0(T), \mu_1(T), \ldots, \mu_N(T), 0, 0, \ldots) \in \Delta$ for any $T \in \mathcal{K}$. Moreover*

$$\|T\|_\Phi = \sup_N \Phi(\mu^N(T)).$$

Note that the sequence $\Phi(\mu^N(T))$ is nondecreasing. Conversely, for any symmetrically normed ideal J there exists a unique symmetrically norming function Φ such that the restriction of $\| \;\|_J$ to the ideal of finite-rank operators is $\| \;\|_\Phi$:

$$\|T\|_J = \|T\|_\Phi \quad \forall T \text{ of finite rank.}$$

This, however, does not imply that $J = J_\Phi$, and this nuance played a crucial role in Section 3 of this chapter. For any symmetrically norming function Φ, let J_Φ^0 be the closure in J_Φ of the ideal of finite-rank operators. Then one can assert that

$$J_\Phi^0 \subset J$$

for any symmetrically normed ideal J.

It is, however, not true, in general, that $J_\Phi^0 = J_\Phi$. In Section 3 we saw, for instance, that:

$$\mathcal{L}_0^{(p,\infty)} \neq \mathcal{L}^{(p,\infty)} \quad \forall p \in [1,\infty[.$$

In fact, one can show (cf. [Goh-K] Corollary 6.1) that $J_\Phi = J_\Phi^0$ iff J_Φ is a separable Banach space.

Theorem 5. [Goh-K] *Any separable symmetrically normed ideal J is of the form J_Φ^0 for a symmetrically norming function Φ, uniquely determined by the equality $\| \ \|_J = \| \ \|_\Phi$.*

For a given symmetrically norming function Φ one defines the dual norm Φ' by

$$\Phi'(\xi) = \sup\left\{\sum_{i=1}^{\infty} \xi_i\, \eta_i \ ; \ \eta \in \Delta\, ,\ \Phi(\eta) \le 1\right\}.$$

Then, provided $J_\Phi \neq \mathcal{K}$, one has the duality

$$(J_\Phi^0)^* = J_{\Phi'}.$$

This duality is given by the pairing

$$\langle A, B\rangle = \operatorname{Trace}(AB) \in \mathbb{C},$$

and uses as an essential ingredient the inequality (cf. [Goh-K])

$$\sum_{n=0}^{N} \mu_n(AB) \le \sum_{n=0}^{N} \mu_n(A)\, \mu_n(B) \quad \forall A, B \in \mathcal{K}.$$

As an example, let (π_n) be a decreasing sequence of positive real numbers, $\pi_n \to 0$ when $n \to \infty$, and let

$$\Phi_\pi(\xi) = \sup_n \ \Big(\sum_{j=1}^{n} \xi_j\Big) \Big/ \Big(\sum_{j=1}^{n} \pi_j\Big) \quad \forall \xi \in \Delta.$$

Then Φ_π is a symmetrically norming function whose dual is

$$\Phi'_\pi(\xi) = \sum_{n=1}^{\infty} \xi_n\, \pi_n \quad \forall \xi \in \Delta.$$

Let us now list a number of symmetrically normed ideals with the corresponding norms.

$$\mathcal{L}^{(p,q)}\ , \ \|T\|_{(p,q)} = \left(\sum_{N=1}^{\infty} N^{(1/p-1)q-1}\ \sigma_N(T)^q\right)^{1/q}.$$

We assume here $p \in\ ]1,\infty[$ and $q \in [1,\infty]$. For $q < \infty$ one has $\mathcal{L}^{(p,q)} = \mathcal{L}_0^{(p,q)}$, i.e. the finite-rank operators are dense in $\mathcal{L}^{(p,q)}$, but for $q = \infty$ this is no longer the case, and one has the dualities

$$\left(\mathcal{L}_0^{(p,\infty)}\right)^* = \mathcal{L}^{(p',1)}, \quad 1/p + 1/p' = 1$$

$$\left(\mathcal{L}^{(p',1)}\right)^* = \mathcal{L}^{(p,\infty)}.$$

For $p, q \in]1, \infty[$ one has the duality

$$\left(\mathcal{L}^{(p,q)}\right)^* = \mathcal{L}^{(p',q')}, \quad 1/p + 1/p' = 1\,, \; 1/q + 1/q' = 1.$$

For $p = 1$ and $q = \infty$ one lets

$$\|T\|_{(1,\infty)} = \sup_{N \geq 2} \frac{\sigma_N(T)}{\log N},$$

and we denote the corresponding normed ideals by $\mathcal{L}^{(1,\infty)}$ and $\mathcal{L}_0^{(1,\infty)}$:

$$\mathcal{L}^{(1,\infty)} = \{T \in \mathcal{K}\,;\, \sigma_N(T) = O(\log N)\}$$

$$\mathcal{L}_0^{(1,\infty)} = \{T \in \mathcal{K}\,;\, \sigma_N(T) = o(\log N)\}.$$

The dual space of $\mathcal{L}_0^{(1,\infty)}$ is the Macaev ideal

$$\mathcal{L}^{(\infty,1)} = \{T \in \mathcal{K}\,;\, \sum_{n=1}^{\infty} n^{-1}\, \mu_n(T) < \infty\}.$$

The dual space of $\mathcal{L}^{(\infty,1)}$ is $\mathcal{L}^{(1,\infty)}$.

The natural norm on $\mathcal{L}^{(\infty,1)}$ is given by $\sum_{N=1}^{\infty} N^{-2}\, \sigma_N(T)$, but it is clearly equivalent to the other norm given by $\sum_{n=1}^{\infty} n^{-1}\, \mu_n(T)$.

Finally we let, as in Section 8, $\mathrm{Li}(\mathcal{H})$ be the symmetrically normed ideal

$$\mathrm{Li}(\mathcal{H}) = \{T \in \mathcal{K}\,;\, \mu_n(T) = O((\log n)^{-1})\}$$

with the symmetric norm

$$\|T\|_{\mathrm{Li}} = \sup_{N>1}\, \mathrm{Li}(N)^{-1}\, \sigma_N(T).$$

One checks that $\mathrm{Li}_0(\mathcal{H}) = \{T \in \mathcal{K}\,;\, \mu_n(T) = o(\log n)^{-1}\}$. We shall show that the ideal $\mathrm{Li}(\mathcal{H})$ plays with respect to the Macaev ideal $\mathcal{L}^{(\infty,1)}$ the same role as the ideal $\mathcal{L}^{(p,\infty)}$ plays with respect to $\mathcal{L}^{(p,1)}$ for $p < \infty$ (cf. Section 2). One has:

Lemma 6. *Let $f \in C_c^\infty(\mathbb{R})$ be a smooth even function on $\mathbb{R}$ with compact support and let D be a selfadjoint invertible unbounded operator in a Hilbert space $\mathcal{H}$. There exists a finite constant $c = c_f$ such that, for any $\varepsilon > 0$,*

$$\|[f(\varepsilon D), a]\|_{(\infty,1)} \leq c\|[D,a]\|\; \|D^{-1}\|_{\mathrm{Li}}$$

for any $a \in \mathcal{L}(\mathcal{H})$ such that $[D,a]$ is bounded.

The proof is the same as that of Lemma 2.13. As in that lemma one easily gets for any operator a,

$$\|[f(\varepsilon D), a]\| \leq \varepsilon \|\widehat{f'}\|_1\; \|[D,a]\|.$$

Moreover, if Supp $f \subset [-k, k]$, the rank of $f(\varepsilon D)$ is bounded above by the number of eigenvalues of $|D|^{-1}$ larger than εk^{-1}. There exists a universal constant $\lambda < \infty$ such that

$$\mu_n(D^{-1}) \leq \lambda(\log n)^{-1} \|D^{-1}\|_{\text{Li}} \text{ for } n > 1,$$

and it follows that

$$\log(\text{rank}(f(\varepsilon D))) \leq \varepsilon^{-1} k\lambda \|D^{-1}\|_{\text{Li}}.$$

The operator $[f(\varepsilon D), a] = T$ has rank $N \leq 2 \, \text{rank}(f(\varepsilon D))$ and it is of norm $\leq \varepsilon \|\widehat{f'}\|_1 \, \|[D, a]\|$. Thus, its Macaev norm

$$\sum_{n=1}^{\infty} n^{-1} \mu_n(T)$$

is at most $(\log N) \, \varepsilon \|\widehat{f'}\|_1 \, \|[D, a]\|$, which by the above estimate concludes the proof.

IV. Appendix D. The Chern Character of Deformations of Algebras

The main theme of this chapter was the construction of the Chern character in K-homology:

$$\text{ch}_* : K^*(\mathcal{A}) \to HC^*(\mathcal{A})$$

where we used Fredholm modules $(\mathcal{H}, F)$ as the basic cycles for K-homology. We have seen in Chapter II Appendix B a variant of the Kasparov bivariant theory based on deformations of algebras. If we specialize E-theory to E-homology, $E(A, \mathbb{C})$, we obtain as our basic cycles the deformations of an algebra A to the algebra $\mathcal{K}$ of compact operators in Hilbert space. (To be more careful one would have to stabilise both A and $\mathcal{K}$ and replace them by SA and $S\mathcal{K}$ but we shall ignore this point.) It is thus natural to extend the construction of the Chern character ch_*, done up to now in K-homology terms, to the case of deformations.

This has been done in [Co-F-S] for closed $*$-products, i.e. for deformations of the algebra of functions on symplectic manifolds. We shall describe this construction in this appendix. It is thus restricted to the special case $\mathcal{A}$ *commutative*, but should extend to the general case using the results of [Ger-S].

We let W be a symplectic smooth manifold. We let ω be the closed symplectic 2-form on W, and let $\widetilde{\omega} \in C^\infty(W, \wedge^2 T)$ be the 2-vector field on W dual to ω. The Poisson bracket on $C^\infty(W)$ is given by:

$$P(f, g) = \{f, g\} = i(\widetilde{\omega})(df \wedge dg) \quad \forall f, g \in C^\infty(W).$$

By construction, P is a Hochschild 2-cocycle:

$$P \in Z^2(\mathcal{A}, \mathcal{A})$$

where $\mathcal{A} = C_c^\infty(W)$ and $\mathcal{A}$ is viewed as a bimodule over $\mathcal{A}$. For each n we let $C^n(\mathcal{A},\mathcal{A})$ and $C^n(\mathcal{A},\mathcal{A}^*)$ be the spaces of cochains continuous in the C^∞ topology.

Definition 1. *A $*$-product is an associative bilinear product on the space $\mathcal{A}[[\nu]]$ of formal power series over $\mathcal{A}$, of the form*

$$f * g = \sum_{r,k,j=0}^{\infty} \nu^{r+k+j}\, C_r(f_k, g_j)$$

for any

$$f = \sum \nu^k f_k\,,\ g = \sum \nu^j g_j \ \text{ in } \mathcal{A}[[\nu]]$$

where $C_r \in C^2(\mathcal{A},\mathcal{A})$ for all r, and $C_0(f,g) = fg$, $C_1 = P$.

Such $*$-products exist on arbitrary symplectic manifolds ([Wil-L]). We shall now consider a more restricted class of deformations, the idea being that if the above formal deformation corresponds to a deformation of $\mathcal{A}$ to an algebra of compact operators in Hilbert space, the canonical trace on $\mathcal{K}$ should yield a trace on the deformed algebra given by

$$\tau(f) = \int f\, \omega^\ell, \quad \ell = \frac{1}{2}\dim(\mathrm{W}).$$

By construction, $\tau(f)$ takes its values in $\mathbb{C}[[\nu]]$, and we shall in fact only need the first ℓ terms of its expansion:

Definition 2. [Co-F-S] *A $*$-product is* closed *if*

$$\int C_r(f,g)\omega^\ell = \int C_r(g,f)\omega^\ell \quad \forall r \le \ell\,;\ f,g \in C_c^\infty(W).$$

This is equivalent to the requirement that

$$\tau_\ell(f) = \text{ Coefficient of } \nu^\ell \text{ in } \int f\,\omega^\ell$$

defines a *trace* on the algebra $\mathcal{A}[[\nu]]$.

If the condition of Definition 2 holds for any value of r, we shall say that the $*$-product is strongly closed.

Theorem 3. [Om-M-Y] *On every symplectic manifold W there exists at least one closed $*$-product.*

The Moyal $*$-product on $W = \mathbb{R}^{2\ell}$ is defined by

$$C_r = (r!)^{-1}\, P^r,$$

using the r-th powers of the bidifferential operator P. It corresponds to the Weyl pseudodifferential calculus and is strongly closed.

Similarly, for $W = T^*M$, the cotangent space of a Riemannian manifold, the $*$-product given by the pseudodifferential calculus with an asymptotic parameter (cf. [Wi$_2$]) is strongly closed.

Finally, note that it is easy to give examples of star products which are not closed. Indeed, using the symplectic measure $\int \cdot\, \omega^\ell$ we get a natural $\mathcal{A}$-bimodule map from $\mathcal{A}$ to $\mathcal{A}^*$:

$$f \in \mathcal{A} \to \tilde{f} \in \mathcal{A}^* \,,\; \langle \tilde{f}, g\rangle = \int fg\, \omega^\ell,$$

and a corresponding map:

$$C \in C^n(\mathcal{A},\mathcal{A}) \mapsto \tilde{C} \in C^n(\mathcal{A},\mathcal{A}^*)\,,$$
$$\tilde{C}(f^0, f^1, \ldots, f^n) = \int_W C(f^1,\ldots,f^n) f^0\, \omega^\ell \quad \forall f^j \in C_c^\infty(W).$$

Lemma 4. *For a closed $*$-product one has $B\tilde{C}_2 = 0$.*

Here $B : C^2(\mathcal{A},\mathcal{A}^*) \to C^1(\mathcal{A},\mathcal{A}^*)$ is the B operator of cyclic cohomology (Chapter III Section 1).

Proof. One has $(B_0\tilde{C}_2)(f,g) = \int C_2(f,g)\omega^\ell$, and $AB_0\tilde{C}_2 = 0$ if and only if $\int C_2(f,g)\omega^\ell = \int C_2(g,f)\omega^\ell \;\; \forall f,g \in \mathcal{A}$. In fact, one can see that, proceeding step by step from the knowledge of the C_k, $k \le t$, defining a star product closed up to order t, the obstruction to the existence of C_{t+1} yielding a $*$-product closed to order $t+1$ is given by

$$(\ker b \cap \ker B)/b(\ker B) = HC^3(\mathcal{A}).$$

We refer to [Co-F-S] for this point, as well as for the discussion of equivalence classes of closed $*$-products.

Let us now pass to the construction of the Chern character of the closed $*$-products, as an element of the cyclic cohomology:

$$\mathrm{ch}_*((C_\ell)) \in HC^*(\mathcal{A}).$$

Theorem 5. [Co-F-S] *Let W be a symplectic manifold of dimension 2ℓ, and let $* = \sum \nu^r\, C_r$ be a closed $*$-product. Then the following equality defines the components of a 2ℓ-dimensional cyclic cocycle in the (b,B) bicomplex on $\mathcal{A} = C_c^\infty(W)$:*

$$\varphi_{2k}(f^0,\ldots,f^{2k}) = \tau_\ell(f^0 * \theta(f^1,f^2) * \cdots * \theta(f^{2k-1},f^{2k}))$$

$\forall f^j \in \mathcal{A}$, $k = 0,1,\ldots,\ell$, where $\theta(f,g) = f * g - fg$ for any $f,g \in \mathcal{A}$, and τ_ℓ is, as above, the coefficient of ν^ℓ in $\int f\, \omega^\ell \in \mathbb{C}[[\nu]]$ for $f \in \mathcal{A}[[\nu]]$.

Thus, the φ_{2k} satisfy $b\varphi_{2k} + B\varphi_{2k+2} = 0 \;\; \forall k = 0,\ldots,\ell$. They vanish, by construction, for $k > \ell$. They also vanish for $0 \le 2k < \ell$, so that the relevant components are the φ_{2k} for $\ell \le 2k \le 2\ell$.

The component $\varphi_{2\ell}$ is always given by

$$\varphi_{2\ell}(f^0,\dots,f^{2\ell}) = \int f^0\, df^1 \wedge df^2 \wedge \cdots \wedge df^{2\ell} \quad \forall f^j \in \mathcal{A}.$$

A simple computation in the 4-dimensional case, $\dim W = 4$, i.e. $\ell = 2$, shows that φ_2 does not vanish in general, and is given by:

$$\varphi_2(f^0,f^1,f^2) = \int f^0\, C_2(f^1,f^2)\omega^2 \quad \forall f^0,f^1,f^2 \in \mathcal{A},$$

i.e. $\varphi_2 = \tilde{C}_2$.

This example shows that a closed $*$-product does not, in general, come from a quantization, i.e. from a deformation of $\mathcal{A}$ to the algebra of compact operators in Hilbert space. Indeed, a necessary condition for that to hold is the *integrality* of the character:

$$\langle \varphi, K_0(\mathcal{A})\rangle \subset \mathbb{Z}.$$

Since in the above example one can add to $\tilde{C}_2$ an arbitrary cyclic 2-cocycle, the above integrality condition does not hold in general.

As another example let M be a compact Riemannian manifold, $W = T^*M$ its cotangent space, and $*$ the $*$-product associated to the pseudodifferential calculus ([Wi]). It is a closed $*$-product coming from the deformation of $C_c^\infty(W)$ to the algebra of compact operators given by the tangent groupoid of M (Chapter II). In particular, it is integral, computes the Atiyah-Singer index map (Chapter II Section 5), and using invariant theory ([At-B-P], [Gi_1]) one computes its character:

$$\mathrm{ch}_*(C_c^\infty(T^*M), *) = \mathrm{Td}(T^*M)$$

as the Todd genus of the symplectic manifold T^*M ([Co-F-S]) viewed as a de Rham current on T^*M.

V

OPERATOR ALGEBRAS

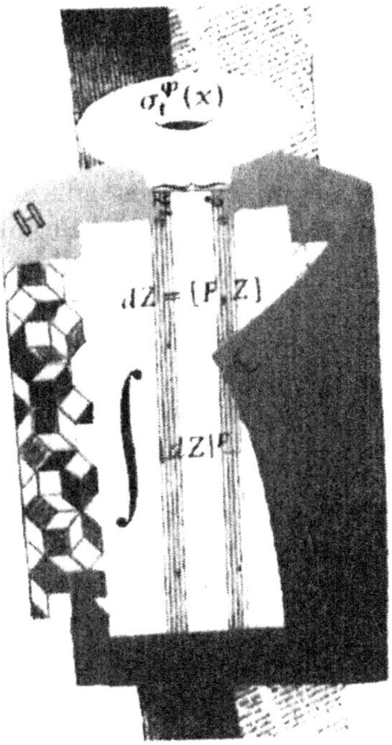

The theory of operator algebras, or von Neumann algebras (according to the terminology of J. Dieudonné), is the noncommutative analogue of measure theory. By the same token, this theory is an essential ingredient in the analysis of noncommutative spaces, which we have already described above. It is remarkable that one can describe for such spaces the associated von Neumann algebra, as well as the most general states and normal weights on that algebra, in terms of random operators. This gives the general theory presented below an inexhaustible source of concrete and explicit examples, in which the general theorems can display their full force.

In contrast with the commutative case, where the only interesting measure space, up to isomorphism, is the Lebesgue space, the noncommutative case is far more complex. Nevertheless, we now have a complete classification of the amenable von Neumann algebras, a class that is described by numerous equivalent properties (cf. Sections 7 and 9).

In a certain sense, the theory of von Neumann algebras is a chapter of linear algebra in infinite dimensions, i.e. in infinite-dimensional separable Hilbert space. It is therefore necessary at the outset to have some familiarity with the general methods of functional analysis. Perhaps the first step consists in understanding the distinction between a von Neumann algebra, which is a $*$-subalgebra of the algebra $\mathcal{L}(\mathfrak{H})$ of operators on a Hilbert space $\mathfrak{H}$ that is *closed for the weak topology*, and a C^*-algebra, or stellar algebra, which is a *norm-closed* $*$-subalgebra of $\mathcal{L}(\mathfrak{H})$. Of course every von Neumann algebra is, in particular, a C^*-algebra, but it is not a very interesting C^*-algebra since it is not *separable for the norm*. (For X compact, $C(X)$ is separable $\iff$ X is metrizable.)

V.1. The Papers of Murray and von Neumann

Let $\mathfrak{H}$ be a complex Hilbert space and let $\mathcal{L}(\mathfrak{H})$ be the C^*-algebra of bounded operators from $\mathfrak{H}$ into $\mathfrak{H}$. This is a Banach algebra equipped with the norm

$$\|T\| = \sup_{\|\xi\| \leq 1} \|T\xi\|$$

and with the involution $T \mapsto T^*$ defined by

$$\langle T^*\xi, \eta \rangle = \langle \xi, T\eta \rangle \quad \forall \xi, \eta \in \mathfrak{H}.$$

Even if $\mathfrak{H}$ has countable dimension, the Banach space $\mathcal{L}(\mathfrak{H})$ is not separable. There exists a unique closed linear subspace of the dual $\mathcal{L}(\mathfrak{H})^*$ whose dual is $\mathcal{L}(\mathfrak{H})$; this is the space of linear forms on $\mathcal{L}(\mathfrak{H})$ that may be written

$$L(T) = \text{Trace}\,(\rho T) \quad \forall T \in \mathcal{L}(\mathfrak{H}),$$

where ρ is a trace-class operator, which means that $|\rho| = (\rho^*\rho)^{1/2}$ satisfies

$$\sum \langle |\rho| \xi_i, \xi_i \rangle = \text{Trace}\,|\rho| < \infty$$

for every orthonormal basis (ξ_i) of $\mathfrak{H}$. The norm of the linear form L is equal to $\text{Trace}\,|\rho|$, and, equipped with this norm, the space

$$\mathcal{L}(\mathfrak{H})_* = \{\rho \in \mathcal{L}(\mathfrak{H});\ \text{Trace}|\rho| < \infty\}$$

is a Banach space, called the *predual* of the Banach space $\mathcal{L}(\mathfrak{H})$.

When $\mathfrak{H}$ has countable dimension, the Banach space $\mathcal{L}(\mathfrak{H})_*$ is separable; the duality between $\mathcal{L}(\mathfrak{H})_*$ and $\mathcal{L}(\mathfrak{H})$ is exactly analogous to the duality between $\ell^1(A)$ and $\ell^\infty(A)$, where A is a set. In particular, whenever $\mathfrak{H}$ is infinite-dimensional, the space $\mathcal{L}(\mathfrak{H})_*$ is not reflexive and the topology $\sigma(\mathcal{L}(\mathfrak{H}), \mathcal{L}(\mathfrak{H})_*)$, called the *ultraweak topology* on $\mathcal{L}(\mathfrak{H})$ [Di$_2$], is not the same as the norm topology of $\mathcal{L}(\mathfrak{H})$. Thus, it is much more restrictive for a subspace of $\mathcal{L}(\mathfrak{H})$ to be closed for $\sigma(\mathcal{L}(\mathfrak{H}), \mathcal{L}(\mathfrak{H})_*)$ than for the norm topology.

This distinction between the two topologies is essential for the sequel. Let M be a $*$-subalgebra of $\mathcal{L}(\mathfrak{H})$ containing the identity operator 1. The following conditions are equivalent:

1) *Topological condition:* M is $\sigma(\mathcal{L}(\mathfrak{H}), \mathcal{L}(\mathfrak{H})_*)$-closed.
2) *Algebraic condition:* M is equal to the commutant $(M')'$ of its commutant M'. (The commutant of a subset S of $\mathcal{L}(\mathfrak{H})$ is defined by the equality $S' = \{T \in \mathcal{L}(\mathfrak{H});\ TA = AT\ \forall A \in S\}$.)

The equivalence between these two properties is von Neumann's double commutant theorem.

Definition 1. *A* von Neumann algebra *on $\mathfrak{H}$ is a $*$-subalgebra of $\mathcal{L}(\mathfrak{H})$ containing the identity operator and satisfying the above equivalent conditions.*

We cite some immediate consequences of the definition.

Let $S \subset \mathcal{L}(\mathfrak{H})$ be a subset such that $S = S^* = \{T^*;\ T \in S\}$; then the commutant S' of S is a von Neumann algebra.

Let M be a von Neumann algebra on $\mathfrak{H}$ and let $M_1 = \{T \in M;\ \|T\| \le 1\}$ be the unit ball of M; then, since M_1 is $\sigma(\mathcal{L}(\mathfrak{H}), \mathcal{L}(\mathfrak{H})_*)$-closed in the unit ball of the dual of $\mathcal{L}(\mathfrak{H})_*$, it is *compact* in the topology $\sigma(M_1, \mathcal{L}(\mathfrak{H})_*)$. In particular, M is the dual of a Banach space M_*, which is in fact the unique (closed) linear subspace of M^* whose dual is M ([S$_3$]).

One introduces, often with the name of weak topology, the topology on $\mathcal{L}(\mathfrak{H})$ arising from the duality with the subspace of $\mathcal{L}(\mathfrak{H})_*$ formed by the operators of finite rank, i.e. the topology characterized by

$$T_\alpha \to T \iff \langle T_\alpha \xi, \eta \rangle \to \langle T\xi, \eta \rangle \quad \forall \xi, \eta \in \mathfrak{H}.$$

This topology is coarser than $\sigma(\mathcal{L}(\mathfrak{H}), \mathcal{L}(\mathfrak{H})_*)$, and, since the space of operators of finite rank is norm-dense in $\mathcal{L}(\mathfrak{H})_*$, it coincides with $\sigma(\mathcal{L}(\mathfrak{H}), \mathcal{L}(\mathfrak{H})_*)$ on the bounded subsets of $\mathcal{L}(\mathfrak{H})$. However, the term "weak topology" is not a good one; what we have in fact is the topology of pointwise weak convergence.

The von Neumann algebras are the $*$-subalgebras of $\mathcal{L}(\mathfrak{H})$ containing 1 that are closed for the topology of pointwise weak convergence, since every commutant S', $S \subset \mathcal{L}(\mathfrak{H})$, has this property.

1.α Examples of von Neumann algebras

1. Abelian von Neumann algebras. The description of this simple example will allow us to give the abstract form of the spectral theorem and the Borel functional calculus. Let $(X, \mathcal{B}, \mu)$ be a standard Borel space equipped with a probability measure μ, and let $\pi(L^\infty(X,\mu))$ be the algebra of operators from $L^2(X,\mu)$ to $L^2(X,\mu)$ defined by

$$\pi(f)g = fg \quad (f \in L^\infty, g \in L^2).$$

Then $M = \pi(L^\infty)$ is a commutative von Neumann algebra, and in fact $M = M'$. The predual of M is the space $L^1(X,\mu)$. Let $x \mapsto n(x)$ be a Borel function from X into $\{1, 2, \ldots, \infty\}$ and let $(\tilde{X}, p)$ be the Borel covering of X defined by

$$\tilde{X} = \{(x,j) \in X \times \mathbb{N};\ 1 \le j \le n(x)\}, \quad p(x,j) = x.$$

The "multiplicity" function n is associated with the representation π_n of $L^\infty(X,\mu)$ on $L^2(\tilde{X},\tilde{\mu})$, where $\tilde{\mu}$ is given by $\int f(x,j)d\tilde{\mu} = \int \sum_j f(x,j)d\mu$, defined by

$$\pi_n(f)g = (f \circ p)g.$$

The image $M = \pi_n(L^\infty(X,\mu))$ is a commutative von Neumann algebra on the space $\mathfrak{H} = L^2(\tilde{X},\tilde{\mu})$, and if $n \not\equiv 1$ then $M' \not\subset M$.

Definition 2. *Let M_i be a von Neumann algebra on $\mathfrak{H}_i$ $(i = 1, 2)$; we say that M_1 is* spatially isomorphic *to M_2 if there exists a unitary $U : \mathfrak{H}_1 \to \mathfrak{H}_2$ such that $UM_1U^* = M_2$.*

If $\mathfrak{H}$ is separable then every commutative von Neumann algebra on $\mathfrak{H}$ is spatially isomorphic to $\pi_n(L^\infty(X,\mu))$ for a suitable space (X,μ) and function n.

Suppose then that $T \in \mathcal{L}(\mathfrak{H})$ is a normal operator ($TT^* = T^*T$); let M be the von Neumann algebra generated by T. One can take $(X,\mathcal{B})$ to be the spectrum $K \subset \mathbb{C}$ of T, and the measure class μ given by the spectral measure of T. The mapping that associates, to the restriction of a polynomial function $\sum a_{ij} z^i \bar{z}^j$ to K, the operator $\sum a_{ij} T^i T^{*j}$, may be extended to an isomorphism π of $L^\infty(K,\mu)$ onto M with $\pi(z) = T$. Thus, for every bounded Borel function f on $\operatorname{Sp} T$, $f(T)$ has meaning and we have

$$(\lambda_1 f_1 + \lambda_2 f_2)(T) = \lambda_1 f_1(T) + \lambda_2 f_2(T),$$
$$(f_1 f_2)(T) = f_1(T) f_2(T),$$
$$(f \circ g)(T) = f(g(T)).$$

The function $x \to n(x)$ of K into $\{1,\ldots,\infty\}$ is unique modulo μ. It expresses the multiplicity of the point x in the spectrum K. The theorem on the bicommutant shows that every operator that doubly commutes with T is a bounded Borel function of T. Finally, let N be a von Neumann algebra, not necessarily commutative, and let $T \in N$; then $T = T_1 + iT_2$, $T_j = T_j^*$; since T_j is normal and every $f(T_j)$, for f a Borel function, is also in N, we see that N is generated by the projections that it contains.

2. The commutant of a unitary representation. Let G be a group and let π be a unitary representation of G on a Hilbert space $\mathfrak{H}_\pi$ (or, more generally, let $\mathcal{A}$ be a $*$-algebra and let π be a nondegenerate $*$-representation of $\mathcal{A}$). The commutant

$$R(\pi) = \{T \in \mathcal{L}(\mathfrak{H}_\pi);\ T\pi(g) = \pi(g)T\ \forall g \in G\}$$

is, by construction, a von Neumann algebra.

The interest in $R(\pi)$ comes from the following proposition:

Proposition 3. a) *Let $E \subset \mathfrak{H}$ be a closed subspace and P the corresponding projection; then*

$$E \text{ reduces } \pi \iff P \in R(\pi).$$

b) *Let E_1 and E_2 be two closed subspaces reducing π; then the reduced representations π^{E_j} are equivalent if and only if*

$$P_1 \sim P_2 \text{ in } R(\pi),$$

i.e. there is a $U \in R(\pi)$ such that $U^*U = P_1$, $UU^* = P_2$.

c) *π^{E_1} is disjoint from π^{E_2} if and only if there exists a projection P in the center of $R(\pi)$ such that $PP_1 = P_1$ and $(1-P)P_2 = P_2$.*

It is natural to say that the representation π is *isotypic* if it does not have two disjoint subrepresentations. This is equivalent to saying that $R(\pi)$ is a factor in the following sense:

Definition 4. *A* factor *M is a von Neumann algebra whose center reduces to the scalars* $\mathbb{C}$.

Another immediate corollary of the proposition and of the Borel functional calculus is the following:

For π *to be irreducible, it is necessary and sufficient that* $R(\pi) = \mathbb{C}$.

3. Finite-dimensional von Neumann algebras. Let M be a finite-dimensional von Neumann algebra. Forgetting the Hilbert space on which it is represented, regard it as a semisimple algebra over the algebraically closed field $\mathbb{C}$. It is then the direct sum of a finite number of matrix algebras:

$$M = \bigoplus_{k=1}^{q} M_{n_k}(\mathbb{C}).$$

Here the sign "=" corresponds to the following definition:

Definition 5. *Let* M_i *be a von Neumann algebra on the space* $\mathfrak{H}_i$ $(i = 1,2)$. *We say that* M_1 *is* algebraically isomorphic *to* M_2 *if there exists an algebra isomorphism* θ *of* M_1 *onto* M_2 *such that* $\theta(x^*) = \theta(x)^*$ $(\forall x \in M_1)$.

Let G be a group and let π be a unitary representation of G on a finite-dimensional Hilbert space $\mathfrak{H}_\pi$. Then $R(\pi)$ is finite-dimensional and to the decomposition $M = \bigoplus_{k=1}^{q} M_{n_k}(\mathbb{C})$ there corresponds the decomposition of π into isotypic components π_k of multiplicity n_k.

4. The action of a discrete group on a manifold. Let V be a manifold and let Γ be a discrete group acting on V by diffeomorphisms. We make the following hypothesis:

$(*)$ *For every* $g \in \Gamma$, $g \neq 1$, *the set* $\{x \in V;\, gx = x\}$ *is negligible.*

Let X be the set V/Γ and let $p : V \to X$ be the quotient mapping. Let $\mathfrak{H}_\alpha = \ell^2(p^{-1}(\alpha))$ for all $\alpha \in X$. By construction, $\mathfrak{H}_\alpha$ has for orthonormal basis the e_x, for $x \in V$ such that $p(x) = \alpha$.

Definition 6. *A* random operator $A = (A_\alpha)_{\alpha \in X}$ *is a family of bounded operators* $A_\alpha \in \mathcal{L}(\mathfrak{H}_\alpha)$ *such that the function*

$$a(x,y) = \langle A_{p(x)} e_x, e_y \rangle \quad \forall x,y \in V,\ p(x) = p(y)$$

is measurable on $V \times V$.

We write $\|A\| = \operatorname{ess\,sup} \|A_\alpha\|$ and call A *bounded* if $\|A\| < +\infty$.

Proposition 7. *Let* M *be the* $*$*-algebra of equivalence classes (modulo equality almost everywhere) of bounded random operators, equipped with the operations*

$$(A+B)_\alpha = A_\alpha + B_\alpha, \quad (A^*)_\alpha = (A_\alpha)^*, \quad (AB)_\alpha = A_\alpha B_\alpha.$$

Then M is a von Neumann algebra (that is, a $*$-algebra algebraically isomorphic to a von Neumann algebra on a Hilbert space).

Moreover, the center of M may be identified with the commutative algebra $L^\infty(X)$, where X is equipped with the image of the Lebesgue class.

Note that Γ acts ergodically on V iff $L^\infty(X) = \mathbb{C}$.

1.β Reduction theory

This theory was developed by von Neumann in 1939, but first published in 1949 [N_2].

Let $\mathfrak{H}$ be a separable Hilbert space and let $\mathfrak{F}$ be the set of all factors in $\mathcal{L}(\mathfrak{H})$. There exists a Borel structure on $\mathfrak{F}$ that makes it a standard Borel space. Let $(X, \mathcal{B})$ be a standard Borel space, μ a probability measure on $(X, \mathcal{B})$, and $t \mapsto M(t)$ a Borel mapping of X into $\mathfrak{F}$. Let M be the C^*-algebra whose elements $x \in M$ are the bounded Borel sections $t \mapsto x(t) \in M(t)$, identified if they are equal μ-almost everywhere, equipped with the obvious operations and the norm

$$\|x\| = \operatorname{ess\,sup} \|x(t)\|.$$

One shows that the C^*-algebra M is a von Neumann algebra (i.e. is algebraically isomorphic to a von Neumann algebra on a suitable space), for example by considering the action of M on the space

$$L^2(X, \mu) \otimes \mathfrak{H} = L^2(X, \mu, \mathfrak{H})$$

defined by

$$(\pi(x)\xi)(t) = x(t)\xi(t) \quad (\forall \xi \in L^2(X, \mu, \mathfrak{H})).$$

The construction of M is summarized by writing

$$M = \int_X M(t) d\mu(t),$$

and M is said to be the *direct integral* of the family $(M(t))_{t \in X}$ with respect to the measure μ.

Theorem 8. *Let M be a von Neumann algebra on a separable Hilbert space. Then M is algebraically isomorphic to a direct integral of factors*

$$\int_X M(t) d\mu(t).$$

This theorem of von Neumann shows that the factors already contain what is original in all of the von Neumann algebras: they suffice to reconstruct every von Neumann algebra as a "generalized direct sum" of factors.

Let G be a group and let π be a unitary representation of G on a separable Hilbert space. To the decomposition of $R(\pi)$ as a direct integral of factors, there corresponds the decomposition of π as a direct integral of isotypic representations.

1.y Comparison of subrepresentations, comparison of projections and the relative dimension function

Let G be a group and let π be a unitary representation of G on the Hilbert space $\mathfrak{H}_\pi$. Suppose π is isotypic; it is then natural to expect, as in finite dimensions, that π is a multiple of an irreducible representation π^E that is a subrepresentation of π. In the correspondence between subrepresentations of π and projections $P \in R(\pi)$, it is easy to see that the irreducible representations correspond to the minimal projections of $R(\pi)$:

π has an irreducible subrepresentation $\Longleftrightarrow$ the factor $R(\pi)$ has a minimal projection $\Longleftrightarrow$ there exist a Hilbert space $\mathfrak{H}_1$ and an isomorphism of $\mathcal{L}(\mathfrak{H}_1)$ onto $R(\pi)$.

Every isotypic representation with an irreducible subrepresentation is a multiple of this subrepresentation. For every factor M on $\mathfrak{H}$ having a minimal projection, $\mathfrak{H}$ can be factored as a tensor product, $\mathfrak{H} = \mathfrak{H}_1 \otimes \mathfrak{H}_2$, in such a way that

$$M = \{T \otimes 1;\ \ T \text{ operating on } \mathfrak{H}_1\}.$$

However, there exist groups G having an isotypic representation π with no irreducible subrepresentation. This phenomenon does not occur if G is a real semisimple Lie group and π is continuous, or more simply if G is compact and π is continuous. However, it does occur for the regular representation of numerous discrete groups: Let Γ be a denumerable discrete group and let Δ be the union of the finite conjugacy classes of Γ. Then Δ is a normal subgroup of Γ; suppose $\Delta = \{1\}$. Let λ' be the right regular representation of Γ; then $\mathfrak{H}_{\lambda'} = \ell^2(\Gamma)$ is the Hilbert space with $(\varepsilon_g)_{g\in\Gamma}$ as orthonormal basis, and one sets

$$\lambda(g)\varepsilon_k = \varepsilon_{gk}, \quad \lambda'(g)\varepsilon_k = \varepsilon_{kg^{-1}}.$$

One shows that the von Neumann algebra $R(\lambda')$ is generated by the operators $\lambda(g)$ $(g \in \Gamma)$, and that the vector ε_1 is cyclic and separating for $R(\lambda')$: $R(\lambda')\varepsilon_1$ and $R(\lambda')'\varepsilon_1$ are dense in $\mathfrak{H}$. The coordinates of $T\varepsilon_1$, where $T \in R(\lambda') \cap R(\lambda')'$, are constant on the conjugacy classes of Γ; since $\Delta = \{1\}$, it follows that $T\varepsilon_1 \in \mathbb{C}\varepsilon_1$ and, since ε_1 is separating, $T \in \mathbb{C}$.

Thus, $R(\lambda')$ is a factor and λ' is isotypic. To see that $R(\lambda')$ is not isomorphic to a factor $\mathcal{L}(\mathfrak{H})$, one makes use of the following concept:

Definition 9. *A trace τ on a factor M is a positive linear form such that $\tau(AB) = \tau(BA)$ for all $A, B \in M$.*

There exists a nonzero trace on $\mathcal{L}(\mathfrak{H})$ only if $\mathfrak{H}$ is finite-dimensional, as one sees by verifying that if $\mathfrak{H}$ is infinite-dimensional then every element of $\mathcal{L}(\mathfrak{H})$ is a sum of commutators. Now, one can define τ on $R(\lambda')$ by

$$\tau(A) = \langle A\varepsilon_1, \varepsilon_1\rangle \quad (\forall A \in R(\lambda')).$$

The property $\tau(AB) = \tau(BA)$ may be verified for A and B of the form $\lambda(g)$ $(g \in \Gamma)$ and may be deduced, in general, by bilinearity and continuity. Since

$\tau(1) = 1$, a nonzero trace on the infinite-dimensional factor $R(\lambda')$ has been constructed.

Thus, there exist infinite-dimensional factors that have no minimal projection.

Let M be a factor. As in Proposition 3, translation of the concepts of equivalent representations and the direct sum of representations into the language of projections yields the following concepts:

1) For a projection P, $P \in M$, one denotes by $[P]$ the equivalence class of P for the relation $P_1 \sim P_2$, which holds if and only if there exists $U \in M$ with $U^*U = P_1$, $UU^* = P_2$.
2) For P_1 and P_2 such that $P_1P_2 = 0$, one denotes by $[P_1] + [P_2]$ the class of $P_1 + P_2$; it depends only on the classes of P_1 and P_2.

The hypothesis that M is a factor implies that the set of classes of projections is totally ordered for the relation $[P_1] \le [P_2]$ defined by the condition that $P_1' \le P_2'$ for suitable representatives. This totally ordered set has a partially defined law of composition making it possible to give meaning to an equality such as $[P] = \frac{n}{m}[Q]$ with n and m positive integers.

Definition 10. *A projection $P \in M$ is said to be* finite *if $Q \sim P$ and $Q \le P$ imply $Q = P$.*

This property depends only on the class of P. In the language of representations, one would adopt the following definition: a representation π is finite if every subrepresentation π^E equivalent to π is equal to π. If π is not finite, then it contains an infinite number of subrepresentations equivalent to π, and conversely.

Theorem 11. (Murray and von Neumann) *Let M be a factor with separable predual. There exists a mapping D of the set of projections of M into $\overline{\mathbb{R}}_+ = [0, +\infty]$, unique up to a scalar factor $\lambda > 0$, such that:*

a) $P_1 \sim P_2 \iff D(P_1) = D(P_2)$.

b) $P_1P_2 = 0 \implies D(P_1 + P_2) = D(P_1) + D(P_2)$.

c) *P finite* $\iff D(P) < +\infty$.

Moreover, up to a normalization, the range of D is one of the following subsets of $\overline{\mathbb{R}}_+$:

$\{1, \ldots, n\}$,	*in which case M is said to be of type*	I_n
$\{1, \ldots, \infty\}$		I_∞
$[0, 1]$		II_1
$[0, +\infty]$		II_∞
$\{0, +\infty\}$		III.

One sees that, for M to have a minimal projection, it is necessary and sufficient that it be of type I. If M is of type I_n, $n < \infty$, then it is $M_n(\mathbb{C})$ and

$D(P)$ is the usual dimension of the subspace of $\mathbb{C}^n$ onto which the projection $P \in M_n(\mathbb{C})$ projects. If $n = \infty$ then $M = \mathcal{L}(\mathfrak{H})$ with $\mathfrak{H}$ infinite-dimensional and separable, and $D(P) =$ dimension of the range of P.

What is remarkable in the II_1 case is the appearance of dimensions with arbitrary values in $[0, 1]$.

A factor M is said to be finite if it contains no subfactor of type I_∞. This is equivalent to saying that M is in one of the cases I_n $(n < \infty)$ or II_1. In particular, if M is infinite (not finite) then there exists no trace τ on M such that $\tau(1) = 1$.

One of the remarkable results of the first papers of Murray and von Neumann is the converse:

Theorem 12. (Murray and von Neumann) *If M is a factor of type II_1, then there exists a unique trace τ on M such that $\tau(1) = 1$.*

Moreover, $\tau \in M_*$. One proof, due to F. J. Yeadon [Y], reduces this theorem to a powerful result of Ryll-Nardzewski: in every $\sigma(X, X^*)$-compact convex subset of a Banach space X, there exists a point that is left fixed by every affine isometry of the convex set. One applies this result by showing that if $\varphi \in M_*$ satisfies $\|\varphi\| = \varphi(1) = 1$, then the closed convex hull K of the orbit of φ under the action of the inner automorphisms of M (transported to M_*) is $\sigma(M_*, M)$-compact.

One then obtains the existence of $\tau \in M_*$, $\tau(1) = 1$, such that $\tau(uxu^*) = \tau(x)$ for u unitary in M and $x \in M$.

1.δ Algebraic isomorphism and spatial isomorphism

Let M_1 and M_2 be two von Neumann algebras, and let θ be a $*$-algebra isomorphism of M_1 onto M_2. Then θ is isometric (because M_1 and M_2 are C^*-algebras and $\|T\|^2 = \|T^*T\| =$ spectral radius of T^*T) and, since the predual is unique, θ is $\sigma(M_i, M_{i*})$-continuous. If M_i acts on the Hilbert space $\mathfrak{H}_i$ $(i = 1, 2)$, the isomorphism θ need not be spatial, for, even though they are isomorphic, M_1 and M_2 may have non-isomorphic commutants. Let us fix M: suppose M has a separable predual and let us try to describe all of the isomorphisms π of M onto a von Neumann subalgebra of $\mathcal{L}(\mathfrak{H})$, where $\mathfrak{H}$ is a separable Hilbert space. Making use of the reduction theory, we may further simplify matters by restricting attention to the case that M is a factor.

We are thus led to study, up to equivalence, the representations π of M on a Hilbert space $\mathfrak{H}_\pi$ that are continuous when M has the topology $\sigma(M, M_*)$ and $\mathcal{L}(\mathfrak{H}_\pi)$ has the topology of the duality with $\mathcal{L}(\mathfrak{H}_\pi)_*$. Since M is a factor, the commutant $R(\pi) = \pi(M)'$ is also a factor and π is isotypic. It follows, by forming $\pi_1 \oplus \pi_2$, that two representations π_1 and π_2 are never disjoint and that every representation π of M is a subrepresentation of an infinite representation ρ that is fixed once and for all. Thus, by Proposition 3 the representations of M are classified by equivalence classes of projections $P \in R(\rho)$, i.e. using Theorem 11, by the dimension function D of $R(\rho)$. This result may be extended to the case that M is no longer a factor. Every representation π of M (continuous for $\sigma(M, M_*)$ and $\sigma(\mathcal{L}(\mathfrak{H}_\pi), \mathcal{L}(\mathfrak{H}_\pi)_*))$ is a subrepresentation of a representation ρ that is *properly infinite* (in the sense that the commutant $R(\rho)$ contains a subfactor of type I_∞) and *faithful*[1] and which is chosen arbitrarily, for example by starting with an isomorphism α of M onto a von Neumann subalgebra of $\mathcal{L}(\mathfrak{H})$ and forming $\rho = \alpha \oplus \alpha \oplus \cdots$.

Thus, once M is known algebraically, there is no serious problem in determining all of the isomorphisms of M onto a von Neumann subalgebra of $\mathcal{L}(\mathfrak{H})$ (see Section 10). The real problem is that of classifying the von Neumann algebras *up to algebraic isomorphism.*

1.ε The first two examples of type II_1 factors, the hyperfinite factor and the property Γ

Recall that if Γ is a countable discrete group all of whose conjugacy classes are infinite and if λ' is its right regular representation, then the von Neumann algebra $R(\lambda')$ is a factor of type II_1, denoted $R(\Gamma)$.

In *On rings of operators. IV* [Mur-N$_3$], Murray and von Neumann showed that all of the factors $R(\Gamma)$ are isomorphic for Γ locally finite (i.e. the union of an increasing directed family of finite subgroups), and that if Γ is the free group with two generators then one obtains a factor not isomorphic to any of these. Let N be a factor of type II_1 and τ the unique trace ($\tau(1) = 1$) of N; one defines the Hilbert-Schmidt norm on N by

$$\|x\|_2 = \tau(x^*x)^{1/2}.$$

This is the analogue of the norm $\left(\sum |a_{ij}|^2\right)^{1/2}$ of a matrix (a_{ij}). It is a pre-Hilbert space norm on N and the corresponding metric is denoted d_2.

The result of Murray and von Neumann is then as follows:

Theorem 13. (Murray and von Neumann [Mur-N$_3$]) *There exists, up to isomorphism, one and only one factor N of type* II_1 *having separable predual and such that* $(\forall x_1, \ldots, x_n \in N,\ \forall \epsilon > 0)\ \exists$ *finite-dimensional* $*$*-subalgebra K such that* $d_2(x_j, K) \le \epsilon\ (\forall j)$.

We denote this unique factor by R, called, for obvious reasons, the *hyperfinite* factor.

In their paper, Murray and von Neumann showed that every infinite-dimensional factor contains a copy of R. For Γ a locally finite, denumerable discrete group, one sees easily that $R(\Gamma)$ satisfies the condition of the theorem, hence is isomorphic to R. Choosing Γ suitably, one sees moreover that R satisfies the following condition, called the *property* Γ:

$$(\forall x_1, \ldots, x_n \in R, \forall \epsilon > 0)\ \ \exists \text{ a unitary } u \text{ of } R \text{ such that}$$

$$\tau(u) = 0, \ \ \|x_j u - u x_j\|_2 \le \epsilon \ \ (j = 1, \ldots, n).$$

[1] *The term "faithful representation" is often used instead of "a representation* π *whose kernel reduces to* 0" ($\pi(x) = 0 \Longrightarrow x = 0$).

Murray and von Neumann then proved that if $\Gamma = \mathbb{Z} * \mathbb{Z}$ is the free group with two generators, then the factor $R(\Gamma)$ does not satisfy Γ. We shall return later on to this property, which was for Murray and von Neumann only a technical tool. We quote these authors: "Certain algebraic invariants of factors in the case II_1 are formed, 1) and 2) in §4.6 and the property Γ, of which the first two are probably of greater general significance, but the last one so far has been put to greater practical use." In fact, the invariants 1) and 2) that they mention are:

1) knowing whether N is anti-isomorphic to N (i.e. isomorphic to the opposite algebra N^o, $x \cdot y = yx$ for $x, y \in N^o$);
2) the subgroup $F(N)$ of $\mathbb{R}_+^*$ constructed as follows: let $\tilde{N} = N \otimes K$, where K is a factor of type I_∞. Then, on $\tilde{N}$, the relative dimension function D has range $\overline{\mathbb{R}}_+$, and if $\theta \in \mathrm{Aut}\,\tilde{N}$ then there exists a *unique* positive real number $\lambda = \mathrm{mod}\ \theta$ with $D(\theta(P)) = \lambda D(P)$ ($\forall$ projections P). The group $F(N) = \{\mathrm{mod}\,\theta;\ \theta \in \mathrm{Aut}\,\tilde{N}\}$ is obviously an algebraic invariant of N.

In fact, an example of a type II_1 factor not anti-isomorphic to itself was not obtained until long after ([Co$_2$]), nor was the existence ([Co$_9$]) of a type II_1 factor whose group F is distinct from $\mathbb{R}_+^*$ (the only calculable examples always gave $F = \mathbb{R}_+^*$).

Finally, to conclude this review of the results of Murray and von Neumann, we mention that they had succeeded in exhibiting a factor in the case III but they noted: "The purely infinite case, i.e. the case III, is the most refractory of all and we have, at least for the time being, scarcely any tools to investigate it." (With the notations of Example 4, one can take $V = P_1(\mathbb{R})$ and $\Gamma = PSL(2, \mathbb{Z})$ acting by homographic transformations. Then M is a factor of type III.)

V.2. Representations of C^*-algebras

One of the key theorems in measure theory is the Riesz representation theorem. We state it only for X compact and metrizable (X metrizable $\Longleftrightarrow$ $C(X)$ separable).

Theorem 1. *Let X be a compact metrizable space and let L be a positive linear form on $C(X)$, i.e.*

$$f \in C(X),\ f(x) \geq 0\ \ \forall x \in X \implies L(f) \geq 0.$$

Then there exists a unique positive measure μ on the σ-algebra $\mathcal{B}$ of Borel sets of X (i.e., the σ-algebra generated by the closed sets of X) such that

$$L(f) = \int f \mathrm{d}\mu \quad \forall f \in C(X).$$

In particular, one can construct the Hilbert space $L^2(X, \mathcal{B}, \mu)$ and the representation π of $C(X)$ on L^2 by multiplication. One knows in addition the von

Neumann algebra generated by $\pi(C(X))$: it is precisely the algebra of multiplication by the elements of $L^\infty(X, \mathcal{B}, \mu)$. The σ-additivity property of μ translates to the equality

$$\varphi\left(\sum_{\alpha \in I} e_\alpha\right) = \sum_{\alpha \in I} \varphi(e_\alpha),$$

where φ denotes the natural extension of L to $L^\infty(X, \mathcal{B}, \mu)$ and where $(e_\alpha)_{\alpha \in I}$ is any countable family of pairwise orthogonal projections $e_\alpha \in L^\infty(X, \mathcal{B}, \mu)$.

Now suppose that A is a *noncommutative* C^*-algebra with unit. The above concepts of positive element, positive linear form and σ-additivity have exact analogues, and this is the point of departure for noncommutative integration theory.

Positive elements in a C^*-algebra

Let $\mathfrak{H}$ be a Hilbert space and let $T \in \mathcal{L}(\mathfrak{H})$. The following conditions are equivalent:

a) $T = T^*$ and Spectrum $T \subset [0, +\infty)$.

b) $\langle T\xi, \xi\rangle \geq 0$ for all $\xi \in \mathfrak{H}$.

For a C^*-algebra A with unit and for $x \in A$, the following properties are equivalent:

1) $x = x^*$ and Spectrum $x \subset [0, +\infty)$.

2) $\exists a \in A$ such that $x = a^*a$.

3) $\exists a \in A$ such that $a^* = a$ and $x = a^2$.

4) $\exists \lambda \geq 0$ such that $\|x - \lambda 1\| \leq \lambda$.

We then say that x is positive and we write $x \geq 0$. The condition 4) shows that the set of positive elements is a closed convex cone in A, which is denoted by A^+. If $A = C(X)$ then $A^+ = \{f;\, f(x) \geq 0\ \forall x \in X\}$.

Positive linear forms on a C^*-algebra

Let A be as above and let A^* be its Banach space dual. We say that $L \in A^*$ is positive if $L(x) \geq 0\ \forall x \geq 0$. We denote by A^*_+ the $\sigma(A^*, A)$-closed convex cone of positive linear forms on A. The analogue of the Riesz representation theorem consists in the following construction (Gel'fand, Naimark, Segal).

Since L is positive, Condition 2) shows that

$$L(x^*x) \geq 0 \quad \forall x \in A.$$

It follows that the sesquilinear form $\langle x, y\rangle_L = L(y^*x)$ defines a pre-Hilbert space structure on A. Let $\mathfrak{H}_L$ be the Hilbert space completion and, for $x \in A$, let $\pi_L(x)$ be the left-multiplication operator defined by

$$\pi_L(x)y = xy \ \text{ for all } y \in A.$$

The inequality $L(y^*x^*xy) \leq \|x\|^2 L(y^*y)$ $(\forall y \in A)$, which results from $\|x\|^2 - x^*x \geq 0$, shows that π_L defines a representation of A on the Hilbert space $\mathfrak{H}_L$.

Just as, in the commutative case, the linear form L can be extended to the Borel functions $f \in L^\infty(X, \mathcal{B}, \mu)$, here the linear form L can be extended to a linear form on the von Neumann algebra $\pi_L(A)''$ generated by $\pi_L(A)$, and the extension, denoted by $\overline{L}$, has the following σ-additivity property:

Definition 2. *Let M be a von Neumann algebra on the Hilbert space $\mathfrak{H}$ and let ψ be a positive linear form on M. We say that ψ is* normal *if*

$$\psi\left(\sum_{\alpha \in I} e_\alpha\right) = \sum_{\alpha \in I} \psi(e_\alpha)$$

for every family $(e_\alpha)_{\alpha \in I}$ of pairwise orthogonal projections.

Here $\sum e_\alpha$ denotes the smallest projection that majorizes all of the finite sums $\sum_{i=1}^n e_{\alpha_i}$; it is an element of M. One proves that, for ψ to be normal, it is necessary and sufficient that it come from the predual M_* of M.

Let us return to C^*-algebras. Let A be such an algebra, with unity. The Hahn-Banach theorem, applied to the convex cone A^+ with nonempty interior, shows that the set

$$S = \{\varphi \in A^*_+;\ \varphi(1) = 1\}$$

of *states* of A is a nonempty convex set that separates the points of A. One is therefore assured of the existence of "positive measures" and, by the Gel'fand-Naimark-Segal construction, of the existence of an *isometric* representation of A as a C^*-subalgebra of $\mathcal{L}(\mathfrak{H})$, with $\mathfrak{H}$ a Hilbert space.

Moreover, S is $\sigma(A^*, A)$-compact, hence is the closed convex hull of its set of extremal points, the *pure states*, which are characterized as follows:

$$\varphi \text{ is a pure state} \iff \text{the representation } \pi_\varphi \text{ is irreducible.}$$

In fact, more generally, there is a bijective correspondence between the face of φ in the cone A^*_+ and the set of positive elements of the von Neumann algebra $R(\pi_\varphi)$, given by

$$\psi \in A^*_+ \text{ is associated with } y \in R(\pi_\varphi)^+ \text{ when } \psi(a) = \langle \pi_\varphi(a)1, y1\rangle, \quad \forall a \in A,$$

where $1 \in \mathfrak{H}_\varphi$ is the vector corresponding to the unit element of A.

Thus, every C^*-algebra A has sufficiently many irreducible representations. In the commutative case, the irreducible representations of A coincide with the homomorphisms of A into $\mathbb{C}$. In the noncommutative case, the nature of the relation of equivalence between irreducible representations of A on a fixed Hilbert space determines a privileged class of C^*-algebras; the following theorem of J. Glimm is fundamental:

Theorem 3. (cf. [Di$_3$]) *Let A be a separable C^*-algebra. The following conditions on A are equivalent:*

1) *Every isotypic representation π of A on a Hilbert space $\mathfrak{H}_\pi$ is a multiple of an irreducible representation.*

2) *For any irreducible representation π of A, the image $\pi(A)$ contains the ideal $k(\mathfrak{H}_\pi)$ of compact operators on $\mathfrak{H}_\pi$.*

3) *Let $\mathfrak{H}$ be a separable Hilbert space and let* $\text{Rep}(A, \mathfrak{H})$ *be the Borel space of irreducible representations of A on $\mathfrak{H}$; then its quotient by the relation of equivalence of representations is countably separated.*

4) *If π_1 and π_2 are two irreducible representations of A having the same kernel, then they are equivalent.*

A C^*-algebra satisfying the above equivalent conditions is said to be *post-liminal.*

V.3. The Algebraic Framework for Noncommutative Integration and the Theory of Weights

To be able to take into account positive measures that are not necessarily finite, it is necessary to introduce the noncommutative analogue of infinite positive measures.

The initial data for noncommutative integration is thus a pair (M, φ) consisting of a von Neumann algebra M and a weight φ on M in the following sense ([Com], [Pe$_1$], [H$_1$]):

Definition 1. *A* weight *on a von Neumann algebra M is an additive, positively homogeneous mapping φ of M_+ into $\overline{\mathbb{R}}_+ = [0, +\infty]$. We say that:*

a) *φ is semifinite if $\{x \in M_+;\ \varphi(x) < \infty\}$ is $\sigma(M, M_*)$-total.*

b) *φ is normal if $\varphi(\sup x_\alpha) = \sup \varphi(x_\alpha)$ for every bounded, increasingly directed family of elements of M_+.*

The simplest example of an infinite weight is that of the usual trace for the bounded operators on a Hilbert space $\mathfrak{H}$. Setting $M = \mathcal{L}(\mathfrak{H})$, for every $T \in M_+$ and every orthonormal basis $(\xi_\alpha)_{\alpha \in I}$ of $\mathfrak{H}$ one has

$$\text{Trace}\ T = \sum \langle T\xi_\alpha, \xi_\alpha \rangle = \sup_{0 \le A \le T} \{\text{Trace}\ A;\ A \text{ of finite rank}\}.$$

In fact, the first infinite weights studied were the traces in the following sense:

Definition 2. *A weight on a von Neumann algebra M is called a* trace *if it is invariant under the inner automorphisms of M.*

Thus, every weight on M which is canonically defined is a trace. The analogues of the concepts of convergence almost everywhere and of the L^p-spaces, $p \in [1, \infty]$, of the classical theory were obtained mainly by Dixmier [Di$_1$] and Segal [Se$_1$] for semifinite normal *traces.*

If φ is a trace, the set

$$C_p = \{x \in M;\ \varphi(|x|^p) < \infty\}$$

is a two-sided ideal of M, and $\|x\|_p = \varphi(|x|^p)^{1/p}$ defines a seminorm on C_p. The completion spaces $L^p(M,\varphi)$ have numerous properties that generalize the commutative case and the classical case of pth-power summable operators on a Hilbert space. In particular, if $x \geq 0$ and $\varphi(x) = 0$ imply $x = 0$ (in which case φ is said to be *faithful*), the predual M_* of M may be identified with the space $L^1(M,\varphi)$. Moreover, the intersection $L^2(M,\varphi) \cap L^\infty(M,\varphi)$ is a Hilbert algebra:

Definition 3. A Hilbert algebra *is a $*$-algebra $\mathcal{A}$ equipped with a positive definite pre-Hilbert space inner product such that:*

1) $\langle x,y\rangle = \langle y^*,x^*\rangle$ $(\forall x,y \in \mathcal{A})$.

2) *The representation of $\mathcal{A}$ on $\mathcal{A}$ by left multiplications is bounded, involutive and nondegenerate.*

The condition 1) defines a conjugate-linear isometry J of the Hilbert space completion $\mathfrak{H}$ of $\mathcal{A}$ onto itself. Condition 2) makes it possible to speak of the left regular representation λ of $\mathcal{A}$ on $\mathfrak{H}$ and therefore to associate with it a von Neumann algebra $\lambda(\mathcal{A})''$ on $\mathfrak{H}$. Then:

a) The commutant of $\lambda(\mathcal{A})$ is generated by the algebra of right multiplications $\lambda'(\mathcal{A}) = J\lambda(\mathcal{A})J$.

b) The von Neumann algebra associated with the Hilbert algebra $L^2(M,\varphi) \cap L^\infty(M,\varphi)$ may be identified with M.

c) For every Hilbert algebra $\mathcal{A}$, there exists a faithful semifinite normal trace τ on the von Neumann algebra $\lambda(\mathcal{A})''$ such that

$$\tau(\lambda(y^*)\lambda(x)) = \langle x,y\rangle \quad \forall x,y \in \mathcal{A},$$

and $\mathcal{A}$ is equivalent to the Hilbert algebra associated with τ.

In general, a von Neumann algebra M does not have a faithful semifinite normal trace. For example, if M is a factor, it has a faithful semifinite normal trace if and only if it is not of type III. A von Neumann algebra M is said to be semifinite if it has a faithful semifinite normal trace (when M acts on a separable space, this is equivalent to saying that M is a direct integral of factors none of which is of type III).

For M semifinite, the supplementary tools that come out of the theory of Hilbert algebras make it possible to prove results that are inaccessible in the general case, such as:

Theorem 4. *Let M_i be a von Neumann algebra on $\mathfrak{H}_i$ $(i = 1,2)$. Then the commutant $(M_1 \otimes M_2)'$ is generated by $M_1' \otimes M_2'$.*

For semifinite von Neumann algebras M_1 and M_2, this result is a consequence of the commutation property a) for Hilbert algebras. Similarly, if G is a *unimodular* locally compact group and dg is a Haar measure on G, then the convolution algebra of continuous functions with compact support is a Hilbert algebra, and the commutation result a) implies:

Theorem 5. *The commutant $R(\lambda')$ of the right regular representation λ' of G on $L^2(G, dg)$ is generated by the left regular representation.*

This theorem was proved, for not necessarily unimodular locally compact groups, by J. Dixmier, who introduced the concept of quasi-Hilbert algebra. Theorem 4 was proved, for not necessarily semifinite von Neumann algebras, by M. Tomita in 1967. As a matter of fact, his theory of generalized Hilbert algebras is the foundation for all of the noncommutative integration theory for weights that are not necessarily traces. The Tomita theory owes much to M. Takesaki, who transformed the original, very difficult to decipher article into an accessible text [T$_2$].

The essentials of the Tomita–Takesaki theory can be summarized by the following definition and theorem:

Definition 6. *A* left Hilbert algebra *is a $*$-algebra $\mathcal{A}$, equipped with a positive definite pre-Hilbert space inner product, such that:*

1) *The operator $x \mapsto x^*$ is* closable.

2) *The representation of $\mathcal{A}$ on $\mathcal{A}$ by left multiplication is bounded, involutive and nondegenerate.*

Thus, the only difference from a Hilbert algebra is that the closure S of the operator $x \mapsto x^*$ can have absolute value $|S| \neq 1$. Let

$$\Delta = (\text{adjoint of } S) \circ S$$

be the square of the absolute value of S. Then $S = J\Delta^{1/2}$, where J is an isometric involution, and the fundamental result is as follows [T$_2$]:

Theorem 7. *Let $\mathcal{A}$ be a left Hilbert algebra and let M be the von Neumann algebra generated by the left regular representation of $\mathcal{A}$. Then $JMJ = M'$ and*

$$\Delta^{it} M \Delta^{-it} = M$$

for all $t \in \mathbb{R}$.

Moreover, just as the Hilbert algebras are associated with traces, the left Hilbert algebras are associated with weights ([Com]):

Let $\mathcal{A}$ be a left Hilbert algebra and let M be the associated von Neumann algebra. Then there exists a faithful semifinite normal weight φ on M such that

$$\varphi(\lambda(y^*)\lambda(x)) = \langle x, y\rangle \quad (\forall x, y \in \mathcal{A}).$$

Conversely, let M be a von Neumann algebra and let φ be a faithful semifinite normal weight on M. Then

$$\mathcal{A}_\varphi = \{x \in M;\ \varphi(x^*x) < \infty,\ \varphi(xx^*) < \infty\},$$

equipped with the multiplication of M and the inner product $\langle x, y\rangle = \varphi(y^*x)$, is a left Hilbert algebra. The associated von Neumann algebra may be identified with M and the corresponding weight with φ.

Since every von Neumann algebra has a faithful semifinite normal weight (if M acts on a separable space then in fact it has a faithful normal state), it follows, in particular, that every von Neumann algebra is isomorphic to the von Neumann algebra generated by the left regular representation of a left Hilbert algebra.

It is here that a remarkable discovery of Takesaki and Winnink relates Tomita's theory— more precisely, the one-parameter group of automorphisms of the von Neumann algebra M defined by $\sigma_t(x) = \Delta^{it}x\Delta^{-it}$—to the fundamental Kubo-Martin-Schwinger condition of quantum statistical mechanics which we already described in Section I.2. Of course the automorphism group σ_t is not unique; it depends on the Hilbert algebra $\mathcal{A}$, that is, on the faithful semifinite normal weight φ on M.

Definition 8. *Let φ be a faithful semifinite normal weight on a von Neumann algebra M. The* group of modular automorphisms *is the one-parameter group $(\sigma_t^\varphi)_{t\in\mathbb{R}}$ of automorphisms of M associated with the left Hilbert algebra $\mathcal{A}_\varphi$.*

One then has the following characterization.

Theorem 9. [T$_2$][Winn] *Let M be a von Neumann algebra and let φ be a faithful normal state on M. Then the group $(\sigma_t^\varphi)_{t\in\mathbb{R}}$ of modular automorphisms is the unique one-parameter group of automorphisms of M that satisfies the Kubo-Martin-Schwinger condition for $\beta = -1$. (Equivalently $(\sigma_{-t}^\varphi)_{t\in\mathbb{R}}$ is KMS for $\beta = 1$.)*

This characterization of (σ_t^φ) holds for faithful semifinite normal weights, with the proper reformulation of the KMS condition taking care of domain problems.

V.4. The Factors of Powers, Araki and Woods, and of Krieger

The noncommutative analogue of a probability space is a pair (M, φ), where M is a von Neumann algebra and φ is a faithful normal state on M. The simplest example corresponds to $M = M_n(\mathbb{C})$. Every state φ on M may be written $\varphi = \mathrm{Tr}(\rho\,\cdot)$, where ρ is a positive matrix the sum of whose eigenvalues is 1:

$$\varphi(x) = \mathrm{Tr}(\rho x) \quad \forall x \in M_n(\mathbb{C}).$$

One can therefore suppose that ρ is diagonal, with eigenvalue $\lambda_i > 0$ in the ith row, and with $\lambda_1 \geq \lambda_2 \geq \cdots \geq \lambda_n > 0$. The list of eigenvalues of ρ is an invariant of φ called its eigenvalue list. The group of modular automorphisms of φ is given by

$$\sigma_t^{\varphi}(x) = e^{itH} x e^{-itH}, \quad \text{where } H = \operatorname{Log} \rho.$$

In particular, if e_{ij} denotes the canonical matrix unit, then

$$\sigma_t^{\varphi}(e_{kl}) = \left(\frac{\lambda_k}{\lambda_l}\right)^{it} e_{kl}.$$

The system (M, φ) is analogous to a probability space having a finite number of points. To obtain examples that are more interesting, the simplest procedure consists in carrying out the analogue of the construction of infinite products of measures.

Thus, let $(M_\nu, \varphi_\nu)_{\nu \in \mathbb{N}}$ be a sequence of pairs (matrix algebra, faithful state); let A be the inductive limit of the C^*-algebras

$$A_\nu = M_1 \otimes M_2 \otimes \cdots \otimes M_\nu,$$

where the embedding $A_\nu \subset A_{\nu+1}$ is by means of the mapping $x \mapsto x \otimes 1$. On A, which is a C^*-algebra with unit, one defines a state $\varphi = \bigotimes_{\nu=1}^{\infty} \varphi_\nu$ by the equality

$$\varphi(x_1 \otimes x_2 \otimes \cdots \otimes x_\nu \otimes 1 \otimes \cdots) = \varphi_1(x_1)\varphi_2(x_2) \cdots \varphi_\nu(x_\nu).$$

One then considers the pair (M, φ) = (von Neumann algebra, normal state) associated with the pair (A, φ) as in Section 3 of this chapter. If every φ_ν is faithful then so is φ, and the group of modular automorphisms of (M, φ) is given by

$$\sigma_t^{\varphi}(x_1 \otimes \cdots \otimes x_\nu \otimes 1 \otimes \cdots) = \sigma_t^{\varphi_1}(x_1) \otimes \cdots \otimes \sigma_t^{\varphi_\nu}(x_\nu) \otimes 1 \otimes \cdots.$$

In fact, this construction of von Neumann algebras is due to von Neumann himself; however, one had to wait until 1967 for it to reveal itself to be fundamental. For the thirty years that followed the birth of von Neumann algebras, only three pairwise nonisomorphic factors of type III were known. In 1967 R. T. Powers, who was trained as a physicist, succeeded in showing that if all of the pairs (M_ν, φ_ν) are taken to be equal to the pair $(M_2(\mathbb{C}), \varphi)$, where $\varphi((a_{ij})) = \alpha a_{11} + (1 - \alpha) a_{22}$, then one obtains a family, with continuous parameter $\alpha \in]0, \frac{1}{2}[$, of pairwise nonisomorphic factors of type III: R_λ, $\lambda = \alpha/(1-\alpha) \in]0, 1[$ ([Pow$_1$]).

In terms of the eigenvalue list $(\lambda_{\nu,j})_{j=1,\ldots,n_\nu}$ one has the following criteria:

1) M is type I if and only if

$$\sum_\nu |1 - \lambda_{\nu,1}| < \infty.$$

2) M is type II$_1$ if and only if

$$\sum_{\nu,i} |(n_\nu)^{-1/2} - (\lambda_{\nu,i})^{1/2}|^2 < \infty.$$

3) If $\lambda_{\nu,1} \geq \delta$ for some $\delta > 0$ and all ν, then M is type III if and only if

$$\sum_{\nu,i} \lambda_{\nu,i} \inf\{|(\lambda_{\nu,1}/\lambda_{\nu,i}) - 1|^2, C\} = \infty$$

for some (and hence all) positive C.

After Powers' discovery, H. Araki and E. J. Woods undertook a classification, up to isomorphism, of the factors that are infinite tensor products of matrix algebras [Ar-W]. They showed, in particular, that, from the eigenvalue list of the φ_ν, $(\lambda_{\nu,j})_{j=1,\ldots,n_\nu}$, one can calculate the following two invariants:

$$r_\infty(M) = \{\lambda \in]0,1[;\ M \otimes R_\lambda \text{ isomorphic to } M\},$$
$$\rho(M) = \{\lambda \in]0,1[;\ M \otimes R_\lambda \text{ isomorphic to } R_\lambda\}.$$

The computation of the asymptotic ratio set $r_\infty(M)$ is done as follows. For each finite subset I of the set of indices $\nu \in \mathbb{N}$ one lets $X(I)$ be the product $\prod_{\nu \in I} \{1,\ldots,n_\nu\}$ with probability measure λ given by the product of the $(\lambda_{\nu,j})$.

The asymptotic ratio set of M is the set of all $x \in [0,\infty]$ such that there exists a sequence of mutually disjoint finite index sets $I_n \subset \mathbb{N}$, mutually disjoint subsets K_n^1 and K_n^2 of $X(I_n)$ for each n such that $a \in K_n^1$ implies $\lambda(a) \neq 0$, and a bijection ϕ_n from K_n^1 to K_n^2 satisfying

$$\sum_n \lambda(K_n^1) = \infty$$

and

$$\lim_{n\to\infty} \max_{a \in K_n^1} |x - \lambda(\phi_n(a))/\lambda(a)| = 0.$$

They showed in addition that $r_\infty(M)$ is a closed subgroup of $\mathbb{R}_+^*$ and that the equality $r_\infty(M) = \lambda^{\mathbb{Z}}$ characterizes the Powers factor R_λ among the infinite tensor products of matrix algebras. Moreover, by studying $\rho(M)$, E. J. Woods succeeded in showing the impossibility of classifying the factors by means of real-valued Borel invariants.

On the other hand, W. Krieger had undertaken a systematic study of the factors associated with ergodic theory. I shall begin by explaining his construction of a von Neumann algebra starting from an equivalence relation with denumerable orbits on a standard Borel space $(X, \mathcal{B})$ and a quasi-invariant measure μ. This construction generalizes the first construction of Murray and von Neumann, which was itself inspired by the crossed products of the theory of central simple algebras over a field. In its definitive form, it is due to J. Feldman and C. Moore [Fel-M].

Thus, let $(X, \mathcal{B}, \mu)$ be a standard Borel probability space and let $\mathcal{R} \subset X \times X$ be the graph, assumed to be analytic, of an equivalence relation with denumerable orbits. One assumes that μ is quasi-invariant in the sense that the saturation of a negligible Borel set by $\mathcal{R}$ is again negligible. One then considers the left Hilbert algebra $\mathcal{A}$ of bounded Borel functions from $\mathcal{R}$ into $\mathbb{C}$ such that, for some $n < \infty$, the set $\{j;\ f(i,j) \neq 0\}$ has cardinality $\leq n$ for all i. The inner product is that of $L^2(\mathcal{R}, \tilde{\mu})$, where

$$\int f(i,j) d\tilde{\mu} = \int \sum_j f(i,j) d\mu(i).$$

The convolution product is given by

$$(f * g)(\gamma) = \sum_{\gamma_1 \gamma_2 = \gamma} f(\gamma_1) g(\gamma_2),$$

where, for $\gamma_1, \gamma_2 \in \mathcal{R}$, one sets $\gamma_1 \gamma_2 = \gamma$ when $\gamma_1 = (i_1, j_1)$, $\gamma_2 = (i_2, j_2)$, $j_1 = i_2$ and $\gamma = (i_1, j_2)$.

The left Hilbert algebra $\mathcal{A}$ thus appears as a generalization of the algebra of square matrices. Since it has a unit element (the function f such that $f(i,j) = 0$ if $i \neq j$, and $f(i,i) = 1$ for all i), it yields a pair (M, φ), where M is a von Neumann algebra and φ is a faithful normal state on M. We shall write simply $M = L^\infty(\mathcal{R}, \tilde{\mu})$, where $\mathcal{R}$ denotes the graph of the equivalence relation, equipped with the groupoid law $\gamma_1 \gamma_2 = \gamma$. It is interesting from a heuristic point of view to make $\mathcal{R}$ play (in noncommutative integration) the role played (in classical integration) by the *space* X. The *noncommutativity* is due to the existence of a nontrivial groupoid law on $\mathcal{R}$ (the trivial law on the set X being $x \cdot x = x$ for all $x \in X$). Up to isomorphism, the von Neumann algebra $L^\infty(\mathcal{R}, \tilde{\mu})$ depends only on the class of the measure μ. One then makes the following definition:

Definition 1. *Let $(X_i, \mathcal{B}_i, \mu_i, \mathcal{R}_i)$ $(i = 1, 2)$ be equivalence relations with denumerable orbits, as above; $\mathcal{R}_1$ is said to be* isomorphic *to $\mathcal{R}_2$ if there exists a Borel bijection θ of X_1 onto X_2 such that $\theta(\mu_1)$ is equivalent to μ_2 and, almost everywhere,*

$$\theta(\text{class of } x) = \text{class of } \theta(x).$$

If T is a Borel transformation of the standard Borel space $(X, \mathcal{B})$ and if μ is a measure quasi-invariant under T, then the orbits of T in X define an equivalence relation $\mathcal{R}_T$; T_1 is said to be *weakly equivalent* to T_2 if $\mathcal{R}_{T_1}$ is isomorphic to $\mathcal{R}_{T_2}$.

Theorem 2. [Dy] *Any two ergodic transformations with invariant measure are weakly equivalent.*

Let T be such a transformation. The state φ (associated with the invariant measure) on the von Neumann algebra $L^\infty(\mathcal{R}_T, \mu)$ is a trace, so that $M = L^\infty(\mathcal{R}_T, \mu)$, which is a factor since $\mathcal{R}_T$ is ergodic, is of type II_1. H. Dye showed, moreover, that it is the hyperfinite factor.

Around 1967, W. Krieger undertook a systematic study of the weak equivalence of the transformations $(X, \mathcal{B}, \mu, T)$, where μ is quasi-invariant under T. He introduced two invariants ([Kr$_1$]):

$$r(T) = \{\lambda \in [0, +\infty[;\ \forall \epsilon > 0,\ \forall A \subset X,\ \mu(A) > 0,$$
$$\exists B \subset A,\ \mu(B) > 0, \text{ and } n \in \mathbb{Z} \text{ such that}$$
$$T^n B \subset A \text{ and } \left| \frac{d\mu(T^n x)}{d\mu(x)} - \lambda \right| \leq \epsilon\ \forall x \in B\},$$
$$\rho(T) = \{\lambda \in \mathbb{R}_+^*;\ \exists \nu \sim \mu \text{ with } \frac{d\nu(Tx)}{d\nu(x)} \in \lambda^{\mathbb{Z}}\ \forall x \in X\},$$

and he showed that r and ρ are not only invariants under weak equivalence, but that in fact $r(T)$ coincides with the Araki–Woods invariant

$$r_\infty(M) = \{\lambda;\ M \otimes R_\lambda \text{ isomorphic to } M\},$$

where $M = L^\infty(\mathcal{R}_T, \mu)$, and similarly that $\rho(T) = \rho(M)$.

This is actually a generalization of the results of Araki and Woods. For, let $(M_\nu, \varphi_\nu)_{\nu \in \mathbb{N}}$ be a sequence of pairs (matrix algebra, faithful state), and let $(\lambda_{\nu,j})_{j=1,\ldots,n_\nu}$ be the corresponding eigenvalue lists. Then the infinite tensor product von Neumann algebra $\bigotimes_{\nu=1}^\infty (M_\nu, \varphi_\nu)$ may also be obtained by Krieger's construction, from the following space and equivalence relation:

$$X = \prod_{\nu=1}^{\infty} X_\nu, \text{ where } X_\nu = \{1, \ldots, n_\nu\} \text{ for each } \nu,$$

$\mathcal{B}$ is the σ-algebra generated by the product topology,

$$\mu = \prod_{\nu=1}^{\infty} \mu_\nu, \text{ where } \mu_\nu(j) = \lambda_{\nu,j} \text{ for } j \in X_\nu,$$

$\mathcal{R}$ is the relation $x \sim y \iff x_i = y_i$ for i sufficiently large.

The relation $\mathcal{R}$ is in fact equal to $\mathcal{R}_T$, where T is the transformation that generalizes the operation of adding 1 in the p-adic integers: to calculate Tx, one looks at the first coordinate x_i of x that is not equal to the maximum possible n_i, replaces it by $x_i + 1$, and then one replaces every preceding x_j $(j < i)$ by 1.

In fact, there exist Krieger factors, i.e. factors of the form $L^\infty(\mathcal{R}_T, \mu)$, that are not infinite tensor products of matrix algebras. The culmination of Krieger's theory is the following theorem, proved around 1973 by means of the invariants of factors that we shall discuss later on:

Theorem 3. [Kr$_2$] *Let $(X_i, \mathcal{B}_i, \mu_i, T_i)$ $(i = 1, 2)$ be ergodic transformations, where each μ_i is quasi-invariant and $(X_i, \mathcal{B}_i)$ is a standard Borel space. Then T_1 is weakly equivalent to T_2 (i.e., $\mathcal{R}_{T_1}$ is isomorphic to $\mathcal{R}_{T_2}$) if and only if the factors $L^\infty(\mathcal{R}_{T_i}, \mu_i)$ are isomorphic.*

This result clearly shows the need for recognizing whether an equivalence relation $\mathcal{R}$ with denumerable orbits is of the form $\mathcal{R}_T$ for some Borel transformation T. In particular, when $\mathcal{R}$ arises from the action of a discrete group Γ, Krieger and I showed [Co-Kr] that if Γ is solvable then $\mathcal{R}$ is of the form $\mathcal{R}_T$. The case that Γ is an amenable group was treated by D. Ornstein and B. Weiss [Or-W] and the definitive answer to the problem was obtained in [Co-F-W] by J. Feldman, B. Weiss and myself.

Theorem 4. [Co-F-W] *$\mathcal{R}$ is of the form $\mathcal{R}_T$ if and only if $\mathcal{R}$ is amenable in the sense of Zimmer* [Zi_1].

R. Zimmer's definition translates the amenability of the von Neumann algebra $L^\infty(\mathcal{R},\mu)$ simply (cf. Section 7), so that the above two theorems imply ([Co-F-W]):

Corollary 5. *Any two Cartan subalgebras of an amenable factor $M = L^\infty(\mathcal{R},\mu)$ are conjugate by an automorphism of M.*

Here, by Cartan subalgebra is meant a maximal abelian subalgebra A of M such that:

a) The normalizer of A, i.e. $\{u \in M \,;\, u^*u = uu^* = 1 \,,\, uAu^* = A\}$, generates M (A is then said to be regular).

b) There exists a normal conditional expectation E of M onto A.

Condition b) is automatic if M is of type II_1, thus one sees that:

The hyperfinite factor R has, up to conjugation, only one regular maximal abelian subalgebra.

We note, by a result of V. Jones and myself, for nonamenable equivalence relations $\mathcal{R}_1$ and $\mathcal{R}_2$, that it is false that the isomorphism $L^\infty(\mathcal{R}_1,\mu_1) \approx L^\infty(\mathcal{R}_2,\mu_2)$ implies the isomorphism of the relations $\mathcal{R}_1$ and $\mathcal{R}_2$. Thus, one cannot hope to translate Zimmer's rigidity results (on equivalence relations arising from ergodic actions of semisimple Lie groups [Zi_2]) directly into a rigidity theorem on the corresponding factors. The problem of adapting his proof to the case of factors is a key open problem. (cf. Appendix B ε))

V.5. The Radon-Nikodým Theorem and Factors of Type III_λ

5.α The Radon-Nikodým theorem

The Tomita–Takesaki theory associates, to every faithful semifinite normal weight φ on a von Neumann algebra M, a one-parameter group σ_t^φ of automorphisms of M, the group of modular automorphisms, defined by

$$\sigma_t^\varphi(x) = \Delta_\varphi^{it} x \Delta_\varphi^{-it},$$

where Δ_φ is the modular operator, the square of the absolute value of the involution $x \to x^*$ regarded as an unbounded operator in the space $L^2(M,\varphi)$, the completion of $\{x \in M; \varphi(x^*x) < \infty\}$ for the inner product $\langle x,y\rangle = \varphi(y^*x)$. On the other hand, the theory of Araki and Woods associates to every factor M the two invariants r_∞ and ρ,

$$r_\infty(M) = \{\lambda;\ M \otimes R_\lambda \sim M\},$$
$$\rho(M) = \{\lambda;\ M \otimes R_\lambda \sim R_\lambda\}.$$

Now, for the factors of Araki and Woods, a direct calculation based on their work yields the following equalities:

$$r_\infty(M) = \bigcap \{\text{Spectrum } \Delta_\varphi;\quad \varphi \text{ a faithful normal state on } M\}. \tag{1}$$

$$\rho(M) = \{\exp(2\pi/T_0);\ \exists \text{ a faithful normal state } \varphi \text{ on } M \text{ such that } \sigma_{T_0}^\varphi = 1\}. \tag{2}$$

These two equalities of course suggest the following definitions for an arbitrary factor M:

$$S(M) = \bigcap \{\text{Spectrum } \Delta_\varphi;\quad \varphi \text{ a faithful normal state on } M\}.$$
$$T(M) = \{\text{ possible periods of groups of modular automorphisms of } M\}.$$

Clearly, the first question this raises is whether the equalities $r_\infty = S$ and $\rho = \exp(2\pi/T)$, valid for the factors of Araki and Woods, remain true in general. A closely related question is the problem of calculating the invariants S and T. The above definitions of these invariants show that in order to calculate S and T one must pass in review all of the faithful normal states on M and calculate their groups of modular automorphisms. Now in general, as is clear for the factors of Section 4 of this chapter, a factor is presented with a privileged state or weight for which the calculation of Δ_φ and σ_t^φ is easy. Thus, the problem that poses itself is to study the precise extent to which the group σ^φ depends on φ.

The complete answer to this problem in fact constitutes precisely the non-commutative version of the Radon-Nikodým theorem.

Theorem 1. [Co_4] *Let M be a von Neumann algebra, let φ be a faithful semifinite normal weight on M, and let $\mathcal{U}$ be the unitary group of M equipped with the topology $\sigma(M,M_*)$.*

a) *For every faithful semifinite normal weight* ψ *on* M *there exists a unique continuous mapping* u *of* $\mathbb{R}$ *into* $\mathcal{U}$ *such that*:

$$u_{t+t'} = u_t\sigma_t^\varphi(u_{t'}) \qquad \forall t,t' \in \mathbb{R},$$
$$\sigma_t^\psi(x) = u_t\sigma_t^\varphi(x)u_t^* \qquad \forall t \in \mathbb{R}, x \in M,$$
$$\psi(x) = \varphi(u_{-i/2}^* x u_{-i/2}) \quad \forall x \in M.$$

This is expressed by writing $u_t = (D\psi : D\varphi)_t$.

b) *Conversely, let* $t \mapsto u_t$ *be a continuous mapping of* $\mathbb{R}$ *into* $\mathcal{U}$ *such that*

$$u_{t+t'} = u_t\sigma_t^\varphi(u_{t'}) \quad \forall t,t' \in \mathbb{R}.$$

Then there exists a unique faithful semifinite normal weight ψ *on* M *such that* $(D\psi : D\varphi) = u$.

If M is commutative, $M = L^\infty(X,\mu)$, then φ and ψ are positive measures on X that are equivalent to μ. Therefore there exists a Radon-Nikodým derivative $h : X \to \mathbb{R}_+$. The h^{it} $(t \in \mathbb{R})$ are in $M = L^\infty(X,\mu)$ and $h^{it} = (D\psi : D\varphi)_t$.

If M is semifinite and τ is a faithful semifinite normal trace on M, then there exist positive operators affiliated with M such that $\varphi = \tau(\rho_\varphi \cdot)$, $\psi = \tau(\rho_\psi \cdot)$, and

$$(D\psi : D\varphi)_t = \rho_\psi^{it}\rho_\varphi^{-it}.$$

The property $\sigma_t^\psi(x) = u_t\sigma_t^\varphi(x)u_t^*$ $(\forall x \in M)$ shows that, although the group of modular automorphisms in general varies with φ, its class modulo the inner automorphisms does not vary. We may then ask if it is not altogether trivial. However, an easy argument based on the above theorem shows that

$$T(M) = \{T_0;\ \sigma_{T_0}^\varphi \text{ is an inner automorphism}\}.$$

Moreover, a theorem of J. Dixmier and M. Takesaki [T2] shows that, assuming M_* separable,

$$T(M) \neq \mathbb{R} \iff M \text{ is not semifinite}.$$

One then introduces the group $\operatorname{Out} M = \operatorname{Aut} M / \operatorname{Inn} M$ of classes of automorphisms of M modulo the inner automorphisms and one associates to every von Neumann algebra M a canonical homomorphism of $\mathbb{R}$ into $\operatorname{Out} M$:

$$\delta(t) = \text{class of } \sigma_t^\varphi$$

(this is independent of the choice of φ). In particular, $T(M) = \ker\delta$ is a subgroup of $\mathbb{R}$. In fact, the range of δ is even contained in the *center* of the group $\operatorname{Out} M$.

Moreover, one can calculate $T(M)$ on the basis of a single faithful semifinite normal weight φ on M, since it suffices to determine the t for which σ_t^φ is an inner automorphism. For example, if M is an Araki-Woods factor $M = \bigotimes_{\nu=1}^\infty (M_\nu, \varphi_\nu)$, the group of modular automorphisms for $\varphi = \bigotimes_{\nu=1}^\infty \varphi_\nu$ is known: it is

$$\sigma_t^\varphi = \bigotimes_{\nu=1}^\infty \sigma_t^{\varphi_\nu}.$$

A simple calculation based on the eigenvalue list $(\lambda_{\nu,j})_{j=1,\dots,n_\nu}$ of φ_ν then shows that

$$T_0 \in T(M) \iff \sum_{\nu=1}^{\infty}\Big(1 - \big|\sum_j \lambda_{\nu,j}^{1+iT_0}\big|\Big) < \infty.$$

From this, one infers, for example, that $T(R_\lambda) = \{T_0;\ \lambda^{iT_0} = 1\}$.

When M is a Krieger factor, or more generally when $M = L^\infty(\mathcal{R},\mu)$ for some equivalence relation $\mathcal{R}$ (see Section 4; $\mathcal{R}$ is not necessarily assumed to be of the form $\mathcal{R}_T$), one also verifies by an easy calculation that

$$T(M) = 2\pi / \operatorname{Log}\rho(\mathcal{R}),$$

where ρ is Krieger's invariant, i.e. $\rho(\mathcal{R})$ is the set of all $\lambda > 0$ for which there exists a $\nu \sim \mu$ such that the Radon-Nikodým derivatives $\frac{d\nu(Sx)}{d\nu(x)}$ belong to $\lambda^{\mathbb{Z}}$ for any Borel transformation S that preserves the orbits of $\mathcal{R}$. The latter calculation shows easily that the equality $\rho(M) = \exp(2\pi/T(M))$ does not hold, in general, with the Araki-Woods definition of ρ.

Finally, we emphasize that Theorem 1, the analogue of the Radon-Nikodým theorem, makes it possible, for numerous von Neumann algebras, to describe explicitly *all* the semifinite normal weights and their groups of modular automorphisms.

Thus, let us take up again Example 4 of Section 1 and let us show how to represent every normal weight on the von Neumann M of random operators on $X = V/\Gamma$. Recall that a quadratic form on a Hilbert space $\mathfrak{H}$ is a mapping q of $\mathfrak{H}$ into $[0,+\infty]$ such that:

a) $q(\xi+\eta) + q(\xi-\eta) = 2q(\xi) + 2q(\eta)\quad (\forall \xi,\eta \in \mathfrak{H})$.

b) $q(\lambda\xi) = |\lambda|^2 q(\xi)\quad (\forall \xi \in \mathfrak{H}, \lambda \in \mathbb{C})$.

Assume moreover that $\operatorname{Dom} q = \{\xi \in \mathfrak{H};\ q(\xi) < \infty\}$ is dense in $\mathfrak{H}$ and that q is lower semicontinuous. Then there is a unique unbounded positive operator T such that $q(\xi) = \|T^{1/2}\xi\|^2\ (\forall \xi \in \mathfrak{H})$.

Let V be a manifold with $n = \dim V$, and use the differential of the action of Γ on V to trivialize the fiber $T(V)$ along each orbit of Γ.

Definition 2. *A random form consists in quadratic forms $q_{\alpha,\nu}$ on $\mathfrak{H}_\alpha$, for all $\alpha \in X$ and $\nu \in \wedge^n T_\alpha(X)$, such that:*

1) $q_{\alpha,\lambda\nu} = |\lambda| q_{\alpha,\nu}\quad \forall \lambda \in \mathbb{R}$.

2) *The mapping $x \mapsto q_{p(x),\nu(x)}(\xi_x)$ of V into $\mathbb{R}$ is measurable for all measurable sections ν and ξ.*

Recall that $p : V \to X$ is the canonical projection. The expression $q_{p(x)}(e_x)$ is by construction a *one density* on V which can be integrated (the dependence on x in V is that of a 1-density). One then sets $\int q = \int_V q_{p(x)}(e_x)$. Theorem 1 makes it possible to prove:

Theorem 3. 1) *Let q be a random form and, for every pair (α,ν), let $T_{\alpha,\nu}$ be the positive unbounded operator in $\mathfrak{H}_\alpha$ associated with the quadratic form $q_{\alpha,\nu}$.*

If $\int q = 1$ then we may define a normal state φ_q on M by

$$\varphi_q(A) = \int_V \langle A_{p(x)} T^{1/2}_{p(x)} e_x, T^{1/2}_{p(x)} e_x \rangle$$

for every random operator $A = (A_\alpha)_{\alpha \in X}$.

2) *The mapping $q \mapsto \varphi_q$ is a bijection between the random forms q, with $\int q = 1$, and the normal states on M.*

3) *φ_q is faithful if and only if the operator $T_{\alpha,\nu}$ is nonsingular for $\nu \neq 0$ and almost every $\alpha \in X$.*

4) *For every faithful normal state $\varphi = \varphi_q$ on M, the group of modular automorphisms σ_t^φ is given by*

$$(\sigma_t^\varphi(A))_\alpha = T^{it}_{\alpha,\nu} A_\alpha T^{-it}_{\alpha,\nu} \quad \text{for all } A = (A_\alpha) \in M.$$

5.β The factors of type III_λ

The Radon-Nikodým theorem makes it possible to calculate the invariant $S(M)$ from a single faithful semifinite normal weight φ on M. The centralizer M_φ of φ is defined by the equality

$$M_\varphi = \{x \in M; \quad \sigma_t^\varphi(x) = x \quad \forall t \in \mathbb{R}\}.$$

For every projection $e \neq 0$, $e \in M_\varphi$, a faithful semifinite normal weight φ_e on the reduced von Neumann algebra $eMe = \{x \in M;\ ex = xe = x\}$ is defined by the equality

$$\varphi_e(x) = \varphi(x) \qquad \forall x \in eMe, x \geq 0.$$

One then has the formula

$$S(M) = \bigcap_{e \neq 0} \text{Spectrum } \Delta_{\varphi_e},$$

where e varies over the nonzero projections of M_φ. Moreover, since e commutes with φ, the calculation of $\sigma_t^{\varphi_e}$, and hence of the spectrum of Δ_{φ_e}, is immediate:

$$\sigma_t^{\varphi_e}(x) = \sigma_t^\varphi(x) \qquad \forall x \in eMe.$$

The above formula permits calculating, for example, $S(M)$ for $M = L^\infty(\mathcal{R}, \mu)$ (Section 4). One has the equality

$$S(M) = r(\mathcal{R}),$$

where r is the invariant defined by Krieger as the set of essential values of the Radon-Nikodým derivatives. Therefore, in general, $S(M) \neq r_\infty(M)$. Moreover, there is a much more satisfying interpretation of $S(M)$ as the spectrum of the modular homomorphism δ.

Let us suppose that the predual M_* is separable. Then a one-parameter group $(\alpha_t)_{t \in \mathbb{R}}$ of automorphisms of M is of the form σ_t^φ for some faithful semifinite normal weight φ on M if and only if the class $\varepsilon(\alpha_t)$ of α_t in $\operatorname{Out} M$ is equal to $\delta(t)$ for all t. To put it another way, the set of all of the groups of modular automorphisms for faithful semifinite normal weights on M is precisely the set of multiplicative Borel sections of δ.

For every faithful semifinite normal weight φ, the spectrum of Δ_φ may be identified with the spectrum of σ^φ in the following sense:

$$\text{Spectrum } \sigma^\varphi = \{\lambda \in \text{ dual group of } \mathbb{R};\ \hat{f}(\lambda) = 0 \text{ for all } f \in L^1(\mathbb{R}) \text{ such that } \int f(t)\sigma_t^\varphi dt = 0\}.$$

The spectrum of σ^φ can also be defined in terms of the supports of the distributions $(\sigma^\varphi(x))^\wedge$ which are Fourier transforms of the functions $t \mapsto \sigma_t^\varphi(x)$. One obtains in this way distributions with values in M, and Spectrum σ^φ is the closure of the union of the supports of the $(\sigma^\varphi(x))^\wedge)$.

We also note that in the above formula, $\mathbb{R}_+^*$ is identified with the dual group of $\mathbb{R}$ by the equality $\langle\lambda, t\rangle = \lambda^{it}$ $(\lambda \in \mathbb{R}_+^*, t \in \mathbb{R})$. The precise equality is then

$$\operatorname{Sp} \Delta_\varphi \cap \mathbb{R}_+^* = \text{Spectrum } \sigma^\varphi.$$

Therefore

$$\mathbb{R}_+^* \cap S(M) = \bigcap_{\varepsilon \circ \alpha = \delta} \text{Spectrum } \alpha,$$

and this formula shows that when M is a factor, $\mathbb{R}_+^* \cap S(M)$ is a subgroup of $\mathbb{R}_+^*$. Moreover, $0 \in S(M) \iff M$ is of type III, and one has the following cases:

$$\begin{aligned}
&\mathrm{III}_0 \quad S(M) = \{0, 1\},\\
&\mathrm{III}_\lambda \quad \lambda \in \,]0,1[: \quad S(M) = \lambda^{\mathbb{Z}} \cup \{0\},\\
&\mathrm{III}_1 \quad S(M) = [0, +\infty[.
\end{aligned}$$

In a certain sense, the above λ expresses the distance between M and the semifinite factors; in fact, λ is related in a monotone and one-to-one fashion to a quantity that measures the obstruction to the existence of a trace on M

$$d(M) = \text{diameter}(S/\operatorname{Inn} M),$$

where S denotes the metric space of normal states on M (with the metric $d(\varphi_1, \varphi_2) = \|\varphi_1 - \varphi_2\|$) and where $\operatorname{Inn} M$ acts on S by $\varphi \mapsto u\varphi u^*$ for u unitary in M. For M of type III_1, $d(M) = 0$ ([Co-St$_1$]); thus one cannot distinguish between two states of a factor of type III_1 by means of a property that is closed and invariant under inner automorphisms.

Let us now give a heuristic interpretation of $S(M)$. We first return to the origins of the terminology "modular automorphisms". The primary example of a left Hilbert algebra is the convolution algebra of continuous functions with compact support on a locally compact group G. Let dg be a left Haar measure on G. The modulus of the group is then the homomorphism δ_G of G into $\mathbb{R}_+^*$ associated with the right action of G on dg. Moreover, the modular operator of the left Hilbert algebra is the operator of multiplication by the function δ_G

in the space $L^2(G, dg)$, thus its spectrum is the closure of the range of δ_G. One can then, from a heuristic point of view, always interpret the invariant $S(M)$ for a factor M as "the image of the modulus of M".

The factors of type III_λ, $\lambda \in]0,1[$, are characterized by the equality $S(M) = \lambda^{\mathbb{Z}} \cup \{0\}$; now, it is easy to see that if G is a locally compact group and the image of the modulus δ_G of G is $\lambda^{\mathbb{Z}}$, then G is the semidirect product of the unimodular group $H = \ker \delta_G$ by an automorphism $\alpha \in \mathrm{Aut}\, H$ that multiplies every Haar measure on H by λ. Conversely, every pair (H, α), where H is unimodular and α multiplies every Haar measure on H by λ, yields by semidirect product a locally compact group $G = H \times_\alpha \mathbb{Z}$ with $\delta_G(G) = \lambda^{\mathbb{Z}}$. This analogy is confirmed by the following theorem:

Theorem 4. [Co$_4$] *Let* $\lambda \in]0,1[$.

a) *Let M be a factor of type* III_λ. *There exist a factor N of type* II_∞ *and a* $\theta \in \mathrm{Aut}\, N$ *multiplying every trace of N by λ (one then writes* mod $\theta = \lambda$) *such that M is isomorphic to the crossed product of N by θ.*

b) *Let N be a factor of type* II_∞ *and let* $\theta \in \mathrm{Aut}\, N$ *with* mod $\theta = \lambda$. *Then the crossed product of N by θ is a factor of type* III_λ.

c) *Two pairs* (N_i, θ_i) $(i = 1, 2)$ *yield isomorphic factors if and only if there exists an isomorphism σ of N_1 onto N_2 such that the classes of* $\sigma\theta_1\sigma^{-1}$ *and* θ_2, *modulo the inner automorphisms of* N_2, *are the same.*

Before studying the implications of this theorem for the problem of classifying factors, let us note some important information concerning the general theory of factors of type III_λ. The concept of trace is replaced by the following:

Definition 5. *A* generalized trace φ *on a factor M of type* III_λ, $\lambda \in]0,1[$, *is a faithful semifinite normal weight φ such that* $\mathrm{Sp}\, \Delta_\varphi = S(M)$ *and* $\varphi(1) = +\infty$.

One proves ([Co$_4$]) the existence of generalized traces on M by studying the relations between the invariants $T(M)$ and $S(M)$ and by showing that, except when $S(M) = \{0, 1\}$, the invariant $T(M)$ is determined by $S(M)$; whereas, in the case III_0, $T(M)$ can be any denumerable subgroup of $\mathbb{R}$, not necessarily closed. Moreover, the following uniqueness theorem holds: if φ_1 and φ_2 are two generalized traces on M, then there exists an *inner* automorphism α such that φ_2 is proportional to $\varphi_1 \circ \alpha$.

The von Neumann algebra of type II_∞, the N of the theorem, is none other than the centralizer M_φ of a generalized trace φ; its position in M is *unique up to inner automorphisms* and it can be characterized as a maximal semifinite subalgebra [Co$_4$].

Theorem 4 shows that the problem of classifying the factors of type III_λ reduces to:

1) classifying the factors of type II_∞;
2) given a factor N of type II_∞, determining the conjugacy classes of θ in $\mathrm{Out}\, N = \mathrm{Aut}\, N / \mathrm{Inn}\, N$ such that mod $\theta = \lambda$.

These two problems are the principal motivations for Sections 6 and 7 below, Problem 2 being subsumed under noncommutative ergodic theory.

V.6. Noncommutative Ergodic Theory

Let $(X, \mathcal{B}, \mu)$ be a standard Borel space equipped with a probability measure μ, and let T be a Borel transformation of $(X, \mathcal{B})$ that leaves μ invariant. Let $M = L^\infty(X, \mathcal{B}, \mu)$ and let φ be the state associated with μ. Then T determines an automorphism of M that preserves φ, by the equality

$$\theta(f) = f \circ T^{-1}.$$

Conversely, every automorphism of M that preserves φ can be obtained in this way. Thus, classical ergodic theory is, after translation, the same thing as the study, up to conjugacy, of the automorphisms of M that fix φ. In fact, one of the justifications for the theory is that all of the triples $(X, \mathcal{B}, \mu)$ (hence all of the pairs (M, φ)) with $\mu(\{x\}) = 0$ for all $x \in X$ are isomorphic. Thus, to each different construction of such a triple there will correspond a family of automorphisms of $(X, \mathcal{B}, \mu)$ and the problem is to compare them. Similarly, in the framework of noncommutative integration theory, there exist numerous different constructions of the hyperfinite factor R, for example as the regular representation of a locally finite discrete group (Section 1), as the infinite tensor product of the pairs $(M_n(\mathbb{C}), \tau_n)$ with τ_n the trace normalized by $\tau_n(1) = 1$, or again by the theorem of H. Dye (Section 4). To each of these constructions, there correspond automorphisms of R. Indeed, R can also be constructed from the canonical anti-commutation relations on a real Hilbert space E, thus obtaining an injective homomorphism of the orthogonal group of E into $\operatorname{Aut} R$. It follows that $\operatorname{Out} R$ in fact contains every separable locally compact group.

Of course, one does not want to distinguish between two automorphisms of the form θ and $\sigma\theta\sigma^{-1}$, where $\sigma \in \operatorname{Aut} R$. Let us adopt the following general definitions:

Definition 1. *Let M be a von Neumann algebra and let $\theta_1, \theta_2 \in \operatorname{Aut} M$ be two automorphisms of M:*

a) *θ_1 and θ_2 are said to be* conjugate *if there exists a $\sigma \in \operatorname{Aut} M$ such that $\theta_2 = \sigma\theta_1\sigma^{-1}$.*

b) *θ_1 and θ_2 are said to be* outer conjugate *if there exists a $\sigma \in \operatorname{Aut} M$ such that $\theta_2 = \sigma\theta_1\sigma^{-1}$ modulo* $\operatorname{Inn} M$.

When M is commutative, the two definitions coincide because $\operatorname{Inn} M = \{1\}$. In the general case, we have two problems: conjugacy and outer conjugacy. I shall begin with results that extend important results of classical ergodic theory to the noncommutative case. I shall then discuss some specifically noncommutative phenomena; the reader interested only in the automorphisms of R can pass directly to Theorem 14 below.

6.α Rokhlin's theorem

Let $(X, \mathcal{B}, \mu)$ be a standard Borel probability space and let T be a Borel transformation of $(X, \mathcal{B})$ that preserves μ. Then there exists an essentially unique partition of X, $X = \bigcup_{i=1}^{\infty} X_i$, where each X_i is invariant under T, and where:

1) For every $i > 0$, the restriction of T to X_i is periodic of period i and

$$\operatorname{card}\{T^j x\} = i \quad \forall x \in X_i.$$

2) For $i = 0$, the restriction of T to X_0 is *aperiodic*, i.e.

$$\operatorname{card}\{T^j x\} = \infty \quad \forall x \in X_0.$$

Rokhlin's theorem may be stated as follows. Let T be an aperiodic transformation of $(X, \mathcal{B}, \mu)$. For every $\epsilon > 0$ and $n > 0$, there exists a Borel set E such that $E, T(E), \ldots, T^{n-1}(E)$ are pairwise disjoint and

$$\mu\Big(X \setminus \bigcup_{j=0}^{n-1} T^j(E)\Big) < \epsilon.$$

Now let (N, τ) be a pair (von Neumann algebra, faithful normal trace normalized by $\tau(1) = 1$) and let θ be an automorphism of N that preserves τ. Then there exists a partition of unity in N, $\sum_{j=0}^{\infty} e_j = 1$, where each e_j is a projection in the *center* of N that is invariant under θ, and where:

a) For every $j > 0$, the restriction of θ to the reduced von Neumann algebra N_{e_j} satisfies the condition that θ^j is inner and, for $k < j$ and for every nonzero projection $e \leq e_j$, there exists a nonzero projection $f \leq e$ such that

$$\|f\theta^k(f)\| \leq \epsilon.$$

b) For $j = 0$, the restriction of θ to the von Neumann algebra N_{e_0} is aperiodic: for every $k > 0$, every nonzero projection $e \leq e_0$ and every $\epsilon > 0$, there exists a nonzero projection $f \leq e$ with

$$\|f\theta^k(f)\| \leq \epsilon.$$

As in the commutative case, this decomposition is unique. When $e_0 = 1$, θ is said to be *aperiodic*.

Theorem 2. [Co$_6$] *Let (N, τ) be a pair (von Neumann algebra, normalized faithful normal trace) and let θ be an aperiodic automorphism of N that preserves τ. For every integer $n > 0$ and every $\epsilon > 0$, there exists a partition of unity $\sum_{j=0}^{n-1} E_j = 1$ in N, where the E_j are projections, such that*

$$\|\theta(E_j) - E_{j+1}\|_2 \leq \epsilon \quad (j = 0, 1, \ldots, n-1, E_n = E_0).$$

Here, as in Section 1, we write $\|x\|_2 = (\tau(|x|^2))^{1/2}$ for all $x \in N$.

6.β Entropy ([Co$_{27}$], [Co-St$_2$], [Co-N-T])

Let $(X, \mathcal{B}, \mu, T)$ be as above and let $\mathcal{P}$ be a Borel partition of X. One uses the term *the entropy of $\mathcal{P}$ relative to T* for the scalar $h(T, \mathcal{P})$ that asymptotically counts $1/n$ times the logarithm of the number of elements of the partition generated by $\mathcal{P}, T\mathcal{P}, \ldots, T^{n-1}\mathcal{P}$:

$$h(T, \mathcal{P}) = \lim_{n\to\infty} \frac{1}{n} h(\mathcal{P} \vee T\mathcal{P} \vee \ldots \vee T^{n-1}\mathcal{P}),$$

where, for a partition $Q = (q_j)_{j\in\{1,\ldots,k\}}$,

$$h(Q) = \sum_{j=1}^{k} \eta(\mu(q_j))$$

with $\eta(t) = -t \log t$ for $t \in]0, 1]$ and $\eta(0) = 0$.

The *entropy of* T is then defined to be the supremum of the $h(T, \mathcal{P})$. It is an invariant, calculable thanks to the Kolmogorov–Sinai theorem: $h(T) = h(T, \mathcal{P})$ for every partition $\mathcal{P}$ for which the $T^j\mathcal{P}$ generate the σ-algebra $\mathcal{B}$. In particular, let us consider a Bernoulli shift: the translation by 1 in $\prod_{\nu\in\mathbb{Z}}(X_\nu, \mu_\nu)$, where the X_ν are all equal to $\{1, \ldots, p\}$ and all of the μ_ν are given by the same measure $j \in \{1, \ldots, p\} \to \lambda_j$. Then $h(T) = \sum_{j=1}^{p} \eta(\lambda_j)$.

The Bernoulli shifts have an analogue in the noncommutative case; let us take the simplest, associated with an integer p. One considers the infinite tensor product $\bigotimes_{\nu\in\mathbb{Z}}(M_\nu, \varphi_\nu)$, where, for all ν, $M_\nu = M_p(\mathbb{C})$ is the algebra of all $p \times p$ matrices and φ_ν is the normalized trace. The shift S_p then yields an automorphism of the pair (R, τ), where R is the hyperfinite factor and τ is its normalized trace. One can pose the following question: Are the S_p pairwise conjugate? This problem led me, together with E. Størmer, to the following generalization of entropy and of the Kolmogorov–Sinai theorem, which enabled us to distinguish the automorphisms S_p up to conjugacy.

These first results were limited to the *unimodular* case, where the measure space $(X, \mathcal{B}, \mu)$ is replaced by a von Neumann algebra M with a finite normal trace τ, and where the automorphism $\theta \in \operatorname{Aut} M$ preserves τ. The role of the finite partitions $Q = (q_j)$ of the space X is played by the finite-dimensional subalgebras $K \subset M$, but at the very outset one encounters an entirely new difficulty: in the noncommutative case it is, in general, false that the subalgebra of M generated by two finite-dimensional subalgebras K_1 and K_2 is again finite-dimensional. (For example, the von Neumann algebra $R(\Gamma)$ of the group $\Gamma = \mathbb{Z}_2 * \mathbb{Z}_3 = PSL(2, \mathbb{Z})$ is generated by two subalgebras $K_1 = \mathbb{C}(\mathbb{Z}_2), K_2 = \mathbb{C}(\mathbb{Z}_3)$ of dimensions 2 and 3.) It is then impossible to make use of the concept of a composite partition $\mathcal{P}_1 \vee \mathcal{P}_2$ as in the commutative case. The key to resolving this difficulty came from the noncommutative theory of information, and in particular from the solution by E. Lieb [Lie] of the Wigner–Yanase–Dyson conjecture on the concavity in the variable $\rho \in M_+$ of the functional

$$I_p(\rho, T) = \tau([\rho^p, T]^*[\rho^{1-p}, T]), \quad \text{where } p \in]0, 1[, T \in M.$$

An immediate corollary of E. Lieb's result is the convexity of the functional

$$S(\rho_1,\rho_2) = \tau(\rho_1(\log\rho_1 - \log\rho_2)) \quad (\rho_j \in M^+).$$

This enabled E. Størmer and me to define a function $H(K_1,\ldots,K_n) \in [0,\infty)$, where the K_j are arbitrary finite-dimensional subalgebras of M, that plays the role of $H(K_1 \vee K_2 \vee \cdots \vee K_n)$, that is, it has the following properties:

a) $H(N_1,\ldots,N_k) \le H(P_1,\ldots,P_k)$ if $N_j \subset P_j$ for $j = 1,\ldots,k$.

b) $H(N_1,\ldots,N_k,N_{k+1},\ldots,N_p) \le H(N_1,\ldots,N_k) + H(N_{k+1},\ldots,N_p)$.

c) $P_1,\ldots,P_n \subset P \Longrightarrow H(P_1,\ldots,P_n,P_{n+1},\ldots,P_m) \le H(P,P_{n+1},\ldots,P_m)$.

d) For every family of minimal projections $e_\alpha \in N$ such that $\sum e_\alpha = 1$, one has $H(N) = \sum \eta(\tau(e_\alpha))$.

e) If $(N_1 \cup N_2 \cup \cdots \cup N_k)''$ is generated by pairwise commuting von Neumann subalgebras $P_j \subset N_j$, then $H(N_1,\ldots,N_k) = H((N_1 \cup \cdots \cup N_k)'')$.

The functional H is moreover continuous in the sense that it satisfies the inequality

f) $H(N_1,\ldots,N_k) \le H(P_1,\ldots,P_k) + \sum_j H(N_j|P_j)$,

where the relative entropy functional $H(N|P)$ satisfies

$\forall\epsilon > 0, n \in \mathbb{N},\ \exists\delta > 0$ such that
$(\dim N = n$ and $(\forall x \in N, \|x\| \le 1)\ \exists y \in P \ni \|x - y\|_2 \le \delta) \Longrightarrow H(N|P) < \epsilon.$

It is this stability of $H(N_1,\ldots,N_k)$ under perturbation of the N_j that would not be satisfied by the functional $H((N_1 \cup N_2 \cup \cdots \cup N_k)'')$.

It is then easy to define, for every automorphism θ of the pair (M,τ), the quantities

$$H(N,\theta) = \lim_{k\to\infty} \frac{1}{k} H(N,\theta(N),\ldots,\theta^k(N))$$
$$H(\theta) = \sup_N H(N,\theta).$$

Moreover, if M is hyperfinite then we have the following analogue of the Kolmogorov-Sinai theorem:

Theorem 3. *Let N_k be an increasing sequence of finite-dimensional subalgebras of M, with $\bigcup N_k$ weakly dense in M. Then*

$$H(\theta) = \sup_k H(N_k,\theta)$$

for every automorphism θ of the pair (M,τ).

Using Properties d) and e), one easily obtains:

Corollary 4. *The shifts S_n of the hyperfinite factor R are pairwise non-conjugate and $H(S_n) = \log n$.*

I refer the reader to [Co-St$_2$] and [Pi-Po] for other entropy calculations. As in the classical case, the relative entropy $H(N|P)$ still has meaning for certain infinite-dimensional pairs (N, P) and plays an important role in [Pi-Po].

I have deliberately omitted giving the explicit definition of the functional H ([Co-St$_2$]), since it only attains its conceptual form with the solution of the general case (*non-unimodular*) given in [Co$_{27}$] and developed in [Co-N-T], which I shall now describe.

Let M be a not necessarily finite von Neumann algebra. The functional $S(\rho_1, \rho_2)$ defined above still has meaning ([Ar$_1$]). It is called the relative entropy. However, ρ_1 and ρ_2 are no longer positive elements of M but are *positive linear forms* $\varphi_1, \varphi_2 \in M_*$. The convexity of this functional $S(\varphi_1, \varphi_2)$ remains true in full generality and has a simple proof ([Ko$_1$], [Pu-W]).

The key concept that makes it possible to treat the general non-unimodular case is that of the *entropy defect* of a completely positive mapping.

Definition 5. *Let A and B be C^*-algebras. A linear mapping $T : A \to B$ is said to be* completely positive *if, for every n, the mapping $1 \otimes T : M_n(A) \to M_n(B)$ is positive.*

I refer the reader to [Arv$_2$] for further information. The principal interest of this concept is that, although the category of C^*-algebras and $*$-homomorphisms has relatively few morphisms, in any case not enough to connect the general C^*-algebras to the *commutative* C^*-algebras, the category of C^*-algebras and *completely positive mappings* is much more flexible, while not being too different from the former thanks to the Stinespring-Kasparov theorem ([Arv$_2$], [Kas$_2$]). Thus, for example, if B is a commutative unital C^*-algebra and $X = \mathrm{Sp}(B)$ is the spectrum (a compact space) of B, then a completely positive mapping T, with $T(1) = 1$, of a C^*-algebra A into B is just a weakly continuous mapping of X into the space $S(A)$ of states of A, $S(A) \subset A^*$. If A is a finite-dimensional C^*-algebra and φ is a state on A, there corresponds to φ a list (λ_i) of positive real numbers, called the eigenvalue list of φ, obtained by setting $\lambda_i = \varphi(e_i)$, where the e_i are minimal projections of A with sum 1, belonging to the centralizer A_φ of φ in A. This list with multiplicity does not depend on the choice of the e_i, and $\sum \lambda_i = 1$. One sets

$$S(\varphi) = \sum \eta(\lambda_i).$$

Let A and B be finite-dimensional C^*-algebras, with B commutative, let μ be a state on B, i.e. a probability measure on $X = \mathrm{Sp}(B)$, and let T be a completely positive mapping of A into B such that $T(1) = 1$.

Definition 6. *The* entropy defect *of T is defined to be the scalar*

$$s_\mu(T) = S(\mu) - \int_X S(T^*\mu, T_x^*)\,d\mu(x).$$

Here $T^*\mu = \mu \circ T$, T_x^* is the image under T^* of the pure state on B associated with $x \in X = \mathrm{Sp}(B)$ and $S(\cdot,\cdot)$ is as above the relative entropy.

One proves that $s_\mu(T) \geq 0$. This number measures the *information loss* due to the translation T of the quantum system (A, φ), where $\varphi = T^*\mu = \mu \circ T$, into a classical system (X, μ). The typical case in which this information loss $s_\mu(T)$ is zero is the one in which T is the conditional expectation of A onto a maximal abelian subalgebra B of the centralizer of φ.

Proposition 7. a) *Let $T_1, T_2 : A \to B$ be completely positive, with $\mu \circ T_i = \varphi$. Then, for every $\lambda \in [0,1]$,*

$$s_\mu(\lambda T_1 + (1-\lambda)T_2) \geq \lambda s_\mu(T_1) + (1-\lambda)s_\mu(T_2).$$

b) *If the state T_x^* is pure for every $x \in X = \mathrm{Sp}(B)$, then*

$$s_\mu(T) = S(\mu) - S(T^*\mu).$$

c) *If $T_1 : A_1 \to A$ is completely positive and unital, then*

$$s_\mu(T \circ T_1) \geq s_\mu(T).$$

d) *For every subalgebra $B_1 \subset B$, let E_1 be the conditional expectation of B onto B_1 associated with μ, and let $T_{B_1} = E_1 \circ T$. Then*

$$s_\mu(T_{B_1 \vee B_2}) \leq s_\mu(T_{B_1}) + s_\mu(T_{B_2})$$

for every pair B_1, B_2 of subalgebras of B, where $B_1 \vee B_2$ denotes the subalgebra generated by B_1 and B_2.

It is property d) that plays a determining role. Let M be a von Neumann algebra and let φ be a faithful normal state on M. Proposition 7 makes it possible to define the functional $H(N_1, \ldots, N_k)$ in full generality, where the N_k are finite-dimensional subalgebras of M. Since this functional depends on the choice of φ, it will be denoted H_φ. The formula that gives $H_\varphi(N_1, \ldots, N_k)$ is the same as that of [Co27] but is clarified conceptually by the idea of entropy defect.

One defines $H_\varphi(N_1, \ldots, N_k)$ to be the supremum of the quantities

$$S(\mu | \vee B_j) - \sum_j s_\mu(T_j), \qquad (*)$$

where (X, μ) is a finite probability space, $B = C(X)$, T is a completely positive mapping of A into B such that $T^*\mu = \varphi$, the B_j are subalgebras of B, and $s_\mu(T_j)$ is the entropy defect of the mapping $T_j = (E_j \circ T)|N_j$ of N_j into B_j.

In other words, one optimizes a *commutative translation* of the situation $(M, \varphi, (N_j))$ that one compares with the commutative situation $(B, \mu, (B_j))$. In the commutative case, the natural quantity is the quantity of information or entropy $S(\mu | \vee B_j)$ of the partition generated by the B_j. However, because of the loss of information in the translation, one must subtract the quantity $\sum_j s_\mu(T_j)$, which yields the formula $(*)$.

It is remarkable that in a great many cases one can effectively calculate the maximum $H_\varphi(N_1,\ldots,N_k)$ of the quantity $(*)$ over all possible commutative translations. This maximum $H_\varphi(N_1,\ldots,N_k)$ is *finite* and satisfies properties analogous to the properties a), b), c), d) and e) listed above ([Co-N-T]). Of course it coincides with the functional H of [Co-St$_2$] when φ is a trace.

In particular, if $\theta \in \operatorname{Aut} M$ is an automorphism that preserves φ, then the limit

$$H_\varphi(N,\theta) = \lim_{k\to\infty} \frac{1}{k} H_\varphi(N,\theta(N),\ldots,\theta^k(N))$$

exists and is finite for every finite-dimensional subalgebra $N \subset M$, and, with $H_\varphi(\theta) = \sup_N H_\varphi(N,\theta)$, we have the following analogue of Theorem 3:

Theorem 8. *Let M be a von Neumann algebra, φ a faithful normal state on M, and $\theta \in \operatorname{Aut} M$ such that $\varphi \circ \theta = \varphi$. Suppose that there exists an increasing sequence $N_k \subset N_{k+1}$ of finite-dimensional subalgebras of M whose union is weakly dense in M (M is then said to be hyperfinite). Then*

$$H_\varphi(\theta) = \lim_{k\to\infty} H_\varphi(N_k,\theta).$$

We thus have available, in the noncommutative case, the analogue of the Kolmogorov-Sinai theory, and one of the most interesting open questions is to use it in quantum statistical mechanics in the same way that the Kolmogorov-Sinai entropy is used in the formalism of thermodynamics ([Rue$_1$]).

The theory of entropy of the automorphisms $\theta \in \operatorname{Aut} M$ of a von Neumann algebra that preserve a state φ on M enables one to formulate the following variational problem. Let A be a nuclear C^*-algebra and let $\theta \in \operatorname{Aut} A$ be an automorphism of A. For every selfadjoint element $V = V^* \in A$, by analogy with the classical case [Rue$_1$] one defines, for fixed $\beta \in [0,+\infty)$,

$$P_\beta(V) = \sup_{\varphi\circ\theta=\varphi} (H_\varphi(\theta) - \beta\varphi(V)),$$

where the supremum is taken over the compact convex set of states φ on A that are invariant under θ.

Suppose that the automorphism θ is sufficiently asymptotically abelian for the following equality to define a derivation δ generating a one-parameter group σ_t of automorphisms of A:

$$\delta(x) = \sum_{n\in\mathbb{Z}} [\theta^n(V), x].$$

The general problem is then as follows:

Problem. *Compare the θ-invariant states on A such that $H_\varphi(\theta) - \beta\varphi(V) = P_\beta(V)$ with the states that are β-KMS for σ_t.*

Let us now take up the discussion of the *simplifications* brought about by noncommutativity; we shall see, for example, that all of the automorphisms S_p are outer conjugate.

6.γ Approximately inner automorphisms

Let M be a von Neumann algebra and let M_* be its predual. The action of $\operatorname{Aut} M$ on M_*, equipped with the norm topology, is equicontinuous; it follows that the topology of pointwise convergence in norm in M_* makes $\operatorname{Aut} M$ a *topological group*.

In what follows, when I speak of $\operatorname{Aut} M$ as a topological group, I shall always be referring to this topology. To convince oneself that this is the right structure on $\operatorname{Aut} M$, it suffices to observe that if M_* is separable, then the topological group $\operatorname{Aut} M$ is *Polish*.

In general, the group $\operatorname{Inn} M \subset \operatorname{Aut} M$ is not closed; for example, for the hyperfinite factor R, $\overline{\operatorname{Inn} R} = \operatorname{Aut} R$. More precisely, assuming M finite, in order that $\operatorname{Inn} M$ be closed in $\operatorname{Aut} M$ it is necessary and sufficient that M not satisfy the property Γ of Section 1 ([Co_3], [S_2]).

When M is a factor of type II_1, the approximately inner automorphisms of M are characterized by the following equivalence:

Theorem 9. [Co_5] *Let N be a factor of type II_1 with separable predual, acting on the Hilbert space $\mathfrak{H} = L^2(N,\tau)$, where τ denotes the normalized trace of N. For an automorphism $\theta \in \operatorname{Aut} N$, the following conditions are equivalent:*

a) $\theta \in \overline{\operatorname{Inn} N}$, *i.e., θ is approximately inner;*

b) $\|\sum_{i=1}^n \theta(a_i)b_i\| = \|\sum_{i=1}^n a_i b_i\|$ *for $a_1,\dots,a_n \in N$ and $b_1,\dots,b_n \in N'$ (the commutant of N).*

Condition b) shows that there then exists an automorphism α of the C^*-algebra $C^*(N,N')$ generated by N and N', such that $\alpha(a) = \theta(a)$ $(\forall a \in N)$ and $\alpha(b) = b$ $(\forall b \in N')$.

Corollary 10. *Let $(N_i)_{i=1,2}$ be factors of type II_1 with separable predual, and let $(\theta_i)_{i=1,2}$ be automorphisms of the N_i. Then*

$$\theta_1 \otimes \theta_2 \text{ approximately inner} \iff \theta_1 \text{ and } \theta_2 \text{ approximately inner.}$$

6.δ Centrally trivial automorphisms

Let N be a factor of type II_1 with separable predual, τ its normalized trace, and θ an approximately inner automorphism of N, $\theta \in \overline{\operatorname{Inn} N}$. Then there exists a sequence $(u_k)_{k\in\mathbb{N}}$ of unitaries of N such that, for every $x \in N$,

$$\theta(x) = \lim_{k\to\infty} u_k x u_k^*$$

for the topology of $L^2(N,\tau)$: $\|x-y\|_2 = (\tau(|x-y|^2))^{1/2}$.

This property is translated into an equality

$$\theta(x) = uxu^* \quad \forall x \in N$$

by introducing a von Neumann algebra containing N in the following way:

Definition 11. *For every ultrafilter $\omega \in \beta\mathbb{N}\setminus\mathbb{N}$ let N^ω be the ultraproduct, N^ω = the von Neumann algebra $\ell^\infty(\mathbb{N}, N)$ divided by the ideal of sequences $(x_n)_{n\in\mathbb{N}}$ such that $\lim_{n\to\omega} \|x_n\|_2 = 0$.*

One proves that this ultraproduct is a finite von Neumann algebra ([Duf], [Ve]) even though, in general, the above-mentioned ideal is not $\sigma(\ell^\infty, \ell^\infty_*)$-closed. Moreover, N may be canonically embedded in the ultraproduct N^ω by associating to $x \in N$ the constant sequence $(x)_{n\in\mathbb{N}}$. The sequence of unitaries $(u_k)_{k\in\mathbb{N}}$ defines a unitary $u \in N^\omega$ and, of course, $\theta(x) = uxu^*$ for all $x \in N$. This equality determines u uniquely, modulo the unitary group of a von Neumann subalgebra of N^ω that plays a crucial role in the sequel:

Definition 12. *Let N and ω be as above; the* asymptotic centralizer *of N for ω is defined to be the commutant N_ω of N in N^ω:*

$$N_\omega = \{y \in N^\omega;\ \ yx = xy\ \forall x \in N\}.$$

The construction of N^ω (resp. N_ω) is functorial, so that every automorphism θ of N defines an automorphism θ^ω (resp. θ_ω) of N^ω (resp. N_ω). As above, let $\theta \in \overline{\operatorname{Inn} N}$ and $u \in N^\omega$ unitary be such that

$$\theta(x) = uxu^* \quad \forall x \in N.$$

The question that arises is then the following: Can u be chosen so that $\theta^\omega(u) = u$? One can multiply u by a unitary v of N_ω without changing the equality $\theta(x) = uxu^*$ $(x \in N)$. Thus, setting $w = u^*\theta^\omega(u)$, the problem is to find $v \in N_\omega$ with $v^*\theta_\omega(v) = w$. By construction, w is a unitary of N_ω. Thus, the problem is to characterize the unitaries of N_ω of the form $v^*\theta_\omega(v)$, $v \in N_\omega$. Rokhlin's theorem for noncommutative ergodic theory yields a complete answer to this problem in the following form.

1) The partition of unity of the center of N_ω associated with the automorphism θ_ω of N_ω is formed by a single $e_j = 1$, and θ_ω^k is outer for $k < j$ and is equal to 1 for $k = j$.

2) For a unitary $w \in N_\omega$ to be of the form $v^*\theta_\omega(v)$, $v \in N_\omega$, it is necessary and sufficient that $w\theta_\omega(w)\cdots\theta_\omega^{j-1}(w) = 1$.

Moreover, the integer j depends only on θ and not on the choice of $\omega \in \beta\mathbb{N}\setminus\mathbb{N}$. It is denoted by $p_a(\theta)$ and called the asymptotic period of θ. This is the period of θ modulo the normal subgroup $\operatorname{Ct} N$ of $\operatorname{Aut} N$ consisting of the automorphisms of N that are *centrally trivial* in the following sense:

$\theta \in \operatorname{Ct} N$ if and only if $\theta_\omega = 1$ for some $\omega \in \beta\mathbb{N}\setminus\mathbb{N}$, or equivalently for every $\omega \in \beta\mathbb{N}\setminus\mathbb{N}$.

Given this definition, let us return to the above problem. The problem now is to know whether $w\theta_\omega(w)\cdots\theta_\omega^{j-1}(w) = 1$ when $j = p_a(\theta)$ and $w = u^*\theta^\omega(u)$. This reduces to asking whether $(\theta^\omega)^j(u) = u$. Now, the period of

θ^{ω} is the same as that of θ, and the problem is to compare it with $p_a(\theta)$; one has

$$\theta^{p_a} = 1,\ \theta \in \overline{\operatorname{Inn} N} \implies \exists \text{ unitaries } (u_n)_{n\in\mathbb{N}} \text{ of } N \text{ such that}$$

$$\theta(u_n) - u_n \to 0 \text{ as } n \to \infty,\ \ \theta(x) = \lim_{n\to\infty} u_n x u_n^* \ \ (\forall x \in N).$$

In particular, if $p_a = 0$ then the condition is satisfied. This is the main motivation for trying to determine the group $\operatorname{Ct} N$ in general. The following theorem may be deduced from [Co$_5$].

Theorem 13. *Let N be a factor of type II_1 with separable predual, acting on $\mathfrak{H} = L^2(N,\tau)$, and let $\theta \in \operatorname{Aut} N$, U the unitary on $L^2(N,\tau)$ associated with θ (the construction of L^2 is functorial). Let $p = p_a(\theta)$ be the asymptotic period of θ and let $\lambda \in \mathbb{C}$, $|\lambda| = 1$.*

Then, in order that $\lambda^p = 1$, it is necessary and sufficient that there exist an automorphism α_λ of the C^*-algebra generated by N, N' and U, such that

$$\alpha_\lambda(U) = \lambda U,\ \ \alpha_\lambda(A) = A \qquad \forall A \in C^*(N,N').$$

The above theorem shows that $\theta_1 \otimes \theta_2 \in \operatorname{Ct}(N_1 \otimes N_2)$ if and only if θ_1 and θ_2 are centrally trivial. Another interesting characterization of $\operatorname{Ct} N$, for N a factor of type II_1 such that $\varepsilon(\overline{\operatorname{Inn} N})$ is noncommutative, where ε is the quotient mapping $\operatorname{Aut} N \to \operatorname{Out} N$, is as follows ([Co$_6$]):

$\varepsilon(\operatorname{Ct} N)$ is the commutant of $\varepsilon(\overline{\operatorname{Inn} N})$ in $\operatorname{Out} N$.

It suffices to know $\varepsilon(\operatorname{Ct} N)$ in order to know $\operatorname{Ct} N$, since $\operatorname{Inn} N \subset \operatorname{Ct} N$ always. If N is the hyperfinite factor R, then $\operatorname{Ct} R = \operatorname{Inn} R$.

6.ε The obstruction $\gamma(\theta)$

Let M be a factor and let $\theta \in \operatorname{Aut} M$. Let $p_0(\theta) \in \mathbb{N}$ be the period of θ modulo inner automorphisms:

$$\theta^j \in \operatorname{Inn} M \iff j \in p_0\mathbb{Z}.$$

This is an invariant of θ under outer conjugation. Suppose $p_0 \neq 0$ and let us try to find a θ' outer conjugate to θ such that $\theta'^{p_0} = 1$. We have a homomorphism of $\mathbb{Z}/p_0\mathbb{Z}$ into $\operatorname{Out} M$ and the problem is to lift it into $\operatorname{Aut} M$. Since the center of the unitary group $\mathcal{U}$ of M is equal to the torus $\mathbb{T} = \{z \in \mathbb{C};\ |z| = 1\}$, the obstruction associated with this problem is an element of $H^3(\mathbb{Z}/p_0\mathbb{Z}, \mathbb{T})$, where the action of $\mathbb{Z}/p_0\mathbb{Z}$ on $\mathbb{T}$ is trivial. This obstruction $\gamma(\theta)$ is in fact the p_0th root of 1 in $\mathbb{C}$ characterized by the equality

$$u \in \mathcal{U},\ \theta^{p_0}(x) = uxu^* \ (\forall x \in M) \implies \theta(u) = \gamma u.$$

The important point is then the existence of automorphisms θ, of factors like the hyperfinite factor, whose obstruction $\gamma(\theta)$ is $\neq 1$. One can easily convince oneself of this existence by the following example. Let us start with

$(X, \mathcal{B}, \mu, (F_t)_{t\in\mathbb{R}})$, where $(X, \mathcal{B}, \mu)$ is a standard Borel probability space and $(F_t)_{t\in\mathbb{R}}$ is a (Borel) one-parameter group of Borel transformations preserving the measure μ. Assume that each F_t, $t \neq 0$, is ergodic (for example one could take a Bernoulli flow [Sc]). Then the crossed product R of $L^\infty(X, \mathcal{B}, \mu)$ by the automorphism associated with F_1 is the hyperfinite factor. The von Neumann algebra $L^\infty(X, \mathcal{B}, \mu)$ is contained in R and the unitary $U \in R$ corresponding to F_1 satisfies:

1) $UfU^* = f \circ F_1 \quad \forall f \in L^\infty(X, \mathcal{B}, \mu)$.
2) $L^\infty(X, \mathcal{B}, \mu)$ and U generate R.

Since F_t, $t \in \mathbb{R}$, commutes with F_1, it defines an automorphism θ_t of R such that

$$\theta_t(U) = U \text{ and } \theta_t(f) = f \circ F_t \quad (f \in L^\infty(X, \mathcal{B}, \mu)).$$

Moreover, for each complex number λ of absolute value 1, let σ_λ be the automorphism of R such that

$$\sigma_\lambda(f) = f \quad \forall f \in L^\infty(X, \mathcal{B}, \mu), \quad \text{and } \sigma_\lambda(U) = \lambda U.$$

By construction, θ and σ commute with each other and $\theta_1(x) = UxU^*$ ($\forall x \in R$). Set $\alpha = \theta_{1/p}\sigma_\gamma$, where $p \in \mathbb{N}$. Then $\alpha^p = \theta_1\sigma_{\gamma^p}$, so that if $\gamma^p = 1$ then $\alpha^p(x) = UxU^*$ ($\forall x \in R$) and $\alpha(U) = \gamma U$. It follows that $p_0(\alpha) = p$ and $\gamma(\alpha) = \gamma$.

A striking feature of this invariant $\gamma(\theta)$ is that it is a complex number $\gamma \neq \bar{\gamma}$ in general. In particular, if one lets θ act not on M but on M^c, the complex conjugate factor, obtained by replacing λx ($\lambda \in \mathbb{C}, x \in M$) by $\bar{\lambda}x$, one gets $\gamma(\theta^c) = \bar{\gamma}(\theta)$. In fact, this is the first invariant that is sensitive to the Galois automorphism $z \mapsto \bar{z}$ of $\mathbb{C}$ over $\mathbb{R}$. This allowed me to construct a factor of type III or of type II_1 ([Co_2]) not anti-isomorphic to itself (see Section 1; the algebra M^c is always isomorphic to M^o via the mapping $x \mapsto x^*$).

6.ζ The list of automorphisms of R up to outer conjugacy

For the hyperfinite factor R, $\overline{\mathrm{Inn}\, R} = \mathrm{Aut}\, R$ and $\mathrm{Ct}\, R = \mathrm{Inn}\, R$. In particular, $p_a(\theta) = p_0(\theta)$ ($\forall \theta \in \mathrm{Aut}R$). We thus have at our disposal two invariants of outer conjugacy, the integer $p_0(\theta)$ and the p_0th root of 1, $\gamma(\theta)$, equal to 1 if $p_0(\theta) = 0$. On the other hand, we have observed the existence of an automorphism of R having a pair (p_0, γ) of invariants given *a priori*.

Theorem 14. [Co_6] *Let θ_1 and θ_2 be two automorphisms of R. Then θ_1 and θ_2 are outer conjugate if and only if*

$$p_0(\theta_1) = p_0(\theta_2), \quad \gamma(\theta_1) = \gamma(\theta_2).$$

For $p_0 = p \neq 0$ and $\gamma \in \mathbb{C}$, $\gamma^p = 1$, there in fact exists an automorphism s_p^γ of R, unique up to conjugacy, having as invariants $p_0(s_p^\gamma) = p$ and $\gamma(s_p^\gamma) = \gamma$, and whose period is the smallest possible compatible with these conditions, that is, equal to $p \times$ (order of γ). In particular, all the outer symmetries $\theta \in \operatorname{Aut} R$ of R, $\theta^2 = 1$, $\theta \notin \operatorname{Inn} R$, are pairwise conjugate. The simplest realization of the symmetry s_2^1 consists in taking the automorphism of $R \otimes R$ that transforms $x \otimes y$ into $y \otimes x$ for all $x, y \in R$. For $p_0 = 0$, there exists, up to outer conjugacy, a unique aperiodic automorphism $\theta \in \operatorname{Aut} R$ (i.e. with $p_0(\theta) = 0$). In particular, all the Bernoulli shifts S_p, although pairwise distinguished up to conjugacy by the entropy, are outer conjugate.

Corollary 15. *The group* Out R *is a simple group with a denumerable number of conjugacy classes.*

In fact, a result more general than the above theorem shows exactly the role played by the equalities $\overline{\operatorname{Inn} R} = \operatorname{Aut} R$ and $\operatorname{Ct} R = \operatorname{Inn} R$. One first shows, for every factor M with separable predual M_*, the following equivalence:

$$\overline{\operatorname{Inn} M} / \operatorname{Inn} M \text{ is nonabelian} \iff M \text{ is isomorphic to } M \otimes R.$$

In this case, let $\theta \in \overline{\operatorname{Inn} M}$. In order that θ be outer conjugate to the automorphism $1 \otimes s_p^\gamma$ of $M \otimes R$ for suitable p and γ, it is necessary and sufficient that $p_0(\theta) = p_a(\theta)$. Moreover, for $\theta \in \operatorname{Aut} M$ to be outer conjugate to $\theta \otimes s_q^1 \in \operatorname{Aut}(M \otimes R)$, it is necessary and sufficient that q divide the asymptotic period $p_a(\theta)$. In particular, every automorphism θ of M is outer conjugate to $\theta \otimes 1$. These results demonstrate the interest of the invariant

$$\chi(M) = \left(\overline{\operatorname{Inn}\ M} \cap \operatorname{Ct}\ M\right) / \operatorname{Inn}\ M,$$

which made it possible ([Co_2]) to show the existence of a factor of type II_1 not anti-isomorphic to itself.

Among other developments on the subject, I mention the following. Jones [Jone_1] has completely classified (up to conjugacy) the actions of arbitrary finite groups on the factor R, by introducing invariants of a cohomological nature generalizing $\gamma(\theta)$ (more elaborate in the general case than in the cyclic case). Next, A. Ocneanu [O_1] succeeded in classifying the outer actions of amenable discrete groups, up to outer conjugacy, by using the techniques of tilings (of amenable groups) introduced by D. Ornstein and B. Weiss [Or-W]. A counterexample due to V. Jones shows that one cannot hope to classify the actions of nonamenable groups. Finally, V. Jones and T. Giordano [Gio-J] on the one hand, and E. Størmer [Stor] on the other, have shown that R possesses, up to conjugacy, just one involutive anti-automorphism.

Let us now apply the above results to the Araki-Woods factor of type II_∞.

6.η Automorphisms of the Araki-Woods factor $R_{0,1}$ of type II_∞

The tensor product $R_{0,1}$ of the hyperfinite factor R by a factor of type I_∞ is the unique Araki-Woods factor of type II_∞ ([Ar-W]). Recall that, for every automorphism θ of a factor N of type II_∞, one writes mod θ for the unique $\lambda \in \mathbb{R}_+^*$ such that $\tau \circ \theta = \lambda\tau$ for every trace τ on N. For $N = R_{0,1}$,

$$\overline{\mathrm{Inn}\, R_{0,1}} = \text{Kernel of mod} = \{\theta;\ \mathrm{mod}\,\theta = 1\}.$$

Moreover, $\mathrm{Ct}\, R_{0,1} = \overline{\mathrm{Inn}\, R_{0,1}}$. From this, one deduces:

Theorem 16. [Co$_6$] a) *Let θ_1 and θ_2 be two automorphisms of $R_{0,1}$. In order that θ_1 be outer conjugate to θ_2, it is necessary and sufficient that*

$$\mathrm{mod}\,\theta_1 = \mathrm{mod}\,\theta_2,\quad p_0(\theta_1) = p_0(\theta_2),\quad \gamma(\theta_1) = \gamma(\theta_2).$$

b) *The following are the only relations between* mod, p_0 *and* γ:

$$\mathrm{mod}\,\theta \neq 1 \implies p_0 = 0,\ \gamma = 1;$$
$$p_0 = 0 \implies \gamma = 1.$$

Whereas the case mod $\theta = 1$ reduces to the case treated above, for every $\lambda \neq 1$ it follows from [Co-T] and part a) of the above theorem that all of the automorphisms $\theta \in \mathrm{Aut}\, R_{0,1}$ with mod $\theta = \lambda$ are *conjugate* (not just outer conjugate). This is a remarkable phenomenon; in effect, for λ an *integer*, one can describe the nature of θ precisely as a shift on an infinite tensor product of $\lambda \times \lambda$ matrix algebras. Thus, when mod $\theta = \lambda$, there exists a $\lambda \times \lambda$ matrix algebra K in $R_{0,1}$ such that:

1) the $\theta^j(K)$ commute pairwise;
2) the $\theta^j(K)$ generate the von Neumann algebra $R_{0,1}$, and this property remains true whenever θ is multiplied by an automorphism of modulus 1.

To summarize, we have arrived at the answer to Problem 2 of the section on factors of type III_λ, and we may conclude that for each $\lambda \in]0,1[$ there is one and only one factor of type III_λ whose associated factor of type II_∞ is $R_{0,1}$. One verifies directly that, for the Powers factor R_λ, the associated factor of type II_∞ is $R_{0,1}$. This shows the interest of the following subproblem of Problem 1) of the section on the III_λ's: characterize the factor $R_{0,1}$ among the factors of type II_∞. One makes the following definition:

Definition 17. *A von Neumann algebra M with separable predual is said to be* hyperfinite *if it is generated by an increasing sequence of finite-dimensional subalgebras.*

(Equivalently ([El-W]) one can require that every finite subset of M be approximable by a finite-dimensional subalgebra.)

It is immediate that $R_{0,1}$, and more generally every Araki-Woods, or even every Krieger factor, is hyperfinite. The above problem can then be reformulated in the following way:

Question 18. *Is $R_{0,1}$ the only hyperfinite factor of type* II_∞?

This problem amounts to knowing whether the commutant of a factor with this approximation property also has it. Around 1967, V. Ya. Golodets offered a proof of this; unfortunately, it contained an irreparable error. However, in another article, the same author used his result to infer that a crossed product by an abelian group does not affect the above approximation property. Although based on an unproven hypothesis, his arguments [Gol] nevertheless show that if a factor M of type III_λ is hyperfinite, then so is the associated factor of type II_∞. This reinforced considerably the interest of the above question, which will be answered in Section V.9.

To conclude this section, let us note that if the factor N of type II_∞ is no longer assumed to be isomorphic to $R_{0,1}$, for $\lambda \in]0,1[$ there is, in general, an infinite number of conjugacy classes in Out N of automorphisms θ of modulus λ ([Co$_3$], [Ph]). To each of these classes there will correspond a factor of type III_λ, and the corresponding factors will be pairwise nonisomorphic.

V.7. Amenable von Neumann Algebras

In this Section I shall review the properties having to do with the approximation of a von Neumann algebra M by finite-dimensional algebras. I shall show that in fact they all define the same class of von Neumann algebras.

7.α Approximation by finite-dimensional algebras

By definition, a von Neumann algebra M is hyperfinite if it is generated by an increasing sequence of finite-dimensional subalgebras. An important reason for the interest in this class is the following result, due to O. Maréchal and based on the proof of Glimm's theorem.

Theorem 1. [Mar] *Let A be a* non-postliminal *separable C^*-algebra. Then, for every hyperfinite von Neumann algebra M having no nonzero finite trace, there exists a state $\varphi \in A^*$ such that the von Neumann algebra generated by $\pi_\varphi(A)$* (Section 2) *is isomorphic to M.*

Thus, a noncommutative integration theory that does not restrict itself to postliminal C^*-algebras, that is, to von Neumann algebras of type I, necessarily involves all the hyperfinite von Neumann algebras. Conversely, for the C^*-algebra A that is the infinite tensor product of 2×2 matrix algebras, it is immediate that all the von Neumann algebras generated by A are hyperfinite.

7.β The properties P of Schwartz, E of Hakeda and Tomiyama, and injectivity

Until 1963, only two examples of nonisomorphic factors of type II were known (of course with separable predual). Specifically, the property Γ distinguished the hyperfinite factor R from the factor Z generated by the regular representation of the free group on two generators. In 1963, J. T. Schwartz introduced a property enabling him to distinguish R from $Z \otimes R$, both of which have property Γ. This property P of R is based on the amenability of a locally finite group. In fact, Schwartz showed that the following conditions on a discrete group Γ are equivalent:

1) Γ is amenable, i.e. there exists a state Φ on $\ell^\infty(\Gamma)$ invariant under translations.
2) $M = R(\Gamma)$ acting on $\mathfrak{H} = \ell^2(\Gamma)$ has the following property P: For every $T \in \mathcal{L}(\mathfrak{H})$, there exists an element of M in the $\sigma(\mathcal{L}(\mathfrak{H}), \mathcal{L}(\mathfrak{H})_*)$-closed convex hull of the uTu^*, u unitary in M'.

Moreover, for every von Neumann algebra M on $\mathfrak{H}$ satisfying 2), he constructed a linear projection P of norm 1 from $\mathcal{L}(\mathfrak{H})$ onto M, satisfying $P(aTb) = aP(T)b$ for all $a, b \in M$, $T \in \mathcal{L}(\mathfrak{H})$.

It is straightforward to verify that every hyperfinite von Neumann algebra satisfies Schwartz's property P. A remarkable result of J. Tomiyama shows that, for every projection P of norm 1 of a von Neumann algebra N onto a von Neumann subalgebra M, the following condition holds automatically [To]:

$$P(aTb) = aP(T)b \quad \forall a, b \in M, T \in N.$$

In [Hak-T], Hakeda and Tomiyama defined a property weaker in appearance than Schwartz's property P:

Definition 2. *A von Neumann algebra M on a Hilbert space $\mathfrak{H}$ is said to have* **property** *E if there exists a projection of norm 1 from the Banach space $\mathcal{L}(\mathfrak{H})$ onto the Banach space $M \subset \mathcal{L}(\mathfrak{H})$.*

Of course $P \Longrightarrow E$; also, the above-mentioned theorem of Tomiyama shows that E plays the same role as P in characterizing the amenability of discrete groups Γ by means of a property of $R(\Gamma)$. Moreover, the property E does not depend on the Hilbert space $\mathfrak{H}$ on which M acts, and, by a theorem of W. Arveson [Arv$_2$], it characterizes the injective objects in the category (von Neumann algebras, completely positive mappings). For this reason, the von Neumann algebras satisfying property E are also called *injective.*

Property E is not descriptive: it says, a priori, very little about the von Neumann algebra M, but it does have remarkable stability properties:

1) If M is an injective von Neumann algebra acting on a Hilbert space $\mathfrak{H}$, then the commutant M' of M is injective.
2) If $(M_\alpha)_{\alpha \in I}$ is a decreasing directed family of injective von Neumann algebras, then $\bigcap_{\alpha \in I} M_\alpha$ is injective.

3) If $(M_\alpha)_{\alpha\in I}$ is an increasing directed family of injective von Neumann algebras, then the closure of $\bigcup_{\alpha\in I} M_\alpha$ is also injective.

4) Let M be a von Neumann algebra with separable predual and let $M = \int_X M(t)d\mu(t)$ be a disintegration of M into factors $M(t)$; then M is injective $\Longleftrightarrow$ $M(t)$ is injective for almost all $t \in X$.

5) Let M be a von Neumann algebra, N a von Neumann subalgebra, and $\mathcal{G}$ a subgroup of the normalizer of N in M. Suppose that N and $\mathcal{G}$ generate M, N is injective and $\mathcal{G}$ is amenable as a discrete group; then M is injective.

6) Let M be an injective von Neumann algebra and Γ an amenable discrete group acting on M by automorphisms; then

$$N = M^\Gamma = \{x \in M;\ \ gx = x\ \ \forall g \in \Gamma\}$$

is injective.

Properties 2 and 3 show that if $\mathfrak{H}$ is a separable Hilbert space, then the monotone class generated by the von Neumann algebras of type I contains only injective von Neumann algebras. Property 4, essentially, reduces the problem of classifying these algebras to the case of injective factors. Property 5 shows that every amenable group of unitaries generates an injective von Neumann algebra. Finally, it is easy to see from Property 6 that, for a factor M of type III_λ, $\lambda \in]0,1[$, to be injective, it is necessary and sufficient that the associated factor of type II_∞ be injective.

Among the most important injective von Neumann algebras, we cite:

a) The crossed product of an abelian von Neumann algebra by an amenable locally compact group.

b) The commutant von Neumann algebra of any continuous unitary representation of a connected locally compact group.

c) The von Neumann algebra generated by any representation of a *nuclear* C^*-algebra (see below for the definition).

7.γ Semidiscrete von Neumann algebras

Let M be a factor of type I on a Hilbert space $\mathfrak{H}$. It gives a decomposition of $\mathfrak{H}$ as a tensor product $\mathfrak{H} = \mathfrak{H}_1 \otimes \mathfrak{H}_2$ in such a way that $M = \mathcal{L}(\mathfrak{H}_1) \otimes 1$ and $M' = 1 \otimes \mathcal{L}(\mathfrak{H}_2)$. One then recovers $\mathcal{L}(\mathfrak{H})$ as the tensor product of M and M'. In one of Murray and von Neumann's first articles, they showed that for every factor M on $\mathfrak{H}$, the homomorphism η of the *algebraic* tensor product

$$M \odot M' = \left\{ \sum_{i=1}^n a_i \otimes b_i;\ \ a_i \in M,\ b_i \in M' \right\}$$

into $\mathcal{L}(\mathfrak{H})$, defined by

$$\eta(\sum_{i=1}^n a_i \otimes b_i) = \sum_{i=1}^n a_i b_i \in \mathcal{L}(\mathfrak{H}),$$

is *injective* and has $\sigma(\mathcal{L}(\mathfrak{H}), \mathcal{L}(\mathfrak{H})_*)$-dense range.

In [E-L], E. Effros and C. Lance succeeded in pushing the analysis much further by studying η from a *metric* point of view. Let A (resp. B) be a unital C^*-algebra acting on a Hilbert space $\mathfrak{H}_A$ (resp. $\mathfrak{H}_B$). Consider the norm on the algebraic tensor product $A \odot B$ that arises from its action on $\mathfrak{H}_A \otimes \mathfrak{H}_B$. This norm on $A \odot B$ makes the completion a C^*-algebra, is independent of the choice of the (faithful) representations of A on $\mathfrak{H}_A$ and B on $\mathfrak{H}_B$, and is characterized by a highly useful theorem of M. Takesaki as the *smallest* norm on $A \odot B$ for which the completion is a C^*-algebra. It is denoted $\| \ \|_{\min}$, and one writes $A \otimes_{\min} B$ for the completion C^*-algebra ([T$_1$]). The C^*-algebra A is said to be *nuclear* if $\| \ \|_{\min}$ is the only pre-C^* norm on $A \odot B$, for every B.

Effros and Lance succeeded in characterizing the factors M for which the above mapping η is *isometric*, by means of a property that is a strengthening of the metric approximation property for the predual M_* of M. The predual M_* is not only an ordered space (for the cone M_*^+) but is *matricially ordered*, in the sense that the tensor product vector space $M_* \otimes M_n(\mathbb{C})$ is ordered for every n, as the predual of $M \otimes M_n(\mathbb{C})$. A completely positive mapping T of M_* into M_* is by definition a linear mapping such that $T \otimes 1_{M_n}$ is positive for every n. The result of Effros and Lance [E-L] is then as follows:

Theorem 3. *Let M be a factor operating on a Hilbert space $\mathfrak{H}$. In order that $\eta : M \otimes_{\min} M' \to \mathcal{L}(\mathfrak{H})$ be isometric, it is necessary and sufficient that the identity mapping on M_* be the pointwise limit in norm of completely positive mappings of finite rank.*

One then defines a *semi-discrete* von Neumann algebra by the above approximation property of its predual M_*. We note that when the factor M is such that neither M nor M' is of type I_∞ or II_∞, and $\mathfrak{H}$ is separable, a corollary of Takesaki's theorem on the min-norm enables one to prove:

Corollary 4. *A factor M acting on a separable Hilbert space $\mathfrak{H}$, such that neither M nor M' is of type I_∞ or II_∞, is semi-discrete if and only if the C^*-algebra $C^*(M, M')$ on $\mathfrak{H}$ generated by M and M' is simple (i.e., has no nontrivial two-sided ideal).*

This corollary is very important when it is combined with the following characterization of the type II_1 factors that do *not* have the property Γ:

Theorem 5. [Co$_5$] *Let M be a factor of type II_1 with separable predual, acting on $L^2(M, \tau) = \mathfrak{H}$, and let $C^*(M, M')$ be the C^*-algebra generated by M and M'. Then:*

M does not have property Γ $\iff$ $C^*(M, M')$ contains the ideal $k(\mathfrak{H})$ of compact operators.

An example of a type II_1 factor for which $C^*(M, M')$ contains the ideal $k(\mathfrak{H})$ was given by C. Akemann and P. Østrand in [Ak-Ø]. Thus, combined with the corollary, the theorem shows that every semi-discrete factor of type II_1 has the property Γ. In fact, Effros and Lance proved in their paper [E-L] that every Araki–Woods factor is semi-discrete. They also proved the implication

$$\text{semi-discrete} \Longrightarrow \text{injective}.$$

The relations between the various properties pertaining to approximation by finite-dimensional algebras can be summarized in the form of a diagram:

$$\begin{array}{ccc} \text{hyperfinite} & & \\ \Downarrow & & \\ \text{Schwartz's property } P & & \\ \Downarrow & & \\ \text{Hakeda–Tomiyama property } E & \Longleftarrow & \text{semi-discrete} \\ = \text{injectivity} & & \end{array}$$

Happily, the situation is in fact remarkably simple:

Theorem 6. $[\mathrm{Co}_5]$ *Let $\mathfrak{H}$ be a separable Hilbert space. For a von Neumann algebra acting on $\mathfrak{H}$, the above four properties are equivalent.*

I defer to Section 9 the description of the corollaries of this theorem relative to the classification problem. Let us begin with a problem of terminology: we surely have at our disposal the right class of von Neumann algebras for the noncommutative theory of integration. Indeed, by O. Maréchal's theorem, this theory should cover at least the hyperfinite case, and, by the results of Effros and Lance, the case of "property E" suffices to account for all the von Neumann algebras associated with nuclear C^*-algebras. In $[\mathrm{Co}_5]$, the term "injective von Neumann algebras" was adopted to denote the above class. Among the advantages of this choice there is above all the simplicity of the definition by means of the property E. However, this terminology has the disadvantage that it is suggestive neither of the fact that it is concerned with an approximation property, nor of the analogy with the amenability of discrete groups. The solution therefore seems to be to choose the term "amenable von Neumann algebra", which is fortunately justified by the equivalence between the above four properties, and a fifth:

Definition 7. *A von Neumann algebra M is said to be* amenable *if, for every normal dual Banach bimodule X over M, the derivations of M with coefficients in X are all inner.*

I refer the reader to the papers of Johnson, Kadison and Ringrose, who established the foundations of the cohomology of von Neumann algebras with coefficients in Banach bimodules ($[\mathrm{Jo}_1]$, $[\mathrm{Jo}_2]$, [Jo-K-R]).

Having accepted the term *amenable* to denote our class of von Neumann algebras, we then have the following corollary, an easy consequence of [E-L] and $[\mathrm{Co}_5]$:

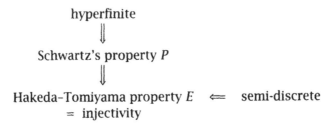

Corollary 8. *Let A be a separable C^*-algebra. Then A is nuclear if and only if for every state φ on A the von Neumann algebra $\pi_\varphi(A)''$ generated by $\pi_\varphi(A)$ is amenable.*

In [Jo_1], B. Johnson introduced the concept of amenable C^*-algebra by the analogue of Definition 7 with the qualifier *normal* omitted. U. Haagerup, by virtue of his remarkable proof of Grothendieck's inequality for arbitrary C^*-algebras (which completes the work of Grothendieck and Pisier), succeeded in showing that a C^*-algebra is amenable if and only if it is nuclear ([Co_8], [H_2], [H_3]).

V.8. The Flow of Weights: mod(M)

In this section we shall describe in detail the main invariant of type III factors, the flow of weights. This invariant emerged from the solution ([Co_4]) of the problem of existence of hyperfinite factors not isomorphic to infinite tensor products of type I factors and from the discrete decomposition of factors of type III_0 (Subsection α) below). In its final form it is due to M. Takesaki [T_3].

8.α The discrete decomposition of factors of type III_0

We have seen above (Section 5 Theorem 4) that any factor M of type III_λ can be uniquely decomposed as the crossed product $M = N \rtimes_\theta \mathbb{Z}$ of a factor N of type II_∞ by an automorphism $\theta \in \mathrm{Aut}\, N$, $\mathrm{mod}(\theta) = \lambda$. We shall now describe an analogous decomposition for arbitrary factors of type III_0 ([Co_4]). There are two nuances with the III_λ case. The first is that N will no longer be a factor but a semifinite type II_∞ von Neumann algebra. The second is that in order to state the uniqueness part of the decomposition we need to introduce the ergodic theoretic notion of induced automorphism θ_e, where $\theta \in \mathrm{Aut}\, N$ and $e \in Z(N)$ is an idempotent in the center of N.

Proposition 1. *Let Z be a commutative von Neumann algebra with no minimal projection, and $\theta \in \mathrm{Aut}\, Z$ be an ergodic automorphism. Then for any nonzero projection $e \in Z$ there exists a unique sequence of projections $e_n \in Z$ such that*

$$e = \sum_{n=1}^{\infty} e_n \,, \quad \theta^n(e_n) \le e \,, \quad e\theta^k(e_n) = 0 \quad \text{for} \quad k = 1, \dots, n-1 .$$

Moreover, one has $e = \sum_{n=1}^{\infty} \theta^n(e_n)$.

We refer, for instance, to [Str] for the simple proof. For brevity a commutative von Neumann algebra is called diffuse if it does not contain any nonzero minimal projection.

Definition 2. *Let N be a von Neumann algebra with diffuse center $Z(N)$ and $\theta \in \mathrm{Aut}\, N$ be ergodic on the center of N. Then for any projection $e \in Z(N)$ we* *let θ_e be the automorphism of the reduced von Neumann algebra N_e determined by*

$$\theta_e(x) = \sum_{n=1}^{\infty} \theta_n(xe_n) \qquad \forall x \in N_e$$

where (e_n) is the sequence of projections $e_n \in Z(N)$ of Proposition 1.

We say that θ_e is the *automorphism induced* by θ on N_e. Next let (N, τ) be a von Neumann algebra with semifinite faithful normal trace τ. We shall say that an automorphism $\theta \in \operatorname{Aut} N$ is a *contraction* if there exists $\lambda < 1$ such that

$$\tau \circ \theta \leq \lambda \, \tau. \tag{1}$$

We can now state the existence and uniqueness of the discrete decomposition of arbitrary factors of type III_0.

Theorem 3. [Co4] *Let M be a factor of type III_0.*

1) *There exist a type II_∞ von Neumann algebra N and a contraction $\theta \in \operatorname{Aut} N$ such that the following isomorphism holds:*

$$M = N \rtimes_\theta \mathbb{Z}$$

2) *Let (N_j, θ_j) be as in 1) with $N_j \rtimes_{\theta_j} \mathbb{Z}$ isomorphic to M; then there exist nonzero projections $e_j \in Z(N_j)$ such that the induced automorphisms θ_{j,e_j} of N_{e_j} are conjugate.*

This theorem has, due to the discreteness of the group $\mathbb{Z}$ involved in the crossed product, the same analytical power as the structure Theorem 5.4 for factors of type III_λ.

The uniqueness statement 2) was originally formulated in terms of outer conjugacy in [Co4] but improved to conjugacy in [Co-T].

8.β Continuous decomposition of type III factors

The above Theorem 3, together with Section V.4, leaves untouched the understanding of the structure of factors of type III_1. This structure was elucidated by M. Takesaki [T3] as a corollary of the following general *continuous decomposition* of type III von Neumann algebras.

Theorem 4. [T3] *Let M be a type III von Neumann algebra.*

1) *There exists a type II_∞ von Neumann algebra N with semifinite normal trace τ and a one-parameter group of automorphisms $(\theta_\lambda)_{\lambda \in \mathbb{R}_+^*}$, so that $\theta_\lambda \in \operatorname{Aut} N$ with $\tau \circ \theta_\lambda = \lambda\tau \quad \forall \lambda \in \mathbb{R}_+^*$, such that the following isomorphism holds:*

$$M = N \rtimes_\theta \mathbb{R}_+^*.$$

2) *The above decomposition is unique up to conjugacy.*

The uniqueness part was originally formulated in terms of outer conjugacy, but improved to conjugacy in [Co-T]. The great virtue of this theorem is its generality and the simplicity of its proof, which is a direct corollary of the biduality theorem for crossed products (Theorem 4 of Appendix A) and of the Radon-Nikodým Theorem 5.1. The von Neumann algebra N is the crossed product

$$N = M \rtimes_{\sigma_t^\varphi} \mathbb{R} \tag{2}$$

of M by an arbitrary modular automorphism group σ^φ and by Theorem 5.1 is independent of the choice of φ. The action on N of the Pontryagin dual group $\mathbb{R}_+^*$ of $\mathbb{R}$ yields the one-parameter group $(\theta_\lambda)_{\lambda \in \mathbb{R}_+^*}$ of automorphisms of N.

Definition 5. *The* flow of weights $W(M)$ *is the restriction of the action of* $(\theta_\lambda)_{\lambda \in \mathbb{R}_+^*}$ *to the center* $Z = Z(N)$.

Stated like this, it is not clear that the flow of weights is defined functorially. We shall give below in Subsection γ) a functorial definition which will justify the name as well. The invariants $S(M)$ and $T(M)$ of a factor are easy to reformulate in terms of the flow of weights; one has

$$S(M) \cap \mathbb{R}_+^* = \{\lambda \in \mathbb{R}_+^* \,;\, \theta_\lambda = \mathrm{id}\} \tag{3}$$

$$T(M) = \{T \in \mathbb{R} \,;\, \exists u \in Z \,,\, u \neq 0 \,,\, \theta_\lambda(u) = \lambda^{iT} u \quad \forall \lambda \in \mathbb{R}_+^*\}. \tag{4}$$

It is also straightforward that

$$M \text{ is a factor} \iff \theta_\lambda \text{ is ergodic on } Z. \tag{5}$$

Thus, one gets, in particular, the following analogue of Theorem 5.4 for factors of type III_1.

Corollary 6. [T3] *Let* M *be a factor of type* III_1.

1) *There exists a factor* N *of type* II_∞, *and a one-parameter group of automorphisms* $\theta_\lambda \in \operatorname{Aut} N$ *with* $\operatorname{mod} \theta_\lambda = \lambda \quad \forall \lambda \in \mathbb{R}_+^*$, *such that* $M = N \rtimes_\theta \mathbb{R}_+^*$.

2) *The above decomposition is unique up to conjugacy.*

With more work one can also derive Theorem 5.4 and Theorem 3 above from the general Theorem 4.

8.γ Functorial definition of the flow of weights

Let M be a von Neumann algebra and φ a semifinite normal weight on M. The support $e = s(\varphi)$ is the unique projection $e \in M$ such that: α) $\varphi | M_e$ is faithful; β) $\varphi(x) = \varphi(exe) \quad \forall x \in M^+$.

Definition 7. *Two semifinite normal weights* φ *and* ψ *are* equivalent *if there exists a partial isometry* u, *with* $u^*u = s(\varphi)$ *and* $uu^* = s(\psi)$ *such that*

$$\psi(x) = \varphi(u^* x u) \qquad \forall x \in M^+.$$

For type III factors the equivalence of projections yields trivial results, and the equivalence of weights is the correct substitute. We shall say that a weight φ on a properly infinite von Neumann algebra M is of *infinite multiplicity* if φ is equivalent to the tensor product $\varphi \otimes \mathrm{Tr}$ of the weight φ by the canonical semifinite faithful normal trace on the type I_∞ factor. There is a slight ambiguity in the definition since one has to give an isomorphism $M \simeq M \otimes F$ where F is the type I_∞ factor, but since $\mathrm{Aut}\, F = \mathrm{Inn}\, F$ one can just use any isomorphism $F \simeq F \otimes F$.

We can now state the analogue of the comparison of projections of Murray and von Neumann.

Theorem 8. [Co-T] *Let M be a properly infinite von Neumann algebra, for instance, a type* III *factor.*

a) *Let Σ be the set of equivalence classes of semifinite normal weights of infinite multiplicity endowed with the operation*

$$(\text{Class}\,\varphi) \vee (\text{Class}\,\psi) = \text{Class}(\varphi \oplus \psi)$$

where $\varphi \oplus \psi = \varphi + \psi$ if $s(\varphi)\, s(\psi) = 0$. Then Σ is the Boolean algebra of σ-finite projections in a unique commutative von Neumann algebra Z.

b) *The action of $\mathbb{R}_+^*$ by multiplication*

$$\lambda(\text{Class}\,\varphi) = \text{Class}(\lambda\varphi) \qquad \forall \lambda \in \mathbb{R}_+^*$$

determines a one-parameter group of automorphisms of Z whose continuous part is canonically isomorphic to the flow of weights of Definition 5.

To understand the theorem one has to remember that a commutative von Neumann algebra Z is uniquely determined by the Boolean algebra $\sigma(Z)$ of projections $e \in Z$ with the operation:

$$e, f \mapsto e \vee f = e + f - ef.$$

In 1) one obtains a very large commutative von Neumann algebra Z whose σ-finite projections $e \in Z$ exactly classify the semifinite normal weights of infinite multiplicity on M. For each $\lambda \in \mathbb{R}_+^*$ the formula 2) defines an automorphism θ_λ of Z. The continuous part of this action of $\mathbb{R}_+^*$ on Z is given by the following von Neumann subalgebra:

$$W = \{x \in Z\,;\ \lambda \mapsto \theta_\lambda(x) \text{ is strongly continuous}\}. \tag{6}$$

One has, moreover, a very simple characterization of exactly those φ for which $(\text{Class}\,\varphi) \in W$.

Proposition 9. [Co-T] *Let φ be a semifinite normal weight of infinite multiplicity. Then* Class $\varphi \in W$ *iff the modular automorphism group σ_t^φ of φ is* integrable, *i.e. iff $\{x \in M\,;\ \int_{-\infty}^{\infty} \sigma_t^\varphi(x^*x)dt \in M\}$ is weakly dense in M.*

There exists moreover a largest integrable weight, the *dominant weight*, unique up to equivalence, which was already used in [Co_4] in the proof of the converse of the Radon-Nikodým theorem. In the type III_1 case this weight plays the same role as the generalized trace of Section 5. It is characterized, among semifinite normal weights of infinite multiplicity, by the invariance $\varphi \sim \lambda\varphi \quad \forall \lambda \in \mathbb{R}_+^*$.

The construction of the flow of weights given in Theorem 8 is obviously functorial. An idempotent $e \in W$ is just an equivalence class of weights, and the action of $\mathbb{R}_+^*$ is just multiplication of φ by $\lambda \in \mathbb{R}_+^*$.

This allows one, in particular, to extend to the type III case the definition of the module $\mathrm{mod}(\alpha)$ of automorphisms $\alpha \in \mathrm{Aut}(M)$.

Definition 10. *Let M be a type* III *factor, then for $\alpha \in \mathrm{Aut}\, M$ the module* $\mathrm{mod}(\alpha)$ *is the automorphism of the flow of weights determined by the equality*

$$\mathrm{mod}(\alpha)\ (\mathrm{Class}\ \varphi) = \mathrm{Class}(\varphi \circ \alpha^{-1}).$$

This allows one to obtain the analogue in the type III case of the classification of automorphisms of Section 6 ([Co-T]) .

It also suggests the notation mod for the functor from von Neumann algebras to flows given by Theorem 8 b).

8.δ Virtual groups and the flow of weights as modular spectrum

The theory of induced representations of locally compact groups led G. Mackey ([M_3]) to a natural generalization of the notion of closed subgroup which has great heuristic value and was further developed by A. Ramsay [Ra] and R. Zimmer [Zi_1]. The nontransitive imprimitivity systems suggest the following definition:

Definition 11. *Let G be a locally compact abelian group. Then a* virtual subgroup *of G is given by an ergodic action α of G on a commutative von Neumann algebra A.*

Given an ordinary *closed* subgroup H of G one lets $A = L^\infty(G/H)$ and α be the action of G by left translations. The group H is then recovered as the isotropy group. The virtue of this notion is that many concepts extend from the special case of closed subgroups to arbitrary virtual subgroups.

For instance, if H_1, H_2 are two closed subgroups of G then the closure $H = \overline{H_1H_2}$ of H_1H_2 is obtained as the isotropy group for the action $\alpha_1 \otimes 1$ of G on the following commutative von Neumann algebra:

$$A = (A_1 \otimes A_2)^G = \{z \in A_1 \otimes A_2\ ;\ \alpha_1(g) \otimes \alpha_2(g^{-1})(z) = z \quad \forall g \in G\} \qquad (7)$$

where $A_j = L^\infty(G/H_j)$, on which G acts by left translations α_j. Obviously the construction (7) works for virtual subgroups as well, and the special case of closed subgroups gives its heuristic meaning.

Now the invariant $S(M)$ for factors of type III is a closed subgroup of $\mathbb{R}_+^*$ (Section 5), but there is no general formula for $S(M_1 \otimes M_2)$ in terms of $S(M_1)$ and $S(M_2)$. Indeed, one can construct factors of type III_0 whose tensor product is of arbitrary type III_λ. This difficulty is completely removed if one uses the above idea of Mackey. Indeed, the flow of weights $\mathrm{mod}(M)$ of a type III factor is, by construction, an ergodic action of $\mathbb{R}_+^*$ on a commutative von Neumann algebra and hence

$$\mathrm{mod}(M) \text{ is a virtual subgroup of } \mathbb{R}_+^*. \tag{8}$$

Moreover, the relation (3) between $S(M)$ and $\mathrm{mod}(M)$ shows that except in the III_0 case, $\mathrm{mod}(M)$ is an ordinary closed subgroup of $\mathbb{R}_+^*$ and is equal to $S(M)$.

Thanks to (7) we can define the closure of the product of two virtual subgroups and one has

Theorem 12. [Co-T] *Let M_1, M_2 be two type* III *factors; then*

$$\mathrm{mod}(M_1 \otimes M_2) = \overline{\mathrm{mod}(M_1)\ \mathrm{mod}(M_2)}.$$

In other words, the flow of weights of $M_1 \otimes M_2$ is obtained by formula (7) from the flow of weights $\mathrm{mod}(M_j)$.

Thus, unlike the invariant S, the invariant mod is compatible with tensor products.

As a further example of the suggestive power of the idea of virtual subgroups, let us note that for general locally compact groups G with closed subgroup H one has:

$$\begin{gathered} H \text{ is of finite covolume in } G \Longleftrightarrow \\ \text{the action of } G \text{ on } A = L^\infty(G/H) \text{ has a finite invariant measure.} \end{gathered} \tag{9}$$

(We mean, of course, in the measure class of the Haar measure.)

This suggests saying that a virtual subgroup has "finite covolume" when the corresponding action has a finite invariant measure. This notion plays an important role in R. Zimmer's work ([Zi_2]).

In our context it allows us to restate the main result of Section III.6 as follows:

Let (V, F) be a codimension 1-foliation with nonvanishing Godbillon-Vey class; then the virtual subgroup $\mathrm{mod}(M)$, $M = W^*(V, F)$, is of finite covolume in $\mathbb{R}_+^*$.

As a final example let us mention the computation of the flow of weights $\mathrm{mod}(M)$ for factors associated to ergodic equivalence relations $\mathcal{R}$ with countable orbits (cf. Section 4) on a measure space (X, μ). Let us, as in Section 4, endow $\mathcal{R}$, the graph of the equivalence relation, with the obvious groupoid law and with the measure $\tilde{\mu}$ which plays the role of a left Haar measure on this groupoid. Let us then consider the homomorphism $\delta : \mathcal{R} \to \mathbb{R}_+^*$ given by the lack of right invariance of the left Haar measure $\tilde{\mu}$, i.e. by:

$$\delta(x,y) = \frac{d\mu(y)}{d\mu(x)} \qquad \forall (x,y) \in \mathcal{R}.$$

This is well defined a.e. modulo μ.

The closure of the range $\delta(\mathcal{R})$ of this homomorphism is well defined as a virtual subgroup of $\mathbb{R}_+^*$ (cf. [Ra]) and one has:

Proposition 13. [Kr$_2$] [Sau$_2$] *Let $\mathcal{R}$ be an ergodic equivalence relation, $M = L^\infty(\mathcal{R}, \tilde{\mu})$ be the associated factor (Definition 4.1). Then the flow of weights of M is given by*

$$\operatorname{mod}(M) = \delta(\mathcal{R}).$$

We have already seen this in a different guise when we described in Chapter I the flow of weights, mod(M), for the von Neumann algebra $W(V,F)$ of a foliation. The closure of the range of a homomorphism δ from a measured groupoid $\mathcal{R}$ to an abelian locally compact group G is defined in general from the action of G on the commutative von Neumann algebra of functions $f \in L^\infty(\mathcal{R} \times G)$ which are invariant under the action of $\mathcal{R}$ on $\mathcal{R} \times G$ given by

$$\gamma(\gamma', h) = (\gamma\gamma', \delta(\gamma)h).$$

V.9. The Classification of Amenable Factors

In this section we shall give the complete classification of amenable von Neumann algebras. All von Neumann algebras here are assumed to have separable predual. By the reduction theory, any amenable von Neumann algebra M can be written as a direct integral, $M = \int M(t)\, d\mu(t)$, of amenable factors. We shall now give the complete list, type by type, of amenable factors. (We ignore the trivial type I case.)

9.α Factors of type II_1

As a corollary of Theorems 1.13 and 7.6 one has

Theorem 1. [Co$_5$] *Any amenable factor of type* II_1 *is isomorphic to the Murray-von Neumann hyperfinite factor R.*

I refer to [Co$_5$], [H$_4$] and [Po$_1$] for simplifications of the original proof.

Now any von Neumann subalgebra M of R is automatically finite (it has a finite faithful normal trace) and amenable (using the canonical normal conditional expectation $E : R \to M$ given by the orthogonal projection on $M \subset L^2(R)$). Thus, one gets:

Corollary 2. [Co_5] *Let M be a von Neumann subalgebra of R; then M is isomorphic to a direct sum of tensor products of the form $M_n(\mathbb{C}) \otimes C_n$ and $R \otimes C_0$, where the C_j are commutative von Neumann algebras.*

In particular, all *subfactors* of R are either finite dimensional or isomorphic to R.

It is straightforward that a discrete group Γ is amenable iff the von Neumann algebra $R(\Gamma)$ of the left regular representation of Γ is amenable. This is a necessary condition for coherence of the terminology. One can then analyse $R(\Gamma)$ as follows:

Corollary 3. [Co_5] *Let Γ be an amenable discrete group and $R(\Gamma)$ the von Neumann algebra generated by its left regular representation in $\ell^2(\Gamma)$. Then $R(\Gamma)$ is a direct sum of tensor products of the form $M_n(\mathbb{C}) \otimes C_n$ and $R \otimes C_0$ where the C_j are commutative von Neumann algebras.*

When all nontrivial conjugacy classes of Γ are infinite one has $R(\Gamma) \simeq R$. As we shall see in Appendix B the analogy between the notion of amenability for discrete groups and for type II_1 factors goes quite far (cf. Theorem 21). We shall now describe the analogue of the invariant means of discrete amenable groups.

Theorem 4. [Co_5] *Let N be a factor in a separable Hilbert space $\mathcal{H}$. Then, unless N is finite-dimensional,*

N is isomorphic to $R \Longleftrightarrow$ there exists a state Φ on $\mathcal{L}(\mathcal{H})$ such that $\Phi(xT) = \Phi(Tx) \quad \forall x \in N\,,\ T \in \mathcal{L}(\mathcal{H})$.

Such a state cannot be normal. By definition a *hypertrace* is a state Φ on $\mathcal{L}(\mathcal{H})$ satisfying the above condition. Even though such states are not normal they can be obtained by nice formulas in some examples. For instance, if we let D be the analogue of the Dirac operator on the noncommutative 2-torus $\mathbb{T}^2_\theta$, $\theta \notin \mathbb{Q}$ (cf. Chapter VI.3.11) and consider $(\mathcal{H}, D)$ as a K-cycle over the C^*-algebra A_θ, the following equality defines a hypertrace Φ_ω on the weak closure $R = A_\theta''$ of A_θ in $\mathcal{H}$:

$$\Phi_\omega(T) = \mathrm{Tr}_\omega(TD^{-2}) \qquad \forall T \in \mathcal{L}(\mathcal{H}) \tag{1}$$

where Tr_ω is the Dixmier trace for a given choice of ω (Chapter IV). One can also use a result of G. Mokobodzki ([Me]) to make Φ_ω universally measurable and commuting with integrals: $\Phi_\omega\left(\int T_\alpha d\mu(\alpha)\right) = \int \Phi_\omega(T_\alpha)\, d\mu(\alpha)$.

In the case of discrete groups Γ, amenability is equivalent to a condition involving finite subsets, the Følner condition:

$\forall g_1, \ldots, g_n \in \Gamma\,,\ \forall \varepsilon > 0\,,\ \exists$ finite non-empty subset F of Γ such that

$$\|1_F - g_i\, 1_F\|_2 \le \varepsilon \|1_F\|_2 \qquad \forall i = 1, \ldots, n$$

where 1_F is the characteristic function of F and $\| \ \|_2$ is the norm in $\ell^2(\Gamma)$.

In a very similar way the existence of a hypertrace on a factor N in $\mathcal{H}$ is characterized by the following condition:

$\forall x_1,\ldots,x_n \in N$, $\forall \varepsilon > 0$, $\exists$ a finite dimensional projection P on $\mathcal{H}$ such that

$$\|x_i P - P x_i\|_{HS} \leq \varepsilon \|P\|_{HS} \tag{2}$$

where $\|T\|_{HS} = (\mathrm{Trace}\,(T^*T))^{1/2}$ is the Hilbert-Schmidt norm of operators.

9.β Factors of type II$_\infty$

Any factor M of type II$_\infty$ is the tensor product $N \otimes F$ of a factor of type II$_1$ by the factor F of type I$_\infty$. Moreover, if M is amenable so is N; thus Theorem 1 implies:

Theorem 5. [Co$_5$] *There exists up to isomorphism only one amenable factor of type* II$_\infty$, *namely* $R_{0,1} = R \otimes F$.

The notation $R_{0,1}$ comes from the Araki-Woods notation for the only ITPFI of type II$_\infty$. We thus obtain the answer to Question 6.18:

Corollary 6. [Co$_5$] *Any hyperfinite factor of type* II$_\infty$ *is isomorphic to* $R_{0,1}$.

Note that the proof of this corollary is very indirect; hyperfiniteness is only used through the apparently much weaker amenability which passes to the II$_1$ factor N if $M = N \otimes F$. There is no direct proof of the hyperfiniteness of N.

We have already given, in Chapter I, examples of foliations (V,F) such as the Kronecker foliation, whose associated von Neumann algebra $W(V,F)$ is the hyperfinite factor of type II$_\infty$.

As another large class of occurrences of this factor one has:

Corollary 7. [Co$_5$] *Let G be a connected separable locally compact group and let λ be the left regular representation of G in $L^2(G)$. Then $R(\lambda_G)$ is a direct integral of factors which are either of type* I *or isomorphic to* $R_{0,1}$.

This result follows from Theorem 5 and a result of Dixmier and Pukanszky showing that no factor of type III occurs in the decomposition of $R(\lambda_G)$.

9.γ Factors of type III_λ, $\lambda \in]0,1[$

Let M be a factor of type III_λ, $\lambda \in]0,1[$, and let $M = N \rtimes_\theta \mathbb{Z}$ be the discrete decomposition of M (Theorem 5.4). Then if M is amenable so is N, as one sees using the natural normal conditional expectation, $E(\sum a_n U^n) = a_0 \in N$, from M to N. Thus, since N is a factor of type II_∞, it is by Theorem 5 isomorphic to $R_{0,1}$ and $\theta \in \mathrm{Aut}(R_{0,1})$ satisfies $\mathrm{mod}(\theta) = \lambda$. By the results of noncommutative ergodic theory (Theorem 6.16) one knows that θ is unique up to outer conjugacy, which thus implies:

Theorem 8. [Co5] *Let $\lambda \in]0,1[$. There exists up to isomorphism only one amenable factor of type III_λ, the Powers factor R_λ.*

We refer to Section 4 for the description of R_λ as an ITPFI. As above the analogue of Theorem 8 holds with amenable replaced by hyperfinite: R_λ is the only hyperfinite factor of type III_λ.

These factors R_λ describe the measure theory of the simplest fractals, namely the self-similar Cantor sets of Chapter IV Section 3 Example 23. Let us first recall that given a pair (A,φ) consisting of a C^*-algebra A and a state φ on A there is a canonically associated von Neumann algebra: the weak closure A'' of A in the GNS representation (cf. Section 2). Let $K \subset [0,1]$ be the self-similar Cantor set of IV.3.23, and let $C(K)$ act by multiplication operators in $\ell^2(D)$ as in IV.3.21, where $D \subset K$ is the denumerable set of endpoints of K. A self-similarity σ of K is the restriction to $\mathrm{Domain}(\sigma) = K \cap J$, where J is a closed interval of an affine transformation $x \mapsto \sigma(x) = ax + b$ of $\mathbb{R}$ which preserves K, i.e.

$$\sigma(x) \in K \qquad \forall x \in \mathrm{Domain}(\sigma). \tag{3}$$

By construction, σ preserves D and thus can be represented in $\ell^2(D)$ as a partial isometry u_σ such that

$$\begin{aligned} u_\sigma(\varepsilon_b) &= 0 \quad \text{if} \quad b \notin \mathrm{Domain}(\sigma) \\ u_\sigma(\varepsilon_b) &= \varepsilon_{\sigma(b)} \quad \text{if} \quad b \in \mathrm{Domain}(\sigma) \end{aligned} \tag{4}$$

where $(\varepsilon_b)_{b\in D}$ is the canonical orthonormal basis of $\ell^2(D)$.

We let A be the C^*-algebra in $\mathcal{H} = \ell^2(D)$ generated by $C(K)$ and the self-similarities u_σ. It can be described in many equivalent ways as in Section II.2. Then let p be the Hausdorff dimension of K and let φ be the positive linear form on the C^*-algebra A given by

$$\varphi(a) = \mathrm{Tr}_\omega(a|dx|^p) \qquad \forall a \in A \tag{5}$$

with the notation of IV.3.23. Recall that the restriction of φ to $C(K)$ is, up to a constant, the Hausdorff measure Λ_p on K.

Proposition 9. *The factor A'' associated to the pair (A,φ) is the hyperfinite factor R_{ρ^p}, of type III_{ρ^p}, where ρ is the similarity ratio of K.*

With the notation of IV.3.23 one has $\rho^p = \frac{1}{q+1}$.

The analogue of the construction of Section 11, with the global field $\mathbb{Q}$ replaced by the field of rational fractions over a finite field $\mathbb{F}_q$, yields the factors $R_{(q^{-\beta})}$ for any $\beta \in]0,1]$.

Finally, we have already given in Chapter I examples of foliations (V,F) with $W(V,F) = R_\lambda$ for arbitrary values of $\lambda \in]0,1[$.

9.δ Factors of type III_0

The analysis of amenable factors of type III_0 combines the discrete decomposition (Theorem 8.3) with Krieger's theorem (Theorem 4.3). Of course, any Krieger factor is hyperfinite and amenable; conversely one has:

Theorem 10. [Co_5] *Any amenable (or hyperfinite) factor of type* III_0 *is a Krieger factor.*

Moreover, such factors are in one-to-one correspondence with ergodic flows by the remarkable result of W. Krieger:

Theorem 11. [Kr_2] 1) *Two Krieger factors M_1, M_2 of type* III_0 *are isomorphic iff their flows of weights are isomorphic:*

$$M_1 \simeq M_2 \Longleftrightarrow \mathrm{mod}(M_1) \simeq \mathrm{mod}(M_2).$$

2) *For any ergodic intransitive flow there exists a unique Krieger factor M with this flow as flow of weights.*

Thus, in other words, the natural extension of the invariant S as a virtual subgroup of $\mathbb{R}_+^*$ is a complete invariant.

Previously to this last result the discrete decomposition was already used in [Co_4], together with an ergodic theory result of W. Krieger, to exhibit a factor M of type III_0 which is hyperfinite but is not isomorphic to any ITPFI (Araki-Woods factor). By now we have the following characterization of ITPFI among hyperfinite factors:

Theorem 12. [Co-W] *Let M be a type* III *hyperfinite factor. Then M is isomorphic to an* ITPFI *iff its flow of weights* $\mathrm{mod}(M)$ *is approximately transitive.*

We need to define approximate transitivity for an action α of a locally compact group G on a commutative von Neumann algebra Z:

Definition 13. *The action α of G on Z is approximately transitive if for any normal states $\mu_1, \dots, \mu_n$ on Z and $\varepsilon > 0$ there exists a normal state μ on Z such that the distance, in norm, of μ_j to the convex hull of $G\mu$ is less than ε for any $j = 1, \dots, n$.*

One can show ([Co-W]) that any measure-preserving flow with nonzero entropy is not approximately transitive. It follows that a factor of type III_0, which is hyperfinite with this flow as its flow of weights, cannot be an ITPFI.

Corollary 14. *There exist type* III_0 *hyperfinite non* ITPFI *factors.*

Let us stress finally that Theorem 11 shows the equivalence of two classification problems, and this in a functorial manner. It had been shown by E.J. Woods, using the theory of ITPFI, that the set of isomorphism classes of ITPFI is not countably separated, and thus that such factors cannot be classified by a countable family of real-valued invariants. We have seen in Chapter I how to construct foliations (V,F) such that $W(V,F)$ is hyperfinite of type III_0.

9.ε Factors of type III_1

Let M be a factor of type III_1 and let $M = N \rtimes_\theta \mathbb{R}_+^*$ be its continuous decomposition (Theorem 8.6). Then N is a factor of type II_∞ and if M is amenable so is N, so that by Theorem 5, the factor N is isomorphic to $R_{0,1}$. We do not, however, have a direct proof of the analogue of the noncommutative ergodic theory result (Theorem 6.16) for flows. The difficulty is that flows are incompatible with the use of ultraproducts. Making use of discrete crossed products instead, to reduce from III_1 to III_λ, $\lambda \in]0,1[$, I showed (cf. [Co_{21}]) that the uniqueness (up to isomorphism) of amenable factors of type III_1 would follow provided one could prove the following:

"Let M be a factor of type III_1. Then for any pair of normal states $\varphi \neq \psi$ on M there exists a bounded sequence (x_n) in M such that $\|[\varphi, x_n]\| \to 0$ and $\|[\psi, x_n]\| \not\to 0$ when $n \to \infty$."

This property is easy to prove for the Araki-Woods factor of type III_1, R_∞. In 1983 U. Haagerup was able to prove the above property for arbitrary amenable factors of type III_1, making use of the continuous decomposition and of the relative commutant theorem ([Co-T]). This result concludes the classification of amenable (or equivalently hyperfinite) factors:

Theorem 15. [H_5] *There exists up to isomorphism only one amenable factor of type* III_1*, the factor* R_∞ *of Araki and Woods.*

Here are some examples of occurrences of this factor.

1) We have already seen in Chapter I that the Anosov foliation (V,F) of the unit-sphere bundle V of a Riemann surface of genus > 1 gives the factor $R_\infty = W(V,F)$.

2) The von Neumann algebra $\mathcal{U}(\mathcal{O})$ generated by bosonic quantum fields with support in any local region $\mathcal{O}$, as in Chapter IV Section 9 Proposition 12, is isomorphic to R_∞. This result, in the free field case, is due to H. Araki. The explicit computation of the modular automorphism group for the vacuum state restricted to $\mathcal{U}(\mathcal{O})$ has been done only in some very special cases and remains a challenging open question.

3) We have seen that the factors R_λ of type III_λ describe the measure theory of the self-similar Cantor sets. Similarly, the factor R_∞ describes the measure theory of the quasi-Fuchsian circles of Chapter IV Theorem 3.17. One lets A be the C^*-algebra crossed product of the algebra of continuous functions on the quasi-circle by the action of the quasi-Fuchsian group Γ. The positive linear form φ is given by the formula (with Tr_ω the Dixmier trace)

$$\varphi(a) = \mathrm{Tr}_\omega(a|dZ|^p) \qquad \forall a \in A.$$

The factor obtained is of type III_1 and isomorphic to R_∞.

4) We shall see in Section 11 that the notion of module in basic number theory [We$_1$] is intimately related to our functor mod, and that the factor R_∞ appears naturally in the statistics of prime numbers.

V.10. Subfactors of Type II_1 Factors

10.α Index of subfactors

V. Jones first extended the classification (Theorem 6.14) of actions of finite cyclic groups on the hyperfinite factor R to arbitrary finite groups G [Jone$_1$]. There is in this situation a Galois-type correspondence between G and the fixed point von Neumann algebra $R^G = \{x \in R \,;\, gx = x \quad \forall g \in G\}$. Jones then went on and investigated arbitrary subfactors N of R satisfying the following *finiteness* condition:

Definition 1. *Let $N \subset M$ be a subfactor of a type II_1 factor M; then N is of* finite index *if the commutant of N in $L^2(M)$ is finite (i.e. of type II_1).*

Here $L^2(M)$ is the Hilbert space completion of M for the inner product $\langle x, y\rangle = \tau(x^*y)$, where $\tau = \mathrm{Tr}_M$ is the unique tracial state on M. It is a bimodule over M with the action by left and right multiplication (cf. Appendix B α) for the description of the identity correspondence in general). By a result of Pimsner and Popa [Pi-Po], N has finite index in M iff M viewed as a left N-module is finite and projective.

The Murray and von Neumann dimension function (Theorem 1.11) yields for any II_1 factor N and any normal representation π of N in a Hilbert space $\mathfrak{H}$, a real number $\dim_N(\mathfrak{H}, \pi)$ which satisfies the following conditions.

1) $\dim_N(\mathfrak{H}, \pi) \in [0, +\infty]$, $\dim_N(L^2(N)) = 1$.
2) $\dim_N(\mathfrak{H}, \pi) = \dim_N(\mathfrak{H}', \pi')$ if and only if the representations π and π' are equivalent.
3) $\dim_N(\bigoplus_{n=1}^\infty(\mathfrak{H}_n, \pi_n)) = \sum_{n=1}^\infty \dim(\mathfrak{H}_n, \pi_n)$.
4) If $e \in \pi(N)'$ is a projection and π_e is the restriction of π to the space $e\mathfrak{H}$, then

$$\dim_N(e\mathfrak{H}, \pi_e) = \mathrm{Tr}_{\pi(N)'}(e) \cdot \dim_N(\mathfrak{H}, \pi).$$

5) $\dim_N(\mathfrak{H},\pi)\cdot\dim_{N'}(\mathfrak{H})=1$, where N' denotes the commutant of $\pi(N)$.

The index of a subfactor N is defined as follows:

Definition 2. *Let $N\subset M$ be a subfactor of a type II_1 factor M; then the index $[M:N]$ of N in M is the multiplicity*

$$\dim_N(L^2(M)).$$

This index satisfies the following easy properties:

Proposition 3. a) *Let N be a subfactor of M of finite index. For every representation $(\mathfrak{H},\pi)$ of M of finite multiplicity, the restriction π_N of π to N has finite multiplicity, and*

$$\dim_N(\mathfrak{H},\pi_N)=[M:N]\cdot\dim_M(\mathfrak{H},\pi).$$

b) *Let N and M be as in* a) *and let P be a subfactor of N of finite index. Then P is a subfactor of M of finite index, and*

$$[M:P]=[M:N][N:P].$$

c) *If $N,M,\mathfrak{H}$ and π are as in* a), *then the commutant $\pi(M)'$ is a subfactor of $\pi(N)'$ of finite index, and*

$$[\pi(N)':\pi(M)']=[M:N].$$

Property a) is immediate; b) follows from a), and c) follows from Property 5) of the dimension function $\dim_M$.

The above example $R^G\subset R$ of fixed points of finite group actions, as well as the inclusions

$$R(\Gamma_1)\subset R(\Gamma_2)$$

of the type II_1 factors associated to discrete groups $\Gamma_1\subset\Gamma_2$ with infinite conjugacy classes, yield only integral values for the index.

One obtains non-integral values for the index in a rather trivial manner as follows:

Let M be a factor of type II_1 and $e\in M$ a projection. We denote by M_e the factor $M_e=\{x\in M;\ xe=ex=x\}$. In order that M_e be isomorphic to M_{1-e}, it is necessary and sufficient that the positive real number $\lambda_0/(1-\lambda_0)$, where $\lambda_0=\mathrm{Tr}_M(e)$, belong to the group $F(M)$ (Section 1). Assuming this to be the case, let θ be an isomorphism $\theta: M_e\to M_{1-e}$ and let

$$N=\{x+\theta(x);\ x\in M_e\}.$$

By construction, N is a subfactor of M, and a straightforward calculation of $\dim_N(L^2(M))$ shows that

$$[M:N]=\frac{1}{\lambda_0}+\frac{1}{1-\lambda_0}.$$

The group $F(R)$ of the hyperfinite factor R is equal to $\mathbb{R}_+^*$. Thus, the above construction yields, for every real number $\alpha \geq 4$, the existence of a subfactor N of R with $[R : N] = \alpha$.

All this shows that the set $\Sigma \subset [1, \infty[$ of values of the index of subfactors satisfies

$$\{1, 2, 3\} \cup [4, +\infty[\subset \Sigma.$$

The first important result of V. Jones is the following:

Theorem 4. [Jone$_2$] *Let Σ be the subset of $\mathbb{R}^+$ consisting of the values of the index $[M : N]$, where M and N are factors of type* II_1, $N \subset M$. *Then*

$$\Sigma = \{4\cos^2 \frac{\pi}{n};\ n \in \mathbb{N}, n \geq 3\} \cup [4, +\infty).$$

We shall briefly describe the basic construction due to V. Jones. It extends to this situation the idea of iterated crossed products, which was already crucial in the original construction of the automorphisms s_p^γ of Theorem 6.14. Here we do not have a group G but only the inclusion $N \subset M$ of its fixed point algebra N in M. First, the analogue of the inclusion of M in its crossed product by G is the inclusion of finite factors

$$M \subset JN'J\ ,\ N' = \text{commutant of } N \text{ in } L^2(M) \tag{1}$$

where J is the canonical antilinear isometric involution in $L^2(M)$,

$$Jx = x^* \qquad \forall x \in L^2(M). \tag{2}$$

One has $M' = JMJ$ (cf. Section 3) so that the inclusion in (1) holds. Moreover, since the inclusion in (1) of type II_1 factors is of the same type as the original inclusion $N \subset M$ (it also has the same index by 3 c), the operation

$$N \subset M \mapsto M \subset M_1 = JN'J \tag{3}$$

can be iterated. It yields an inductive system M_k of type II_1 factors such that

$$M_k \subset M_{k+1},\quad [M_{k+1} : M_k] = [M : N] \quad \forall k \tag{4}$$

and one can endow the inductive limit M_∞ with its unique tracial state, which we denote by τ.

The crucial fact now is that in the inclusion in (1) one has a canonical projection $e_1 \in JN'J = M_1$ given by

$$e_1 \text{ is the orthogonal projection of } L^2(M) \text{ on the closure } \overline{N} = \overline{\{x1\ ;\ x \in N\}}. \tag{5}$$

By construction, $Je_1J = e_1$ because $N^* = N$. Moreover, e_1 is the probabilistic conditional expectation of M on N and, in particular, is an N-bimodule map from M to N so that ([To])

$$e_1 \in N'\ ,\ e_1 \in JN'J. \tag{6}$$

Using the bicommutant theorem, one then checks (cf. [Jone$_2$]) that

$$M_1 \text{ is generated by } M \text{ and } e_1. \tag{7}$$

Now when the construction (1) is iterated we get a sequence of projections $e_n \in M_n$ and (cf. [Jone$_2$]) they satisfy the following relations:

Lemma 5. *Let τ be the unique tracial state on M_∞. The (e_n) form a sequence of projections $e_m = e_m^* = e_m^2 \in M_\infty$ such that:*

α) $e_i\, e_j = e_j\, e_i \quad \forall i,j \in \mathbb{N}\,,\ |i-j| > 1$.

β) $e_{m+1}\, e_m\, e_{m+1} = [M:N]^{-1}\, e_{m+1}$ and $e_m\, e_{m+1}\, e_m = [M:N]^{-1}\, e_m$ for all $m \geq 1$.

γ) $\tau(x\, e_{m+1}) = [M:N]^{-1}\, \tau(x)$ for any element x of the algebra generated by $e_1, \dots, e_m$.

The relations α) and β) imply that the algebra A_m generated by $e_1, \dots, e_m$ is a finite-dimensional C^*-algebra. The proof of Theorem 4 ([Jone$_2$]) is based on the analysis of the C^*-algebra inductive limit of the A_m's. The resulting C^*-algebra $A(t)$ depends only upon $t = [M:N]^{-1}$ and is nontrivial only for $t^{-1} \in \Sigma$ with Σ as in Theorem 4. The first example of non-integral index corresponds to $n = 5$; the index is then the square of the golden ratio $\frac{1+\sqrt{5}}{2} = 2\cos\frac{\pi}{5}$, and the C^*-algebra generated by the projections e_i is canonically isomorphic to the C^*-algebra associated with the parameter space for the Penrose tilings (cf. Chapter II).

I refer the reader to [Goo] for more detailed information. By analyzing the relative commutant of N in M_n, V. Jones also introduces an invariant, consisting of a Dynkin diagram, which is finer than the index $[M:N]$, and is canonically associated with a subfactor $N \subset M$.

The classification of the subfactors of index $4\cos^2\frac{\pi}{n}$ of the hyperfinite factor was carried out by A. Ocneanu and S. Popa ([O$_2$], [Po$_2$]). Finally, V. Jones' discovery of new polynomial invariants for knots [Jone$_3$], based on his analysis of the subfactors, has had a remarkable impact on low-dimensional topology. However, it lies beyond the scope of this book, and we shall just explain below the relation with the Hecke algebras.

10.β Positive Markov traces on Hecke algebras

We shall explain in this section the link between the relations 5 α) and β) and the Hecke algebras of bi-invariant functions on $SL_n(\mathbb{F}_q)$, where $\mathbb{F}_q$ is the finite field with q elements. We need a general definition which will be used again in Section 11.

Definition 6. *Let $\Gamma_0 \subset \Gamma$ be a subgroup of a discrete group Γ such that the left action of Γ on Γ/Γ_0 has only finite orbits. Then the Hecke algebra $\mathcal{H}(\Gamma,\Gamma_0)$ is the convolution algebra of Γ_0-bi-invariant functions on Γ with finite support in $\Gamma_0\backslash\Gamma/\Gamma_0$.*

More explicitly, the convolution $f * f'$ of two such functions is given by

$$(f * f')(\gamma) = \sum_{\Gamma_0\backslash\Gamma} f(\gamma\gamma_1^{-1})\, f'(\gamma_1) \qquad \forall \gamma \in \Gamma. \tag{8}$$

When Γ is finite one can view $\mathcal{H}(\Gamma,\Gamma_0)$ as the subalgebra of Γ_0-bi-invariant elements in the group ring $\mathbb{C}\Gamma$.

Let q be a power of a prime and $\mathbb{F}_q$ the finite field with q elements. Let $\Gamma = SL_n(\mathbb{F}_q)$ and $\Gamma_0 \subset \Gamma$ be the Borel subgroup given by upper triangular matrices. The Bruhat decomposition $\Gamma = \bigcup_{w\in S_n} \Gamma_0\, w\, \Gamma_0$, where S_n is the permutation group, gives a natural basis $(t_w)_{w\in S_n}$ for $\mathcal{H}_n(q) = \mathcal{H}(\Gamma,\Gamma_0)$, where

$$t_w(g) = 1 \text{ if } g \in \Gamma_0\, w\, \Gamma_0, \qquad t_w(g) = 0 \text{ if } g \notin \Gamma_0\, w\, \Gamma_0. \tag{9}$$

Moreover, one checks, using $t_i = t_{\sigma_i}$ where σ_i is the transposition $(i, i+1)$:

Proposition 7. *The algebra $\mathcal{H}_n(q) = \mathcal{H}(\Gamma,\Gamma_0)$ is generated by the t_i, $i = 1,\ldots,n-1$, and admits the following presentation:*

a) $(t_i + 1)(t_i - q) = 0\,,\ i = 1,\ldots,n-1$

b) $t_i\, t_j = t_j\, t_i \quad \forall i,j\,,\ |i-j| > 1$

c) $t_{i+1}\, t_i\, t_{i+1} = t_i\, t_{i+1}\, t_i \quad \forall i = 1,\ldots,n-2.$

For any integer $n \geq 1$ and complex number $q \in \mathbb{C}$, let $\mathcal{H}_n(q)$ be the algebra over $\mathbb{C}$ with generators t_i, $i = 1,\ldots,n-1$ and the presentation a), b) and c). We also allow the value $n = \infty$, and let $\mathcal{H}_\infty(q)$ be the inductive limit of the $(\mathcal{H}_n(q), \rho_{nm})$, where $\rho_{nm}(t_{j,n}) = t_{j,m}$ with obvious notation.

For $q = 1$, $\mathcal{H}_n(q)$ is the group ring $\mathbb{C}S_n$ of the symmetric group. For $q \neq 0$ and $q^n \neq 1$, $\mathcal{H}_n(q)$ is isomorphic to $\mathbb{C}S_n$; but, for $q^n = 1$, $q \neq 1$, or $q = 0$ the algebra $\mathcal{H}_n(q)$ is not semisimple.

For $q \neq -1$ let $e_j = \frac{1}{q+1}\, t_j$; then the e_j, $j \geq 1$, generate $\mathcal{H}_\infty(q)$ and the presentation a) b) c) becomes, with $t = (2 + q + q^{-1})^{-1}$

$$\begin{aligned} &\text{a}')\quad e_j^2 = e_j \qquad \forall j \geq 1 \\ &\text{b}')\quad e_i\, e_j = e_j\, e_i \qquad \text{if } |i-j| > 1 \\ &\text{c}')\quad e_{i+1}\, e_i\, e_{i+1} - t\, e_{i+1} = e_i\, e_{i+1}\, e_i - t\, e_i \qquad \forall i \geq 1. \end{aligned} \tag{10}$$

The relations α) and β) of Lemma 5 are stronger since, while a′) and b′) are unchanged c′) is replaced by the stronger

$$\text{c}'')\quad e_{i+1}\, e_i\, e_{i+1} - t\, e_{i+1} = e_i\, e_{i+1}\, e_i - t\, e_i = 0. \tag{11}$$

The Jones construction thus gives:

Theorem 8. [Jone$_4$] *Assume $q \in [1,\infty[\, \cup \{e^{i2\pi/m}; m = 3,4,\ldots\}$ and also set $t = (2+q+q^{-1})^{-1}$. Let $\mathcal{H}_\infty(q)$ be the above algebra with the unique involution $*$ such that $e_j^* = e_j \quad \forall j \geq 1$.*

1) *There exists a unique trace* τ *on* $\mathcal{H}_\infty(q)$ *such that* $\tau(1) = 1$ *and such that for any* n, $\tau(x\, e_{n+1}) = t\, \tau(x) \quad \forall x \in \mathcal{H}_n(q)$;

2) *The trace* τ *is positive* ($\tau(x^*x) \geq 0 \quad \forall x \in \mathcal{H}_\infty(q)$) *and the weak closure of* $\mathcal{H}_\infty(q)$ *in the GNS representation is the hyperfinite factor* R.

3) *The von Neumann subalgebra of* $\mathcal{H}_\infty(q)''$ *generated by the* e_j, $j > 1$, *is a subfactor* $N \subset R$, $[R : N] = t^{-1}$.

There are two refinements, due to A. Ocneanu and H. Wenzl [Wen]. First consider for $q + q^{-1} \in \mathbb{R}$ the involutive algebra $(\mathcal{H}_\infty(q), *)$ as defined in Theorem 8. Then this involutive algebra admits nontrivial Hilbert space representations iff $q \in [1, \infty[\cup \{e^{2\pi i/m}\ ;\ m = 3, 4, \ldots\}$. Next, with q as above and $z \in \mathbb{C}$ there exists a unique normalized trace $\phi_{q,z}$ on $\mathcal{H}_\infty(q)$ satisfying the Markov property

$$\phi(x\, e_{m+1}) = z\, \phi(x) \qquad \forall x \in \mathcal{H}_n(q). \tag{12}$$

Moreover, with q as in Theorem 8, the trace $\phi_{q,z}$ on the involutive algebra $(\mathcal{H}_\infty(q), *)$, is positive iff $z \in [0,1]$ and the pair (q, z) satisfies one of the following conditions:

$$\begin{aligned} & q \geq 1\ ,\ (1+q)^{-1} \leq z \leq q(1+q)^{-1} \\ & q \geq 1 \text{ and } \exists k \in \mathbb{Z}\ ,\ k \neq 0\ ,\ z = \frac{1 - q^{k+1}}{(1+q)(1-q^k)} \\ & q = e^{2\pi i/m} \text{ and } \exists k \in \mathbb{Z}\ ,\ |k| < m - 1\ ,\ k \neq 0 \\ & \qquad \text{with } z = \frac{1 - q^{k+1}}{(1+q)(1-q^k)} \qquad t = (2 + q + q^{-1})^{-1}. \end{aligned} \tag{13}$$

As in Theorem 8, one obtains for each allowed value of (q, z) a subfactor of the hyperfinite factor R, whose index has been computed by H. Wenzl [Wen].

V.11. Hecke Algebras, Type III Factors and Statistical Theory of Prime Numbers

In this section we shall discuss an example of a quantum statistical mechanical system, arising from the theory of prime numbers, which exhibits a phase transition with spontaneous symmetry breaking [Bos-C]. The original motivation for these results comes from the work of B. Julia [J] (cf. also [Spe]).

11.α Description of the system and its phase transition

Let us first recall our discussion of quantum statistical mechanics of Chapter I Section 2. Thus, a quantum statistical system is given by

α) The C^*-algebra of observables A.

β) The time evolution $(\sigma_t)_{t \in \mathbb{R}}$, which is a one-parameter group of automorphisms of A.

An equilibrium or KMS state at inverse temperature β is a state φ on A which fulfills the KMS_β condition, I.2 and V.3, i.e. for any $x, y \in A$, there exists a bounded holomorphic function $F_{x,y}(z)$, continuous on the closed strip $0 \leq \text{im}\, z \leq \beta$, such that

$$F_{x,y}(t) = \varphi(x\, \sigma_t(y)) \quad \forall t \in \mathbb{R}$$

$$F_{x,y}(t + i\beta) = \varphi(\sigma_t(y)x) \quad \forall t \in \mathbb{R}.$$

In the simplest case, where $A = M_N(\mathbb{C})$ is the algebra of $N \times N$ matrices, any one-parameter group of automorphisms $(\sigma_t)_{t\in\mathbb{R}}$ of A is of the form

$$\sigma_t(x) = e^{itH}\, x\, e^{-itH} \quad \forall x \in A\, ,\ t \in \mathbb{R}$$

for some selfadjoint element $H = H^* \in A$. Then for any $\beta \in [0, \infty[$, one has a *unique* KMS$_\beta$ state for σ_t, and it is given by the formula

$$\varphi_\beta(x) = \frac{\text{Trace}\,(e^{-\beta H}\, x)}{\text{Trace}(e^{-\beta H})} \quad \forall x \in A.$$

Note here that H is only defined up to an additive constant by σ_t, so that the normalization factor, Trace $(e^{-\beta H})$, cannot be recovered from σ_t. However, the following formula holds:

$$\text{Log Trace}\,(e^{-\beta H}) = \underset{\varphi}{\text{Sup}}\,(S(\varphi) - \beta\varphi(H))$$

where φ varies over all states on A and $S(\varphi)$ is the entropy of the state,

$$S(\varphi) = -\text{Trace}\,(\rho \,\text{Log}\, \rho) \ \text{ for } \ \varphi = \text{Trace}\,(\rho\,\cdot).$$

In a slightly more involved situation, that of *systems without interaction*, it is still true that for any $\beta \in [0, \infty[$ there exists a unique KMS$_\beta$ state. More precisely, one immediately has the following:

Proposition 1. *Let $A = \bigotimes_{\nu\in I} A_\nu$ be an infinite tensor product of matrix algebras $A_\nu = M_{n_\nu}(\mathbb{C})$, and let $\sigma_t = \bigotimes_{\nu\in I} \sigma_t^\nu$ be a product time evolution. Then for any $\beta \geq 0$, there exists a unique KMS$_\beta$ state φ_β for (A, σ_t), and one has $\varphi_\beta = \bigotimes_\nu \varphi_{\beta,\nu}$, where $\varphi_{\beta,\nu}$ is the unique KMS$_\beta$ state for (A_ν, σ_t^ν).*

For interesting systems *with interaction* one expects, in general, that for *large* temperature, i.e. small β, the disorder will be predominant so that there will exist only one KMS$_\beta$ state. For small enough temperature some order should set in and allow for the existence of various thermodynamical phases, i.e. of various KMS$_\beta$ states. It is a very important general fact of the C^*-algebraic formulation of quantum statistical mechanics that for a given β every KMS$_\beta$ state decomposes *uniquely* as a statistical superposition of *extreme* KMS$_\beta$ states:

Proposition 2. [Br-R] [H] *Let (A, σ_t) be a C^*-dynamical system and $\beta \in [0, \infty[$. Then the space of* KMS$_\beta$ *states is a compact convex Choquet simplex.*

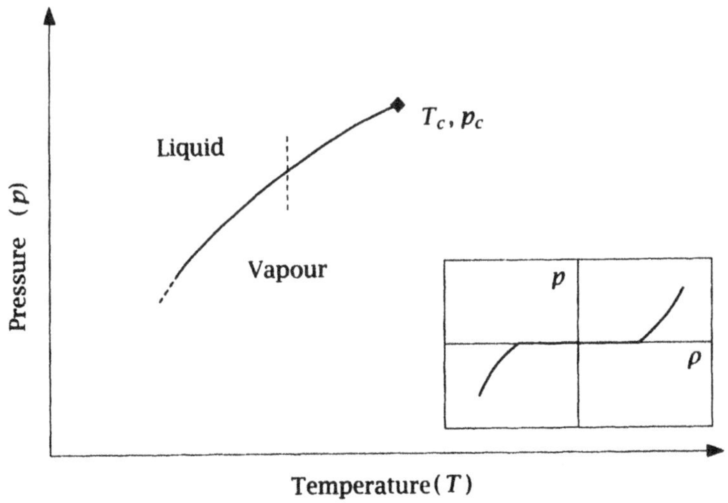

Figure V.1.

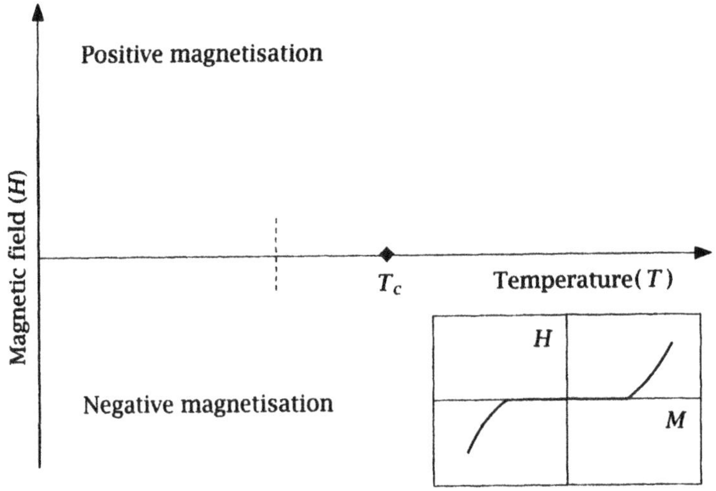

Figure V.2.

For a careful discussion of the link between extreme KMS_β states and thermodynamical phases we refer the reader to [H].

As a simple (classical) example illustrating the coexistence of phases at small temperature one can think of the phase diagram for water and vapor (Figure 1), or better yet for the ferromagnet (Figure 2). In the latter example, when the temperature T is larger than the *critical* temperature T_c of the order of $10^3 K$, the disorder dominates, while for $T < T_c$ the individual magnets tend to align with each other, which in the classical 3-dimensional set-up yields a set of thermodynamical homogeneous phases parametrized by the 2-dimensional sphere of directions in 3-space.

This example serves to illustrate the phenomenon of *spontaneous symmetry breaking*: The group $SO(3)$ of rotations in $\mathbb{R}^3$ is a symmetry group of the dynamical system one starts with, and for large T, $T > T_c$, the equilibrium state is unique and hence invariant under rotation. For small T however, $T < T_c$, the group $SO(3)$ acts nontrivially on the set of thermodynamical phases and the choice of an equilibrium state breaks the symmetry.

The C^*-algebraic formulation of this is straightforward. One has a compact group G of automorphisms of the C^*-algebra A which commutes with the time evolution:

$$\alpha_g \in \operatorname{Aut} A\,,\ \forall g \in G\,,\ \alpha_g \sigma_t = \sigma_t \alpha_g \quad \forall t \in \mathbb{R}.$$

Such a group obviously acts on the compact convex space of KMS_β states, and hence on its extreme points.

We shall now describe (the precise motivation will be explained below) a C^*-dynamical system intimately related to the distribution of prime numbers and exhibiting the above behaviour of spontaneous symmetry breaking.

The C^*-algebra A is a *Hecke algebra*, which contains the algebra of the usual Hecke operators of number theory [Ser$_1$], i.e. those related to Hecke correspondences for lattices in $\mathbb{C}$ [Ser$_1$]. The latter algebra is commutative and is, essentially, the algebra of composition of double cosets

$$\gamma \in GL(2,\mathbb{Z})\backslash GL(2,\mathbb{Q})/GL(2,\mathbb{Z}).$$

Recall that given a discrete group Γ and a subgroup Γ_0 which is *almost normal*, so that

"The orbits of Γ_0 acting on the left on Γ/Γ_0 are *finite*"

one defines the Hecke algebra $\mathcal{H}(\Gamma,\Gamma_0)$ as the convolution algebra of ($\mathbb{C}$-valued for our purposes) functions with finite support on $\Gamma_0\backslash\Gamma/\Gamma_0$. More specifically, given two such functions $f, f' \in \mathcal{H}(\Gamma,\Gamma_0)$, their convolution is

$$(f * f')(\gamma) = \sum_{\Gamma_0\backslash\Gamma} f(\gamma\gamma_1^{-1})\, f'(\gamma_1) \quad \forall \gamma \in \Gamma.$$

In this formula f and f' are viewed as Γ_0-bi-invariant functions on Γ with finite support in $\Gamma_0\backslash\Gamma/\Gamma_0$.

To complete $\mathcal{H}$ to a C^*-algebra we just close it in norm in the following *regular representation* of $\mathcal{H}$ in $\ell^2(\Gamma_0\backslash\Gamma)$ (cf. [Bi]).

Proposition 3. *Let $\Gamma_0 \subset \Gamma$ be an almost normal subgroup of the discrete group Γ. Then the following defines an (involutive) representation λ of $\mathcal{H}(\Gamma, \Gamma_0)$ in $\ell^2(\Gamma_0 \backslash \Gamma)$:*

$$(\lambda(f)\xi)(\gamma) = \sum_{\Gamma_0 \backslash \Gamma} f(\gamma\gamma_1^{-1})\, \xi(\gamma_1) \quad \forall \gamma \in \Gamma_0 \backslash \Gamma\,,\ \forall f \in \mathcal{H}.$$

One checks that $\lambda(f)$ is bounded for any $f \in \mathcal{H}$. The involution on $\mathcal{H}$ such that

$$\lambda(f^*) = \lambda(f)^* \quad \forall f \in \mathcal{H}$$

is given by the equality

$$f^*(\gamma) = \overline{f(\gamma^{-1})} \quad \forall \gamma \in \Gamma_0 \backslash \Gamma / \Gamma_0.$$

Thus, we let A be the C^*-algebra norm closure of $\mathcal{H}(\Gamma, \Gamma_0)$ in $\ell^2(\Gamma_0 \backslash \Gamma)$. A good notation for it, compatible with the discrete group case (Chapter II Section 4) is

$$\overline{\mathcal{H}} = C_r^*(\Gamma, \Gamma_0).$$

Let us now define the one-parameter group of automorphisms $\sigma_t \in \operatorname{Aut} A$. We first need to introduce notation. Since each Γ_0 orbit on Γ/Γ_0 is finite, we shall let, for $\gamma \in \Gamma$

$$L(\gamma) = \text{cardinality of } \Gamma_0(\gamma\Gamma_0) \text{ in } \Gamma/\Gamma_0$$
$$R(\gamma) = \text{cardinality of } (\Gamma_0\gamma)\Gamma_0 \text{ in } \Gamma_0 \backslash \Gamma.$$

Thus, by construction, $L(\gamma) \in \mathbb{N}^*$, $R(\gamma) \in \mathbb{N}^*$, $R(\gamma) = L(\gamma^{-1})$, and L and R are both Γ_0-bi-invariant functions.

Proposition 4. *Let $\Gamma_0 \subset \Gamma$ be an almost normal subgroup of the discrete group Γ. There exists a unique one-parameter group of automorphisms $\sigma_t \in \operatorname{Aut}(C_r^*(\Gamma, \Gamma_0))$ such that*

$$(\sigma_t(f))(\gamma) = \left(\frac{L(\gamma)}{R(\gamma)}\right)^{-it} f(\gamma) \quad \forall \gamma \in \Gamma_0 \backslash \Gamma / \Gamma_0.$$

In fact, σ_{-t} is the restriction of the modular automorphism group σ_t^φ for the state on $M = \lambda(\mathcal{H})''$ given by the unit vector corresponding to the coset $\Gamma_0 \in \Gamma_0 \backslash \Gamma$.

Let us now consider the Hecke algebra $\mathcal{H}$ for the groups

$$\Gamma = P_{\mathbb{Q}}\,,\ \Gamma_0 = P_{\mathbb{Z}}$$

where P is the group of 2×2 matrices $P = \left\{\begin{bmatrix} 1 & b \\ 0 & a \end{bmatrix} ;\ a \text{ invertible}\right\}$.

One checks that $P_{\mathbb{Z}}$ is almost normal in $P_{\mathbb{Q}}$.

We shall now describe the phase transition with spontaneous symmetry breaking for the dynamical system corresponding to $\Gamma = P_{\mathbb{Q}}$ and $\Gamma_0 = P_{\mathbb{Z}}$.

Let us denote by ψ_β the following function on the group $\mathbb{Q}/\mathbb{Z}$. Given $n = \frac{a}{b} \in \mathbb{Q}/\mathbb{Z}$, with $a, b \in \mathbb{Z}$, with a relatively prime to $b > 0$, one lets $b = \prod p^{k_p}$ be the prime factor decomposition of b and one sets

$$\psi_\beta(n) = \prod p^{-k_p\beta}(1 - p^{\beta-1})(1 - p^{-1})^{-1}.$$

Theorem 5. [Bos-C] *Let (A, σ_t) be the C^*-dynamical system associated to the almost normal subgroup $P_{\mathbb{Z}}$ of $P_{\mathbb{Q}}$. Then:*

a) *For $0 < \beta \le 1$ there exists a unique* KMS_β *state φ_β. Its restriction to $C^*(\mathbb{Q}/\mathbb{Z}) \subset C_r^*(\Gamma, \Gamma_0)$ (through the inclusion $\mathbb{Q}/\mathbb{Z} \subset \Gamma_0\backslash\Gamma/\Gamma_0$ of the unipotent subgroup $\left\{\begin{bmatrix} 1 & n \\ 0 & 1 \end{bmatrix}\right\} \subset P$) is given by the above function of positive type ψ_β on $\mathbb{Q}/\mathbb{Z}$. Each φ_β is a factor state and the associated factor is the hyperfinite factor of type III_1, that is R_∞.*

b) *For $\beta > 1$ the* KMS_β *states form a simplex whose extreme points $\varphi_{\beta,\chi}$ are parametrized by imbeddings $\chi : K \to \mathbb{C}$ of the field $K = \overline{\mathbb{Q}}^{\mathrm{ab}}$ (the field of roots of unity) in $\mathbb{C}$, and whose restrictions to $C^*(\mathbb{Q}/\mathbb{Z})$ are given by the formula*

$$\varphi_{\beta,\chi}(\gamma) = \zeta(\beta)^{-1} \sum_{n=1}^{\infty} n^{-\beta} \chi(\gamma)^n.$$

These states are type I_∞ factor states.

The normalization factor is the inverse of the Riemann ζ-function evaluated at β.

In other words the critical temperature here is $T_c = 1$, and at low temperature ($\beta > 1$) the phases of the system are parametrized by all possible embeddings of $K = \overline{\mathbb{Q}}^{\mathrm{ab}}$ in the field of complex numbers.

As we shall see below, the Galois group $G = [K : \mathbb{Q}]$ does act naturally as automorphisms $G \subset \mathrm{Aut}(C_r^*(\Gamma, \Gamma_0))$ commuting with the time evolution, and spontaneous symmetry breaking occurs for $\beta > 1$.

We shall now explain how the above C^*-dynamical system is related to the distribution of prime numbers and to class field theory.

11.β Bosonic second quantization and prime numbers as a subset of $\mathbb{R}$

It is a saying attributed to E. Nelson that first quantization is a mystery while second quantization is a *functor*. In the bosonic case this functor S, from the category of Hilbert spaces to itself, assigns to every Hilbert space $\mathcal{H}$ the new Hilbert space $S\mathcal{H}$ given by:

$$S\mathcal{H} = \bigoplus_{n=0}^{\infty} S^n\mathcal{H}$$

where $S^n\mathcal{H}$ is the nth symmetric power of $\mathcal{H}$ endowed with the inner product

$$\langle \xi_1 \cdots \xi_n, \eta_1 \cdots \eta_n \rangle = \sum_{\sigma} \prod_{i=1}^{n} \langle \xi_i, \eta_{\sigma(i)} \rangle \quad \forall \xi_j, \eta_j \in \mathcal{H}.$$

Given an operator T on $\mathcal{H}$ (or more generally $T : \mathcal{H}_1 \to \mathcal{H}_2$), the operator $\mathbf{S}T$ on $\mathbf{S}\mathcal{H}$ is given by

$$(\mathbf{S}T)(\xi_1 \cdots \xi_n) = (T\xi_1)(T\xi_2) \cdots (T\xi_n) \quad \forall \xi_i \in \mathcal{H}.$$

Even if T is bounded, $\mathbf{S}T$ is not bounded in general but if T is selfadjoint so is $\mathbf{S}T$. Thus, we shall work with such operators. One has the formula

$$\text{Trace}\,(\mathbf{S}T) = \frac{1}{\det(1 - T)} \tag{$*$}$$

which makes good sense if $\|T\| < 1$ and $T \in \mathcal{L}^1(\mathcal{H})$. The problem we shall now consider is the following: Give a simple characterization of selfadjoint operators T in $\mathcal{H}$ whose spectrum is the subset $\mathcal{P} \subset \mathbb{R}$ formed of all prime numbers, each with multiplicity one:

$$\mathcal{P} = \{2, 3, 5, 7, 11, 13, 17, \dots\} \subset \mathbb{R}.$$

The corresponding problem for the set $\mathbb{N} = \{0, 1, 2, 3, \dots\}$ of natural numbers, or $\mathbb{N}^* = \mathbb{N}\backslash\{0\}$, is easier, and was solved in Dirac's paper [Dir] which inaugurated quantum field theory. In that case the solution is simply that there exists an operator a such that

$$aa^* - a^*a = 1 \;,\; a^*a = T.$$

(For $\mathbb{N}^*$ one requires that aa^* be equal to T.)

Let us now state the result for the subset $\mathcal{P} \subset \mathbb{R}$:

Lemma 6. [Bos-C] *Let T be a selfadjoint operator in a Hilbert space $\mathcal{H}$; then, counting multiplicities*

$$\text{Spectrum } T = \mathcal{P} \Longleftrightarrow \text{Spectrum } \mathbf{S}T = \mathbb{N}^*.$$

Proof. Let us first assume that Spectrum $\mathbf{S}T = \mathbb{N}^*$. Then, as quite generally $\text{Spec}\,T \subset \text{Spec}\,\mathbf{S}T$, using the inclusion $\mathcal{H} \subset \mathbf{S}\mathcal{H}$ we see that $\Sigma = \text{Spec}\,T \subset \mathbb{N}^*$. Let us show that $\mathcal{P} \subset \Sigma$. Indeed, let $p \in \mathcal{P}$ not be in Σ. Then because $\Sigma \subset \mathbb{N}^*$ one has $p \notin \Sigma^n$ for any n (with $\Sigma^n = \{k_1 k_2 \cdots k_n; k_j \in \Sigma\}$). This shows that $p \notin \text{Spec}(\mathbf{S}T) = \bigcup \Sigma^n$, whence a contradiction. Thus, $\mathcal{P} \subset \Sigma$. If $k \in \Sigma\backslash\mathcal{P}$ then as $k \in \mathcal{P}^n$ for some $n > 1$, this would mean that k is not a simple eigenvalue for $\mathbf{S}T$. Thus $\mathcal{P} = \Sigma$. The converse is obvious from Euclid's unique factorization theorem, but we shall fix the corresponding notation: we let $\mathcal{H}_1 = \ell^2(\mathcal{P})$ be the Hilbert space with basis $(\varepsilon_p)_{p\in\mathcal{P}}$, and we identify it with the one-particle subspace of $\mathbf{S}\mathcal{H}_1 = \ell^2(\mathbb{N}^*)$, the Hilbert space of square integrable sequences of complex numbers, with canonical basis the ε_n, $n \in \mathbb{N}^*$. We shall denote by T the operator

$$T : \ell^2(\mathcal{P}) \to \ell^2(\mathcal{P}) \; ; \; T\varepsilon_p = p\varepsilon_p \quad \forall p \in \mathcal{P}$$

and by $\mathbf{S}T$ the corresponding operator

$$\mathbf{S}T : \ell^2(\mathbb{N}^*) \to \ell^2(\mathbb{N}^*) \; ; \; (\mathbf{S}T)\varepsilon_n = n\varepsilon_n \quad \forall n \in \mathbb{N}^*.$$

We shall let $H = \log(\mathbf{S}T)$. It is the generator of a one-parameter unitary group $U_t = \exp(itH) = (\mathbf{S}T^{it})$, whose role is made clear by the following special case of formula $(*)$, which is the Euler product formula for the Riemann ζ function:

$$\text{For Re } s > 1; \zeta(s) = \text{Trace } (\mathbf{S}T)^s = \frac{1}{\det(1 - T^s)}.$$

The meaning of Lemma 6 is that the subset $\mathcal{P} \subset \mathbb{R}$ has a neat definition provided one is ready to use the formalism of *bosonic quantum field theory.* That formalism includes *the algebra* of creation and annihilation operators, respectively $a^*(\xi)$ and $a(\eta)$, for $\xi, \eta \in \mathcal{H}$, given by

$$a^*(\xi)\xi_1 \cdots \xi_n = \xi\xi_1 \cdots \xi_n \quad \forall \xi_j \in \mathcal{H}$$

$$a(\eta) = (a^*(\eta))^*.$$

It also includes the *time evolution,* in Heisenberg's picture, given by

$$\sigma_t(x) = U_t \, x \, U_t^* = e^{itH} \, x \, e^{-itH} \quad \forall t \in \mathbb{R}.$$

In our case the corresponding C^*-algebra in $\mathbf{S}\mathcal{H} = \ell^2(\mathbb{N}^*)$ and time evolution are given by the following:

Proposition 7. a) *For each $p \in \mathcal{P}$ let μ_p be the isometry in $\ell^2(\mathbb{N}^*)$ given by the polar decomposition of the creation operator associated to the unit vector $\varepsilon_p \in \mathcal{H}$. The C^*-algebra $C^*(\mathbb{N}^*)$ generated by the μ_p's is the same as that generated by the isometrics μ_n*

$$\mu_n \, \varepsilon_k = \varepsilon_{kn} \quad \forall n \in \mathbb{N}^* \;\; k \in \mathbb{N}^*.$$

b) *This C^*-algebra is the infinite tensor product*

$$C^*(\mathbb{N}^*) = \bigotimes_{p \in \mathcal{P}} \tau_p$$

where each τ_p is the C^*-algebra generated by μ_p, and is the Toeplitz C^*-algebra.

c) *The equality $\sigma_t(x) = e^{itH} \, x \, e^{-itH}$, $\forall x \in C^*(\mathbb{N}^*)$, $t \in \mathbb{R}$, where $H = \log(\mathbf{S}T)$, defines a one-parameter group of automorphisms of $C^*(\mathbb{N}^*)$ given as*

$$\sigma_t = \bigotimes_{p \in \mathcal{P}} \sigma_{t,p}; \sigma_{t,p}(\mu_p) = p^{it} \, \mu_p \quad \forall t \in \mathbb{R}.$$

We recall that the Toeplitz C^*-algebra $\mathcal{T}$ is the C^*-algebra with generator u and presentation relation $u^*u = 1$. If u is any non-unitary isometry in a (separable) Hilbert space the smallest C^*-algebra containing u is isomorphic to $\mathcal{T}$. This C^*-algebra is nuclear so that the finite tensor products $\bigotimes_{p \leq n} \mathcal{T}_p$ are unambiguously defined. Their inductive limit $\bigotimes_{p \in \mathcal{P}} \mathcal{T}_p$ is the C^*-algebra $C^*(\mathbb{N}^*)$.

The C^*-dynamical system thus obtained is not very interesting because it is without interaction (7 c). Nevertheless the corresponding unique KMS_β states will be useful later and are given by the following corollary of Proposition 1 and of the Araki-Woods classification of ITPFI (Section 4).

Proposition 8. [Bos-C] a) *For every $\beta > 0$, there exists a unique* KMS_β *state on* $(C^*(\mathbb{N}^*), \sigma_t)$. *It is the infinite tensor product*

$$\varphi_\beta = \bigotimes_{p \in \mathcal{P}} \varphi_{\beta,p}$$

where $\varphi_{\beta,p}$ is the unique KMS_β *state on the Toeplitz algebra for $\sigma_{t,p}$. Its eigenvalue list is*

$$\{(1 - p^{-\beta})p^{-n\beta}\,;\ n \in \mathbb{N}\}.$$

b) *For $\beta > 1$, the state φ_β is of type I_∞ and is given by*

$$\varphi_\beta(x) = \zeta(\beta)^{-1}\ \mathrm{Trace}(e^{-\beta H}\, x) \quad \forall x \in C^*(\mathbb{N}^*).$$

c) *For $\beta = 1$, the state φ_β is a factor state of type III_1 given by*

$$\varphi_\beta(x) = \mathrm{Tr}_\omega(e^{-H}\, x) \quad \forall x \in C^*(\mathbb{N}^*)$$

where Tr_ω *is the Dixmier trace.*

d) *For $0 < \beta \leq 1$, φ_β is a factor state of type III_1 and the associated factor is the factor R_∞ of Araki-Woods.*

Statement d) for $\beta = 1$ is due to B. Blackadar [Bl$_4$]. We refer to Chapter IV for the definition of the Dixmier trace, whose general properties make it clear that the equality c) defines a KMS_1 state.

11.γ Products of trees and the noncommutative Hecke algebra

In this section we shall relate the C^*-dynamical system $(C^*(\mathbb{N}^*), \sigma_t)$, of Section β) with basic number theory notions ([We]) and get to the Hecke dynamical system of Theorem 5.

Let P be the $ax + b$ group, i.e. the group of triangular 2×2 matrices of the form $\begin{bmatrix} 1 & b \\ 0 & a \end{bmatrix}$, with a invertible. We view it as an algebraic group, i.e. as a functor $A \mapsto P_A$ from abelian rings to groups. It plays an important role in the elementary classification of locally compact (commutative and non discrete) fields (cf. [We]). Indeed, given such a field K, the group $G = P_K$ is a *locally* *compact* group, and as such it has a module: $\delta : G \to \mathbb{R}_+^*$, obtained from the lack of invariance of a left Haar measure dg on G under right translations:

$$d(gk) = \delta(k)dg \quad \forall k \in G.$$

Or equivalently $d(g^{-1}) = \delta(g)^{-1}dg$ as measures on G.

This module $\delta : P_K \to \mathbb{R}_+^*$ is 1 on the additive group, and its restriction to the multiplicative group (extended by 0 on $K\backslash K^*$) yields a proper continuous multiplicative map

$$\text{mod}_K : K \to \mathbb{R}_+$$

such that the open sets $\{k \in K; \text{mod}_K(k) < \varepsilon\}$ form a basis of neighborhoods of 0 (cf. [We]). The image of δ is a closed subgroup of $\mathbb{R}_+^*$ and, except for the case of the Archimedean fields $\mathbb{R}$ or $\mathbb{C}$, this closed subgroup is discrete and equal to $\lambda^{\mathbb{Z}}$ for some $\lambda \in]0,1[$ whose inverse $q = \lambda^{-1}$ is called the *module* of K. The function $\text{mod}_K(x - y) = d(x, y)$ is then an ultrametric distance giving the topology of K ([We]). Given $x \in K$, the integer $\nu(x)$ such that $\text{mod}_K(x) = q^{-\nu(x)}$ is called the valuation of x.

Proposition 9. (cf. [We]) *Let K be a non-discrete commutative locally compact field, $K \neq \mathbb{R}$ or $\mathbb{C}$. Then there exists a prime p such that* $\text{mod}_K(p) < 1$. *Call R, $R^\times$ and P the subsets of K respectively given by*

$$R = \{x \in K\ ;\ \text{mod}_K(x) \leq 1\}\ ,\ R^\times = \{x \in K\ ;\ \text{mod}_K(x) = 1\},$$

$$J = \{x \in K\ ;\ \text{mod}_K(x) < 1\}.$$

Then K is ultrametric; R is the unique maximal compact subring of K; $R^\times$ is the group of invertible elements of R; J is the unique maximal ideal of R, and there is $\pi \in J$ such that $J = \pi R = R\pi$. Moreover, the residue field $k = R/J$ is a finite field of characteristic p; if q is the number of its elements, the image of $K^\times$ in $\mathbb{R}_+^*$ under mod_K *is the subgroup of $\mathbb{R}_+^*$ generated by q; and* $\text{mod}_K(\pi) = q^{-1}$.

As a basic example the field $\mathbb{Q}_p$ of p-adic numbers is defined for any prime number p as the completion of the field $\mathbb{Q}$ of rational numbers for the distance function

$$d(x, y) = |x - y|_p$$

where for $x \in \mathbb{Q}$, $x = p^n \frac{a}{b}$ (with n, a, b integers and a, b relatively prime to p) one sets

$$|x|_p = p^{-n}.$$

The maximal subring R of $K = \mathbb{Q}_p$ is the ring $\mathbb{Z}_p$ of p-adic integers, and the residue field $k = R/J$ is the finite field $\mathbb{F}_p$.

One obtains in this way, together with the inclusion $\mathbb{Q} \subset \mathbb{R}$, every inclusion $\mathbb{Q} \subset K$ of the field of rational numbers as a dense subfield of a local field K. Such inclusions, (or rather in general, equivalence classes of completions) are called *places* and, to distinguish the real place $\mathbb{Q} \subset \mathbb{R}$ from the others, the latter are called *finite places*.

By putting together the inclusions of $\mathbb{Q}$ in its completions $\mathbb{Q}_v = K$ parametrized by the places of $\mathbb{Q}$ one gets a single inclusion of $\mathbb{Q}$ in the *locally compact* commutative ring of *adèles*, which is the restricted product of the fields $\mathbb{Q}_v$. More specifically this ring is the product $\mathbb{R} \times \mathcal{A}$, where the ring $\mathcal{A}$ of *finite* adèles is obtained as follows:

a) Elements x of $\mathcal{A}$ are arbitrary families (x_p), $x_p \in \mathbb{Q}_p$ such that $x_p \in \mathbb{Z}_p$ for all but a finite number of p.

b) $(x+y)_p = x_p + y_p$ and $(xy)_p = x_p y_p$ define the addition and product in $\mathcal{A}$.

c) Finally, $\mathcal{A}$ has a unique topology as a locally compact ring such that the subring $\mathcal{R} = \prod \mathbb{Z}_p$ is open and closed and inherits its compact product topology.

We shall now relate the C^*-dynamical system $(C^*(\mathbb{N}^*), \sigma_t)$ of Subsection β) with the locally compact ring $\mathcal{A}$ of finite adèles.

We just need to recall that given a non-unimodular locally compact group G one has a natural one-parameter group of automorphisms σ_t of $C^*(G)$ given by the formula, valid say on $L^1(G)$,

$$(\sigma_t(f))(g) = \delta(g)^{-it} f(g) \quad \forall g \in G \, , \; t \in \mathbb{R}. \qquad (*)$$

The group σ_{-t} is the modular automorphism group of the Plancherel weight on $C^*(G)$.

Proposition 10. *Let $\mathcal{A}$ be the ring of finite adèles over $\mathbb{Q}$, and $\mathcal{R}$ its maximal open compact subring. Let G be the locally compact group $G = P_{\mathcal{A}}$, and $e \in C^*(G)$ the characteristic function of the open and compact subgroup $P_{\mathcal{R}} \subset P_{\mathcal{A}}$. Then:*

1) One has $e = e^* = e^2$, and the reduced C^*-algebra $C^*(G)_e = \{x \in C^*(G) \, ; \, ex = xe = x\}$ is canonically isomorphic to the C^*-algebra $C^*(\mathbb{N}^*)$ of Subsection β).

2) One has $\sigma_t(e) = e \quad \forall t \in \mathbb{R}$, and the restriction of σ_t to the reduced C^*-algebra $C^*(\mathbb{N}^*)$ is the time evolution of Subsection β).

We think of the characteristic function of $P_{\mathcal{R}}$ as an element of $L^1(G, dg) \subset C^*(G)$, with dg the unique left Haar measure which gives measure 1 to $P_{\mathcal{R}}$. The group G is solvable and hence amenable, so that there is no distinction between $C^*(G)$ and the reduced C^*-algebra $C_r^*(G)$. The proof of Proposition 10 is not difficult because it reduces immediately to a *local* statement, namely: If $K = \mathbb{Q}_p$ and $R \subset K$ is the maximal compact subring, the C^*-dynamical system $(C^*(P_K), \sigma_t)$ given by $(*)$, reduced by the projection e_p = characteristic function of P_R, is isomorphic to the Toeplitz C^*-algebra τ_p, with the time evolution $\sigma_{t,p}$ of Proposition 7 c).

We then take for $\mu_p \in C^*(P_{\mathbb{Q}_p})_{e_p}$ the isometry given by the L^1 function

$$\mu_p\left(\begin{bmatrix} 1 & b \\ 0 & a \end{bmatrix}\right) = 1 \text{ if } b \in R \, , \; |a|_p = p^{-1}, \text{ and equal to 0 otherwise.}$$

The C^*-dynamical system $(C^*(P_{\mathcal{A}}), \sigma_t)$ of Proposition 10 is without interaction, exactly as is $(C^*(\mathbb{N}^*), \sigma_t)$ (Proposition 8); there is an exact analogue of Proposition 8, which states the existence and uniqueness (up to scale) of KMS_β *weights* on the above system. One needs to use weights because one is dealing with non-unital C^*-algebras. On the technical side, such weights have to be semicontinuous and semifinite (for the norm topology) (cf. [Com]). It is, however, instructive to work out the explicit formula for those KMS_β weights. Using the natural isomorphism of the Pontryagin dual $\widehat{\mathcal{A}}$ of the additive group $\mathcal{A}$ with itself, one gets an isomorphism

$$C^*(P_{\mathcal{A}}) = C_0(\mathcal{A}) \rtimes \mathcal{A}^*$$

where the multiplicative group $\mathcal{A}^*$ acts by homotheties on the locally compact space $\mathcal{A}$. The KMS_β weight on $(C^*(P_{\mathcal{A}}), \sigma_t)$ is then the weight dual to the measure μ_β on $\mathcal{A}$

$$\mu_\beta(f) = \zeta(\beta)^{-1} \int_{\mathcal{A}^*} |j|^\beta f(j) d^*j.$$

Here d^*j is the Haar measure on the multiplicative group $\mathcal{A}^*$, $j \mapsto |j|$ is the module, and the formula makes sense as such for $\beta > 1$, and by analytic continuation for $0 < \beta < 1$ ([We]).

It is clear that to obtain a C^*-dynamical system *with interaction* we need to use not only the locally compact ring $\mathcal{A}$ but also the fundamental inclusion

$$\mathbb{Q} \subset \mathcal{A}.$$

We shall use the corresponding inclusion $P_{\mathbb{Q}} \subset P_{\mathcal{A}} = G$ in the action of $P_{\mathcal{A}}$ on the C^*-module $\mathcal{E} = C^*(G)e$ over $C^*(\mathbb{N}^*)$ given by the isomorphism of Proposition 10: $C^*(G)_e = C^*(\mathbb{N}^*)$. Indeed, given any C^*-algebra B and selfadjoint projection $e \in B$, the space $\mathcal{E} = Be = \{x \in B \,;\, xe = x\}$ is in a natural way a right C^*-module over the reduced C^*-algebra $B_e = \{x \in B \,;\, ex = xe = x\}$. One thus lets

$$\langle \xi, \eta \rangle = \xi^* \eta \in B_e \ , \quad \forall \xi, \eta \in \mathcal{E} = Be$$

$$\xi a \in \mathcal{E} \quad \forall \xi \in \mathcal{E} \, , \, a \in B_e.$$

This C^*-module has moreover a natural left B-module structure, which is given by $(b, \xi) \mapsto b\xi \in \mathcal{E}$, $\forall b \in B$, $\xi \in \mathcal{E}$.

In our case $\mathcal{E} = C^*(G)e$ is a space of functions on G which are invariant under right multiplication by elements of $P_{\mathcal{R}} \subset G$ or, in other words, it is a space of functions on the homogeneous space

$$\Delta = G/P_{\mathcal{R}}.$$

This space Δ is, by construction, the restricted product of the spaces

$$\Delta_p = P_{\mathbb{Q}_p}/P_{\mathbb{Z}_p}$$

relative to the base point given by $P_{\mathbb{Z}_p}$.

Proposition 11. *The homogeneous space $\Delta_p = P_{\mathbb{Q}_p}/P_{\mathbb{Z}_p}$ over the group $P_{\mathbb{Q}_p} \subset GL(2,\mathbb{Q}_p)$ is naturally isomorphic to the (set of vertices of the) tree of $SL(2,\mathbb{Q}_p)$. The group $P_{\mathbb{Q}_p}$ acts by isometries of Δ_p and preserves a point at ∞.*

Let us recall (cf. [Ser$_2$]) that the tree of $SL(2,K)$, where K is a local field, is defined in terms of equivalence classes of lattices in a two-dimensional vector space V over K. With the notation of Proposition 9, a lattice $L \subset V$ is an R-submodule of V which is of finite type and generates V as a vector space. The multiplicative group K^* operates on the set of lattices by $(L,x) \mapsto xL$ for $x \in K^*$, and one lets T be the set of orbits of this action of K^*. Given a lattice $L \subset V$ and a class $\Lambda' \in T$ there exists a unique representative $L' \in \Lambda'$ such that $L' \subset L$ and $L' \not\subset \pi L$, with π given by Proposition 9. Then $L/L' = R/\pi^n R$ and the integer n, which depends only upon the classes of L and L', defines a distance d on T, by

$$d(\text{class of } L,\ \text{class of } L') = n.$$

Using the set of pairs with mutual distance equal to 1 to define a 1-dimensional simplicial complex, one gets a tree, the tree of $SL(2,K)$, and the above distance is the length of the unique path joining two elements of this tree (cf. [Ser$_2$]). The group $GL(V)$ of automorphisms of the vector space V acts on the set of lattices by

$$(L,g) \mapsto gL \quad \forall g \in GL(V),$$

and, since this action commutes with that of K^*, it gives an action, by isometries, of $GL(V)$ on the tree T. Let us identify V with K^2, $GL(V)$ with $GL(2,K)$, and consider P_K as a subgroup of $GL(2,K)$: $P_K = \left\{\begin{bmatrix} 1 & b \\ 0 & a \end{bmatrix} ;\ a \in K^*\ ,\ b \in K\right\}$. Let L_0 be the lattice $R^2 \subset K^2$. Then one checks that P_K acts transitively on T and that the stabilizer of the class of L_0 is P_R. We thus get a canonical identification $T = P_K/P_R$. Taking $K = \mathbb{Q}_p$ yields the conclusion.

Proposition 12. 1) *The homogeneous space $G/P_{\mathcal{R}} = \Delta$ is canonically isomorphic to the restricted product of the trees T_p with base point $*$, and the action of G on Δ is simplicial.*

2) *The subgroup $P_{\mathbb{Q}} \subset P_{\mathcal{A}} = G$ acts transitively on Δ, and the isotropy subgroup of the base point $*$ is $P_{\mathbb{Z}}$.*

We can thus identify Δ with $P_{\mathbb{Q}}/P_{\mathbb{Z}}$, and we shall now get the Hecke algebra of Theorem 5 from the commutant of the action of $P_{\mathbb{Q}}$ in the space of functions on Δ. We need for that purpose first to obtain Hilbert spaces, and the construction will follow from the following general lemma applied to the C^*-module $\mathcal{E} = C^*(G)e$ over $C^*(\mathbb{N}^*)$ and the time evolution σ_t of $C^*(\mathbb{N}^*)$.

Lemma 13. *Let C be a unital C^*-algebra, $\mathcal{E}$ a C^*-module over C, $(\sigma_t)_{t\in\mathbb{R}}$ a one-parameter group of automorphisms of C, $\beta \in]0,\infty[$, φ_β a KMS$_\beta$ state on C, and $\mathcal{H}_{\varphi_\beta}$ the Hilbert space of the GNS construction for φ_β.*

a) Let $\mathcal{H}_\beta$ be the completion of $\mathcal{E}$ for the inner product given by

$$\langle \xi, \eta \rangle_\beta = \varphi_\beta(\langle \xi, \eta \rangle) \quad \forall \xi, \eta \in \mathcal{E}.$$

Then the action of the endomorphisms, $\mathrm{End}_C(\mathcal{E})$, *on $\mathcal{E}$ extends by continuity to $\mathcal{H}_\beta$.*

b) There exists a unique representation ρ of C^o, the opposite algebra of C, in $\mathcal{H}_\beta$ such that for any $\xi \in \mathcal{H}_\beta$ and $a \in C$ in the domain of $\sigma_{i\beta/2}$ one has

$$\rho(a)\xi = \xi\sigma_{i\beta/2}(a).$$

This representation commutes with the left action of $\mathrm{End}_C(\mathcal{E})$.

The Hilbert space $\mathcal{H}_\beta$ is the tensor product:

$$\mathcal{H}_\beta = \mathcal{E} \otimes_C \mathcal{H}_{\varphi_\beta}$$

so that the first assertion follows (Chapter II Appendix A). The second assertion also follows, using $\mathcal{H}_{\varphi_\beta}$ as a left Hilbert algebra.

We apply this lemma with $C = C^*(\mathbb{N}^*)$, $\mathcal{E} = C^*(G)e$, and $\sigma_t \in \mathrm{Aut}\, C$ given by the time evolution (Proposition 7 c) of $C^*(\mathbb{N}^*)$. As $\mathcal{E}$ is a space of functions on Δ, so is each of the $\mathcal{H}_\beta$, and for each $\alpha \in \Delta$ we let ε_α be the characteristic function of $\{\alpha\} \subset \Delta$. The vectors ε_α, $\alpha \in \Delta$, are of unit length in each $\mathcal{H}_\beta$ and always span a dense subspace of $\mathcal{H}_\beta$. For $\beta = 1$ they give an orthonormal basis, so that $\mathcal{H}_1 = \ell^2(\Delta)$. In general, the inner product is uniquely determined by the following function of positive type on $P_{\mathbb{Q}}$, with $* =$ base point of Δ,

$$\psi_\beta(g) = \langle g\varepsilon_*, \varepsilon_* \rangle_\beta$$

and the computation of ψ_β yields the following:

1) $\psi_\beta(g) = 0$ if $g \notin N = \left\{ \begin{bmatrix} 1 & n \\ 0 & 1 \end{bmatrix} ; n \in \mathbb{Q} \right\}$

2) $\psi_\beta\left(\begin{bmatrix} 1 & n \\ 0 & 1 \end{bmatrix}\right) = \prod p^{-k_p\beta}\,(1 - p^{\beta-1})(1 - p^{-1})^{-1}$

where $b = \prod p^{k_p}$ is the prime factor decomposition of the denominator $b > 0$ of the irreducible fraction $n = \frac{a}{b}$.

The commutant of $P_{\mathbb{Q}}$ in $\mathcal{H}_\beta$ is given by a right action ρ_β of the Hecke algebra $A = C_r^*(P_{\mathbb{Q}}, P_{\mathbb{Z}})$ of Theorem 5 as follows:

Proposition 14. *Let $\beta \in \,]0, \infty[$.*

1) The following equality defines a faithful unitary representation of $A^o = C_r^*(P_{\mathbb{Q}}, P_{\mathbb{Z}})^o$ in $\mathcal{H}_\beta$:

$$\rho_\beta(f)\varepsilon_\alpha = \sum_{\alpha' \in \alpha \circ \gamma} \left(\frac{L(\gamma)}{R(\gamma)}\right)^{\beta/2} f(\gamma)\, \varepsilon_{\alpha'} \quad \forall f \in \mathcal{H}.$$

2) For each β, $\rho_\beta(A^o)$ generates the commutant of $P_{\mathbb{Q}}$ in $\mathcal{H}_\beta$.

Thus, the Hecke algebra $A = C_r^*(P_{\mathbb{Q}}, P_{\mathbb{Z}})$ appears uniquely as the norm closure, in each $\mathcal{H}_\beta$, of the algebra of operators in $\mathcal{H}_\beta$ which commute with $P_{\mathbb{Q}}$ and preserve the linear span of the ε_α, $\alpha \in \Delta$. The latter algebra is normalized by any $g \in P_{\mathcal{A}} = G$, and it follows that the group $C = P_{\mathcal{A}}/\overline{P}_{\mathbb{Q}} = \mathcal{A}^*/\mathbb{Q}^*$ acts by automorphisms on A. This action is independent of the real parameter β and will be denoted by

$$g \in C \mapsto \theta_g \in \operatorname{Aut} A.$$

The compact group C is the idèle class group [We]. By construction, the fixed point algebra $A^C = \{a \in A\,;\ \theta_g(a) = a \quad \forall g \in C\}$ is isomorphic to $C^*(\mathbb{N}^*)$. It is this action of C on A which is the symmetry group of the dynamical system (A, σ_t) of Theorem 5, with spontaneous symmetry breaking for $\beta = 1$.

V. Appendix A. Crossed Products of von Neumann Algebras

Let M be a von Neumann algebra, G a locally compact group and $\alpha : G \to \operatorname{Aut} M$ a continuous action of G on M. The group $\operatorname{Aut} M$ is endowed with the topology of pointwise norm convergence in the predual M_* of M. Thus, for each $\varphi \in M_*$ the map $g \to \alpha_g(\varphi)$ is norm continuous. Let dg be a left Haar measure on G and λ be the left regular representation of G in the Hilbert space $L^2(G)$,

$$(\lambda(g)\xi)(h) = \xi(g^{-1}\, h) \qquad \forall g,h \in G\,,\ \xi \in L^2(G). \tag{1}$$

The action α of G on M is encoded by the homomorphism $\tilde{\alpha} : M \to M \otimes L^\infty(G)$ given by

$$\tilde{\alpha}(x) = (\alpha_g^{-1}(x))_{g\in G} \in L^\infty(G, M) = M \otimes L^\infty(G) \tag{2}$$

which satisfies the equivariance condition

$$\tilde{\alpha} \circ \alpha_g(x) = \lambda(g)\, \tilde{\alpha}(x)\, \lambda(g)^{-1} \qquad \forall g \in G\,,\ x \in M. \tag{3}$$

Definition 1. *The crossed product $M \rtimes_\alpha G$ is the von Neumann subalgebra of $M \otimes \mathcal{L}(L^2(G))$ generated by $\tilde{\alpha}(M)$ and $1 \otimes \lambda(G)$.*

The equality (3) shows that the crossed product contains as a weakly dense subalgebra the finite sums

$$\sum \tilde{\alpha}(x_g)\, \lambda(g) \qquad x_g \in M.$$

When the group G is *discrete* any element of the crossed product can be uniquely written as a formal sum

$$\sum \tilde{\alpha}(x_g)\, \lambda(g)$$

where the x_g are uniquely determined as matrix elements in the natural basis of $\ell^2(G)$ (cf. [Di$_2$]). This is no longer the case when the group G is not discrete, and a good description of the generic element of $M \rtimes_\alpha G$ is lacking in general.

Definition 2. *Let α and α' be two actions of G on M. Then α and α' are said to be outer equivalent if there exists a strongly continuous map $g \mapsto u_g$ from G to the unitary group $\mathcal{U}(M)$ such that*

1) $u_{g_1 g_2} = u_{g_1}\ \alpha_{g_1}(u_{g_2}) \qquad \forall g_1, g_2 \in G$

2) $\alpha'_g(x) = u_g\ \alpha_g(x)\ u_g^* \qquad \forall x \in M\ ,\ g \in G.$

The crossed products $M \rtimes_\alpha G$ and $M \rtimes_{\alpha'} G$ are then canonically isomorphic using the unitary $u \in M \otimes L^\infty(G)$, $u = (u_{g^{-1}})_{g \in G}$. One has

$$u\ \tilde{\alpha}(x)\ u^* = \widetilde{\alpha'}(x) \qquad \forall x \in M$$

$$\lambda(g)\ u\ \lambda(g)^{-1} = u\ \tilde{\alpha}(u_g) \qquad \forall g \in G. \tag{4}$$

When G is abelian let $\hat{G}$ be the Pontryagin dual of G and let $\langle g, g' \rangle$ for $g \in G$, $g' \in \hat{G}$, be the canonical pairing with values in $\mathbb{T} = \{z \in \mathbb{C}\ ;\ |z| = 1\}$.

Proposition 3. [T₃] *Let G be an abelian locally compact group and α an action of G on a von Neumann algebra M. The following equality defines a canonical action $\hat{\alpha}$ of the Pontryagin dual $\hat{G}$ of G on the crossed product $M \rtimes_\alpha G$:*

$$\hat{\alpha}_{g'}(y) = (1 \otimes v_{g'})\ y\ (1 \otimes v_{g'}^*) \qquad \forall y \in M \rtimes_\alpha G\ ,\ g' \in \hat{G}$$

where $(v_{g'}\ \xi)(g) = \langle g, g' \rangle\ \xi(g) \quad \forall \xi \in L^2(G)$.

By construction, $\hat{\alpha}$ fixes pointwise the von Neumann subalgebra M ($= \tilde{\alpha}(M)$) of $M \rtimes_\alpha G$ and multiplies $1 \otimes \lambda(g)$ by $\langle g, g' \rangle \in \mathbb{T}$.

The central result of the theory of crossed products of von Neumann algebras is the following:

Theorem 4. [T₃] *Let G be an abelian locally compact group and α an action of G on a von Neumann algebra M. There exists a canonical isomorphism θ of the double crossed product $(M \rtimes_\alpha G) \rtimes_{\hat{\alpha}} \hat{G}$ with $M \otimes \mathcal{L}(L^2(G))$ which transforms the double dual action $\hat{\hat{\alpha}}$ into an action outer equivalent to the action $\alpha \otimes 1$ of G on $M \otimes \mathcal{L}(L^2(G))$.*

This theorem has since then been extended to nonabelian groups and Hopf-von Neumann algebras (cf. [Str] [En-S]).

V. Appendix B. Correspondences

We shall describe in this appendix a third natural notion of morphism between von Neumann algebras. We have already met two notions:

1) $\rho : M \to N$ is a normal involutive algebra homomorphism.

2) $T : M \to N$ is a completely positive normal linear map.

While the first is the most obvious notion of morphism and basic concepts were introduced for automorphisms (Section 6), the second notion plays a key role in Section 7 and in the entropy theory (Section 6). In the commutative case, $M = L^\infty(X, \mu_X)$ and $N = L^\infty(Y, \mu_Y)$, notion 1 corresponds to a non-singular map $\psi : Y \to X$, while notion 2 corresponds to a non-singular map $y \to p_y$ from Y to positive measures on X.

We shall introduce a third notion of morphism, intimately related to those above.

Definition 1. *Let M and N be von Neumann algebras. A correspondence from M to N is a Hilbert space $\mathcal{H}$ which is an N-M-bimodule.*

Thus, more explicitly, we are given commuting normal $*$-representations π_N of N and π_{M^0} of M^0 in $\mathcal{H}$, where M^0 is the opposite von Neumann algebra. To make the notation lighter we shall write, whenever no confusion can arise,

$$\pi_N(y)\, \pi_{M^0}(x^0)\, \xi = y\, \xi\, x \qquad \forall \xi \in \mathcal{H}\,,\ y \in N\,,\ x \in M. \tag{1}$$

To justify the terminology (one could simply call $\mathcal{H}$ an N-M-bimodule) we shall first consider the commutative case.

Let (X, μ_X) and (Y, μ_Y) be standard measure spaces, $M = L^\infty(X, \mu_X)$ and $N = L^\infty(Y, \mu_Y)$. Then the most general correspondence $\mathcal{H}$ between M and N is given by a measure class μ on $X \times Y$ with projections $\mathrm{pr}_X(\mu)$, $\mathrm{pr}_Y(\mu)$ absolutely continuous with respect to μ_X and μ_Y, and an integer-valued μ-measurable function $n(s,t)$ $((s,t) \in X \times Y)$. The Hilbert space $\mathcal{H}$ is equal to $\int H_{(s,t)}\, d\mu(s,t)$ where $H_{(s,t)}$ is a Hilbert space of dimension $n(s,t) \in \{0, 1, \ldots, \infty\}$ while the structure of bimodule is given by

$$(g\, \xi\, f)(s,t) = g(t)\, f(s)\, \xi(s,t) \qquad \forall f \in M\,,\ g \in N\,,\ \xi \in \mathcal{H}. \tag{2}$$

In general the measure μ is not absolutely continuous with respect to $\mu_X \times \mu_Y$; this measure represents the graph of the correspondence, while the function n represents the multiplicity of the correspondence.

If in the above example we take $(X, \mu_X) = (Y, \mu_Y)$ and μ equal to the image of $\mu_X = \mu_Y$ on the diagonal $\Delta = \{(x,x); x \in X\}$, while $n(s,s) = 1 \quad \forall s \in X$, we get the *identity* as a correspondence from M to $N = M$. The Hilbert space $\mathcal{H}$ is equal to $L^2(X, \mu_X)$ and the bimodule structure corresponds to the standard representation of M.

We shall now describe with special care the identity correspondence $L^2(M)$ for an arbitrary von Neumann algebra M.

B.α Half densities and the identity correspondence

Let M be a von Neumann algebra, M_*^+ the positive cone of the predual of M. Recall that any $\varphi \in M_*^+$ has a well defined support $e = s(\varphi)$, $e = e^*$, $e^2 = e$, $e \in M$, and that the modular automorphism group σ_t^φ is a one-parameter group of automorphisms of the reduced von Neumann algebra $M_e = \{x \in M; ex = xe = x\}$. We shall introduce the notation

$$\sigma_t^\varphi(x) = \varphi^{it}\, x\, \varphi^{-it} \qquad \forall x \in M_e\,,\ \forall t \in \mathbb{R} \tag{3}$$

and we shall continue to use it for imaginary values of t and elements of M_e for which $\sigma_t^\varphi(x)$ exists by analytic continuation. Consider now a monomial of the form

$$\delta = a_1\, \varphi_1^{z_1}\, a_2\, \varphi_2^{z_2} \cdots a_n\, \varphi_n^{z_n}\, a_{n+1}$$

where $a_i \in M$, $\varphi_i \in M_*^+$ and $z_i \in \mathbb{C}$, $\mathrm{Re}\, z_i > 0$.

We shall give to such a monomial the degree $\sum z_i \in \mathbb{C}$. This degree plays the same role as the degree of densities in differential geometry.

When the degree of δ is equal to 1 we define the integral or expectation value $\langle\delta\rangle$ of δ using the rules

$$\langle a_1\, \psi^{z_1}\, a_2\, \psi^{z_2} \cdots a_n\, \psi^{z_n}\, a_{n+1}\rangle = F(z_1, \dots, z_n) \tag{4}$$

for ψ faithful, $a_j \in M$, $\sum z_i = 1$, where the function F is the unique bounded holomorphic function in the tube $\{(z_i) \in \mathbb{C}^n; \mathrm{Re}\, z_i > 0, \sum z_i = 1\}$, with boundary values given for $t_1, t_2, \dots, t_{n-1} \in \mathbb{R}$ by

$$F(it_1, it_2, \dots, it_{n-1}, 1 - i \sum t_j) = \psi(a_1\, \sigma_{t_1}^\psi(a_2)\, \sigma_{t_1+t_2}^\psi(a_3) \cdots). \tag{5}$$

In order to pass from δ to a monomial of the form (4) one uses the rule

$$\varphi^z = (D\varphi : D\psi)_{-iz}\, \psi^z \tag{6}$$

where $(D\varphi : D\psi)_z$ is the value at z of the Radon-Nikodým derivative (Section 5). Note that this operator is bounded if $\varphi \le \psi$ and $\mathrm{Re}\ z \le \frac{1}{2}$. It follows from [Co$_4$] and [H$_0$] that the above rules are consistent and that the expectation value $\langle\delta\rangle$ is well defined, independently of the choice of ψ, say with $\psi \ge \sum \varphi_i$, and $\mathrm{Re}\ z_i \le \frac{1}{2}$, which can always be assumed. The outcome is that one can manipulate the above monomials, or linear combinations of such monomials of the same degree α, exactly like densities of degree α in differential geometry. The commutation rules are given by (3) and (6).

The expectation value $\langle\delta\rangle$ satisfies

$$\langle a\delta\rangle = \langle\delta a\rangle \qquad \forall a \in M. \tag{7}$$

In particular, a 1-density defines a normal linear form on M given by $a \in M \mapsto \langle a\delta\rangle \in \mathbb{C}$.

There is a natural involution $\delta \mapsto \delta^*$:

$$(a_1\, \varphi_1^{z_1}\, a_2\, \varphi_2^{z_2} \cdots a_n\, \varphi_n^{z_n}\, a_{n+1})^* = a_{n+1}^*\, \varphi_n^{\bar z_n}\, a_n^* \cdots \varphi_1^{\bar z_1}\, a_1^* \tag{8}$$

which replaces the degree α by $\overline{\alpha}$ and is antilinear. Let us consider the sesquilinear form on $\frac{1}{2}$-densities given by

$$\langle \delta_1, \delta_2 \rangle = \langle \delta_1 \delta_2^* \rangle. \tag{9}$$

One shows that this sesquilinear form is positive and moreover that

$$\langle \delta_1 \delta_2 \rangle = \langle \delta_2 \delta_1 \rangle \qquad \forall \delta_1, \delta_2 \ \tfrac{1}{2}\text{-densities}. \tag{10}$$

It follows immediately that the Hilbert space $L^2(M)$ thus obtained is a normal bimodule for the action

$$(x, \delta, y) \mapsto x\, \delta\, y \qquad \forall x, y \in M\,,\ \delta \in L^2(M). \tag{11}$$

Moreover, one can show that any element of $L^2(M)$ admits a canonical left polar decomposition of the form

$$\delta = u\, \varphi^{1/2} \tag{12}$$

where u is a partial isometry with initial support $u^*u = s(\varphi)$, $\varphi \in M_*^+$ (cf. [Co$_1$] [H$_0$]).

The canonical involution J of $L^2(M)$, given by

$$J\delta = \delta^* \qquad \forall \delta \in L^2(M) \tag{13}$$

is isometric by (10), and it obviously exchanges the left and right actions of M on $L^2(M)$.

Finally, $L^2(M)$ is endowed with a natural positive cone $L^2(M)^+$ whose elements are the $\frac{1}{2}$-densities of the form $\varphi^{1/2}$, $\varphi \in M_*^+$ (cf. [Co$_1$] for a characterization of the self-dual cones in Hilbert space obtained in this manner).

Definition 2. *Let M be a von Neumann algebra. The* identity correspondence *between M and M is the canonical bimodule $L^2(M)$ of $\frac{1}{2}$-densities on M.*

To describe this bimodule, one may also use an auxiliary faithful (semifinite normal) weight ν. Then the Hilbert space $L^2(M, \nu)$, the completion of $\mathrm{Dom}_{1/2}(\nu) = \{x \in M; \nu(x^*x) < \infty\}$ with the obvious pre-Hilbert-space structure, is naturally equipped with:

A normal $*$-representation π_ν of M by left multiplication.

An isometric antilinear involution J_ν such that

$$J_\nu\, \pi_\nu(M)\, J_\nu = \pi_\nu(M)' \qquad (\text{commutant of } \pi_\nu(M)).$$

Then the equality $\pi_\nu(x^o) = J_\nu\, \pi_\nu(x)^*\, J_\nu$ defines a normal $*$-representation of M^o in $L^2(M, \nu)$ which hence becomes an M-M-bimodule.

The Hilbert space $L^2(M, \nu)$ comes equipped with a natural self-dual cone $L^2(M, \nu)^+$, whose elements are in bijection with the positive cone M_*^+ of the predual of M by

$$\xi \in L^2(M, \nu)^+ \mapsto \omega_{\xi,\xi} \in M_*^+ \qquad ([\mathrm{H}_0] \text{ Lemma 2.10})$$

where $\omega_{\xi,\xi}(x) = \langle \pi_\nu(x)\, \xi, \xi \rangle \quad \forall x \in M$.

It follows immediately that there exists a unique unitary equivalence U of the bimodule $L^2(M,\nu)$ with $L^2(M)$ such that

$$U\xi = (\omega_{\xi,\xi})^{1/2} \qquad \forall \xi \in L^2(M,\nu)^+. \tag{14}$$

Since $L^2(M,\nu)$ is the completion of $\mathrm{Dom}_{1/2}(\nu) = \{x \in M \,;\, \nu(x^*x) < \infty\}$, we let η_ν be the canonical map from $\mathrm{Dom}_{1/2}(\nu)$ to $L^2(M,\nu)$. Then the isometry U allows one to extend the notation $\varphi^{1/2}$, $\varphi \in M_*^+$, to weights as

$$x\,\nu^{1/2} = U\,\eta_\nu(x) \qquad \forall x \in \mathrm{Dom}_{1/2}(\nu). \tag{15}$$

B.β Correspondences and $*$-homomorphisms

Let M and N be von Neumann algebras. Let ρ be a normal $*$-homomorphism of M in N. We do not assume that $\rho(1) = 1$, so $\rho(1) = e$ is a projection, and the Hilbert space $L^2(\rho) = \{\xi \in L^2(N) \,;\, \xi e = \xi\}$ is an N-M-bimodule with

$$\pi_N(y)\, \pi_{M^0}(x^0)\, \xi = y\, \xi\, \rho(x) \qquad \forall y \in N\,,\ x \in M. \tag{16}$$

To avoid useless complications we shall assume that both M and N have separable preduals, and that $\mathcal{H}$ is separable.

Proposition 3. *Assume that N is properly infinite.*

a) *Every correspondence $\mathcal{H}$ between M and N is equivalent to an $L^2(\rho)$.*

b) *The intertwining operators from $L^2(\rho_1)$ to $L^2(\rho_2)$ are the elements y of $\rho_2(1)\, N\, \rho_1(1)$ such that*

$$\rho_2(x)\, y = y\, \rho_1(x) \qquad \forall x \in M.$$

Proof. a) As N is properly infinite, the representation π_N of N in $\mathcal{H}$ is subequivalent to the standard representation of N in $L^2(N)$. Thus, we can assume that $\mathcal{H} = L^2(N)e$, where e is a projection, $e \in N$, and that $\pi_N(y)\xi = y\xi \quad \forall y \in N\,,\ \xi \in L^2(N)e$. The commutant of $\pi_N(N)$ is then the algebra of right multiplications in $L^2(N)e$ by elements of eNe, so π_{M^0} determines a normal $*$-homomorphism ρ, $\rho(1) = e$ of M in N_e, such that

$$\pi_{M^0}(x^0)\, \xi = \xi\, \rho(x) \qquad \forall x \in M\,,\ \forall \xi \in L^2(N)e.$$

b) With the obvious notation, the intertwining operators from π_N^1 to π_N^2 correspond to the elements of $\rho_2(1)\, N\, \rho_1(1)$, and the intertwining condition with respect to the action of M is exactly

$$y\, \rho_1(x) = \rho_2(x)\, y \qquad \forall x \in M.$$

Corollary 4. *Let θ_1 and $\theta_2 \in \mathrm{Aut}\,N$; then $L^2(\theta_1)$ is equivalent to $L^2(\theta_2)$ iff $\varepsilon(\theta_1) = \varepsilon(\theta_2)$ in $\mathrm{Out}\,N = \mathrm{Aut}\,N/\mathrm{Inn}\,N$.*

If N is not properly infinite, Proposition 3 does not hold (in general there need not be any nonzero $*$-homomorphism of M into N while there is always the coarse correspondence (Example 5 a) between M and N). This, however, will not create any difficulty since, letting F be the factor of type I_∞ of all bounded operators in $\ell^2(\mathbb{N})$, the von Neumann algebra $\tilde{N} = N \otimes F$ is properly infinite, and replacing N by $\tilde{N}$ does not affect the correspondences from M to N. (Let $\mathcal{H}$ be a correspondence from M to N, then $\mathcal{H} \otimes \ell^2$ is in an obvious way a correspondence from M to $\tilde{N}$. Conversely, let $e = 1 \otimes e_{11} \in \tilde{N}$, where $(e_{ij})_{i,j\in\mathbb{N}}$ is the canonical system of matrix units in F; then if $\tilde{\mathcal{H}}$ is a correspondence from M to $\tilde{N}$ the space $e\,\tilde{\mathcal{H}}$ is a correspondence from M to $\tilde{N}_e = N$.)

Let $\mathcal{H}$ be a correspondence from M to N, and let M_1 and N_1 be von Neumann subalgebras of M and N, respectively. It is clear that by restriction of the bimodule structure of $\mathcal{H}$ we obtain a correspondence from M_1 to N_1. This operation of restriction does not look so natural from the first point of view using normal involutive algebraic homomorphisms, so even though it is equivalent to the correspondence point of view (Proposition 3) it is important to keep both of them.

We shall now describe several examples of correspondences for which the associated homomorphism ρ is less natural.

Examples 5. a) In the commutative case, $M = L^\infty(X,\mu_X)$, $N = L^\infty(Y,\mu_Y)$, we can consider the coarse correspondence which associates to each $x \in X$ an arbitrary point of Y. This means that we take the measure $\mu = \mu_X \times \mu_Y$ and the multiplicity function $n(s,t) = 1 \quad \forall (s,t) \in X \times Y$.

This construction extends immediately to an arbitrary pair of von Neumann algebras M and N. The coarse correspondence is then the obvious bimodule of Hilbert-Schmidt operators from $L^2(M)$ to $L^2(N)$:

$$T \in \mathcal{L}^2(L^2(M), L^2(N))$$
$$\pi_N(y)\pi_{M^o}(x^o)T = y\,T\,x \qquad \forall y \in N\,,\ x \in M\,,\ T \in \mathcal{L}^2. \tag{17}$$

Equivalently, one can use $L^2(M) \otimes L^2(N)$ with

$$y(\xi \otimes \eta)x = \xi x \otimes y\eta \qquad \forall x \in M\,,\ y \in N.$$

b) Let Γ be a countable group acting freely by nonsingular transformations of the measure space (X,μ_X); then the restriction of the identity correspondence of M to $L^\infty(X,\mu_X) \subset M = L^\infty(X,\mu_X) \rtimes \Gamma$ (the crossed product by Γ) is the graph in $X \times X$, with its natural measure class, of the equivalence relation $x \sim y$ iff $\exists g \in \Gamma$, $gx = y$.

c) Let $M = N \rtimes_\alpha G$ be the crossed product of a von Neumann algebra N by an action α of the locally compact group G (cf. Appendix A). Then every unitary representation π of G in a Hilbert space H_π defines canonically a correspondence from M to M as follows:

$$\mathcal{H}_\pi = L^2(M) \otimes H_\pi \text{ ; the right action } \pi_{M^o}(x^o)\,,\ x \in M \text{ is}$$

$$\pi_{M^o}(x^o)(\xi \otimes \eta) = \xi x \otimes \eta \qquad \forall \xi \in L^2(M)\,,\ \eta \in H_\pi, \tag{18}$$

the left action π_M of M is given, using $M = N \rtimes_\alpha G$, by

$$\pi_M(x)(\xi \otimes \eta) = x\,\xi \otimes \eta \qquad \forall x \in N\,,\ \xi \in L^2(M)\,,\ \eta \in H_\pi$$

$$\pi_M(g)(\xi \otimes \eta) = g\,\xi \otimes \pi(g)\eta \qquad \forall g \in G\,,\ \xi \in L^2(M)\,,\ \eta \in H_\pi.$$

B.γ Coefficients of correspondences and completely positive maps

A correspondence from the von Neumann algebra M to N is by definition a representation π of the C^*-algebra $N \otimes_{\max} M^o$ which is *binormal*, i.e. whose restrictions to both M^o and N are normal representations. The coefficients of such representations, i.e. the functionals

$$z \in N \otimes_{\max} M^o \mapsto \langle \pi(z)\xi, \xi\rangle \in \mathbb{C}$$

are exactly the *binormal* states [E-L]. We refer to [E-L] for the definition of the C^*-algebra $N \otimes_{\rm bin} M^o$ corresponding to such states. We shall now relate these coefficients to completely positive normal maps from M to N. We let ν be a faithful semifinite normal weight on N and use the following:

Proposition 6. a) *Let $\mathcal{H}$ be a normal left N-module. Then*

$$D(\mathcal{H}, \nu) = \{\xi \in \mathcal{H}\ ;\ \exists c < \infty\,,\ \|y\xi\| \le c\ \nu(y^*y)^{1/2} \quad \forall y \in \mathrm{Dom}_{1/2}(\nu)\}$$

is dense in $\mathcal{H}$.

b) *The equality $I_\nu(\varphi) = \nu^{-1/2}\ \varphi\ \nu^{-1/2}$, $\forall \varphi \in \mathrm{Face}(\nu) = \{\varphi \in M_*^+\ ;\ \varphi \le \lambda\nu$ for some $\lambda > 0\}$ defines a completely positive linear map I_ν from the face of ν in N_* to the domain of ν in M. Its inverse $x \mapsto \nu^{1/2}\ x\ \nu^{1/2}$ $\forall x \in \mathrm{Dom}(\nu)$ is also completely positive.*

The proof is straightforward (cf. [Co_{22}] [E-L]).

Definition 7. *With $N, \mathcal{H}$ and ν as in Proposition 6 a, a vector $\xi \in \mathcal{H}$ is called ν-bounded if $\xi \in D(\mathcal{H}, \nu)$.*

We can now state the precise relation between coefficients of correspondences and completely positive normal maps.

Proposition 8. *Let M and N be von Neumann algebras, and ν a faithful semifinite normal weight on N.*

a) *Let $\mathcal{H}$ be a correspondence from M to N, and ξ a ν-bounded vector (Definition 7). Then there exists a unique completely positive map P from M to N such that for any $x \in M$, $y \in N$, one has*

$$\langle y\ \xi\ x, \xi\rangle = \langle y\ \nu^{1/2}\ P(x)\ \nu^{1/2}\rangle.$$

b) *Let P be a completely positive normal map from M to N such that also $\nu(P(1)) < \infty$. Then there exists a unique pair $(\mathcal{H}, \xi)$ where $\mathcal{H}$ is a correspondence from M to N, $\xi \in \mathcal{H}$ and:*

α) $N\xi M$ is dense in $\mathcal{H}$

β) ξ is a ν-bounded vector and for any $x \in M$ and $y \in N$, $\nu(y^*y) < \infty$ one has

$$\langle y\, \xi\, x, \xi \rangle = \langle y\, \nu^{1/2}\, P(x)\, \nu^{1/2} \rangle.$$

Proof. Both statements are immediate from Proposition 7 and the GNS construction for binormal states.

Corollary 9. *If N is properly infinite and $P : M \to N$ is a completely positive normal map, there exists a normal $*$-homomorphism $\rho : M \to N$ and a partial isometry $v \in N$, $v^*v \leq \rho(1)$, $vv^* =$ Support $P(1)$ with $P(x) = P(1)^{1/2}\, v\, \rho(x)\, v^*\, P(1)^{1/2}$.*

This follows from Proposition 3.

We shall introduce the following notation for the coefficients:

Notation 10. *Let M, N, ν and $\mathcal{H}$ be as in Proposition 8 a. Then for any ν-bounded vectors $\xi_1, \xi_2 \in \mathcal{H}$ we let $(\xi_1, \xi_2)_\nu$ be the unique normal map P from M to N such that*

$$\langle y\, \xi_1\, x, \xi_2 \rangle = \langle y\, \nu^{1/2}\, P(x)\, \nu^{1/2} \rangle \qquad \forall x \in M\,,\ y \in N.$$

For $a \in M$ we have $(\xi_1\, a, \xi_2)_\nu\,(x) = (\xi_1, \xi_2)_\nu(ax)$ and $(\xi_1, \xi_2\, a)_\nu\,(x) = (\xi_1, \xi_2)_\nu(xa^*)$ for any $x \in M$.

Lemma 11. a) *Let $b \in N$ be such that $t \mapsto \sigma_t^\nu(b) \in N$ extends analytically from $t \in \mathbb{R}$ to $\operatorname{im} t \in \left[0, \frac{1}{2}\right]$. Then for any $\xi_1 \in D(\mathcal{H}, \nu)$ one has $b\, \xi_1 \in D(\mathcal{H}, \nu)$ and $(b\, \xi_1, \xi_2)_\nu(x) = \sigma_{i/2}^\nu(b)(\xi_1, \xi_2)_\nu(x)$ $\forall x \in M$ (and also $(\xi_1, b\, \xi_2)_\nu(x) = (\xi_1, \xi_2)_\nu(x)\, (\sigma_{i/2}^\nu(b))^*$ $\forall x \in M$).*

b) *Let ν' be another weight on N, with $\nu' \geq \lambda\nu$ for some $\lambda > 0$; then $D(\mathcal{H}, \nu) \subset D(\mathcal{H}, \nu')$, the Radon-Nikodým derivative $(D\nu' : D\nu)_t \in \mathbb{R}$ extends analytically from $t \in \mathbb{R}$ to $\operatorname{im} t \in \left[-\frac{1}{2}, 0\right]$, and with $b = (D\nu' : D\nu)_{-i/2}$ one has for any $\xi_1, \xi_2 \in D(\mathcal{H}, \nu)$, $x \in M$,*

$$(\xi_1, \xi_2)_{\nu'}(x) = b^*\, (\xi_1, \xi_2)_\nu(x)\, b.$$

B.δ Composition of correspondences

In the previous sections we related correspondences with the two classical notions 1) and 2) of morphisms for von Neumann algebras. While for 1) and 2) the composition of morphisms is the obvious one, defining *in a canonical manner* the composition of correspondences requires more care and will be dealt with in detail below to avoid any confusion.

Let M_1, N and M_2 be three von Neumann algebras, $\mathcal{H}_1$ a correspondence from M_1 to N and $\mathcal{H}_2$ a correspondence from N to M_2. Thus, $\mathcal{H}_1$ is, in particular, a left N-module and $\mathcal{H}_2$ a right N-module, and we shall construct canonically a tensor product

$$\mathcal{H} = \mathcal{H}_2 \otimes_N \mathcal{H}_1 \tag{19}$$

which will be naturally a correspondence from M_1 to M_2. The subtle point is that, while $\mathcal{H}$ will be defined like any tensor product as the linear span of simple tensors satisfying simple algebraic relations, the basic tensors generating $\mathcal{H}$ are not just of the form $\xi_2 \otimes \xi_1$, $\xi_j \in \mathcal{H}_j$ but rather

$$\xi_2 \otimes \nu^{-1/2}\, \xi_1\,,\ \xi_j \in \mathcal{H}_j\,,\ \nu \in N_*^+,\quad \nu \text{ faithful.} \tag{20}$$

In fact, it is quite useful also to allow ν to be a semifinite faithful normal weight; but this will introduce no difficulty and will not change the tensor product.

Here we have *three* variables ξ_1, ξ_2 and ν, and thus besides the obvious bilinearity of (20) in ξ_1, ξ_2, and the simplification by N

$$\xi_2\, x \otimes \nu^{-1/2}\, \xi_1 = \xi_2 \otimes \nu^{-1/2}(\nu^{1/2}\, x\, \nu^{-1/2})\xi_1\,,\ \forall x \in N \cap \operatorname{Dom} \sigma^{\nu}_{-i/2} \tag{21}$$

(which uses, of course, our notation 3), we shall have the following relation which eliminates the choice of ν:

$$\xi_2 \otimes (\nu')^{-1/2}\, \xi_1 = \xi_2 \otimes \nu^{-1/2}(\nu^{1/2}\, \nu'^{-1/2})\xi_1, \tag{22}$$

whenever one has $\nu \le \nu'$, so that $\nu^{1/2}\, \nu'^{-1/2} = (D\nu : D\nu')_{-i/2} \in N$. It is clear that relation (22) allows, using $\nu' = \nu_1 + \nu_2$ for $\nu_j \in N_*^+$, comparison of simple tensors (20) for different faithful $\nu_j \in N_*^+$. The case of weights will not be more difficult.

To proceed carefully we shall first of all fix ν and construct a Hilbert space generated by the simple tensors (20) and satisfying the relation (21). We shall then use (22) to eliminate the dependence on ν.

Let us define a sesquilinear form on the algebraic tensor product:

$$\mathcal{H}_2 \odot D(\mathcal{H}_1, \nu) \tag{23}$$

of $\mathcal{H}_2$ by the dense subspace of ν-bounded vectors in $\mathcal{H}_1$ (Definition 7) by the equality

$$\langle \xi_2 \otimes \nu^{-1/2}\, \xi_1\,,\ \eta_2 \otimes \nu^{-1/2}\, \eta_1 \rangle = \varphi_2(\nu^{-1/2}\, \varphi_1\, \nu^{-1/2}) \tag{24}$$

where $\varphi_j \in N_*$ are given by

$$\varphi_1(y) = \langle y\, \xi_1, \eta_1 \rangle \qquad \forall y \in N \tag{25}$$

$$\varphi_2(y) = \langle \xi_2\, y, \eta_2 \rangle \qquad \forall y \in N. \tag{26}$$

We have used, *as a notation*, $\xi_2 \otimes \nu^{-1/2}\, \xi_1$ instead of $\xi_2 \otimes \xi_1$.

Proposition 12. a) *The equality* (24) *defines a positive sesquilinear form and the relation* (21) *holds in the associated Hilbert space* $\mathcal{H}_2 \otimes_\nu \mathcal{H}_1$.

b) *Let* $\nu \le \nu'$ *be faithful semifinite normal weights on* N*; then with* $b = (D\nu : D\nu')_{-1/2}$ *an isometry* $\mathcal{H}_2 \otimes_{\nu'} \mathcal{H}_1 \xrightarrow{V} \mathcal{H}_2 \otimes_\nu \mathcal{H}_1$ *is defined by*

$$V(\xi_2 \otimes (\nu')^{-1/2}\, \xi_1) = \xi_2 \otimes \nu^{-1/2}\, b\, \xi_1 \qquad \forall \xi_2 \in \mathcal{H}_2\,,\ \xi_1 \in D(\mathcal{H}, \nu_1').$$

Proof. a) The positivity follows from the complete positivity of the map $I_\nu = \nu^{-1/2} \cdot \nu^{-1/2}$ from N_* to N (cf. Proposition 6). To prove (21) one checks using (24) that for $x \in \operatorname{Dom} \sigma^\nu_{-i/2}$ one has

$$\langle \xi_2\, x \otimes \nu^{-1/2}\, \xi_1\,,\ \eta_2 \otimes \nu^{-1/2}\, \eta_1 \rangle = \langle \xi_2 \otimes \nu^{-1/2}(\nu^{1/2}\, x\, \nu^{-1/2})\xi_1\,,\ \eta_2 \otimes \nu^{-1/2}\, \eta_1 \rangle$$

for any $\xi_2, \eta_2 \in \mathcal{H}_2$, $\xi_1, \eta_1 \in D(\mathcal{H}_1, \nu)$.

b) follows from the equality $I_\nu(\varphi) = b\, I_{\nu'}(\varphi) b^*$ proven in Lemma 11 b) above. This shows that V is an isometry, and the faithfulness of ν shows that it has dense range.

Theorem 13. 1) *Let* N *be a von Neumann algebra,* $\mathcal{H}_1$ *a normal left module and* $\mathcal{H}_2$ *a normal right module over* N. *There is a canonical Hilbert space* $\mathcal{H} = \mathcal{H}_2 \otimes_N \mathcal{H}_1$ *generated by the* $\xi_2 \otimes \nu^{-1/2}\, \xi_1$*;* $\xi_2 \in \mathcal{H}_2$, $\nu \in N_*^+$, *faithful, and* $\xi_1 \in D(\mathcal{H}, \nu_1)$ *and satisfying the relations* (21), (22) *and* (24).

2) *Let* $\mathcal{H}_1$ *(resp.* $\mathcal{H}_2$*) be a correspondence from* M_1 *to* N *(resp. from* N *to* M_2*) and* $\mathcal{H} = \mathcal{H}_1 \otimes_N \mathcal{H}_2$ *as in* 1). *Then the following equality defines a correspondence between* M_1 *and* M_2*:*

$$\begin{aligned} x_2(\xi_2 \otimes \nu^{-1/2}\, \xi_1)x_1 &= (x_2\, \xi_2) \otimes \nu^{-1/2}(\xi_1\, x_1) \\ &\forall x_j \in M_j\,,\ \forall \nu \in N_*^+\,,\ \forall \xi_2 \in \mathcal{H}_2\,,\ \forall \xi_1 \in D(\mathcal{H}_1, \nu). \end{aligned}$$

Proof. 1) follows from Proposition 12.

2) It is enough to check that if both x_j are unitaries the formula 2) defines an isometry in $\mathcal{H}$, which is easy using (24).

Remark 14. Using Proposition 12 we see that $\xi_2 \otimes \nu^{-1/2}\, \xi_1$ is still well defined as an element of $\mathcal{H}$ when ν is a faithful semifinite normal weight and $\xi_1 \in D(\mathcal{H}, \nu)$.

Definition 15. *The correspondence* $\mathcal{H}_2 \otimes_N \mathcal{H}_1$ *from* M_1 *to* M_2 *is the* composition *of the correspondences* $\mathcal{H}_2$ *and* $\mathcal{H}_1$.

We can easily compute the coefficients of the composition:

Proposition 16. *Let $\mathcal{H}_1, \mathcal{H}_2$ and $\mathcal{H} = \mathcal{H}_2 \otimes_N \mathcal{H}_1$ be as above: Let ν be a faithful semifinite normal weight on N and let ν_2 be a faithful semifinite normal weight on M_2. Then for any $\xi_1, \eta_1 \in D(\mathcal{H}_1, \nu)$ and $\xi_2, \eta_2 \in D(\mathcal{H}_2, \nu_2)$ one has*

$$(\xi_2 \otimes \nu^{-1/2}\, \xi_1\,,\, \eta_2 \otimes \nu^{-1/2}\, \eta_1)_{\nu_2} = (\xi_2, \eta_2)_{\nu_2} \circ (\xi_1, \eta_1)_\nu .$$

In other words the coefficients of the composition are obtained by composing the coefficients of the correspondences.

The proof of Proposition 16 is immediate using (24).

Let us now relate the composition of correspondences with the composition of $*$-homomorphisms. Thus, we let $\rho_1 : M_1 \to N$ and $\rho_2 : N \to M_2$ be $*$-homomorphisms and $L^2(\rho_i)$ the associated correspondences.

Proposition 17. *The correspondence $L^2(\rho_2) \otimes_N L^2(\rho_1)$ is canonically equivalent to $L^2(\rho_2 \circ \rho_1)$.*

This follows from a slightly more general fact: Let $\mathcal{H}_2$ be any correspondence from N to M_2. Then consider the following M_2-M_1-bimodule:

$$\mathcal{H} = \mathcal{H}_2\, \rho_1(1)\,, \quad \pi(x_2 \otimes x_1^0)\xi = x_2\, \xi\, \rho_1(x_1) \qquad \forall x_j \in M_j . \tag{27}$$

One then has:

Lemma 18. *The bimodule $\mathcal{H}_2\, \rho_1(1)$ is canonically equivalent to $\mathcal{H}_2 \otimes_N L^2(\rho_1)$.*

Proof. One checks that the following formula gives the desired equivalence, recalling the construction (Subsection α) of $L^2(N)$ as half-densities, so that $\nu^{-1/2}\, \xi \in N$ for any ν-bounded element of $L^2(N)$ viewed as a left N-module.

$$V(\xi_2 \otimes \nu^{-1/2}\, \xi_1) = \xi_2(\nu^{-1/2}\, \xi_1) \qquad \forall \xi_2 \in \mathcal{H}_2\,,\ \xi_1 \in D(\mathcal{H}_1, \nu).$$

Using the above Propositions 3 and 17 one checks that the composition of correspondences is associative; but this can be done directly as in Theorem 13 with a specific equivalence. We shall end this section with the basic notion of the contragredient of a correspondence.

Definition 19. *Let $\mathcal{H}$ be a correspondence from M to N; then its contragredient $\overline{\mathcal{H}}$ is the correspondence from N to M given by*

$$x\, \bar{\xi}\, y = \overline{(y^*\, \xi\, x^*)} \qquad \forall \xi \in \mathcal{H}\,,\ x \in M\,,\ y \in N$$

with $\overline{\mathcal{H}}$ the conjugate Hilbert space of $\mathcal{H}$.

We have used the canonical antilinear isometry $\xi \mapsto \bar{\xi}$ from $\mathcal{H}$ to $\overline{\mathcal{H}}$.

Example. Let $\rho : M \to N$ be a $*$-isomorphism; then one checks that $\overline{L^2(\rho)} = L^2(\rho^{-1})$.

The meaning of $\overline{L^2(\rho)}$ for a $*$-homomorphism is far less obvious.

Theorem 20. *Let $\mathcal{H}_1$ be a correspondence from M_1 to N and $\mathcal{H}_2$ a correspondence from N to M_2; then one has a canonical equivalence of correspondences from M_2 to M_1:*

$$\overline{(\mathcal{H}_2 \otimes_N \mathcal{H}_1)} = \overline{\mathcal{H}_1} \otimes_N \overline{\mathcal{H}_2}.$$

Proof. Let ν be a faithful normal semifinite weight on N. The meaning of the theorem is that in the construction of $\mathcal{H}_1 \otimes_N \mathcal{H}_2$ of Theorem 13 we could equally have used expressions of the form $\xi_2\, \nu^{-1/2} \otimes \xi_1$, $\xi_2 \in D(\mathcal{H}, \nu)$, $\xi_1 \in \mathcal{H}_1$ and defined the inner product using the following analogue of (24):

$$\langle \xi_2\, \nu^{-1/2} \otimes \xi_1\, ,\, \eta_2\, \nu^{-1/2} \otimes \eta_1 \rangle = \varphi_1(\nu^{-1/2}\, \varphi_2\, \nu^{-1/2}) \tag{28}$$

with $\varphi_j \in N_*$ given by (25) (26).

One checks indeed that the following map defines the desired unitary equivalence:

$$V(\xi_2\, \nu^{-1/2} \otimes \xi_1) = \xi_2 \otimes \nu^{-1/2}\, \xi_1 \qquad \forall \xi_j \in D(\mathcal{H}_j, \nu).$$

B.ε Correspondences, hyperfiniteness and property T

In this section we specialize to factors N of type II_1 and exploit the following analogy between correspondences from N to N and representations of discrete groups. First, since such correspondences are exactly binormal representations of the C^*-algebra $N \otimes_{\max} N^{\mathrm{o}}$ all classical notions of representation theory are available, such as unitary equivalence, irreducibility, intertwining operators, etc.

We also have a natural topology on any set of equivalence classes of correspondences from N to N using, as in representation theory (cf. [Di$_3$]), the coefficients to define it.

Finally, by Subsection δ) we have the analogue of the tensor product of representations, but, of course, there is a fundamental difference since the product $\mathcal{H}_1 \otimes_N \mathcal{H}_2$ is no longer commutative (and $\mathrm{Out}\, N$ is not commutative in general).

The identity correspondence is a unit for this product and is the analogue of the trivial representation of discrete groups. The coarse correspondence from N to N (cf. Example 5 a) has the same absorbing property for the product as the regular representation of discrete groups and is the analogue of the latter in general. With this small dictionary at hand we can now translate two key notions of the representation theory of discrete groups: amenability and the property T of Kazhdan and see what they give.

Recall that a (discrete) group Γ is said to be *amenable* if the trivial representation is weakly contained in the regular representation. Here we have:

Theorem 21. *Let N be a factor of type* II_1. *Then the identity correspondence is weakly contained in the coarse one iff N is hyperfinite.*

Proof. First recall (Example 5 c) that if $R(\Gamma) = N$ is the von Neumann algebra of the left regular representation of a discrete group Γ we have a natural functor from representations of Γ to correspondences from N to N. Since this functor is obviously continuous, it is enough to write $R = R(\Gamma)$ for some amenable discrete group Γ with infinite conjugacy classes to check the property for the hyperfinite factor. Conversely, if the identity correspondence of N is weakly contained in the coarse one it is immediate that N is semidiscrete using Theorem 7.3. Thus, the conclusion follows from Theorem 7.6.

Next, recall that a (discrete) group Γ is said to have Kazhdan's property T if the trivial representation is isolated.

Theorem 22. [Co-J] *Let Γ be a discrete group with infinite conjugacy classes. Then Γ has property T iff the identity correspondence of the II_1 factor $R(\Gamma)$ is isolated.*

Using the above functor it is trivial that if Γ does not have property T then the identity correspondence of $R(\Gamma)$ is not isolated. Conversely, if Γ has property T one shows that the identity correspondence is isolated using the following functor from correspondences (from $R(\Gamma)$ to $R(\Gamma)$) to representations of Γ:

To the $R(\Gamma)$-bimodule $\mathcal{H}$ one assigns the unitary representation π of Γ in $\mathcal{H}$ given by

$$\pi(g)\,\xi = g\,\xi\,g^{-1} \qquad \forall g \in \Gamma\,,\ \xi \in \mathcal{H}.$$

One checks easily that π contains the trivial representation of Γ iff $\mathcal{H}$ contains the identity correspondence. This gives the desired result.

The great interest of a "property T" factor as in Theorem 22 is that its outer automorphism group Out N is countable as well as its fundamental group $F(N)$ ([Co_9]) leading to the first example $F(N) \neq \mathbb{R}_+^*$.

The factors with property T enjoy remarkable rigidity properties. For example, there does not exist a nontrivial sequence $N_n \subset N_{n+1}$ of subfactors with $\bigcup N_n$ dense in N. The main problem is as follows:

Problem 1. *Show that if Γ_1 and Γ_2 are not isomorphic, then $R(\Gamma_1)$ and $R(\Gamma_2)$ are not isomorphic.*

Of course, here Γ_1 and Γ_2 are rigid in the sense defined above (for amenable groups, the situation if radically opposite; see Section 9). The only results known are: 1) if Γ_1 is rigid, in the above sense, and if Γ_2 is discrete in $SL(2,\mathbb{R})$, then there does not exist any homomorphism of $R(\Gamma_1)$ into $R(\Gamma_2)$; 2) one can distinguish countably many discrete subgroups Γ of $Sp(n,\mathbb{R})$ by their associated factors $R(\Gamma)$ [C-H].

Problem 2. *Determine the fundamental group of $R(\Gamma)$ for Γ rigid.*

VI

THE METRIC ASPECT OF NONCOMMUTATIVE GEOMETRY

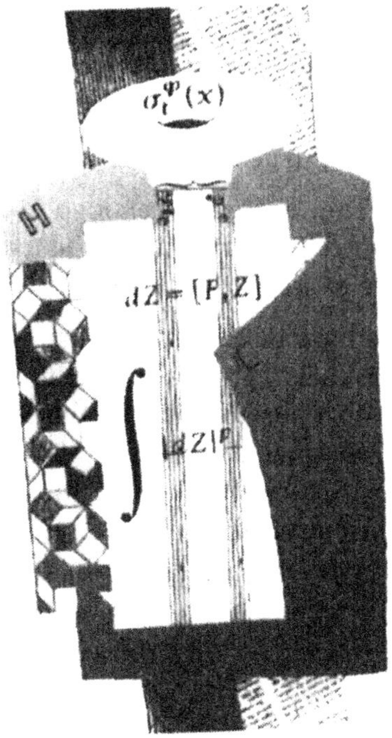

The geometric spaces of Gauss and Riemann are defined as manifolds in which the metric is given by the formula

$$\delta(p,q) = \text{ infimum of length of paths } \gamma \text{ from } p \text{ to } q \tag{1}$$

where the length of a path γ is computed as the integral of the square root of a quadratic form in the differential of the path

$$\text{Length of } \gamma = \int_p^q (g_{\mu\nu}\, dx^\mu \;\; dx^\nu)^{1/2} \tag{2}$$

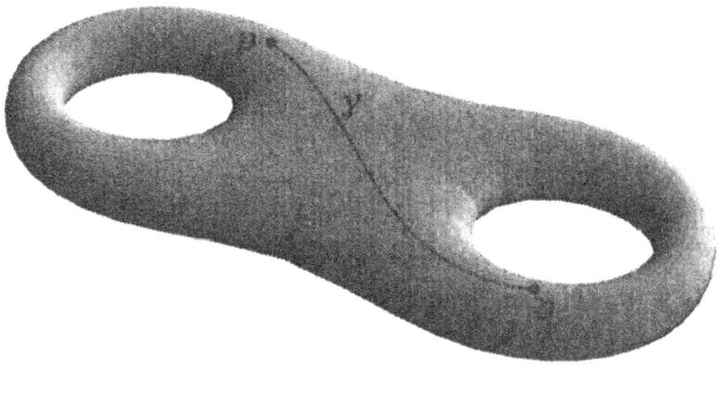

Figure VI.1.

These geometric spaces form a relevant class of metric spaces, inasmuch as:

α) They are general enough to include numerous examples ranging from non-Euclidean geometries through surfaces embedded in $\mathbb{R}^3$ to space like hypersurfaces in general relativity.

β) They are special enough to deserve the name "geometry", since, being determined by local data, all the tools of differential and integral calculus are available to analyse them.

We have developed in Chapter IV a differential and integral calculus of "infinitesimals", given a Fredholm module $(\mathcal{H}, F)$ over the algebra $\mathcal{A}$ of coordinates on a possibly noncommutative space X. The Fredholm module $(\mathcal{H}, F)$ over $\mathcal{A}$ specifies the calculus on X but not the metric structure. For instance, the construction of $(\mathcal{H}, F)$ in the manifold case (Section IV.4) only used the conformal structure. In fact, in the example of Section IV.3, where $X = \mathbb{S}^1$ and $(\mathcal{H}, F)$ is the Hilbert transform, the quantum differential expression

$$dZ\, d\overline{Z} = [F, Z][F, \overline{Z}] \quad , \quad Z : \mathbb{S}^1 \to \mathbb{C} \tag{3}$$

where Z is the boundary value of a univalent map, yields an infinitesimal unit of length intimately tied up with the metric on $Z(\mathbb{S}^1)$ induced by the usual Riemannian metric $dz d\overline{z}$ of $\mathbb{C}$. If we vary Z, even the *dimension* of $\mathbb{S}^1$ for the "metric" (3) will change (cf. Section IV.3).

Let $\mathcal{A}$ be an involutive algebra and $(\mathcal{H}, F)$ a Fredholm module over $\mathcal{A}$. To define a "unit of length" in the corresponding space X, we shall consider an operator of the form

$$G = \sum_1^q (dx^\mu)^* \, g_{\mu\nu} (dx^\nu) \tag{4}$$

where $dx = [F, x]$ for any $x \in \mathcal{A}$, the x^μ are elements of $\mathcal{A}$ and where $g = (g_{\mu\nu})_{\mu,\nu=1,\ldots,q}$ is a positive element of the matrix algebra $M_q(\mathcal{A})$.

We want to think of G as the ds^2 of Riemannian geometry. It is by construction a positive "infinitesimal", i.e. a positive compact operator on $\mathcal{H}$. The unit of length is its positive square root

$$ds = G^{1/2}. \tag{5}$$

To measure distances in the possibly noncommutative space X we first replace the points $p, q \in X$ by the corresponding pure states φ, ψ on the C^*-algebra closure of $\mathcal{A}$

$$\varphi(f) = f(p)\, , \; \psi(f) = f(q) \quad \forall f \in \mathcal{A}. \tag{6}$$

We then dualise the basic formula (1) as follows

$$\text{dist}(p, q) = \sup \{ |f(p) - f(q)| \, ; \, f \in \mathcal{A} \, , \, \|df/ds\| \le 1 \} \tag{7}$$

which only involves p, q through the associated pure states (6). Since we are in the noncommutative set up we need to deal with the ambiguity in the order of the terms in an expression such as df/ds which can be either $df(ds)^{-1}$ or $(ds)^{-1}\, df$ or $(ds)^{-\alpha}\, df(ds)^{-(1-\alpha)}$ for instance. Instead of handling this problem directly we shall assume that G commutes with F, i.e. that $dG = 0$, a condition similar to the Kähler condition, and introduce the following selfadjoint operator

$$D = FG^{-1/2} = F\, ds^{-1}, \tag{8}$$

whose existence assumes that G is nonsingular, i.e. $\ker G = 0$. We shall then formulate (7) as follows

$$d(p,q) = \sup\{|f(p) - f(q)|\,;\; f \in \mathcal{A}\,,\; \|[D,f]\| \le 1\}. \tag{9}$$

Now the operator F is by construction the sign of D, while G is obtained from D by the formula

$$G = D^{-2}. \tag{10}$$

Thus it is more economical to take as our basic data the triple $(\mathcal{A}, \mathcal{H}, D)$ consisting of a Hilbert space $\mathcal{H}$, an involutive algebra $\mathcal{A}$ of operators on $\mathcal{H}$ and an unbounded selfadjoint operator D on $\mathcal{H}$. The conditions satisfied by such triples are

$$\begin{aligned} &[D,a] \quad &&\text{is bounded for any} \quad a \in \mathcal{A} \\ &(D-\lambda)^{-1} \quad &&\text{is compact for any} \quad \lambda \notin \mathbb{R} \end{aligned} \tag{11}$$

and were already formalised in Chapter IV Definition 2.11. In the present chapter we shall begin a systematic investigation of those geometric spaces. Besides Riemannian manifolds (see below) and spaces of non integral Hausdorff dimension (Section IV.3) the following are examples of geometric spaces described by our data:

a) Discrete spaces

b) Duals of discrete subgroups of Lie groups

c) Configuration space in supersymmetric quantum field theory.

We shall deal with Example a in Section 3 below. We have described already in great detail the triples $(\mathcal{A}, \mathcal{H}, D)$ corresponding to b) and c) in Section IV.9.

Our first task in this chapter will be to show that the Riemannian spaces are special cases of the above notion of geometric spaces. This will be done using an elliptic differential operator of order one, the Dirac operator (or the signature operator in the non-spin case). We shall first see that formula (9) applied to the triple (algebra of functions, Hilbert space of spinors, Dirac operator) readily gives back the geodesic distance (1) on the Riemannian manifold. Our next task will be to develop the analogue of the Lagrangian formulation of electrodynamics involving matter fields and gauge bosons for our more general geometric spaces. This will be done using the tools of the quantized calculus developed in Chapter IV Section 2. As mentioned above the commutator $[D, f]$, $f \in \mathcal{A}$ will play the role of the differential quotient df/ds. As a central result we shall prove the inequality between the second Chern number of a "vector bundle" and the minimum of the Yang-Mills action on vector potentials. We shall see that our new notion of geometric space treats on an equal footing the continuum and the discrete, while the action for electrodynamics on the simplest mixture of continuum and discrete–the product of 4-dimensional continuum by a discrete 2-point space–gives the Glashow-Weinberg-Salam model for leptons. The notion of manifold in noncommutative geometry will be reached only after an understanding of Poincaré duality, i.e., that the K-homology cycle $(\mathfrak{H}, D)$ yields the fundamental class of the space under consideration. The notion of manifold obtained is directly inspired by the work of D. Sullivan [Su$_7$] who discovered the basic role played by the K-homology fundamental class of a manifold.

The main example of a space to which all these considerations will be applied is Euclidean space-time in physics, i.e., space-time but with imaginary time. What we shall give is a geometric interpretation of the now experimentally confirmed effective low-energy model of particle physics, namely the Glashow-Weinberg-Salam standard model. This model is a gauge theory model with gauge group $U(1) \times SU(2) \times SU(3)$ and a pair of complex Higgs fields providing masses by the symmetry-breaking mechanism. We interpret this model geometrically as a pure gauge theory, i.e. electrodynamics, but on a more elaborate space-time $E' = E \times F$, the product of ordinary Euclidean space-time by a finite space F. The geometry of this finite space is specified by a pair $(\mathfrak{H}, D)$ as above, where $\mathfrak{H}$ is finite-dimensional and the selfadjoint operator D encodes the nine fermion masses and the four Kobayashi–Maskawa mixing parameters of the standard model.

The values of the hypercharges do not have to be fitted artificially to their experimental values but come out right from a simple unimodularity condition on the space E'.

Our analysis is limited to the classical context and does not at the moment address the questions related to renormalization, such as the existence of relations between coupling constants or the naturalness problem. Nevertheless, our more geometric and conceptual interpretation of the standard model gives a clear indication that particle physics is not so much a long list of elementary particles as the unveiling of the fine geometric structure of space-time.

VI.1. Riemannian Manifolds and the Dirac Operator

Let M be a compact Riemannian spin manifold, and let $D = \partial_M$ be the corresponding Dirac operator (cf. [Gi$_1$]). Thus, D is an unbounded selfadjoint operator acting in the Hilbert space $\mathfrak{H}$ of L^2-spinors on the manifold M.

We shall give four formulas below that show how to reconstruct the *metric space* (M, d), where d is the geodesic distance, the *volume measure* dv on M, the space of *gauge potentials*, and, finally, the *Yang-Mills* action functional, from the purely operator-theoretic data

$$(\mathcal{A}, \mathfrak{H}, D),$$

where D is the Dirac operator on the Hilbert space $\mathfrak{H}$ and where $\mathcal{A}$ is the abelian von Neumann algebra of multiplication by bounded measurable functions on M.

Thus, $\mathcal{A}$ is an abelian von Neumann algebra on $\mathfrak{H}$, and knowing the pair $(\mathfrak{H}, \mathcal{A})$ yields essentially no information (cf. Chapter V) except for the multiplicity, which is here the constant $2^{[d/2]}$, where $d = \dim M$. Similarly, the mere knowledge of the operator D on $\mathfrak{H}$ is equivalent to giving its list of eigenvalues $(\lambda_n)_{n\in\mathbb{N}}$, $\lambda_n \in \mathbb{R}$, and is an impractical point of departure for reconstructing M. The growth of these eigenvalues, i.e., the behavior of $|\lambda_n|$ as $n \to \infty$, is again governed by the dimension d of M, namely, $|\lambda_n| \sim Cn^{1/d}$ as $n \to \infty$.

What is relevant is the triple $(\mathcal{A}, \mathfrak{H}, D)$. Elements of $\mathcal{A}$ other than the constants do not commute with D, and the *boundedness* of the commutator $[D, a]$ already implies the following regularity condition on a measurable function a:

Lemma 1. *If a is a bounded measurable function on M then the densely defined operator $[D, a]$ is bounded if and only if a is almost everywhere equal to a Lipschitz function f, $|f(p) - f(q)| \leq Cd(p, q)$ $(\forall p, q \in M)$.*

Here, d is the geodesic distance in M. The operator $[D, a]$ should be viewed in effect as a quadratic form

$$\xi, \eta \mapsto \langle a\xi, D\eta\rangle - \langle D\xi, a^*\eta\rangle,$$

which is well-defined for ξ and η in the domain of D; its boundedness means an inequality

$$|\langle a\xi, D\eta\rangle - \langle D\xi, a^*\eta\rangle| \leq c\|\xi\|\|\eta\| \qquad \forall \xi, \eta \in \operatorname{dom} D.$$

The proof of the lemma follows immediately from results in [Fe].

Now, every Lipschitz function on M is continuous and the algebra of Lipschitz functions is norm-dense in the algebra of continuous functions on M; it follows that the C^*-algebra $C(M)$ of continuous functions on M is identical to the *norm closure* A in $\mathcal{L}(\mathfrak{H})$ of $\mathbf{A} = \{a \in \mathcal{A};\ [D, a] \text{ is bounded}\}$.

By Gel'fand's theorem (Chapter II), we recover the compact topological space M as the spectrum of A. Thus, a point p of M is a $*$-homomorphism $\rho : A \to \mathbb{C}$,

$$\rho(ab) = \rho(a)\rho(b) \quad \forall a, b \in A.$$

Any such homomorphism ρ is given by evaluation of a at p for some point $p \in M$,

$$\rho(a) = a(p) \in \mathbb{C}.$$

All of this is still qualitative; we now come to the first interesting formula, giving us a natural distance function, which turns out in this case to be the geodesic distance

Formula 1. *For any pair of points $p, q \in M$, their geodesic distance is given by the formula*

$$d(p,q) = \sup\{|a(p) - a(q)|;\ a \in \mathcal{A},\ \|[D,a]\| \leq 1\}.$$

The proof is straightforward, but it is relevant to go through it to see what is involved. The operator $[D, a]$, which by Lemma 1 is bounded iff a is Lipschitz, is then given by the Clifford multiplication $i^{-1}\gamma(da)$ by the gradient da of a. This gradient is ([Fe]) a bounded measurable section of the cotangent bundle T^*M of M, and we have

$$\|[D,a]\| = \operatorname{ess\,sup} \|da\| = \text{the Lipschitz norm of } a.$$

It follows at once that the right-hand side of Formula 1 is less than or equal to the geodesic distance $d(p, q)$. However, fixing the point p and considering the function $a(q) = d(q, p)$, one checks that a is Lipschitz with constant 1, so that $\|[D, a]\| \leq 1$, which yields the desired equality. Note that Formula 1 is in essence dual to the original formula

$$d(p,q) = \text{infimum of the length of paths } \gamma \text{ from } p \text{ to } q, \qquad (*)$$

in the sense that, instead of involving arcs, namely copies of $\mathbb{R}$ inside the manifold M, it involves *functions* a, that is, maps from M to $\mathbb{R}$ (or to $\mathbb{C}$).

This is an essential point for us since, in the case of discrete spaces or of noncommutative spaces X, there are no interesting arcs in X but there are plenty of *functions*, namely, the elements $a \in \mathcal{A}$ of the defining algebra. We note at once that the right-hand side of Formula 1 is meaningful in that general context and it defines a metric on the space of *states* of the C^*-algebra A, the norm closure of $\mathbf{A} = \{a \in \mathcal{A};\ [D, a] \text{ is bounded}\}$

$$d(\varphi, \psi) = \sup\{|\varphi(a) - \psi(a)|;\ \|[D,a]\| \leq 1\}.$$

Finally, we also note that, although both Formula 1 and the formula $(*)$ give the same result for Riemannian manifolds, they are of quite different nature if we try to use them in actual measurements of distances. The formula $(*)$ uses the idealized notion of a path, and quantum mechanics teaches us that there is nothing like "the path followed by a particle". Thus, for measurements of very small distances, it is more natural to use wave functions and Formula 1.

We have now recovered from our original data $(\mathcal{A}, \mathfrak{H}, D)$ the metric space (M, d), where d is the geodesic distance. Let us now deal with the tools of differential and integral calculus, the first obvious example being the measure given by the volume form

$$f \mapsto \int_M f dv,$$

where, in local coordinates $x^\mu, g_{\mu\nu}$, we have

$$dv = \left(\det(g_{\mu\nu})\right)^{1/2} |dx^1 \wedge \cdots \wedge dx^n|.$$

This takes us to our second formula, which is nothing more than a restatement of H. Weyl's theorem about the asymptotic behavior of elliptic differential operators ([Gu], [Gi$_1$]). It does, however, involve a new tool, *the Dixmier trace* Tr_ω (cf. Chapter IV.2), which, unlike asymptotic expansions, makes sense in full generality in our context and is the correct operator-theoretic substitute for integration:

Formula 2. *For every* $f \in A$, *we have* $\int_M f dv = c(d)\ \mathrm{Tr}_\omega(f|D|^{-d})$, *where* $d = \dim M$, $c(d) = 2^{(d-[d/2])}\ \pi^{d/2}\ \Gamma\left(\frac{d}{2}+1\right)$.

By convention we let D^{-1} be equal to 0 on the finite-dimensional subspace $\ker D$.

Let us refer to Section IV.2 for the detailed definition and properties of the Dixmier trace Tr_ω. We can interpret the right-hand side of the equality as the limit of the sequence

$$\frac{1}{\log N} \sum_{j=0}^{N} \lambda_j,$$

where the λ_j are the eigenvalues of the compact operator $f|D|^{-d}$, or, equivalently, as the residue, at the point $s = 1$, of the function

$$\zeta(s) = \mathrm{Trace}(f|D|^{-ds}) \quad (\mathrm{Re}\ s > 1).$$

The crucial fact for us is that the Dixmier trace makes sense independently of the context of pseudodifferential operators and that all properties of the integral $\int_M f dv$, such as positivity, finiteness, covariance, etc., become obvious corollaries of the general properties of the Dixmier trace:

A) *Positivity*: $\mathrm{Tr}_\omega(T) \geq 0$ if T is a positive operator.

B) *Finiteness*: $\mathrm{Tr}_\omega(T) < \infty$ if the eigenvalues of $|T|$ satisfy $\sum_0^N \mu_n(T) = O(\log N)$.

C) *Covariance*: $\mathrm{Tr}_\omega(UTU^*) = \mathrm{Tr}_\omega(T)$ for every unitary U.

D) *Vanishing*: $\mathrm{Tr}_\omega(T) = 0$ if T is of trace class.

Property D is the counterpart of locality in our framework; it shows that the Dixmier trace of an operator is unaffected by a finite-rank perturbation, and allows many identities to hold, as we have seen in Chapter IV.

Now, setting up the integral of functions, i.e., the Riemannian volume form, is a good indication but quite far from the full story. In particular, many distinct Riemannian metrics yield the same volume form. Since our aim is to investigate physical space-time at the scale of elementary particle physics, we shall now make a deliberate choice: instead of focusing on the intrinsic Riemannian curvature, which would drive us towards general relativity, we shall concentrate on the measurement (using $(\mathfrak{H}, D)$) of the curvature of connections on vector bundles, and on the Yang-Mills functional, which takes us to the theory of matter fields. This line is of course easier since it does not involve derivatives of the $g_{\mu\nu}$.

Let us state our aim clearly: to recover the Yang–Mills functional on connections on vector bundles, making use of only the following data (IV.2.11):

Definition 2. *A K-cycle $(\mathfrak{H}, D)$, over an algebra $\mathcal{A}$ with involution $*$, consists of a $*$-representation of $\mathcal{A}$ on a Hilbert space $\mathfrak{H}$ together with an unbounded selfadjoint operator D with compact resolvent, such that $[D, a]$ is bounded for every $a \in \mathcal{A}$.*

We shall assume that $\mathcal{A}$ is unital and that the unit $1 \in \mathcal{A}$ acts as the identity on $\mathfrak{H}$ (cf. Remark 12 for the nonunital case).

If the eigenvalues λ_n of $|D|$ are of the order of $n^{1/d}$ as $n \to \infty$, we say that the K-cycle is (d, ∞)-summable (cf. Section 2 of Chapter IV). On the algebra of functions on a compact Riemannian spin manifold, the Dirac operator determines a K-cycle that is (d, ∞)-summable, where $d = \dim M$. Finer regularity of functions, such as infinite differentiability, is easily expressed using the domains of powers of the derivation δ, $\delta(a) = [|D|, a]$.

We shall not be too specific about the choice of regularity; our discussion applies to any degree of regularity higher than Lipschitz.

The value of the following construction is that it will also apply when the $*$-algebra $\mathcal{A}$ is noncommutative, or when D is no longer the Dirac operator (cf. Section 3). The reader can have in mind both the Riemannian case and the slightly more involved case where the algebra $\mathcal{A}$ is the $*$-algebra of matrices of functions on a Riemannian manifold, just to bear in mind that the notion of exterior product no longer makes sense over such an algebra.

We shall begin with the notion of connection on the trivial bundle, i.e., the case of "electrodynamics", and define vector potentials and the Yang-Mills action in that case. We shall then treat the general case of arbitrary Hermitian bundles, i.e., in algebraic terms, of arbitrary Hermitian, finitely generated projective modules over $\mathcal{A}$.

We wish to define k-forms over $\mathcal{A}$ as operators on $\mathfrak{H}$ of the form

$$\omega = \sum a_0^j [D, a_1^j] \cdots [D, a_k^j],$$

where the a_i^j are elements of $\mathcal{A}$ represented as operators on $\mathfrak{H}$. This idea arises because, although the operator D fails to be invariant under the representation on $\mathfrak{H}$ of the unitary group $\mathcal{U}$ of $\mathcal{A}$,

$$\mathcal{U} = \{u \in \mathcal{A};\ u^*u = uu^* = 1\},$$

the following equality shows that the failure of invariance is governed by a 1-form in the above sense: by $\omega_u = u[D, u^*]$, that is,

$$uDu^* = D + \omega_u.$$

Note that ω_u is selfadjoint as an operator on $\mathfrak{H}$. Thus, it is natural to adopt the following definition:

Definition 3. *A vector potential V is a selfadjoint element of the space of 1-forms $\sum a_0^j[D, a_1^j]$, where $a_k^j \in \mathcal{A}$.*

One can immediately check that in the basic example of the Dirac operator on a spin Riemannian manifold, a vector potential in the above sense is exactly a 1-form ω on the manifold M and that this form is imaginary, the corresponding operator in the space of spinors being given by the Clifford multiplication:

$$V = i^{-1}\gamma(\omega) \quad (i = \sqrt{-1}).$$

The action of the unitary group $\mathcal{U}$ on vector potentials is such that it replaces the operator $D + V$ by $u(D + V)u^*$; thus it is given by the algebraic formula

$$\gamma_u(V) = u[D, u^*] + uVu^* \quad (u \in \mathcal{U}).$$

We now need only define the curvature or field strength θ for a vector potential, and use the analogue of the above Formula 2 to integrate the square of θ: the formula

$$\mathrm{YM}(V) = \mathrm{Tr}_\omega(\theta^2 |D|^{-d})$$

should give us the Yang-Mills action.

The formula for θ should be of the form $\theta = \mathrm{d}V + V^2$; the only difficulty is in defining properly the "differential" $\mathrm{d}V$ of a vector potential, as an operator on $\mathfrak{H}$.

Let us examine what happens; the naive formulation is

$$\text{If } V = \sum a_0^j[D, a_1^j] \text{ then } \mathrm{d}V = \sum [D, a_0^j][D, a_1^j].$$

Before we point out what the difficulty is, let us check that if we replace V by $\gamma_u(V)$, where

$$\gamma_u(V) = u[D, u^*] + \sum u a_0^j[D, a_1^j]u^*,$$

then the curvature is transformed covariantly:

$$\mathrm{d}(\gamma_u(V)) + \gamma_u(V)^2 = u(\mathrm{d}V + V^2)u^*.$$

As this computation is instructive, we shall carry it out in detail. First, in order to write $\gamma_u(V)$ in the same form as V, we use the equality

$$[D, a_1^j]u^* = [D, a_1^j u^*] - a_1^j [D, u^*].$$

Thus, $\gamma_u(V) = u[D, u^*] + \sum u a_0^j [D, a_1^j u^*] - \sum u a_0^j a_1^j [D, u^*]$, and we have

$$d\gamma_u(V) = [D, u][D, u^*] + \sum [D, u a_0^j][D, a_1^j u^*] - \sum [D, u a_0^j a_1^j][D, u^*].$$

We now claim that the following operators on $\mathfrak{H}$ are indeed equal:

$$d\gamma_u(V) + \gamma_u(V)^2 = u(dV + V^2)u^*.$$

We start with the left-hand side; it is equal to

$$\begin{aligned} &d\gamma_u(V) + (u[D, u^*] + uVu^*)^2 \\ &\quad = d\gamma_u(V) + u[D, u^*]u[D, u^*] + u[D, u^*]uVu^* \\ &\qquad + uVu^*u[D, u^*] + uV^2u^* \\ &\quad = d\gamma_u(V) - [D, u][D, u^*] - [D, u]Vu^* + uV[D, u^*] + uV^2u^* \\ &\quad = \sum [D, u a_0^j][D, a_1^j u^*] - \sum [D, u a_0^j a_1^j][D, u^*] \\ &\qquad - [D, u]Vu^* + uV[D, u^*] + uV^2u^* \\ &\quad = udVu^* + uV^2u^*, \end{aligned}$$

where the last equality follows from

$$\begin{aligned} &\sum [D, u]a_0^j [D, a_1^j u^*] - \sum [D, u]a_0^j a_1^j [D, u^*] = [D, u]Vu^*, \\ &\sum u[D, a_0^j][D, a_1^j u^*] - \sum u[D, a_0^j]a_1^j [D, u^*] = udVu^*, \\ &\sum u a_0^j [D, a_1^j][D, u^*] = uV[D, u^*]. \end{aligned}$$

The difficulty that we overlooked is the following: the same vector potential V might be written in several ways as $V = \sum a_0^j [D, a_1^j]$, so that the definition of dV as

$$dV = \sum [D, a_0^j][D, a_1^j]$$

is ambiguous.

To understand the nature of the problem, let us introduce some algebraic notation. We let $\Omega^*\mathcal{A}$ be the reduced universal differential graded algebra over $\mathcal{A}$ (Chapter III.1). It is by definition equal to $\mathcal{A}$ in degree 0 and is generated by symbols da ($a \in \mathcal{A}$) of degree 1 with the following presentation:

α) $d(ab) = (da)b + adb \quad (\forall a, b \in \mathcal{A})$,

β) $d1 = 0$.

One can check that $\Omega^1\mathcal{A}$ is isomorphic as an $\mathcal{A}$-bimodule to the kernel $\ker(m)$ of the multiplication mapping $m : \mathcal{A} \otimes \mathcal{A} \to \mathcal{A}$, the isomorphism being given by the mapping

$$\sum a_i \otimes b_i \in \ker(m) \mapsto \sum a_i db_i \in \Omega^1\mathcal{A}.$$

The involution $*$ of $\mathcal{A}$ extends uniquely to an involution on Ω^* with the rule

$$(da)^* = -da^*.$$

The differential d on $\Omega^*\mathcal{A}$ is defined *unambiguously* by

$$d(a^0 da^1 \cdots da^n) = da^0 da^1 \cdots da^n \quad \forall a^j \in \mathcal{A},$$

and it satisfies the relations

$$d^2\omega = 0 \quad \forall \omega \in \Omega^*\mathcal{A},$$
$$d(\omega_1\omega_2) = (d\omega_1)\omega_2 + (-1)^{\partial\omega_1}\omega_1 d\omega_2 \quad \forall \omega_j \in \Omega^*\mathcal{A}.$$

Proposition 4. 1) *The following equality defines a $*$-representation π of the reduced universal algebra $\Omega^*(\mathcal{A})$ on $\mathfrak{H}$:*

$$\pi(a^0 da^1 \cdots da^n) = a^0[D, a^1] \cdots [D, a^n] \quad \forall a^j \in \mathcal{A}.$$

2) *Let $J_0 = \ker \pi \subset \Omega^*$ be the graded two-sided ideal of Ω^* given by $J_0^{(k)} = \{\omega \in \Omega^k;\ \pi(\omega) = 0\}$; then $J = J_0 + dJ_0$ is a graded differential two-sided ideal of $\Omega^*(\mathcal{A})$.*

The first statement is obvious; let us discuss the second. By construction, J_0 is a two-sided ideal but it is not, in general, a *differential* ideal, i.e., if $\omega \in \Omega^k(\mathcal{A})$ and $\pi(\omega) = 0$, one does not in general have $\pi(d\omega) = 0$. This is exactly the reason why the above definition of $\sum [D, a_0^j][D, a_1^j]$ as the differential of $\sum a_0^j[D, a_1^j]$ was ambiguous.

Let us show, however, that $J = J_0 + dJ_0$ is still a two-sided ideal. Since $d^2 = 0$ it is obvious that J is then a differential ideal. Let $\omega \in J^{(k)}$ be a homogeneous element of J; then ω is of the form $\omega = \omega_1 + d\omega_2$, where $\omega_1 \in J_0 \cap \Omega^k$, $\omega_2 \in J_0 \cap \Omega^{k-1}$. Let $\omega' \in \Omega^{k'}$, and let us show that $\omega\omega' \in J^{(k+k')}$. We have

$$\begin{aligned} \omega\omega' &= \omega_1\omega' + (d\omega_2)\omega' = \omega_1\omega' + d(\omega_2\omega') - (-1)^{k-1}\omega_2 d\omega' \\ &= (\omega_1\omega' + (-1)^k\omega_2 d\omega') + d(\omega_2\omega'). \end{aligned}$$

But, the first term belongs to $J_0 \cap \Omega^{k+k'}$ and $\omega_2\omega' \in J_0 \cap \Omega^{k+k'-1}$.

Using 2) of Proposition 4, we can now introduce the graded differential algebra

$$\Omega_D^* = \Omega^*(\mathcal{A})/J.$$

Let us first investigate Ω_D^0, Ω_D^1 and Ω_D^2.

We have $J \cap \Omega^0 = J_0 \cap \Omega^0 = \{0\}$ provided that we assume, as we shall, that $\mathcal{A}$ is a *subalgebra* of $\mathcal{L}(\mathfrak{H})$. Thus, $\Omega_D^0 = \mathcal{A}$.

Next, $J \cap \Omega^1 = J_0 \cap \Omega^1 + d(J_0 \cap \Omega^0) = J_0 \cap \Omega^1$; thus Ω_D^1 is the quotient of Ω^1 by the kernel of π, and it is thus exactly the $\mathcal{A}$-bimodule $\pi(\Omega^1)$ of operators ω of the form

$$\omega = \sum a_j^0[D, a_j^1] \quad (a_j^k \in \mathcal{A}).$$

Finally, $J \cap \Omega^2 = J_0 \cap \Omega^2 + \mathrm{d}(J_0 \cap \Omega^1)$ and the representation π gives an isomorphism

$$\Omega_D^2 \cong \pi(\Omega^2)/\pi(\mathrm{d}(J_0 \cap \Omega^1)). \qquad (*)$$

More precisely, this means that we can view an element ω of Ω_D^2 as a class of elements ρ of the form

$$\rho = \sum a_j^0 [D, a_j^1][D, a_j^2] \quad (a_j^k \in \mathcal{A})$$

modulo the sub-bimodule of elements of the form

$$\rho_0 = \sum [D, b_j^0][D, b_j^1] \quad ; \quad b_j^k \in \mathcal{A}, \ \sum b_j^0 [D, b_j^1] = 0.$$

It is now clear that since we work modulo this subspace $\pi(\mathrm{d}(J_0 \cap \Omega^1))$, the question of ambiguity in the definition of $\mathrm{d}\omega$ for $\omega \in \pi(\Omega^1)$ no longer arises.

The equality $(*)$ makes sense for all k,

$$\Omega_D^k \cong \pi(\Omega^k)/\pi(\mathrm{d}(J_0 \cap \Omega^{k-1})), \qquad (*)$$

and allows us to define the following inner product on Ω_D^k: for each k let $\mathfrak{H}_k$ be the Hilbert space completion of $\pi(\Omega^k)$ with the inner product

$$\langle T_1, T_2 \rangle_k = \mathrm{Tr}_\omega(T_2^* T_1 |D|^{-d}) \quad \forall T_j \in \pi(\Omega^k).$$

Let P be the orthogonal projection of $\mathfrak{H}_k$ onto the orthogonal complement of the subspace $\pi(\mathrm{d}(J_0 \cap \Omega^{k-1}))$. By construction, the inner product $\langle P\omega_1, \omega_2 \rangle = \langle P\omega_1, P\omega_2 \rangle$ for $\omega_j \in \pi(\Omega^k)$ depends only on their classes in Ω_D^k. We denote by Λ^k the Hilbert space completion of Ω_D^k for this inner product; it is, of course, equal to $P\mathfrak{H}_k$.

Proposition 5. 1) *The actions of $\mathcal{A}$ on Λ^k by left and right multiplication define commuting unitary representations of $\mathcal{A}$ on Λ^k.*

2) *The functional* $\mathrm{YM}(V) = \langle \mathrm{d}V + V^2, \mathrm{d}V + V^2 \rangle$ *is positive, quartic and invariant under gauge transformations,*

$$\gamma_u(V) = u\mathrm{d}u^* + uVu^* \quad \forall u \in \mathcal{U}(\mathcal{A}).$$

3) *The functional* $I(\alpha) = \mathrm{Tr}_\omega(\theta^2 |D|^{-d})$, with $\theta = \pi(\mathrm{d}\alpha + \alpha^2)$, *is positive, quartic and gauge invariant on* $\{\alpha \in \Omega^1(\mathcal{A});\ \alpha = \alpha^*\}$.

4) *One has* $\mathrm{YM}(V) = \inf\{I(\alpha);\ \pi(\alpha) = V\}$.

Let us say a few words about the easy proof. First, the left and right actions of $\mathcal{A}$ on $\mathfrak{H}_k$ are unitary. The unitarity of the right action of $\mathcal{A}$ follows from the equality $\mathrm{Tr}_\omega(Ta|D|^{-d}) = \mathrm{Tr}_\omega(aT|D|^{-d}) \quad \forall T \in \mathcal{L}(\mathcal{H}), a \in \mathcal{A}$. Since $\pi(\mathrm{d}(J_0 \cap \Omega^{k-1})) \subset \pi(\Omega^k)$ is a sub-bimodule of $\pi(\Omega^k)$ it follows that P is a bimodule morphism:

$$P(a\xi b) = aP(\xi)b \quad \forall a, b \in \mathcal{A},\ \xi \in \mathfrak{H}_k.$$

Thus 1) follows. As for 2), one merely notes that by the above calculation, with dV now unambiguous, $\theta = dV + V^2$ is covariant under gauge transformations, whence the result. For 3), one again uses the above calculation to show that $d\alpha + \alpha^2$ transforms covariantly under gauge transformations.

Finally, 4) follows from the property of the orthogonal projection P: as an element of Λ^2, $dV + V^2$ is equal to $P(\pi(d\alpha + \alpha^2))$ for any α with $\pi(\alpha) = V$, and since the ambiguity in $\pi(d\alpha)$ is exactly $\pi(d(J_0 \cap \Omega^1))$ one gets 4).

Stated in simpler terms, the meaning of Proposition 5 is that the ambiguity that we met above in the definition of the operator curvature $\theta = dV + V^2$ can be ignored by taking the infimum

$$\mathrm{YM}(V) = \inf \mathrm{Tr}_\omega(\theta^2 |D|^{-d})$$

over all possibilities for $\theta = dV + V^2$, $dV = \sum [D, a_j^0][D, a_j^1]$ being ambiguous. The action obtained is nevertheless quartic by 2).

We shall now check that in the case of Riemannian manifolds with the Dirac K-cycle, the graded differential algebra Ω_D^* is canonically isomorphic to the *de Rham algebra of ordinary forms on M with their canonical pre-Hilbert space structure.* The whole point is that Propositions 4 and 5 give us these concepts in far greater generality, and the formula in 4) will allow extending to this generality, in the case $d = 4$, the inequality between the topological action and the Yang–Mills action YM (cf. Section 2). We refer the skeptical reader to the examples of Section 3.

We now specialize to the Riemannian case, where $\mathcal{A}$ is the algebra of functions (with some regularity) on the compact spin manifold M, and $D = \partial_M$ is the Dirac operator in the Hilbert space $L^2(M, S)$ of spinors. We let C be the bundle over M whose fiber at each $p \in M$ is the complexified Clifford algebra $\mathrm{Cliff}_{\mathbb{C}}(T_p^*(M))$ of the cotangent space at $p \in M$. Any bounded measurable section ρ of C defines a bounded operator $\gamma(\rho)$ on $\mathfrak{H} = L^2(M, S)$. For any $f^0, \ldots, f^n \in \mathcal{A}$ one has

$$\pi(f^0 df^1 \cdots df^n) = i^{-n} \gamma(f^0 d_c f^1 \cdot d_c f^2 \cdots d_c f^n),$$

where the usual differential $d_c f$ is regarded as a section of C, and $\cdot$ denotes the product in C.

For each $p \in M$, the Clifford algebra C_p has a $\mathbb{Z}/2$ grading given by the parity of the number of terms ξ_j, $\xi_j \in T_p^*(M)$ in a product $\xi_1 \cdot \xi_2 \cdots \xi_n$, and a filtration, where $C_p^{(k)}$ is the subspace spanned by products of $n \leq k$ elements of $T_p^*(M)$. The associated graded algebra is canonically isomorphic to the complexified exterior algebra $\Lambda_{\mathbb{C}}(T_p^*(M))$ and $\sigma_k : C^{(k)} \to \Lambda_{\mathbb{C}}^k(T_p^*)$ is the quotient mapping.

Using the canonical inner product on C given by the trace in the spinor representation, one can also identify $\Lambda_{\mathbb{C}}^k$ with the orthogonal complement of $C^{(k-1)}$ in $C^{(k)}$; equivalently, if we let C^k be the subspace of $C^{(k)}$of elements of the same parity as k, then $\Lambda_{\mathbb{C}}^k = C^k \ominus C^{k-2}$.

The differential algebra Ω_D^* is determined by the following lemma:

Lemma 6. *Let $(\mathfrak{H}, D)$ be the Dirac K-cycle on the algebra $\mathcal{A}$ of functions on M and let $k \in \mathbb{N}$. Then a pair T_1, T_2 of operators on $\mathfrak{H}$ is of the form $T_1 = \pi(\omega)$, $T_2 = \pi(d\omega)$ for some $\omega \in \Omega^k(\mathcal{A})$ if and only if there exist sections ρ_1 and ρ_2 of C^k and C^{k+1} such that*

$$T_j = \gamma(\rho_j) \quad (j = 1, 2), \quad \mathrm{d}_c \sigma_k(\rho_1) = i\sigma_{k+1}(\rho_2).$$

Here $\sigma_k(\rho_1)$ is an ordinary k-form on M and d_c is the classical differential. Note that for $k > \dim M$ one has $\sigma_k(\rho) = 0$. The proof is straightforward.

We can now easily determine the graded differential algebra Ω_D^*. First, let us identify $\pi(\Omega^k)$ with the space of sections of C^k; Lemma 6 then shows that

$$\pi(\mathrm{d}(J_0 \cap \Omega^{k-1})) = \ker \sigma_k.$$

(If ρ is a section of C^k with $\sigma_k(\rho) = 0$ then the pair of sections $\rho_1 = 0$, $\rho_2 = \rho$ of C^{k-1} and C^k fulfills the condition of Lemma 6, so that $\rho = \pi(\mathrm{d}\omega)$ for some ω, $\pi(\omega) = 0$.) Thus, σ_k is an isomorphism, $\Omega_D^k \cong$ sections of $\wedge_{\mathbb{C}}^k(T^*)$, which, again by Lemma 6, commutes with the differential. We can then state:

Formula 3. *The mapping $a^0 \mathrm{d}a^1 \cdots \mathrm{d}a^n \mapsto a^0 \mathrm{d}_c a^1 \cdots \mathrm{d}_c a^n$ for $a^j \in \mathcal{A}$ extends to a canonical isomorphism of the differential graded algebra Ω_D^* with the de Rham algebra of differential forms on M. Under this isomorphism, the inner product on Ω_D^k is $c(d)^{-1}$ times the Riemannian inner product of k-forms*

$$\langle \omega, \omega' \rangle = \int_M \omega \wedge *\omega'.$$

The last equality follows from the computation of the Dixmier trace for the operator on $\mathfrak{H} = L^2(M, S)$ associated with a section ρ of the bundle C of Clifford algebras (cf. Section IV.2):

$$\mathrm{Tr}_\omega(\rho|D|^{-d}) = 2^{-d}\Gamma\left(\frac{d}{2} + 1\right)^{-1} \pi^{-d/2} \int_M \mathrm{trace}(\rho(p))\mathrm{d}\nu(p).$$

As an immediate corollary of Formula 3, we have

$$\mathrm{YM}(V) = c(d)^{-1} \int \|\mathrm{d}V\|^2 \mathrm{d}\nu.$$

Let us now extend the definition of the action YM to connections on arbitrary Hermitian vector bundles.

First of all, we need to express in algebraic terms — i.e., using only the involutive algebra $\mathcal{A}$ — the notion of a Hermitian vector bundle over M. A vector bundle E is entirely characterized by the right $\mathcal{A}$-module $\mathcal{E}$ of sections of E with the same regularity as the elements of $\mathcal{A}$; the local triviality of E and the finite-dimensionality of its fibers translate algebraically into the statement that $\mathcal{E}$ is a direct summand of a free module $\mathcal{A}^N$ for some finite N, or, in fancier terms, that $\mathcal{E}$ is a finitely generated projective module over $\mathcal{A}$.

The Hermitian structure on E, that is, the inner product $\langle \xi, \eta \rangle_p$ on each fiber E_p, yields a sesquilinear map

$$\langle\ ,\ \rangle : \mathcal{E} \times \mathcal{E} \to \mathcal{A},$$

given by $\langle \xi, \eta \rangle(p) = \langle \xi(p), \eta(p) \rangle_p$. The mapping $\langle\ ,\ \rangle$ satisfies the following conditions:

1) $\langle \xi a, \eta b \rangle = a^* \langle \xi, \eta \rangle b \quad (\forall \xi, \eta \in \mathcal{E},\ a, b \in \mathcal{A})$,
2) $\langle \xi, \xi \rangle \geq 0 \quad (\forall \xi \in \mathcal{E})$,
3) $\mathcal{E}$ is self-dual for $\langle\ ,\ \rangle$.

Thus, the Hermitian vector bundles over M correspond bijectively to the Hermitian, finitely generated projective modules over $\mathcal{A}$ in the following sense:

Definition 7. *Let $\mathcal{A}$ be a unital $*$-algebra and let $\mathcal{E}$ be a finitely generated projective module over $\mathcal{A}$. A* Hermitian structure *on $\mathcal{E}$ is given by a sesquilinear mapping $\langle\ ,\ \rangle : \mathcal{E} \times \mathcal{E} \to \mathcal{A}$ satisfying the above conditions* 1, 2 *and* 3.

We shall use this notion only in the case where $\mathcal{A}$ is a subalgebra stable under the holomorphic functional calculus in a C^*-algebra, in which case all reasonable notions of positivity coincide in $\mathcal{A}$.

In this case, all Hermitian structures on a given finitely generated projective module $\mathcal{E}$ over $\mathcal{A}$ are isomorphic to each other and are thus obtained as follows: one writes $\mathcal{E}$ as a direct summand $\mathcal{E} = e\mathcal{A}^N$ of a free module $\mathcal{E}_0 = \mathcal{A}^N$, where the idempotent $e \in M_N(\mathcal{A})$ is *self-adjoint*, and one then restricts to $\mathcal{E}$ the Hermitian structure on $\mathcal{A}^N$ given by

$$\langle \xi, \eta \rangle = \sum \xi_i^* \eta_i \in \mathcal{A} \quad \forall \xi = (\xi_i), \eta = (\eta_i) \in \mathcal{A}^N.$$

The algebra $\mathrm{End}_{\mathcal{A}}(\mathcal{E})$ of endomorphisms of a Hermitian, finitely generated projective module $\mathcal{E}$ has a natural involution, given by

$$\langle T^* \xi, \eta \rangle = \langle \xi, T\eta \rangle \quad \forall \xi, \eta \in \mathcal{E}.$$

With this involution, $\mathrm{End}_{\mathcal{A}}(\mathcal{E})$ is isomorphic to the reduced $*$-algebra $eM_N(\mathcal{A})e$.

As above, we now let $(\mathfrak{H}, D)$ be a K-cycle over $\mathcal{A}$, and Ω_D^1 the $\mathcal{A}$-bimodule of operators in $\mathfrak{H}$ of the form $V = \sum a_i[D, b_i]$, with $a_i, b_i \in \mathcal{A}$.

Definition 8. *Let $\mathcal{E}$ be a Hermitian, finitely generated projective module over $\mathcal{A}$. A* connection *on $\mathcal{E}$ is given by a linear mapping $\nabla : \mathcal{E} \to \mathcal{E} \otimes_{\mathcal{A}} \Omega_D^1$ such that*

$$\nabla(\xi a) = (\nabla \xi)a + \xi \otimes \mathrm{d}a \quad \forall \xi \in \mathcal{E},\ a \in \mathcal{A}.$$

A connection ∇ is compatible *(with the metric) if and only if*

$$\langle \xi, \nabla \eta \rangle - \langle \nabla \xi, \eta \rangle = \mathrm{d}\langle \xi, \eta \rangle \quad \forall \xi, \eta \in \mathcal{E}.$$

The last equality has a clear meaning in Ω_D^1. In the computations, one should remember that $(\mathrm{d}a)^* = -\mathrm{d}a^*$ $(\forall a \in \mathcal{A})$, and if $\nabla \xi = \sum \xi_i \otimes \omega_i$, $\omega_i \in \Omega_D^1$, then $\langle \nabla \xi, \eta \rangle = \sum \omega_i^* \langle \xi_i, \eta \rangle$.

Such connections always exist; for $\mathcal{E}$ expressed as $e\mathcal{A}^N$ as above, one may take ∇ as the "Grassmannian" connection

$$(\nabla_0\xi) = e\eta, \text{ where } \eta_j = d\xi_j.$$

Two compatible connections ∇ and ∇' on $\mathcal{E}$ can only differ by an element $\Gamma \in \mathrm{Hom}_{\mathcal{A}}(\mathcal{E}, \mathcal{E} \otimes_{\mathcal{A}} \Omega_D^1)$.

As in Proposition 5, we shall now give two equivalent definitions of the action functional $\mathrm{YM}(\nabla)$ on the affine space $C(\mathcal{E})$ of compatible connections.

The group $\mathcal{U}(\mathcal{E})$ of unitary automorphisms of $\mathcal{E}$, $\mathcal{U}(\mathcal{E}) = \{u \in \mathrm{End}_{\mathcal{A}}(\mathcal{E});\ uu^* = u^*u = 1\}$, acts by conjugation $\gamma_u(\nabla) = u\nabla u^*$ on the space $C(\mathcal{E})$. To define the curvature θ of a connection ∇, one first extends ∇ to a unique linear mapping $\tilde{\nabla}$ from $\tilde{\mathcal{E}}$ to $\tilde{\mathcal{E}}$, $\tilde{\mathcal{E}} = \mathcal{E} \otimes_{\mathcal{A}} \Omega_D^*$, such that

$$\tilde{\nabla}(\xi \otimes \omega) = (\nabla\xi)\omega + \xi \otimes d\omega \quad \forall \xi \in \mathcal{E},\ \omega \in \Omega_D^*,$$

and one checks that this mapping satisfies

$$\tilde{\nabla}(\eta\omega) = (\tilde{\nabla}\eta)\omega + (-1)^{\partial\eta}\eta d\omega$$

for every homogeneous $\eta \in \tilde{\mathcal{E}}$ and $\omega \in \Omega_D^*$. It then follows that $\theta = \tilde{\nabla}^2$ is an endomorphism of the right Ω_D^*-module $\tilde{\mathcal{E}}$; it is determined by its restriction to $\mathcal{E}$, again denoted θ,

$$\theta \in \mathrm{Hom}_{\mathcal{A}}(\mathcal{E}, \mathcal{E} \otimes_{\mathcal{A}} \Omega_D^2).$$

Next, using the inner product on Ω_D^2 and the Hermitian structure on $\mathcal{E}$, one has a natural inner product on

$$\mathrm{Hom}_{\mathcal{A}}(\mathcal{E}, \mathcal{E} \otimes_{\mathcal{A}} \Omega_D^2).$$

Using this, we make the following definition.

Definition 9. $\mathrm{YM}(\nabla) = \langle\theta, \theta\rangle$.

By construction, this action is gauge invariant, positive and quartic. It is moreover obvious from the above Formula 3 in the case of the Dirac K-cycle on a Riemannian spin manifold that one has:

Formula 4. *Let M be a Riemannian spin manifold with its Dirac K-cycle $(\mathfrak{H}, D)$. Then, the notion of connection* (Definition 8) *is the usual one, and*

$$\mathrm{YM}(\nabla) = c(d)^{-1} \int_M \|\theta\|_{\mathrm{HS}}^2 d\nu,$$

where θ is the usual curvature of ∇.

Thus, we recover in this case the usual Yang-Mills action. For computational purposes, and also to see the curvature as an operator in $\mathfrak{H}$, we shall now mention the easy adaptation of Proposition 5, 4) to the general case.

First of all, any compatible connection in the sense of Definition 8 is the composition with π of a *universal compatible connection,*

Definition 10. *Let $\mathcal{E}$ be a Hermitian, finitely generated projective module over $\mathcal{A}$. Then a* universal compatible connection *on $\mathcal{E}$ is a linear mapping ∇ of $\mathcal{E}$ to $\mathcal{E} \otimes_{\mathcal{A}} \Omega^1$ such that:*

a) $\nabla(\xi a) = (\nabla\xi)a + \xi \otimes da \quad (\forall \xi \in \mathcal{E},\ a \in \mathcal{A})$,

b) $\langle \xi, \nabla\eta\rangle - \langle \nabla\xi, \eta\rangle = d\langle \xi, \eta\rangle \quad (\forall \xi, \eta \in \mathcal{E})$.

To see the surjectivity of the mapping $\pi : CC(\mathcal{E}) \to C(\mathcal{E})$, where $CC(\mathcal{E})$ is the space of universal compatible connections, it is enough to check that the special Grassmannian connection ∇_0 is of this form and that π is a surjection of Ω^1 onto $\pi(\Omega^1)$. Next (cf. Chapter III Section 3), a universal compatible connection extends uniquely as a linear mapping

$$\tilde{\nabla} : \mathcal{E} \otimes_{\mathcal{A}} \Omega^* \to \mathcal{E} \otimes_{\mathcal{A}} \Omega^*$$

such that $\tilde{\nabla}$ is equal to ∇ on $\mathcal{E} \otimes 1$ and such that

$$\tilde{\nabla}(\eta\omega) = (\tilde{\nabla}\eta)\omega + (-1)^{\deg \eta}\eta d\omega$$

for every homogeneous η in $\mathcal{E} \otimes_{\mathcal{A}} \Omega^*$ and $\omega \in \Omega^*$.

The curvature $\theta = \tilde{\nabla}^2$ is then an endomorphism of the induced module $\tilde{\mathcal{E}} = \mathcal{E} \otimes_{\mathcal{A}} \Omega^*$ over Ω^*, and $\pi(\theta)$ makes sense as a bounded operator in the Hilbert space $\mathcal{E} \otimes_{\mathcal{A}} \mathfrak{H}$, as does the following operator D_∇ (cf. Section 3):

$$D_\nabla(\xi \otimes \eta) = \xi \otimes D\eta + ((1 \otimes \pi)\nabla\xi)\eta \qquad \forall \xi \in \mathcal{E},\ \eta \in \mathfrak{H};$$

the analogue of the action I of Proposition 5 3) is then given by

$$I(\nabla) = \mathrm{Tr}_\omega(\pi(\theta)^2 |D_\nabla|^{-d}).$$

One then proves in the same way that

Proposition 11. *For any compatible connection $\nabla \in C(\mathcal{E})$, one has*

$$\mathrm{YM}(\nabla) = \inf\{I(\nabla_1);\ \pi(\nabla_1) = \nabla\}.$$

Remark 12. All the above considerations apply to noncompact spaces as well, i.e. the $*$-algebra $\mathcal{A}$ is no longer unital. The summability hypothesis $(D-\lambda)^{-1} \in \mathcal{L}^{(d,\infty)}$ $\forall \lambda \notin \mathbb{R}$ is being replaced by $a(D-\lambda)^{-1} \in \mathcal{L}^{(d,\infty)}$ $\forall \lambda \notin \mathbb{R}, a \in \mathcal{A}$.

VI.2. Positivity in Hochschild Cohomology and Inequalities for the Yang–Mills Action

The residue theorem of Section IV.2 is a basic result that allows one to express the Hochschild cohomology class of an (n, ∞)-summable K-cycle by means of the Dixmier trace of suitable products of commutators. However, the Dixmier trace enjoys the fundamental property of being *positive*:

$$\mathrm{Tr}_\omega(T) \geq 0 \quad \forall T \in \mathcal{L}^{(1,\infty)}(\mathfrak{H}),\ T \geq 0.$$

This shows that in the even case, the Hochschild class of an (n, ∞)-summable K-cycle $(\mathfrak{H}, D, \gamma)$ in fact has a *positive* representative, given by the following positive Hochschild cocycle:

$$(*)$$
$$\psi_\omega(a^0, \dots, a^n)$$
$$= (-1)^{n/2}\Gamma\left(\tfrac{n}{2}+1\right)\mathrm{Tr}_\omega((1+\gamma)a^0[D,a^1]\cdots[D,a^n]D^{-n}) \quad \forall a^0, \dots, a^n \in \mathcal{A}.$$

By definition (cf. [Co-C]) a Hochschild cocycle ψ on a $*$-algebra $\mathcal{A}$ is *positive* if it has *even* dimension $n = 2m$ and the following equality defines a positive sesquilinear form on the vector space $\mathcal{A}^{\otimes(m+1)}$:

$$\langle a^0 \otimes a^1 \otimes \cdots \otimes a^m, b^0 \otimes b^1 \otimes \cdots \otimes b^m\rangle = \psi(b^{0*}a^0, a^1, \dots, a^m, b^{m*}, \dots, b^{1*})$$

for any $a^j, b^j \in \mathcal{A}$. In general the positive Hochschild cocycles form a convex cone

$$Z^n_+(\mathcal{A}, \mathcal{A}^*) \subset Z^n(\mathcal{A}, \mathcal{A}^*)$$

in the vector space Z^n of Hochschild cocycles on $\mathcal{A}$.

To familiarize ourselves with the notion of positivity, we shall consider the example where $\mathcal{A}$ is the $*$-algebra of smooth functions on a compact manifold M (and of course we only consider continuous multilinear forms on $\mathcal{A}$).

For $n = 0$, the space $Z^0 = Z^0(\mathcal{A}, \mathcal{A}^*)$ is the space of 0-dimensional currents on M, and Z^0_+ is the cone of *positive measures* in the usual sense.

For $n = 2$, a Hochschild cocycle *class* C is characterized by a 2-dimensional de Rham current C which is obtained directly by antisymmetrization from any cocycle $\varphi \in C$:

$$\langle \mathrm{C}, f^0 df^1 \wedge df^2\rangle = \frac{1}{2}(\varphi(f^0, f^1, f^2) - \varphi(f^0, f^2, f^1)) \quad \forall f^j \in \mathcal{A}.$$

In the class C there is a unique element Φ_C that is skew-symmetric in the last two variables; it is given by

$$\Phi_C(f^0, f^1, f^2) = \langle \mathrm{C}, f^0 df^1 \wedge df^2\rangle \quad \forall f^j \in \mathcal{A}.$$

The mapping $C \mapsto \Phi_C$ is the natural cross-section that one usually uses to identify de Rham currents with Hochschild cocycles rather than Hochschild cohomology classes. It does, however, have one bad feature: the cocycle $\lambda\Phi_C$,

$\lambda \in \mathbb{C}$, is *never* positive. (Otherwise the equality $L_f(g) = \langle \mathbf{C}, g df \wedge d\bar{f} \rangle$ would define a positive measure; but since $L_{\bar{f}} = -L_f$ one has $L_f = 0$ and $\mathbf{C} = 0$.) This shows that if one sticks to this canonical representative, the notion of positivity remains hidden. Let M be a 2-dimensional oriented compact manifold, and take for the class C the class of the de Rham current $\mathbf{C}$

$$\langle \mathbf{C}, f^0\, df^1 \wedge df^2 \rangle = \frac{-1}{2\pi i} \int_M f^0\, df^1 \wedge df^2 \qquad \forall f^j \in C^\infty(M).$$

As we shall see, the positive representatives, $\varphi \in Z_+^2 \cap C$, of this class will correspond to conformal structures g on M. More precisely, since $Z_+^2 \cap C$ is *convex*, the correspondence will be established between conformal structures on M and the *extreme points* of $Z_+^2 \cap C$ (cf. Figure 2).

Thus, let g be a conformal structure on M or equivalently, since M is oriented, a *complex* structure. Then, to the Lelong notion of positive current ([Le]) there corresponds the positivity in our sense of the following Hochschild 2-cocycle:

$$\varphi_g(f^0, f^1, f^2) = \frac{i}{\pi} \int_M f^0 \partial f^1 \wedge \bar{\partial} f^2,$$

where ∂ and $\bar{\partial}$ are inherited from the complex structure.

An immediate check shows that the antisymmetrization of φ_g is $\frac{1}{(2\pi i)}[M]$, where $i = \sqrt{-1}$,

$$\frac{1}{2}(\varphi_g(f^0, f^1, f^2) - \varphi_g(f^0, f^2, f^1)) = \frac{-1}{2\pi i} \int_M f^0 df^1 \wedge df^2.$$

The positivity of φ_g corresponds to the Dirichlet Hilbert space structure on the space of forms of type $(1,0)$. It is also clear that the mapping $g \mapsto \varphi_g$ is an injection, since one can read off from φ_g what it means for a function f to be holomorphic in a given small open set $U \subset M$.

Indeed, from the positive inner product on $\mathcal{A} \otimes \mathcal{A}$ associated with φ_g one reconstructs the $\mathcal{A}$-bimodule of L^2 forms of type $(1,0)$ as well as the complex differentiation ∂:

$$\mathcal{A} \xrightarrow{\partial} L^2(M, \Lambda^{(1,0)}).$$

Each φ_g is an extreme point of the convex set $Z_+^2 \cap C$, and, conversely, the exposed points of this convex set can be determined as follows: for any element of the *dual cone* $(Z_+^2)^\wedge$ of Z_+^2, of the form

$$G = \sum_{\mu,\nu=1}^{d} g_{\mu\nu} dx^\mu (dx^\nu)^* \in \Omega^2(\mathcal{A}),$$

where $g_{\mu\nu}$ is a *positive* element of $M_d(\mathcal{A})$, one can show, assuming a suitable condition of nondegeneracy, that the linear form

$$\langle G, \varphi \rangle = \sum \varphi(g_{\mu\nu}, x^\mu, (x^\nu)^*)$$

attains its minimum at a unique point in $Z_+^2 \cap C$, and that this point is equal to φ_g, where g is the conformal structure on M associated with the classical Riemannian metric

$$g = \sum g_{\mu\nu} \mathrm{d}x^\mu (\mathrm{d}x^\nu)^*.$$

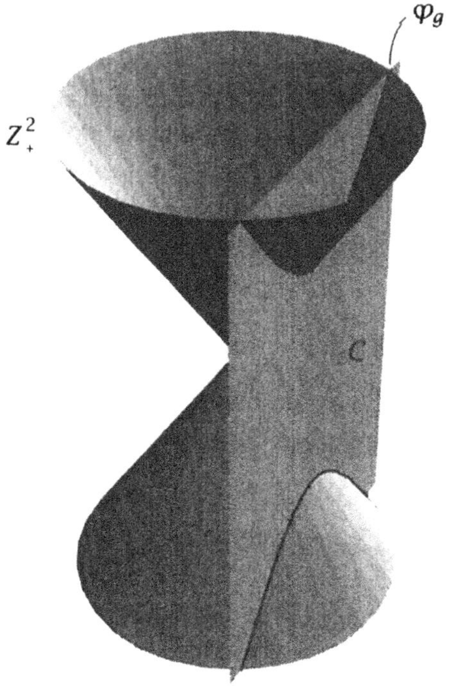

Figure VI.2. Conformal structures and extreme points of $Z_+^2 \cap C$

This allows us to understand the complex structures on M as the solutions of a variational problem involving the fundamental class of M and *positivity* in Hochschild cohomology. This problem is by no means restricted in its formulation to the *commutative* case, but it requires the notion of fundamental class in cyclic cohomology. It can be taken as a starting point for developing complex geometry in the noncommutative case.

Finally, let us note that the cocycle φ_g can be expressed in terms of the Dirac operator D for g on M:

$$\varphi_g(f^0, f^1, f^2) = -\mathrm{Tr}_\omega((1+\gamma) f^0 [D, f^1][D, f^2] D^{-2}) \quad \forall f^j \in \mathcal{A},$$

where the $\mathbb{Z}/2$ grading γ of the spinor bundle is provided by the orientation of M and the choice of spin structure is irrelevant.

This shows that, for $n = 2$, the formula $(*)$ for a positive representative of the Hochschild class of a $(2, \infty)$-summable K-cycle is the relevant construction of positive Hochschild cocycles.

We shall now concentrate on the case $n = 4$. Let us first state an easy proposition for the case of Riemannian spin manifolds of dimension 4:

Proposition 1. *Let M be a 4-dimensional compact Riemannian spin manifold and let D be the corresponding Dirac operator in the Hilbert space $\mathfrak{H} = L^2(M, S)$ with the $\mathbb{Z}/2$-grading γ. Then the Hochschild class of the following positive Hochschild 4-cocycle on $\mathcal{A} = C^\infty(M)$ is $(2\pi i)^{-2}$ times the fundamental class of M:*

$$\psi_\omega(f^0, \dots, f^4) = 2\mathrm{Tr}_\omega((1 + \gamma)f^0[D, f^1] \cdots [D, f^4]D^{-4}).$$

The cocycle ψ_ω is independent of ω and of the spin structure on M; it depends only on the conformal structure of M and is equal to

$$\frac{1}{16\pi^2}\int_M \mathrm{tr}((1 + \gamma)f^0 df^1 \cdot df^2 \cdot df^3 \cdot df^4) dv = \varphi_g(f^0, \dots, f^4) \quad \forall f^j \in \mathcal{A}.$$

In the last expression, the trace tr is the natural trace on the Clifford algebra C_x at every point $x \in M$, the differentials df^j are regarded as sections of the Clifford algebra bundle, and dv is the Riemannian volume element on M.

We shall now consider the general case of a $(4, \infty)$-summable K-cycle denoted $(\mathfrak{H}, D, \gamma)$ on a $*$-algebra $\mathcal{A}$ and show, using the *positivity* of ψ_ω, how to extend the usual inequality $|c_2(E) - \frac{1}{2}c_1(E)^2| \le \mathrm{YM}(\nabla)$ between Chern classes of vector bundles E and the value of the Yang-Mills action on arbitrary compatible connections ∇ on E.

The action functional $\mathrm{YM}(\nabla)$ of Section 1 Definition 9 has the following homogeneity property

If D is replaced by λD , $\lambda > 0$, then $\mathrm{YM}(\nabla)$ is replaced by λ^{4-d} $\mathrm{YM}(\nabla)$.

This shows that we can hope for an inequality relating it to topological invariants of finite projective modules only if $d = 4$, which we shall assume from now on.

Let $\mathcal{E}$ be a Hermitian finite projective module over $\mathcal{A}$ and, as in Section 1, let π be the natural surjection

$$\pi : CC(\mathcal{E}) \to C(\mathcal{E})$$

from the space of universal compatible connections $CC(\mathcal{E})$ to the space $C(\mathcal{E})$ of compatible connections.

Of course, as in Section 1, the group $\mathcal{U}(\mathcal{E})$ of unitary endomorphisms of $\mathcal{E}$ acts by gauge transformations $\gamma_u(\nabla) = u\nabla u^*$ on the space of universal compatible connections and the curvature θ makes sense, as an element of $\mathrm{Hom}_{\mathcal{A}}(\mathcal{E}, \mathcal{E} \otimes_{\mathcal{A}} \Omega^2(\mathcal{A}))$, and is covariant under gauge transformations. The mapping $\nabla \mapsto \pi(\nabla)$ is covariant and the two curvatures are related by $\theta_{\pi \cdot \nabla} = \pi(\theta_\nabla)$.

We shall now assume that the K-cycle $(\mathfrak{H}, D)$ is *even* (i.e., we are given a $\mathbb{Z}/2$-grading γ) and that it is $(4,\infty)$-summable. We now describe two traces on $\Omega(\mathcal{A})$. The first is positive and yields the action $\mathrm{YM}(\nabla)$ of Section 1; the second is *closed* and yields topological invariants of finitely generated projective modules $\mathcal{E}$.

Lemma 2. *If $(\mathfrak{H}, D, \gamma)$ is a $(4,\infty)$-summable K-cycle over $\mathcal{A}$, then the following equalities define traces on $\Omega^{\mathrm{even}}(\mathcal{A})$:*

1) $\tau(a^0 da^1 \cdots da^4) = \mathrm{Tr}_\omega(a^0[D,a^1]\cdots[D,a^4]D^{-4}) \quad \forall a^j \in \mathcal{A}$;

2) $\Phi(a^0 da^1 \cdots da^4) = \mathrm{Tr}_\omega(\gamma a^0[D,a^1]\cdots[D,a^4]D^{-4}) \quad \forall a^j \in \mathcal{A}$.

In fact, both are traces on $\Omega^*(\mathcal{A})$ but Φ is a $\mathbb{Z}/2$-graded trace, i.e.,

$$\Phi(\omega_2\omega_1) = (-1)^{\partial\omega_1\partial\omega_2}\Phi(\omega_1\omega_2) \quad \forall \omega_j \in \Omega^*(\mathcal{A}).$$

Let $\mathcal{E}$ be a Hermitian, finitely generated projective module over $\mathcal{A}$, and extend τ to a unique trace $\tilde{\tau}$ on the endomorphisms of the induced module $\mathcal{E}\otimes_{\mathcal{A}}\Omega^*(\mathcal{A})$; then with the notation of Proposition 1.11 one has

$$I(\nabla) = \tilde{\tau}(\theta^2) \qquad \forall \nabla \in CC(\mathcal{E}).$$

Since the operator $(1 \pm \gamma)$ is positive and commutes with $\pi(\Omega^{\mathrm{even}}(\mathcal{A}))$, we get the general inequality

$$\tilde{\tau}(\theta^2) \geq |\tilde{\Phi}(\theta^2)| \quad \forall \nabla \in CC(\mathcal{E}).$$

It remains to understand the topological meaning of the term $\tilde{\Phi}(\theta^2)$. This will follow from the pairing between K-theory and cyclic cohomology as expressed in terms of connections and curvature (cf. Chapter III Section 3) but requires the following additional hypothesis. We know that if B is the natural boundary operator

$$B : H^4(\mathcal{A}, \mathcal{A}^*) \to HC^3(\mathcal{A}),$$

then $B\Phi = 0$, simply because, by the residue formula (Chapter IV Section 2), the Hochschild class of Φ is (assuming $H^*(\mathcal{A},\mathcal{A}^*)$ to be Hausdorff) the same as the Hochschild class of a cyclic cocycle, the character of $(\mathfrak{H}, D, \gamma)$. However, we shall need the following hypothesis:

Hypothesis 3. $B\Phi = 0$ *as a cochain.*

We shall of course have to check the hypothesis in the relevant examples, but note already that in the case of the Dirac operator on a Riemannian manifold M of dimension $d = 4$, we have

$$\Phi(f^0, \ldots, f^4) = -\frac{1}{8\pi^2}\int f^0 df^1 \wedge \cdots \wedge df^4,$$

which is a cyclic cocycle and therefore satisfies Hypothesis 3.

Note also that, in general, since $B_0\Phi$ is already cyclic,

$$B_0\Phi(a^0,a^1,a^2,a^3) = \Phi(1,a^0,a^1,a^2,a^3) = \mathrm{Tr}_\omega(\gamma[D,a^0]\cdots[D,a^4]D^{-4});$$

the condition $B\Phi = 0$ means in fact that $B_0\Phi = 0$, i.e., that Φ is a cyclic cocycle. The corresponding cyclic cohomology class

$$[\Phi] \in HC^4(\mathcal{A})$$

is not, in general, the same as the cyclic cohomology class $\mathrm{ch}_*(\mathfrak{H},D,\gamma)$ of the Chern character of the K-cycle, but differs from it by lower-dimensional classes, i.e., by $S(HC^2(\mathcal{A}))$.

We may now conclude using 1.11:

Theorem 4. *Let $\mathcal{A}$ be a $*$-algebra, $(\mathfrak{H},D,\gamma)$ a $(4,\infty)$-summable K-cycle over $\mathcal{A}$, and φ_ω the Hochschild cocycle*

$$\varphi_\omega(a^0,\ldots,a^4) = 2\mathrm{Tr}_\omega(\gamma a^0[D,a^1]\cdots[D,a^4]D^{-4}).$$

Assume that $B\varphi_\omega = 0$. Then, for any Hermitian, finitely generated projective module $\mathcal{E}$ over $\mathcal{A}$, we have

$$|\langle[\mathcal{E}],\varphi_\omega\rangle| \leq \mathrm{YM}(\nabla) \quad \forall\nabla \in C(\mathcal{E}).$$

The left-hand side is the pairing between K-theory and cyclic cohomology (Chapter III Section 3). In the case of the Dirac operator on a compact spin manifold M, the cocycle φ_ω is $(2\pi i)^{-2}$ times the fundamental homology class of M and the action $\mathrm{YM}(\nabla)$ is $(8\pi^2)^{-1}$ times the usual Yang-Mills action. Thus, the above inequality is the usual one: $|c_2(E) - \frac{1}{2}c_1(E)^2| \leq \mathrm{YM}(\nabla)$ for any compatible connection ∇ on a Hermitian vector bundle E on M.

VI.3. Product of the Continuum by the Discrete and the Symmetry Breaking Mechanism

We have shown how to extend, to our context of finitely summable K-cycles $(\mathfrak{H},D)$ over an algebra $\mathcal{A}$, the concepts of gauge potentials and Yang-Mills action, as well as the way in which this action is related to a topological action in the case of dimension 4. In this section we shall give several examples of computations of this action. We first briefly recall its definition and use the opportunity to add to it a fermionic part.

We are given a $*$-algebra $\mathcal{A}$ and a (d,∞)-summable K-cycle $(\mathfrak{H},D)$ over $\mathcal{A}$. This gives us a representation on $\mathfrak{H}$ of the reduced universal differential algebra $\Omega^*\mathcal{A}$:

$$\pi(a^0da^1\cdots da^k) = a^0[D,a^1]\cdots[D,a^k] \quad \forall a^j \in \mathcal{A},$$

which defines a quotient differential graded algebra

$$\Omega_D^*(\mathcal{A}) = \Omega^*(\mathcal{A})/J, \quad J = J_0 + dJ_0, \quad J_0^{(k)} = \Omega^k \cap \ker \pi.$$

A compatible connection ∇ on a Hermitian, finitely generated projective module $\mathcal{E}$ over $\mathcal{A}$ is given by a linear mapping

$$\nabla : \mathcal{E} \to \mathcal{E} \otimes_{\mathcal{A}} \Omega_D^1$$

which satisfies the Leibniz rule and is compatible with the inner product. The affine space $C(\mathcal{E})$ of such connections is acted on by the unitary group $\mathcal{U}(\mathcal{E})$ of the $*$-algebra of endomorphisms $\mathrm{End}_{\mathcal{A}}(\mathcal{E})$. This action transforms the curvature $\theta = \nabla^2$ of such connections covariantly, and

$$\mathrm{YM}(\nabla) = \mathrm{Tr}_\omega(\pi(\theta)^2 |D|^{-d})$$

is a gauge invariant quartic positive action on $C(\mathcal{E})$ (cf. Section 1).

In the case of the trivial module $\mathcal{E} = \mathcal{A}$ (with the right action of $\mathcal{A}$ on itself), a vector potential is a selfadjoint element V of Ω_D^1, and the following expression is also gauge invariant

$$\langle \psi, (D + \pi(V))\psi \rangle \quad \psi \in \mathfrak{H}, \ V \in \Omega_D^1$$

where the unitary group $\mathcal{U} = \mathcal{U}(\mathcal{E}) = \mathcal{U}(\mathcal{A})$ acts on $\mathfrak{H}$ by restriction of the action of $\mathcal{A}$, whereas it acts on vector potentials by gauge transformations:

$$\gamma_u(V) = u\,d(u^*) + uVu^*, \quad u \in \mathcal{U}, \ V \in \Omega_D^1.$$

This is the fermionic action that we want to add to the action $\mathrm{YM}(\nabla)$; it extends to the case of arbitrary Hermitian, finitely generated projective modules $\mathcal{E}$ over $\mathcal{A}$, by means of the next lemma.

Lemma 1. *Let $\mathcal{A}$, $\mathcal{E}$, and $(\mathfrak{H}, D)$ be as above. Then:*

1) *The tensor product $\mathcal{E} \otimes_{\mathcal{A}} \mathfrak{H}$ is a Hilbert space with inner product given by*

$$\langle \xi_1 \otimes \eta_1, \xi_2 \otimes \eta_2 \rangle = \langle \eta_1, (\xi_1, \xi_2)\eta_2 \rangle \quad \forall \xi_j \in \mathcal{E}, \ \eta_j \in \mathfrak{H}.$$

2) *For any compatible connection ∇, the following equality defines a selfadjoint operator D_∇ in the above Hilbert space:*

$$D_\nabla(\xi \otimes \eta) = \xi \otimes D\eta + ((1 \otimes \pi)\nabla\xi)\eta \quad \forall \xi \in \mathcal{E}, \ \eta \in \mathfrak{H}.$$

Thus, the fermionic action is now given by

$$\langle \psi, D_\nabla \psi \rangle, \qquad \in \mathcal{E} \otimes_{\mathcal{A}} \mathfrak{H}, \ \nabla \in C(\mathcal{E}),$$

and one checks that it is invariant under gauge transformations by elements of $\mathcal{U}(\mathcal{E})$.

The total action is

$$\mathcal{L}(\nabla, \psi) = \lambda \, \mathrm{YM}(\nabla) + \langle \psi, D_\nabla \psi \rangle,$$

where λ is a coupling constant.

We shall compute it in three cases: a) the discrete case of a 2-point space; b) the product case of a 4-dimensional manifold by case a); c) the irrational rotation algebra A_θ.

In order to see what the relevant concepts are in the 0-dimensional case a), we first need to discuss product spaces briefly. We are given two triples

$$(\mathcal{A}_1, \mathfrak{H}_1, D_1),\ (\mathcal{A}_2, \mathfrak{H}_2, D_2)$$

and we assume that one of them is *even*, i.e., that we are given a $\mathbb{Z}/2$-grading, say γ_1, on $\mathfrak{H}_1$. The product is then given by the triple $(\mathcal{A}, \mathfrak{H}, D)$, where

$$\mathcal{A} = \mathcal{A}_1 \otimes \mathcal{A}_2,\ \ \mathfrak{H} = \mathfrak{H}_1 \otimes \mathfrak{H}_2,\ \ D = D_1 \otimes 1 + \gamma_1 \otimes D_2.$$

This corresponds to the external product of K-cycles. There is an obvious notion of external product of Hermitian finitely generated projective modules over the $\mathcal{A}_j$.

Next, the formula $D^2 = D_1^2 \otimes 1 + 1 \otimes D_2^2$, which follows from the anti-commutation of D_1 with γ_1, shows that dimensions add up, that is, if D_j is (p_j, ∞)-summable then D is (p, ∞)-summable $p = p_1 + p_2$; moreover, once the limiting procedure Lim_ω is fixed, one can show that if one of the two terms is convergent then $\forall T_j \in \mathcal{L}(\mathfrak{H}_j)$,

$$\frac{\Gamma(\frac{p}{2}+1)}{\Gamma(\frac{p_1}{2}+1)\,\Gamma(\frac{p_2}{2}+1)}\,\mathrm{Tr}_\omega((T_1 \otimes T_2)|D|^{-p}) = \mathrm{Tr}_\omega(T_1|D_1|^{-p_1})\mathrm{Tr}_\omega(T_2|D_2|^{-p_2}) \quad .$$

All of this is true provided that $p_j \geq 1$, but in the case we are interested in (Example b)) we have $p_1 = 4$ and $p_2 = 0$. The corresponding formula turns out to be

$$\mathrm{Tr}_\omega((T_1 \otimes T_2)|D|^{-p}) = \mathrm{Tr}_\omega(T_1|D_1|^{-p})\mathrm{Trace}(T_2),$$

where Trace is the ordinary trace. To understand how this occurs, one can use the following general equality, assuming that $|D|^{-1} \in \mathcal{L}^{(p,\infty)}$:

$$\mathrm{Lim}_\omega\left(\frac{1}{\lambda}\mathrm{Trace}(Te^{-\lambda^{-2/p}D^2})\right) = \Gamma\left(\frac{p}{2}+1\right)\ \mathrm{Tr}_\omega(T|D|^{-p})$$

for all $T \in \mathcal{L}(\mathfrak{H})$.

Thus, the 0-dimensional analogue of the action $\mathrm{YM}(\nabla)$ is just given by $\mathrm{Trace}(\pi(\theta)^2)$.

2. Example a). The space we are dealing with has *two points* a and b. Thus, the algebra $\mathcal{A}$ is just the direct sum $\mathbb{C} \oplus \mathbb{C}$ of two copies of $\mathbb{C}$. An element $f \in \mathcal{A}$ is given by two complex numbers $f(a), f(b) \in \mathbb{C}$. Let $(\mathfrak{H}, D, \gamma)$ be a 0-dimensional K-cycle over $\mathcal{A}$; then $\mathfrak{H}$ is *finite-dimensional* and the representation of $\mathcal{A}$ in $\mathfrak{H}$ corresponds to a decomposition of $\mathfrak{H}$ as a direct sum $\mathfrak{H} = \mathfrak{H}_a + \mathfrak{H}_b$, with the action of $\mathcal{A}$ given by

$$f \in \mathcal{A} \mapsto \begin{bmatrix} f(a) & 0 \\ 0 & f(b) \end{bmatrix}.$$

If we write D as a 2×2 matrix in this decomposition,

$$D = \begin{bmatrix} D_{aa} & D_{ab} \\ D_{ba} & D_{bb} \end{bmatrix},$$

we can ignore the diagonal elements since they commute exactly with the action of $\mathcal{A}$. We shall thus take D to be of the form

$$D = \begin{bmatrix} 0 & D_{ab} \\ D_{ba} & 0 \end{bmatrix},$$

where $D_{ba} = D_{ab}^*$ and D_{ba} is a linear mapping from $\mathfrak{H}_a$ to $\mathfrak{H}_b$. We shall denote this linear mapping by M and take for γ the $\mathbb{Z}/2$-grading given by the matrix

$\begin{bmatrix} 1 & 0 \\ 0 & -1 \end{bmatrix} = \gamma$. We thus have

$$\mathcal{A} = \mathbb{C} \oplus \mathbb{C}, \quad \mathfrak{H} = \mathfrak{H}_a \oplus \mathfrak{H}_b, \quad D = \begin{bmatrix} 0 & M^* \\ M & 0 \end{bmatrix}, \quad \gamma = \begin{bmatrix} 1 & 0 \\ 0 & -1 \end{bmatrix}.$$

Let us first compute the metric on the space $X = \{a, b\}$, given by Formula 1 of Section 1. Given $f \in \mathcal{A}$, we have

$$[D, f] = \left[\begin{bmatrix} 0 & M^* \\ M & 0 \end{bmatrix}, \begin{bmatrix} f(a) & 0 \\ 0 & f(b) \end{bmatrix} \right]$$

$$= \begin{bmatrix} 0 & M^*(f(b) - f(a)) \\ -M(f(b) - f(a)) & 0 \end{bmatrix} = (f(b) - f(a)) \begin{bmatrix} 0 & M^* \\ -M & 0 \end{bmatrix}.$$

Thus, the norm of this commutator is $|f(b) - f(a)|\lambda$, where λ is the largest eigenvalue $\|M\|$ of $|M|$. Therefore

$$d(a, b) = \sup\{|f(a) - f(b)|;\ \|[D, f]\| \le 1\} = 1/\lambda.$$

Let us now determine the space of gauge potentials, the curvature and the action in two cases.

Case α): $\mathcal{E} = \mathcal{A}$ (i.e., the trivial bundle over X)

The space $\Omega^1(\mathcal{A})$ of universal 1-forms over $\mathcal{A}$ is given by the kernel of the multiplication $m : \mathcal{A} \otimes \mathcal{A} \to \mathcal{A}$, $m(f \otimes g) = fg$. These are functions on $X \times X$ that vanish on the diagonal. Thus, $\Omega^1(\mathcal{A})$ is a 2-dimensional space; if $e \in \mathcal{A}$ is the idempotent $e(a) = 1$, $e(b) = 0$, this space has as basis

$$ede, \ (1 - e)de,$$

so that every element of $\Omega^1(\mathcal{A})$ is of the form $\lambda ede + \mu(1 - e)d(1 - e)$. The differential $d : \mathcal{A} \to \Omega^1(\mathcal{A})$ is the finite difference

$$df = (\Delta f)ede - (\Delta f)(1 - e)d(1 - e), \quad \Delta f = f(a) - f(b);$$

it is a derivation with values in the bimodule $\Omega^1(\mathcal{A})$, which *fails to be commutative since $f\omega \neq \omega f$* for $\omega \in \Omega^1$, $f \in \mathcal{A}$.

Also, if $M \neq 0$ then the representation $\pi : \Omega^*(\mathcal{A}) \to \mathcal{L}(\mathfrak{H})$ is injective on $\Omega^1(\mathcal{A})$, so that $\Omega^1(\mathcal{A}) = \Omega_D^1(\mathcal{A})$. We have

$$\pi(\lambda e\mathrm{d}e + \mu(1-e)\mathrm{d}e) = \begin{bmatrix} 0 & -\lambda M^* \\ \mu M & 0 \end{bmatrix} \in \mathcal{L}(\mathfrak{H}).$$

A vector potential is given by a selfadjoint element of Ω_D^1, i.e., by a single complex number Φ, with

$$\pi(V) = \begin{bmatrix} 0 & \overline{\Phi} M^* \\ \Phi M & 0 \end{bmatrix}.$$

Since $V = -\overline{\Phi} e\mathrm{d}e + \Phi(1-e)\mathrm{d}e$, its curvature is

$$\theta = \mathrm{d}V + V^2 = -\overline{\Phi}\mathrm{d}e\mathrm{d}e - \Phi\mathrm{d}e\mathrm{d}e + (\overline{\Phi}e\mathrm{d}e - \Phi(1-e)\mathrm{d}e)^2,$$

and, using the equalities $e\mathrm{d}e(1-e) = e\mathrm{d}e$, $e(\mathrm{d}e)e = 0$, $(1-e)\mathrm{d}e(1-e) = 0$, we have

$$\theta = -(\Phi + \overline{\Phi})\mathrm{d}e\mathrm{d}e - (\Phi\overline{\Phi})\mathrm{d}e\mathrm{d}e.$$

Under the representation π, we have $\pi(\mathrm{d}e) = \begin{bmatrix} 0 & -M^* \\ M & 0 \end{bmatrix}$ and $\pi(\mathrm{d}e\mathrm{d}e) = \begin{bmatrix} -M^*M & 0 \\ 0 & -MM^* \end{bmatrix}$. This yields the formula for the Yang-Mills action

$$\mathrm{YM}(V) = 2(|\Phi + 1|^2 - 1)^2 \mathrm{Trace}((M^*M)^2),$$

where Φ is an arbitrary complex number. The action of the gauge group $\mathcal{U} = U(1) \times U(1)$ on the space of vector potentials, i.e., on Φ, is given by

$$\gamma_u(V) = u\mathrm{d}u^* + uVu^*;$$

for $u = u_a e + u_b(1-e)$, this gives

$$\begin{aligned}\gamma_u(V) =& (u_a e + u_b(1-e))(\overline{u}_a\mathrm{d}e - \overline{u}_b\mathrm{d}e) \\ &+ (u_a e + u_b(1-e))(-\overline{\Phi}e\mathrm{d}e + \Phi(1-e)\mathrm{d}e)(\overline{u}_a e + \overline{u}_b(1-e)) \\ =& e\mathrm{d}e + u_b\overline{u}_a(1-e)\mathrm{d}e - u_a\overline{u}_b e\mathrm{d}e - (1-e)\mathrm{d}e \\ &- u_a\overline{u}_b\overline{\Phi}e\mathrm{d}e + u_b\overline{u}_a\Phi(1-e)\mathrm{d}e,\end{aligned}$$

which, on the variable $1 + \Phi$, just means multiplication by $u_b\overline{u}_a$.

In this very simple case, our action $\mathrm{YM}(V)$ reproduces the usual situation of broken symmetries (Figure 3); it has a non-unique minimum, $|\Phi + 1| = 1$, which is acted upon nontrivially by the gauge group.

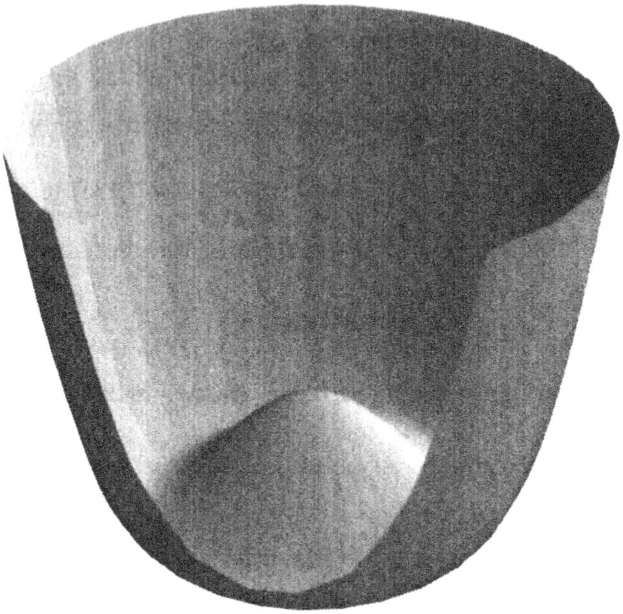

Figure VI.3. The potential $YM(V)$

The fermionic action is in this case given by

$$\langle \psi, (D + \pi(V))\psi \rangle,$$

where the operator $D + \pi(V)$ is equal to

$$\begin{bmatrix} 0 & M^* \\ M & 0 \end{bmatrix} + \begin{bmatrix} 0 & \overline{\Phi} M^* \\ \Phi M & 0 \end{bmatrix} = \begin{bmatrix} 0 & (1 + \overline{\Phi})M^* \\ (1 + \Phi)M & 0 \end{bmatrix},$$

which is a term of Yukawa type coupling the fields $(1 + \Phi)$ and ψ.

Case β): Let us take for $\mathcal{E}$ the nontrivial bundle over $X = \{a, b\}$ with fibers of dimensions n_a and n_b, respectively, over a and b. This bundle is nontrivial if and only if $n_a \neq n_b$; we shall consider the simplest case $n_a = 2$, $n_b = 1$. The finitely generated projective module $\mathcal{E}$ of sections is of the form

$$\mathcal{E} = f\mathcal{A}^2,$$

where the idempotent $f \in M_2(\mathcal{A})$ is given by the formula

$$f = \begin{bmatrix} (1,1) & 0 \\ 0 & (1,0) \end{bmatrix} = \begin{bmatrix} 1 & 0 \\ 0 & e \end{bmatrix}$$

in terms of the notation of α).

To the idempotent f there corresponds a particular compatible connection on $\mathcal{E}$, given by $\nabla_0\xi = f d\xi$ with the obvious notation. An arbitrary compatible connection on $\mathcal{E}$ has the form

$$\nabla\xi = \nabla_0\xi + \rho\xi,$$

where $\rho = \rho^*$ is a selfadjoint element of $M_2(\Omega_D^1(\mathcal{A}))$ such that $f\rho = \rho f = \rho$. If we write ρ as a matrix,

$$\rho = \begin{bmatrix} \rho_{11} & \rho_{12} \\ \rho_{21} & \rho_{22} \end{bmatrix},$$

these conditions read:

$$e\rho_{21} = \rho_{21},\ e\rho_{22} = \rho_{22} = \rho_{22}e,\ \rho_{12}e = \rho_{12};$$

thus we get

$$\rho_{11} = -\overline{\Phi}_1 ede + \Phi_1(1-e)de,\ \rho_{21} = \overline{\Phi}_2 ede,\ \rho_{12} = \rho_{21}^*,\ \rho_{22} = 0,$$

where Φ_1 and Φ_2 are arbitrary complex numbers.

The curvature θ is given by

$$\begin{aligned}\theta &= f d f d f + f d\rho f + \rho^2 \\ &= \begin{bmatrix} 0 & 0 \\ 0 & edede \end{bmatrix} + \begin{bmatrix} d\rho_{11} & (d\rho_{12})e \\ ed\rho_{21} & 0 \end{bmatrix} + \begin{bmatrix} \rho_{11}\rho_{11} + \rho_{12}\rho_{21} & \rho_{11}\rho_{12} \\ \rho_{21}\rho_{11} & \rho_{21}\rho_{12} \end{bmatrix}.\end{aligned}$$

An easy calculation gives the action $\mathrm{YM}(\nabla)$ in terms of the variables Φ_1, Φ_2:

$$\mathrm{YM}(\nabla) = \left(1 + 2(1 - (|\Phi_1 + 1|^2 + |\Phi_2|^2))^2\right)\mathrm{Tr}((M^*M)^2).$$

It is, by construction, invariant under the gauge group $U(1) \times U(2)$. What we learn in Example β) rather than in α) is that the choice of vacuum corresponds to a choice of connection minimizing the action, and in case β) there is really no preferred choice of ∇_0, the point 0 of the space of vector potentials (case α)) having no intrinsic meaning. In fact, the space of connections realizing the minimum of the action YM is a 3-sphere

$$\{(\Phi_1, \Phi_2) \in \mathbb{C}^2;\ |\Phi_1 + 1|^2 + |\Phi_2|^2 = 1\}$$

whose elements have the following meaning. Let E_a (resp. E_b) be the fiber of our Hermitian bundle over the point a (resp. b) of X; then $\dim E_a = 2$, $\dim E_b = 1$. As we saw above, the differential $d : \mathcal{A} \to \Omega^1(\mathcal{A})$ is the finite difference. One way to extend it to the bundle E is to use an isometry $u : E_b \to E_a$ and the formula

$$(\Delta\xi)_a = \xi_a - u\xi_b,\ (\Delta\xi)_b = \xi_b - u^*\xi_a.$$

All minimal connections ∇ are of the form

$$\nabla\xi = (\Delta\xi)_a \otimes ede + (\Delta\xi)_b \otimes (1-e)d(1-e).$$

Since the minimum of $\mathrm{YM}(\nabla)$ is > 0 we also see that the bundle E is not flat; it does not admit any compatible connection with vanishing curvature. Since the dimension of the space X is 0, the action YM has of course no topological meaning. However, we shall now return to the 4-dimensional case and work out the case of the product space in detail.

3. Example b). (4-dimensional Riemannian manifold V) $\times$ (2-point space X).

Let us fix the notation: V is a compact Riemannian spin 4-manifold, $\mathcal{A}_1$ the algebra of functions on V and $(\mathfrak{H}_1, D_1, \gamma_5)$ the Dirac K-cycle on $\mathcal{A}_1$, with its canonical $\mathbb{Z}/2$-grading γ_5 given by the orientation, let $\mathcal{A}_2$, $\mathfrak{H}_2$, D_2 be as in Example a) above, that is, $\mathcal{A}_2 = \mathbb{C} \oplus \mathbb{C}$, $\mathfrak{H}_2$ is the direct sum $\mathfrak{H}_{2,a} \oplus \mathfrak{H}_{2,b}$, and D_2 is given by the matrix

$$D_2 = \begin{bmatrix} 0 & M^* \\ M & 0 \end{bmatrix}.$$

Let $\mathcal{A} = \mathcal{A}_1 \otimes \mathcal{A}_2$, $\mathfrak{H} = \mathfrak{H}_1 \otimes \mathfrak{H}_2$ and $D = D_1 \otimes 1 + \gamma_5 \otimes D_2$.

The algebra $\mathcal{A}$ is commutative; it is the algebra of complex-valued functions on the space $Y = V \times X$, which is the union of two copies of the manifold V: $Y = V_a \cup V_b$.

Let us first compute the metric on Y associated with the K-cycle $(\mathfrak{H}, D)$:

$$d(p,q) = \sup_{f \in \mathcal{A}} \{|f(p) - f(q)|;\ \|[D,f]\| \le 1\}.$$

To the decomposition $Y = V_a \cup V_b$ there corresponds a decomposition of $\mathcal{A}$ as $\mathcal{A}_a \oplus \mathcal{A}_b$, so that every $f \in \mathcal{A}$ is a pair (f_a, f_b) of functions on V. Also, to the decomposition

$$\mathfrak{H}_2 = \mathfrak{H}_{2,a} \oplus \mathfrak{H}_{2,b},$$

there corresponds a decomposition $\mathfrak{H} = \mathfrak{H}_a \oplus \mathfrak{H}_b$, in which the action of $f = (f_a, f_b) \in \mathcal{A}$ is diagonal:

$$f \mapsto \begin{bmatrix} f_a & 0 \\ 0 & f_b \end{bmatrix} \in \mathcal{L}(\mathfrak{H}).$$

In this decomposition the operator D becomes

$$D = \begin{bmatrix} \partial_V \otimes 1 & \gamma_5 \otimes M^* \\ \gamma_5 \otimes M & \partial_V \otimes 1 \end{bmatrix},$$

where ∂_V is the Dirac operator on V and γ_5 is the $\mathbb{Z}/2$-grading of its spinor bundle.

This gives us the formula for the "differential" of a function $f \in \mathcal{A}$:

$$[D,f] = \begin{bmatrix} i^{-1}\gamma(df_a) \otimes 1 & (f_b - f_a)\gamma_5 \otimes M^* \\ (f_a - f_b)\gamma_5 \otimes M & i^{-1}\gamma(df_b) \otimes 1 \end{bmatrix}.$$

The differential $[D,f]$ thus contains three parts:

α) the usual differential df_a of the restriction of f to the copy V_a of V;

β) the usual differential df_b of the restriction of f to the copy V_b of V;

γ) the finite difference $\Delta f = f(p_a) - f(p_b)$, where p_a and p_b are the points of V_a and V_b above a given point p of V.

The norm of the operator $[D, f]$ can be computed easily: if λ is the norm of M, i.e., the largest eigenvalue of $|M| = (M^*M)^{1/2}$, then

$$\|[D, f]\| = \operatorname*{ess\,sup}_{p \in V} \begin{bmatrix} \|\mathrm{d}f_a(p)\| & -i\lambda(\Delta f)(p) \\ i\lambda(\Delta f)(p) & \|\mathrm{d}f_b(p)\| \end{bmatrix},$$

where $\|\mathrm{d}f_a(p)\|$ is the length of the gradient of f_a at $p \in V_a$.

We thus obtain:

Proposition 4. 1) *The restriction of the metric d on $V_a \cup V_b$ to each copy (V_a or V_b) of V is the Riemannian geodesic distance of V.*

2) *For each point $p = p_a$ of V_a, the distance $d(p_a, V_b) = \mathrm{Inf}\{d(p_a, q);\ q \in V_b\}$ is equal to λ^{-1} and is attained at the unique point p_b.*

Now recall that, given a metric space (Y, d) and two subsets Y_1, Y_2 of Y, their Hausdorff distance $d(Y_1, Y_2)$ is given by

$$d(Y_1, Y_2) = \sup\{d(x, Y_2), x \in Y_1;\ d(x, Y_1), x \in Y_2\}.$$

Thus, the metric d on $V_a \cup V_b = Y$ is clearly related to the following definition of the Gromov distance between two metric spaces (X_1, d_1) and (X_2, d_2):

Definition 5. [Gro$_1$] *Let (X_1, d_1) and (X_2, d_2) be two metric spaces. Then, given $\epsilon > 0$, the* Gromov distance *$\delta(X_1, X_2)$ is smaller than ϵ if and only if there exists a metric d on $X = X_1 \cup X_2$ such that $d|X_j = d_j$, and $d(X_1, X_2) < \epsilon$.*

Thus, we see that if we try to take different Riemannian metrics g_a and g_b on the two copies, V_a and V_b, of V, say by letting

$$D = \begin{bmatrix} D_a \otimes 1 & \gamma \otimes M^* \\ \gamma \otimes M & D_b \otimes 1 \end{bmatrix},$$

then Proposition 4 fails unless the Gromov distance between g_a and g_b is less than $\epsilon = 1/\lambda$, $\lambda = \|M\|$.

Let us now pass to the computation of the $\mathcal{A}$-bimodule Ω_D^1 of 1-forms over the space Y. The above computation of $[D, f] = \pi(\mathrm{d}f)$ for $f \in \mathcal{A}$ shows that an element α of the $\mathcal{A}$-bimodule $\Omega_D^1 = \pi(\Omega^1)$ is given by:

α) an ordinary differential form ω_a on V_a;

β) an ordinary differential form ω_b on V_b;

γ) a pair of complex-valued functions δ_a, δ_b on V.

The corresponding operator in $\mathfrak{H}$ is given by

$$\begin{bmatrix} i^{-1}\gamma(\omega_a) \otimes 1 & \delta_a \gamma_5 \otimes M^* \\ \delta_b \gamma_5 \otimes M & i^{-1}\gamma(\omega_b) \otimes 1 \end{bmatrix} = \alpha;$$

the bimodule structure over $\mathcal{A}$ is given, with obvious notation, by

$$(f_a, f_b)(\omega_a, \omega_b, \delta_a, \delta_b) = (f_a\omega_a, f_b\omega_b, f_a\delta_a, f_b\delta_b),$$
$$(\omega_a, \omega_b, \delta_a, \delta_b)(f_a, f_b) = (f_a\omega_a, f_b\omega_b, f_b\delta_a, f_a\delta_b).$$

The involution $*$ is given by $(\omega_a, \omega_b, \delta_a, \delta_b)^* = (-\overline{\omega}_a, -\overline{\omega}_b, \overline{\delta}_b, \overline{\delta}_a)$.

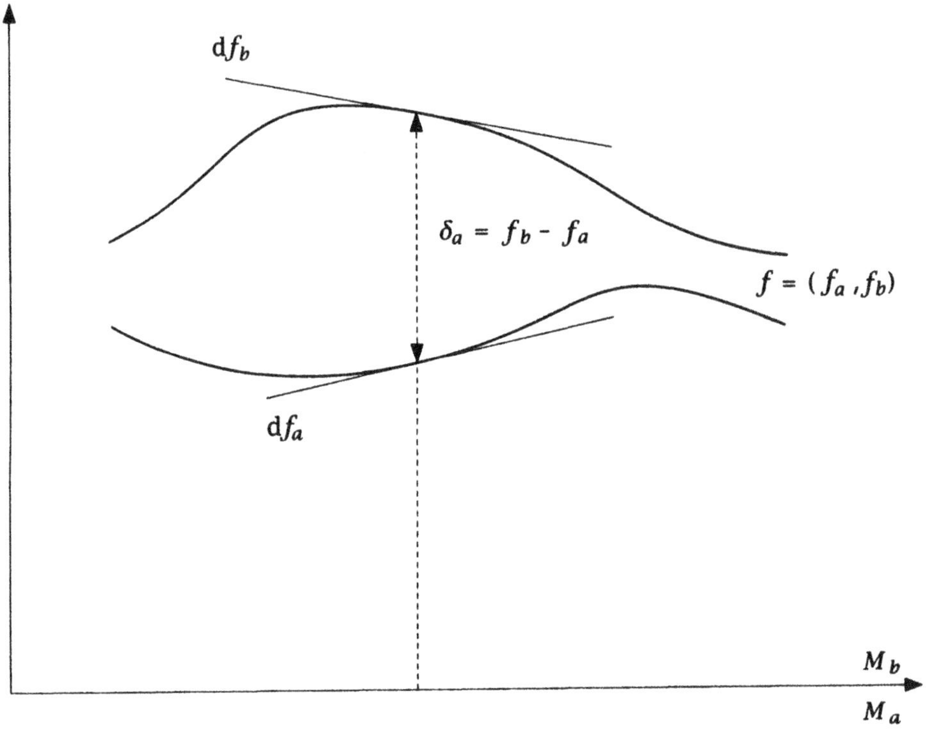

Figure VI.4. Differential of a function on a double-space

The terms δ_a and δ_b correspond to the bimodule of finite differences on passing from one copy V_a to the other copy V_b of V. Note that even though $\mathcal{A}$ is commutative, this bimodule is *not commutative*; for, if it were commutative then the finite difference would fail to be a derivation. With the above notation, the differential $f \in \mathcal{A} \mapsto \pi(df)$ reads as follows:

$$f = (f_a, f_b) \mapsto (df_a, df_b, f_b - f_a, f_a - f_b) \in \Omega_D^1.$$

When we project on V, the bimodule Ω_D^1 can be viewed as a 10-dimensional bundle over V, given by two copies of the complexified cotangent bundle, and a trivial 2-dimensional bundle

$$T_p^*(V)_{\mathbb{C}} \oplus T_p^*(V)_{\mathbb{C}} \oplus \mathbb{C} \oplus \mathbb{C};$$

however, one must keep in mind the nontrivial bimodule structure in the last two terms. Figure IV.4 illustrates the situation.

As in the case of the Dirac operator on Riemannian manifolds (Section 1, Lemma 6), let us compute the pairs of operators of the form $\pi(\rho) = T_1$, $\pi(d\rho) = T_2$ for $\rho \in \Omega^1(\mathcal{A})$. Given $\rho = \sum f_j dg_j \in \Omega^1(\mathcal{A})$, with $f_j, g_j \in \mathcal{A}$, we have

$$\pi(\rho) = \begin{bmatrix} i^{-1}\gamma(\omega_a) \otimes 1 & \delta_a \gamma_5 \otimes M^* \\ \delta_b \gamma_5 \otimes M & i^{-1}\gamma(\omega_b) \otimes 1 \end{bmatrix},$$

where $\omega_a = \sum f_{ja} dg_{ja}$, $\omega_b = \sum f_{jb} dg_{jb}$ and

$$\delta_a = \sum f_{ja}(g_{jb} - g_{ja}), \quad \delta_b = \sum f_{jb}(g_{ja} - g_{jb}).$$

We have $\pi(d\rho) = \sum \pi(df_j)\pi(dg_j)$, which gives the 2×2 matrix

$$\pi(d\rho) = \begin{bmatrix} -\gamma(\xi_a) \otimes 1 + (\delta_a + \delta_b) \otimes M^*M & \gamma_5 i^{-1}\gamma(\eta_a) \otimes M^* \\ \gamma_5 i^{-1}\gamma(\eta_b) \otimes M & -\gamma(\xi_b) \otimes 1 + (\delta_a + \delta_b) \otimes MM^* \end{bmatrix},$$

where $\xi_a = \sum df_{ja} dg_{ja}$ and $\xi_b = \sum df_{jb} dg_{jb}$ are sections of the Clifford algebra bundle C^2 over V, whereas

$$\eta_b = \sum ((f_{ja} - f_{jb}) dg_{ja} - (g_{ja} - g_{jb}) df_{jb}),$$
$$\eta_a = \sum ((f_{jb} - f_{ja}) dg_{jb} - (g_{jb} - g_{ja}) df_{ja}).$$

Using the equalities

$$d\delta_a = \sum (f_{ja}(dg_{jb} - dg_{ja}) + (g_{jb} - g_{ja}) df_{ja}),$$
$$d\delta_b = \sum (f_{jb}(dg_{ja} - dg_{jb}) + (g_{ja} - g_{jb}) df_{jb}),$$
$$\omega_a = \sum f_{ja} dg_{ja}, \quad \omega_b = \sum f_{jb} dg_{jb},$$

we can rewrite η_a and η_b as

$$\eta_a = \omega_b - d\delta_a - \omega_a, \quad \eta_b = \omega_a - d\delta_b - \omega_b.$$

As in the Riemannian case (Lemma 6 of Section 1), the sections ξ_a and ξ_b of C^2 are arbitrary except for $\sigma_2(\xi_a) = d\omega_a$ and $\sigma_2(\xi_b) = d\omega_b$. This shows that the subspace $\pi(d(J_0 \cap \Omega^1))$ of $\pi(\Omega^2)$ is the space of 2×2 matrices of operators of the form

$$T = \begin{bmatrix} \gamma(\xi_a) \otimes 1 & 0 \\ 0 & \gamma(\xi_b) \otimes 1 \end{bmatrix},$$

where ξ_a and ξ_b are sections of C^0, i.e., are just arbitrary scalar-valued functions on V, so that $\gamma(\xi_a) = \xi_a$, $\gamma(\xi_b) = \xi_b$.

A general element of $\pi(\Omega^2)$ is a 2×2 matrix of operators of the form

$$T = \begin{bmatrix} -\gamma(\alpha_a) \otimes 1 + h_a \otimes M^*M & \gamma_5 i^{-1}\gamma(\beta_a) \otimes M^* \\ \gamma_5 i^{-1}\gamma(\beta_b) \otimes M & -\gamma(\alpha_b) \otimes 1 + h_b \otimes MM^* \end{bmatrix},$$

where α_a, α_b are arbitrary sections of C^2, h_a, h_b are arbitrary functions on V and β_a, β_b are arbitrary sections of C^1 (i.e., 1-forms).

Lemma 6. *Assume that M^*M is not a scalar multiple of the identity matrix. Then, an element of Ω_D^2 is given by:*

1) *a pair of ordinary 2-forms α_a, α_b on V;*
2) *a pair of ordinary 1-forms β_a, β_b on V;*
3) *a pair of scalar functions h_a, h_b on V.*

The hypothesis $M^*M \neq \lambda$ is important since otherwise the functions h_a, h_b are eliminated by $\pi(\mathrm{d}(J_0 \cap \Omega^1))$.

Using the above computation of $\pi(\mathrm{d}\rho)$ we can, moreover, compute the sextuple $(\alpha_a, \alpha_b, \beta_a, \beta_b, h_a, h_b)$; then for the differential $\mathrm{d}\omega$ of an element $\omega = (\omega_a, \omega_b, \delta_a, \delta_b)$ of Ω_D^1, we get:

1) $\alpha_a = \mathrm{d}\omega_a$, $\alpha_b = \mathrm{d}\omega_b$;
2) $\beta_a = \omega_b - \omega_a - \mathrm{d}\delta_a$, $\beta_b = \omega_a - \omega_b - \mathrm{d}\delta_b$;
3) $h_a = \delta_a + \delta_b$, $h_b = \delta_a + \delta_b$.

Thus, we see that the differential $\mathrm{d}\omega \in \Omega_D^2$ involves the differential terms $\mathrm{d}\omega_a$, $\mathrm{d}\omega_b$, $\mathrm{d}\delta_a$ and $\mathrm{d}\delta_b$ as well as the finite-difference terms $\omega_a - \omega_b$ and $\delta_a + \delta_b$, but in combinations such as $\omega_b - \omega_a - \mathrm{d}\delta_a$ imposed by $\mathrm{d}(\mathrm{d}f) = 0$. Let us compute the product $\omega\omega' \in \Omega_D^2$ of two elements $\omega = (\omega_a, \omega_b, \delta_a, \delta_b)$, $\omega' = (\omega'_a, \omega'_b, \delta'_a, \delta'_b)$ of Ω_D^1; we get:

1) $\alpha_a = \omega_a \wedge \omega'_a$, $\alpha_b = \omega_b \wedge \omega'_b$;
2) $\beta_a = \delta_a \omega'_b - \delta'_a \omega_a$, $\beta_b = \delta_b \omega'_a - \delta'_b \omega_b$;
3) $h_a = \delta_a \delta'_b$, $h_b = \delta_b \delta'_a$.

The next step is to determine the inner product on the space Ω_D^2 of 2-forms given in Section 1. By definition, we take the orthogonal complement of $\pi(\mathrm{d}(J_0 \cap \Omega^1))$ in $\pi(\Omega^2)$ with the inner product $\langle T_1, T_2 \rangle = \mathrm{Tr}_\omega(T_1^* T_2 |D|^{-4})$. An easy calculation then gives:

Lemma 7. *Let $\lambda(M^*M)$ be the orthogonal projection (for the Hilbert-Schmidt scalar product) of the matrix M^*M onto the scalar matrices $\lambda\,\mathrm{id}$. Then, the squared norm of an element $(\alpha_a, \alpha_b, \beta_a, \beta_b, h_a, h_b)$ of Ω_D^2 is given by $(8\pi^2)^{-1}$ times*

$$\int_V (N_a \|\alpha_a\|^2 + N_b \|\alpha_b\|^2)\mathrm{d}v + \mathrm{tr}(M^*M) \int_V (\|\beta_a\|^2 + \|\beta_b\|^2)\mathrm{d}v$$
$$+ \mathrm{tr}\Big((M^*M - \lambda(M^*M))^2\Big) \times \int_V (\|h_a\|^2 + \|h_b\|^2)\mathrm{d}v,$$

where $N_a = \dim \mathfrak{H}_{2,a}$ and $N_b = \dim \mathfrak{H}_{2,b}$.

We are now ready to compute the action $\mathrm{YM}(\nabla)$. We shall take the Hermitian bundle on $Y = V_a \cup V_b$ that has complex fiber $\mathbb{C}^2$ on the copy V_a of V, and is trivial with one-dimensional fiber $\mathbb{C}$ on the copy V_b of V. In other words, we consider the product of the example of Section 1 by the example a), β) on the space X. From the above description of Ω_D^1, we see that if ∇ is a compatible connection on E then it is given by a triple:

α) an ordinary compatible connection ∇_a on the restriction of E to V_a;

β) an ordinary compatible connection ∇_b on the restriction of E to V_b;

γ) a section u on V of the bundle $\mathrm{Hom}(E_b, E_a)$ of linear mappings from the fiber $E_{b,p}$ to the fiber $E_{a,p}$.

Both α) and β) have the obvious meaning, while γ) prescribes the value of the finite-difference operation on sections ξ of E. At the point p_a, this finite difference is

$$(\Delta\xi)(p_a) = \xi(p_a) - u_p\xi(p_b) \in E_{p_a} = \mathbb{C}^2,$$

whereas at the point p_b it is given by

$$(\Delta\xi)(p_b) = \xi(p_b) - u_p^*\xi(p_a) \in E_{p_b} = \mathbb{C}.$$

Of course, the choice of u is given by a pair ϕ_1, ϕ_2 of complex scalar fields on V, namely the two components of $u(1)$ for the basis of $\mathbb{C}^2$ (cf. Example a), β)).

2) *For each point $p = p_a$ of V_a, the distance $d(p_a, V_b) = \mathrm{Inf}\{d(p_a, q);\ q \in V_b\}$ is equal to λ^{-1} and is attained at the unique point p_b.*

The gauge group $\mathcal{U} = \mathrm{End}_{\mathcal{A}}(\mathcal{E})$ is the group of unitary endomorphisms of the bundle E over $Y = V_a \cup V_b$, or, equivalently, the group

$$\mathcal{U} = \mathrm{Map}(V, U(1) \times U(2)).$$

Its actions on the $U(2)$ connection ∇_a and on the $U(1)$ connection ∇_b are the obvious ones, and the action on a field $u \in \mathrm{Hom}(E_b, E_a)$ is given by composition.

Let us be more explicit in the description of ∇ as a linear mapping from the space $\mathcal{E}$ of sections of E to $\mathcal{E} \otimes_{\mathcal{A}} \Omega_D^1$. An element of $\mathcal{E} \otimes_{\mathcal{A}} \Omega_D^1$ is given by:

1) an ordinary differential form $\omega = (\omega_a, \omega_b)$ on $Y = V_a \cup V_b$, with coefficients in E,

2) a section $\delta = (\delta_a, \delta_b)$ on $Y = V_a \cup V_b$ of the bundle E.

The mapping ∇ is then given by

$$\nabla(\xi_a, \xi_b) = (\nabla_a\xi_a, \nabla_b\xi_b), (\xi_a - u\xi_b), (\xi_b - u^*\xi_a)$$

for any section $\xi = (\xi_a, \xi_b)$ of E.

Since the restriction of E to V_a is trivial with fiber $\mathbb{C}^2$, we may as well describe ∇_a by a 2×2 matrix $\begin{bmatrix} \omega_{11}^a & \omega_{12}^a \\ \omega_{21}^a & \omega_{22}^a \end{bmatrix}$ of 1-forms on V that is skew-adjoint. Similarly, ∇_b is given by a single skew-adjoint 1-form $[\omega_{11}^b]$, and u by a pair of complex fields $(1 + \varphi_1, \varphi_2) = u(1)$.

With these notations, the connection ∇ is given by the equality

$$\nabla\xi = f d\xi + \rho\xi \in \mathcal{E} \otimes_{\mathcal{A}} \Omega_D^1 \quad \forall \xi \in \mathcal{E},$$

where $\mathcal{E} = f\mathcal{A}^2$, $f \in M_2(\mathcal{A})$ being the idempotent $f = \begin{bmatrix} 1 & 0 \\ 0 & e \end{bmatrix}$, $e = (0,1) \in \mathcal{A}$, and where $\rho \in M_2(\Omega_D^1)$ is the 2×2 matrix whose entries are the following elements of Ω_D^1:

$$\rho_{11} = (\omega_{11}^a, \omega_{11}^b, \varphi_1, \overline{\varphi}_1),$$
$$\rho_{12} = (\omega_{12}^a, 0, 0, \overline{\varphi}_2),$$
$$\rho_{21} = (\omega_{21}^a, 0, \varphi_2, 0),$$
$$\rho_{22} = (\omega_{22}^a, 0, 0, 0),$$

or, equivalently,

$$\rho = \left[\begin{bmatrix} \omega_{11}^a & \omega_{12}^a \\ \omega_{21}^a & \omega_{22}^a \end{bmatrix}, \begin{bmatrix} \omega_{11}^b & 0 \\ 0 & 0 \end{bmatrix}, \begin{bmatrix} \varphi_1 & 0 \\ \varphi_2 & 0 \end{bmatrix}, \begin{bmatrix} \overline{\varphi}_1 & \overline{\varphi}_2 \\ 0 & 0 \end{bmatrix}\right].$$

The curvature θ is then the following element of $fM_2(\Omega_D^2)f$:

$$\theta = f\mathrm{d}f\mathrm{d}f + f\mathrm{d}\rho f + \rho^2,$$

which is easily determined using the above computation of

$$\mathrm{d} : \Omega_D^1 \to \Omega_D^2, \quad \wedge : \Omega_D^1 \times \Omega_D^1 \to \Omega_D^2.$$

As we saw in Lemma 6, elements of Ω_D^2 have a differential degree and a finite-difference degree (α, β) adding up to 2. Let us thus begin with terms in θ of bi-degree $(2,0)$. To compute them we just use the formulas 1) following Lemma 6:

$$\alpha_a = \mathrm{d}\omega_a, \ \alpha_b = \mathrm{d}\omega_b; \ \ \alpha_a = \omega_a \wedge \omega'_a, \ \alpha_b = \omega_b \wedge \omega'_b.$$

We thus see that the component $\theta^{(2,0)}$ of bi-degree $(2,0)$ is the following 2×2 matrix of 2-forms on $V_a \cup V_b$:

$$\theta_a^{(2,0)} = \mathrm{d}\omega^a + \omega^a \wedge \omega^a, \ \ \theta_b^{(2,0)} = \mathrm{d}\omega^b + \omega^b \wedge \omega^b = \begin{bmatrix} \mathrm{d}\omega_{11}^b & 0 \\ 0 & 0 \end{bmatrix}.$$

Next, we look at the component $\theta^{(1,1)}$ of bi-degree $(1,1)$ and use the formulas 2):

$$\beta_a = \omega_b - \omega_a - \mathrm{d}\delta_a, \qquad \beta_a = \delta_a \omega'_b - \omega_a \delta'_a,$$

$$\beta_b = \omega_a - \omega_b - \mathrm{d}\delta_b, \qquad \beta_b = \delta_b \omega'_a - \omega_b \delta'_b.$$

Thus, $\theta^{(1,1)}$ is the following 2×2 matrix of 1-forms on $V_a \cup V_b$:

$$\begin{aligned}\theta_a^{(1,1)} &= \left(\begin{bmatrix} \omega_{11}^b & 0 \\ 0 & 0 \end{bmatrix} - \begin{bmatrix} \omega_{11}^a & \omega_{12}^a \\ \omega_{21}^a & \omega_{22}^a \end{bmatrix} - \begin{bmatrix} \mathrm{d}\varphi_1 & 0 \\ \mathrm{d}\varphi_2 & 0 \end{bmatrix}\right)\begin{bmatrix} 1 & 0 \\ 0 & 0 \end{bmatrix} \\ &\quad + \begin{bmatrix} \varphi_1 & 0 \\ \varphi_2 & 0 \end{bmatrix}\begin{bmatrix} \omega_{11}^b & 0 \\ 0 & 0 \end{bmatrix} - \begin{bmatrix} \omega_{11}^a & \omega_{12}^a \\ \omega_{21}^a & \omega_{22}^a \end{bmatrix}\begin{bmatrix} \varphi_1 & 0 \\ \varphi_2 & 0 \end{bmatrix} \\ &= \begin{bmatrix} -\mathrm{d}\varphi_1 - (\omega_{11}^a - \omega_{11}^b)(\varphi_1 + 1) - \omega_{12}^a \varphi_2 & 0 \\ -\mathrm{d}\varphi_2 - \omega_{21}^a(\varphi_1 + 1) - (\omega_{22}^a - \omega_{11}^b)\varphi_2 & 0 \end{bmatrix}.\end{aligned}$$

Similarly, we have

$$\theta_b^{(1,1)} = \begin{bmatrix} 1 & 0 \\ 0 & 0 \end{bmatrix}\left(\begin{bmatrix} \omega_{11}^a & \omega_{12}^a \\ \omega_{21}^a & \omega_{22}^a \end{bmatrix} - \begin{bmatrix} \omega_{11}^b & 0 \\ 0 & 0 \end{bmatrix} - \begin{bmatrix} d\overline{\varphi}_1 & d\overline{\varphi}_2 \\ 0 & 0 \end{bmatrix}\right)$$
$$+ \begin{bmatrix} \overline{\varphi}_1 & \overline{\varphi}_2 \\ 0 & 0 \end{bmatrix}\begin{bmatrix} \omega_{11}^a & \omega_{12}^a \\ \omega_{21}^a & \omega_{22}^a \end{bmatrix} - \begin{bmatrix} \omega_{11}^b & 0 \\ 0 & 0 \end{bmatrix}\begin{bmatrix} \overline{\varphi}_1 & \overline{\varphi}_2 \\ 0 & 0 \end{bmatrix}$$
$$= \begin{bmatrix} d\overline{\varphi}_1 + (\omega_{11}^a - \omega_{11}^b)(\overline{\varphi}_1 + 1) + \omega_{21}^a \overline{\varphi}_2 & -d\overline{\varphi}_2 + \omega_{12}^a(\overline{\varphi}_1 + 1) + (\omega_{22}^a - \omega_{11}^b)\varphi_2 \\ 0 & 0 \end{bmatrix}.$$

Finally, we have to compute the component $\theta^{(0,2)}$; we use the formulas 3):

$$h_a = \delta_a + \delta_b, \qquad h_a = \delta_a \delta_b',$$

$$h_a = \delta_a + \delta_b, \qquad h_b = \delta_b \delta_a'.$$

We then have

$$\theta_a^{(0,2)} = \begin{bmatrix} \varphi_1 & 0 \\ \varphi_2 & 0 \end{bmatrix} + \begin{bmatrix} \overline{\varphi}_1 & \overline{\varphi}_2 \\ 0 & 0 \end{bmatrix} + \begin{bmatrix} \varphi_1 & 0 \\ \varphi_2 & 0 \end{bmatrix}\begin{bmatrix} \overline{\varphi}_1 & \overline{\varphi}_2 \\ 0 & 0 \end{bmatrix}$$
$$= \begin{bmatrix} \varphi_1 + \overline{\varphi}_1 + \varphi_1\overline{\varphi}_1 & \overline{\varphi}_2(1+\varphi_1) \\ \varphi_2(1+\overline{\varphi}_1) & 0 \end{bmatrix},$$

$$\theta_b^{(0,2)} = \begin{bmatrix} 1 & 0 \\ 0 & 0 \end{bmatrix} + \begin{bmatrix} 1 & 0 \\ 0 & 0 \end{bmatrix}\begin{bmatrix} \varphi_1 + \overline{\varphi}_1 & \overline{\varphi}_2 \\ \varphi_2 & 0 \end{bmatrix}\begin{bmatrix} 1 & 0 \\ 0 & 0 \end{bmatrix}$$
$$+ \begin{bmatrix} \overline{\varphi}_1 & \overline{\varphi}_2 \\ 0 & 0 \end{bmatrix}\begin{bmatrix} \varphi_1 & 0 \\ \varphi_2 & 0 \end{bmatrix}$$
$$= \begin{bmatrix} 1 + \varphi_1 + \overline{\varphi}_1 + \overline{\varphi}_1\varphi_1 + \overline{\varphi}_2\varphi_2 & 0 \\ 0 & 0 \end{bmatrix}.$$

Thus,

$$\mathrm{YM}(\nabla) = I_2 + I_1 + I_0,$$

where each I_j is the integral over M of a Lagrangian density given by the following formulas.

First, for I_2:

$$|d\omega^a + \omega^a \wedge \omega^a|^2 N_a + |d\omega^b|^2 N_b,$$

where $N_a = \dim \mathfrak{H}_{2,a}$, $N_b = \dim \mathfrak{H}_{2,b}$ and the norms are the squared norms for the curvatures of the connections ∇^a and ∇^b, respectively.

Next, for I_1:

$$2\left|\nabla\begin{pmatrix} 1+\varphi_1 \\ \varphi_2 \end{pmatrix}\right|^2 \operatorname{tr}(M^*M),$$

where ∇ is the covariant differentiation of a pair of scalar fields, given by

$$d + \begin{bmatrix} \omega_{11}^a - \omega_{11}^b & \omega_{12}^a \\ \omega_{21}^a & \omega_{22}^a - \omega_{11}^b \end{bmatrix}.$$

Finally, for I_0:

$$\left(1 + 2(1 - (|1+\varphi_1|^2 + |\varphi_2|^2))^2\right)\mathrm{Tr}\left((\lambda^\perp(M^*M))^2\right),$$

where $\lambda^\perp$ is the orthogonal projection in the Hilbert-Schmidt space of matrices onto the orthogonal complement of the scalar multiples of the identity. These terms are obtained, with the right coefficients, from the computation of the Hilbert space norm on Ω_D^2 in Lemma 7.

The fermionic action is even easier to compute. We have

$$\langle \psi, D_\nabla \psi\rangle = J_0 + J_1,$$

where $\psi \in \mathcal{E}\otimes_{\mathcal{A}}\mathfrak{H}$, $\gamma\psi = \psi$, is given by a pair of left-handed sections of $S\otimes\mathfrak{H}_{2,a}$ denoted by $\begin{bmatrix}\psi_1^a\\ \psi_2^a\end{bmatrix}$, and a right-handed section of $S\otimes\mathfrak{H}_{2,b}$ denoted by ψ^b. Both J_0 and J_1 are given by Lagrangian densities:

$$J_0 : \overline{\psi}^a(\partial + i^{-1}\gamma(\omega^a))\psi^a + \overline{\psi}^b(\partial + i^{-1}\gamma(\omega^b))\psi^b,$$
$$J_1 : \overline{\psi}_b M[(1+\varphi_1), \varphi_2]\psi_a + \text{h.c.}$$

We can now make the point concerning this example b): modulo a few nuances that we shall deal with, the five terms of our action

$$I_0 + I_1 + I_2 + J_0 + J_1$$

are the five terms of the Glashow-Weinberg-Salam unification of electromagnetic and weak forces for N generations of leptons (where $N = N_a = N_b$ is the dimension of $\mathfrak{H}_{2,a}$ and $\mathfrak{H}_{2,b}$).

Let us describe for instance, from [Ell], and using the conventional notation of physics, the five constituents of the G.W.S. Lagrangian, which we write directly in the Euclidean (i.e., imaginary time) framework. For each constituent, we give the corresponding fields and Lagrangian:

$$\mathcal{L} = \mathcal{L}_G + \mathcal{L}_f + \mathcal{L}_\Phi + \mathcal{L}_Y + \mathcal{L}_V,$$

where

1) $\mathcal{L}_G$: The *pure gauge boson* part is just

$$\mathcal{L}_G = \frac{1}{4}(G_{\mu\nu a}G_a^{\mu\nu}) + \frac{1}{4}(F_{\mu\nu}F^{\mu\nu}),$$

where $G_{\mu\nu a} = \partial_\mu W_{\nu a} - \partial_\nu W_{\mu a} + g\varepsilon_{abc}W_{\mu b}W_{\nu c}$ and $F_{\mu\nu} = \partial_\mu B_\nu - \partial_\nu B_\mu$ are the field strength tensors of an $SU(2)$ gauge field $W_{\mu a}$ and a $U(1)$ gauge field B_μ. (Einstein summation over repeated indices is used here.)

2) $\mathcal{L}_f$: The *fermion kinetic term* has the form

$$\mathcal{L}_f = -\sum\left[\bar{f}_L\gamma^\mu\left(\partial_\mu + ig\frac{\tau_a}{2}W_{\mu a} + ig'\frac{Y_L}{2}B_\mu\right)f_L + \bar{f}_R\gamma^\mu\left(\partial_\mu + ig'\frac{Y_R}{2}B_\mu\right)f_R\right],$$

where the f_L (resp. f_R) are the left-handed (resp. right-handed) fermion fields, which for leptons and for each generation are given by a pair, i.e., an isodoublet, of left-handed spinors (such as $\begin{bmatrix} \nu_L \\ e_L \end{bmatrix}$), and a singlet ($e_R$), i.e., a right-handed spinor.

We shall return later to the hypercharges Y_L and Y_R, which for leptons are given by $Y_L = -1$, $Y_R = -2$.

3) $\mathcal{L}_\Phi$: The *kinetic terms for the Higgs fields* are

$$\mathcal{L}_\Phi = -\left|\left(\partial_\mu + ig\frac{\tau_a}{2}W_{\mu a} + i\frac{g'}{2}B_\mu\right)\Phi\right|^2,$$

where $\Phi = \begin{bmatrix} \Phi_1 \\ \Phi_2 \end{bmatrix}$ is an $SU(2)$ doublet of complex scalar fields Φ_1 and Φ_2 with hypercharge $Y_\Phi = 1$.

4) $\mathcal{L}_Y$: The *Yukawa coupling* of Higgs fields with fermions is

$$\mathcal{L}_Y = -\sum [H_{ff'}(\overline{f}_L \cdot \Phi)f'_R + H^*_{ff'}\overline{f'}_R(\Phi^+ \cdot f_L)],$$

where $H_{ff'}$ is a general coupling matrix in the space of different families.

5) $\mathcal{L}_V$: The *Higgs self-interaction* is the potential

$$\mathcal{L}_V = \mu^2(\Phi^+\Phi) - \frac{1}{2}\lambda(\Phi^+\Phi)^2,$$

where $\lambda > 0$ and $\mu^2 > 0$ are scalars.

All of these terms are deeply rooted in both experimental and theoretical physics, but we postpone an elaboration of this point to the complete model invoking quarks and strong interactions as well. For the moment we shall establish a dictionary, or change-of-variables, between our action and the Glashow–Weinberg–Salam action.

The first obvious nuance between the two actions is that our action involves a $U(1)$ and a $U(2)$ gauge field while the G.W.S. action involves a $U(1)$ and an $SU(2)$ gauge field (however, cf. [Wein] for an interesting perspective on this point). We shall thus, *in an artificial manner*, reduce our theory to $U(1) \times SU(2)$ by imposing on the connections $\nabla^a = d + \omega^a$ and $\nabla^b = d + \omega^b$ the following condition:

$$\text{Ad hoc condition:} \quad \text{tr}(\omega^a) = \omega^b.$$

Let us now spell out the dictionary.

Noncommutative geometry	Classical field theory
vector $\psi \in \mathcal{E} \otimes_{\mathcal{A}} \mathfrak{H}$, $\gamma\psi = \psi$	chiral fermion f
differential components of connection ω^a, ω^b	pure gauge bosons W, B
finite-difference component of connection $(1 + \delta^a)$, δ^b	Higgs field Φ
I_2	$\mathcal{L}_G$
I_1	$\mathcal{L}_\Phi$
I_0	$\mathcal{L}_V$
J_0	$\mathcal{L}_f$
J_1	$\mathcal{L}_Y$

It is moreover straightforward, using the above "ad hoc condition", to work out the change of variables from our fields ψ, ω, δ to the fields f, W, B, Φ which gives the equality

$$g^{-2}\mathrm{YM}(\nabla) + \langle \psi, D_\nabla \psi \rangle = \mathcal{L}(f, W, B, \Phi),$$

where the right-hand side is a special case of the G.W.S. Lagrangian, with a few constraints. These relations are of limited use for three reasons. The first is that the model incorporates neither the quarks nor the strong interaction, the second is the artificial nature of the "ad hoc condition", and the third and most important is that, due to renormalization, the coupling constants such as g, g', μ, λ, $H_{ff'}$ that appear in the G.W.S. model are all functions of an effective energy Λ to which, for the moment, we can give no preferred value.

In Section 5 we shall remove the first two defects by explaining how to incorporate quarks and strong interactions, as well as how the hypercharges of physics occur, from a conceptual point of view.

8. Example c). The noncommutative torus.

In this example we shall treat the noncommutative torus (cf. Chapter IV Section 6) from a metric point of view, and show how the classical solutions of the extrema of the Yang-Mills action modulo gauge transformations allow us to recover an ordinary torus.

Thus, let us begin with the following $*$-algebra $\mathcal{A}_\theta$, where $\theta \in [0, 1]$ is an irrational number and $\lambda = e(\theta) = \exp 2\pi i\theta$:

$$\mathcal{A}_\theta = \left\{ \sum a_{n,m} U^n V^m ;\ a \in S(\mathbb{Z}^2) \right\},$$

where $S(\mathbb{Z}^2)$ is the vector space of sequences $(a_{n,m})_{n,m\in\mathbb{Z}}$ that decay faster than the inverse of any polynomial in (n, m). The product in $\mathcal{A}_\theta$ is specified by the relation

$$VU = \lambda UV, \quad \lambda = e(\theta) = \exp 2\pi i\theta,$$

and the involution by $U^* = U^{-1}$, $V^* = V^{-1}$.

In Chapter III we computed the cyclic cohomology of this algebra; thus, $HC^0(\mathcal{A}_\theta)$ is 1-dimensional and is generated by the unique trace τ_0 of $\mathcal{A}_\theta$,

$$\tau_0(\sum a_{n,m}U^nV^m) = a_{0,0} \in \mathbb{C},$$

whereas besides $S\tau_0 \in HC^2$ the cyclic cohomology $HC^2(\mathcal{A}_\theta)$ is generated by the cyclic 2-cocycle

$$\tau_2(a^0,a^1,a^2) = -2\pi i \sum_{\substack{n_0+n_1+n_2=0 \\ m_0+m_1+m_2=0}} (n_1m_2 - n_2m_1)\, a^0_{n_0,m_0}a^1_{n_1,m_1}a^2_{n_2,m_2}.$$

The above considerations only involve the smooth algebra $\mathcal{A}_\theta$; we shall now fix the metric. To that end we take the element G of $\Omega_2^+(\mathcal{A}_\theta)$ defined by

$$G = (\mathrm{d}U)(\mathrm{d}U)^* + (\mathrm{d}V)(\mathrm{d}V)^*,$$

and we solve the following extremum problem (which, in the commutative case, gave us the conformal structure on a Riemann surface):

Lemma 9. *On the intersection of the cyclic cohomology class $\tau_2 + b(\ker B)$ with the positive cone Z_+^2 in Hochschild cohomology, the functional G defined by*

$$\varphi \in Z^2 \mapsto \langle G,\varphi\rangle = \varphi(1,U,U^*) + \varphi(1,V,V^*)$$

reaches its minimum at a unique point φ_2 given by

$$\varphi_2(a^0,a^1,a^2) = 2\pi \sum_{\substack{n_0+n_1+n_2=0 \\ m_0+m_1+m_2=0}} (n_1 - im_1)(-n_2 - im_2)\, a^0_{n_0,m_0}a^1_{n_1,m_1}a^2_{n_2,m_2}.$$

We shall now use the noncommutative analogue of a conformal structure, i.e., the positive cocycle φ_2 together with the trace τ_0, to construct the analogue of the Dirac operator for $\mathcal{A}_\theta$, that is, we shall obtain a $(2,\infty)$-summable K-cycle $(\mathfrak{H}, D)$ on $\mathcal{A}_\theta$. The Hilbert space $\mathfrak{H}$ is the direct sum $\mathfrak{H} = \mathfrak{H}^+ \oplus \mathfrak{H}^-$ of the Hilbert space $\mathfrak{H}^- = L^2(\mathcal{A}_\theta, \tau_0)$ of the G.N.S. construction of τ_0, and a Hilbert space $\mathfrak{H}^+$ of forms of type $(1,0)$ on the noncommutative torus which is obtained canonically from φ_2 as follows:

Proposition 10. *Let $\mathcal{A}$ be a $*$-algebra and let $\varphi_2 \in Z_+^2(\mathcal{A},\mathcal{A}^*)$ be a positive Hochschild 2-cocycle on $\mathcal{A}$. Let $\mathfrak{H}^+$ be the Hilbert space completion of $\Omega^1(\mathcal{A})$ equipped with the inner product*

$$\langle a^0\mathrm{d}a^1, b^0\mathrm{d}b^1\rangle = \varphi_2(b^{0*}a^0, a^1, b^{1*}).$$

Then the actions of $\mathcal{A}$ on $\mathfrak{H}^+$ by left and right multiplications are unitary.

They are automatically bounded if $\mathcal{A}$ is a pre-C^*-algebra.

Thus, $\mathfrak{H}^+$ is a bimodule over $\mathcal{A}$ (i.e., a correspondence in the sense of Chapter V) and the differential $\mathrm{d} : \mathcal{A} \to \Omega^1(\mathcal{A})$ gives a derivation which, for reasons that will become clear, we shall denote by $\partial : \mathcal{A} \to \mathfrak{H}^+$.

In our specific example, the computation is straightforward and gives $\mathfrak{H}^+ = L^2(\mathcal{A}_\theta, \tau_0)$ as an $\mathcal{A}_\theta$-bimodule and $\partial : \mathcal{A} \to \mathfrak{H}^+$ given by $\partial = \frac{1}{\sqrt{2\pi}}(\delta_1 - i\delta_2)$, where δ_1, δ_2 are the canonical derivations of $\mathcal{A}$. One can immediately check the following:

Proposition 11. *Let* $\mathcal{A} = \mathcal{A}_\theta$ *act on the left on both* $\mathfrak{H}^- = L^2(\mathcal{A}_\theta, \tau_0)$ *and* $\mathfrak{H}^+$ *(associated with* φ_2 *by Proposition* 10*). Then, the operator*

$$D = \begin{bmatrix} 0 & \partial \\ \partial^* & 0 \end{bmatrix}$$

in $\mathfrak{H} = \mathfrak{H}^+ \oplus \mathfrak{H}^-$ *defines a* $(2, \infty)$*-summable K-cycle over* $\mathcal{A}_\theta$.

The $\mathbb{Z}/2$*-grading* γ *is given by the matrix* $\gamma = \begin{bmatrix} 1 & 0 \\ 0 & -1 \end{bmatrix}$.

We have thus arrived at the desired *quantization* of the cocycle φ_2. Further calculation yields the formula

$$-\mathrm{Tr}_\omega((1+\gamma)a^0[D, a^1][D, a^2]D^{-2}) = \varphi_2(a^0, a^1, a^2). \qquad (*)$$

The point is that if we had started not with τ_2 but with a multiple, $\lambda\tau_2$, where $\lambda > 0$, this would have replaced φ_2 in Lemma 9 by $\lambda\varphi_2$ and would thus have modified the inner product of $\mathfrak{H}^+$, multiplying it by λ. This is equivalent to the replacement of ∂ by $\lambda^{1/2}\partial$ keeping the inner product of $\mathfrak{H}^+$ fixed, in view of the equality

$$\langle a^0 \partial a^1, \partial a^2 \rangle = \varphi_2(a^0, a^1, a^{2*}).$$

Thus, this replacement of τ_2 by $\lambda\tau_2$ replaces D by $\lambda^{1/2}D$ and clearly leaves unchanged the left-hand side of formula $(*)$. This shows that there is a unique normalization of φ_2 for which $(*)$ holds. In fact, this normalization is dictated by the integrality of the pairing of φ_2 with $K_0(\mathcal{A}_\theta)$:

$$\langle \varphi_2, K_0(\mathcal{A}_\theta) \rangle = \mathbb{Z}.$$

The same independence of the choice of trace τ_0 applies to the homogeneous formula $(*)$, but of course the value of the trace

$$a \in \mathcal{A}_\theta \mapsto \mathrm{Tr}_\omega(aD^{-2})$$

does depend on the scaling of D.

Note, finally, that we have found a completely canonical procedure for constructing the K-cycle $(\mathfrak{H}, D, \gamma)$ over $\mathcal{A}_\theta$ from the fundamental class in cyclic cohomology, i.e., the choice of orientation, and the formal positive element

$$G = \mathrm{d}U(\mathrm{d}U)^* + \mathrm{d}V(\mathrm{d}V)^* \in \Omega^2_+(\mathcal{A}_\theta).$$

This possibility of going backwards from cyclic cohomology to K-homology is at the moment limited to the 2-dimensional situation.

We shall now use the $(2,\infty)$-summable K-cycle $(\mathfrak{H}, D, \gamma)$ over $\mathcal{A}_\theta$ and compute the action $\mathrm{YM}(\nabla)$ on connections ∇ on Hermitian finitely generated projective modules over $\mathcal{A}_\theta$. Our first task will be to determine the differential graded algebra $\Omega_D^*(\mathcal{A}_\theta)$. We thus need the analogue of Lemma 6 of Section 1. By construction, the Hilbert space $\mathfrak{H}$ is the direct sum of two copies of the left regular representation λ of $\mathcal{A}_\theta$ in $L^2(\mathcal{A}_\theta, \tau_0)$, and we have:

Lemma 12. 1) *If $k \geq 1$ then $\pi(\Omega^k) \subset \mathcal{L}(\mathfrak{H})$ is the $\mathcal{A}_\theta$-bimodule of all 2×2 matrices with entries in $\lambda(\mathcal{A}_\theta)$ and of degree $(-1)^k$ for the $\mathbb{Z}/2$-grading defined by $\gamma = \begin{bmatrix} 1 & 0 \\ 0 & -1 \end{bmatrix}$.*

2) *Let $a_{ij} \in \mathcal{A}_\theta$ and consider the elements*

$$\alpha = \begin{bmatrix} 0 & \lambda(a_{12}) \\ \lambda(a_{21}) & 0 \end{bmatrix} \in \pi(\Omega^1), \quad \beta = \begin{bmatrix} \lambda(a_{11}) & 0 \\ 0 & \lambda(a_{22}) \end{bmatrix} \in \pi(\Omega^2).$$

Then there exists an $\omega \in \Omega^1(\mathcal{A}_\theta)$ with $\alpha = \pi(\omega)$ and $\beta = \pi(d\omega)$ if and only if $a_{22} - a_{11} = \partial^* a_{12} - \partial a_{21}$.

This is not difficult to verify: first, the image under π of the subspace $\{aU^{-1}dU + bV^{-1}dV;\ a, b \in \mathcal{A}_\theta\}$ of $\Omega^1(\mathcal{A}_\theta)$ already includes all elements of the form

$$\begin{bmatrix} 0 & \lambda(a_{12}) \\ \lambda(a_{21}) & 0 \end{bmatrix}$$

with $a_{ij} \in \mathcal{A}_\theta$. This shows that $\pi(\Omega^1)$ has the desired form, and so does $\pi(\Omega^k)$ since it contains the kth power of $\pi(\Omega^1)$.

To verify (2), one uses the commutation $\partial\partial^* = \partial^*\partial$ as well as the equalities

$$\pi(\omega) = 0, \quad \pi(d\omega) = \begin{bmatrix} \lambda(a) & 0 \\ 0 & \lambda(a) \end{bmatrix},$$

where $\omega \in \Omega^1(\mathcal{A}_\theta)$ is given by

$$\omega = \frac{1}{2\pi} a\left(U^{-1}dU - \frac{1}{2}U^{-2}d(U^2)\right) \quad (a \in \mathcal{A}_\theta).$$

It is straightforward to compute the Dixmier trace $\mathrm{Tr}_\omega(TD^{-2})$, where $T = (\lambda(a_{ij}))$ is a 2×2 matrix of elements $\lambda(a_{ij})$ $(a_{ij} \in \mathcal{A}_\theta)$, since $\mathcal{A}_\theta$ (and hence $M_2(\mathcal{A}_\theta)$) has a unique normalized trace τ_0. Thus, we have

$$\mathrm{Tr}_\omega\left((\lambda(a_{ij}))D^{-2}\right) = \tau_0(a_{11}) + \tau_0(a_{22}).$$

With the notation of Section 1, Proposition 5, we thus see that, as an $\mathcal{A}_\theta$-bimodule, the Hilbert space completion $\mathfrak{H}_2$ of $\pi(\Omega^2)$ for the inner product given by the Dixmier trace may be identified with two copies of $L^2(\mathcal{A}_\theta, \tau_0)$, each viewed as an $\mathcal{A}_\theta$-bimodule.

The projection P is the projection onto the orthogonal complement of the elements $\pi(d\omega)$ with $\omega \in \Omega^1$ and $\pi(\omega) = 0$. Thus, P is the projection on the orthogonal complement of the pairs (a_{11}, a_{22}) with $a_{jj} \in \mathcal{A}_\theta$, $a_{11} = a_{22}$. This shows that we can identify the $\mathcal{A}_\theta$-bimodule $L^2(\mathcal{A}_\theta, \tau_0)$ with Ω_D^2 by the map $a \mapsto \begin{bmatrix} \lambda(a) & 0 \\ 0 & -\lambda(a) \end{bmatrix}$. Similarly, we identify the $\mathcal{A}_\theta$-bimodule Ω_D^1 with a sum of two copies of $L^2(\mathcal{A}_\theta, \tau_0)$ using the map $(a_1, a_2) \mapsto \begin{bmatrix} 0 & \lambda(a_1 - ia_2) \\ -\lambda(a_1 + ia_2) & 0 \end{bmatrix}$.

Proposition 13. *With the above identifications of $\mathcal{A}_\theta$-bimodules the differentials $d : \mathcal{A}_\theta \to \Omega^1$, $d : \Omega^1 \to \Omega^2$ and the product: $\Omega^1 \times \Omega^1 \to \Omega^2$ are given by:* $da = (\delta_1 a, \delta_2 a)$, $d(a_1, a_2) = \delta_2(a_1) - \delta_1(a_2)$,

$$(a_1, a_2)(b_1, b_2) = a_1 b_2 - a_2 b_1 \qquad \forall a, a_j, b_j \in \mathcal{A}_\theta.$$

The proof is straightforward. This proposition shows that the notion of compatible connection $\nabla \in C(\mathcal{E})$ on a Hermitian, finitely generated projective module $\mathcal{E}$ over $\mathcal{A}_\theta$, as defined in Section 1, is identical with the notion of compatible connection that we introduced with M. Rieffel in [Co-R] and which depended on the natural parallelizable structure of $\mathcal{A}_\theta$. Moreover, up to normalization, the two actions YM—the first as defined in Section 1, the second as defined in [Co-R]—do coincide. We can thus exploit the results of [Co-R] and state:

Theorem 14. *Let $\mathcal{E}$ be an arbitrary Hermitian, finitely generated projective module over $\mathcal{A}_\theta$ and let d be the largest integer such that $\mathcal{E} = \Lambda^d$ is a multiple of a finitely generated projective module Λ. Then the moduli space of the equivalence classes under $\mathcal{U}(\mathcal{E})$ of the compatible connections ∇ on $\mathcal{E}$ that minimize the action $\mathrm{YM}(\nabla)$ is homeomorphic to $(\mathbb{T}^2)^d / S_d$, the quotient of the dth power of a 2-torus by the action of the symmetric group S_d.*

This shows that even though we started with the irrational rotation algebra, a fairly irrational or singular datum, the Yang-Mills problem takes us back to a fairly regular situation.

We should again stress that the only data used in setting up the Yang-Mills problem are 1) the $*$-algebra $\mathcal{A}_\theta$, and 2) the formal "metric"

$$dU dU^* + dV dV^* \in \Omega_+^2(\mathcal{A}_\theta).$$

We shall now give a thorough description of all the finitely generated projective modules over $\mathcal{A}_\theta$ and of the actual connections that minimize the action YM.

Let us recall from Chapter III Theorem 3.14 that the isomorphism classes of finitely generated projective modules over $\mathcal{A}_\theta$ are parametrized by the intersection of the lattice $\mathbb{Z}^2$ with the half-space

$$\{(x, y);\ x - \theta y > 0\},$$

i.e.,

$$\{(p,q) \in \mathbb{Z}^2;\ p - \theta q > 0\}.$$

The mapping that associates with a finitely generated projective module $\mathcal{E}$ the *positive* real number $p - \theta q$ is the pairing with the trace τ_0 that is the generator of $HC^0(\mathcal{A}_\theta)$. It thus plays the role of the *dimension*, and indeed it is the Murray-von Neumann dimension over the II_1 factor R, the weak closure of $\mathcal{A}_\theta$ on the Hilbert space $\mathfrak{H}$ of the K-cycle $(\mathfrak{H}, D)$.

The mapping that assigns to $\mathcal{E}$ the integer $q \in \mathbb{Z}$ plays the role of the first Chern class and is given by the pairing with the fundamental class $\tau_2 \in HC^2(\mathcal{A}_\theta)$ in cyclic cohomology. It will be relevant to compare it below with the action YM.

Thus, let $(p,q) \in \mathbb{Z}^2$, $q > 0$, be a pair of relatively prime integers ($p = 0$ being allowed). We recall (Section III.3) that a finitely generated projective module $\mathcal{E}$ over $\mathcal{A}_\theta$ is obtained as follows. Let $S(\mathbb{R})$ be the usual Schwartz space of complex-valued functions on the real line and define two operators V_1 and V_2 on $S(\mathbb{R})$ by

$$(V_1\xi)(s) = \xi(s-\epsilon),\quad (V_2\xi)(s) = e(s)\xi(s) \quad (s \in \mathbb{R},\ \xi \in S(\mathbb{R})),$$

where $\epsilon = \frac{p}{q} - \theta$ and $e(s) = \exp 2\pi i s$ $(\forall s \in \mathbb{R})$. Of course,

$$V_2V_1 = e(\epsilon)V_1V_2.$$

Next, let K be a finite-dimensional Hilbert space and let w_1, w_2 be unitary operators on K such that

$$w_2w_1 = \bar{e}(p/q)w_1w_2,\quad w_1^q = w_2^q = 1.$$

We make $\mathcal{E}$ into a right $\mathcal{A}_\theta$-module by defining

$$\xi U = (V_1 \otimes w_1)\xi,\quad \xi V = (V_2 \otimes w_2)\xi \quad (\forall \xi \in S(\mathbb{R}) \otimes K = \mathcal{E}).$$

By Theorem 14 of Chapter III Section 3 the above modules, together with the free modules $\mathcal{A}_\theta^p$, give us the complete list of finitely generated projective modules $\mathcal{E}$ over $\mathcal{A}_\theta$. Since, given such an $\mathcal{E}$, all the Hermitian structures on $\mathcal{E}$ are pairwise isomorphic, we need only describe one of them on $S(\mathbb{R}) \otimes K$.

We view $\mathcal{E}$ as the space $S(\mathbb{R}, K)$ of K-valued Schwartz functions on $\mathbb{R}$ and we define the Hermitian metric, i.e., the $\mathcal{A}_\theta$-valued inner product, by the formula

$$\langle \xi, \eta \rangle_{\mathcal{A}_\theta}(m,n) = \int_{-\infty}^{\infty} \langle w_2^n w_1^m \xi(s - m\epsilon), \eta(s) \rangle \bar{e}(ns)ds,$$

where we have identified the elements a of $\mathcal{A}_\theta$ with sequences $f(m,n)$ of rapid decay on $\mathbb{Z}^2$:

$$a = \sum f(m,n)U^mV^n.$$

It is not difficult to check that this inner product defines a Hermitian structure on $\mathcal{E}$.

All of the above constructions have a clear geometric origin if one remembers that the algebra $\mathcal{A}_\theta$ corresponds to the noncommutative space of leaves of the irrational Kronecker foliation of the 2-torus (cf. Chapter II). Then the closed geodesics of the 2-torus are transverse to the foliation and yield the above description of *Hermitian* finitely generated projective modules. However, we do not want at present to deviate from our canonical route from $\mathcal{A}_\theta$ and the metric $dUdU^* + dVdV^*$ to the Yang-Mills problem under discussion.

We are now ready to describe the connections ∇ that minimize the action YM. By Proposition 13, 2), giving a connection ∇ on a finitely generated projective module $\mathcal{E}$ is equivalent to giving the two covariant differentials ∇_j $(j = 1, 2)$ such that

$$\nabla_j(\xi a) = (\nabla_j \xi)a + \xi\delta_j(a) \quad \forall \xi \in \mathcal{E}, a \in \mathcal{A}_\theta, j = 1, 2$$

(∇_j is a linear mapping from $\mathcal{E}$ to $\mathcal{E}$). Also, by Proposition 13, the curvature is, under the identification of Ω_D^2 with $L^2(\mathcal{A}_\theta, \tau_0)$, equal to the endomorphism $\nabla_1\nabla_2 - \nabla_2\nabla_1$ of $\mathcal{E}$. We shall thus say (as in [Co-R]) that the curvature is *constant* if the endomorphism $\nabla_1\nabla_2 - \nabla_2\nabla_1$ of $\mathcal{E}$ is a scalar multiple of the identity.

Let $\mathcal{E}$ be the Hermitian, finitely generated projective module $S(\mathbb{R}, K)$ as above; we define a connection ∇ on $\mathcal{E}$ by

$$(\nabla_1\xi)(s) = 2\pi i(s/\epsilon)\xi(s), \quad (\nabla_2\xi)(s) = (d\xi/ds)(s) \quad \forall s \in \mathbb{R},$$

where $\epsilon = \frac{p}{q} - \theta$ as above. Straightforward calculations show that this is indeed a connection and that it is compatible with the Hermitian metric defined above.

Moreover, one checks that the curvature $\theta = \nabla_1\nabla_2 - \nabla_2\nabla_1$ is *constant* and is given by

$$\theta = -\frac{2\pi i}{\epsilon} \text{ id}.$$

All of this remains true, with the same value of the constant curvature, for the connections ∇^σ obtained from the formulas

$$(\nabla_1^\sigma \xi)(s) = 2\pi i(\frac{s}{\epsilon})\xi(s) + 2\pi i\sigma_1\xi(s),$$

$$(\nabla_2^\sigma \xi)(s) = (\frac{d\xi}{ds})(s) + 2\pi i\sigma_2\xi(s),$$

where σ_1 and σ_2 are commuting selfadjoint operators in K that commute with w_1 and w_2. We then have:

Theorem 15. [Co-R] 1) *All the above connections* ∇^σ *are compatible, have constant curvature* $-2\pi i/\epsilon$ *and minimize the action* YM(∇).

2) *Every compatible connection* ∇ *that minimizes the action* YM *has constant curvature* $-2\pi i/\epsilon$ *and is gauge equivalent under* $\mathcal{U}(\mathcal{E})$ *to a connection of the form* ∇^σ.

To conclude this section we note that with $\mathcal{E}$ as above, both the dimension (i.e., the Murray-von Neumann dimension $p - \theta q$) and the curvature constant

$$\frac{\theta}{2\pi i} = -\frac{1}{\epsilon}, \quad \text{with } \epsilon = \frac{p}{q} - \theta,$$

are irrational numbers, but their product is *an integer*, which gives the pairing between $\mathcal{E}$ viewed as an element of $K_0(\mathcal{A}_\theta)$ with the cyclic cohomology fundamental class τ_2 of $\mathcal{A}_\theta$ (cf. Chapter III).

VI.4. The Notion of Manifold in Noncommutative Geometry

Let X be a noncommutative space and $\mathcal{A}$ the corresponding $*$-algebra. We saw above that giving a K-cycle $(\mathfrak{H}, D)$ over $\mathcal{A}$ yields a metric $d(\varphi, \psi)$ on the state space of $\mathcal{A}$, permits constructing a differential graded algebra Ω_D^*, and, in the (d, ∞)- summable case, recovers integration in terms of the Dixmier trace. The nontriviality of the K-homology class, i.e., of the stable homotopy class, of the K-cycle $(\mathfrak{H}, D)$ played a crucial role in the residue theorem (Chapter IV.2) in ensuring the nonvanishing of the Dixmier trace on the d-dimensional forms Ω_D^d.

In this section we shall first expound classical results from the theory of manifolds and their characteristic classes, in particular, those of D. Sullivan, which exhibit the central role played by K-homology. We shall then explain how to formulate Poincaré duality in K-homology in the noncommutative context. This will allow us to get closer to the concept of a *manifold* in noncommutative geometry, and we shall see how ordinary manifolds, the noncommutative tori and, later, Euclidean space-time as determined by the $U(1) \times SU(2) \times SU(3)$ standard model, fit in with this algebraic notion of manifold.

4.α The classical notion of manifold

A d-dimensional closed topological manifold X is a compact space locally homeomorphic to open sets in Euclidean space of dimension d. Such local homeomorphisms are called charts. If two charts overlap in the manifold one obtains an overlap homeomorphism between open subsets of Euclidean space. A smooth (resp. $PL \ldots$) structure on X is given by a covering by charts so that all overlap homeomorphisms are smooth (resp. $PL \ldots$). By definition a PL homeomorphism is simply a homeomorphism which is piecewise affine.

Smooth manifolds can be triangulated and the resulting PL structure up to equivalence is uniquely determined by the original smooth structure. We can thus write

$$\text{Smooth} \Longrightarrow PL \Longrightarrow \text{Top}. \tag{1}$$

The above three notions of Smooth, PL and Topological manifolds are compared using the respective notions of tangent bundles. A smooth manifold X possesses a tangent bundle TX which is a *real vector bundle* over X. The stable isomorphism class of TX in the real K-theory of X is classified by the homotopy class of a map

$$X \to BO. \tag{2}$$

Similarly, a PL (resp. Top) manifold possesses a tangent bundle which is no longer a vector bundle but rather a suitable neighborhood of the diagonal in

$X \times X$ for which the projection $(x, y) \mapsto x$ on X defines a PL (resp. Top) bundle. Such bundles are stably classified by the homotopy class of a natural map

$$X \to BPL \qquad (\text{resp. } B\,\text{Top}). \tag{3}$$

The implication (1) yields natural maps:

$$BO \to BPL \to B\,\text{Top} \tag{4}$$

and the nuances between the three above kinds of manifolds are governed by the ability to lift up to homotopy the classifying maps (3) for the tangent bundles. (In dimension 4 this statement has to be made unstably to go from Top to PL.) It follows, for instance, that every PL manifold of dimension $d \leq 7$ possesses a compatible smooth structure. Also for $d \geq 5$, a topological manifold X^d admits a PL structure iff a single topological obstruction $\delta \in H^4(X, \mathbb{Z}/2)$ vanishes.

For $d = 4$ one has Smooth $= PL$ but topological manifolds only sometimes possess smooth structures (and when they do they are not unique up to equivalence) as follows from the work of Donaldson and Freedman.

The KO-orientation of a manifold.

Any finite simplicial complex can be embedded in Euclidean space and has the homotopy type of a manifold with boundary. The homotopy types of these manifolds with boundary are thus rather arbitrary. For closed manifolds this is no longer true and we shall now discuss this point.

Let X be a closed oriented manifold. Then the orientation class $\mu_X \in H_n(X, \mathbb{Z}) = \mathbb{Z}$ defines a natural isomorphism

$$a \in H^i(X) \mapsto a \cap \mu_X \in H_{n-i}(X) \tag{5}$$

which is called the *Poincaré duality* isomorphism. This continues to hold for any space Y homotopy equivalent to X since homology and cohomology are invariant under homotopy.

Conversely, let X be a finite simplicial complex which satisfies Poincaré duality (5) for a suitable class μ_X; then X is called a Poincaré complex. If one assumes that X is simply connected ($\pi_1(X) = \{e\}$), then ([Mi$_4$]) there exists a spherical fibration $E \xrightarrow{p} X$ over X (the fibers $p^{-1}(b)$, $b \in X$ have the homotopy type of a sphere), which is unique up to fibre homotopy equivalence, and which plays the role of the stable tangent bundle when X is homotopy equivalent to a manifold. Moreover, in the simply connected case and with $d = \dim X \geq 5$, the problem of finding a PL manifold in the homotopy type of X is the same as that of promoting this spherical fibration to a PL bundle. There are, in general, obstructions for doing that, but a key result of D. Sullivan [ICM, Nice 1970] asserts that after tensoring the relevant abelian obstruction groups by $\mathbb{Z}\left[\frac{1}{2}\right]$, a PL bundle is the same thing as a spherical fibration together with a KO-orientation. This shows first that the characteristic feature of the homotopy type of a PL manifold is to possess a KO-orientation

$$\nu_X \in KO_*(X) \tag{6}$$

which defines a Poincaré duality isomorphism in real K-theory, after tensoring by $\mathbb{Z}\left[\frac{1}{2}\right]$:

$$a \in KO^*(X)_{1/2} \mapsto a \cap \nu_X \in KO_*(X)_{1/2}. \tag{7}$$

Moreover, it was shown that this element $\nu_X \in KO_*(X)$ describes all the invariants of the PL manifolds in a given homotopy type, provided the latter is simply connected and all relevant abelian obstruction groups are tensored by $\mathbb{Z}\left[\frac{1}{2}\right]$. Among these invariants are the rational Pontryagin classes of the manifold. For smooth manifolds they are the Pontryagin classes of the tangent vector bundle, but, in general, they are obtained from the Chern character of the KO-orientation class ν_X. These classes continue to make sense for topological manifolds and are known to be homeomorphism invariants thanks to the work of S. Novikov.

We can thus assert that, in the simply connected case, a closed manifold is, in a rather deep sense, more or less the same thing as a homotopy type X satisfying Poincaré duality in ordinary homology together with a preferred element $\nu_X \in KO_*(X)$ which induces Poincaré duality in KO-theory tensored by $\mathbb{Z}\left[\frac{1}{2}\right]$. In the non-simply-connected case one has to take into account the equivariance with respect to the fundamental group $\pi_1(X) = \Gamma$ acting on the universal cover $\widetilde{X}$.

4.β Bivariant K-theory and Poincaré duality

The main property of (the stable homotopy class of) the fundamental class in K-homology is that it gives a *Poincaré duality* that exchanges K-homology with K-theory. In the noncommutative case this is a little more involved than in the commutative case, so we shall first review how Poincaré duality is formulated in bivariant K-theory ([Kas$_6$], [Co-S]).

Kasparov's bivariant K-theory ([Kas$_1$]) is a bifunctor KK from the category C^*-Alg of C^*-algebras (with morphisms given by algebra $*$-homomorphisms) to the category of abelian groups. The abelian group associated with a pair (A,B) of C^*-algebras is denoted $KK(A,B)$; it is covariant in B and contravariant in A, so that, for instance, to a morphism $\rho : A_1 \to A_2$ there corresponds a map $\rho^* : KK(A_2,B) \to KK(A_1,B)$. This bifunctor has the following properties:

1) For $A = \mathbb{C}$, $KK(\mathbb{C},B)$ is naturally isomorphic to the K-theory group $K_0(B)$.
2) For $B = \mathbb{C}$, $KK(A,\mathbb{C})$ is the K-homology of A; in particular, to every K-cycle on a $*$-algebra $\mathcal{A}$ there corresponds an element of $KK(A,\mathbb{C})$, where A is the norm closure of $\mathcal{A}$ in $\mathcal{L}(\mathfrak{H})$.
3) $KK(A,B)$ is homotopy invariant. This means that the morphisms ρ^* and ρ_* of abelian groups associated with a morphism ρ of C^*-algebras depend only on the homotopy class of ρ.
4) A bilinear, associative *intersection product* is defined, given C^*-algebras A_1, A_2, B_1, B_2 and D:

$$KK(A_1, B_1 \otimes D) \otimes_D KK(D \otimes A_2, B_2) \longrightarrow KK(A_1 \otimes A_2, B_1 \otimes B_2).$$

We refer the reader to Chapter IV Appendix A for the properties of the intersection product. One way to remember these properties is to think of the elements of $KK(A,B)$ as (homotopy classes of) generalized morphisms from A to B, the intersection product then being composition. For any C^*-algebra A the intersection product provides the abelian group $KK(A,A)$ with a ring structure, whose unit element will be denoted 1_A.

Now let A and B be C^*-algebras and let us assume that we have two elements

$$\alpha \in KK(A \otimes B, \mathbb{C}), \quad \beta \in KK(\mathbb{C}, A \otimes B)$$

such that

$$\begin{aligned} \beta \otimes_A \alpha &= 1_B \in KK(B,B), \\ \beta \otimes_B \alpha &= 1_A \in KK(A,A). \end{aligned} \tag{$*$}$$

It then follows from the general properties of the intersection product that there are canonical isomorphisms

$$\begin{aligned} K_*(A) = KK(\mathbb{C}, A) &\cong KK(B, \mathbb{C}) = K^*(B), \\ K_*(B) = KK(\mathbb{C}, B) &\cong KK(A, \mathbb{C}) = K^*(A) \end{aligned}$$

that exchange the K-theory of A with the K-homology of B.

More explicitly, the map from $K_*(A)$ to $K^*(B)$ is given by the intersection product with α:

$$x \in K_*(A) = KK(\mathbb{C}, A) \to x \otimes_A \alpha \in KK(B, \mathbb{C}) = K^*(B).$$

The inverse map from $K^*(B) = KK(B, \mathbb{C})$ to $K_*(A)$ is given by the intersection product with β:

$$y \in K^*(B) = KK(B, \mathbb{C}) \to \beta \otimes_B y \in KK(\mathbb{C}, A) = K_*(A).$$

More generally, for any pair of C^*-algebras C and D, we have canonical isomorphisms

$$\begin{aligned} KK(C, A \otimes D) &\cong KK(C \otimes B, D), \\ KK(C, B \otimes D) &\cong KK(C \otimes A, D), \end{aligned}$$

which show that the above pair α, β establishes a duality between A and B with arbitrary coefficients.

In [Kas$_5$] and [Co-S] an example of this duality was worked out with $A = C(V)$ and $B = C_0(TV)$, where V is a compact manifold, A is the C^*-algebra of continuous functions on V, and TV is the total space of the tangent bundle of V. The differentiable structure of V then provides, through the pseudodifferential calculus, the desired elements $\alpha \in KK(A \otimes B, \mathbb{C})$, $\beta \in KK(\mathbb{C}, A \otimes B)$ fulfilling the above condition $(*)$.

The Thom isomorphism for vector bundles ([Kas$_1$]) provides a natural KK-equivalence (i.e., an isomorphism in the category of C^*-algebras with $KK(A,B)$ as the morphisms from A to B)

$$C_0(TV) \cong C_V,$$

where C_V is the C^*-algebra of continuous sections of the bundle over V of Clifford algebras $C_p = \text{Cliff}_{\mathbb{C}}(T_p(V))$. We thus get, using the KK-equivalence, a natural duality between $A = C(V)$ and $B = C_V$. We shall now describe in greater detail the corresponding elements

$$\alpha \in KK(C(V) \otimes C_V, \mathbb{C}), \quad \beta \in KK(\mathbb{C}, C(V) \otimes C_V).$$

Since, as a rule, K-homology is always more difficult than K-theory, we shall concentrate on the description of α. The description of β is much simpler.

We shall describe α as a very specific K-cycle on $C(V) \otimes C_V$: we let $\mathfrak{H}$ be the Hilbert space of square-integrable differential forms on V, where V is equipped with a Riemannian metric g:

$$\mathfrak{H} = L^2(V, \wedge_{\mathbb{C}}^* T^*V).$$

We let $D = \mathrm{d} + \mathrm{d}^*$ be the selfadjoint operator in $\mathfrak{H}$ given by the sum of the exterior differential d with its adjoint d^*. The action of $C(V)$ on $\mathfrak{H}$ is the obvious one, by multiplication. For the action of C_V we have the following:

Lemma 1. *Let $(\mathfrak{H}, D)$ be the K-cycle over $C(V)$ given above. Then the commutant on $\mathfrak{H}$ of the algebra generated by $C(V)$ and the $[D, f]$ ($f \in C(V)$, $\|[D, f]\| < \infty$) is canonically isomorphic to the algebra of bounded measurable sections of the bundle C of Clifford algebras.*

Indeed, the commutant of $C(V)$ on $\mathfrak{H}$ is the algebra of bounded measurable sections of the bundle $\text{End}(\wedge_{\mathbb{C}}^* T^*V)$ of endomorphisms of $\wedge_{\mathbb{C}}^* T^*V$, so that it is enough to compute for each $p \in V$ the commutant of the algebra generated by the operators $\gamma(\xi)$, $\xi \in T_p^*(V)$, where

$$\gamma(\xi)\eta = \xi \wedge \eta + i_\xi \eta \quad \forall \eta \in \wedge_{\mathbb{C}}^* T_p^*(V).$$

However, γ defines a representation of the Clifford algebra C_p on the Hilbert space $\wedge_{\mathbb{C}} T_p^*(V)$ with $1 \in \wedge_{\mathbb{C}} T_p^*(V)$ as cyclic and separating vector, so that its commutant is also given by a canonical representation of C_p, given explicitly by the formula

$$\gamma_p'(\xi)\eta = (-1)^{\partial\eta}(\xi \wedge \eta - i_\xi \eta) \quad \forall \eta \in \wedge_{\mathbb{C}}^* T_p^*(V).$$

This K-cycle $(\mathfrak{H}, D)$ over $C(V) \otimes C_V$ defines the fundamental class of V in K-homology, $\alpha \in KK(C(V) \otimes C_V, \mathbb{C})$, and yields, for instance, the construction of the Dirac operator with coefficients in a Clifford bundle ([Gi$_1$]) as the natural mapping

$$K_*(C_V) \to K^*(C(V)).$$

The K-theory class $\beta \in KK(\mathbb{C}, C(V) \otimes C_V)$ is easier to describe; it is just the family, parametrized by $p \in V$, of Bott elements $\beta_p \in K_*(C_V)$ obtained from the Bott periodicity applied to a small disk centered at $p \in V$.

In general, C_V is not Morita equivalent to $C(V)$. Giving a Spin^c structure on V determines such a Morita equivalence and thus permits replacing α and β by equivalent elements

$$\alpha \in KK(C(V) \otimes C(V), \mathbb{C}), \quad \beta \in KK(\mathbb{C}, C(V) \otimes C(V)).$$

This time α is given by the Dirac K-cycle on V and the two representations of $C(V)$ on $\mathfrak{H}$ are identical, thus yielding the diagonal representation of $C(V) \otimes C(V)$ on $\mathfrak{H}$. This is a very special feature of the *commutative* case: if A is abelian then every A-module is in a trivial way an A-bimodule, since one then has the diagonal homomorphism $A \otimes A \to A$. In general, as we saw above, the fundamental class in K-homology involves an algebra A, its Poincaré dual B and (A, B)-bimodules.

We shall now review the construction in Section 3, Example c) of the K-cycle $(\mathfrak{H}, D, \gamma)$ on the $*$-algebra $\mathcal{A}_\theta$ and find out that the $\mathcal{A}_\theta$-module $\mathfrak{H}$ is in fact in a canonical way an $\mathcal{A}_\theta$-bimodule, thus yielding the desired fundamental class for the noncommutative torus:

$$[(\mathfrak{H}, D, \gamma)] = \alpha \in KK(A_\theta \otimes A_\theta^{o}, \mathbb{C}),$$

where A_θ, the norm closure of $\mathcal{A}_\theta$, is the irrational rotation C^*-algebra.

Recall how we constructed $\mathfrak{H}$: we have $\mathfrak{H} = \mathfrak{H}^+ \oplus \mathfrak{H}^-$, where $\mathfrak{H}^-$ is the Hilbert space of the G.N.S. construction relative to the unique normalized trace τ_0, and $\mathfrak{H}^+$ is given by Proposition 10 of Section 3. On $\mathfrak{H}^-$ we have natural actions of $\mathcal{A}_\theta$ on both sides (left and right) on the G.N.S. representation relative to a trace. On $\mathfrak{H}^+$ Proposition 10 of Section 3 shows that we also have a natural *bimodule* structure on $\mathcal{A}_\theta$. We thus see that $\mathfrak{H}$ is naturally an $\mathcal{A}_\theta \otimes \mathcal{A}_\theta^{o}$-module and it remains to verify that the operator $D = \begin{bmatrix} 0 & \partial \\ \partial^* & 0 \end{bmatrix}$ is still a K-cycle for $\mathcal{A}_\theta \otimes \mathcal{A}_\theta^{o}$. In fact, we have the following analogue of Lemma 1:

Lemma 2. a) *For any $a, b \in \mathcal{A}_\theta$ the commutator $[D, a \otimes b^{o}]$ is bounded.*

b) *The commutant on $\mathfrak{H}$ of the algebra generated by the left action of $\mathcal{A}_\theta$ and the commutators $[D, a]$ $(a \in \mathcal{A}_\theta)$ is the weak closure $R = (\mathcal{A}_\theta^{o})''$ of the right action of $\mathcal{A}_\theta$ on $\mathfrak{H}$.*

The assertion a) is clear. To see b), just use Lemma 3.12, which shows that (with the notation of that lemma) the $*$-algebra generated by $\mathcal{A}_\theta$ and the $[D, a]$ $(a \in \mathcal{A}_\theta)$ is the left action of $M_2(\mathcal{A}_\theta)$ on $\mathfrak{H}$.

This is a good point at which to comment a little on noncommutative measure theory in this example; the von Neumann algebra $(\mathcal{A}_\theta^{o})''$ is the hyperfinite factor R of type II_1 (Chapter V), and the Dixmier trace gives the following formula for a *hypertrace* on R:

$$\varphi(T) = \mathrm{Tr}_\omega(TD^{-2}) \quad \forall T \in \mathcal{L}(\mathfrak{H})$$

(cf. Chapter V).

The $\mathbb{Z}/2$-grading $\gamma = \begin{bmatrix} 1 & 0 \\ 0 & -1 \end{bmatrix}$ on $\mathfrak{H} = \mathfrak{H}^+ \oplus \mathfrak{H}^-$ then turns $(\mathfrak{H}, D, \gamma)$ into an element $\alpha \in KK(A_\theta \otimes A_\theta^{\rm o}, \mathbb{C})$.

We leave it to the reader to determine a K-theory class $\beta \in K_0(A_\theta \otimes A_\theta^{\rm o}) = KK(\mathbb{C}, A_\theta \otimes A_\theta^{\rm o})$ such that the pair (α, β) satisfies the rules $(*)$ of Poincaré duality.

The lesson that we want to draw from this section is that the K-homology fundamental class of a noncommutative space is given by a *bimodule*, not just a module, $(\mathfrak{H}, D)$.

We should of course stress that we are interested in the actual K-cycle $(\mathfrak{H}, D)$ over $\mathcal{A} \otimes \mathcal{B}$ and not only in its stable homotopy class.

4.γ Poincaré duality and cyclic cohomology

The above discussion involved Poincaré duality in K-theory; we shall now turn to cohomology. We showed in Section 1 how to associate to a K-cycle $(\mathfrak{H}, D)$ on a $*$-algebra $\mathcal{A}$ a differential graded algebra Ω_D^* which, in the case of the Dirac K-cycle on a Riemannian manifold, gives the de Rham differential algebra of ordinary forms. Now, Ω_D^* makes sense in general, and, as does every differential graded algebra, it has a cohomology ring $H^*(\Omega_D^*)$ which, in general, will fail to be graded commutative. We shall now explain how in the finitely summable case, using the Dixmier trace, one has natural maps

A) $\Omega_D^k(\mathcal{A}) \to H^{d-k}(\mathcal{B}, \mathcal{B}^*)$,

B) $H^k(\Omega_D^*(\mathcal{A})) \to H^{d-k}(\mathcal{B})$,

that relate a pair $\mathcal{A}$, $\mathcal{B}$ of Poincaré dual algebras, where $H^*(\mathcal{B}, \mathcal{B}^*)$ and $H^*(\mathcal{B})$ are, respectively, the Hochschild and the periodic cyclic cohomologies of the algebra $\mathcal{B}$.

Note that unlike $H^*(\Omega_D^*(\mathcal{A}))$, the periodic cyclic cohomology $H^*(\mathcal{B})$ of an algebra $\mathcal{B}$ *does not*, in general, have a natural ring structure. In fact, even in the commutative case, say if $\mathcal{B}$ is the algebra of Lipschitz functions on a simplicial complex X, the cyclic *cohomology* $H^*(\mathcal{B})$ of $\mathcal{B}$ is the *homology* of X, which, unless X is a manifold, has no natural ring structure.

Thus, let $(\mathfrak{H}, D, \gamma)$ be an even (d, ∞)-summable K-cycle over $\mathcal{A} \otimes \mathcal{B}$. In order to construct these mappings we shall need the following conditions, which, thanks to the above Lemmas 1 and 2, are fulfilled in the examples.

The order 1 condition on D: $[[D, a], b] = 0 \quad \forall a \in \mathcal{A}, b \in \mathcal{B}$

Note that since $[a, b] = 0$ $(\forall a \in \mathcal{A}, b \in \mathcal{B})$, this condition is symmetric in $\mathcal{A}$ and $\mathcal{B}$. Note also that given the K-cycle $(\mathfrak{H}, D)$ on $\mathcal{A}$, there is a largest algebra $\mathcal{B}$ fulfilling the above condition; namely, if M is the von Neumann algebra commutant of $\mathcal{A} \cup \{[D, a];\ a \in \mathcal{A}\}$ then

$$\mathcal{B} = \{x \in M;\ [D, x] \text{ is bounded}\}.$$

The second condition is the one we already met in Theorem 4 of Section 2; namely, we want to assume that $B\Phi_\omega = 0$, where Φ_ω is the Hochschild cocycle on $\mathcal{A} \otimes \mathcal{B}$ given by

$$\mathrm{Tr}_\omega(\gamma x^0[D,x^1]\cdots[D,x^n]|D|^{-n}) = \Phi_\omega(x^0,\ldots,x^n).$$

The closedness condition: $\mathrm{Tr}_\omega(\gamma[D,x^1]\cdots[D,x^n]|D|^{-n}) = 0 \quad \forall x^j \in \mathcal{A}\otimes\mathcal{B}$

This condition is fulfilled and easy to check in the examples of Riemannian manifolds with the Dirac K-cycle, or in the $C(V)\otimes C_V$ example discussed above. Let us verify it in the case of the noncommutative torus with $\mathcal{A} = \mathcal{A}_\theta$ and $\mathcal{B} = \mathcal{A}_\theta$. We have to show that

$$\mathrm{Tr}_\omega(\gamma[D,a^1\otimes b^1][D,a^2\otimes b^2]D^{-2}) = 0 \quad \forall a^j, b^j \in \mathcal{A}_\theta,$$

the action of $a\otimes b$ on $\mathfrak{H}$ being given by the 2×2 matrix

$$\pi(a\otimes b) = \begin{bmatrix} \lambda(a)\lambda'(b) & 0 \\ 0 & \lambda(a)\lambda'(b)\end{bmatrix}, \qquad a\in\mathcal{A}_\theta,\ b\in\mathcal{A}_\theta^0,$$

where λ (resp. λ') is the left (resp. right) regular representation of $\mathcal{A}_\theta$ on $L^2(\mathcal{A}_\theta,\tau_0)$ (cf. Lemma 12 of Section 3). Since $D = \begin{bmatrix} 0 & \partial \\ \partial^* & 0\end{bmatrix}$, we have

$$[D,\pi(a\otimes b)] = \begin{bmatrix} 0 & \lambda(\partial a)\lambda'(b)+\lambda(a)\lambda'(\partial b) \\ \lambda(\partial^* a)\lambda'(b)+\lambda(a)\lambda'(\partial^* b) & 0\end{bmatrix}$$

just using the derivation rule for both $\partial = \delta_1 - i\delta_2$ and $\partial^* = -\delta_1 - i\delta_2$. Next, for any 2×2 matrix $X_{ij} = \lambda(a_{ij})\lambda'(b_{ij})$ with $a_{ij}, b_{ij} \in \mathcal{A}_\theta$, one verifies that, by the uniqueness of the trace on $\mathcal{A}_\theta$,

$$\mathrm{Tr}_\omega([X_{ij}]D^{-2}) = \tau_0(a_{11})\tau_0(b_{11}) + \tau_0(a_{22})\tau_0(b_{22}).$$

Now $[D,a^1\otimes b^1][D,a^2\otimes b^2] = [X_{ij}]$, with

$$X_{11} = (\lambda(\partial a^1)\lambda'(b^1)+\lambda(a^1)\lambda'(\partial b^1))(\lambda(\partial^* a^2)\lambda'(b^2)+\lambda(a^2)\lambda'(\partial^* b^2)),$$
$$X_{22} = (\lambda(\partial^* a^1)\lambda'(b^1)+\lambda(a^1)\lambda'(\partial^* b^1))(\lambda(\partial a^2)\lambda'(b^2)+\lambda(a^2)\lambda'(\partial b^2)).$$

Thus, $\mathrm{Tr}_\omega(\gamma[D,a^1\otimes b^1][D,a^2\otimes b^2]D^{-2})$ is the sum of 8 terms

$$\begin{aligned}&\tau_0(\partial a^1\partial^* a^2)\tau_0(b^1b^2) + \tau_0((\partial a^1)a^2)\tau_0(b^1\partial^* b^2)\\ &+\tau_0(a^1\partial^* a^2)\tau_0((\partial b^1)b^2) + \tau_0(a^1a^2)\tau_0(\partial b^1\partial^* b^2)\\ &-\tau_0(\partial^* a^1\partial a^2)\tau_0(b^1b^2) - \tau_0(\partial^* a^1 a^2)\tau_0(b^1\partial b^2)\\ &-\tau_0(a^1\partial a^2)\tau_0((\partial^* b^1)b^2) - \tau_0(a^1a^2)\tau_0(\partial^* b^1\partial b^2).\end{aligned}$$

One sees that these terms add up to 0, thanks to the following relations:

α) $\tau_0(\partial a^1\partial^* a^2) = \tau_0(\partial^* a^1\partial a^2)$ (using transposition and the commutation of ∂ and ∂^*). Similarly, $\tau_0(\partial b^1\partial^* b^2) = \tau_0(\partial^* b^1\partial b^2)$.

β) $\tau_0((\partial a^1)a^2) = -\tau_0(a^1\partial a^2)$ (transposition), $\tau_0(b^1\partial^* b^2) = -\tau_0((\partial^* b^1)b^2)$.

We are now ready to construct the mappings A and B in general.

Lemma 3. *Let* $(\mathfrak{H}, D, \gamma)$ *be a* (d,∞)*-summable* $(\mathcal{A},\mathcal{B})$*-bimodule satisfying the order* 1 *condition. Then:*

1) *For every* $k \le d$ *and* $\alpha \in \Omega^k(\mathcal{A})$, *a Hochschild cocycle* $C_\alpha \in Z^{d-k}(\mathcal{B},\mathcal{B}^*)$ *is defined by*

$$C_\alpha(b^0,\dots,b^{d-k}) = \mathrm{Tr}_\omega(\gamma\pi(\alpha)b^0[D,b^1]\cdots[D,b^{d-k}]|D|^{-d}) \quad \forall b^j \in \mathcal{B}.$$

2) *If the closedness condition is satisfied, then* C_α *depends only on the class of* α *in* $\Omega_D^k(\mathcal{A})$, *and we have*

$$B_0 C_\alpha = (-1)^k C_{\mathrm{d}\alpha}.$$

Here B_0 is the cyclic cohomology operator

$$(B_0\varphi)(b^0,\dots,b^{q-1}) = \varphi(1,b^0,\dots,b^{q-1}) + (-1)^q\varphi(b^0,\dots,b^{q-1},1).$$

To verify 1), simply note that $\pi(\alpha)$ belongs to the commutant of $\mathcal{B}$; this is sufficient because when computing the coboundary

$$bC_\alpha(b^0,\dots,b^{d-k+1}),$$

one gets only two terms

$$\mathrm{Tr}_\omega(\gamma\pi(\alpha)b^0[D,b^1]\cdots[D,b^{d-k}]b^{d-k+1}|D|^{-d}) \\ - \mathrm{Tr}_\omega(\gamma\pi(\alpha)b^{d-k+1}b^0[D,b^1]\cdots[D,b^{d-k}]|D|^{-d}),$$

and these terms cancel due to the commutation of b^{d-k+1} with $|D|^{-d}$ modulo trace class operators.

To verify 2), let us consider the differential graded algebra $\Omega_D^*(\mathcal{A}\otimes\mathcal{B})$. First, we have $[\mathrm{d}a, b] = 0$ ($\forall a \in \mathcal{A}, b \in \mathcal{B}$) and similarly $[\mathrm{d}b, a] = 0$, using the order 1 condition. Next, for $\alpha \in \Omega_D^d(\mathcal{A}\otimes\mathcal{B})$, the value of $\mathrm{Tr}_\omega(\gamma\pi(\alpha_0)|D|^{-d})$, where $\alpha_0 \in \Omega^d(\mathcal{A}\otimes\mathcal{B})$ is any representative of α, is well-defined. Indeed, it depends only on $\pi(\alpha_0)$, and if $\alpha_0 \in \mathrm{d}J_0^{d-1}$ then it vanishes by the closedness condition. We shall denote this value by $\int\alpha$, and we note that it gives a closed, graded trace on the differential algebra $\Omega_D^*(\mathcal{A}\otimes\mathcal{B})$.

Using the natural homomorphism $\Omega_D^*(\mathcal{A}) \to \Omega_D^*(\mathcal{A}\otimes\mathcal{B})$, the quotient of $\Omega^*(\mathcal{A}) \to \Omega^*(\mathcal{A}\otimes\mathcal{B})$, we see that C_α only depends on the class $\alpha \in \Omega_D^k(\mathcal{A})$. By construction, the Hochschild cocycle $C_\alpha(b^0,\dots,b^{d-k})$ vanishes if any b^j, $j \ge 1$, is one. Thus,

$$\begin{aligned}(B_0C_\alpha)(b^0,\dots,b^{d-k-1}) &= \int \alpha\,\mathrm{d}b^0\mathrm{d}b^1\cdots\mathrm{d}b^{d-k-1} \\ &= (-1)^k\int(\mathrm{d}\alpha)b^0\mathrm{d}b^1\cdots\mathrm{d}b^{d-k-1} \\ &= (-1)^k C_{\mathrm{d}\alpha}(b^0,\dots,b^{d-k-1}).\end{aligned}$$

Proposition 4. *Let $(\mathfrak{H}, D, \gamma)$ be an $(\mathcal{A}, \mathcal{B})$-bimodule which is (d, ∞)-summable and satisfies both the order 1 and the closedness condition. Then:*

1) *For $0 \leq k \leq d$, the mapping $\alpha \to C_\alpha$ is well-defined from $\Omega_D^k(\mathcal{A})$ to the Hochschild cocycles $Z^{d-k}(\mathcal{B}, \mathcal{B}^*)$.*

2) *The image under C of $\ker d \subset \Omega_D^k(\mathcal{A})$ is contained in $Z_\lambda^{d-k}(\mathcal{B})$, i.e., C_α is a cyclic cocycle if $d\alpha = 0$ in $\Omega_D^{k+1}(\mathcal{A})$.*

3) *The image under C of $\operatorname{im} d \subset \Omega_D^k(\mathcal{A})$ is contained in $\operatorname{im} B$, where $B : H^{d-k+1}(\mathcal{B}, \mathcal{B}^*) \to HC^{d-k}(\mathcal{B})$ is the cyclic cohomology operation.*

4) *C defines a mapping from $H^*(\Omega_D^*(\mathcal{A}))$ to periodic cyclic cohomology $H^*(\mathcal{B})$ that is compatible with the natural filtrations.*

The first assertion follows from Lemma 3. To prove 2), just note that by Lemma 3, if $d\alpha = 0$ then $B_0 C_\alpha = 0$, but since C_α is a Hochschild cocycle it then follows that C_α is also a cyclic cocycle. To get 3), note that if $\alpha = d\beta$ then first of all, $d\alpha = 0$, so that C_α is a cyclic cocycle by (2). Then, by 2) of Lemma 3, we have $C_\alpha = (-1)^{k-1} B_0 C_\beta$ and, since C_α is cyclic, $AC_\alpha = (d-k+1)C_\alpha$, where A is cyclic antisymmetrization; thus $C_\alpha = \frac{(-1)^{k-1}}{d-k+1} BC_\beta$ belongs to the range of B.

Thus, C is a well-defined mapping from $H^k(\Omega_D^*)$ to $HC^{d-k}(B)/\operatorname{im} B$, and the assertion 4) follows.

4.δ Bivector potentials on an $(\mathcal{A}, \mathcal{B})$-bimodule $(\mathfrak{H}, D, \gamma)$

Let $\mathcal{A}, \mathcal{B}$ be a pair of $*$-algebras and let $(\mathfrak{H}, D, \gamma)$ be an even K-cycle on $\mathcal{A} \otimes \mathcal{B}^o$ that satisfies the order 1 condition:

$$[[D, a], b] = 0 \quad (\forall a \in \mathcal{A},\ b \in \mathcal{B}^o). \tag{$*$}$$

Since $(\mathfrak{H}, D, \gamma)$ is a K-cycle over $\mathcal{A} \otimes \mathcal{B}^o = C$, the entire construction of Section 1 applies, in particular, the concepts of a vector potential V and its Yang-Mills action $\mathrm{YM}(V)$ when $(\mathfrak{H}, D, \gamma)$ is (d, ∞)-summable. We only want to remark here that if one considers the group $G = \mathcal{U}_\mathcal{A} \times \mathcal{U}_\mathcal{B}$ which is the product of the unitary group of $\mathcal{A}$ by that of $\mathcal{B}$, then there is a natural affine subspace $\mathcal{V}$ of the space $\mathcal{V}_{\mathcal{A} \otimes \mathcal{B}^o}$ of vector potentials for $\mathcal{A} \otimes \mathcal{B}^o$, which satisfies the following conditions:

α) $\mathcal{V}$ is invariant under the affine action of $G = \mathcal{U}_\mathcal{A} \times \mathcal{U}_\mathcal{B}$.

β) For every $V \in \mathcal{V}$ the operator $D+V$ also satisfies the order 1 condition $(*)$.

Proposition 5. *If $\mathcal{V} = \mathcal{V}_\mathcal{A} + \mathcal{V}_{\mathcal{B}^o}$ is the subspace of $\mathcal{V}_{\mathcal{A} \otimes \mathcal{B}^o}$ of sums of vector potentials relative to $\mathcal{A}$ and $\mathcal{B}^o$, then $\mathcal{V}$ satisfies the above conditions α) and β).*

The action of the unitary group of $\mathcal{A} \otimes \mathcal{B}^o$ on vector potentials is determined (cf. Section 1) by the equality $g(D+V)g^* = D + \gamma_g(V)$. Let us specialize it to elements $g = uv \in \mathcal{U}_\mathcal{A} \times \mathcal{U}_\mathcal{B}$ and verify α). Let $V = V^a + V^b \in \mathcal{V}_\mathcal{A} + \mathcal{V}_\mathcal{B}$. Then

$$uv(D + V^a + V^b)v^* u^* = uvDv^* u^* + uV^a u^* + vV^b v^*,$$

since, by the order 1 condition, every element V^a of $\mathcal{V}^a$ (resp. V^b of $\mathcal{V}^b$) commutes with $\mathcal{B}$ (resp. $\mathcal{A}$). Next,

$$u v D v^* u^* = u(D + v[D, v^*])u^* = D + u[D, u^*] + v[D, v^*],$$

using again the order 1 condition. Thus, we get

$$\gamma_{uv}(V^a + V^b) = \gamma_u(V^a) + \gamma_v(V^b),$$

which shows that the action of the gauge group is the product action. One checks β) similarly.

VI.5. The Standard $U(1) \times SU(2) \times SU(3)$ Model

In this section, we shall start from the main point of the computation of our Yang-Mills functional in Example b) of Section 3 (referred to briefly as Example 3b)), i.e., in the case of the product of a continuum by a discrete 2-point space. The point is that we recovered the Glashow-Weinberg-Salam model for leptons, with the five different pieces of its Lagrangian, from this simple modification of the (4-dimensional) continuum. The question we shall answer in the present section is the following: Can one, by a similar procedure, incorporate the quarks as well as strong interactions?

Before embarking on this problem, some preparation is required to explain better what our aim is. First, there is at present (1993) no question that the standard model of electro-weak and strong interactions is a remarkably successful phenomenological model of particle physics. Since I did not take any part in its elaboration, I shall refrain from a survey of the experimental roots of this model or of the long history of its elaboration. I refer the reader to the beautiful book of A. Pais [P] or to the more technical papers [Ell]. This seems an important prerequisite for a mathematician reader of the present section, who might otherwise underestimate the depth of the physical roots of the model.

Next, by the work of 't Hooft [Hoo] this model is renormalizable ([Bo-S], [F-S]), a necessary requirement for applying the only known perturbative recipe for quantizing the theory. It nevertheless has problems, such as the naturalness problem [Ell], which make specialists doubt that it is really of fundamental significance, thus leading them to look for alternative routes, grand unification, technicolor... These alternate routes all share a common feature: they deny any fundamental significance to the Higgs boson.

Our contribution does not throw any new light on the above theoretical problems of the standard model, since it is limited to the *classical level*. However, it specifies very precisely which modification of the continuum, in fact its replacement by a product with a *finite space*, entails that the Lagrangian of quantum electrodynamics becomes the Lagrangian of the standard model. As we shall see, the geometry of the finite set will be, as advocated above, specified by its *Dirac operator*, and this will be an operator in a finite-dimensional

Hilbert space encoding both the masses of the elementary particles and the Kobayashi-Maskawa mixing parameters.

Once the structure of this finite space F is given, we merely apply our general action to the space (continuum) $\times F$ to get the standard model action. In many ways, our contribution should be regarded as an *interpretation*, of a geometric nature, of all the intricacies of the most accurate phenomenological model of high-energy physics; if it makes the model more intelligible to a mathematical audience, then our purpose will in some small measure be achieved. It does undoubtedly confirm that high-energy (i.e., small-distance) physics is in fact unveiling the fine structure of space-time. Finally, it gives a status to the Higgs boson as just another gauge field, but corresponding to a finite difference rather than a differential.

5.α The dictionary

We shall first need to refine our dictionary between the language of noncommutative geometry and that of particle physics.

In the basic data of noncommutative geometry, $(\mathcal{A}, \mathfrak{H}, D)$, the Hilbert space $\mathfrak{H}$ and the operator D have a straightforward translations (cf. Example 3b)):

$$\mathfrak{H} = \text{the Hilbert space of Euclidean fermions,}$$
$$D = \text{the inverse of the Euclidean propagator of fermions.}$$

The algebra $\mathcal{A}$ is related to the gauge group of local gauge transformations, but the relation deserves to be spelled out at the mathematical level. It is given by

$$*\text{-algebra } \mathcal{A} \to \text{ unitary group } \mathcal{U}(\mathcal{A}).$$

We need to make some comments on this functor $\mathcal{U}$ from $*$-algebras to groups.

Firstly, a similar functor, namely $\mathcal{A} \to GL(\mathcal{A})$, which replaces an algebra $\mathcal{A}$ by its group of invertible elements, plays a fundamental role in algebraic K-theory ([Mi$_1$]).

When the above functor $\mathcal{U}$ is applied to $M_n(\mathcal{A})$, the algebra of $n \times n$ matrices over $\mathcal{A}$, and one retains the corresponding inclusion of $U(n) = \mathcal{U}(M_n(\mathbb{C})) \subset G = \mathcal{U}(M_n(\mathcal{A}))$, one recovers uniquely the algebra $\mathcal{A}$ from the pair of groups $U(n) \subset G$, provided that $n > 2$. Indeed, at the Lie algebra level one knows the inclusion of the complexified Lie algebra of $U(n)$, i.e., the Lie algebra of matrices, in the complexified Lie algebra of $\mathcal{U}(M_n(\mathcal{A}))$, i.e., again the Lie algebra $M_n(\mathcal{A})$ with bracket $[a, b] = ab - ba$. Let e_{ij}, $i, j = 1, \ldots, n$, be the usual matrix units in $M_n(\mathbb{C})$, and $H(\lambda)$ the diagonal matrix $\sum_{j=1}^{n} \lambda_j \, e_{jj}$ for $\lambda_j \in \mathbb{C}$. Then for $i \neq j$ the subspace of $M_n(\mathcal{A})$

$$E_{ij} = \left\{x; [H(\lambda), x] = (\lambda_i - \lambda_j)\, x \quad , \quad \forall \lambda_j \in \mathbb{C}\right\}$$

is equal to $\mathcal{A}\, e_{ij}$ and these subspaces are pairwise isomorphic using the maps $x \mapsto [x, e_{k\ell}]$. Thus the algebra structure on $\mathcal{A}$ is uniquely recovered from the equality

$$[a\, e_{ij}, b\, e_{jk}] = ab\, e_{ik} \quad i \neq k\,,\ a, b \in \mathcal{A}.$$

This shows that the replacement of an algebra by its associated groups loses very little information, at least in the nonabelian case. Giving the algebra $\mathcal{A}$ from which the group $G = \mathcal{U}(\mathcal{A})$ comes singles out a very narrow class of representations of G. Indeed, there is a natural mapping

$$\mathrm{Rep}(\mathcal{A}) \xrightarrow{\mathrm{Res}} \mathrm{Rep}(G)$$

which associates to every unitary representation π of $\mathcal{A}$ on a Hilbert space $\mathfrak{H}_\pi$ its restriction to the unitary group $\mathcal{U}(\mathcal{A})$. This mapping is, by construction, compatible with direct sums, but—and this is an essential point—while group representations $\rho_1, \rho_2 \in \mathrm{Rep}\, G$ can be tensored, yielding a representation $\rho = \rho_1 \otimes \rho_2$ of G on $\mathfrak{H}_1 \otimes \mathfrak{H}_2$, there is *no corresponding operation* for representations of an (involutive) algebra.

In particular, the above mapping Res : $\mathrm{Rep}(\mathcal{A}) \to \mathrm{Rep}(\mathcal{U}(\mathcal{A}))$ is, in general, *far from surjective.* Just take the simplest example, $\mathcal{A} = M_n(\mathbb{C})$ so that $G = U(n)$, and the range of Res consists of the multiples of the fundamental representation φ of $U(n)$ in $\mathbb{C}^n$. In other words, to assume that a representation ρ of G comes from a representation of $\mathcal{A}$ by restriction is a very restrictive condition.

Now, the group G does have a clear significance as the group of local gauge transformations, and we shall take it as a characteristic of its representation in the *one-particle* space of Euclidean fermions that this representation of $\mathcal{U}(\mathcal{A})$ comes by restriction from a representation of $\mathcal{A}$.

Notice also that it is only through the group $G = \mathcal{U}(\mathcal{A})$ and the corresponding algebra $\mathcal{A}$ that Euclidean space-time enters the picture, since in quantum field theory the space-time points merely play the role of indices or labels except in their occurrence in the construction of the group of *local* gauge transformations. Thus, the last piece of the dictionary is:

$*$-algebra $\mathcal{A}$ $\to$ $\mathcal{U}(\mathcal{A})$ = Group of local gauge transformations on Euclidean space-time.

All of this works perfectly well for the Glashow-Weinberg-Salam model for leptons, as shown in Example 3b), and we shall see how the incorporation of quarks and strong interactions will require the setup of Poincaré duality and *bimodules* of Section 4. We first need to recall the detailed description of the standard model.

5.β The standard model

Just as for the Glashow-Weinberg-Salam model for leptons, the Lagrangian of the standard model contains five different terms,

$$\mathcal{L} = \mathcal{L}_G + \mathcal{L}_f + \mathcal{L}_\varphi + \mathcal{L}_Y + \mathcal{L}_V,$$

which we now describe together with the field content of the theory. (As before, we shall use the Einstein convention of summation over repeated indices.)

1) The pure gauge boson part $\mathcal{L}_G$

$$\mathcal{L}_G = \frac{1}{4}(G_{\mu\nu a}G_a^{\mu\nu}) + \frac{1}{4}(F_{\mu\nu}F^{\mu\nu}) + \frac{1}{4}(H_{\mu\nu b}H_b^{\mu\nu}),$$

where $G_{\mu\nu a}$ is the field strength tensor of an $SU(2)$ gauge field $W_{\mu a}$, F_μ is the field strength tensor of a $U(1)$ gauge field B_μ, and $H_{\mu\nu b}$ is the field strength tensor of an $SU(3)$ gauge field $V_{\mu b}$. This last gauge field, the *gluon field*, is the carrier of the strong force; the gauge group $SU(3)$ is the *color group*, and is thus the essential new ingredient. The respective coupling constants for the fields W, B, and V will be denoted g, g', and g'', consistent with the previous notation.

2) The fermion kinetic term $\mathcal{L}_f$

To the leptonic terms

$$-\sum_f [\bar{f}_L \gamma^\mu (\partial_\mu + ig\frac{\tau_a}{2}W_{\mu a} + ig'\frac{Y_L}{2}B_\mu)f_L + \bar{f}_R\gamma^\mu(\partial_\mu + ig'\frac{Y_R}{2}B_\mu)f_R],$$

one adds the following similar terms involving the quarks:

$$-\sum_f [\bar{f}_L \gamma^\mu (\partial_\mu + ig\frac{\tau_a}{2}W_{\mu a} + ig'\frac{Y_L}{2}B_\mu + ig''\lambda_b V_{\mu b})f_L$$
$$+ \bar{f}_R\gamma^\mu(\partial_\mu + ig'\frac{Y_R}{2}B_\mu + ig''\lambda_b V_{\mu b})f_R].$$

For each of the three generations of quarks $\begin{bmatrix} u \\ d \end{bmatrix}$, $\begin{bmatrix} c \\ s \end{bmatrix}$, and $\begin{bmatrix} t \\ b \end{bmatrix}$ one has a left-handed isodoublet (such as $\begin{bmatrix} u_L \\ d_L \end{bmatrix}$), two right-handed $SU(2)$ singlets (such as $\begin{bmatrix} u_R \\ d_R \end{bmatrix}$), and each quark field appears in 3 colors so that, for instance, there are three u_R fields: u_R^r, u_R^y, u_R^b. All of these quark fields are thus in the fundamental representation **3** of $SU(3)$.

The hypercharges Y_L and Y_R are identical for different generations and are given by the following table:

	e, μ, τ	ν_e, ν_μ, ν_τ	u, c, t	d, s, b
Y_L	-1	-1	$\frac{1}{3}$	$\frac{1}{3}$
Y_R	-2		$\frac{4}{3}$	$-\frac{2}{3}$

These numbers are not explained by theory but are set by hand so as to get the correct electromagnetic charges Q_{em} from the formulas

$$2Q_{\text{em}} = Y_L + 2I_3, \quad 2Q_{\text{em}} = Y_R,$$

where I_3 is the third generator of the weak isospin group $SU(2)$.

3) The kinetic terms for the Higgs fields

$$\mathcal{L}_\varphi = -\left|\left(\partial_\mu + ig\frac{\tau_a}{2}W_{\mu a} + i\frac{g'}{2}B_\mu\right)\varphi\right|^2,$$

where $\varphi = \begin{bmatrix} \varphi_1 \\ \varphi_2 \end{bmatrix}$ is an $SU(2)$ doublet of complex scalar fields with hypercharge $Y_\varphi = 1$. This term is *exactly the same* as in the G.W.S. model for leptons.

4) The Yukawa coupling of Higgs fields with fermions

$$\mathcal{L}_Y = -\sum_{f,f'}[H_{ff'}\,\overline{f}_L \cdot \varphi\, f'_R + H^*_{ff'}\overline{f'_R}(\varphi^* \cdot f_L)],$$

where $H_{ff'}$ is a general coupling matrix in the space of different fermions, about which we must now be more explicit. First, there is no $H_{ff'} \neq 0$ between leptons and quarks, so that $\mathcal{L}_Y$ is a sum of a leptonic and a quark part. Since there is no right-handed neutrino in this model, the leptonic part can always be put into the form

$$\mathcal{L}_{Y,\text{lepton}} = -G_e(\overline{L}_e \cdot \varphi)e_R - G_\mu(\overline{L}_\mu \cdot \varphi)\mu_R - G_\tau(\overline{L}_\tau \cdot \varphi)\tau_R + \text{h.c.},$$

where L_e is the isodoublet $\begin{bmatrix} \nu_{e,L} \\ e_L \end{bmatrix}$, and similarly for the other generations. The coupling constants G_e, G_μ, and G_τ provide the lepton masses through the Higgs vacuum contribution.

The quark Yukawa coupling is more complicated owing to new terms which provide the masses of the up particles, and to the mixing angles. We have three new terms. The first is of the form

$$G\overline{L}u_R\tilde{\varphi}, \qquad (*)$$

where the isodoublet $L = \begin{bmatrix} u_L \\ q_L \end{bmatrix}$ is obtained from a left-handed up quark and a mixing q_L of left-handed down quarks (taken from the three families); the two others have a similar structure but with the up quark replaced by the charm and top quarks respectively. Also, $\tilde{\varphi}$ needs to have the same isospin but opposite hypercharge to the Higgs doublet φ and is given by

$$\tilde{\varphi} = J\varphi^*, \quad J = \begin{bmatrix} 0 & 1 \\ -1 & 0 \end{bmatrix}. \qquad (**)$$

We refer to [Ell] for more details on this point, to which we shall return later.

5) The Higgs self-interaction

$$\mathcal{L}_V = \mu^2 \varphi^+ \varphi - \frac{1}{2}\lambda(\varphi^+ \varphi)^2$$

has exactly the same form as in the previous case.

Thus, we see that there are, essentially, three novel features of the complete standard model with respect to the leptonic case:

A) The new gauge symmetry: *color*, with gluons responsible for the strong interaction.

B) The new values $\frac{1}{3}$, $\frac{4}{3}$, $-\frac{2}{3}$ of the hypercharge for quarks.

C) The new Yukawa coupling terms $(*)$.

We shall now briefly explain how these new features motivate a corresponding modification of Example 3b), which led us above to the G.W.S. model for leptons. First, our model will still be a *product* of an ordinary Euclidean continuum by a finite space.

In Example 3b), for the algebra $\mathcal{A}$ of functions on the finite space, we took the algebra $\mathbb{C}_a \oplus \mathbb{C}_b$. But since we then considered a bundle on $\{a, b\}$ with fiber $\mathbb{C}^2$ over a and $\mathbb{C}$ over b, we could have in an equivalent fashion taken $\mathcal{A} = M_2(\mathbb{C}) \oplus \mathbb{C}$ and then dealt with vector potentials, instead of connections on vector bundles. Let us see how C) leads to replacing $\mathcal{A} = M_2(\mathbb{C}) \oplus \mathbb{C}$ by $\mathcal{A} = \mathbb{H} \oplus \mathbb{C}$, where $\mathbb{H}$ is the Hamiltonian algebra of quaternions. The point is simply that the equation $(**)$ which relates φ and $\tilde{\varphi}$ is the same as the unitary equivalence $\mathbf{2} \sim \overline{\mathbf{2}}$ of the fundamental representation $\mathbf{2}$ of $SU(2)$ with the complex-conjugate or contragradient representation $\overline{\mathbf{2}}$, i.e., we have

$$g \in U(2), \quad JgJ^{-1} = \overline{g} \iff g \in SU(2).$$

Let us simply remark that $x \in M_2(\mathbb{C})$, $JxJ^{-1} = \overline{x}$ defines an algebra, the quaternion algebra $\mathbb{H}$.

Next, let us see how A) leads us to the formalism of bimodules and Poincaré duality of Section 4. Indeed, let us look at any isodoublet of the form $\begin{bmatrix} u_L \\ d_L \end{bmatrix}$ of left-handed quarks. It appears in 3 colors,

$$\begin{matrix} u_L^r & u_L^y & u_L^b \\ d_L^r & d_L^y & d_L^b \end{matrix},$$

which makes it clear that the corresponding representation of $SU(2) \times SU(3)$ is the external tensor product $\mathbf{2}_{SU(2)} \otimes \mathbf{3}_{SU(3)}$ of their fundamental representations. It is easy to convince oneself that even if one neglects the nuance between $U(n)$ and $SU(n)$ in general, there is no way to obtain such groups and representations from a single algebra and its unitary group. The solution that we found, namely to take $(\mathcal{A}, \mathcal{B})$-bimodules, with $\mathcal{B} = \mathbb{C} \oplus M_3(\mathbb{C})$ (and $\mathcal{A} = \mathbb{C} \oplus \mathbb{H}$ as above) is in fact already suggested by the following picture in a paper of J. Ellis [Ell], very close to that of the diagonal $\Delta \subset X \times X$ in Poincaré duality:

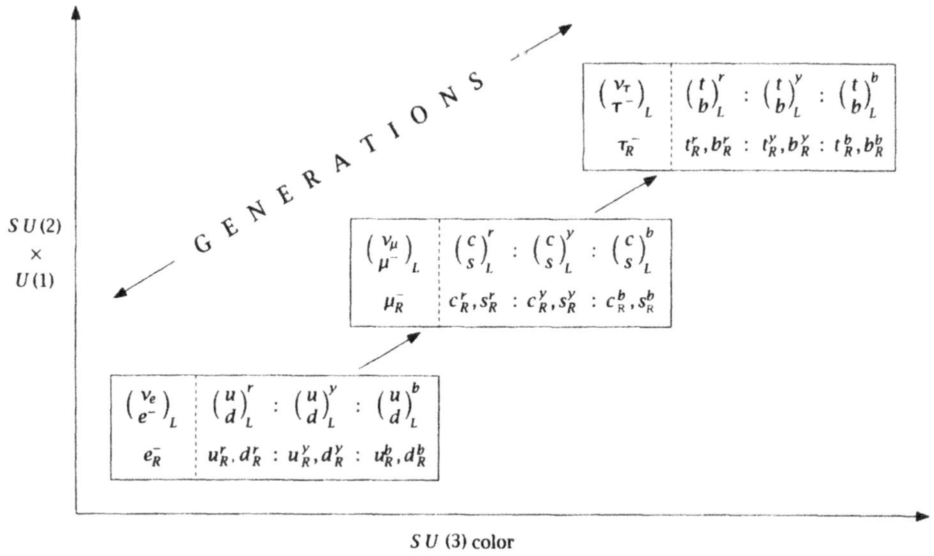

Figure VI.5. The apparent "generation" or "family" structure of fundamental fermions. The horizontal axis corresponds to $SU(3)$ color properties, the vertical axis to $SU(2) \times U(1)$ representation contents

We just refine it by taking algebras—$\mathbb{C} \oplus \mathbb{H}$ for the y-axis, $\mathbb{C} \oplus M_3(\mathbb{C})$ for the x-axis—instead of groups, which allows us to account better for the leptons (by using the $\mathbb{C}$ of $\mathbb{C} \oplus M_3(\mathbb{C})$).

Finally, we shall get a conceptual understanding of the numbers B) from a general unimodularity condition that makes sense in noncommutative geometry, but we need not anticipate that point.

We are now ready to describe in detail the geometric structure of the finite space F which, once crossed by $\mathbb{R}^4$, gives the standard model.

5.γ Geometric structure of the finite space F

This structure is given by an $(\mathcal{A}, \mathcal{B})$-module $(\mathfrak{H}, D, \gamma)$, where $\mathcal{A}$ is the $*$-algebra $\mathbb{C} \oplus \mathbb{H}$ while $\mathcal{B}$ is the $*$-algebra $\mathbb{C} \oplus M_3(\mathbb{C})$. Unlike $\mathcal{B}$, the algebra $\mathcal{A}$ is only an algebra over $\mathbb{R}$. The $*$-representations π of $\mathcal{A}$ on a finite-dimensional Hilbert space are characterized (up to unitary equivalence) by three multiplicities: n_+,

n_-, and m, where $\mathfrak{H}_\pi = \mathbb{C}^{n_+} \oplus \mathbb{C}^{n_-} \oplus \mathbb{C}^{2m}$; if $a = (\lambda, q) \in \mathcal{A} = \mathbb{C} \oplus \mathbb{H}$, then $\pi(a)$ is the block diagonal matrix

$$\pi(a) = (\lambda \otimes \mathrm{id}_{n^+}) \oplus (\overline{\lambda} \otimes \mathrm{id}_{n^-}) \oplus \left(\begin{bmatrix} \alpha & \beta \\ -\overline{\beta} & \overline{\alpha} \end{bmatrix} \otimes \mathrm{id}_m \right),$$

where the quaternion q is $q = \alpha + \beta \mathrm{j}$ with $\alpha, \beta \in \mathbb{C} \subset \mathbb{H}$. The representation of the complex $*$-algebra $\mathcal{B}$ on $\mathfrak{H}$ gives a decomposition

$$\mathfrak{H} = \mathfrak{H}_0 \oplus (\mathfrak{H}_1 \otimes \mathbb{C}^3)$$

in which $\mathcal{B}$ acts by $\pi(b) = b_0 \oplus (1 \otimes b_1)$ for $b = (b_0, b_1) \in \mathbb{C} \oplus M_3(\mathbb{C})$; thus the commuting representation of $\mathcal{A}$ is given by *a pair* π_0, π_1 of representations on $\mathfrak{H}_0$ and $\mathfrak{H}_1$. The $(\mathcal{A}, \mathcal{B})$-bimodule $\mathfrak{H}$ is thus completely described by six multiplicities: (n_+^0, n_-^0, m^0) for π_0 and (n_+^1, n_-^1, m^1) for π_1. We shall take these to be of the form

$$(n_+^0, n_-^0, m^0) = N(1,0,1), \quad (n_+^1, n_-^1, m^1) = N(1,1,1)$$

(where N will eventually be the number of generations $N = 3$). We shall take the $\mathbb{Z}/2$-grading γ in $\mathfrak{H}$ to be given, as in Example 3b), by the element $\gamma = (1, -1)$ of the center of $\mathcal{A}$. Finally, we shall take for D the most general selfadjoint operator in $\mathfrak{H}$ that anticommutes with γ ($D\gamma = -\gamma D$) and commutes with $\mathbb{C} \otimes \mathcal{B}$, where $\mathbb{C} \subset \mathcal{A}$ is the diagonal subalgebra $\{(\lambda, \lambda);\ \lambda \in \mathbb{C}\}$. (As we shall see, D encodes both the masses of the fermions and the Kobayashi-Maskawa mixing parameters.) It follows that the action of $\mathcal{A}$ and the operator D in $\mathfrak{H}_0$ (resp. $\mathfrak{H}_1$) have the following general form (with $q = \alpha + \beta \mathrm{j} \in \mathbb{H}$):

$$\pi_0(f, q) = \begin{bmatrix} f & 0 & 0 \\ 0 & \alpha & \beta \\ 0 & -\overline{\beta} & \overline{\alpha} \end{bmatrix}, \quad D_0 = \begin{bmatrix} 0 & M_e^* & 0 \\ M_e & 0 & 0 \\ 0 & 0 & 0 \end{bmatrix},$$

$$\pi_1(f, q) = \begin{bmatrix} f & 0 & 0 & 0 \\ 0 & \overline{f} & 0 & 0 \\ 0 & 0 & \alpha & \beta \\ 0 & 0 & -\overline{\beta} & \overline{\alpha} \end{bmatrix}, \quad D_1 = \begin{bmatrix} 0 & 0 & M_d^* & 0 \\ 0 & 0 & 0 & M_u^* \\ M_d & 0 & 0 & 0 \\ 0 & M_u & 0 & 0 \end{bmatrix},$$

where M_e, M_u, M_d are arbitrary $N \times N$ complex matrices.

Since π_0 is a degenerate case ($M_u = 0$) of π_1, we just restrict to π_1 in order to determine $\Omega_D^1(\mathcal{A})$.

A straightforward computation gives $\pi_1(\sum a_j da_j')$ with $a_j, a_j' \in \mathcal{A}$, $a_j = (\lambda_j, q_j)$, $q_j = \alpha_j + \beta_j \mathrm{j}$, $q_j' = \alpha_j' + \beta_j' \mathrm{j}$; we have

$$\pi_1\left(\sum a_j da_j'\right) = \begin{bmatrix} 0 & X \\ Y & 0 \end{bmatrix},$$

where X and Y are the matrices

$$X = \begin{bmatrix} M_d^* \varphi_1 & M_d^* \varphi_2 \\ -M_u^* \overline{\varphi}_2 & M_u^* \overline{\varphi}_1 \end{bmatrix}, \quad Y = \begin{bmatrix} M_d \varphi_1' & M_u \varphi_2' \\ -M_d \overline{\varphi_2'} & M_u \overline{\varphi_1'} \end{bmatrix},$$

with

$$\varphi_1 = \sum \lambda_i(\alpha_i' - \lambda_i'), \quad \varphi_2 = \sum \lambda_i \beta_i',$$
$$\varphi_1' = \sum \alpha_i(\lambda_i' - \alpha_i') + \beta_i \overline{\beta_i'}, \quad \varphi_2' = \sum -\alpha_i \beta_i' + \beta_i(\overline{\lambda_i'} - \overline{\alpha_i'}).$$

It follows that $\Omega_D^1(\mathcal{A}) = \mathbb{H} \oplus \mathbb{H}$ with the $\mathcal{A}$-bimodule structure given by

$$(\lambda, q)(q_1, q_2) = (\lambda q_1, q q_2) \quad \forall q_1, q_2 \in \mathbb{H},$$
$$(q_1, q_2)(\lambda, q) = (q_1 q, q_2 \lambda) \quad \forall \lambda \in \mathbb{C},\ q \in \mathbb{H},$$

and the differential d again being the *finite difference*:

$$\mathrm{d}(\lambda, q) = (q - \lambda, \lambda - q) \in \mathbb{H} \oplus \mathbb{H}$$

(just set $q_1 = \varphi_1 + \varphi_2 \mathrm{j}$, $q_2 = \varphi_1' + \varphi_2' \mathrm{j}$ with the above φ's).

Finally, the involution $*$ on $\Omega_D^1(\mathcal{A})$ is given by

$$(q_1, q_2)^* = (\overline{q}_2, \overline{q}_1) \quad \forall q_j \in \mathbb{H}.$$

The space $\mathcal{U}$ of vector potentials is thus naturally isomorphic to $\mathbb{H}$, and a similar computation shows that $\Omega_D^2(\mathcal{A}) = \mathbb{H} \oplus \mathbb{H}$ with the $\mathcal{A}$-bimodule structure

$$(\lambda, q)(q_1, q_2)(\lambda', q') = (\lambda q_1 \lambda', q q_2 q') \qquad \forall \lambda, \lambda' \in \mathbb{C},\ q, q' \in \mathbb{H};$$

the product $\Omega_D^1 \times \Omega_D^1 \to \Omega_D^2$ is given by

$$(q_1, q_2) \wedge (q_1', q_2') = (q_1 q_2', q_2 q_1'),$$

and the differential $\mathrm{d} : \Omega_D^1 \to \Omega_D^2$ by

$$\mathrm{d}(q_1, q_2) = (q_1 + q_2, q_1 + q_2).$$

Thus, the curvature θ of a vector potential $V = (q, \overline{q})$ is

$$\theta = \mathrm{d}V + V^2 = (q + q^* + q q^*, q + q^* + q^* q) = (|1 + q|^2 - 1)(1, 1),$$

where $q \mapsto |q|$ denotes the norm of quaternions.

We thus see that the action $\mathrm{YM}(V) = \mathrm{Trace}(\pi(\theta)^2)$ (we are in the 0-dimensional case) is the same symmetry-breaking quartic potential for a pair of complex numbers as in Example 3b).

The detailed expression for the Hilbert space norm on $\Omega_D^2(\mathcal{A}) = \mathbb{H} \oplus \mathbb{H}$ is given, for $\omega = (q_1, q_2)$, $q_j = \alpha_j + \beta_j \mathrm{j}$, by

$$\|\omega\|^2 = \lambda_1 |\alpha_1|^2 + \mu_1 |\beta_1|^2 + \lambda_2(|q_2|^2),$$

where

$$\lambda_1 = \mathrm{Trace}(|M_e|^4) + 3\mathrm{Trace}(|M_d|^4 + |M_u|^4),$$
$$\mu_1 = 6\,\mathrm{Trace}(|M_d|^2 |M_u|^2),$$
$$\lambda_2 = \frac{1}{2}\mathrm{Trace}(|M_e|^4 + 3(|M_d|^4 + |M_u|^4 + 2|M_d|^2 |M_u|^2)).$$

Finally, we shall investigate what freedom we have in the choice of the selfadjoint operators D_0, D_1 on $\mathfrak{H}_0$, $\mathfrak{H}_1$ in the above example. Two pairs $(\mathfrak{H}_j, D_j)$ and $(\mathfrak{H}'_j, D'_j)$ give identical results if there exist unitaries $U_j : \mathfrak{H}_j \to \mathfrak{H}'_j$ such that:

α) $U_j D_j U_j^* = D_j \ \ (j = 1,2)$,

β) $U_j \pi_j(a) U_j^* = \pi'_j(a) \ \ \forall a \in \mathcal{A},\ j = 1,2.$

Making use of this freedom, we can assume that D_0 is diagonal in $\mathfrak{H}_0$ and has positive eigenvalues e_1, e_2, e_3. Thus, the situation for D_0 is described by these 3 positive numbers.

For $\mathfrak{H}_1$, a general element of the commutant of $\pi_1(\mathcal{A})$ is of the form

$$U_1 = \begin{bmatrix} V_1 & 0 & 0 & 0 \\ 0 & V_2 & 0 & 0 \\ 0 & 0 & V_3 & 0 \\ 0 & 0 & 0 & V_3 \end{bmatrix},$$

where the V_j are unitary operators when U_1 is unitary.

Conjugating D_1 by U_1 replaces M_d and M_u, respectively, by

$$M'_d = V_1 M_d V_3^*, \quad M'_u = V_2 M_u V_3^*.$$

We thus see that we can assume that both M_u and M_d are positive matrices and that one of them, say M_u, is diagonal.

The invariants are thus the eigenvalues of M_u and M_d, i.e., a total of 6 positive numbers, and the pair of maximal abelian subalgebras generated by M_u and M_d. Since any pair $\mathcal{A}_1$, $\mathcal{A}_2$ of maximal abelian subalgebras of $M_3(\mathbb{C})$ are conjugate by a unitary W, $W\mathcal{A}_1 W^* = \mathcal{A}_2$, which is given modulo the unitary groups $\mathcal{U}(\mathcal{A}_j)$, there remain 4 parameters with which to specify W so that $W M_d W^*$ is also diagonal. Such a W corresponds to the Kobayashi-Maskawa mixing matrix in the standard model.

5.δ Geometric structure of the standard model

We shall show in this section how the standard model is obtained from the product geometry of the usual 4-dimensional continuum by the above finite geometry F. Thus, we let V be a 4-dimensional spin manifold and $(L^2, \partial_V, \gamma_5)$ its Dirac K-cycle. The product geometry is, according to the general rule for forming products, described by the algebras

$$\mathcal{A} = C^\infty(V) \otimes (\mathbb{C} \oplus \mathbb{H}), \quad \mathcal{B} = C^\infty(V) \otimes (\mathbb{C} \oplus M_3(\mathbb{C})).$$

The Hilbert space $H = L^2(V,S) \otimes \mathfrak{H}_F$, where $\mathfrak{H}_F$ is described in γ) above, i.e., $\mathfrak{H}_F = \mathfrak{H}_0 \oplus (\mathfrak{H}_1 \otimes \mathbb{C}^3)$. There is a corresponding decomposition $H = H_0 \oplus (H_1 \otimes \mathbb{C}^3)$, with corresponding representations π_j of $\mathcal{A}$ on H_j.

Then $D = \partial_V \otimes 1 + \gamma_5 \otimes D_F$, where D_F is as above. This gives a decomposition $D = D_0 \oplus (D_1 \otimes 1)$, where, according to γ), we take M_e, M_u and M_d to be *positive* matrices:

$$D_0 = \begin{bmatrix} \partial_V \otimes 1 & \gamma_5 \otimes M_e & 0 \\ \gamma_5 \otimes M_e & \partial_V \otimes 1 & 0 \\ 0 & 0 & \partial_V \otimes 1 \end{bmatrix},$$

$$D_1 = \begin{bmatrix} \partial_V \otimes 1 & 0 & \gamma_5 \otimes M_d & 0 \\ 0 & \partial_V \otimes 1 & 0 & \gamma_5 \otimes M_u \\ \gamma_5 \otimes M_d & 0 & \partial_V \otimes 1 & 0 \\ 0 & \gamma_5 \otimes M_u & 0 & \partial_V \otimes 1 \end{bmatrix}.$$

We shall first restrict attention to the algebra $\mathcal{A}$, the case of $\mathcal{B}$ being easier. Note that $\mathcal{A} = C^\infty(V,\mathbb{C}) \oplus C^\infty(V,\mathbb{H})$, so that every $a \in \mathcal{A}$ is given by a pair (f,q) consisting of a $\mathbb{C}$-valued function f on V and an $\mathbb{H}$-valued function q on V.

Let us first compute $\Omega_D^1(\mathcal{A})$. Given $\rho = \sum a_s da'_s \in \Omega^1(\mathcal{A})$, with $a_s, a'_s \in \mathcal{A}$, we have $a_s = (f_s, q_s)$, $a'_s = (f'_s, q'_s)$, where f_s, f'_s are complex-valued functions on V and q_s, q'_s are $\mathbb{H}$-valued functions on V, of the form

$$q_s = \alpha_s + \beta_s \mathrm{j}, \quad q'_s = \alpha'_s + \beta'_s \mathrm{j}.$$

Then

$$\pi_1(\rho) = \begin{bmatrix} i^{-1}\gamma(A) \otimes 1 & 0 & \varphi_1\gamma_5 \otimes M_d & \varphi_2\gamma_5 \otimes M_d \\ 0 & i^{-1}\gamma(\overline{A}) \otimes 1 & -\overline{\varphi}_2\gamma_5 \otimes M_u & \overline{\varphi}_1\gamma_5 \otimes M_u \\ \varphi'_1\gamma_5 \otimes M_d & \varphi'_2\gamma_5 \otimes M_u & i^{-1}\gamma(W_1) \otimes 1 & i^{-1}\gamma(W_2) \otimes 1 \\ -\overline{\varphi'_2}\gamma_5 \otimes M_d & \overline{\varphi'_1}\gamma_5 \otimes M_u & i\gamma(\overline{W}_2) \otimes 1 & i^{-1}\gamma(\overline{W}_1) \otimes 1 \end{bmatrix},$$

where $A = \sum f_s df'_s$ is a $\mathbb{C}$-valued 1-form on V, and $W_1 + W_2\mathrm{j} = W = \sum q_s dq'_s$ is an $\mathbb{H}$-valued 1-form on V (cf. [At$_4$]).

Also, φ_j and φ'_j are complex-valued functions on V given by the same formulas as above for the finite geometry, namely,

$$\varphi_1 = \sum f_s(\alpha'_s - f'_s), \quad \varphi_2 = \sum f_s \beta'_s,$$
$$\varphi'_1 = \sum (\alpha_s(f'_s - \alpha'_s) + \beta_s\overline{\beta'_s}), \quad \varphi'_2 = \sum (\beta_s(\overline{f'_s} - \overline{\alpha'_s}) - \alpha_s\beta'_s).$$

This means that the pair (q,q') of $\mathbb{H}$-valued functions, given by $q = \varphi_1 + \varphi_2\mathrm{j}$, $q' = \varphi'_1 + \varphi'_2\mathrm{j}$, satisfies

$$(q,q') = \sum a_s \Delta a'_s,$$

where $\Delta a'_s = (q'_s - \lambda'_s, \lambda'_s - q'_s)$ is the finite-difference operation, while the $\mathcal{A}$-bimodule structure on the space of $\mathbb{H} \oplus \mathbb{H}$-valued functions (q_1, q_2) is given by

$$(f,q)(q_1,q_2) = (fq_1, qq_2), \quad (q_1,q_2)(f,q) = (q_1 q, q_2 f) \quad \text{for all } (f,q) \in \mathcal{A}.$$

This shows that $\Omega_D^1(\mathcal{A})$ is the direct sum of two subspaces:

$$\Omega_D^1(\mathcal{A}) = \Omega_D^{(1,0)} \oplus \Omega_D^{(0,1)},$$

where:

$\Omega_D^{(1,0)}$, the subspace of elements of *differential* type, is the space of pairs (A, W) consisting of a $\mathbb{C}$-valued 1-form A on V and an $\mathbb{H}$-valued 1-form W on V;

$\Omega_D^{(0,1)}$, the subspace of elements of *finite-difference* type, is the space of pairs (q_1, q_2) of $\mathbb{H}$-valued functions on V, with the above $\mathcal{A}$-bimodule structure.

The geometric picture is that of two copies V_R and V_L of V, with $\mathbb{C}$-valued functions on V_R, and $\mathbb{H}$-valued functions on V_L. More generally, the differential forms on V_R are $\mathbb{C}$-valued, whereas on V_L they are $\mathbb{H}$-valued, exactly as in Atiyah's book [At$_4$]. Of course the finite difference mixes both sides, so they are not independent.

Given an element $a = (f, q)$ of $\mathcal{A}$, the element $\mathrm{d}a$ of Ω_D^1 has a *differential* component $(\mathrm{d}f, \mathrm{d}q)$, given by the $\mathbb{C}$-valued 1-form $\mathrm{d}f$ and the $\mathbb{H}$-valued 1-form $\mathrm{d}q$; and a *finite-difference* component $(q - f, f - q)$.

The involution $*$ on Ω_D^1 is given by

$$((A, W), (q_1, q_2))^* = ((-\overline{A}, -\overline{W}), (\overline{q}_2, \overline{q}_1)),$$

so that a vector potential is given by:

α) an ordinary $U(1)$ vector potential on V;

β) an $SU(2)$ vector potential on V (cf. [At$_4$]);

γ) a pair $q = \alpha + \beta \mathrm{j}$ of complex scalar fields on V.

The next step is to compute $\Omega_D^2(\mathcal{A})$ as well as the product

$$\Omega_D^1 \times \Omega_D^1 \to \Omega_D^2$$

and the differential $\mathrm{d} : \Omega_D^1 \to \Omega_D^2$.

We shall first state the result. It holds provided a certain nondegeneracy condition is satisfied, namely, that the following matrices are not scalar multiples of the identity matrix:

$$M_d^2 \text{ or } \frac{1}{2}(M_d^2 + M_u^2), \quad M_d^2 \text{ or } M_e^2 \text{ or } M_u^2.$$

The result then is that an element of $\Omega_D^2(\mathcal{A})$ has:

1) a component of type $(2,0)$ given by a pair (F, G) consisting of a $\mathbb{C}$-valued 2-form F and an $\mathbb{H}$-valued 2-form G on V;

2) a component of type $(1,1)$ given by a pair (ω_1, ω_2) of quaternionic 1-forms ω_1, ω_2 on V;

3) a component of type $(0,2)$ given by a pair (q_1, q_2) of quaternionic functions on V.

Moreover, with the obvious notation, the following formulas hold:

$$\begin{aligned}&\mathrm{d}((A,W),(q_1,q_2))\\&\quad=\{(\mathrm{d}A,\mathrm{d}W),(\mathrm{d}q_1+A-W,\mathrm{d}q_2+W-A),(q_1+q_2,q_1+q_2)\},\\&((A,W),(q_1,q_2)\cdot((A',W'),(q_1',q_2'))\\&\quad=\{(A\wedge A',W\wedge W'),(Aq_1'-q_1W',Wq_2'-q_2A',(q_1q_2',q_2q_1')\};\end{aligned}$$

they show that we are dealing with the graded tensor product of the differential algebra of V (the de Rham algebra) by the differential algebra Ω_D of the finite space F.

In order to compute the Hilbert space norm on $\Omega_D^2(\mathcal{A})$, we have to write down explicitly the class in $\pi(\Omega^2(\mathcal{A}))/\pi(J)$ associated with an arbitrary element $((F,G),(\omega_1,\omega_2),(q_1,q_2))$ of $\Omega_D^2(\mathcal{A})$. We shall first write what happens for π_1; the case of π_0 is a degenerate case obtained by taking $M_u=0$. The subspace $\pi(J)$ only interacts with the elements of degree 0 in the Clifford algebra, and any element of $\pi(J)$ is given by 5 complex-valued functions α, β, γ, Y, and Z on V, whose representation in $\mathfrak{H}_1$ is given by:

$$\begin{bmatrix}\alpha\otimes 1 & 0 & 0 & 0\\ 0 & \overline{\alpha}\otimes 1 & 0 & 0\\ 0 & 0 & T_{11} & T_{12}\\ 0 & 0 & T_{21} & T_{22}\end{bmatrix},$$

where

$$T=\begin{bmatrix}\beta & \gamma\\ -\overline{\gamma} & \overline{\beta}\end{bmatrix}\otimes 1+\begin{bmatrix}Y & Z\\ \overline{Z} & -\overline{Y}\end{bmatrix}\otimes\frac{1}{2}(M_d^2-M_u^2).$$

In the Hilbert space $\mathfrak{H}_0$, the same element is represented by

$$\begin{bmatrix}\alpha\otimes 1 & 0 & 0\\ 0 & T'_{11} & T'_{12}\\ 0 & T'_{21} & T'_{22}\end{bmatrix},$$

with

$$T'=\begin{bmatrix}\beta & \gamma\\ -\overline{\gamma} & \overline{\beta}\end{bmatrix}\otimes 1+\begin{bmatrix}Y & Z\\ \overline{Z} & -\overline{Y}\end{bmatrix}\otimes\frac{1}{2}M_e^2.$$

The elements (F,G) and (ω_1,ω_2) of degree $(2,0)$ and $(1,1)$ have canonical representatives given by the following expressions, where $\omega_k=\alpha_k+\beta_k\mathrm{j}$ and α_k,β_k are complex-valued 1-forms on V:

$$\begin{bmatrix}i^{-2}\gamma(F)\otimes 1 & 0 & 0 & 0\\ 0 & i^{-2}\gamma(\overline{F})\otimes 1 & 0 & 0\\ 0 & 0 & & \\ & & & i^{-2}\gamma(G)\otimes 1\\ 0 & 0 & & \end{bmatrix}\text{for }\mathfrak{H}_1,$$

$$\begin{bmatrix} i^{-2}\gamma(F)\otimes 1 & 0 & 0 \\ 0 & & \\ & i^{-2}\gamma(G)\otimes 1 & \\ 0 & & \end{bmatrix} \quad \text{for } \mathfrak{H}_0,$$

$$\begin{bmatrix} 0 & 0 & i^{-1}\gamma(\alpha_1)\gamma_5\otimes M_d & i^{-1}\gamma(\beta_1)\gamma_5\otimes M_d \\ 0 & 0 & i\gamma(\overline{\beta}_1)\gamma_5\otimes M_u & i^{-1}\gamma(\overline{\alpha}_1)\gamma_5\otimes M_u \\ i^{-1}\gamma(\alpha_2)\gamma_5\otimes M_d & i^{-1}\gamma(\beta_2)\gamma_5\otimes M_u & 0 & 0 \\ i\gamma(\overline{\beta}_2)\gamma_5\otimes M_d & i^{-1}\gamma(\overline{\alpha}_2)\gamma_5\otimes M_u & 0 & 0 \end{bmatrix} \quad \text{for } \mathfrak{H}_1,$$

$$\begin{bmatrix} 0 & i^{-1}\gamma(\alpha_1)\gamma_5\otimes M_e & i^{-1}\gamma(\beta_1)\gamma_5\otimes M_e \\ i^{-1}\gamma(\alpha_2)\gamma_5\otimes M_e & 0 & 0 \\ i\gamma(\overline{\beta}_2)\gamma_5\otimes M_e & 0 & 0 \end{bmatrix} \quad \text{for } \mathfrak{H}_0.$$

The component (q_1, q_2) of degree $(0,2)$, $q_i = \alpha_i + \beta_i j$, has the representative modulo $\pi(J)$

$$\begin{bmatrix} \alpha_1 M_d^2 & \beta_1 M_d M_u & 0 & 0 \\ -\overline{\beta}_1 M_u M_d & \overline{\alpha}_1 M_u^2 & 0 & 0 \\ 0 & 0 & & \\ & & q_2\otimes \frac{1}{2}(M_d^2 + M_u^2) & \\ 0 & 0 & & \end{bmatrix},$$

$$\begin{bmatrix} \alpha_1 M_e^2 & 0 & 0 \\ 0 & & \\ & q_2 \otimes \frac{1}{2}M_e^2 & \\ 0 & & \end{bmatrix}.$$

The nondegeneracy condition ensures that the various terms such as $\alpha_1 M_d^2$ do not disappear modulo $\pi(J)$. It is now straightforward, as in Section 3, to compute the action. One gets the five terms of the $U(1)\times SU(2)$ part of the standard model, with the new Yukawa coupling terms C). But before we discuss the coefficients with which they arise we need to show how to reduce the gauge group $\mathcal{U}(\mathcal{A})\times\mathcal{U}(\mathcal{B})$ of the theory to the global gauge group $U(1)\times SU(2)\times SU(3)$ and obtain the intricate table of hypercharges of the standard model. Note that we did not make the straightforward calculation of vector potentials and action for the algebra $\mathcal{B}$, which yield a pure gauge action with group $U(1)\times U(3)$.

5.ε Unimodularity condition and hypercharges

We shall now see how, from a general condition of unimodularity valid in the general context of noncommutative geometry, one obtains the intricate table of hypercharges of elementary particles:

	e,μ,τ	$\nu_e\nu_\mu\nu_\tau$	u,c,t	d,s,b
Y_L	-1	-1	$\frac{1}{3}$	$\frac{1}{3}$
Y_R	-2		$\frac{4}{3}$	$-\frac{2}{3}$

(Note that since we are dealing with the Lie algebra of $U(1)$, this table is only determined up to a common scale.)

To obtain these values and at the same time obtain the global gauge group $U(1)\times SU(2)\times SU(3)$, we shall simply replace the local gauge group $\mathcal{U}(\mathcal{A})\times\mathcal{U}(\mathcal{B})$ by its unimodular subgroup $S\mathcal{U}$ *relative to* $\mathcal{A}$.

In our context, the notion of *determinant* of a unitary makes sense provided that a *trace* τ is given. More precisely, by [Harp-S], given a C^*-algebra C and a *self-adjoint* trace τ on C (i.e., $\tau(x^*)=\overline{\tau(x)}$ for all $x\in C$), one obtains the phase of the determinant of a unitary u as follows:

$$\mathrm{Phase}_\tau(u)=\frac{1}{2\pi i}\int_0^1\tau(u(t)'u(t)^{-1})dt,$$

where $u(t)$ is a smooth path of unitaries joining u to 1. Thus, this phase is only defined in the connected component $\mathcal{U}_0(C)$ of the identity, and it is ambiguous, by the image $\langle\tau,K_0(C)\rangle$ of $K_0(C)$ under the trace τ, which is a countable subgroup of $\mathbb{R}$.

The condition $\mathrm{Phase}_\tau(u)=0$ is well-defined and gives a normal subgroup of $\mathcal{U}(C)$. We let $S_\tau\mathcal{U}(C)$ be the connected component of its identity element.

Now let $\mathcal{A}$ and $\mathcal{B}$ be $*$-algebras and let $(\mathfrak{H},D)$ be a (d,∞)-summable bimodule over $\mathcal{A},\mathcal{B}$. We shall apply the above considerations to the C^*-algebra C on $\mathfrak{H}$ generated by $\mathcal{A}$ and $\mathcal{B}$ and to the family of self-adjoint traces τ on C given by the selfadjoint elements $\rho=\rho^*$ of the *center* of $\mathcal{A}$, so

$$\tau_\rho(x)=\mathrm{Tr}_\omega(\rho x D^{-d})\quad\forall x\in C.$$

We thus get a normal subgroup $S_{\mathcal{A}}(C)$ of the unitary group of C by intersecting all of the S_τ, $\tau=\tau_\rho$ as above. Since $\mathcal{U}(\mathcal{A})\times\mathcal{U}(\mathcal{B})$ is a subgroup of $\mathcal{U}(C)$, its intersection with $S_{\mathcal{A}}(C)$ gives a normal subgroup $S(\mathcal{A},\mathcal{B})$ of $\mathcal{U}(\mathcal{A})\times\mathcal{U}(\mathcal{B})$. We shall see what $S(\mathcal{A},\mathcal{B})$ is, in simple examples, but our main point now is:

Theorem 1. *Let $(\mathcal{A},\mathcal{B},\mathcal{H},D)$ be the product geometry of a 4-dimensional Riemannian spin manifold V by the finite geometry F. Then the group $S(\mathcal{A},\mathcal{B})$ is equal to* $\mathrm{Map}(V,U(1)\times SU(2)\times SU(3))$ *and its representation in $\mathcal{H}$ is, for the $U(1)$ factor, given by the above table of hypercharges.*

By construction, $\mathcal{A} = C^\infty(V) \otimes \mathcal{A}_F$, $\mathcal{B} = C^\infty(V) \otimes \mathcal{B}_F$, $\mathcal{H} = L^2(V,S) \otimes \mathfrak{H}_F$, $D = \partial_V \otimes 1 + \gamma_5 \otimes D_F$, and it follows by a straightforward argument that

$$S(\mathcal{A},\mathcal{B}) = \mathrm{Map}(V, S(\mathcal{A}_F,\mathcal{B}_F)),$$

where $S(\mathcal{A}_F,\mathcal{B}_F)$ is defined as above, but using the ordinary trace in the finite-dimensional space $\mathfrak{H}_F$ instead of the Dixmier trace. Thus, we need only compute the group $S(\mathcal{A}_F,\mathcal{B}_F)$ over a point, and its representation in $\mathfrak{H}_F$. Now, every selfadjoint element ρ of the center $Z(\mathcal{A}_F)$ is of the form $\rho = \lambda_1 e + \lambda_2(1-e)$, where the λ_j are real numbers, $e = (0,1) \in \mathbb{C} \oplus \mathbb{H}$ and $1 - e = (1,0)$. It follows easily that

$$S(\mathcal{A}_F,\mathcal{B}_F) = (\mathcal{U}(\mathcal{A}_F) \times \mathcal{U}(\mathcal{B}_F)) \cap (SU(e\mathfrak{H}_F) \times SU((1-e)\mathfrak{H}_F)).$$

In other words, the unimodularity condition means that the action is unimodular on both $e\mathfrak{H}_F$ and $(1-e)\mathfrak{H}_F$. Let, then, U be an element of $\mathcal{U}(\mathcal{A}_F) \times \mathcal{U}(\mathcal{B}_F)$. It is given by a quadruple

$$U = ((\lambda, q), (u, v));\quad \lambda \in U(1), q \in SU(2), u \in U(1), v \in U(3).$$

We have $\mathfrak{H}_F = \mathfrak{H}_0 \oplus (\mathfrak{H}_1 \otimes \mathbb{C}^3)$ and, with the notation of γ), the action of U on $\mathfrak{H}_F$ is given by

$$\pi_0(\lambda, q)u \oplus (\pi_1(\lambda, q) \otimes v).$$

This operator restricts to both $e\mathfrak{H}_F$ and $(1-e)\mathfrak{H}_F$, and we have to compute the determinants of these restrictions. With N generations, we get

$$\det(U_e) = u^{2N} \times (\det(v))^{2N}, \quad \det(U_{1-e}) = (\lambda u)^N \times (\det v)^{2N},$$

hence the unimodularity condition means, independently of N, that

$$\lambda = u, \quad \det v = u^{-1}. \qquad (*)$$

It follows that $S(\mathcal{A}_F,\mathcal{B}_F) = U(1) \times SU(2) \times SU(3)$. Let us compute the table of hypercharges, taking u as generator. Since $\lambda = u$, for $\mathfrak{H}_0$ we get

$$\pi_0(\lambda,q)u = \begin{bmatrix} \lambda u & 0 \\ 0 & qu \end{bmatrix} = \begin{bmatrix} u^2 & 0 \\ 0 & qu \end{bmatrix},$$

which gives the hypercharge 2 to e_R, μ_R, τ_R and 1 to e_L, μ_L, τ_L and ν_e, ν_μ, ν_τ. For $\mathfrak{H}_1$ we get $v = v_0 u^{-1/3}$ with $v_0 \in SU(3)$ and $u^{1/3}$ a cube root of u, so that

$$\pi_1(\lambda,q)u^{-1/3} = \begin{bmatrix} \lambda u^{-1/3} & 0 & 0 & 0 \\ 0 & \lambda^{-1}u^{-1/3} & 0 & 0 \\ 0 & 0 & \multicolumn{2}{c}{\multirow{2}{*}{$qu^{-1/3}$}} \\ 0 & 0 & & \end{bmatrix},$$

which gives the hypercharge $\frac{2}{3}$ to d_R, s_R, b_R; $-\frac{4}{3}$ to u_R, c_R, t_R; and $-\frac{1}{3}$ for the left-handed quarks. An overall sign is of course irrelevant (change u to u^{-1}), so we get the desired table.

Remark. 1) *Let $\mathcal{A}$ be the algebra of functions on a 4-dimensional spin manifold V, let $(\mathfrak{H}, D)$ be the sum of N copies of the Dirac K-cycle (L^2, ∂_V) on V, and $\mathcal{B}$ the commutant of $\mathcal{A} \cup [D, \mathcal{A}]$ on $\mathfrak{H}$. Then the group $S(\mathcal{A}, \mathcal{B})$ is the group* $\mathrm{Map}(V, SU(N))$ *of local gauge transformations associated with the global gauge group $SU(N)$.*

2) *To the above reduction of the gauge group there corresponds a similar reduction of bivector potentials A, which in the case of Theorem 1 means that A is traceless in the spaces $e\mathfrak{H}_F$ and $(1-e)\mathfrak{H}_F$, and in the case of Remark 1 yields the corresponding $SU(N)$ pure gauge theory.*

We are now ready to compare our theory with the standard model of electro-weak and strong interactions. We first note that the bimodule $(\mathfrak{H}, D)$ over $\mathcal{A}, \mathcal{B}$ of Theorem 1 is *not* irreducible, i.e., the commutant of the algebra generated by $\mathcal{A}$, $\mathcal{B}$ and D does not reduce to $\mathbb{C}$. With generic M_e, M_u, M_d, one can check that this commutant is $\mathbb{C}^4$ so that $(\mathfrak{H}, D)$ splits as a direct sum of 4 irreducible bimodules, three for the three lepton generations, i.e., for $\mathfrak{H}_0$, and one for $\mathfrak{H}_1$ which is irreducible.

Theorem 2. *Let $(\mathcal{A}, \mathcal{B}, \mathfrak{H}, D)$ be the product geometry of a 4-dimensional spin Riemannian manifold by the finite geometry F. Let $S(\mathcal{A}, \mathcal{B})$ be the reduced gauge group and $\mathcal{V}$ the corresponding space of bivector potentials. Then the following action gives the standard model with its 18 free parameters:*

$$\mathrm{Tr}_\omega((\lambda_{\mathcal{A}}\theta_{\mathcal{A}}^2 + \lambda_{\mathcal{B}}\theta_{\mathcal{B}}^2)D^{-4}) + \langle \psi, D_V \psi \rangle,$$

where $\lambda_{\mathcal{A}}$, $\lambda_{\mathcal{B}}$ belong to the commutant of the bimodule and $\theta_{\mathcal{A}}$, $\theta_{\mathcal{B}}$ are the respective curvatures.

The proof is straightforward, given Theorem 1 and the above computations.

Let us now note that, at the classical level which we have not left so far, there is a natural 17-dimensional subspace given by the following restriction:

$$\lambda_{\mathcal{A}}, \lambda_{\mathcal{B}} \text{ belong to the algebra generated by } \mathcal{A} \text{ and } \mathcal{B}.$$

The corresponding classical relation can only have a heuristic value. In the limit of large top mass m_t, i.e., neglecting the other fermion masses, one gets the following:

$$m_t \geq \sqrt{3}\, m_W \quad \text{if } \lambda_{\mathcal{A}} > 0; \tag{1}$$

$$m_H = m_t \left(3 \frac{x^2 + 8x + 14}{x^2 + 9x + 18} \right)^{1/2}, \quad \frac{1}{4}x + \frac{3}{4} = \left(\frac{m_t}{2m_W} \right)^2. \tag{2}$$

Here x is the ratio between the two eigenvalues of the operator $\lambda_{\mathcal{A}}$, so that the value $x = 1$ is the most natural, leading to $m_t = 2m_W$. None of these relations is a physical prediction, but if they were nearly satisfied by the physical values of m_t and m_H it would show that the change of parametrization in the standard model given by the above theorem is qualitatively a good one at the classical level.

In order to transform our interpretation of the standard model into a predictive theory it is important to solve the following problems:

1) Find a nontrivial finite quantum group of symmetries of the finite space F.

2) Determine the structure of the Clifford algebra of the finite space F, given by the linear map from the space $\mathbb{H} \oplus \mathbb{H}$ of 1-forms into the algebra of endomorphisms of $\mathcal{H}_F$ which to a 1-form $\sum a_i \, db_i$ associates $\sum a_i[D, b_i]$.

BIBLIOGRAPHY

[Ab] H. Abels. Parallelizability of proper actions, global K-slices and maximal compact subgroups. *Math. Ann.* **212** (1974/75), 1-19; MR **51** # 11460.

[Ad-A-K] V.M.Adamjan, D.Z. Arov and M.G. Krein. Analytic properties of the Schmidt pairs of a Hankel operator and the generalized Schur-Takagi problem. *Mat. Sb. (N.S.)* **86(128)** (1971), 34-75; *Math. U.S.S.R. Sb.* **15** (1971), 31-73; MR **45** # 7505.

[Ad-A-K] V.M. Adamjan, D.Z. Arov and M.G. Krein. Infinite Hankel matrices and generalized problems of Carathéodory-Fejér and Riesz. *Funkt. Anal. i Prilozhen.* **2** (1968), 1-19; *Functional Anal. Appl.* **2** (1968), 1-18; MR **38** # 2591.

[Ada] J.F. Adams. Stable homotopy and generalized homology. *Univ. of Chicago Press, Chicago, Ill.*, 1974; MR **53** # 6534.

[Ak-O] C.A. Akemann and P. Østrand. On a tensor product C^*-algebra associated with the free group of two generators. *J. Math. Soc. Japan* **27** (1975), no. 4, 589-599; MR **53** # 3190.

[Ak-W] C.A. Akemann and M.E. Walter. Unbounded negative definite functions. *Canad. J. Math.* **33** (1981), no. 4, 862-871; MR 83b:43009.

[Al] L. Alvarez-Gaumé. Supersymmetry and the Atiyah-Singer index theorem. *Comm. Math. Phys.* **90** (1983), 161-173; MR 85d:58078.

[Ar_1] H. Araki. Relative entropy of states of von Neumann algebras I, II. *Publ. RIMS* **11** (1975/76), no. 3, 809-833; *ibid.* **13** (1977/78), no. 1, 173-192; MR **54** # 13585 and **56** # 12905.

[Ar_2] H. Araki. Relative Hamiltonian for faithful normal states of a von Neumann algebra. *Publ. Res. Inst. Math.* **9** (1973/74), 165-209; MR **49** # 6826.

[Ar_3] H. Araki. Golden-Thompson and Peierls-Bogolubov inequalities for a general von Neumann algebra. *Comm. Math. Phys.* **34** (1973), 167-178; MR **49** # 5864.

[Ar_4] H. Araki. Expansional in Banach algebra. *Ann. Sci. École Norm. Sup.* 4 (1973), 67-84; MR **55** # 8793.

[Ar-W] H. Araki and E.J. Woods. A classification of factors. *Publ. Res. Inst. Math Sci. Ser. A* **4** (1968/69), 51-130; MR **39** # 6087.

[Arn-C-M-P] D. Arnal, J.C. Cortet, P. Molin and C. Pinczon. Covariance and geometrical invariance in $*$-quantization. *J. Math. Phys.* **24** (1983), no. 2, 276-283 MR 84h:58061.

[Arv_1] W. Arveson. Notes on extensions of C^*-algebras. *Duke Math. J.* **44** (1977) no. 2, 329-355; MR **55** # 11056.

[Arv_2] W. Arveson. Subalgebras of C^*-algebras. *Acta Math.* **123** (1969), 141-224 MR **40** # 6274.

[Arv_3] W. Arveson. The harmonic analysis of automorphism groups. *Operator algebras and applications, Part 1, pp. 199-269, Proc. Sympos. Pure Math. 38, Amer. Math. Soc., Providence, R.I.,* 1982; MR 84m:46085.

[At_1] M.F. Atiyah. Elliptic operators and compact groups. *Lecture Notes in Math., 401, Springer, Berlin,* 1974; MR **58** # 2910.

[At_2] M.F. Atiyah. Global theory of elliptic operators. *Proc. Internat. Conf. on Functional Analysis and Related Topics (Tokyo, 1969), pp. 21-30, Univ. of Tokyo Press, Tokyo,* 1970; MR **42** # 1154.

[At_3] M.F. Atiyah. *K*-Theory. *Benjamin, New York,* 1967; MR **36** # 7130.

[At_4] M.F. Atiyah. Geometry of Yang-Mills fields. *Accad. Naz. dei Lincei, Scuola Norm. Sup., Pisa,* 1979; MR 81a:81047.

[At_5] M.F. Atiyah. Elliptic operators, discrete groups and von Neumann algebras. *Analyse et topologie, pp. 43-72, Astérisque No. 32/33, Soc. Math France, Paris,* 1976; MR **54** # 8741.

[At-B-P] M.F. Atiyah, R. Bott and V.K. Patodi. On the heat equation and the index theorem. *Invent. Math.* **19** (1973), 279-330; MR **58** # 31287.

[At-B-S] M.F. Atiyah, R. Bott and A. Shapiro. Clifford modules. *Topology* **3** (1964) *Suppl.* 1, 3-38; MR **29** # 5250.

[$At\text{-}S_1$] M.F. Atiyah and W. Schmid. A new proof of the regularity theorem for invariant eigendistributions on semisimple Lie groups. *To appear.*

[$At\text{-}Sc_2$] M.F. Atiyah and W. Schmid. A geometric construction of the discrete series for semisimple Lie groups. *Invent. Math.* **42** (1977), 1-62; MR **57** # 3310.

[$At\text{-}Si_1$] M.F. Atiyah and I. Singer, The index of elliptic operators, I and III. *Ann. of Math.(2)* **87** (1968) 484-530 and 546-604; MR **38** # 5243 and #5245.

[$At\text{-}Si_2$] M.F. Atiyah and I.M. Singer. The index of elliptic operators, IV. *Ann. of Math.(2)* **93** (1971), 119-138; MR **43** # 5554.

[$At\text{-}Si_3$] M.F. Atiyah and I.M. Singer. The index of elliptic operators on compact manifolds. *Bull. Amer. Math. Soc.* **69** (1963), 422-433; MR **28** # 626.

[B] S. Baaj. Multiplicateurs non bornés. *Thèse, Univ. P. et M. Curie, Paris* 1980.

[B-J] S. Baaj and P. Julg. Théorie bivariante de Kasparov et opérateurs non bornés dans les C^*-modules hilbertiens. *C.R. Acad. Sci. Paris Sér. I* **296** (1983), no. 21, 875-878; MR 84m:46091.

[Ba] H. Bacry. Localizability and space in quantum physics. *Lecture Notes in Physics, 308, Springer, Berlin,* 1988; MR 90a:81001.

[Bau-C_1] P. Baum and A. Connes. Geometric K-theory for Lie groups and foliations. *Preprint, Inst. Hautes Études Sci., Paris,* 1982.

[Bau-C_2] P. Baum and A. Connes. Leafwise homotopy equivalence and rational Pontrjagin classes. *Foliations, pp. 1-14, Adv. Stud. Pure Math., 5, North Holland, Amsterdam,* 1985; MR 88d:58111.

[Bau-C_3] P. Baum and A. Connes. Chern character for discrete groups. *A fête of topology, pp. 163-232, Academic, Boston, Mass.,* 1988; MR 90c:58149.

[Bau-C-H] P. Baum, A. Connes and N. Higson. Classifying space for proper actions and K-theory of group C^*-algebras. *C^*-algebras: 1943–1993, A fifty year celebration, Contemporary Math. 167 (1994), Amer Math. Soc., Providence, RI* 241–291.

[Bau-D] P. Baum and R. Douglas. K-homology and index theory. *Operator algebras and applications, Part 1, pp. 117-173, Proc. Sympos. Pure Math., 38, Amer. Math. Soc., Providence, R.I.,* 1982; MR 84d:58075.

3-F-F-L-S_1] F. Bayen, M. Flato, C. Fronsdal, A. Lichnerowicz and D. Sternheimer. Deformation theory and quantization, I and II. *Ann. Phys.* **111** (1978), 61-110, 111-151; MR **58** # 14737.

3-F-F-L-S_2] F. Bayen, M. Flato, C. Fronsdal, A. Lichnerowicz and D. Sternheimer. Quantum mechanics as a deformation of classical mechanics. *Lett. Math. Phys.* **1** (1975/77), 521-530.

[B-G-M-S] J. Bellissard, D.R. Grempel, F. Martinelli and E. Scoppola. Localization of electrons with spin-orbit or magnetic interactions in a two dimensional crystal. *Preprint, CPT-85/P-1788,* June, 1985.

[Be_1] A.F. Beardon. The Hausdorff dimension of singular sets of properly discontinuous groups. *Amer. J. Math.* **88** (1966), 722-736; MR **33** # 7532.

[Be_2] A.F. Beardon. The exponent of convergence of Poincaré series. *Proc. London Math. Soc. (3)* **18** (1968), 461-483; MR **37** # 2986.

[Bed-H] E. Bedos and P. de la Harpe. Moyennabilité intérieure des groupes: définitions et exemples. *Enseign. Math. (2)* **32** (1986), 139-157; MR 87k:43001.

[Bel_1] J. Bellissard. K-theory of C^*-algebras in solid state physics. *Statistical mechanics and field theory: mathematical aspects, pp. 99-156, Lecture Notes in Phys., 257, Springer, Berlin,* 1986; MR 88e:46053.

[Bel_2] J. Bellissard. Ordinary quantum Hall effect and noncommutative cohomology. *Proc. on localization in disordered systems (Bad Schandau, 1986), Teubner, Leipzig,* 1988.

[B-S] J. Bellissard and B. Simon. Cantor spectrum for the almost Mathieu equation. *J. Functional Anal.* **48** (1982), 408-419; MR 84h:81019.

[Ber_1] L. Bers. Uniformization by Beltrami equations. *Comm. Pure Appl. Math.* **14** (1961), 215-228; MR **24** # A2022.

[Ber_2] L. Bers. Uniformization, moduli and Kleinian groups. *Bull. London Math. Soc.* **4** (1972), 257-300; MR **50** # 595.

[Ber_3] L. Bers. On Hilbert's 22nd problem. *Mathematical developments arising from Hilbert problems, pp. 559-609, Proc. Sympos. Pure Math., 28, Amer. Math. Soc., Providence, R.I.*, 1976; MR **55** # 654.

[Bi] M. Binder. On induced representations of discrete groups. *Proc. Amer. Math. Soc.* **118** (1993), 301-309; MR 93f:22005.

[Bi-K-S] M. Birman, L. Koplienko and M. Solomiak. Estimates of the spectrum of a difference of fractional powers of selfadjoint operators. *Izv. Vysš. Učebn. Zaved. Mat.* **1975**, no. 3(154), 3-10; MR **52** # 6458.

[Bl_1] B. Blackadar. K-theory for operator algebras. *Math. Sci. Res. Inst. Publ., 5, Springer, New York,* 1986; MR 88g:46082.

[Bl_2] B. Blackadar. A simple unital projectionless C^*-algebra. *J. Operator Theory* **5** (1981), 63-71; MR 82h:46076.

[Bl_3] B. Blackadar. A simple C^*-algebra with no nontrivial projections. *Proc. Amer. Math. Soc.* **78** (1980), 504-508; MR 81j:46092.

[Bl_4] B. Blackadar. The regular representation of restricted direct product groups. *J. Functional Anal.* **25** (1977), 267-274; MR **55** # 12860.

[Bla-B] P. Blanc and J.L. Brylinski. Cyclic homology and the Selberg principle. *J. Functional Anal.* **109** (1992), 289-330.

[Bla-W] P. Blanc and D. Wigner. Homology of Lie groups and Poincaré duality. *Lett. Math. Phys.* **7** (1983), 259-270; MR 85g:22023.

[Bo-S] N.N. Bogolyubov and D.V. Shirkov. Quantum fields. *Benjamin, Reading, Mass.*, 1983; MR 85g:81096 .

[Bos_1] J.-B. Bost. Principe d'Oka, K-théorie et systèmes dynamiques non commutatifs. *Invent. Math.* **101** (1990), 261-333; MR 92j:46126.

[Bos_2] J.-B. Bost. Witt group of commutative involutive Banach algebras. *Unpublished.*

[Bos-C] J.-B. Bost and A. Connes. Produits eulériens et facteurs de type III. *C.R. Acad. Sci. Paris Ser. I Math.* **315** (1992), 279-284.

[Bot_1] R. Bott. On characteristic classes in the framework of Gel'fand-Fuchs cohomology. *Analyse et topologie, pp. 113-139, Astérisque No. 32/33, Math. Soc. France, Paris,* 1976; MR **53** # 6582.

[Bot_2] R. Bott. The index theorems for homogeneous differential operators. *Differential and combinatorial topology, pp. 167-186, Princeton Univ. Press, Princeton, N.J.*, 1965; MR **31** # 6246.

[Bow_1] R. Bowen. Equilibrium states and the ergodic theory of Anosov diffeomorphisms. *Lecture Notes in Math., 470, Springer, Berlin,* 1975; MR **56** # 1364.

[Bow_2] R. Bowen. Hausdorff dimension of quasi-circles. *Inst. Hautes Études Sci. Publ. Math. No. 50* (1979), 11-25; MR 81g:57023.

[Br] O. Bratteli. Inductive limits of finite dimensional C^*-algebras. *Trans. Amer. Math. Soc.* **171** (1972), 195-234; MR **47** # 844.

[Br-R] O. Bratteli and D.W. Robinson. Operator algebras and quantum statistical mechanics, Vol. I and II. *Springer, New York,* 1979 and 1981; MR 81a:46070 and 82k:82013.

[Bro_1] R. Brown. From groups to groupoids: a brief survey. *Bull. London Math. Soc.* **19** (1987), 113-134; MR 87m:18009.

[Bro_2] R. Brown. Topological aspects of holonomy groupoids. *UCNW Preprint,* 88-10.

[Bro_3] L.G. Brown. Stable isomorphism of hereditary subalgebras of C^*-algebras. *Pacific J. Math.* **71** (1977), 335-348; MR **56** # 12894.

[B-D-F] L.G. Brown, R.G. Douglas and P.A. Fillmore. Extensions of C^*-algebras and K-homology. *Ann. of Math. (2)* **105** (1977), 265-324; MR **56** # 16399.

[B-G-R] L.G. Brown, P. Green and M.A. Rieffel. Stable isomorphism and strong Morita equivalence of C^*-algebras. *Pacific J. Math.* **71** (1977), 349-363; MR **57** # 3866.

[B-O] T.P. Branson and B. Ørsted. Explicit functional determinants in four dimensions. *Proc. Amer. Math. Soc.* **113** (1991), 669-682; MR 92b:58238.

[Bry] J.-L. Brylinski. Some examples of Hochschild and cyclic homology. *Algebraic groups (Utrecht, 1986), pp. 33-72, Lecture Notes in Math., 1271, Springer, Berlin,* 1987; MR 90e:22018.

[Bry_2] J.-L. Brylinski. Cyclic homology and equivariant theories. *Ann. Inst. Fourier (Grenoble)* **37** (1987), 15-28; MR 89j:55008.

[Bry-N] J.-L. Brylinski and V. Nistor. Cyclic cohomology of smooth discrete groupoids. *Preprint,* 1993.

[Bu] D. Burghelea. The cyclic homology of the group rings. *Comment. Math. Helv.* **60** (1985), 354-365; MR 88e:18007.

[Bur-F] D. Burghelea and Z. Fiedorowicz. Cyclic homology and algebraic K-theory of spaces, II. *Topology* **25** (1986), 303-317; MR 88i:18009b.

[C] A.P. Calderón. Intermediate spaces and interpolation, the complex method. *Stud. Math.* **24** (1964), 113-190; MR **29** # 5097.

[Ca-P] R. Carey and J.D. Pincus. Almost commuting algebras. *K-theory and operator algebras, pp. 19-43, Lecture Notes in Math., 575, Springer, Berlin,* 1977; MR **58** # 23652.

[Car-E] H. Cartan and S. Eilenberg. Homological algebra. *Princeton Univ. Press, Princeton, N.J.,* 1956; MR **17**, 1040.

[Cart] P. Cartier. Homologie cyclique. *Séminaire Bourbaki, Vol. 1983/84, pp. 123-146, Astérisque No. 121/122, Soc. Math. France, Paris,* 1984; MR 86e:18012.

[Ch-E] M. Choi and E. Effros. Separable nuclear C^*-algebras and injectivity. *Duke Math. J.* **43** (1976), 309-322; MR **53** # 8912.

[Ch-E-Y] M. Choi, G. Elliott and N. Yui. Gauss polynomials and the rotation algebra. *Preprint.*

[C-H] M. Cowling and U. Haagerup. Completely bounded multipliers of the Fourier algebra of a simple Lie group of real rank one. *Invent. Math.* **96** (1989), 507-549; MR 90h:22008.

[Com] F. Combes. Poids associé à une algèbre hilbertienne à gauche. *Compos. Math.* **23** (1971), 49-77; MR **44** # 5786.

[Co_1] A. Connes. Caractérisation des espaces vectoriels ordonnés sous-jacents aux algèbres de von Neumann. *Ann. Inst. Fourier (Grenoble)* **24** (1974), 121-155; MR **51** # 13705.

[Co_2] A. Connes. Sur la classification des facteurs de type II. *C.R. Acad. Sci. Paris Ser. A-B* **281** (1975), A13-A15; MR **51** # 13706.

[Co_3] A. Connes. Almost periodic states and factors of type III_1. *J. Functional Anal.* **16** (1974), 415-445; MR **50** # 10840.

[Co_4] A. Connes. Une classification des facteurs de type III. *Ann. Sci. École Norm. Sup. (4)* **6** (1973), 133-252; MR **49** # 5865.

[Co_5] A. Connes. Classification of injective factors. *Ann. of Math. (2)* **104** (1976), 73-115; MR **56** # 12908.

[Co_6] A. Connes. Outer conjugacy classes of automorphisms of factors. *Ann. Sci. École Norm. Sup. (4)* **8** (1975), 383-419; MR **52** # 15031.

[Co_7] A. Connes. The von Neumann algebra of a foliation. *Mathematical problems in theoretical physics, pp. 145-151, Lecture Notes in Phys., 80, Springer, Berlin,* 1978; MR 80i:46057.

[Co_8] A. Connes. On the cohomology of operator algebras. *J. Functional. Anal.* **28** (1978), 248-253; MR **58** # 12407.

[Co_9] A. Connes. A factor of type II_1 with countable fundamental group. *J. Operator Theory* **4** (1980), 151-153; MR 81j:46099.

[Co_{10}] A. Connes. Sur la théorie non commutative de l'intégration. *Algèbres d'opérateurs, pp. 19-143, Lecture Notes in Math., 725, Springer, Berlin,* 1979; MR 81g:46090.

[Co_{11}] A. Connes. A survey of foliations and operator algebras. *Operator algebras and applications, Part 1, pp. 521-628, Proc. Sympos. Pure Math., 38, Amer. Math. Soc, Providence, R.I.,* 1982; MR 84m:58140.

[Co_{12}] A. Connes. Classification des facteurs. *Operator algebras and applications, Part 2, pp. 43-109, Proc. Sympos. Pure Math., 38, Amer. Math. Soc, Providence, R.I.,* 1982; MR 84e:46068.

[Co_{13}] A. Connes. C^*-algèbres et géométrie différentielle. *C.R. Acad. Sci. Paris Sér. A-B* **290** (1980), A599-A604; MR 81c:46053.

[Co_{14}] A. Connes. Cyclic cohomology and the transverse fundamental class of a foliation. *Geometric methods in operator algebras (Kyoto, 1983), pp. 52-144, Pitman Res. Notes in Math., 123, Longman, Harlow,* 1986; MR 88k:58149.

[Co_{15}] A. Connes. Spectral sequence and homology of currents for operator algebras. *Math. Forschungsinst. Oberwolfach Tagungsber., 41/81, Funktionalanalysis und C^*-Algebren, 27-9/3-10,* 1981.

[Co_{16}] A. Connes. The action functional in noncommutative geometry. *Comm. Math. Phys.* **117** (1988), 673-683; MR 91b:58246.

[Co_{17}] A. Connes. Noncommutative differential geometry. Part I: The Chern character in K-homology, *Preprint I.H.E.S.* (M/82/53), 1982. Part II: de Rham homology and noncommutative algebra, *Preprint IHES* (M/83/19), 1983.

[Co_{18}] A. Connes. Noncommutative differential geometry. *Inst. Hautes Études Sci. Publ. Math. No. 62* (1985), 257-360; MR 87i:58162.

[Co_{19}] A. Connes. Feuilletages et algèbres d'opérateurs. *Bourbaki Seminar, Vol. 1979/80, pp. 139-155, Lecture Notes in Math., 842, Springer, Berlin,* 1981; MR 83c:58077.

[Co_{20}] A. Connes. An analogue of the Thom isomorphism for crossed products of a C^*-algebra by an action of $\mathbb{R}$. *Adv. in Math.* **39** (1981), 31-55; MR 82j:46084.

[Co_{21}] A. Connes. Factors of type III_1, property L'_λ and closure of inner automorphisms. *J. Operator Theory* **14** (1985), 189-211; MR 88b:46088.

[Co_{22}] A. Connes. On the spatial theory of von Neumann algebras. *J. Functional Anal.* **35** (1980), 153-164; MR 81g:46083.

[Co_{23}] A. Connes. Cyclic cohomology and noncommutative differential geometry. *Proceedings of the International Congress of Math. (ICM Berkeley, Calif., 1986) Vol.2, pp. 879-889, Amer. Math. Soc., Providence, R.I.,* 1987; MR 89h:58182 .

[Co_{24}] A. Connes. Entire cyclic cohomology of Banach algebras and characters of θ-summable Fredholm modules. *K-Theory* **1** (1988), 519-548; MR 90c:-46094.

[Co_{25}] A. Connes. Compact metric spaces, Fredholm modules and hyperfiniteness. *Ergodic. Theor. and Dynam. Systems* **9** (1989), 207-220; MR 90i:-46124.

[Co_{26}] A. Connes. Cohomologie cyclique et foncteur Ext^n. *C.R. Acad. Sci. Paris Ser. I Math.* **296** (1983), 953-958; MR 86d:18007.

[Co_{27}] A. Connes. Entropie de Kolmogoroff-Sinai et mécanique statistique quantique. *C.R. Acad. Sci. Paris Ser. I Math.* **301** (1985), 1-6; MR 87b:46072.

[Co_{28}] A. Connes. Essay on physics and noncommutative geometry. *The interface of mathematics and particle physics, pp. 9-48, Oxford Univ. Press, New York,* 1990; MR 92g:58007.

[Co_{29}] A. Connes. Géométrie non commutative. *InterEditions, Paris,* 1990; MR 92e:58016 .

[Co_{30}] A. Connes. On the Chern character of θ-summable Fredholm modules. *Comm. Math. Phys.* **139** (1991), 171-181; MR 92i:19003.

[Co_{31}] A. Connes. Noncommutative geometry and physics. *Les Houches, Preprint IHES M/93/32,* 1992.

[Co-C] A. Connes and J. Cuntz. Quasi homomorphismes, cohomologie cyclique et positivité. *Comm. Math. Phys.* **114** (1988), 515-526; MR 89h:46098.

[Co-F] A. Connes and T. Fack. Morse inequalities for measured foliations.

[Co-F-S] A. Connes, M. Flato and D. Sternheimer. Closed star products and cyclic cohomology. *Lett. Math. Phys.* **24** (1992), 1-12; MR 93d:19003.

[Co-F-W] A. Connes, J. Feldman and B. Weiss. An amenable equivalence relation is generated by a single transformation. *Ergodic Theory and Dynamical Systems* **1** (1982), 431-450; MR 84h:46090.

[Co-G-M_1] A. Connes, M. Gromov and H. Moscovici. Conjecture de Novikov et fibrés presque plats. *C.R. Acad. Sci. Paris Ser. I Math.* **310** (1990), 273-277; MR 91e:57041.

[Co-G-M_2] A. Connes, M. Gromov and H. Moscovici. Group cohomology with Lipschitz control and higher signatures. *Geom. Functional Anal.* **3** (1993), 1-78.

[Co-H] A. Connes and N. Higson. Déformations, morphismes asymptotiques et K-théorie bivariante. *C.R. Acad. Sci. Paris Ser. I Math.* **311** (1990), 101-106; MR 91m:46114.

[Co-H-S] A. Connes, U. Haagerup and E. Størmer. Diameters of state spaces of type III factors. *Operator algebras and their connections with topology and ergodic theory (Busteni, 1983), pp. 91-116, Lecture Notes in Math., 1132, Springer, Berlin*, 1985; MR 88b:46089.

[Co-J] A. Connes and V. Jones, Property T for von Neumann algebras. *Bull. London Math. Soc.* **17** (1985), 57-62; MR 86a:46083.

[Co-K] A. Connes and M. Karoubi. Caractère multiplicatif d'un module de Fredholm. *K-Theory* **2** (1988), 431-463; MR 90c:58174.

[Co-Kr] A. Connes and W. Krieger. Measure space automorphisms, the normalizers of their full groups and approximate finiteness. *J. Functional Anal.* **24** (1977), 336-352; MR **56** # 3246.

[Co-L] A. Connes and J. Lott. Particle models and noncommutative geometry. *Nuclear Phys. B* **18B** (1990), *suppl.*, 29-47 (1991); MR 93a:58015.

[Co-M_1] A. Connes and H. Moscovici. Cyclic cohomology, the Novikov conjecture and hyperbolic groups. *Topology* **29** (1990), 345-388; MR 92a:58137.

[Co-M_2] A. Connes and H. Moscovici. Transgression du caractère de Chern et cohomologie cyclique. *C.R. Acad. Sc. Paris Sér. I Math.* **303** (1986), 913-918; MR 88c:58006.

[Co-M_3] A. Connes and H. Moscovici. The L^2-index theorem for homogeneous spaces. *Bull. Amer. Math. Soc. (N.S.)* **1** (1979), 688-690; MR 80g:58046.

[Co-M_4] A. Connes and H. Moscovici. The L^2-index theorem for homogeneous spaces of Lie groups. *Ann. of Math. (2)* **115** (1982), 291-330; MR 84f:58108.

[Co-M_5] A. Connes and H. Moscovici. Conjecture de Novikov et groupes hyperboliques. *C.R. Acad. Sci. Paris Ser. I Math.* **307** (1988), 475-480; MR 90e:57048.

[Co-N-T] A. Connes, H. Narnhofer and W. Thirring. Dynamical entropy of C^*-algebras and von Neumann algebras. *Comm. Math. Phys.* **112** (1987), 691-719; MR 89b:46078.

[Co-R] A. Connes and M. Rieffel. Yang-Mills for noncommutative two tori. *Operator algebras and mathematical physics (Iowa City, Iowa, 1985), pp. 237-266, Contemp. Math. Oper. Algebra. Math. Phys., 62, Amer. Math. Soc., Providence, R.I.*, 1987; MR 88b:58033.

[Co-S] A. Connes and G. Skandalis. The longitudinal index theorem for foliations. *Publ. Res. Inst. Math. Sci. Kyoto* **20** (1984), 1139-1183; MR 87h:58209.

[Co-St_1] A. Connes and E. Størmer. Homogeneity of the state space of factors of type III_1. *J. Functional Anal.* **28** (1978), 187-196; MR **57** #10435.

[Co-St_2] A. Connes and E. Størmer. Entropy for automorphisms of II_1 von Neumann algebras. *Acta Math.* **134** (1975), 289-306; MR **56** # 12906.

[Co-Su] A. Connes and D. Sullivan. Quantized calculus on S^1 and quasi-Fuchsian groups.

[Co-S-T] A. Connes, D. Sullivan and N. Teleman. Quasiconformal mappings, operators on Hilbert space, and local formulae for characteristic classes. *Preprint I.H.E.S.*, M/93/38.

[Co-T] A. Connes and M. Takesaki. The flow of weights on factors of type III. *Tôhoku Math. J.* **29** (1977), 473-575; MR 82a:46069a.

[Co-W] A. Connes and E.J. Woods. A construction of approximately finite-dimensional non-ITPFI factors. *Canad. Math. Bull.* **23** (1980), 227-230; MR 82m:46069.

[Coq-E-S] R. Coquereaux, G. Esposito-Farese and F. Scheck. Noncommutative geometry and graded algebras in electroweak interactions. *Internat. J. Modern Phys. A* 7 (1992), 6555-6593.

[Cu_0] J. Cuntz. Simple C^*-algebras generated by isometries. *Comm. Math. Phys.* **57** (1977), 173-185; MR **57** # 7189.

[Cu_1] J. Cuntz. K-theoretic amenability for discrete groups. *J. Reine Angew. Math.* **344** (1983), 180-195; MR 86e:46064.

[Cu_2] J. Cuntz. A new look at KK-theory. *K-Theory* **1** (1987), 31-51; MR 89a:46142.

[Cu_3] J. Cuntz. Universal extensions and cyclic cohomology. *C.R. Acad. Sci. Paris Ser. I Math.* **309** (1989), 5-8; MR 91a:19002.

[Cu_4] J. Cuntz. K-theory for certain C^*-algebras. *Ann. of Math. (2)* **113** (1981), 181-197; MR 84c:46058.

[Cu_5] J. Cuntz. K-theory for certain C^*-algebras, II. *J. Operator Theory* **5** (1981), 101-108; MR 84k:46053.

[Cu_6] J. Cuntz. A class of C^*-algebras and topological Markov chains, II. *Invent. Math.* **63** (1981), 25-40; MR 82f:46073b.

[Cu_7] J. Cuntz. The K-groups for free products of C^*-algebras. *Operator algebras and applications, Part 1, pp. 81-84, Proc. Sympos. Pure Math., 38, Amer. Math. Soc., Providence, R.I.,* 1982; MR 84f:46078.

[Cu_8] J. Cuntz. Generalized homomorphisms between C^*-algebras and KK-theory. *Dynamics and processes, (Bielefeld, 1981), pp. 31-45, Lecture Notes in Math., 1031, Springer, Berlin,* 1983; MR 85j:46126.

[Cu_9] J. Cuntz. K-theory and C^*-algebras. *Algebraic K-theory, number theory, geometry and analysis (Bielefeld, 1982), pp. 55-79, Lecture Notes in Math., 1046, Springer, Berlin,* 1984; MR 86d:46071.

[Cu-K] J. Cuntz and W. Krieger. A class of C^*-algebras and topological Markov chains. *Invent. Math.* **56** (1980), 251-268; MR 82f:46073a.

[C-Q] J. Cuntz and D. Quillen. Algebra extensions and nonsingularity. *J. Amer. Math. Soc., to appear.*

[C-Q_2] J. Cuntz and D. Quillen. Operators on noncommutative differential forms and cyclic homology. *J. Differential Geometry, to appear.*

[C-Q_3] J. Cuntz and D. Quillen. Cyclic homology and nonsingularity. *J. Amer. Math. Soc., to appear.*

[C-Q_4] J. Cuntz and D. Quillen. On excision in periodic cyclic cohomology, I and II. *C. R. Acad. Sci Paris Sér. I Math.,* **317** (1993), 917–922; **318** (1994), 11–12.

[Cu-S] J. Cuntz and G. Skandalis. Mapping cones and exact sequences in KK-theory. *J. Operator Theory* **15** (1986), 163-180; MR 88b:46099.

[Dad_1] M. Dadarlat. Shape theory and asymptotic morphisms for C^*-algebras, *Duke Math. J.* **73** (1994), no. 3, 687–711.

[Dad_2] M. Dadarlat. A note on asymptotic homomorphisms. *K-Theory, to appear.*

[D-L_1] M. Dadarlat and T.A. Loring. Deformations of topological spaces predicted by E-theory. *Proc. GPOTS,Iowa 1993, to appear.*

[D-L_2] M. Dadarlat and T.A. Loring. K-homology, asymptotic representations and unsuspended E-theory. *J. Functional Anal., to appear.*

[Da] E.B. Davies. Lipschitz continuity of functions of operators in the Schatten classes. *J. London Math. Soc.* **37** (1988), 148-157; MR 89c:47009.

[D-K] C. Delaroche and A.A. Kirillov. Sur les relations entre l'espace dual d'un groupe et la structure de ses sous-groupes fermés. *Séminaire Bourbaki, Exp. 343*, 1967/68.

[De] P. Delorme. 1-cohomologie des représentations unitaires des groupes de Lie semi-simples et résolubles. *Bull. Soc. Math. France* **105** (1977), 281-336; MR **58** # 28272.

[D-H-K] R. Douglas, S. Hurder and J. Kaminker. Cyclic cocycles, renormalization and eta-invariants, *Invent. Math.* **103** (1991), 101-179; MR 91m:58152.

[D-H-R] S. Doplicher, R. Haag and J. Roberts. Local observables and particle statistics, I. *Comm. Math. Phys.* **23** (1971), 199-230; MR **45** # 6316.

[DeW-L] M. De Wilde and P.B. Lecomte. Existence of star-products and of formal deformations of the Poisson-Lie algebra of arbitrary symplectic manifolds. *Lett. Math. Phys.* **7** (1983), 487-496; MR 85j:17021.

[Dir] P.A.M. Dirac. The quantum theory of the emission and absorption of radiation. *Proc. Roy. Soc. London Ser. A* **114** (1927), 243-265.

[Di_1] J. Dixmier. Formes linéaires sur un anneau d'opérateurs. *Bull. Soc. Math. France* **81** (1953), 9-39; MR **15**, 539.

[Di_2] J. Dixmier, Les algèbres d'opérateurs dans l'espace hilbertien. *2nd edition, Gauthier-Villars, Paris*, 1969.

[Di_3] J. Dixmier. Les C^*-algèbres et leurs représentations. *Gauthier-Villars, Paris*, 1964; MR **30** # 1404.

[Di_4] J. Dixmier. Existence de traces non normales. *C.R. Acad. Sci. Paris Ser A-B* **262** (1966), A1107-A1108; MR **33** # 4695.

[Di-D] J. Dixmier and A. Douady. Champs continus d'espaces hilbertiens et de C^*-algèbres. *Bull. Soc. Math. France* **91** (1963), 227-284; MR **29** # 485.

[D-S] S. Donaldson and D. Sullivan. Quasiconformal 4-manifolds. *Acta Math.* **163** (1989), 181-252; MR 91d:57012.

[Don] W.F. Donoghue, Jr. Monotone matrix functions and analytic continuation. *Springer, New York*, 1974; MR **58** #6279.

[Dop] S. Doplicher. Abstract compact group duals. *Operator algebras and quantum field theory (Proc. Internat. Congr. Math. (Kyoto, 1990), pp. 1319-1333, Math. Soc. Japan, Tokyo*, 1991; MR 93e:81065.

[Dou_1] R.G. Douglas. C^*-algebra extensions and K-homology. *Ann. of Math. Stud., 95, Princeton Univ. Press, Princeton, N.J.*, 1980; MR 82c:46082.

[Dou_2] R.G. Douglas. Banach algebra techniques in operator theory. *Pure Appl. Math., 49, Academic, New York*, 1972; MR **50** # 14335.

[Dou-V] R.G. Douglas and D. Voiculescu. On the smoothness of sphere extensions. *J. Operator Theory* **6** (1981), 103-111; MR 83h:46080.

[Dr] V.G. Drinfel'd. Quantum groups. *Proceedings of the International Congress of Math. (Berkeley, Calif., 1986), pp. 798-820, Amer. Math. Soc., Providence, R.I.*, 1987; MR 89f:17017.

[Du-K-M] M. Dubois-Violette, R. Kerner and J. Madore. Noncommutative geometry and new models of gauge theory. *Preprint, Orsay*, 1989.

[Duf] D. McDuff. Central sequences and the hyperfinite factor. *Proc. London Math. Soc. (3)* **21** (1970), 443-461; MR **43** # 6737.

[Dw-K] W.G. Dwyer and D.M. Kan. Normalizing the cyclic modules of Connes. *Comment. Math. Helv.* **60** (1985), 582-600; MR 88d:18009.

[Dy] H.A. Dye. On groups of measure preserving transformations, I and II. *Amer. J. Math.* **81** (1959), 119-159; *ibid.* **85** (1963), 551-576; MR **24** # A1366, **28** # 1275.

[Ehr_1] C. Ehresmann. Oeuvres complètes et commentées. *Imprimerie Evrard, Amiens,* 1982.

[Ehr_2] C. Ehresmann. Catégories topologiques et catégories différentiables. *Colloq. Géom. Diff. Globale (Bruxelles, 1959), pp. 137-150, Centre Belge Rech. Math., Louvain,* 1959; MR **22** # 7148.

[Ehr_3] C. Ehresmann. Sur les catégories différentiables. *Atti Conv. Int. Geom. Diff. (Bologna, 1967), pp. 31-40, Zanichelli, Bologna,* 1970.

[Ehr_4] C. Ehresmann. Structures feuilletées. *Proc. Fifth Canadian Math. Congress (Montréal, 1961), pp. 109-172, Univ. Toronto Press, Toronto,* 1963.

[E-H-S] E.G. Effros, D.E. Handelman and C.L. Shen. Dimension groups and their affine representations. *Amer. J. Math.* **102** (1980), 385-407; MR 83g:46061.

[E-L] E.G. Effros and C. Lance. Tensor products of operator algebras. *Adv. in Math.* **25** (1977), 1-34; MR **56** # 6402.

[El_1] G. Elliott. On the classification of inductive limits of sequences of semisimple finite-dimensional algebras. *J. Algebra* **38** (1976), 29-44; MR **53** # 1279.

[El_2] G.A. Elliott. On totally ordered groups, and K_0. *Ring theory (Waterloo, 1978), pp. 1-49, Lecture Notes in Math., 734, Springer, Berlin,* 1979; MR 81g:06012.

[El-N-N] G. Elliott, T. Natsume and R. Nest. Cyclic cohomology for one-parameter smooth crossed products. *Acta Math.* **160** (1988), 285-305; MR 89h:46093.

[El-W] G. Elliott and E.J. Woods, The equivalence of various definitions for a properly infinite von Neumann algebra to be approximately finite dimensional. *Proc. Amer. Math. Soc.* **60** (1976), 175-178; MR **58** # 23630.

[Ell] J. Ellis. Phenomenology of unified gauge theories. *Les Houches, Session XXXVII (1981), North-Holland,* 1983.

[En-S] M. Enock and J.-M. Schwartz. Une dualité dans les algèbres de von Neumann. *Bull. Soc. Math. France Mem., 44 suppl., Soc. Math. France, Paris,* 1975; MR **56** # 1091.

[F-S] T. Fack and G. Skandalis. Sur les représentations et idéaux de la C^*-algèbre d'un feuilletage. *J. Operator Theory* **8** (1982), 95-129; MR 84d:46101.

[Fa-S] L.D. Faddeev and A.A. Slavnov. Gauge fields. *Frontiers in Phys., 50, Benjamin/Cummings, Reading, Mass.,* 1980; MR 83e:81001.

[Fa-R-T] L.D. Faddeev, N.Y. Reshetikhin and L.A. Takhtajan. Quantization of Lie groups and Lie algebras. *Preprint, LOMI,* 1987.

[Fe] H. Federer. Geometric measure theory. *Grundlehren Math., 153, Springer, New York,* 1969; MR **41** # 1976.

[Fed] B.V. Fedosov. Analytic formulae for the index of elliptic operators. *Trudy Moskov. Mat. Obšč.* **30** (1974), 159-241; MR **54** # 8743.

[F-S$_1$] B.V. Fedosov and M.A. Subin [Shubin]. The index of random elliptic operators and of families of them. *Dokl. Akad. Nauk SSSR* **236** (1977), 812-815; MR **58** # 24397.

[F-S$_2$] B.V. Fedosov and M.A. Subin [Shubin]. The index of random elliptic operators, I and II. *Mat. Sb.* **106** (1978), 108-140; *ibid.* **106** (1978), 455-483; MR **58** # 18610-18611.

[F-T] B.L. Feigin [Feĭgin] and B.L. Tsygan. Additive K-theory. *K-theory, arithmetic and geometry, pp. 67-209, Lecture Notes in Math., 1289, Springer, Berlin,* 1987; MR 89a:18017.

[Fef] C. Fefferman. Characterizations of bounded mean oscillation. *Bull. Amer. Math. Soc.* **77** (1971), 587-588; MR **43** # 6713.

[Fel-M] J. Feldman and C. Moore. Ergodic equivalence relations, cohomology and von Neumann algebras I, II. *Trans. Amer. Math. Soc.* **234** (1977), 289-324, 325-359; MR **58** # 28261.

[Fey] R. Feynman. The reason for antiparticles. *Elementary particles and the laws of physics, pp. 1-59, Cambridge Univ. Press, Cambridge, Mass.,* 1988.

[Fied-L] Z. Fiedorowicz and J.-L. Loday. Crossed simplicial groups and their associated homology. *Trans. Amer. Math. Soc.* **326** (1991), 57-87; MR 91j:18018.

[F-H] J. Fox and P. Haskell. K-amenability for $SU(n,1)$. *J. Functional Anal.* **117** (1993) 279-307; MR 94i:22033.

[F-M-S-S] J. Fröhlich, F. Martinelli, E. Scoppola and T. Spencer. Constructive proof of localization in the Anderson tight binding model. *Comm. Math. Phys.* **101** (1985), 21-46; MR 87a:82047.

[F-S] J. Fröhlich and T. Spencer. Absence of diffusion in the Anderson tight binding model for large disorder or low energy. *Comm. Math. Phys.* **88** (1983), 151-184; MR 85c:82004.

[Fr] J. Fröhlich. Non-perturbative quantum field theory. *Adv. Ser. Math. Phys., 15, World Scientific, River Edge, N.J.,* 1992.

[Ful] W. Fulton. Intersection theory. *Ergebn. der Math. und ihrer Grenzgeb., 2, Springer, Berlin,* 1984; MR 85k:14004.

[Fu] H. Furstenberg. Recurrence in ergodic theory and combinatorial number theory. *Princeton Univ. Press, Princeton, N.J.,* 1981; MR 82j:28010.

[G-F] I.M. Gel'fand and D. Fuchs. Cohomology of the Lie algebra of formal vector fields. *Math. USSR Izv.* **4** (1970), 327-342.

[G-N$_1$] I.M. Gel'fand and M.A. Naimark. On the imbedding of normed rings into the ring of operators in Hilbert space. *Mat. Sb.* **12** (1943), 197-213; MR **5**, 147.

[G-N$_2$] I.M. Gel'fand and M.A. Naimark. Anneaux normés à involution et leurs représentations. *Izv. Akad. Nauk SSSR* **12** (1948), 445-480; MR **10**, 199.

[Ger-S] M. Gerstenhaber and S. Schack. Algebraic cohomology and deformation theory. *Deformation theory of algebras and structures and applications, pp. 11-264, NATO Adv. Sci. Inst. Ser. C: Math. Phys. Sci., 247, Kluwer, Dordrecht,* 1988; MR 90c:16016.

[Ge_1] E. Getzler. Pseudodifferential operators on supermanifolds and the Atiyah-Singer index theorem. *Comm. Math. Phys.* **92** (1983), 163-178; MR 86a:-58104.

[Ge_2] E. Getzler. A short proof of the local Atiyah-Singer index theorem. *Topology* **25** (1986), 111-117; MR 87h:58207.

[Ge_3] E. Getzler. The odd Chern character in cyclic homology and spectral flow. *Topology* **32** (1993), 489-507; MR 87h:58207.

[Ge-J] E. Getzler and J.D.S. Jones. A_∞-algebras and the cyclic bar complex. *Illinois J. Math.* **34** (1990), 256-283; MR 91e:19001.

[Ge-J_2] E. Getzler and J.D.S. Jones. The cyclic homology of crossed product algebras. *To appear in Crelle's Journal Reine Angew. Math.*.

[Ge-S] E. Getzler and A. Szenes. On the Chern character of a theta-summable Fredholm module. *J. Functional Anal.* **84** (1989), 343-357; MR 91g:19007.

[Gi_1] P. Gilkey. Invariance theory, the heat equation and the Atiyah-Singer index theorem. *Math. Lecture Ser., 11, Publish or Perish, Wilmington, Del.*, 1984; MR 86j:58144.

[Gi_2] P. Gilkey. The index theorem and the heat equation. *Math. Lecture Ser., 4, Publish or Perish, Boston, Mass.*, 1974; MR **56** # 16704.

[Gio-J] T. Giordano and V. Jones. Antiautomorphismes involutifs du facteur hyperfini de type II_1. *C.R. Acad. Sci. Sér. A-B* **290** (1980), A29-A31; MR 82b:46079.

[Gl-J] J. Glimm and A. Jaffe. Quantum physics. *Springer, New York*, 1981; MR 83c:81001.

[Go-V] C. Godbillon and J. Vey. Un invariant des feuilletages de codimension 1. *C.R. Acad. Sci. Paris Ser. A-B* **273** (1971), A92-A95; MR **44** #1046.

[Goh-K_1] I.C. Gohberg and M.G. Krein. Introduction to the theory of linear nonselfadjoint operators. *Transl. Math. Monogr., 18, Amer. Math. Soc., Providence, R.I.*, 1969; MR **39** # 7447.

[Goh-K_2] I.C. Gohberg and M.G. Krein. Introduction to the theory of non-selfadjoint operators. *Moscow*, 1985.

[Gol] V. Ya. Golodets. Crossed products of von Neumann algebras. *Uspekhi Mat. Nauk.* **26** (1971) 3-50; MR **53** # 14155; see A. Connes, P. Ghez, R. Lima, D. Testard and E.J. Woods: review of a paper of Golodets.

[Goo] F. Goodman, P. de la Harpe and V.F.R. Jones. Coxeter graphs and towers of algebras. *Math. Sci. Res. Inst. Publ., 14, Springer, New York*, 1989; MR 91c:46082.

[G-P] S.E. Goodman and J.F. Plante. Holonomy and averaging in foliated sets. *J. Diff. Geom.* **14** (1979), 401-407; MR 81m:57020.

[Good] T. Goodwillie. Cyclic homology, derivations, and the free loopspace. *Topology* **24** (1985), 187-215; MR 87c:18009.

[Gr] P. Green. Equivariant K-theory and crossed-product C^*-algebras. *Operator algebras and applications, Part 1, pp. 337-338, Proc. Sympos. Pure Math., 38, Amer. Math. Soc., Providence, R.I.*, 1982; MR 83j:46004a.

[Gro_1] M. Gromov. Groups of polynomial growth and expanding maps. *Inst. Hautes Études Sci. Publ. Math. No.* 53 (1981), 53-73; MR 83b:53053.

[Gro_2] M. Gromov. Volume and bounded cohomology. *Inst. Hautes Études Sci. Publ. Math. No.* 56 (1982), 5-99; MR 84h:53053.

[Gro_3] M. Gromov. Hyperbolic groups. *Essays in group theory, pp. 75-263, Math. Sci. Res. Inst. Publ., 8, Springer, New York,* 1987; MR 89e:20070.

[Gro-L] M. Gromov and H.B. Lawson. Positive scalar curvature and the Dirac operator on complete Riemannian manifolds. *Inst. Hautes Études Sci. Publ. Math. No.* 58 (1983), 83-196 (1984); MR 85g:58082.

[$Grot_1$] A. Grothendieck. Produits tensoriels topologiques et espaces nucléaires. *Mem. Amer. Math. Soc., 16, Amer. Math. Soc., Providence, R.I.,* 1955; MR **17**, 763.

[$Grot_2$] A. Grothendieck. Sur quelques points d'algèbre homologique. *Tôhoku Math. J. (2)* **9** (1957), 119-221; MR **21** # 1328.

[$Grot_3$] A. Grothendieck. Techniques de construction et théorèmes d'existence en géométrie algébrique, III. Preschemas quotients. *Séminaire Bourbaki, 212,* 1960/61.

[Gru-S] B. Grünbaum and G.C. Shephard. Tilings and patterns. *Freeman, New York,* 1989; MR 90a:52027.

[Gu] V.W. Guillemin. A new proof of Weyl's formula on the asymptotic distribution of eigenvalues. *Adv. in Math.* **55** (1985), 131-160; MR 86i:58135.

[Gu-S] V. Guillemin and S. Sternberg. Geometric asymptotics. *Math. Surveys, 14, Amer. Math. Soc., Providence, R.I.,* 1977; MR **58** # 24404.

[H] R. Haag. Local quantum physics. *Springer, Berlin,* 1992; MR 94f:81097.

[H-H-W] R. Haag, N.M. Hugenholtz and M. Winnink. On the equilibrium states in quantum statistical mechanics. *Comm. Math. Phys.* **5** (1967), 215-236; MR **36** # 2366.

[H_0] U. Haagerup. The standard form of von Neumann algebras. *Math. Scand.* **37** (1975), 271-283; MR **53** # 11387.

[H_1] U. Haagerup. Normal weights on W^*-algebras. *J. Functional Anal.* **19** (1975), 302-317; MR **52** # 1338.

[H_2] U. Haagerup. The Grothendieck inequality for bilinear forms on C^*-algebras. *Adv. in Math.* **56** (1985), 93-116; MR 86j:46061.

[H_3] U. Haagerup. All nuclear C^*-algebras are amenable. *Invent. Math.* **74** (1983), 305-319; MR 85g:46074.

[H_4] U. Haagerup. A new proof of the equivalence of injectivity and hyperfiniteness for factors on a separable Hilbert space. *J. Functional Anal.* **62** (1985), 160-201; MR 86k:46091.

[H_5] U. Haagerup. Connes' bicentralizer problem and uniqueness of the injective factor of type III_1. *Acta Math.* **158** (1987), 95-148; MR 88f:46117.

[H_6] U. Haagerup. An example of a nonnuclear C^*-algebra which has the metric approximation property. *Invent. Math.* **50** (1978/79), 279-293; MR 80j:46094.

[H_7] U. Haagerup. Operator-valued weights in von Neumann algebras I. *J. Functional Anal.* **32** (1979), 175-206; MR 81e:46049a.

[Ha_1] A. Haefliger. Differentiable cohomology. *Course given at C.I.M.E.,* 1976.

[Ha_2] A. Haefliger. Variétés feuilletées. *Ann. Scuola Norm. Sup. Pisa (3)* **16** (1962), 367-397; MR **32** #6487.

[Ha_3] A. Haefliger. Groupoïde d'holonomie et classifiants. *Astérisque No.* 116 (1984), 70-97; MR 86c:57026a.

[Hak-T] J. Hakeda and J. Tomiyama. On some extension properties of von Neumann algebras. *Tôhoku Math. J. (2)* **19** (1967), 315-323; MR **36** # 5706, erratum **37**, p. 1470.

[Hal] E.H. Hall. On a new action of the magnet on electric currents. *Amer. J. Math.* **2** (1879), 287-292.

[Han] D. Handelman. Positive matrices and dimension groups affiliated to topological Markov chains. *Operator algebras and applications, Part 1, pp. 191-194, Proc. Sympos. Pure Math., 38, Amer. Math. Soc., Providence, R.I.,* 1982; MR 83j:46004a.

[Har] G.H. Hardy. Divergent series. *Clarendon, Oxford,* 1949; MR **11**, 25.

[$Hari_1$] Harish-Chandra. Harmonic analysis on semisimple Lie groups. *Bull. Amer. Math. Soc.* **76** (1970), 529-551; MR **41** # 1933.

[$Hari_2$] Harish-Chandra. Harmonic analysis on real reductive groups, III. The Maass-Selberg relations and the Plancherel formula. *Ann. of Math. (2)* **104** (1976), 117-201; MR **55** # 12875.

[Harp] P. de la Harpe. Groupes hyperboliques, algèbres d'opérateurs et un théorème de Jolissaint. *C.R. Acad. Sci. Paris Sér. I Math.* **307** (1988), 771-774; MR 90d:22005.

[Harp-S] P. de la Harpe and G. Skandalis. Déterminant associé à une trace sur une algèbre de Banach. *Ann. Inst. Fourier (Grenoble)* **34** (1984), 241-260; MR 87i:46146a.

[Harp-V] P. de la Harpe and A. Valette. La propriété (T) de Kazhdan pour les groupes localement compacts. *Astérisque No. 175, Soc. Math. France, Paris,* 1989; MR 90m:22001.

[He] W. Heisenberg. The physical principles of the quantum theory. *Dover, New York,* 1969.

[Hei-La] J.L. Heitsch and C. Lazarov. Homotopy invariance of foliation Betti numbers. *Invent. Math.* **104** (1991), 321-347; MR 92b:58219.

[$Hel\text{-}H_1$] J. Helton and R. Howe. Integral operators, commutators, traces, index and homology. *Proc. conf. on operator theory, pp. 141-209, Lecture Notes in Math., 345, Springer, Berlin,* 1973; MR **52** # 11652.

[$Hel\text{-}H_2$] J. Helton and R. Howe. Traces of commutators of integral operators. *Acta Math.* **135** (1975), 271-305; MR **55** # 11106.

[Helem] A. Ya. Helemskiĭ. The homology of Banach and topological algebra. *Kluwer, Dordrecht,* 1986; MR 87k:46149.

[Her] M. Herman. Sur la conjugaison différentiable des difféomorphismes du cercle à des rotations. *Inst. Hautes Études Sci. Publ. Math. No.* 49 (1979), 5-233; MR 81h:58039.

[$Hi\text{-}S_1$] M. Hilsum and G. Skandalis. Morphismes K-orientés d'espaces de feuilles et fonctorialité en théorie de Kasparov. *Ann. Sci. École Norm. Sup. (4)* **20** (1987), 325-390; MR 90a:58169.

[$Hi\text{-}S_2$] M. Hilsum and G. Skandalis. Stabilité des C^*-algèbres de feuilletages. *Ann. Inst. Fourier (Grenoble)* **33** (1983), 201-208; MR 85f:58115.

[Hi-S_3] M. Hilsum and G. Skandalis. Invariance par homotopie de la signature à coefficients dans un fibré presque plat. *J. Reine Angew. Math.* **423** (1992), 73-99; MR 93b:46137.

[Hig] N. Higson. Categories of fractions and excision in KK-theory. *J. Pure Appl. Alg* **65** (1990), no. 2, 119–138; MR 91i:19005.

[Hi_1] M. Hilsum. Signature operator on Lipschitz manifolds and unbounded Kasparov bimodules. *Operator algebras and their connections with topology and ergodic theory, pp. 254-288, Lecture Notes Math., 1132, Springer, Berlin,* 1985; MR 87d:58133.

[Hi_2] M. Hilsum. Fonctorialité en K-théorie bivariante pour les variétés lipschitziennes. *K-Theory* **3** (1989), 401-440; MR 91j:19012.

[Ho-K-R] G. Hochschild, B. Kostant and A. Rosenberg. Differential forms on regular affine algebras. *Trans. Amer. Math. Soc.* **102** (1962), 383-408; MR **26** # 167.

[Hoo] G. 't Hooft. Renormalizable Lagrangians for massive Yang-Mills fields. *Nuclear Phys.* **35** (1971), 167-188.

[Hoo-J] C.E. Hood and J.D.S. Jones. Some algebraic properties of cyclic homology groups. *K-Theory* **1** (1987), 361-384; MR 89f:18011.

[Hop] E. Hopf. Ergodic theory and the geodesic flow on surfaces of constant negative curvature. *Bull. Amer. Math Soc.* **77** (1971), 863-877; MR **44** # 1789.

[Hor_1] L. Hörmander. On the index of pseudodifferential operators. *Elliptische Differentialgleichungen, Band II, pp. 127-146, Akademie-Verlag, Berlin,* 1971; MR **58** #31292.

[Hor_2] L. Hörmander. The Weyl calculus of pseudodifferential operators. *Comm. Pure Appl. Math.* **32** (1979), 360-444; MR 80j:47060.

[Hor_3] L. Hörmander. Fourier integral operators, I. *Acta Math.* **127** (1971), 79-183; MR **52** # 9299.

[Hor_4] L. Hörmander. The analysis of linear partial differential operators, Vol. I–IV. *Grundlehren der Math. Wissensch., 274-5, 256-7, Springer, Berlin,* 1983 and 1985; MR 85g:35002a-b and 87d:35002a-b.

[Hor_5] L. Hörmander. Pseudodifferential operators and hypoelliptic equations. *Singular integrals, pp. 138-183, Proc. Sympos. Pure Math., X, Amer. Math. Soc., Providence, R.I.,* 1967; MR **52** #4033.

[Hu-K_1] S. Hurder and A. Katok. Secondary classes and transverse measure theory of a foliation. *Bull. Amer. Math. Soc. (N.S.)* **11** (1984), 347-350; MR 85h:58100.

[Hu-K_2] S. Hurder and A. Katok. Ergodic theory and Weil measures for foliations. *Ann. of Math.* **126** (1987), 221-275; MR 89d:57042.

[Hus] D. Husemoller. Fibre bundles. *McGraw-Hill, New York,* 1966; MR **37** # 4821.

[Ig] K. Igusa. Higher singularities of smooth functions are unnecessary. *Ann. of Math.* **119** (1984), 1-58; MR 85k:57034.

[I-M] T. Iwaniec and G. Martin. Quasiregular mappings in even dimensions, *Mittag-Leffler Inst.* **19** (1989-1990).

[I-W] E. İnönü and E.P. Wigner. On the contraction of groups and their representations. *Proc. Nat. Acad. Sci. USA* **39** (1953), 510-524; MR **14**, 1061.

[J-L-O] A. Jaffe, A. Lesniewski and K. Osterwalder. Quantum K-theory: the Chern character. *Comm. Math. Phys.* **118** (1988), 1-14; MR 90a:58170.

[J-L-W] A. Jaffe, A. Lesniewski and J. Weitsman. Index of a family of Dirac operators on loop space. *Comm. Math. Phys.* **112** (1987), 75-88; MR 88i:58157.

[J-W] S. Janson and T. Wolff. Schatten classes and commutators of singular integral operators. *Ark. Mat.* **20** (1982), 301-310; MR 85a:47026.

[Jo_1] B. Johnson. Cohomology in Banach algebras. *Mem. Amer. Math. Soc., 127, Amer. Math. Soc., Providence, R.I.,* 1972; MR **51** # 11130.

[Jo_2] B. Johnson. Introduction to cohomology in Banach algebras. *Algebras in analysis (Birmingham, 1973), pp. 84-100, Academic, London,* 1975; MR **54** # 5835.

[Jo-K-R] B. Johnson. R.V. Kadison and J. Ringrose, Cohomology of operator algebra, III. Reduction to normal cohomology. *Bull. Soc. Math. France* **100** (1972), 73-96; MR **47** # 7454.

[Jol] P. Jolissaint. Les fonctions à décroissance rapide dans les C^*-algèbres réduites de groupes. *Thèse, Univ. de Genève,* 1987.

[Jol_1] P. Jolissaint. Rapidly decreasing functions in reduced C^*-algebras of groups. *Trans. Amer. Math. Soc.* **317** (1990), 167-196; MR 90d:22006.

[Jol_2] P. Jolissaint. K-theory of reduced C^*-algebras and rapidly decreasing functions on groups. *K-Theory* **2** (1989), 723-735; MR 90j:22004.

[Jon] J.D.S. Jones. Cyclic homology and equivariant homology. *Invent. Math.* **87** (1987), 403-423; MR 88f:18016.

[$Jone_1$] V.F.R. Jones. Actions of finite groups on the hyperfinite type II_1 factor. *Mem. Amer. Math. Soc., 28, Amer. Math. Soc., Providence, R.I.,* 1980; MR 81m:46094.

[$Jone_2$] V.F.R. Jones. Index for subfactors. *Invent. Math.* **72** (1983), 1-25; MR 84d:46097.

[$Jone_3$] V.F.R. Jones. A polynomial invariant for knots via von Neumann algebras. *Bull. Amer. Math. Soc.* **12** (1985), 103-111; MR 86e:57006.

[$Jone_4$] V.F.R. Jones. Braid groups, Hecke algebras and type II_1 factors. *Geometric methods in operator algebras (Kyoto, 1983), pp. 242-273, Pitman Res. Notes in Math. Ser., 123, Longman, New York,* 1986; MR 88k:46069.

[J] B. Julia. Statistical theory of numbers. *Number theory and physics (Les Houches Winter School, 1989), pp. 276-293, Springer Proc. Phys., 47, Springer, Berlin,* 1990; MR 91h:11088.

[Ju] P. Julg. K-théorie du produit croisé d'une C^*-algèbre par un groupe compact. *Thèse de 3e cycle, Université Pierre et Marie Curie, Paris,* 1982.

[Ju-K] P. Julg and G. Kasparov. L'anneau $KK_G(\mathbb{C}, \mathbb{C})$ pour $G = SU(n, 1)$. *C.R. Acad. Sci. Paris Ser. I Math.* **313** (1991), 259-264; MR 92i:19008.

[$Ju-V_1$] P. Julg and A. Valette. K-moyennabilité pour les groupes opérant sur les arbres. *C.R. Acad. Sci. Paris Sér. I Math.* **296** (1983), 977-980; MR 86m:46063.

[$Ju-V_2$] P. Julg and A. Valette. Twisted coboundary operator on a tree and the Selberg principle. *J. Operator Theory* **16** (1986), 285-304; MR 88d:22026.

[Ju-V_3] P. Julg and A. Valette. L'opérateur de co-bord tordu sur un arbre, et le principe de Selberg, II. *J. Operator Theory* **17** (1987), 347-355; MR 88m:-22041.

[K] R.V. Kadison. Irreducible operator algebras. *Proc. Nat. Acad. Sci. USA* **43** (1957), 273-276; MR **19**, 47.

[K-R] R.V. Kadison and J.R. Ringrose. Fundamentals of the theory of operator algebras, Vol. I and II. *Academic, New York,* 1983; MR 85j:46099.

[Ka-Kam-S] D.S. Kahn, J. Kaminker and C. Schochet. Generalized homology theories on compact metric spaces. *Michigan Math. J.* **24** (1977), 203-224; MR **57** # 13921.

[Kam-M] J. Kaminker and J.G. Miller. Homotopy invariance of the analytic index of signature operators over C^*-algebras. *J. Operator Theory* **14** (1985), 113-127; MR 87b:58082.

[Kam-S] J. Kaminker and C. Schochet. Steenrod homology and operator algebras. *Bull. Amer. Math. Soc.* **81** (1975), 431-434; MR **56** # 9287.

[Kar_1] M. Karoubi. Connexions, courbures et classes caractéristiques en K-théorie algébrique. *Current trends in algebraic topology, Part I (London, Ont., 1981), pp. 19-27, CMS Conf. Proc.,2, Amer. Math. Soc., Providence, R.I.,* 1982; MR 84f:57013.

[Kar_2] M. Karoubi. K-theory, an introduction. *Grundlehren der Math. Wissen., 226, Springer, Berlin,* 1978; MR **58** # 7605.

[Kar_3] M. Karoubi. Homologie cyclique et K-théorie. *Astérisque No. 149* (1987); MR 89c:18019.

[Kar_4] M. Karoubi. Foncteurs dérivés et K-théorie. Catégories filtrées. *C. R. Acad. Sci. Paris Ser. A-B* **267** (1968), A328-A331; MR **38** # 2192.

[Kar_5] M. Karoubi. Foncteurs dérivés et K-théorie. *Séminaire Heidelberg-Saarbrücken-Strasbourg sur la K-théorie, pp. 107-186, Lecture Notes in Math., 136, Springer, Berlin,* 1970; MR **42** # 344.

[Kar-V] M. Karoubi and O. Villamayor. K-théorie algébrique et K-théorie topologique, I. *Math. Scand.* **28** (1971), 265-307; MR **47** # 1915.

[Kas_1] G. Kasparov. The operator K-functor and extensions of C^*-algebras. *Izv. Akad. Nauk. SSSR Ser. Mat.* **44** (1980), 571-636; MR 81m:58075.

[Kas_2] G. Kasparov. Hilbert C^*-modules: theorems of Stinespring and Voiculescu. *J. Operator Theory* **4** (1980), 133-150; MR 82b:46074.

[Kas_3] G. Kasparov. Index of invariant elliptic operators, K-theory and representations of Lie groups. *Dokl. Akad. Nauk. SSSR* **268** (1983), 533-537; MR 84g:58100.

[Kas_4] G. Kasparov. Lorentz groups: K-theory of unitary representations and crossed products. *Dokl. Akad. Nauk. SSSR* **275** (1984), 541-545; translated as *Soviet Math. Dokl.* **29** (1984), 256-260; MR 85k:22015.

[Kas_5] G. Kasparov. Equivariant KK-theory and the Novikov conjecture. *Invent. Math.* **91** (1988), 147-201; MR 88j:58123.

[Kas_6] G. Kasparov. K theory, group C^*-algebras and higher signatures. *Conspectus, Chernogolovka,* 1981.

[Kas_7] G. Kasparov. Topological invariants of elliptic operators, I. K-homology. *Math. USSR Izv.* **9** (1975), 751-792; MR **58** # 7603.

[Kas-S] G. Kasparov and G. Skandalis. Groups acting on buildings, operator K-theory and Novikov's conjecture. *K-Theory* **4** (1991), 303-337; MR 92h:-19009.

[Kass] C. Kassel. Cyclic homology, comodules and mixed complexes. *J. Algebra* **107** (1987), 195-216; MR 88k:18019.

[Kass-L] C. Kassel and J.-L. Loday. Extensions centrales d'algèbres de Lie. *Ann. Inst. Fourier (Grenoble)* **32** (1982), 119-142; MR 85g:17004.

[Kast_0] D. Kastler. The C^*-algebras of a free boson field, I. Discussion of the basic facts. *Comm. Math. Phys.* **1** (1965), 14-48; MR **33** # 2197.

[Kast_1] D. Kastler. Cyclic cohomology within the differential envelope. An introduction to Alain Connes' noncommutative differential geometry. *Hermann, Paris,* 1988; MR 89h:18013.

[Kast_2] D. Kastler. Cyclic cocycles from graded KMS functionals. *Comm. Math. Phys.* **121** (1989), 345-350; MR 90e:46068.

[Kast_3] D. Kastler. A detailed account of Alain Connes' version of the standard model in noncommutative differential geometry, I and II. *Rev. Math. Phys.* **5** (1993), 477-532.

[Kast_4] D. Kastler. Introduction to algebraic theory of superselection sectors (space-time dimension = 2, strictly localized morphisms. *The algebraic theory of superselection sectors (Palermo, 1989), pp. 113-214, World Scientific, River Edge, N.J.,* 1990.

[Kast_5] D. Kastler. Entire cyclic cohomology of $\mathbb{Z}/2$-graded Banach algebras. *K-Theory* **2** (1989), 485-509; MR 90e:46067.

[Kat] T. Kato. Continuity of the map $S \mapsto |S|$ for linear operators. *Proc. Japan Acad.* **49** (1973), 157-160; MR **53** # 8943.

[Katz] Y. Katznelson. Smooth mappings of the circle without sigma-finite invariant measures. *Sympos. Math.* **22** (1977), 363-369.

[Kaz_1] D. Kazhdan. On the connection of the dual space of a group with the structure of its closed subgroups. *Funkt. Anal. i Prilozhen.* **1** (1967), 71-74; MR **35** # 288.

[Kha] M. Khalkhali. Algebraic connections, universal bimodules and entire cyclic cohomology. *Comm. Math. Phys., to appear.*

[Ki_1] F. Kittaneh. Inequalities for the Schatten p-norm, II. *Glasgow Math. J.* **29** (1987), 99-104; MR 88d:47027.

[Ki_2] F. Kittaneh. Inequalities for the Schatten p-norm, III. *Comm. Math. Phys.* **104** (1986), 307-310; MR 88j:47009.

[Ki_3] F. Kittaneh. Inequalities for the Schatten p-norm, IV. *Comm. Math. Phys.* **106** (1986), 581-585; MR 88j:47010.

[Kir] E. Kirchberg and G. Vaillant. On C^*-algebras having linear, polynomial and subexponential growth. *Invent. Math.* **108** (1992), 635-652.

[Kl-D-P] K. von Klitzing, G. Dorda and M. Pepper. New method for high-accuracy determination of the fine-structure constant based on quantized hall resistance. *Phys. Rev. Lett.* **45** (1980), 494-497.

[Ko_1] H. Kosaki. Interpolation theory and the Wigner-Yanase-Dyson-Lieb concavity. *Comm. Math. Phys.* **87** (1982/83), 315-329; MR 84h:46101.

[Ko_2] H. Kosaki. On the continuity of the map $\phi \to |\phi|$ from the predual of a W^*-algebra. *J. Functional Anal.* **59** (1984), 123-131; MR 86c:46072.

[Ko_3] H. Kosaki. Unitarily invariant norms under which the map $A \to |A|$ is Lipschitz continuous. *Publ. Res. Inst. Math. Sci.* **28** (1992), 299-313; MR 93d:47043.

[Kos_1] B. Kostant. Graded manifolds, graded Lie theory and prequantization. *Differential geometrical methods in mathematical physics, pp. 177-306, Lecture Notes in Math., 570, Springer, Berlin,* 1977; MR **58** # 28326.

[Kos_2] B. Kostant. On the existence and irreducibility of certain series of representations. *Bull. Amer. Math. Soc.* **75** (1969), 627-642; MR **39** # 7031.

[Kr_1] W. Krieger. On the Araki-Woods asymptotic ratio set and non-singular transformations of a measure space. *Contributions to ergodic theory and probability, pp. 158-177, Lecture Notes in Math., 160, Springer, Berlin,* 1970; MR **54** # 2915.

[Kr_2] W. Krieger. On ergodic flows and the isomorphism of factors. *Math. Ann.* **223** (1976), 19-70; MR **54** # 3430.

[Ku] R. Kubo. Statistical-mechanical theory of irreversible processes, I. General theory and simple applications to magnetic and conduction problems. *J. Phys. Soc. Japan* **12** (1957), 570-586; MR **20** # 4940a.

[Kun] H. Kunz. Quantized currents and topological invariants for electrons in incommensurate potentials. *Phys. Rev. Letters* **57** (1986), 1095-1097; MR 87j:81298.

[La] R.B. Laughlin. Quantized Hall conductivity in two dimensions. *Phys. Rev. B* **23** (1981), 5632-5633.

[Lap] M.L. Lapidus and C. Pomerance. Fonction zêta de Riemann et conjecture de Weyl-Berry pour les tambours fractals. *C.R. Acad. Sci. Paris Sér. I Math.* **310** (1990), 343-348; MR 91d:58248.

[L] H.B. Lawson. Foliations. *Bull. Amer. Math. Soc.* **80** (1974), 369-418; MR **49** # 8031.

[L-M] H.B. Lawson and L. Michelsohn. Spin geometry. *Princeton Math. Ser., 38, Princeton Univ. Press, Princeton, N.J.,* 1989; MR 91g:53001.

[Le] P. Lelong. Fonctions plurisousharmoniques et formes différentiables positives. *Gordon and Breach, Paris,* 1968; MR **39** # 4436.

[Li] A. Lichnerowicz. Déformations d'algèbres associées à une variété symplectique (les $*_\nu$-produits). *Ann. Inst. Fourier (Grenoble)* **32** (1982), 157-209; MR 83k:58095.

[Lie] E.H. Lieb. Convex trace functions and the Wigner-Yanase-Dyson conjecture. *Adv. in Math.* **11** (1973), 267-288; MR **48** #10407.

[Lio-P] J.-L. Lions and J. Peetre. Sur une classe d'espaces d'interpolation. *Inst. Hautes Études Sci. Publ. Math. No.* 19 (1964), 5-68; MR **29** # 2627.

[Lo] J.-L. Loday. Cyclic homology. *Springer, Berlin,* 1992; MR 94a:19004.

[Lo-Q] J.L. Loday and D. Quillen. Cyclic homology and the Lie algebra of matrices. *C.R. Acad. Sci. Paris Sér. I Math.* **296** (1983), 295-297; MR 85d:17010.

[Lo-Q_1] J.L. Loday and D. Quillen. Cyclic homology and the Lie algebra homology of matrices. *Comment. Math. Helv.* **59** (1984), 569-591; MR 86i:17003.

[Lor$_1$] T.A. Loring. Deformations of nonorientable surfaces as torsion E-theory elements. *C.R. Acad. Sci. Paris Sér. I Math.* **316** (1993), 341-346.

[Lor$_2$] T.A. Loring. A test for injectivity for asymptotic morphisms. *Preprint.*

[Lu] G. Luke. Pseudodifferential operators on Hilbert bundles. *J. Differential Equations* **12** (1972), 566-589; MR **49** # 11577.

[Lus] G. Lusztig. Novikov's higher signature and families of elliptic operators. *J. Differential Geometry* **7** (1972), 229-256; MR **48** # 1250.

[Mac] K. Mackenzie. Lie groupoids and Lie algebroids in differential geometry. *London Math. Soc. Ser., 124, Cambridge Univ. Press, Cambridge,* 1987; MR 89g:58225.

[M$_1$] G.W. Mackey. On the analogy between semisimple Lie groups and certain related semi-direct product groups. *Lie groups and their representations, pp. 339-363, Halsted, New York,* 1975; MR **53** # 13478.

[M$_2$] G.W. Mackey. Virtual groups. *Topological dynamics, pp. 335-364, Benjamin, New York,* 1968; MR **39** # 2907.

[M$_3$] G.W. Mackey. Ergodic theory and virtual groups. *Math. Ann.* **166** (1966), 187-207; MR **34** # 1444.

[M$_4$] G.W. Mackey. The theory of unitary group representations. *Univ. of Chicago Press, Chicago, Ill.,* 1976; MR **53** # 686.

[Ma] S. Mac Lane. Homology. *Springer, Berlin,* 1975; MR **50** # 2285.

[Mak] N.G. Makarov. Metric properties of harmonic measure. *Proceedings of the International Congress of Mathematicians (Berkeley, Calif., 1986), Vol. I, pp. 766-776, Amer. Math. Soc., Providence, R.I.,* 1987; MR 89h:30039.

[Man] B. Mandelbrot. Fractals, form, chance and dimension. *Freeman, San Francisco, Calif.,* 1977; MR **57** # 11224.

[Mani$_1$] Yu. I. Manin. Algebraic aspects of non-linear differential equations. *J. Soviet Math.* **11** (1979), 1-122.

[Mani$_2$] Yu. I. Manin. Quantum groups and noncommutative geometry. *Centre Rech. Math., Univ. Montréal, Montréal, Qué.,* 1988; MR 91e:17001.

[Mar] O. Maréchal. Une remarque sur un théorème de Glimm. *Bull. Sci. Math. (2)* **99** (1975), 41-44; MR **55** # 3802.

[Mart-S] P.C. Martin and J. Schwinger. Theory of many-particle systems, I. *Phys. Rev. (2)* **115** (1959), 1342-1373; MR **22** # 588.

[Me] P.A. Meyer. Limites médiales, d'après Mokobodzki. *Séminaire de probabilités VII, (Strasbourg, 1971-1972), pp. 198-204, Lecture Notes in Math., 321, Springer, Berlin,* 1973; MR **53** # 4247.

[Mi$_1$] J. Milnor. Introduction to algebraic K-theory. *Ann. of Math. Stud., 72, Princeton Univ. Press, Princeton, N.J.,* 1971; MR **50** # 2304.

[Mi$_2$] J. Milnor. Topology from the differentiable viewpoint. *Univ. Press of Virginia, Charlottesville, Va.,* 1965; MR **37** # 2239.

[Mi$_3$] J. Milnor. Morse theory. *Ann. of Math. Stud., 51, Princeton University Press, Princeton, N.J.,* 1963; MR **29** #634.

[Mi$_4$] J. Milnor and D. Stasheff. Characteristic classes. *Ann. of Math. Stud., 76, Princeton Univ. Press, Princeton, N.J.,* 1974; MR **55** # 13428.

[Mis_1] A.S. Mishchenko. Infinite-dimensional representations of discrete groups and higher signatures. *Math. SSSR Izv.* **8** (1974), 85-112; MR **50** # 14848.

[Mis_2] A.S. Mishchenko. Homotopy invariants of multiply connected manifolds, I. Rational invariants. *Izv. Akad. Nauk. SSSR Ser. Mat.* **34** (1970), 501-514; MR **42** # 3795.

[Mis_3] A.S. Mishchenko. C^*-algebras and K-theory, *Algebraic topology (Aarhus 1978), pp. 262-274, Lecture Notes in Math., 763, Springer, Berlin,* 1979; MR 81d:58049.

[Mis-F] A.S. Mishchenko and A.T. Fomenko. The index of elliptic operators over C^*-algebras. *Izv. Akad. Nauk SSSR Ser. Mat.* **43** (1979), 831-859; translated as *Math. USSR-Izv.* **15** (1980), 87-112; MR 81i:46075.

[Mis-S] A.S. Mishchenko and Yu P. Solov'jev. Representations of Banach algebras and formulas of Hirzebruch type. *Mat. Sb. (N.S.)* **111(153)** (1980), 209-226; MR 82f:46059.

[Mo-S] C.C. Moore and C. Schochet. Global analysis on foliated spaces. *Math. Sci. Res. Inst. Publ., 9, Springer, New York,* 1988; MR 89h:58184.

[Mos-V] H. Moscovici and A. Verona. Harmonically induced representations of nilpotent Lie groups. *Invent. Math.* **48** (1978), 61-73; MR 80a:22011.

[Mou] P. van Mouche. The coexistence problem for the discrete Mathieu operator. *Comm. Math. Phys.* **122** (1989), 23-33; MR 90e:47027.

[Moy] J.E. Moyal. Quantum mechanics as a statistical theory. *Proc. Cambridge Philos. Soc.* **45** (1949), 99-124; MR **10**, 582.

[Mu_1] N. Mukunda. Expansion of Lie groups and representations of $SL(3,\mathbb{C})$. *J. Math. Phys.* **10** (1969), 897-911; MR **39** # 7887.

[Mu_2] N. Mukunda. Unitary representations of the group $SL(3,C)$ in an $SU(3)$ basis. *J. Math. Phys.* **11** (1970), 1759-1771; MR **41** # 7946.

[Muh-R-W] P. Muhly. J. Renault and D. Williams, Equivalence and isomorphism for groupoid C^*-algebras. *J. Operator Theory* **17** (1987), 3-22; MR 88h:46123.

[Mur-N_1] F.J. Murray and J. von Neumann. On rings of operators. *Ann. of Math.* **37** (1936), 116-229.

[Mur-N_2] F.J. Murray and J. von Neumann. On rings of operators, II. *Trans. Amer. Math. Soc.* **41** (1937), 208-248.

[Mur-N_3] F.J. Murray and J. von Neumann. On rings of operators, IV. *Ann. of Math. (2)* **44** (1943), 716-808; MR **5**, 101.

[Nat_1] T. Natsume. The C^*-algebras of codimension one foliations without holonomy. *Math. Scand.* **56** (1985), 96-104; MR 87b:58066.

[Nat_2] T. Natsume. Topological K-theory for codimension one foliations without holonomy. *Foliations (Tokyo 1983), pp. 15-27, Adv. Stud. Pure Math., 5, North-Holland, Amsterdam,* 1985; MR 88f:57047.

[N_1] J. von Neumann. On ring operators, III. *Ann. of Math.* **41** (1940), 94-161; MR **1**, 146.

N_2] J. von Neumann. On rings operators: reduction theory. *Ann. of Math. (2)* **50** (1949), 401-485; MR **10**, 548.

[Ni] V. Nistor. Group cohomology and the cyclic cohomology of crossed products. *Invent. Math.* **99** (1990), 411-424; MR 91f:46097.

[Ni_2] V. Nistor. Cyclic cohomology of crossed products by algebraic groups. *Invent. Math.* **112** (1993), 615-638.

[Ni_3] V. Nistor. A bivariant Chern character for p-summable quasihomomorphisms. *K-Theory* 5 (1991), 193-211; MR 93d:19009.

[Ni_4] V. Nistor. A bivariant Chern-Connes character. *Ann. of Math.* **138** (1993), 555-590.

[No_1] S.P. Novikov. Magnetic Bloch functions and vector bundles. Typical dispersion laws in their quantum numbers. *Sov. Math. Dokl.* **23** (1981), 298-303; MR 82h:81111.

[No_2] S.P. Novikov. Pontrjagin classes, the fundamental group and some problems of stable algebra. *Essays on topology and related topics, (Mémoires dedicated to G. de Rham), pp. 147-155, Springer, New York,* 1970; MR **42** # 3804.

[$N\text{-}S_1$] S.P. Novikov and M.A. Shubin. Morse inequalities and von Neumann II_1 factors. *Dokl. Akad. Nauk SSSR* **289** (1986), 289-292; MR 88c:58065.

[$N\text{-}S_2$] S.P. Novikov and M.A. Shubin. Morse theory and von Neumann invariants of non-simply connected manifolds. *Uspekhi Mat. Nauk* **41** (1986), 222-223.

[O] C. Ogle. Assembly maps, K-theory and hyperbolic groups. *K-Theory* **6** (1992), 235-265.

[O_1] A. Ocneanu. Actions des groupes moyennables sur les algèbres de von Neumann. *C.R. Acad. Sci. Paris Sér. A-B* **291** (1980), A399-A401; MR 82d:46093.

[O_2] A. Ocneanu. Quantized groups, string algebras and Galois theory for algebras. *Operator algebras and applications, Vol. 2, pp. 119-172, London Math. Soc. Lecture Note Ser., 136, Cambridge Univ. Press, Cambridge,* 1988; MR 91k:46068.

[$Om\text{-}M\text{-}Y_1$] H. Omori, Y. Maeda and A. Yoshioka. Weyl manifolds and deformation quantization. *Adv. in Math.* **85** (1991), 224-255; MR 92d:58071.

[$Om\text{-}M\text{-}Y_2$] H. Omori, Y. Maeda and A. Yoshioka. Existence of a closed star product. *Lett. math. Phys.* **26** (1992) 285-294; MR 94c:58078.

[Or-W] D. Ornstein and B. Weiss. Ergodic theory of amenable group actions, I. The Rohlin lemma. *Bull. Amer. Math. Soc. (N.S.)* **2** (1980), 161-164; MR 80j:28031.

[Ox-U] J. Oxtoby and S. Ulam. Measure-preserving homeomorphisms and metrical transitivity. *Ann. of Math. (2)* **42** (1941), 874-920; MR **3**, 211.

[P] A. Pais. Inward bound of matter and forces in the physical world. *Clarendon, Oxford,* 1986.

[Pa] R.S. Palais. Seminar on the Atiyah-Singer index theorem. *Ann. of Math. Stud., 57, Princeton Univ. Press, Princeton, N.J.,* 1965; MR **33** # 6649.

[Pat_1] S.J. Patterson. The limit set of a Fuchsian group. *Acta Math.* **136** (1976), 241-273; MR **56** # 8841.

[Pat_2] S.J. Patterson. Some examples of Fuchsian groups. *Proc. London Math. Soc.* **39** (1979), 276-298; MR 80j:30070.

[Par] R. Parthasarathy. Dirac operator and the discrete series. *Ann. of Math.* **96** (1972), 1-30; MR **47** # 6945.

[Pas] W. Paschke. Inner product modules over B^*-algebras. *Trans. Amer. Math. Soc.* **182** (1973), 443-468; MR **50** # 8087.

[Pe_1] G. Pedersen. Measure theory for C^*-algebras, I-III. *Math. Scand.* **19** (1966), 131-145, *ibid.* **22** (1968), 63-74, and *ibid.* **25** (1969), 71-93; MR **35** # 3453, **39** # 7444 and **41** # 4263.

[Pe_2] G. Pedersen. C^*-algebras and their automorphism groups. *London Math. Soc. Monographs, 14, Academic, New York,* 1979; MR 81e:46037.

[Pel_1] V.V. Peller. Smooth Hankel operators and their applications. *Dokl. Akad. Nauk SSSR* **252** (1980), 43-48; translated as *Soviet. Math. Dokl.* **21** 3 (1980), 683-688; MR 83g:47030.

[Pel_2] V.V. Peller. Nuclearity of Hankel operators. *Preprint, Steklov Inst. Math., Leningrad,* 1979.

[Pel_3] V.V. Peller. Hankel operators of the class φ_p and their applications (rational approximation, Gaussian processes, majorization problem for operators). *Mat. Sb. (N.S.)* **113(155)** (1980), 538-581; MR 82g:47022.

[Pel_4] V.V. Peller. Vectorial Hankel operators, commutators and related operators of the Schatten-von Neumann class φ_p. *Preprint, Leningrad,* 1981.

[Pel-H] V.V. Peller and S.V. Hruschev. Hankel operators, best approximations, and stationary Gaussian processes, I-III. *Uspekhi Mat. Nauk* **37** (1982), 53-124; translated as *Russian Math. Surveys* **37** (1982), 61-144; MR 84e:47036.

[Pen] M. Penington. K-theory and C^*-algebras of Lie groups and foliations. *PhD thesis, Oxford, Michaelmas,* Term 1983.

[Pen-Pl] M. Penington and R. Plymen. The Dirac operator and the principal series for complex semisimple Lie groups. *J. Functional Anal.* **53** (1983), 269-286; MR 85d:22016.

[Ph] J. Phillips. Automorphisms of full II_1 factors, with applications to factors of type III. *Duke Math. J.* **43** (1976), 375-385; MR **53** # 6337.

[Pi-Po] M. Pimsner and S. Popa. Entropy and index for subfactors. *Ann. Sci. École Norm. Sup. (4)* **19** (1986), 57-106; MR 87m:46120.

[$Pi-V_1$] M. Pimsner and D. Voiculescu. Exact sequences for K-groups and Ext-groups of certain cross-product C^*-algebras. *J. Operator Theory* **4** (1980), 93-118; MR 82c:46074.

[$Pi-V_2$] M. Pimsner and D. Voiculescu. Imbedding the irrational rotation C^*-algebra into an AF-algebra. *J. Operator Theory* **4** (1980), 201-210; MR 82d:46086.

[$Pi-V_3$] M. Pimsner and D. Voiculescu. K-groups of reduced crossed products by free groups. *J. Operator Theory* **8** (1982), 131-156; MR 84d:46092.

[Pl_1] R. Plymen. Strong Morita equivalence, spinors and symplectic spinors. *J. Operator Theory* **16** (1986), 305-324; MR 88d:58112.

[Pl_2] R. Plymen. Reduced C^*-algebra for reductive p-adic groups. *J. Functional Anal.* **88** (1990), 251-266; MR 91a:46064.

[Po_1] S. Popa. A short proof of "injectivity implies hyperfiniteness" for finite von Neumann algebras. *J. Operator Theory* **16** (1986), 261-272; MR 87m:46115.

[Po_2] S. Popa. Classification of subfactors: the reduction to commuting squares. *Invent. Math.* **101** (1990), 19-43; MR 91h:46109.

[Pow_1] R.T. Powers. Representations of uniformly hyperfinite algebras and their associated von Neumann rings. *Ann. of Math. (2)* **86** (1967), 138-171; MR **36** # 1989.

[Pow_2] R.T. Powers. Simplicity of the C^*-algebra associated with the free group on two generators. *Duke Math. J.* **42** (1975), 151-156; MR **51** # 10534.

[Pow-S] R.T. Powers and E. Størmer. Free states of the canonical anticommutation relations. *Comm. Math. Phys.* **16** (1970), 1-33; MR **42** # 4126.

[$Powe_1$] S. Power. Hankel operators on Hilbert space. *Bull. London Math. Soc.* **12** (1980), 422-442; MR 82a:47030.

[$Powe_2$] S. Power. Hankel operators on Hilbert space. *Res. Notes in Math., 64, Pitman, Boston, Mass.*, 1982; MR 84e:47037.

[Pr_1] J. Pradines. Théorie de Lie pour les groupoïdes différentiables. Relations entre propriétés locales et globales. *C.R. Acad. Sci. Paris Sér. A-B* **263** (1966), A907-A910; MR **35** # 4954.

[Pr_2] J. Pradines. Théorie de Lie pour les groupoïdes différentiables. Calcul différentiel dans la catégorie des groupoïdes infinitésimaux. *C.R. Acad. Sci. Paris Sér. A-B* **264** (1967), A245-A248; MR **35** # 7242.

[Pr_3] J. Pradines. Géométrie différentielle au-dessus d'un groupoïde. *C.R. Acad. Sci. Paris Sér. A-B* **266** (1968), A1194-A1196; MR **37** # 6861.

[Pr_4] J. Pradines. Troisième théorème de Lie pour les groupoïdes différentiables. *C.R. Acad. Sci. Paris Sér. A-B* **267** (1968), A21-A23; MR **37** # 6969.

[Pr_5] J. Pradines. How to define the differentiable graph of a singular foliation. *Cahiers Topol. Géom. Different. Catégoriques* **26** (1985), 339-380; MR 87g:57043.

[Pu-W] W. Pusz and S. Woronowicz. Form convex functions and the WYDL and other inequalities. *Lett. Math. Phys.* **2** (1977/78), 505-512; MR 80j:47019.

[Q_1] D. Quillen. Superconnections and the Chern character. *Topology* **24** (1985), 89-95; MR 86m:58010.

[Q_2] D. Quillen. Algebra cochains and cyclic cohomology. *Inst. Hautes Études Sci. Publ. Math. No. 68* (1989), 139-174; MR 90j:18008.

[Q_3] D. Quillen. Chern-Simons forms and cyclic cohomology. *Interface of mathematics and particle physics (Oxford, 1988), pp. 117-134, Oxford Univ. Press, New York,* 1990; MR 92d:19004.

[R] A. Ranicki. The algebraic theory of surgery, I, II. *Proc. London Math. Soc.* **40** (1980), 87-192, 193-283; MR 82f:57024.

[Ra] A. Ramsay. Virtual groups and group actions. *Adv. in Math.* **6** (1971), 253-322; MR **43** # 7590.

[Re-S] M. Reed and B. Simon. Fourier analysis, self-adjointness. *Academic, New York,* 1975.

[Ren_1] J. Renault. A groupoid approach to C^*-algebras. *Lecture Notes in Math., 793, Springer, Berlin,* 1980; MR 82h:46075.

[Ren_2] J. Renault. Représentation des produits croisés d'algèbres de groupoïdes. *J. Operator Theory* **18** (1987), 67-97; MR 89g:46108.

[Ri_1] M.A. Rieffel. C^*-algebras associated with irrational rotations. *Pacific J. Math.* **93** (1981), 415-429; MR 83b:46087.

[Ri_2] M.A. Rieffel. Morita equivalence for C^*-algebras and W^*-algebras. *J. Pure Appl. Algebra* 5 (1974), 51-96; MR **51** # 3912.

[Ri_3] M.A. Rieffel. Deformation quantization of Heisenberg manifolds. *Comm. Math. Phys.* **122** (1989), 531-562; MR 90e:46060.

[Ri_4] M.A. Rieffel. The cancellation theorem for projective modules over irrational rotation C^*-algebras. *Proc. London Math. Soc. (3)* **47** (1983), 285-302; MR 85g:46085.

[Ri_5] M.A. Rieffel. Induced representations of C^*-algebras. *Bull. Amer. Math. Soc.* **78** (1972), 606-609; MR **46** # 677.

[Ri_6] M.A. Rieffel. Dimension and stable rank in the K-theory of C^*-algebras. *Proc. London Math. Soc.* **46** (1983), 301-333; MR 84g:46085.

[Ro] J. Roe. An index theorem on open manifolds, I, II. *J. Diff. Geometry* **27** (1988), 87-113; *ibid.* **27** (1988), 115-136; MR 89a:58102.

[Ro_2] J. Roe. Operator algebras and index theory on non-compact manifolds. *Index theory of elliptic operators, foliations and operator algebras (New Orleans, 1986), pp. 229-249, Contemp. Math., 70, Amer. Math. Soc., Providence, R.I.*, 1988; MR 89i:58139.

[Ros_1] J. Rosenberg. C^*-algebras, positive scalar curvature and the Novikov conjecture. *Inst. Hautes Études Sci. Publ. Math. No. 58* (1983), 197-212; MR 85g:58083.

[Ros_2] J. Rosenberg. Realization of square-integrable representations of unimodular Lie groups on L^2-cohomology spaces. *Trans. Amer. Math. Soc.* **261** (1980), 1-32; MR 81k:22012.

[Ros_3] J. Rosenberg. The role of K-theory in noncommutative algebraic topology. *Operator algebras and K-theory, pp. 155-182, Contemp. Math., 10, Amer. Math. Soc., Providence, R.I.*, 1982; MR 84h:46097.

[Ros_4] J. Rosenberg. Group C^*-algebras and topological invariants. *Operator algebras and group representations, II (Neptun, Romania, 1980), pp. 95-115, Monographs Stud. Math., 18, Pitman, Boston, Mass.*, 1984; MR 85g:22011.

[R-S_1] R. Rochberg and S. Semmes. Nearly weakly orthonormal sequences, singular value estimates, and Calderón-Zygmund operators. *J. Functional Anal.* **86** (1989), 237-306; MR 90k:47047.

[R-S_1] R. Rochberg and S. Semmes. End point results for estimates of singular values of singular integral operators. *Contributions to operator theory and its applications (Mesa, Ariz., 1987), pp. 217-231, Oper. Theory Adv. Appl., 35, Birkhäuser, Basel*, 1988; MR 90j:47069.

[Ru] W. Rudin. Real and complex analysis. *McGraw-Hill, New York*, 1966; MR **35** # 1420.

[Rue_1] D. Ruelle. Thermodynamic formalism. *Addison-Wesley, Reading, Mass.*, 1978; MR 80g:82017.

[Rue_2] D. Ruelle. Statistical mechanics. *Benjamin, New York*, 1969; MR **44** # 6279.

[Rue-S] D. Ruelle and D. Sullivan. Currents, flows and diffeomorphisms. *Topology* **14** (1975), 319-327; MR **54** # 3759.

[S_1] S. Sakai. Automorphisms and tensor products of operator algebras. *Amer. J. Math.* **97** (1975), 889-896; MR **52** # 11610.

[S_2] S. Sakai. On automorphism groups of II_1-factors. *Tohôku Math. J.* **26** (1974), 423-430; MR **52** # 1343.

[S_3] S. Sakai. C^*-algebras and W^*-algebras. *Ergebn. Math. und ihrer Grenzgeb., 60, Springer, New York,* 1971; MR **56** # 1082.

[Sa] D. Sarason. Functions of vanishing mean oscillation. *Trans. Amer. Math. Soc.* **207** (1975), 391-405; MR **51** # 13690.

[Sau] J.L. Sauvageot. Semi-groupe de la chaleur transverse sur la C^*-algèbre d'un feuilletage riemannien. *C.R. Acad. Sci. Paris Sér. I Math* **310** (1990), 531-536; MR 91g:58279.

[Sau_2] J.L. Sauvageot. Sur le type du produit croisé d'une algèbre de von Neumann par un groupe localement compact d'automorphismes. *C.R. Acad. Sci. Paris Sér. A* **278** (1974), 941-944; MR **49** # 5867.

[Sc] P. Shields. The theory of Bernoulli shifts. *Univ. of Chicago Press, Chicago, Ill.,* 1973; MR **56** # 584.

[Sch] L.B. Schweitzer. A short proof that $M_n(A)$ is local if A is local and Frechet. *Internat. J. Math.* **3** (1992), 581-589.

[Sch_1] W. Schmid. On a conjecture of Langlands. *Ann. of Math. (2)* **93** (1971), 1-42; MR **44** # 4149.

[Sch_2] W. Schmid. Some properties of square-integrable representations of semisimple Lie groups. *Ann. of Math. (2)* **102** (1975), 535-564; MR **58** # 28303.

[Sch_3] W. Schmid. L^2-cohomology and the discrete series. *Ann. of Math. (2)* **103** (1976), 375-394; MR **53** # 716.

[Schw] J. Schwartz. Two finite, non-hyperfinite, non-isomorphic factors. *Comm. Pure Appl. Math.* **16** (1963), 19-26; MR **26** # 6812.

[Se_1] I. E. Segal. A noncommutative extension of abstract integration. *Ann. of Math. (2)* **57** (1953), 401-457; MR **14**, 991.

[Se_2] I. E. Segal. Quantized differential forms. *Topology* **7** (1968), 147-172; MR **38** # 1113.

[Se_3] I. E. Segal. Quantization of the de Rham complex. *Global analysis, pp. 205-210, Proc. Sympos. Pure Math., 16, Amer. Math. Soc., Providence, R.I.,* 1970; MR **42** # 1157.

[Se_4] I. E. Segal. A class of operator algebras which are determined by groups. *Duke Math. J.* **18** (1951), 221-265; MR **13**, 534.

[Se_5] I. E. Segal. Foundations of the theory of dynamical systems of infinitely many degrees of freedom, I. *Mat.-Fys. Medd. Danske Vid. Selsk.* **31** (1959), 39 pp.; MR **22** # 3477.

[Se_6] G. Segal. Equivariant K-theory. *Inst. Hautes Études Sci. Publ. Math. No. 34* (1968), 129-151; MR **38** # 2769.

[Se-W] G. Segal and G. Wilson. Loop groups and equations of KDV type. *Inst. Hautes Études Sci. Publ. Math. No. 61* (1985), 5-65; MR 87b:58039.

[Sei] P. Seibt. Cyclic homology of algebras. *World Scientific, Singapore,* 1987; MR 89i:18013.

[Ser_1] J.-P. Serre. Cours d'arithmétique. *Presses Univ. France, Paris,* 1970; MR **41** # 138.

[Ser_2] J.-P. Serre. Arbres, amalgames, SL_2. *Astérisque No. 46, Soc. Math. France, Paris,* 1977; MR **57** #16426.

[Ser_3] J.-P. Serre. Modules projectifs et espaces fibrés à fibre vectorielle. *Séminaire Dubreil-Pisot: algèbre et théorie des nombres* **11** (1957/58); Oeuvres I, pp. 531-543.

[Ser-B] C. Series and R. Bowen. Markov maps associated with Fuchsian groups. *Inst. Hautes Études Sci. Publ. Math. No. 50*(1979), 153-170; MR 81b:58026.

[Sh_1] M. Shubin. Elliptic almost periodic operators and von Neumann algebras. *Funkt. Anal. i Prilozhen.* **9** (1975), 89-90; MR **58** # 23168.

[Sh_2] M. Shubin. The density of states for elliptic operators with almost periodic coefficients. Conference on differential equations and applications (Ruse, Bulgaria, 1975). *Godišnik Visš. Učebn. Zaved. Priložna Mat.* **11** (1975), 209-216 (1977); MR **57** # 13245.

[Sh_3] M. Shubin. Almost periodic pseudodifferential operators and von Neumann algebras. *Trudy Moscov. Mat. Obšč.* **35** (1976), 103-164; MR **58** # 30521.

[Sh_4] M. Shubin. The density of states of self-adjoint elliptic operators with almost periodic coefficients. *Trudy Sem. Petrovsk. No. 3* (1978), 243-275; MR **58** # 17587.

[Sh_5] M. Shubin. Almost periodic functions and partial differential operators. *Uspekhi Mat. Nauk* **33** (1978), 3-47; MR **58** # 30522.

[Sh_6] M. Shubin. Almost periodic elliptic operators. *Proc. of All-Union Conference on Partial Differential Equations (Moscow State Univ., 1976), pp. 245-248, MGU, Moscow,* 1978.

[Sh_7] M. Shubin. Spectral theory and the index of elliptic operators with almost-periodic coefficients. *Uspekhi Mat. Nauk* **34** (1979), 95-135 (in Russian); MR 81f:35090.

[Si] B. Simon. Trace ideals and their applications. *London Math. Soc. Lecture Notes, 35, Cambridge Univ. Press, Cambridge,* 1979; MR 80k:47048.

[Sin_1] I.M. Singer. Future extensions of index theory and elliptic operators. *Prospects in math., pp. 171-185, Ann. of Math. Stud., 70, Princeton Univ. Press, Princeton, N.J.,* 1971; MR **49** # 8061.

[Sin_2] I.M. Singer. Some remarks on operator theory and index theory. *K-theory and operator algebras (Athens, Ga., 1975), pp. 128-138; Lecture Notes in Math., 575, Springer, Berlin,* 1977; MR **57** # 7699.

[Sk_1] G. Skandalis. Some remarks on Kasparov theory. *J. Functional Anal.* **56** (1984), 337-347; MR 86c:46085.

[Sk_2] G. Skandalis. Exact sequences for the Kasparov groups of graded algebras. *Canad. J. Math.* **37** (1985), 193-216; MR 86d:46072.

[Sk_3] G. Skandalis. Une notion de nucléarité en K-théorie. *K-Theory* **1** (1988), 549-573; MR 90b:46131.

[Sk_4] G. Skandalis. Le bifoncteur de Kasparov n'est pas exact. *C.R. Acad. Sci. Paris Sér. I Math.* **313** (1991), 939-941; MR 93b:46136.

[Skl] E.K. Sklyanin. Quantum variant of the method of the inverse scattering problem. *Zap. Nauchn. Sem. Leningrad. Otdel. Math. Inst. Steklov* (LOMI) **95** (1980), 55-128 (in Russian); MR 83a:81057.

[Sp] E.H. Spanier. Algebraic topology. *McGraw-Hill, New York,* 1966; MR **35** # 1007.

[Spe] D. Spector. Supersymmetry and the Möbius inversion function. *Comm. Math. Phys.* **127** (1990) 239-252; MR 91a:81055.

[St] M. Stadler [Macho Stadler]. La conjecture de Baum-Connes pour un feuilletage sans holonomie de codimension un sur une variété fermée. *Publ. Mat.* **33** (1989), 445-457; MR 91b:58247.

[Stei] E. Stein. Singular integrals and differentiability properties of functions. *Princeton Univ. Press, Princeton, N.J.*, 1970; MR **44** # 7280.

[Ste] J. Stern. Le problème de la mesure. *Séminaire Bourbaki, Vol. 1983/84, Exp. 632, pp. 325-346, Astérisque No. 121/122, Soc. Math. France, Paris*, 1985; MR 86f:03085.

[Sto] M.H. Stone. A general theory of spectra, I. *Proc. Nat. Acad. Sci. USA* **26** (1940), 280-283; MR **1**, 338.

[Stor] E. Størmer. Real structure in the hyperfinite factor. *Duke Math. J.* **47** (1980), 145-153; MR 81g:46088.

[Str] S. Strătilă. Modular theory in operator algebras. *Abacus Press, Tunbridge Wells*, 1981; MR 85g:46072.

[Su_1] D. Sullivan. Quasiconformal homeomorphisms in dynamics, topology and geometry. *Proceedings of the International Congress of Math., Vol. 2 (Berkeley, Calif., 1986), pp.1216-1228, Amer. Math. Soc., Providence, R.I.*, 1987; MR 90a:58160.

[Su_2] D. Sullivan. The density at infinity of a discrete group of hyperbolic motions. *Inst. Hautes Études Sci. Publ. Math. No. 50* (1979), 171-202; MR 81b:58031.

[Su_3] D. Sullivan. Cycles for the dynamical study of foliated manifolds and complex manifolds. *Invent. Math.* **36** (1976), 225-255; MR **55** # 6440.

[Su_4] D. Sullivan. Entropy, Hausdorff measures old and new, and limit sets of geometrically finite Kleinian groups. *Acta Math.* **153** (1984), 259-277; MR 86c:58093.

[Su_5] D. Sullivan. On the ergodic theory at infinity of an arbitrary discrete group of hyperbolic motions. *Riemann surfaces and related topics, pp. 465-496, Ann. of Math. Stud., 97, Princeton Univ. Press, Princeton, N.J.*, 1981; MR 83f:58052.

[Su_6] D. Sullivan. Differential forms and the topology of manifolds. *Manifolds (Tokyo, 1973), pp. 37-49, Univ. Tokyo Press, Tokyo*, 1975; MR **51** # 6838.

[Su_7] D. Sullivan. Geometric periodicity and the invariants of manifolds. *Manifolds — Amsterdam Conf. (1970), pp. 44-75, Lecture Notes in Math., 197, Springer, Berlin*, 1971; MR **44** # 2236.

[Su_8] D. Sullivan. Hyperbolic geometry and homeomorphisms. *Geometric topology (Athens, Ga., 1977), pp. 543-555, Academic, New York*, 1979; MR 81m:57012.

[Su_9] D. Sullivan and N. Teleman. An analytic proof of Novikov's theorem on rational Pontrjagin classes. *Inst. Hautes Études Sci. Publ. Math. No. 58* (1983), 79-81; MR 85k:57024.

[Sw] R.G. Swan. Vector bundles and projective modules. *Trans. Amer. Math. Soc.* **105** (1962), 264-277; MR **26** # 785.

[T_0] H. Takai. On a duality for crossed products of C^*-algebras. *J. Functional Anal.* **19** (1975), 25-39; MR **51** # 1413.

[T_1] M. Takesaki. On the cross-norm of the direct product of C^*-algebras. *Tohôku Math. J.* **16** (1964), 111-122.

[T_2] M. Takesaki. Tomita's theory of modular Hilbert algebras and its applications. *Lecture Notes in Math., 128, Springer, New York,* 1970.

[T_3] M. Takesaki. Duality for crossed products and the structure of von Neumann algebras of type III. *Acta Math.* **131** (1973), 249-310; MR **55** # 11068.

[T_4] M. Takesaki. Theory of operator algebras, I. *Springer, New York,* 1979; MR 81e:46038.

[T_5] M. Takesaki. Theory of operator algebras, II. *To appear.*

[Tat] J. Tate. Fourier analysis in number fields, and Hecke's zeta functions. *Algebraic number theory (Brighton, 1965), pp. 305-347, Thompson, Washington, D.C.,* 1967; MR **36** # 121.

[Ta] J.L. Taylor. Topological invariants of the maximal ideal space of a Banach algebra. *Adv. in Math.* **19** (1976), 149-206; MR **53** # 14134.

[Te] N. Teleman. The index of signature operators on Lipschitz manifolds. *Inst. Hautes Études Sci. Publ. Math. No. 58* (1983), 39-78; MR 85f:58112.

[Te_1] N. Teleman. The index theorem for topological manifolds. *Acta Math.* **153** (1984), 117-152; MR 86c:58137.

[Tho] D.J. Thouless. Localization and the two-dimensional Hall effect. *J. Phys. C* **14** (1981), 3475-3480.

[Tho-K-N-dN] D.J. Thouless, M. Kohmoto, M. Nightingale and M. den Nijs. Quantized Hall conductance in two-dimensional periodic potential. *Phys. Rev. Lett.* **49** (1982), 405-408.

[Thu_1] W. Thurston. Geometry and topology of 3-manifolds. *Notes, Princeton Univ., Princeton, N.J.,* 1978.

[Thu_2] W. Thurston. Noncobordant foliations of S^3. *Bull. Amer. Math. Soc.* **78** (1978), 511-514.

[To] J. Tomiyama. On the projection of norm one in W^*-algebras. *Proc. Japan Acad.* **33** (1957), 608-612; MR **20** # 2635.

[Tor] A.M. Torpe. K-theory for the leaf space of foliations by Reeb components. *J. Functional Anal.* **61** (1985), 15-71; MR 86h:46102.

[Ts_1] B.L. Tsygan. Homology of matrix Lie algebras over rings and the Hochschild homology. *Uspekhi Math. Nauk.* **38** (1983), 217-218; MR 85i:17014.

[V_1] A. Valette. K-theory for the reduced C^*-algebra of semisimple Lie groups with real rank 1 and finite center. *Quart. J. Math. Oxford (2)* **35** (1984), 341-359; MR 86j:58145.

[V_2] A. Valette. Minimal projections, integrable representations and property (T). *Arch. Math. (Basel)* **43** (1984), 397-406; MR 86j:22006.

[Va-A] J.L. van Hemmen and T. Ando. An inequality for trace ideals. *Comm. Math. Phys.* **76** (1980), 143-148; MR 81j:47014.

[Ver_1] A. Vershik. A theorem on the lacunary isomorphism of monotonic sequences of partitionings. *Funkt. Anal. i Prilozhen.* **2** (1968), 17-21; MR **39** # 2943.

[Ver_2] A. Vershik. Nonmeasurable decompositions, orbit theory, algebras of operators. *Soviet Math. Dokl.* **12** (1971), 1218-1222.

[Ve] J. Vesterstrom. Quotients of finite W^*-algebras. *J. Functional Anal.* **9** (1972), 322-335; MR **45** # 5774.

[Vey] J. Vey. Déformation du crochet de Poisson sur une variété symplectique. *Comment. Math. Helv.* **50** (1975), 421-454; MR **54** # 8765.

[Vo_1] D. Voiculescu. Some results on norm-ideal perturbations of Hilbert space operators, I, II. *J. Operator Theory* **2** (1979), 3-37; *ibid.* **5** (1981), 77-100; MR 80m:47012 and 83f:47014.

[Vo_2] D. Voiculescu. A noncommutative Weyl-von Neumann theorem. *Rev. Roumaine Math. Pures Appl.* **21** (1976), 97-113; MR **54** # 3427.

[Vo_3] D. Voiculescu. On the existence of quasicentral approximate units relative to normed ideals, Part 1. *J. Functional Anal.* **91** (1990), 1-36; MR 91m:46089.

[W] S. Wang. On the first cohomology group of discrete groups with property (T). *Proc. Amer. Math. Soc.* **42** (1974), 621-624; MR **50** # 7413.

[Wa_1] A. Wassermann. Une démonstration de la conjecture de Connes-Kasparov pour les groupes de Lie linéaires connexes réductifs. *C.R. Acad. Sci. Paris Sér. I Math.* **304** (1987), 559-562; MR 89a:22010.

[Wa_2] A. Wassermann. Ergodic actions of compact groups on operator algebras, III. *Invent. Math.* **93** (1988), 309-354; MR 91e:46093.

[We_1] A. Weil. Basic number theory. *Springer, New York,* 1974; MR **55** # 302.

[We_2] A. Weil. Fonction zêta et distributions. *Séminaire Bourbaki, Exp. 312, June,* 1966.

[Wein] S. Weinberg. Conceptual foundations of the unified theory of weak and electromagnetic interactions. *Nobel Lecture* (Dec. 8, 1979).

[Wei] S. Weinberger. Aspects of the Novikov conjecture. *Geometric and topological invariants of elliptic operators. (Brunswick, Maine, 1988), Contemp. Math., 105, Amer. Math. Soc., Providence, R.I.,* 1990; MR 91a:57020.

[Wen] H. Wenzl. On sequences of projections. *C.R. Math. Rep. Acad. Sci. Canada* **9** (1987), 5-9; MR 88k:46070.

[Wi_1] H. Widom. Families of pseudodifferential operators. *Topics in functional analysis, I, pp. 345-395, Adv. in Math. Supp. Stud., 3, Academic, New York,* 1978; MR 81c:58062.

[Wi_2] H. Widom. A complete symbolic calculus for pseudodifferential operators. *Bull. Sci. Math. (2)* **104** (1980), 19-63; MR 81m:58078.

[Wil-L] M. De Wilde and P.B. Lecomte. Existence of star-products and of formal deformations of the Poisson-Lie algebra of arbitrary symplectic manifolds. *Lett. Math. Phys.* **7** (1983), 487-496; MR 85j:17021.

[Win] E. Winkelnkemper. The graph of a foliation. *Ann. Global Anal. and Geom.* **1** (1983), no. 3, 51-75; MR 85j:57043.

[Winn] M. Winnink. Algebraic aspects of the Kubo-Martin-Schwinger boundary condition. *Cargese Lectures in Physics, 4, pp. 235-255, Gordon and Breach, New York,* 1989.

[Wit] E. Witten. Supersymmetry and Morse theory. *J. Differential Geom.* **17** (1982), 661-692; MR 84b:58111.

[Wo_1] M. Wodzicki. Excision in cyclic homology and in rational algebraic K-theory. *Ann. of Math. (2)* **129** (1989), 591-639; MR 91h:19008.

[Wo_2] M. Wodzicki. Local invariants of spectral asymmetry. *Invent. Math.* **75** (1984), 143-177; MR 85g:58089.

[Wo_3] M. Wodzicki. Noncommutative residue, Part I. Fundamentals. *K-theory, arithmetic and geometry (Moscow, 1984-86), pp 320-399, Lecture Notes Math., 1289, Springer, Berlin,* 1987; MR 90a:58175.

[Woo] R. Wood. Banach algebras and Bott periodicity. *Topology* **4** (1965/66), 371-389; MR **32** # 3062.

[Wor_1] S.L. Woronowicz. Twisted SU(2) group. An example of a noncommutative differential calculus. *Publ. Res. Inst Math. Sci.* **23** (1987), 117-181; MR 88h:46130.

[Wor_2] S.L. Woronowicz. Compact matrix pseudogroups. *Comm. Math. Phys.* **111** (1987), 613-665; MR 88m:46079.

[Y] F.J. Yeadon. A new proof of the existence of a trace in a finite von Neumann algebra. *Bull. Amer. Math. Soc.* **77** (1971), 257-260; MR **42** # 6629.

[Za] J. Zak. Magnetic translation group, II. *Phys. Rev. A* **134** (1964), A1602-A1606; MR **31** # 2031.

[Z] R. Zekri. A new description of Kasparov's theory of C^*-algebra extensions. *J. Functional Anal.* **84** (1989), 441-471; MR 90g:46106.

[Zi_1] R. Zimmer. Hyperfinite factors and amenable ergodic actions. *Invent. Math.* **41** (1977), 23-31; MR **57** # 10438.

[Zi_2] R. Zimmer. Ergodic theory of semisimple groups. *Monographs in Math., 81, Birkhäuser, Boston,* 1984; MR 86j:22014.

[Zi_3] R. Zimmer. On the von Neumann algebra of an ergodic group action. *Proc. Amer. Math. Soc.* **66** (1977), 289-293; MR **57** # 592.

NOTATION AND CONVENTIONS

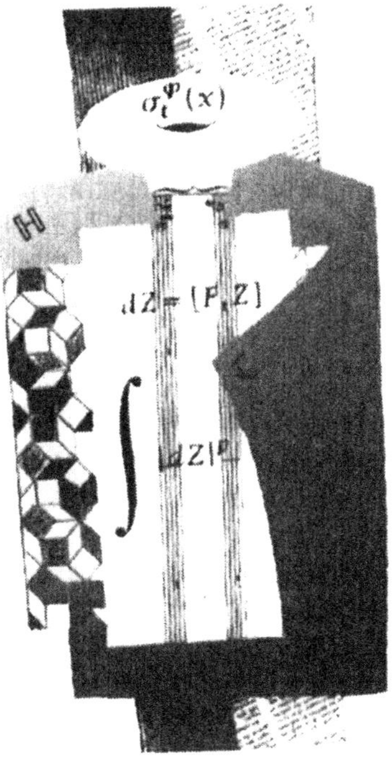

1. For $x \in \mathbb{R}_+$ we let $[x]$ be the integral part of x, i.e. the largest $n \in \mathbb{N}$, $n \leq x$.

2. Let V be a smooth manifold. We let TV (resp. T^*V) be its tangent (resp. cotangent) bundle. For any $\alpha \in \mathbb{C}$ we let $|\wedge|^\alpha V$ be the complex line bundle over V whose fiber at $p \in V$ is the one-dimensional complex vector space of maps

$$\rho : \wedge^{\dim(V)} T_p(V) \to \mathbb{C}$$

such that $\rho(\lambda v) = |\lambda|^\alpha \rho(v) \quad \forall \lambda \in \mathbb{R}$.

An α-density on V is a section of the bundle $|\wedge|^\alpha V$. Any measurable positive 1-density has an integral

$$\int_V \rho \in [0, +\infty].$$

For $\alpha \in [0,1]$, $q = \frac{1}{\alpha} \in [1, \infty]$ we let $L^q(V)$ be the Banach space of (classes of) measurable α-densities on V such that

$$\int |\rho|^q < \infty$$

with the norm $\|\rho\|_q = (\int |\rho|^q)^{1/q}$.

In particular $V \to L^2(V)$ is a functor to the category of Hilbert spaces. (We refer to [Gu-S] for more details.)

3. Let X be a locally compact space. We let $C(X)$ be the linear space of complex-valued continuous functions on X. We use C_c, C_b, and C_0 to indicate, respectively, functions with compact support, bounded functions, and functions vanishing at ∞. We use the notation $C(X, E)$ for continuous sections of a vector bundle E over X. Thus for instance $C_c(X,E)$ is the linear space of continuous sections with compact support of E. When X is a smooth manifold we let $C^\infty(X)$ be the linear subspace of $C(X)$ whose elements are smooth functions. Similarly $C^\infty(X,E)$ is the space of smooth sections of the smooth vector bundle E over X. We let $C^{-\infty}(X,E)$ be the space of generalized sections of E, i.e. (cf. [Gu-S]) the topological dual of the linear space:

$$C_c^\infty(X, E^* \otimes |\wedge| X).$$

4. We let $S(\mathbb{R}^k)$ (resp. $S(\mathbb{Z}^k)$) be the Schwartz space of smooth functions $f(x)$ on $\mathbb{R}^k$ all of whose derivatives decay faster than $|x|^{-n}$ for any n (resp. of sequences $a(s)$ which decay faster than $|s|^{-n}$ for any n).

5. Given an algebra $\mathcal{A}$ over $\mathbb{C}$ we let $Z(\mathcal{A})$ be its center. For $a, b \in \mathcal{A}$ we let $[a,b] = ab - ba$ be their commutator. For $X \subset \mathcal{A}$ we let $X' = \{b \in \mathcal{A}\,;\, ab = ba \quad \forall a \in X\}$ be the commutant of X. For each $n \in \mathbb{N}$ we let $M_n(\mathbb{C})$ be the algebra of $n \times n$ matrices and $M_n(\mathcal{A}) = M_n(\mathbb{C}) \otimes \mathcal{A}$ the algebra of $n \times n$ matrices over $\mathcal{A}$. We let $GL_n(\mathcal{A})$ be the group of invertible elements of $M_n(\mathcal{A})$.

6. Let $\mathcal{H}$ be a Hilbert space, T a bounded operator on $\mathcal{H}$. We let T^* be its adjoint ($\langle T^*\xi, \eta\rangle = \langle \xi, T\eta\rangle$, $\forall \xi, \eta \in \mathcal{H}$) and $|T| = \sqrt{T^*T}$ its absolute value.
We let $\mathcal{L}(\mathcal{H})$ be the $*$-algebra of bounded operators in $\mathcal{H}$.

7. Let G be a small category. We let $G^{(0)}$ be the set of objects of G and $r, s : G \to G^{(0)}$ be the range and source maps. We define, for $x \in G^{(0)}$

$$G^x = \{\gamma \in G\,;\, r(\gamma) = x\} \;,\quad G_x = \{\gamma \in G\,;\, s(\gamma) = x\}.$$

We let BG be the classifying space of G, i.e. the geometric realization of the nerve of G. This continues to make sense when G is a topological category. We refer to [Lo] and to appendix III.A.

8. Let E be a finite-dimensional Euclidean vector space (resp. a Euclidean vector bundle over a base space X). We let $\mathrm{Cliff}(E)$ be the Clifford algebra of E (resp. the bundle of Clifford algebras $\mathrm{Cliff}(E_x)$, $x \in X$).

By definition $\mathrm{Cliff}(E)$ is the quotient of the tensor algebra of E by the ideal generated by the elements $\xi \otimes \xi - \|\xi\|^2\, 1$.

We endow the complexified Clifford algebra $\mathrm{Cliff}_{\mathbb{C}}(E)$ with the unique involution (antilinear anti-automorphism of square 1) such that

$$\xi^* = \xi \quad \forall \xi \in E.$$

When confusion might otherwise arise we denote the product in $\mathrm{Cliff}(E)$ by $\omega_1, \omega_2 \mapsto \omega_1 \cdot \omega_2$.

9. To an orientation of E (as in 8) we associate the element ε, $\varepsilon^2 = 1$, of $\mathrm{Cliff}_{\mathbb{C}}(E)$ given by

$$\varepsilon = i^{-[n/2]}\, \xi_1 \cdots \xi_n \;,\quad n = \dim E$$

where $\xi_1, \ldots, \xi_n$ is an oriented orthonormal basis of E. When n is odd $\varepsilon \in Z(\mathrm{Cliff}_{\mathbb{C}}(E))$.

A spin structure on an oriented Euclidean vector bundle E is a reduction of its structure group $SO(n)$ to its covering group $\mathrm{Spin}(n)$. We let $S = S(E)$ be the associated Hermitian complex vector bundle of spinors.

For $c \in \mathrm{Cliff}_{\mathbb{C}}(E)$ we let $\gamma(c)$ be the corresponding operator in $S(E)$.

When n is even we define the $\mathbb{Z}/2$ grading of $S(E)$ by

$$\gamma = \gamma(\varepsilon).$$

When n is odd we assume that $\gamma(\varepsilon) = 1$.

10. The letters $c, h, \hbar$, and k are used for the standard physical constants (speed of light, Planck's constant, Planck's constant$/2\pi$, and the Boltzmann constant).

11. Let V be a Riemannian Spin manifold and S the Spinor bundle of $E = T^*V$. We let ∂_V or $\not{\partial}_V$ be the Dirac operator on V

$$\partial_V = i^{-1}\, \gamma \circ \nabla$$

where $i = \sqrt{-1}$, ∇ is the Levi-Civita connection

$$\nabla : C^\infty(V, S) \to C^\infty(V, T^* \otimes S)$$

and $\gamma : C^\infty(V, T^* \otimes S) \to C^\infty(V, S)$ is the Clifford multiplication as in (9).

12. We refer to [Mi_4] for the standard theory of characteristic classes and shall use the sign conventions of [Mi_4]. In particular the Chern character of a vector bundle is given by trace $(\exp(\nabla^2/2\pi i))$ in terms of connections and curvature. Aside from the standard multiplicative classes Td, $\hat{A}$, and L associated respectively to the formal power series $x/(1-e^{-x})$, $(x/2)/\sinh(x/2)$, and $x/\tanh(x)$, we shall also use the $\mathcal{L}$-genus ([$\mathrm{At\text{-}Si}_1$]) given by the formal power series $(x/2)/\tanh(x/2)$.

INDEX

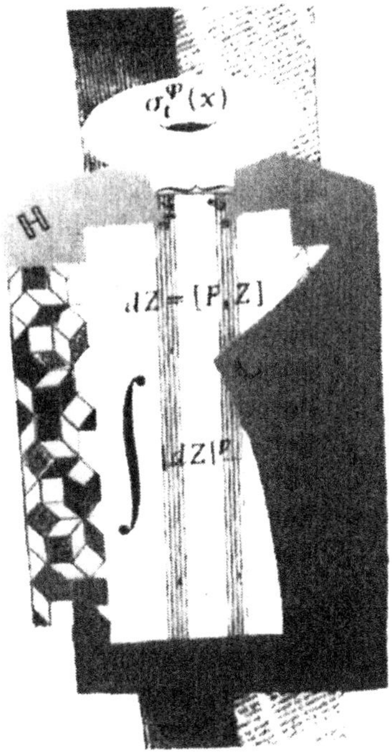

F

This
book is set
in Lucida Bright and
Lucida New Math designed by
Bigelow & Holmes and produced in Type 1
format by Y&Y. Book design by Peter Renz.
PRODUCTION by Associated Ions of Ann Arbor using
TEX implemented as Textures from Blue Sky Research and
running on Apple Macintosh equipment. Macros by Patrick Ion,
beginning with the author's compuscript. Editing and copy editing by
Patrick Ion and Arthur Greenspoon. Corrections, copy editing, and
bibliographic work by Bonita Ion. ORIGINAL FIGURES by Willy
Payet of Villeurbanne, France, with refinements and
additions by Nicholas Ion. Software & devices used
included: Mathematica, Adobe Illustrator,
Adobe Photoshop, StudioCraft and a
Hewlett-Packard scanner. The
FRONT COVER image is by
Alain Connes.

Colophon composed in Apple Chancery by Bigelow & Holmes.

Printed and bound by CPI Group (UK) Ltd, Croydon, CR0 4YY
03/10/2024
01040314-0015